T0317513

Game Theory

Wiley Series in
OPERATIONS RESEARCH AND MANAGEMENT SCIENCE

Founding Series Editor
James J. Cochran

Operations Research and Management Science (ORMS) is a broad, interdisciplinary branch of applied mathematics concerned with improving the quality of decisions and processes and is a major component of the global modern movement towards the use of advanced analytics in industry and scientific research. *The Wiley Series in Operations Research and Management Science* features a broad collection of books that meet the varied needs of researchers, practitioners, policy makers, and students who use or need to improve their use of analytics. Reflecting the wide range of current research within the ORMS community, the Series encompasses application, methodology, and theory and provides coverage of both classical and cutting edge ORMS concepts and developments. Written by recognized international experts in the field, this collection is appropriate for students as well as professionals from private and public sectors including industry, government, and nonprofit organization who are interested in ORMS at a technical level. The Series is comprised of three sections: Decision and Risk Analysis; Optimization Models; and Stochastic Models.

Game Theory

An Introduction

Third Edition

E. N. Barron
Loyola University Chicago
Chicago, USA

Library of Congress Cataloging-in-Publication Data applied for:

Hardback ISBN: 9781394169115

Cover Design: Wiley
Cover Image: © Aaron Foster/Getty images

Set in 9.5/12.5pt STIXTwoText by Straive, Chennai, India

To Christina, Michael, and Anastasia; and Fotini and Michael

Contents

Preface for the Third Edition

I have offended God and mankind because my work didn't reach the quality it should have.
– Leonardo da Vinci

I do not feel obliged to believe that the same God who endowed us with sense and reason would have intended us to forgo their use.
– Galileo Galilei

Philip II of Macedon: "If I invade Lakonia you will be destroyed, never to rise again." The Spartans replied, "If."
– Plutarch

Game theory is the study of how people make decisions in situations where their choices affect the outcome for others. It is a branch of mathematics that has been applied to a wide range of fields, including economics, political science, biology, and computer science. Game theory is important because it helps us to understand how people behave in situations where there is conflict or cooperation. It can be used to predict how people will act in certain situations and to design strategies that will help us to achieve our goals. For example, game theory can be used to understand how businesses compete with each other, how politicians negotiate with each other, and how animals interact with each other in the wild. It can also be used to design new products and services and to create more effective marketing campaigns. Game theory is a complex and fascinating field of study, and it is still evolving. As we learn more about how people make decisions, we can use game theory to improve our lives in many different ways. Game theory is a powerful tool that can be used to improve our understanding of the world and our ability to make decisions. In addition, knowing what is optimal helps us understand what is suboptimal.

I am gratified that my presentation of game theory in the previous editions of the book has been well received among many different groups of students and researchers. I have received suggestions and criticisms from many diverse areas of expertise, including mathematics, economics, and engineering, as well as from professors, students, and practitioners. Since I agree with most of these suggestions, I decided to write an edition incorporating many of the suggestions and taking note of the criticisms. Accordingly, this third edition adds several topics and takes a slightly new approach. The new sections include

- A new section on Bayesian Games connected to games with imperfect information. This topic introduces the concept of Bayesian Equilibrium and Subgame Perfect Bayesian Equilibrium. The idea is to apply Bayes' rule to information sets in order to maximize the expected payoff to a player whose actions emanate from the information set. A Perfect Bayesian Equilibrium gives yet another way to pick out a correct, or desirable, equilibrium in games where there are multiple Nash equilibria.

- A new chapter on repeated games. The question answered here is whether cooperation among the players will occur if the game is played repeatedly. Strategies to elicit cooperation are developed when the probability a game is played again is sufficiently large.

- A new section on the Stable Matching problem. Students applying to medical residency programs are players in this game.

- Many new exercises and examples. Exercises which have been added at the end of most chapters can be used as projects or in-class discussions. I would like to thank Prof. Peter Tingley for many of these exercises. In addition, I would like to thank Peter for his numerous suggestions and motivation for this edition.

Chapter 1 now combines Chapters 1 and 2 from the second edition. Theory is downplayed to a certain extent throughout the book, although there are still two distinct proofs of the von Neumann minimax theorem. The first proof, which is accessible to anyone knowing calculus, is in an appendix of Chapter 1, and the second proof, which is due to Josef Hofbauer [15], is a very clever use of the replicator equations for population games in Chapter 8. Some knowledge of ordinary differential equations is needed, which is covered in most sophomore year undergraduate courses.

I have decided to remove the Maple© software from this edition and concentrate on Mathematica©. All of the Mathematica code snippets have been updated to work on version 13.1. One other point is that most, if not all of the odd-numbered problems in the book have hints, or solutions in the back of the book. Most newly added problems are not so-favored.

When I became aware of errors, I fixed them, since obviously I cannot fix errors I have no knowledge of except inadvertently. It is well known among coders that every attempt to fix bugs introduces new bugs, so I fully expect to have introduced new errors in this edition. There is now definitely too much material to be covered in one semester but the choice of topics after the essential material is up to the instructor. This subject is still one of the few that brings together many topics in an undergraduate math curriculum–calculus, linear algebra, probability, differential equations, and optimization. As such, it is the perfect capstone course as much as it is one of the most important subjects to learn on its own.

I would like to rededicate this book to my wife Christina and to our grandchildren, Jack Mario and Sofia Vittoria.

Chicago, IL   *E. N. Barron*
ebarron@luc.edu
2023

Preface for the Second Edition

This second edition expands the book in several ways. There is a new chapter on extensive games which takes advantage of the open source software package **GAMBIT**[1] to both draw the trees and solve the games. Even simple examples will show that doing this by hand is not really feasible and that's why it was skipped in the first edition. These games are now included in this edition because it provides a significant expansion of the number of game models which the student can think about and implement. It is an important modeling experience and one cannot avoid thinking about the game and the details to get it right.

Many more exercises have been added to the end of most sections. Some material has been expanded upon and some new results have been discussed. For instance, the book now has a section on correlated equilibria and a new section on explicit solutions of three-player cooperative games due to recent results of Leng and Parlar [22]. The use of software makes these topics tractable. Finding a correlated equilibrium depends on solving a linear programming problem which becomes a trivial task with Maple/Mathematica or any linear programming package.

Once again, there is more material in the book than can be covered in one semester, especially if one now wants to include extensive games. Nevertheless all of the important topics can be covered in one semester if one does not get sidetracked into linear programming or economics. The major topics forming the core of the course are zero sum games, nonzero sum games, and cooperative games. A course covering these topics could be completed in one quarter.

The foundation of this class is examples. Every concept is either introduced by or expanded upon and then illustrated with examples. Even though proofs of the main theorems are included, in my own course, I skip virtually all of them and focus on their use. In a more advanced course, one might include proofs and more advanced material. For example, I have included a brief discussion of mixed strategies for continuous games but this is a topic which actually requires knowledge of measure theory to present properly. Even knowing about Stieltjes integrals is beyond the prerequisites of the class. Incidentally, the prerequisites for the book are very elementary probability, calculus, and a basic knowledge of matrices (like multiplying and inverses).

Another new feature of the second edition is the availability of a solution manual which includes solutions to all of the problems in the book. The new edition of the book will contain answers to *odd numbered* problems. Some instructors have indicated that they would prefer to not have all the solutions in the book so that they could assign homework for grades without making up all new problems. I am persuaded by this argument after teaching this course many times.

All the software in the book in both Maple and Mathematica is available for download from my website

www.math.luc.edu/~enb.

My classes in game theory have had a mix of students with majors in mathematics, economics, biology, chemistry, but even French, theology, philosophy, and English. Most of the students had college mathematics prerequisites but some had only taken calculus and probability/statistics in high school. The prerequisites are not a strong impediment for this course.

I have recently begun my course with introducing extensive games almost from the first lecture. In fact, when I introduce the Russian Roulette and 2×2 Nim examples in chapter 1, I take that

---

1 Savani, Rahul and Turocy, Theodore L. (2023). Gambit: The package for computation in game theory, Version 16.1.0. http://www.gambit-project.org.

opportunity to present them in Gambit in a classroom demonstration. From that point on, extensive games are just part of the course and they are intermingled with matrix theory as a way to model a game and come up with the game matrix. In fact demonstrating the construction using Gambit of a few examples in class is enough for students to be able to construct their own models. Exercises on extensive games can be assigned from the first week. In fact the chapter on extensive games does not have to be covered as a separate chapter but used as a source of problems and homework. The only concepts I discuss in that chapter are backward induction and subgame perfect equilibrium which can easily be covered through examples.

A suggested syllabus for this course may be useful:

1. Chapters 1 and 2 (3 weeks)
    (a) Upper and lower values, mixed strategies, introduction to game trees and Gambit.
    (b) Expected payoffs, minimax theorem, graphical method.
    (c) Invertible matrix games, symmetric games, linear programming method.
2. Chapter 3 (2 weeks)
    (a) Nonzero sum two person games, pure and mixed Nash equilibrium.
    (b) Best responses, Equality of Payoffs.
    (c) Calculus method, Lemke–Howson, Correlated Equilibrium.
3. Chapter 4 (1 week)
    (a) Extensive form games
        i. Trees in Gambit, information sets. Examples
        ii. Subgame perfect equilibrium, backward induction. Examples
    (b) Exam 1
4. Chapter 5 (2 weeks)
    (a) Pure Nash equilibrium for games with a continuum of strategies
    (b) Selected examples: Cournot, Stackelberg, Traveler's paradox, Braess' paradox. War of attrition.
5. Chapter 6 (3 weeks)
    (a) Cooperative Games
        i. Characteristic functions, imputations, core, least core
        ii. Nucleolus, Shapley Value.
    (b) Bargaining, Nash solution, Threats
    (c) Exam 2
6. Chapter 7 (1 week)
    (a) Evolutionary Stable strategies
    (b) Population games and Stability
    (c) Review

Naturally, instructors may choose from the many peripheral topics available in the book if they have time, or for assignments as extra credit or projects. I have made no attempt to make the book exhaustive of topics that should be covered and I think that would be impossible in any case. The topics I have chosen I consider to be foundational for all of game theory and within the constraints of the prerequisites of an undergraduate course. For further topics, there are many excellent books on the subject, some of which are listed in the references.

As a final note on software, this class does not require the writing of any programs. All one needs is a basic facility with using software packages. In all cases, solving any of the games in Maple or Mathematica involves looking at the examples and modifying the matrices as warranted. The use of software has not been a deterrent to any student I have had in any game theory class and in fact the class can be designed without the use of any software.

Acknowledgements: I am very grateful to everyone who has contacted me with possible errors in the first edition. I am particularly grateful to Professor Kevin Easley at Truman State, for his many suggestions, comments, and improvements for the book over the time he has been teaching game theory. His comments and the comments from his class were invaluable to me. I am grateful to all of those instructors who have adopted the book for use in their course and I hope that the second edition removes some of the deficiencies in the first and makes the course better for everyone.

As part of an independent project, I assigned my student Zachary Schaefer the problem of writing some very useful Mathematica programs to do various tasks. The projects ranged from setting up the graphs for any appropriate-sized matrix game, solving any game with linear programming by both methods, automating the search for Nash equilibria in a nonzero sum game, and finding the nucleolus and Shapley value for any cooperative game (this last one is a real *tour de force*). All of these projects are available from my website. Zachary did a great job.

I also would like to thank the National Science Foundation for partial support of this project under grant 1008602.

I would be grateful for notification of any errors found.

Chicago, IL
ebarron@luc.edu
2012

E. N. Barron

Preface for the First Edition

> Man is a gaming animal. He must always be trying to get the better in something or other.
>
> –Charles Lamb, *Essays of Elia, 1823*

Why do countries insist on obtaining nuclear weapons? What is a fair allocation of property taxes in a community? Should armies ever be divided, and in what way in order to attack more than one target? How should a rat run to avoid capture by a cat? Why do conspiracies almost always fail? What percentage of offensive plays in football should be passes, and what percentage of defensive plays should be blitzes? How should the assets of a bankrupt company be allocated to the debtors? These are the questions that game theory can answer. Game theory arises in almost every facet of human interaction (and inhuman interaction as well). Either every interaction involves objectives that are directly opposed, or the possibility of cooperation presents itself. Modern game theory is a rich area of mathematics for economics, political science, military science, finance, biological science (because of competing species and evolution), and so on.[1]

This book is intended as a mathematical introduction to the basic theory of games, including noncooperative and cooperative games. The topics build from zero sum matrix games, to nonzero sum, to cooperative games, to population games. Applications are presented to the basic models of competition in economics: Cournot, Bertrand, and Stackelberg models. The theory of auctions is introduced and the theory of duels is a theme example used in both matrix games, nonzero sum games, and games with continuous strategies. Cooperative games are concerned with the distribution of payoffs when players cooperate. Applications of cooperative game theory to scheduling, cost savings, negotiating, bargaining, etc., are introduced and covered in detail.

The prerequisites for this course or book include a year of calculus, and very small parts of linear algebra and probability. For a more mathematical reading of the book, it would be helpful to have a class in advanced calculus, or real analysis. Chapter 7 uses ordinary differential equations. All of these courses are usually completed by the end of the sophomore year, and many can be taken concurrently with this course. Exercises are included at the end of almost every section, and odd numbered problems have solutions at the end of the book. I have also included appendixes on the basics of linear algebra, probability, Maple,[2] and Mathematica,[3] commands for the code discussed in the book using Maple.

One of the unique features of this book is the use of Maple[4] or Mathematica[5] to find the values and strategies of games, both zero and nonzero sum, and noncooperative and cooperative. The major computational impediment to solving a game is the roadblock of solving a linear or nonlinear program. Maple/Mathematica gets rid of those problems and the theories of linear and nonlinear programming do not need to be presented to do the computations. To help present some insight into the basic simplex method which is used in solving matrix games and in finding the nucleolus,

---

1 In an ironic twist, game theory cannot help with most common games, like chess, because of the large number of strategies involved.
2 Trademark of Maplesoft Corporation.
3 Trademark of Wolfram Research Corp.
4 Version 10.0.
5 Version 8.0.

a section on the simplex method specialized to solving matrix games is included. If a reader does not have access to Maple or Mathematica, it is still possible to do most of the problems by hand, or using the free software Gambit,[1] or Gams.[2]

The approach I took in the software in this book is to not reduce the procedure to a canned program in which the student simply enters the matrix and the software does the rest (Gambit does that). To use Maple/Mathematica and the commands to solve any of the games in this book, the student has to know the procedure, that is, what is going on with the game theory part of it, and then invoke the software to do the computations.

My experience with game theory for undergraduates is that students greatly enjoy both the theory and applications, which are so obviously relevant and fun. I hope that instructors who offer this course as either a regular part of the curriculum, or as a topics course, will find that this is a very fun class to teach, and maybe to turn students on to a subject developed mostly in this century and still under hot pursuit. I also like to point out to students that they are studying the work of Nobel Prize winners: Herbert Simon[3] in 1979, John Nash,[4] J. C. Harsanyi[5] and R. Selten[6] in 1994, William Vickrey[7] and James Mirrlees[8] in 1996, and Robert Aumann[9] and Thomas Schelling[10] in 2005. In 2007 the Nobel Prize in economics was awarded to game theorists Roger Myerson,[11] Leonid Hurwicz,[12] and Erik Maskin.[13] In addition, game theory was pretty much invented by John von Neumann,[14] one of the true geniuses of the twentieth century.

Chicago, IL   *E. N. Barron*
2007

---

1 Available from www.gambit.sourceforge.net/.

2 Available from www.gams.com.

3 June 15, 1916–February 9, 2001, a political scientist who founded organizational decision making.

4 See the short biography in Appendix D.

5 May 29, 1920–August 9, 2000, Professor of Economics at University of California, Berkeley, instrumental in equilibrium selection.

6 Born October 5, 1930, Professor Emeritus, University of Bonn, known for his work on bounded rationality.

7 June 21, 1914–October 11, 1996, Professor of Economics at Columbia University, known for his work on auction theory.

8 Born July 5, 1936, Professor Emeritus at University of Cambridge.

9 Born June 8, 1930, Professor at Hebrew University.

10 Born April 14, 1921, Professor in School of Public Policy, University of Maryland.

11 Born March 29, 1951, Professor at University of Chicago.

12 Born August 21, 1917, Regents Professor of Economics Emeritus at the University of Minnesota.

13 Born December 12, 1950, Professor of Social Science at Institute for Advanced Study, Princeton.

14 See a short biography in Appendix D and Reference [25] for a full biography.

Acknowledgments

I am very grateful to Susanne Steitz-Filler, who was my original editor at Wiley. I gratefully acknowledge my new editor at Wiley, Paul Sayer. In addition, I am grateful to all of the support staff of John Wiley & Sons and to the reviewers of this book.

E. N. Barron

Introduction

Mostly for the Instructor

My goal is to present the basic concepts of noncooperative and cooperative game theory and introduce students to a very important application of mathematics. In addition, this course introduces students to an understandable theory created by geniuses in many different fields that even today has a low cost of entry.

My experience is that the students who enroll in game theory are primarily mathematics students interested in applications, with about one-third to one-half of the class majoring in economics or other disciplines (such as biology or biochemistry or physics). The modern economics and operations research curriculum requires more and more mathematics, and game theory is typically a required course in those fields. For economics students with a more mathematical background, this course is set at an ideal level. For mathematics students interested in a graduate degree in something other than mathematics, this course exposes them to another discipline in which they might be interested and that will enable them to further their studies, or simply to learn some fun mathematics. Many students get the impression that applied mathematics is physics or engineering, and this class shows that there are other areas of applications that are very interesting and that open up many other alternatives to a pure math or classical applied mathematics concentration.

Game theory can be divided into two broad classifications: noncooperative and cooperative. The sequence of main topics in this book is as follows:

1. Two-person zero sum matrix games.
2. Nonzero sum games, both bimatrix, and with a continuum of strategies, Nash and correlated equilibria.
3. Cooperative games, covering both the nucleolus concept and the Shapley value.
4. Bargaining with and without threats.
5. Evolution and population games and the merger with stability theory.

This is generally more than enough to fill one semester, but if time permits (which I doubt) or if the instructor would like to cover other topics (duels, auctions, economic growth, evolutionary stable strategies, population games) these are all presented at some level of depth in this book appropriate for the intended audience. Game theory has a lot of branches and these topics are the main branches, but not all of them. Combinatorial game theory is a branch that could be covered in a separate course but it is too different from the topics considered in this book to be consistent. Repeated games and stochastic games are two more important topics that are skipped as too advanced and too far afield.

This book begins with the classical zero sum two-person matrix games, which is a very rich theory with many interesting results. I suggest that the first two chapters be covered in their entirety, although many of the examples can be chosen on the basis of time and the instructor's interest. For classes that are more mathematically oriented, one could cover the proofs given of the von Neumann minimax theorem. The use of linear programming as an algorithmic method for solving matrix games is essential, but one must be careful to avoid getting sucked into spending too much time on the simplex method. Linear programming is a course in itself, and as long as students understand the transformation of a game into a linear program, they get a flavor of the power of the method. It is a little magical when implemented in Mathematica, and I give two ways to do it, but there are reasons for preferring one over the other when doing it by hand.

The generalization to nonzero sum two-person games comes next with the foundational idea of a Nash equilibrium introduced. It is an easy extension from a saddle point of a zero sum game to a Nash equilibrium for nonzero sum. Several methods are used to find the Nash equilibria from the use of calculus to full-blown nonlinear programming. A short introduction to correlated equilibrium is presented with a conversion to a linear programming problem for solution. Again, Mathematica is an essential tool in the solution of these problems. Both linear and nonlinear programs are used in this course as a tool to study game theory, and not as a course to study the tools. I suggest that the entire chapter be covered.

It is essential that the instructor cover at least the main points in Chapters 1–7. Chapter 5 is a generalization of two-person nonzero sum games with a finite number of strategies (basically matrix games) to games with a continuum of strategies. Calculus is the primary method used. The models included in Chapter 5 involve the standard economic models, the theory of duels, which are just games of timing, and the theory of auctions. An entire semester can be spent on this one chapter, so the instructor will probably want to select the applications for her or his own and the class's interest. Students find the economic problems particularly interesting and a very strong motivation for studying both mathematics and economics.

When cooperative games are covered, I present both the theory of the core, leading to the nucleolus, and the very popular Shapley value. Students find the nucleolus extremely computationally challenging because there are usually lots of inequalities to solve and one needs to find a special condition for which the constraints are nonempty. Doing this by hand is not trivial even though the algebra is easy. Once again, Mathematica can be used as a tool to assist in the solution for problems with four or more players, or even three players. In addition, the graphical abilities of software permit a demonstration of the actual shrinking or expanding of the core according to an adjustment of the dissatisfaction parameter. On the other hand, the use of software is not a procedure in which one simply inputs the characteristic function and out pops the answer. A student may use software to assist but not solve the problem.[1] Chapter 6 ends with a presentation of the theory of bargaining in which nonlinear programming is used to solve the bargaining problem following Nash's ideas. In addition, the theory of bargaining with optimal threat strategies is included. This serves also as a review section because the concepts of matrix games are used for safety levels, saddle points, and so on.

The last chapter serves as a basic introduction to evolutionary stable strategies and population games. If you have a lot of biology or biochemistry majors in your class, you might want to make

---

1 On the other hand, the recent result of Leng and Parlar [22] has explicit formulas for the nucleolus of any 3 player game. The software doing the calculation is a black box.

time for this chapter. The second half of the chapter does require an elementary class in ordinary differential equations. The connection between stability, Nash equilibria, and evolutionary stable strategies can be nicely illustrated with the assistance of the Mathematica differential equation packages, circumventing the need for finding the actual solution of the equations by hand. Testing for stability is a calculus method. One possible use of this chapter is for projects or extra credit.

My own experience is that I run out of time with a 14-week semester after Chapter 6. Too many topics need to be skipped, but adjusting the topics in different terms makes the course fresh from semester to semester. Of course, topics can be chosen according to your own and the class's interests. On the other hand, this book is not meant to be exhaustive of the theory of games in any way.

The prerequisites for this course have been kept to a minimum. This course is presented in our mathematics department, but I have had many economics, biology, biochemistry, business, finance, political science, and physics majors. The prerequisites are listed as a class in probability, and a class in linear algebra, but very little of those subjects are actually used in the class. I tell students that if they know how to multiply two matrices together, they should do fine; and the probability aspects are usually nothing more than the definitions. The important prerequisite is really not being afraid of the concepts. I have had many students with a background of only two semesters of calculus, no probability or linear algebra, or only high school mathematics courses. Students range from sophomores to graduate students (but I have even taught this course to freshman honors students). As a minor reference I include appendixes on linear algebra, probability, and the major procedures in the text to Mathematica.

The use of software in this class is also optional, but then it is like learning multivariable calculus without an ability to graph the surfaces. It can be done, but it is more difficult. Why do that when the tool is available? That may be one of the main features of this book, because before the technology was available, this subject had to be presented in a very mathematical way or a very nonmathematical way. I have tried to take the middle road, and it is not a soft class. On the other hand, the new Chapter 4 on extensive games absolutely requires the use of GAMBIT because the interesting models are too complex to do by hand. I feel that this is appropriate considering that GAMBIT is free and runs on all machines. It is also possible for an instructor to skip this chapter entirely if so desired.

There are at least two important websites in game theory. The first is

gametheory.net,

which is a repository for all kinds of game theory stuff. I especially like the notes by T. Ferguson at UCLA, H. Moulin at Rice University, W. Bialis at SUNY Buffalo, and Y. Peres, at the University of California, Berkeley. The second site is

www.gambit.sourceforge.net,

which contains the extremely useful open source software **GAMBIT,** which students may download and install on their own computers. Gambit is a game-solving program that will find the Nash equilibria of all N-person matrix games with any number of strategies. It may also be used to solve any zero sum game by entering the matrix appropriately. Students love it. Finally, if a user has Mathematica, there is a cooperative game solver available from the Wolfram website, known as **TuGames**, written by Holgar Meinhardt, that can be installed as a Mathematica package. TuGames can solve any characteristic function cooperative game, and much more. There is also a MATLAB package written by Professor Meinhardt for solving cooperative games.

I would be grateful for any notification of errors, and I will post errata on my website

$$\boxed{\textbf{www.math.luc.edu/}{\sim}\textbf{enb}.}$$

I will end this introduction with an intriguing quote that Professor Avner Friedman included in his book *Differential Games*, which he had me read for my graduate studies. The quote is from Lord Byron: "There are two pleasures for your choosing; the first is winning and the second is losing." Is losing really a pleasure? I can answer this now. The answer is no.

1

Matrix Two-Person Games

If you must play, decide upon three things at the start: the rules of the game, the stakes, and the quitting time.

–Chinese proverb

Everyone has a plan until you get hit.
–Mike Tyson, Heavyweight Boxing Champ, 1986–1990, 1996

You can't always get what you want.

–Mothers everywhere

Be sober, be vigilant; because your adversary the devil, as a roaring lion, walketh about, seeking whom he may devour.

–1 Peter 5:8

1.1   What Is Game Theory?

We should start by making sure we understand what this subject is about. Game theory as a branch of **optimization theory** is lots of things, not all of which are covered in this book. Game theory studied here is roughly the math/science of how to make decisions in situations in which

- You might have incomplete information, or the situation may involve some randomness you don't control.
- There may be other players whose goals are different from yours. These can be **opponents** whose aims are opposite yours, or collaborators whose aims are consistent with yours (as in **cooperative games**), or relevant actors whose aims may simply differ from yours.
- You are trying to maximize some kind of outcome to yourself called your **utility** or **payoff**—this could be a probability of winning, an amount of money, your happiness—really anything.

Game Theory: An Introduction, Third Edition. E. N. Barron.
© 2024 John Wiley & Sons, Inc. Published 2024 by John Wiley & Sons, Inc.

Figure 1.1 The Card Players by Paul Cezanne. Everett Collection/Shutterstock.

When random elements are part of the game, like the dealing of cards or the chances a dictator will attack, you can't predict exactly what will happen, but you try to make things work out as best as you can on average. There are many examples of games in this category. To name just a few:

- **Economic**: Companies trying to make decisions such as setting production limits, who likely have only partial information on the effects of their choices, and where there may be competitors whose choices are not known.
- **Military**: It seems pretty clear that strategic decisions such as troop deployments in a combat zone involve both uncertainty and opponents whose actions are not known.
- **Actual** games: Some standard games played for fun or profit fall into this category. Probably the best example is poker, which we will spend some time thinking about. Of course the full game of poker is way too complicated to analyze, and we will only do so for very specialized and simple games whose rules will be specified precisely.

We will study many examples of all of these, and many have very important and impactful consequences in the real world.

Another main branch of game theory is combinatorial game theory. This would include games like chess, go, life, etc.—in theory you could work out every possibility and find the absolute perfect moves, but it is just too complicated. But, using game theory, modern computers can defeat any grand master in chess and any combinatorial game for which computer programmers have already tackled the problem. We will touch on a few games in this category of combinatorial games, but it is not the focus of this book. If you picked this up to be a better chess player read a chess book. The game theory in this book has a lot to say about poker, however, and some professional poker players use and have developed game theory for poker.

1.2 Motivating Examples

In this section we will present three simple games to provide some insight into why game theory is interesting and important and has attracted the attention of many scientists (and gamblers).

1.2.1 Three Card Poker

Here is a very simplified model of a game of poker. The rules are as follows.

- Three possible cards are involved in this game: Ace, Queen, and King. There are two players Actor, known as player I, and Responder, known as player II. They each must ante $1 to play the game.
- Actor will be dealt either an Ace (A) or a queen (Q), the two possibilities being equally likely. Actor knows the card dealt but Responder does not.
- Responder will always be dealt a King (K). Both players know this.
- Actor may choose to pass (equivalently, fold) or bet. If Actor chooses to pass, she loses $1 no matter what card she has.
- If Actor bets, then Responder must choose to call or fold. If Responder folds then Actor wins the pot, i.e., Actor wins $1 from Responder. If Responder calls, the highest card wins the pot, which now contains $2 from each player. So if Actor has an Ace, Responder wins $2, and if Actor has a Queen, Responder loses $2.

You might think the strategy for Actor should be to fold with a Queen and bet with an Ace, but this may not be the case: Actor may want to bluff by betting on a Queen at least some of the time, and indeed we see poker players bluffing in real life. It will be very helpful to draw a diagram, called a game tree, of each player's choices and the resulting payoffs. These game trees will be studied extensively in Chapter 3.

It is important to note that in Figure 1.2 a dotted line connecting two nodes means that the two nodes are in the same **information set**, labeled 2:1 in this case. What that means is that player II, Responder, knows that player I, Actor, has made a bet but player II does not know what card she holds, i.e., player I has arrived at the information set 2:1 at one of the two nodes, but player II doesn't know which one. **This models information available to player II** and is a very important part of game theory.

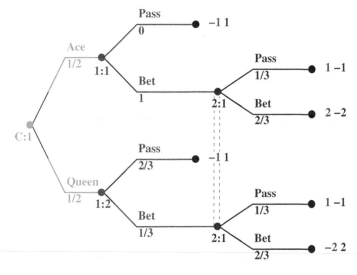

Figure 1.2 A simple poker game.

We cannot (easily) figure out exactly what should happen from the tree. Since Responder does not immediately know if Actor has Ace or Queen, she doesn't know how to respond to a bet. If Actor has Ace she wants to fold, but if Actor has a Queen, she prefers to call. Because of that, Actor does not immediately know what to do if she has a Queen. If Responder is planning to fold, Actor should bet and win $1 instead of losing $1. But if Responder would call a bet, Actor should fold, and accept the loss of only $1 instead of losing $2. Notice that the numbers at the end of the branches represent the payoffs to each player if each branch is followed.

To analyze the game, we convert the tree to a matrix. To do that, we need to understand what the possible strategies are. **A strategy for a player is a plan for what to do starting from the beginning of the game to the last possible move.** You can imagine then why strategies can be so complicated for a full poker game like 5 card stud, or a game of chess. Here is a set of strategies for our game that works for Actor.

PP : pass on Ace and pass on Queen
PB : pass on Ace and bet on Queen
BP : bet on Ace and pass on Queen
BB : bet on Ace and bet on King

These strategies take into account the fact that Actor must have a plan on what to do if she gets an Ace or a Queen.

For Responder it is simpler: She only has to make a choice if there is a bet since she plays second, and there are only two choices, call or fold. Her strategies are

c : if there is a bet, call
f : if there is a bet, fold

We can now calculate the **expected payoff** to Actor for each pair of strategies, getting a description of this as a matrix game, with Actor playing rows:

$$
\begin{array}{c|cc}
 & c & f \\
\hline
PP & -1 & -1 \\
PB & -\frac{3}{2} & 0 \\
BP & \frac{1}{2} & 0 \\
BB & 0 & 1
\end{array}.
$$

The expected payoff is a calculation to see how much a player can expect to win or lose on each play on average if the game is played many times. This is a consequence of what mathematicians call the **law of averages** to account for payoffs depending on chance outcomes. To see how this is calculated for our problem, take for example BP versus c. Then

- If Actor has an Ace she bets and Responder calls; Actor has the higher card so she wins $2.
- If Actor has a Queen, she passes; so she loses $1.

Each of these possibilities happens with probability $\frac{1}{2}$, so the expected outcome is

$$
E(BP, c) = \frac{1}{2}2 + \frac{1}{2}(-1) = \frac{1}{2}.
$$

Similarly, PB versus c means Actor folds with an Ace and bets with a Queen, while Responder always bets. The expected payoff to Actor is $E(PB, c) = \frac{1}{2}(-1) + \frac{1}{2}(-2) = -\frac{3}{2}$. The rest of the entries you can verify.

What would each player do? Remember that the numbers in the matrix represent the expected payoff to Actor. Responder obtains the negative of the payoffs to Actor. This is the hallmark of a zero-sum game.

For Actor, it is clear that *PP* would be a bad strategy since Actor can always do better by playing *BB* or *BP*. This means we could essentially drop *PP* from consideration. For the same reason *PB* should be dropped because Actor would be foolish to play it instead of *BP* or *BB*. This means the matrix can actually be reduced to

$$\begin{array}{c|cc} & c & f \\ \hline BP & \frac{1}{2} & 0 \\ BB & 0 & 1 \end{array}.$$

Now it is not so clear what to do for either player. If Actor decides to play *BP*, then Responder would play *f*; knowing that, Actor would instead play *BB* but then Responder would play *c*. This reasoning never ends until game theory enters the stage.

Let's assume that Actor is not going to always do the same thing but will randomize between *BP* and *BB*. After all, Actor and Responder would be exceedingly bad poker players if they always did the same thing. What does randomize mean? It means that Actor will play *BP* with probability, say *p*, and then *BB* with probability $1 - p$. Naturally, we then have to calculate Actor's expected payoff. We're going to do it first assuming Responder plays *c* and then if Responder plays *f*.

$$E_c(p) = \tfrac{1}{2}p + 0(1 - p) = \tfrac{p}{2}, \text{ and } E_f(p) = 0p + 1(1 - p) = 1 - p.$$

A way to think about this now is Actor will choose *p* so that it doesn't matter whether Responder plays *c* or *f*. This would say Actor is **indifferent** to Responder's choice and it would give the same expected payoff to Actor. This requires

$$\tfrac{p}{2} = 1 - p \implies p = \tfrac{2}{3}.$$

Voila! Actor should play *BP* exactly $\frac{2}{3}$ of the time and *BB*, $\frac{1}{3}$ of the time. Using similar reasoning for Responder, she should call $\frac{2}{3}$ of the time and fold $\frac{1}{3}$ of the time. The **indifference principle**, also called the **equality of payoffs principle**, for Responder says that Responder would choose the probability of playing *c*, namely *q* so that Responder is indifferent between Actor playing *BP* or *BB*. This would lead to the equations

$$q\tfrac{1}{2} = (1 - q) \cdot 1 \implies q = \tfrac{2}{3}.$$

(It is a coincidence here that these probabilities are the same for the two players. This does not always have to be the case.) Thus, in terms of the 4×2 matrix game, the strategies for Actor is to never play *PP* or *PB* and play *BP* with probability $\frac{2}{3}$ and *BB* with probability $\frac{1}{3}$.

But this is a kind of strange way of saying it: if you are playing, and you are looking at a card, what you need to know is, how often should I bet?

- If Actor has a Queen: she should play *BP* exactly $\frac{2}{3}$ of the time, so that two thirds of the time she should pass. Actor should also play *BB* $\frac{1}{3}$ of the time, so bet $\frac{1}{3}$ of the time with a Queen.
- If Actor has an Ace she should bet no matter what.

The upshot is: Actor always bets with the Ace, and bets $\frac{1}{3}$ of the time with a Queen. For Responder, if there is a bet, call with probability $\frac{2}{3}$ and fold with probability $\frac{1}{3}$. This could be implemented, say by flipping an unbalanced coin.

Now we see a very interesting phenomenon that is well known to poker players. Namely, the optimal strategy for Actor has her betting one-third of the time when she has a losing card (Queen). We expected some type of result like this because of the incentive for Actor to bet even when she has a Queen. **Bluffing with positive probability is a part of an optimal strategy when done in the right proportion.**

The **indifference principle** gave us a way of finding the best strategies to play for each player. The reason it works is that when a player doesn't care which strategy the opponent will play, since she gets the same payoff either way, then the player uses a strategy that immunizes against the choices of the opponent. We will see later that this will mean that if either player deviates from the indifference strategy, then the opponent will do better.

Another point to be made about this example is you can see why a complete game of poker with five cards and many rounds of betting, raising, calling, folding, etc., is beyond analysis. Strategies are just too complicated for a complete poker analysis at least with classical computers.

1.2.2 Simplified Baseball

In this focused example of what happens between a pitcher and a batter in baseball, we consider that the pitcher can throw one of two pitches. The batter must guess which of the two pitches he will be faced with. In major league baseball, if the batter guesses wrong, it is almost impossible to hit the ball.

The pitcher, who we label as player I or PI, can throw

i) A fastball
ii) A changeup.

The batter, who we label as player II or PII can guess

i) A fastball is coming
ii) A changeup is coming

Note that batter is forced to make one of these guesses; they can't just wait and see what happens because the pitch comes in so fast. The choices are essentially made simultaneously. Once the players have made their choices, the pitcher pitches and the batter tries to hit the ball. What matters is how likely it is that the batter gets a hit. We will first look at things from the pitcher's point of view, so we record the **probability of the batter being out.** This probability, of course, depends on the two choices the players made, and let's say it depends as follows:

• If Pitcher throws fastball and Batter guesses fastball: 0.6
• If Pitcher throws fastball and Batter guesses changeup: 0.9
• If Pitcher throws changeup, and Batter guesses fastball: 0.8
• If Pitcher throws changeup, and Batter guesses changeup: 0.4

The pitcher's goal is just to maximize the probability of an out, and the batter's goal is to maximize the probability of a hit, which is the same as minimizing the probability of an out. We will arrange this as a matrix. The rows represent the pitcher's possible choices: Fastball (F) and change-up (C). The columns represent the batter's possible choices: Guess fastball (GF) and Guess changeup (GC). The entries are the payoff to the pitcher, which in this case is the probability of an out:

	GF	GC
F	0.6	0.9
C	0.8	0.4

The question is, what should the players do? Well, if players are forced to make firm decisions, we can't figure it out. Someone always wants to change. For instance,

• If Pitcher is playing F and Batter is playing GF, then Pitcher would like to change to C, since her payoff goes up from 0.4 to 0.6.

• If Pitcher is playing C and Batter is playing GF, then Batter would like to change to GC—then the outcome switches form 0.8 to 0.4, meaning Batter is less likely to be out.

You can check the other two cases.

What does this mean? We analyze this by having Pitcher choose the pitch to throw first and then letting Batter respond. The possibilities work out as

• If Pitcher plays F. Then Batter could play GF with outcome 0.6 or GC with an outcome 0.9. Hitter wants a low probability of an out, so she plays GF. GF is PII's best response to PI's choice of F.
• If Pitcher plays C. Then Batter could play GF with outcome 0.8 or GC with outcome 0.4. Hitter wants a low probability of an out, so she plays GC. GC is PII's best response to PI's choice of C.

The best Pitcher can do is play F and Batter plays GF, with outcome 0.6. Naturally, in real life, if a pitcher always threw a fastball, every batter would always play GF and the pitcher would be out of the game in no time. But there are other possible strategies for a pitcher. The pitcher can play randomly! Let's use the most obvious random strategy: Throwing a fastball or changeup each with probability 0.5, maybe by flipping a fair coin to decide. Then

• If Batter plays GF, the expected outcome to Pitcher is $0.5 \times 0.6 + 0.5 \times 0.8 = 0.7$
• If Batter plays GC, the expected outcome to Pitcher is $0.5 \times 0.9 + 0.5 \times 0.4 = 0.65$

Of course Batter then would play GC, because she is happier with the outcome 0.65. Well, 0.65 is better than 0.6, so adding randomness helped the Pitcher! And actually we should have known that from the real world: in baseball, pitchers rarely throw the same pitch all the time; they mix them up and keep the batter guessing.

Now the question is, is Pitcher using the best strategy now? I would argue no: Batter is responding by always guessing changeup, so Pitcher can do better by throwing fastball a bit more often. But, exactly how much more often? Well, again we should look back to how we solved the poker example: we should figure out how Batter should respond to every possible Pitcher strategy, and the Pitcher chooses the strategy that works out best when Batter always uses this response.

Figure 1.3 is a graphical representation of the payoffs to each player as a function of p.

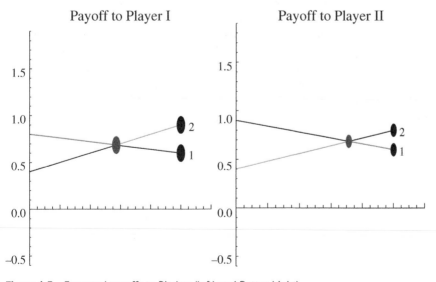

Figure 1.3 Expected payoffs to Pitcher (left) and Batter (right).

The horizontal axis is for the variable p =the probability Pitcher, PI, throws a fastball. So, if $p = 0.3$, this would mean throw a fastball with probability 0.3, and a changeup with probability 0.7. Equivalently, throw a fastball 30% of the time, and a changeup 70% of the time. The vertical axis is the payoff to Pitcher (for the left graph) depending on which line is chosen by Batter and which p is chosen by Pitcher.

If Batter is playing GF, then the payoff to Pitcher is a function of this p:

$$E_{GF}(p) = 0.6p + 0.8(1 - p) = 0.8 - 0.2p.$$

This is line labeled 1 in Figure 1.3 for the payoff to PI. Similarly, if Batter is playing GC, then the payoff is a different function of p:

$$E_{GC}(p) = 0.9p + 0.4(1 - p) = 0.4 + 0.5p.$$

This is line labeled 2 in Figure 1.3 for the payoff to PI. Depending on what Batter does, the payoff to Pitcher is somewhere on either line 1 or line 2, depending on Pitcher's choice of p.

We are forcing Pitcher to choose p first, so Batter will always choose the **lower** of the two lines above that p because Batter wants Pitcher to get the lowest possible payoff. Knowing that, the best thing Pitcher can do is choose the value of p at the intersection of the two lines, which is the value of p for which it doesn't matter what Batter does. This is the value of p for which Pitcher is indifferent to Batter's choices. The indifference principle shows up again.

To find this exact value of p, notice that, at that point, the two possible outcomes must be the same for Batter. This lets us find p by solving an equation!

$$E_{GF}(p) = E_{GC}(p) \Leftrightarrow 0.8 - 0.2p = 0.4 + 0.5p \Leftrightarrow p = \frac{4}{7}.$$

The expected payoff to Pitcher using $p = \frac{4}{7}$ is then $E_{GF}(p) = E_{GC}(p) = 0.6857$. This is clearly better than the expected payoff of 0.65 if $p = 0.5$.

We could then do a similar calculation for PII using the graph on the right of the two lines representing the expected payoff to Batter on the right of the figure, and get that they should play GF with probability $\frac{5}{7}$ and GC with probability $\frac{2}{7}$. Notice that PI wants to get the highest point on the lower of the two lines, while PII wants to get the lowest point on the higher of the two lines. It is then automatic to calculate the expected payoff to batter will be 0.6857. It is not a coincidence that the expected payoff to each player is the same when both players are playing their best strategies.

One of the great applications of game theory to sports is that a team could have an analysis like this for every pitcher and every batter and determine the best strategy for every at bat. In addition, it is possible to see what happens if, say, the pitcher improves his pitches by any amount and then analyze the batter's best strategy to the improvement. This is the content of sports analytics and there is no reason why a similar analysis can't be applied to football, basketball, or any sport.

Here's another example in which it would be foolish to always play the same thing. It will be clear that *randomization of strategies* must be included and is an essential element of games. The simple game of showing fingers illustrates this basic principle.

Example 1.1 *Evens or Odds:*
In this game, each player decides to show one, two, or three fingers. If the total number of fingers shown is even, player I wins +1 and player II loses −1. If the total number of fingers is odd, player I

loses −1, and player II wins +1. The strategies in this game are simple: deciding how many fingers to show. We may represent the payoff matrix as follows:

Evens	Odds		
I/II	1	2	3
1	1	−1	1
2	−1	1	−1
3	1	−1	1

The row player here and throughout this book will always want to maximize his payoff, while the column player wants to **minimize** the payoff to the row player, so that her own payoff is maximized (because it is a zero or constant sum game). The rows are called the **pure strategies** for player I, and the columns are called the **pure strategies** for player II.

The following question arises: How should each player decide what number of fingers to show? If the row player **always** chooses the same row, say, one finger, then player II can **always** win by showing two fingers. No one would be stupid enough to play like that. So what do we do? There is no obvious strategy that will always guarantee a win for either player.

If a player always plays the same strategy, the opposing player can win the game. It seems that the only alternative is for the players to mix up their strategies and play some rows and columns sometimes and other rows and columns at other times. Another way to put it is that the only way an opponent can be prevented from learning about your strategy is if you yourself do not know exactly what pure strategy you will use. That only can happen if you choose a strategy randomly.

Our next example is a simple illustration of strategies in a combinatorial game.

1.2.3 2 × 2 NIM

The game works as follows: Four pennies are set out in two piles of two pennies each. Player I chooses a pile and then decides to remove one or two pennies from the pile chosen. Then player II chooses a pile with at least one penny and decides how many pennies to remove. Then player I starts the second round with the same rules. When both piles have no pennies, the game ends and the loser is the player who removed the last penny. The loser pays the winner one dollar. Figure 1.4 shows all the possibilities for how this came can progress. When the game is drawn as a tree representing the successive moves, just as we did in the poker example, it is called a game in **extensive form**.

It turns out that we can also represent this using a matrix, like we did with the example in Section 1.2.1, we just need to be very careful about what a strategy should be. As discussed in Section 1.2.1, a strategy should be a complete plan for what to do for the whole game, made before the game starts. Here are the possible strategies for each player in 2 × 2 Nim:

Strategies for player I
(1) Play (1,2) then, if at (0,2) play (0,1)
(2) Play (1,2) then if at (0,2) play (0,0)
(3) Play (0,2)

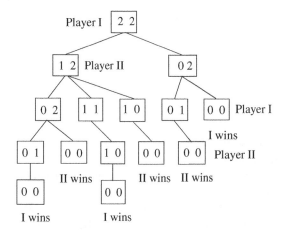

Figure 1.4 2 × 2 Nim tree.

Strategies for II are more involved:

Strategies for player II
(1) If at (1,2) → (0,2); if at (0,2) → (0,1)
(2) If at (1,2) → (1,1); if at (0,2) → (0,1)
(3) If at (1, 2) → (1, 0); if at (0, 2) → (0, 1)
(4) If at (1, 2) → (0, 2); if at (0, 2) → (0, 0)
(5) If at (1, 2) → (1, 1); if at (0, 2) → (0, 0)
(6) If at (1, 2) → (1, 0); if at (0, 2) → (0, 0)

Playing strategies for player I against player II results in the payoff matrix for the game, with the entries representing the payoffs to player I.

Player I/Player II	1	2	3	4	5	6
1	1	1	−1	1	1	−1
2	−1	1	−1	−1	1	−1
3	−1	−1	−1	1	1	1

From here it is pretty clear that PII should play strategy 3, and win the dollar no matter what player I does. There will be no need for randomization and, despite the setup being more complicated, this works out more simply than the baseball example.

This game is not very interesting because there is always a winning strategy for player II and it is pretty clear what it is (The former, but not the latter, is pretty much true for any combinatorial game). Why would player I ever want to play this game? There are actually many games like this (tic-tac-toe is an obvious example) that are not very interesting because their outcome (the winner and the payoff) is determined as long as the players play optimally. Chess is not so obvious an example because the number of strategies is so vast that the game cannot, or has not, been analyzed in this way.

One of the main points in this example is the complexity of the strategies. In a game even as simple as tic-tac-toe, the number of strategies is fairly large, and in a game like chess you can forget about writing them all down.

The Nim example is called a **combinatorial game** because both players know every move, and there is no element of chance involved. A separate branch of game theory considers only combinatorial games but this takes us too far afield and the theory of these games is not considered in this book.[1]

1.3 Mathematical Setup

We are now ready to start making things more precise.

1.3.1 Definition of a Matrix Game

For now our games will have only two players, who we call player I (PI) and player II (PII). The game works by

- PI makes a choice between n possible strategies, labeled $1, \ldots, n$.
- PII simultaneously makes a choice between m strategies, labeled $1, \ldots, m$,
- Once the pair of strategies, i for PI and j for PII is chosen, PI gets a payoff of a_{ij}, and PII gets a payoff of $-a_{ij}$.

The game is determined by the numbers a_{ij}, and we can arrange these numbers in a matrix. These numbers are called the **payoffs to player I** and the matrix is called the **payoff** or **game matrix**. This is a zero sum game, because no matter what strategies are chosen, the sum of the two payoffs to the two players is 0. In a **zero sum game**, whatever one player wins the other loses, and we only need to record the payout to I. The result is

	Player II			
Player I	Strategy 1	Strategy 2	$\cdots$	Strategy m
Strategy 1	a_{11}	a_{12}	$\cdots$	a_{1m}
Strategy 2	a_{21}	a_{22}		a_{2m}
$\vdots$	$\vdots$	$\vdots$	$\vdots$	$\vdots$
Strategy n	a_{n1}	a_{n2}	$\cdots$	a_{nm}

By agreement we place player I as the row player and player II as the column player. We also agree that the numbers in the matrix represent the payoff to player I. In a zero sum game the payoffs to player II would be the negative of those in the matrix so we don't have to record both of those. Of course, if player I has some payoff which is negative, then player II would have a positive payoff.

Summarizing, a **two-person zero sum game** in matrix form means that there is a matrix $A = (a_{ij}), i = 1, \ldots, n, j = 1, \ldots, m$ of real numbers so that if player I, the row player, chooses to play row i and player II, the column player, chooses to play column j, then the payoff to player I is a_{ij} and the payoff to player II is $-a_{ij}$. Both players want to choose strategies that will maximize their individual payoffs. In the matrix, player I will always be the row player and she wants to maximize, while player II will be the column player and she wants to minimize the payoff to player I.

A **constant sum matrix game**, meaning the sum of the two payoffs for each player is the same no matter what strategies are chosen. Constant sum games can also be reduced to the zero sum case: if the sum of the payoffs is C then the game is equivalent to a new game where you start by giving PI the payoff C, then play a new game where PII's payoff is the negative of PI's payoff. So for our purposes constant sum games will work out the same as zero sum games.

---

1 The interested reader can see the book *Winning Ways for your Mathematical Plays* (Academic Press, 1982) by Elwyn R. Berlekamp, John H. Conway, and Richard K. Guy.

For example, the simplified baseball game in Section 1.2.2 is governed by the matrix $A = \begin{bmatrix} 0.6 & 0.9 \\ 0.8 & 0.4 \end{bmatrix}$. Pitcher's payoff was the probability of an out and Batter's payoff was the probability of a hit, and these always add up to 1, not to zero. This is an example of a constant sum game with $C = 1$.

In this section we will focus on finding the solution of a game, and on defining what we mean when we say players are playing their best strategies. The starting point here is that we force I to play first. Then it is possible to figure out how each player should play. We will then turn it around and force II to play first, and see what happens there.

We look at a game with matrix $A = (a_{ij})$ from player I's viewpoint. Player I assumes that player II is playing her best, so II chooses a column j so as to

Minimize a_{ij} over $j = 1, \ldots, n$

for any given row i. Then player I can guarantee that she can choose the row i that will maximize this. So **player I** can **guarantee** that in the **worst possible situation** she can get at least

$$v^- \equiv \max_{i=1,\ldots,n} \min_{j=1,\ldots,m} a_{ij},$$

and we call this number v^- the **lower value of the game**. It is also called player I's gain floor.

Next, consider the game from II's perspective. Player II assumes that player I is playing her best, so that I will choose a row so as to

Maximize a_{ij} over $i = 1, \ldots, n$

for any given column $j = 1, \ldots, m$. **Player II** can therefore choose her column j so as to **guarantee** a loss of no more than

$$v^+ \equiv \min_{j=1,\ldots,m} \max_{i=1,\ldots,n} a_{ij},$$

and we call this number v^+ the **upper value of the game**. It is also called player II's loss ceiling.

In summary, v^- represents the least amount that player I can be guaranteed to receive and v^+ represents the largest amount that player II can guarantee can be lost. This description makes it clear that we should always have $v^- \leq v^+$.

Here is how to find the upper and lower values for any given matrix in a two-person zero sum game with a finite number of strategies for each player.

For each row of the game matrix, find the minimum payoff in each column and write it in a new additional last column. Then the lower value is the largest number in that last column, that is, the maximum over rows of the minimum over columns. Similarly, in each column find the maximum of the payoffs (written in the last row). The upper value is the smallest of those numbers in the last row. Here's the procedure:

$$
\begin{array}{cccc|l}
a_{11} & a_{12} & \cdots & a_{1m} & \longrightarrow \min_j a_{1j} \\
a_{21} & a_{22} & \cdots & a_{2m} & \longrightarrow \min_j a_{2j} \\
\vdots & \vdots & \cdots & \vdots & \\
a_{n1} & a_{n2} & \cdots & a_{nm} & \longrightarrow \min_j a_{nj} \\
\hline
\downarrow & \downarrow & \cdots & \downarrow & \\
\max_i a_{i1} & \max_i a_{i2} & \cdots & \max_i a_{im} & v^- = \text{largest min} \\
& & & & v^+ = \text{smallest max}
\end{array}
$$

Here is the precise definition of the value of a game in pure strategies.

Definition 1.3.1 *A matrix game with matrix $A_{n\times m} = (a_{ij})$ has the lower value* lower value

$$v^- \equiv \max_{i=1,\ldots,n} \min_{j=1,\ldots,m} a_{ij}.$$

and the upper value

$$v^+ \equiv \min_{j=1,\ldots,m} \max_{i=1,\ldots,n} a_{ij},$$

The **game has a value in pure strategies** *if $v^- = v^+$, and we write it as $v = v(A) = v^+ = v^-$.*

One way to look at the value of a game is as a handicap. This means that if the value v is positive, player I should pay player II the amount v in order to make it a **fair game**, with $v = 0$. If $v < 0$, then player II should pay player I the amount $-v$ in order to even things out for player I before the game begins. If a game did have $v^+ = v^-$ it would mean that the players should always play a particular row and column. These games, which may be fairly complicated are not that interesting. Think of Baseball, or Poker, when it is known exactly what the players will do.

Example 1.2 Let's see if there is a value for the examples Poker, Baseball, and 2×2 Nim. First for NIM,

$$
\begin{array}{cccccc|l}
1 & 1 & -1 & 1 & 1 & -1 & \longrightarrow \quad \min = -1 \\
-1 & 1 & -1 & -1 & 1 & -1 & \longrightarrow \quad \min = -1 \\
-1 & -1 & -1 & 1 & 1 & 1 & \longrightarrow \quad \min = -1 \\
\hline
\downarrow & \downarrow & \downarrow & \downarrow & \downarrow & \downarrow & \boxed{v^- = -1} \\
\end{array}
$$

$$\max = 1 \quad \max = 1 \quad \max = -1 \quad \max = 1 \quad \max = 1 \quad \max = 1 \quad \boxed{v^+ = -1}$$

We see that $v^- =$ largest min $= -1$ and $v^+ =$ smallest max $= -1$. This says that $v^+ = v^- = -1$, and so 2×2 NIM has value $v = -1$.

For Poker, our matrix was

	c	f
PP	0	0
PB	$-\frac{1}{2}$	1
BP	$\frac{1}{2}$	0
BB	0	1

.

We calculate $v^- = \max\{0, -\frac{1}{2}, 0, 0\} = 0$, and $v^+ = \min\{\frac{1}{2}, 1\} = \frac{1}{2}$. This game does not have a value in pure strategies.

For Baseball we had the matrix

	GF	GC
F	0.6	0.9
C	0.8	0.4

and we calculate $v^- = \max\{0.6, 0.4\} = 0.6, v^+ = \min\{0.8, 0.9\} = 0.8$. This game also does not have a value in pure strategies.

We have mentioned that the most that player I can be guaranteed to win should be less than (or equal to) the most that player II can be guaranteed to lose, (i.e., $v^- \le v^+$), Here is a quick verification of this fact.

Claim: $v^- \leq v^+$. For any column j we know that for any fixed row i, $\min_j a_{ij} \leq a_{ij}$, and so taking the max of both sides over rows, we obtain

$$v^- = \max_i \min_j a_{ij} \leq \max_i a_{ij}.$$

This is true for any column $j = 1, \ldots, m$. The left side is just a number (i.e., v^-) independent of i as well as j, and it is smaller than the right side for any j. But this means that $v^- \leq \min_j \max_i a_{ij} = v^+$, and we are done.

1.3.2   Saddle Points: What It Means to be Optimal

Now here is a precise definition of a (pure) saddle point involving only the payoffs, which basically tells the players what to do in order to obtain the value of the game when $v^+ = v^-$.

Definition 1.3.2   *A saddle point in pure strategies is a pair of strategies row i^* for PI and column j^* for PII such that, if the players choose (i^*, j^*), then neither one wants to change to another row or column. Equivalently, $a_{i^*j^*}$ is the max in its column and min in its row. Written mathematically,*

$$a_{ij^*} \leq a_{i^*j^*} \leq a_{i^*j}, \text{ for all rows } i = 1, \ldots, n \text{ and columns } j = 1, \ldots, m. \tag{1.3.1}$$

In words, (i^*, j^*) is a saddle point if when player I deviates from row i^*, but II still plays j^*, then player I will get less. Also, if player II deviates from column j^* but I sticks with i^*, then player I will do better.

You can spot a **saddle point in a matrix** (if there is one) as the entry which is **simultaneously the smallest in a row and largest in the column.** A matrix may have none, one, or more than one saddle point. here's a condition which guarantees at least one saddle.

Lemma 1.3.3   *A game will have a saddle point in pure strategies if and only if*

$$v^- = \max_i \min_j a_{ij} = \min_j \max_i a_{ij} = v^+. \tag{1.3.2}$$

Proof: If (1.3.1) is true, then

$$v^+ = \min_j \max_i a_{ij} \leq \max_i a_{ij^*} \leq a_{i^*j^*} \leq \min_j a_{i^*j} \leq \max_i \min_j a_{ij} = v^-.$$

But $v^- \leq v^+$ always, and so we have equality throughout and $v = v^+ = v^- = a_{i^*j^*}$.
  On the other hand, if (1.3.2) holds, then

$$v^+ = \min_j \max_i a_{ij} = \max_i \min_j a_{ij} = v^-.$$

Let j^* be such that $v^+ = \max_i a_{ij^*}$ and i^* such that $v^- = \min_j a_{i^*j}$. Then

$$a_{i^*j} \geq v^- = v^+ \geq a_{ij^*}, \text{ for any } i = 1, \ldots, n, \ j = 1, \ldots, m.$$

In addition, taking $j = j^*$ on the left, and $i = i^*$ on the right, gives $a_{i^*j^*} = v^+ = v^-$. This satisfies the condition for (i^*, j^*) to be a saddle point. ∎

When a saddle point exists in pure strategies (which will happen if $v^+ = v^-$), (1.3.1) says that if any player deviates from playing her part of the saddle, then the other player can take advantage and improve his payoff. In this sense, each part of a saddle is a **best response** to the other.

In the game of NIM we know there is a saddle point and it occurs at column 3, but any row. This shows that saddle points are not necessarily unique.

We now know that $v^+ \geq v^-$ is always true. We also know how to play if $v^+ = v^-$. The issue is what do we do if $v^+ > v^-$. That will be a topic we study in the next section.

Problems

1.1 There are 100 bankers lined up in each of 100 rows. Pick the richest banker in each row. Javier is the poorest of those. Pick the poorest banker in each column. Raoul is the richest of those. Who is richer: Javier or Raoul?

1.2 Suppose that two companies are both thinking about introducing competing products into the marketplace. They choose the time to introduce the product, and their choices are 1 month, 2 months, or 3 months. The payoffs correspond to market share.

I/II	1	2	3
1	0.5	0.6	0.8
2	0.4	0.5	0.9
3	0.2	0.4	0.5

For instance, if player I introduces the product in 3 months and player II introduces it in 2 months, then it will turn out that player I will get 40% of the market. The companies want to introduce the product in order to maximize their market share. This is also a constant sum game. Find v^+, v^- and a saddle point if any.

1.3 In a Nim game start with 4 pennies. Each player may take 1 or 2 pennies from the pile. Suppose player I moves first. The game ends when there are no pennies left and the player who took the last penny pays 1 to the other player.
   (a) Draw the game as we did in 2×2 Nim.
   (b) Write down all the strategies for each player and then the game matrix.
   (c) Find v^+, v^-. Would you rather be player I or player II?

1.4 In the game rock–paper–scissors both players select one of these objects simultaneously. The rules are as follows: paper beats rock, rock beats scissors, and scissors beats paper. The losing player pays the winner $1 after each choice of object. If both choose the same object the payoff is 0.
   (a) What is the game matrix?
   (b) Find v^+ and v^- and determine whether a saddle point exists in pure strategies, and if so, find it.

1.5 Each of two players must choose a number between 1 and 5. If a player's choice = opposing player's choice +1, she loses $2; if a player's choice $\geq$ opposing player's choice +2, she wins $1. If both players choose the same number the game is a draw.
   (a) What is the game matrix?

(b) Find v^+ and v^- and determine whether a saddle point exists in pure strategies, and if so, find it.

1.6 Each player displays either one or two fingers and simultaneously guesses how many fingers the opposing player will show. If both players guess either correctly or incorrectly, the game is a draw. If only one guesses correctly, he wins an amount equal to the total number of fingers shown by both players. Each pure strategy has two components: the number of fingers to show, the number of fingers to guess. Find the game matrix, v^+, v^-, and optimal pure strategies if they exist.

1.7 Let x be an unknown number and consider the matrices

$$A = \begin{bmatrix} 0 & x \\ 1 & 2 \end{bmatrix}, \quad B = \begin{bmatrix} 2 & 1 \\ x & 0 \end{bmatrix}.$$

Show that no matter what x is, each matrix has a pure saddle point.

1.8 If we have a game with matrix A and we modify the game by adding a constant C to every element of A, call the new matrix $A + C$, is it true that $v^+(A + C) = v^+(A) + C$?
   (a) If it happens that $v^-(A + C) = v^+(A + C)$, will it be true that $v^-(A) = v^+(A)$, and conversely?
   (b) What can you say about the optimal pure strategies for $A + C$ compared to the game for just A?

1.9 Consider the square game matrix $A = (a_{ij})$ where $a_{ij} = i - j$ with $i = 1, 2, \ldots, n$, and $j = 1, 2, \ldots, n$. Show that A has a saddle point in pure strategies. Find them and find $v(A)$.

1.10 Player I chooses 1, 2, or 3 and player II guesses which number I has chosen. The payoff to I is |I's number − II's guess|. Find the game matrix. Find v^- and v^+.

1.11 In a baseball example player I, the batter, expects the pitcher (player II) to throw a fastball (F), a slider(S), or a curveball(C). This is the game matrix.

I/II	F	C	S
F	0.30	0.25	0.20
C	0.26	0.33	0.28
S	0.28	0.30	0.33

Find v^+ and v^-.

1.12 In a football game the offense has two strategies: run or pass. The defense also has two strategies: defend against the run, or defend against the pass. A possible game matrix is

$$A = \begin{bmatrix} 3 & 6 \\ x & 0 \end{bmatrix}.$$

This is the game matrix with the offense as the row player I. The numbers represent the number of yards gained on each play. The first row is run, the second is pass. The first column is defend the run and the second column is defend the pass. Assuming that $x > 0$, find the value of x so that this game has a saddle point in pure strategies.

1.4   Mixed Strategies

In the Baseball example and the Poker example we found that they do not have a saddle point in pure strategies. This means that if a player always plays a row (or column) all the time, the other player can take advantage of it. The only way to avoid this is if a player randomizes her plays. This implies that not only does the opponent not know which row (or column) will be played with certainty, but the same will be true of the player who is choosing the row (or column). The players will be mixing up their choices of rows and columns, i.e., they will choose mixed strategies.

1.4.1   Definition of Mixed Strategies

Definition 1.4.1   *A **mixed strategy** is a vector $X = (x_1, \ldots, x_n)$ for player I and $Y = (y_1, \ldots, y_m)$ for player II, where*

$$x_i \geq 0, \quad \sum_{i=1}^{n} x_i = 1 \ \text{ and } y_j \geq 0, \quad \sum_{j=1}^{m} y_j = 1.$$

The components x_i represent the probability that row i will be used by player I, so $x_i = Prob$ (I uses row i), and y_j the probability column j will be used by player II, that is, $y_j = Prob(II$ uses row j). The set of mixed strategies with k components is denoted by

$$S_k \equiv \left\{ (z_1, z_2, \ldots, z_k) \mid z_i \geq 0, i = 1, 2, \ldots, k, \sum_{i=1}^{k} z_i = 1 \right\}.$$

In this terminology, a mixed strategy for player I is any element $X \in S_n$ and for player II any element $Y \in S_m$.

If player I uses the mixed strategy $X = (x_1, \ldots, x_n) \in S_n$ then she will use row i on each play of the game with probability x_i. Every pure strategy is also a mixed strategy by choosing all the probability to be concentrated at the row or column that the player wants to always play. For example, if player I wants to always play row 3, then the mixed strategy she would choose is $X = (0, 0, 1, 0, \ldots, 0)$. Therefore, allowing the players to choose mixed strategies permits many more choices, and the mixed strategies make it possible to mix up the pure strategies used. The set of mixed strategies contains the set of all pure strategies in this sense, and it is a generalization of the idea of strategy.

Now, if the players use mixed strategies the payoff can be calculated only in the **expected** sense. That means the game payoff will represent what each player can expect to receive and will actually receive on average only if the game is played many, many times. More precisely, we calculate as follows.

Definition 1.4.2   *Given a choice of mixed strategy $X \in S_n$ for player I and $Y \in S_m$ for player II, the* **expected payoff** *to player I of the game is*

$$E(X, Y) = \sum_{i=1}^{n} \sum_{j=1}^{m} a_{ij} \ Prob(I \text{ uses } i \text{ and } II \text{ uses } j)$$

$$= \sum_{i=1}^{n} \sum_{j=1}^{m} a_{ij} \ Prob(I \text{ uses } i) Prob(II \text{ uses } j)$$

$$= \sum_{i=1}^{n} \sum_{j=1}^{m} x_i a_{ij} y_j = X \, A \, Y^T.$$

In a zero sum two-person game the expected payoff to player II would be $-E(X, Y)$. *The independent choice of strategy by each player justifies the fact that*

$$\text{Prob}(I \text{ uses } i \text{ and } II \text{ uses } j) = \text{Prob}(I \text{ uses } i)\text{Prob}(II \text{ uses } j).$$

Notation 1.4.3 *For an* $n \times m$ *matrix* $A = (a_{ij})$ *we denote the jth column vector of A by* A_j *and the ith row vector of A by* $_iA$. *So*

$$A_j = \begin{bmatrix} a_{1j} \\ a_{2j} \\ \vdots \\ a_{nj} \end{bmatrix} \text{ and } _iA = \left(a_{i1}, a_{i2}, \dots, a_{im} \right).$$

If player I decides to use the pure strategy $X = (0, \dots, 0, 1, 0, \dots, 0)$ *with row i used 100% of the time and player II uses the mixed strategy Y, we denote the expected payoff by* $E(i, Y) = {}_iA \cdot Y^T$. *Similarly, if player II decides to use the pure strategy* $Y = (0, \dots, 0, 1, 0, \dots, 0)$ *with column j used 100% of the time, we denote the expected payoff by* $E(X, j) = XA_j$. *We may also write*

$$E(i, Y) = {}_iA \cdot Y^T = \sum_{j=1}^{m} a_{ij}y_j, \quad E(X, j) = X A_j = \sum_{i=1}^{n} x_i a_{ij}, \text{ and } E(i, j) = a_{ij}.$$

Notice too that

$$E(X, Y) = \sum_{i=1}^{n} x_i E(i, Y) = \sum_{j=1}^{m} E(X, j)y_j.$$

In $E(i, Y)$ *we say that row i is played against Y and in* $E(X, j)$ *we say X is played against column j.*

1.4.2 Optimal Mixed Strategies

In the matrix zero sum game the goals are that player I wants to maximize her expected payoff and player II wants to minimize the expected payoff to I using mixed strategies.

We may define the **upper and lower values of the mixed game** as

$$v^+ = \min_{Y \in S_m} \max_{X \in S_n} XAY^T, \text{ and } v^- = \max_{X \in S_n} \min_{Y \in S_m} XAY^T.$$

We will see shortly, however, that this is really not needed because it is always true that $v^+ = v^-$ when we allow mixed strategies. Of course, we have seen that this is not true when we permit only pure strategies.

We can define what we mean by a saddle point in mixed strategies.

Definition 1.4.4 *A* **saddle point in mixed strategies** *is a pair* (X^*, Y^*) *of probability vectors* $X^* \in S_n, Y^* \in S_m$, *which satisfies*

$$E(X, Y^*) \le E(X^*, Y^*) \le E(X^*, Y), \quad \forall (X \in S_n, Y \in S_m). \tag{1.4.1}$$

If player I decides to use a strategy other than X^* *but player II still uses* Y^*, *then I receives an expected payoff smaller than that obtainable by sticking with* X^*. *A similar statement holds for player II. So* (X^*, Y^*) *is an* **equilibrium** *in this sense.*

We will see in the next section that a saddle point in mixed strategies always exists. That's the good news! But first we deal with the more practical question of actually finding these saddle points. And before that comes an even simpler question: how do you check that a pair of mixed strategies (X^*, Y^*) really is a saddle point, once you think you've found it? This is an issue because Definition 1.4.4 would require you to check the outcome of infinitely many possible strategies, since the sets S_n, S_m are infinite. The following theorem says it is enough to check only the pure strategies.

Theorem 1.4.5 *A pair of mixed strategies X^*, Y^* for a zero sum matrix game is a saddle point if and only if*

$$E(i, Y^*) \le E(X^*, Y^*) \le E(X^*, j)$$

for all $i = 1, 2, \ldots, n, j = 1, 2, \ldots, m$.

Proof: If (X^*, Y^*) is a saddle point, then the inequalities hold by substituting $X = (0, 0, \ldots, 1, 0, 0)$, $Y = (0, 1, 0, \ldots, 0)$ with 1 in the ith place for X and 1 in the jth place for Y, into (1.4.1). To show the other direction, assume the inequalities of the theorem hold. We must show that (1.4.1) holds for all (mixed) X and Y. So, choose a strategy $X = (x_1, \ldots, x_n)$. Then

$$E(X, Y^*) = \sum_{i=1}^{n} x_i E(i, Y^*) \le \sum_{i=1}^{n} x_i E(X^*, Y^*) = E(X^*, Y^*),$$

where the middle inequality holds by the condition in the theorem, and the last equality because $\sum_{i=1}^{n} x_i = 1$ since $X = (x_1, \ldots, x_n)$ is a probability vector,
    Now choose a strategy $Y = (y_1, \ldots, y_m)$. Then

$$E(X^*, Y) = \sum_{j=1}^{m} y_j E(X^*, j) \ge \sum_{j=1}^{m} y_j E(X^*, Y^*) = E(X^*, Y^*),$$

where again the middle inequality holds by the condition in the theorem, and the last equality because $\sum_{j=1}^{m} y_j = 1$ since $Y = (y_1, \ldots, y_m)$ is a probability vector. ∎

Example 1.3   Consider the matrix game defined by

$$A = \begin{bmatrix} -1 & 3 & 2 \\ 3 & 1 & 1 \\ 2 & -2 & 1 \end{bmatrix}.$$

We claim that $X* = (\frac{2}{5}, \frac{3}{5}, 0), Y^* = (\frac{1}{5}, 0, \frac{4}{5})$ is a saddle point in mixed strategies. We don't know how to find this yet but we can check that it is a saddle point by verifying the inequalities in the theorem. First,

$$E(X^*, Y^*) = [\tfrac{2}{5}, \tfrac{3}{5}, 0] \begin{bmatrix} -1 & 3 & 2 \\ 3 & 1 & 1 \\ 2 & -2 & 1 \end{bmatrix} \begin{bmatrix} \tfrac{1}{5} \\ 0 \\ \tfrac{4}{5} \end{bmatrix} = \frac{7}{5}.$$

Now check what happens if each player switches to a pure strategy:

$$E(1, Y^*) = \tfrac{7}{5}, E(2, Y^*) = \tfrac{7}{5}, E(3, Y^*) = \tfrac{6}{5}. \text{ All} \le \tfrac{7}{5}$$
$$\implies \text{PI does not want to change from } X^*.$$
$$E(X^*, 1) = \tfrac{7}{5}, E(X^*, 2) = \tfrac{9}{5}, E(X^*, 3) = \tfrac{7}{5}. \text{ All} \ge \tfrac{7}{5}$$
$$\implies \text{PII does not want to change from } Y^*.$$

In fact, we can do this calculation in a somewhat easier way by calculating X^*A:

$$\left[\frac{2}{5}, \frac{3}{5}, 0\right] \begin{bmatrix} -1 & 3 & 2 \\ 3 & 1 & 1 \\ 2 & -2 & 1 \end{bmatrix} = \left[\frac{7}{5}, \frac{9}{5}, \frac{7}{5}\right],$$

Against X^* player II can only choose between these three values, so PII can't get lower than $\frac{7}{5}$. Similarly,

$$\begin{bmatrix} -1 & 3 & 2 \\ 3 & 1 & 1 \\ 2 & -2 & 1 \end{bmatrix} \begin{bmatrix} \frac{1}{5} \\ 0 \\ \frac{4}{5} \end{bmatrix} = \begin{bmatrix} \frac{7}{5} \\ \frac{7}{5} \\ \frac{6}{5} \end{bmatrix}.$$

Against Y^*, player I can only choose between these three values, and PI can't get more than $\frac{7}{5}$. This says (X^*, Y^*) is indeed a saddle point in mixed strategies.

Some of the observations in the previous example are important enough that we should state them:

Theorem 1.4.6 *Let $A = (a_{ij})$ be an $n \times m$ game with value $v(A)$. Let $X^* \in S_n$ be a strategy for player I and $Y^* \in S_m$ be a strategy for player II. Then*

1. $v(A) \geq \min_j E(X^*, j)$.
2. $v(A) \leq \max_i E(i, Y^*)$.
3. *If $\min_j E(X^*, j) = \max_i E(i, Y^*)$, then (X^*, Y^*) is a saddle point.*
4. *if v is any number and (X^*, Y^*) are strategies satisfying $E(i, Y^*) \leq v \leq E(X^*, j)$ for all i, j, then $v = value(A)$ and (X^*, Y^*) is a saddle point.*

The last statement is not obvious because it depends on a result that is the major result of game theory, namely the von Neumann minimax theorem. In fact, up to this point we have assumed saddle points exist and we are working toward finding them under this assumption. But, we have to prove they exist and moreover that if they exist then the game must have a value $v(A)$ in mixed strategies. All this is left for later in this chapter.

The next corollary says that only one player really needs to play mixed strategies in order to calculate $v(A)$. It also says that $v(A)$ is always between the pure strategy lower and upper values v^- and v^+.

Corollary 1.4.7

$$v(A) = \min_{Y \in S_m} \max_{1 \leq i \leq n} E(i, Y) = \max_{X \in S_n} \min_{1 \leq j \leq m} E(X, j).$$

In addition, the value of the game with mixed strategies is always between the lower and upper values with pure strategies, i.e., $v^- = \max_i \min_j a_{ij} \leq v(A) \leq \min_j \max_i a_{ij} = v^+$.

Be aware of the fact that not only are the min and max in the corollary being switched (and there is no reason to think this can be done in general) but also the sets over which the min and max are taken are changing. The second part of the corollary is immediate from the first part since

$$v(A) = \min_{Y \in S_m} \max_{1 \leq i \leq n} E(i, Y) \leq \min_{1 \leq j \leq m} \max_{1 \leq i \leq n} a_{ij} = v^+$$

and

$$v(A) = \max_{X \in S_n} \min_{1 \le j \le m} E(X, j) \ge \max_{1 \le i \le n} \min_{1 \le j \le m} a_{ij} = v^-.$$

Remember that $\min_{S_m} \le \min_j$ because mixed strategies S_m always contain all pure strategies. Similarly, $\max_i \le \max_{S_n}$. In the next example we will see how to use the last statement of the theorem.

Example 1.4 Evens and Odds: In the game of evens or odds each player shows 1, 2, or 3 fingers. If the total number of fingers shown is even, player I is paid 1 by player II. If the total number is odd, player I pays player II 1. The game matrix is

I/II	1	2	3
1	1	−1	1
2	−1	1	−1
3	1	−1	1

We calculate that $v^- = -1$ and $v^+ = +1$, so this game does not have a saddle point using only pure strategies. But it does have a value and saddle point using mixed strategies. Suppose that v is the value of this game and $(X^* = (x_1, x_2, x_3), Y^* = (y_1, y_2, y_3))$ is a saddle point. According to Theorem 1.4.6 these quantities should satisfy

$$E(i, Y^*) = {}_iA\, Y^{*T} = \sum_{j=1}^{3} a_{ij} y_j \le v \le E(X^*, j) = X^* A_j = \sum_{i=1}^{3} x_i a_{ij},$$

$$i = 1, 2, 3,\ j = 1, 2, 3.$$

Using the values from the matrix, we have the system of inequalities

$$y_1 - y_2 + y_3 \le v, -y_1 + y_2 - y_3 \le v, \text{ and } y_1 - y_2 + y_3 \le v,$$
$$x_1 - x_2 + x_3 \ge v, -x_1 + x_2 - x_3 \ge v, \text{ and } x_1 - x_2 + x_3 \ge v.$$

Let's go through finding the strategy X^*. We are looking for numbers x_1, x_2, x_3, and v satisfying $x_1 - x_2 + x_3 \ge v, -x_1 + x_2 - x_3 \ge v$, as well as $x_1 + x_2 + x_3 = 1$ and $x_i \ge 0, i = 1, 2, 3$. But then $x_1 = 1 - x_2 - x_3$, and we reduce to finding x_1, x_2, and v so that

$$1 - 2x_2 \ge v \text{ and } -1 + 2x_2 \ge v \implies -v \ge 1 - 2x_2 \ge v \implies v \le 0.$$

Similarly, looking for $y_1, y_2, y_3 = 1 - y_1 - y_2$ we get

$$1 - 2y_2 \le v \text{ and } -1 + 2y_2 \le v \implies v \ge 1 - 2y_2 \ge -v \implies v \ge 0.$$

This means we must have $v = 0$. Then $x_2 = y_2 = \frac{1}{2}$. This would force $x_1 + x_3 = \frac{1}{2}$ as well.

Instead of substituting for x_1, substitute $x_2 = 1 - x_1 - x_3$ hoping to be able to find x_1 or x_3. You would see that we would once again get $x_1 + x_3 = \frac{1}{2}$. Something is going on with x_1 and x_3, and we don't seem to have enough information to find them. But we can see from the matrix that it doesn't matter whether player I shows one or three fingers! The payoffs in all cases are the same. This means that row 3 (or row 1) is a redundant strategy and we might as well drop it. (We can say the same for column 1 or column 3.) If we drop row 3 we perform the same set of calculations but we quickly find that $x_2 = \frac{1}{2} = x_1$. Now we have our candidates for the saddle points and value, namely, $v = 0, X^* = \left(\frac{1}{2}, \frac{1}{2}, 0\right)$ and also, in a similar way $Y^* = \left(\frac{1}{2}, \frac{1}{2}, 0\right)$. Check that with these candidates the inequalities of Theorem 1.4.6 are satisfied and so they are the actual value and saddle.

However, it is important to remember that with all three rows and columns, the theorem does not give a single characterization of the saddle point. Indeed, there are an infinite number of saddle points, $X^* = (x_1, \frac{1}{2}, \frac{1}{2} - x_1), 0 \le x_1 \le \frac{1}{2}$ and $Y^* = (y_1, \frac{1}{2}, \frac{1}{2} - y_1), 0 \le y_1 \le \frac{1}{2}$. This makes sense because whether player I or player II shows 1 or 3 fingers doesn't really matter as long as 2 fingers is shown 50% of the time. Nevertheless, there is only one value for this, or any matrix game, and it is $v = 0$ in the game of odds and evens.

Later we will see that the theorem gives a method for solving any matrix game if we pair it up with another theory, namely, linear programming, which is a way to optimize a linear function over a set with linear constraints. Linear programming will accommodate the more difficult problem of solving a system of inequalities.

The next theorem give us a way of solving games algebraically without having to solve inequalities.

Theorem 1.4.8 **(Equilibrium Theorem)** *If Y is optimal for II and $y_j > 0$, then $E(X,j) = value(A)$ for any optimal mixed strategy X for I. Similarly, if X is optimal for I and $x_i > 0$, then $E(i, Y) = value(A)$ for any optimal Y for II. In symbols,*

$$y_j > 0 \implies E(X,j) = v(A), \text{ and } x_i > 0 \implies E(i, Y) = v(A).$$

Thus, if any optimal mixed strategy for a player has a strictly positive probability of using a row or a column, then that row or column played against any optimal opponent strategy will yield the value. Secondly, if there are two or more rows, say i and k, with positive probability of being played, then $E(i, Y) = E(k, Y)$ since they are both equal to $v(A)$. A similar statement holds for columns. This would say that player II is indifferent to player I playing row i or row k. Therefore, this theorem is also called the **indifference principle**. We have already seen how useful this is in Section 1.2.1.

Proof: If it happens that (X^*, Y^*) are optimal and there is a component of $X^* = (x_1, \ldots, x_k^*, \ldots, x_n)$, say, $x_k^* > 0$ but $E(k, Y^*) < v(A)$, then multiplying both sides of $E(k, Y^*) < v(A)$ by x_k^* yields $x_k^* E(k, Y^*) < x_k^* v(A)$. Now, it is always true that for any row $i = 1, 2, \ldots, n$,

$$E(i, Y^*) \le v(A), \text{ which implies that } x_i E(i, Y^*) \le x_i v(A).$$

But then, by adding, we get

$$\sum_{i=1, i \ne k}^{n} x_i E(i, Y^*) + x_k^* E(k, Y^*) = \sum_{i=1}^{n} x_i E(i, Y^*) < \sum_{i=1}^{n} x_i v(A) = v(A).$$

We see that, under the assumption $E(k, Y^*) < v(A)$, we have

$$v(A) = E(X^*, Y^*) = \sum_{i=1}^{n} \sum_{j=1}^{m} x_i a_{ij} y_j = \sum_{i=1}^{n} x_i E(i, Y^*) < v(A),$$

which is a contradiction. But this means that if $x_k^* > 0$ we must have $E(k, Y^*) = v(A)$.

Similarly, suppose $E(X^*, j) > v(A)$ where $Y^* = (y_1^*, \ldots, y_j^*, \ldots, y_m^*), y_j^* > 0$. Then

$$v(A) = E(X^*, Y^*) = \sum_{\ell=1}^{m} E(X^*, \ell) y_\ell^* > v(A) y_j^* + \sum_{\ell=1, \ell \ne j}^{m} v(A) y_\ell^* = v(A)$$

again a contradiction. Hence $E(X^*, j) > v(A) \implies y_j^* = 0$. ∎

Example 1.5 For an illustration of the use of the theorem let's consider the matrix $A = \begin{bmatrix} 3 & 2 \\ -2 & 3 \end{bmatrix}$.
If we assume $Y = (y_1, y_2)$ with both $y_j > 0$, then we know

$$E(X, 1) = 3x_1 - 2x_2 = v \text{ and } E(X, 2) = 2x_1 + 3x_2 = v.$$

This says $3x_1 - 2x_2 = 2x_1 + 3x_2$. Since $X = (x_1, x_2)$ is a strategy, $x_1 + x_2 = 1$ so that $x_1 = \frac{5}{6}, x_2 = \frac{1}{6}, v = \frac{13}{6}$. Similarly,

$$3y_1 + 2y_2 = -2y_1 + 3y_2 \implies y_1 = \frac{1}{6}, y_2 = \frac{5}{6}.$$

The next remark summarizes several important and useful results to find optimal strategies and the value of a game.

Remark 1.4.9 *Properties of Optimal Strategies*

1. *A number v is the value of the game and (X, Y) is a saddle point if and only if $E(i, Y) \leq v \leq E(X, j), i = 1, \ldots, n, j = 1, \ldots, m$.*
2. *If X is a strategy for player I and $value(A) \leq E(X, j), j = 1, \ldots, m$, then X is optimal for player I.*
3. *If Y is a strategy for player II and $value(A) \geq E(i, Y), i = 1, \ldots, m$, then Y is optimal for player II.*
4. *If X is any optimal strategy for player I and $E(X, j) > value(A)$ for some column j, then for any optimal strategy Y for player II, we must have $y_j = 0$. Player II would never use column j in any optimal strategy for player II. Similarly, if Y is any optimal strategy for player II and $E(i, Y) < value(A)$, then any optimal strategy X for player I must have $x_i = 0$. If row i for player I gives a payoff when played against an optimal strategy for player II strictly below the value of the game, then player I would never use that row in any optimal strategy for player I. In symbols, if $(X = (x_i), Y = (y_j))$ is optimal, then*

$$E(X, j) > v(A) \implies y_j = 0, \text{ and } E(i, Y) < v(A) \implies x_i = 0.$$

5. *If player I has more than one optimal strategy, then player I's set of optimal strategies is a convex, closed, and bounded set. Also, if player II has more than one optimal strategy, then player II's set of optimal strategies is a convex, closed, and bounded set.*

1.4.3 Best Response Strategies

If you are playing a game and you determine, in one way or another, that your opponent is using a particular strategy, or is assumed to use a particular strategy, then what should you do? To be specific, suppose that you are player I and you know, or simply assume, that player II is using the mixed strategy Y, optimal or not for player II. In this case you should play the mixed strategy X that maximizes $E(X, Y)$. This strategy that you use would be a **best response** to the use of Y by player II. The best response strategy to Y may not be the same as what you would use if you knew that player II were playing optimally; that is, it may not be a part of a saddle point. Here is the precise definition.

Definition 1.4.10 *A mixed strategy X^* for player I is a **best response strategy** to the strategy Y for player II if it satisfies*

$$\max_{X \in S_n} E(X, Y) = \sum_{i=1}^{n} \sum_{j=1}^{m} x_i^* a_{ij} y_j = E(X^*, Y).$$

A mixed strategy Y^* for player II is a best response strategy to the strategy X for player I if it satisfies

$$\min_{Y \in S_n} E(X, Y) = \sum_{i=1}^{n} \sum_{j=1}^{m} x_i a_{ij} y_j^* = E(X, Y^*).$$

Incidentally, if (X^*, Y^*) is a saddle point of the game, then X^* is the best response to Y^*, and vice versa. Unfortunately, knowing this doesn't provide a good way to calculate X^* and Y^* because they are **both** unknown at the start.

Example 1.6 Consider the 3×3 game

$$A = \begin{bmatrix} 1 & 1 & 1 \\ 1 & 2 & 0 \\ 1 & 0 & 2 \end{bmatrix}.$$

The saddle point is $X^* = \left(0, \frac{1}{2}, \frac{1}{2}\right) = Y^*$ and $v(A) = 1$. Now suppose that player II, for some reason, thinks she can do better by playing $Y = \left(\frac{1}{4}, \frac{1}{4}, \frac{1}{2}\right)$. What is an optimal response strategy for player I? Let $X = (x_1, x_2, 1 - x_1 - x_2)$. Calculate

$$E(X, Y) = XAY^T = -\frac{x_1}{4} - \frac{x_2}{2} + \frac{5}{4}.$$

We want to maximize this as a function of x_1, x_2 with the constraints $0 \leq x_1, x_2 \leq 1$. We see right away that $E(X, Y)$ is maximized by taking $x_1 = x_2 = 0$ and then necessarily $x_3 = 1$. Hence, the best response strategy for player I if player II uses $Y = \left(\frac{1}{4}, \frac{1}{4}, \frac{1}{2}\right)$ is $X^* = (0, 0, 1)$. Then, using this strategy, the expected payoff to I is $E(X^*, Y) = \frac{5}{4}$, which is larger than the value of the game $v(A) = 1$. So that is how player I should play if player II decides to deviate from the optimal Y. This shows that any deviation from a saddle could result in a better payoff for the opposing player. However, if one player knows that the other player will not use her part of the saddle, then the best response may not be the strategy used in the saddle. In other words, if (X^*, Y^*) is a saddle point, the best response to $Y \neq Y^*$ may not be X^*, but some other X, even though it will be the case that $E(X^*, Y) \geq E(X^*, Y^*)$.

Because $E(X, Y)$ is linear in each strategy when the other strategy is fixed, the best response strategy for player I will usually be a pure strategy. For instance, if Y is given, then $E(X, Y) = ax_1 + bx_2 + cx_3$, for some values a, b, c that will depend on Y and the matrix. The maximum payoff is then achieved by looking at the largest of a, b, c, and taking $x_i = 1$ for the x multiplying the largest of a, b, c, and the remaining values of $x_j = 0$. How do we know that? Well, to show you what to do in general, we will show that

$$\max\{ax_1 + bx_2 + cx_3 \mid x_1 + x_2 + x_3 = 1, \ x_1, x_2, x_3 \geq 0\} \tag{1.4.2}$$
$$= \max\{a, b, c\}.$$

Suppose that $\max\{a, b, c\} = c$ for definiteness. Now, by taking $x_1 = 0, x_2 = 0, x_3 = 1$, we get

$$\max\{ax_1 + bx_2 + cx_3 \mid x_1 + x_2 + x_3 = 1, \ x_1, x_2, x_3 \geq 0\}$$
$$\geq a \cdot 0 + b \cdot 0 + c \cdot 1 = c.$$

On the other hand, since $x_1 + x_2 + x_3 = 1$, we see that

$$ax_1 + bx_2 + c(1 - x_1 - x_2) = x_1(a - c) + x_2(b - c) + c \leq c,$$

since $a - c < 0, b - c < 0$ and $x_1, x_2 \geq 0$. So, we conclude that

$$c \geq \max\{ax_1 + bx_2 + cx_3 \mid x_1 + x_2 + x_3 = 1, \ x_1, x_2, x_3 \geq 0\} \geq c,$$

and this establishes (1.4.2). This shows that $X^* = (0, 0, 1)$ is a best response to Y.

It is possible to get a mixed strategy best response but only if some or all of the coefficients a, b, c are equal. For instance, if $b = c$, then

$$\max\{ax_1 + bx_2 + cx_3 \mid x_1 + x_2 + x_3 = 1, \ x_1, x_2, x_3 \geq 0\} = \max\{a, c\}.$$

To see this, suppose that $\max\{a, c\} = c$. We compute

$$\begin{aligned}
\max\{ax_1 &+ bx_2 + cx_3 \mid x_1 + x_2 + x_3 = 1, \ x_1, x_2, x_3 \geq 0\} \\
&= \max\{ax_1 + c(x_2 + x_3) \mid x_1 + x_2 + x_3 = 1\} \\
&= \max\{ax_1 + c(1 - x_1) \mid 0 \leq x_1 \leq 1\} \\
&= \max\{x_1(a - c) + c \mid 0 \leq x_1 \leq 1\} \\
&= c.
\end{aligned}$$

This maximum is achieved at $X^* = (0, x_2, x_3)$ for any $x_2 + x_3 = 1, x_2 \geq 0, x_3 \geq 0$, and we see that we can get a mixed strategy as a best response.

In general, **if one of the strategies, say, Y is given and known**, then

$$\max_{X \in S_n} \sum_{i=1}^{n} x_i \left(\sum_{j=1}^{m} a_{ij} y_j \right) = \max_{1 \leq i \leq n} \left(\sum_{j=1}^{m} a_{ij} y_j \right).$$

In other words,

$$\max_{X \in S_n} E(X, Y) = \max_{1 \leq i \leq n} E(i, Y).$$

We proved this in the proof of Theorem 1.4.6.

Best response strategies are frequently used when we assume that the opposing player is Nature or some nebulous player that we think may be trying to oppose us, like the market in an investment game, or the weather. The next example helps to make this more precise.

Example 1.7 Suppose that player I has some money to invest with three options: stock (S), bonds (B), or certificates of deposit (CDs). The rate of return depends on the state of the market for each of these investments. Stock is considered risky, bonds have less risk than stock, and CDs are riskless. The market can be in one of three states: good (G), neutral (N), or bad (B), depending on factors such as the direction of interest rates, the state of the economy, prospects for future growth. Here is a possible game matrix in which the numbers represent the annual rate of return to the investor who is the row player:

I/II	G	N	B
S	12	8	−5
B	4	4	6
CD	5	5	5

The column player is the market. This game does not have a saddle in pure strategies. (Why?) If player I assumes that the market is the opponent with the goal of minimizing the investor's rate of return, then we may look at this as a two-person zero sum game. On the other hand, if the investor thinks that the market may be in any one of the three states with equal likelihood, then the market will play the strategy $Y = \left(\frac{1}{3}, \frac{1}{3}, \frac{1}{3} \right)$, and then the investor must choose how to respond to that; that is, the investor seeks an X^* for which $E(X^*, Y) = \max_{X \in S_3} E(X, Y)$, where $Y = \left(\frac{1}{3}, \frac{1}{3}, \frac{1}{3} \right)$. Of course, the investor uses the same procedure no matter which Y she thinks the market will use. The investor can use this game to compare what will happen under various Y strategies.

This problem may actually be solved fairly easily. If we assume that the market is an opponent in a game then the value of the game is $v(A) = 5$, and there are many optimal strategies, one of which is $X^* = (0, 0, 1)$, $Y^* = (0, \frac{1}{2}, \frac{1}{2})$. If instead $Y = (\frac{1}{3}, \frac{1}{3}, \frac{1}{3})$, then the best response for player I is $X = (0, 0, 1)$, with payoff to I equal to 5. If $Y = (\frac{2}{3}, 0, \frac{1}{3})$, the best response is $X = (1, 0, 0)$, that is, invest in the stock if there is a $\frac{2}{3}$ probability of a good market. The payoff then is $\frac{19}{3} > 5$.

It may seem odd that the best response strategy in a zero sum two person game is usually a pure strategy, but it can be explained easily with a simple example. Suppose that someone is flipping a coin that is not fair—say heads comes up 75% of the time. On each toss you have to guess whether it will come up heads or tails. How often should you announce heads in order to maximize the percentage of time you guess correctly? If you think it is 75% of the time, then you will be correct $75 \times 75 = 56.25\%$ of the time! If you say heads **all the time**, you will be correct 75% of the time, and that is the best you can do.

Here is a simple game with an application of game theory to theology!

Example 1.8 Blaise Pascal constructed a game to show that belief in God is the only rational strategy. The model assumes that you are player I, and you have two possible strategies: believe or don't believe. The opponent is taken to be God, who either plays God exists, or God doesn't exist. (Remember, God can do anything without contradiction.) Here is the matrix, assuming $\alpha, \beta, \gamma > 0$:

You/God	God exists	God doesn't exist
Believe	α	$-\beta$
Don't believe	$-\gamma$	0

If you believe and God doesn't exist, then you receive $-\beta$ because you have foregone evil pleasures in the belief God exists. If you don't believe and God exists, then you pay a price $-\gamma$. Pascal argued that this would be a big price. If you believe and God exists, then you receive the amount α, from God, and Pascal argued this would be a very large amount of spiritual currency.

To solve the game, we first calculate v^+ and v^-. Clearly $v^+ = 0$ and $v^- = \max(-\beta, -\gamma) < 0$, so this game does not have a saddle point in pure strategies unless $\beta = 0$ or $\gamma = 0$. If there is no loss or gain to you if you play don't believe, then that is what you should do, and God should play **not exist**. In this case the value of the game is zero.

Assuming that none of α, β, or γ is zero, we will have a mixed strategy saddle. Let $Y = (y, 1 - y)$ be an optimal strategy for God. Then it must be true that

$$E(1, Y) = \alpha y - \beta(1 - y) = v(A) = -\gamma y = E(2, Y).$$

These are the two equations for a mixed strategy from the fact we are assuming $0 < y < 1$. Solving, we get that $y = \beta / (\alpha + \beta + \gamma)$. The optimal strategy for God is

$$Y = \left(\frac{\beta}{\alpha + \beta + \gamma}, \frac{\alpha + \gamma}{\alpha + \beta + \gamma} \right)$$

and the value of the game to you is

$$v(A) = \frac{-\gamma\beta}{\alpha + \beta + \gamma} < 0.$$

Your optimal strategy $X = (x, 1 - x)$ must satisfy

$$E(X, 1) = \alpha x - \gamma(1 - x) = -\beta x = E(X, 2) \implies$$

$$x = \frac{\gamma}{\alpha + \beta + \gamma}, \text{ and } X = \left(\frac{\gamma}{\alpha + \beta + \gamma}, \frac{\alpha + \beta}{\alpha + \beta + \gamma} \right).$$

Pascal argued that if γ, the penalty to you if you don't believe and God exists is loss of eternal life, represented by a very large number. In this case, the percent of time you play believe, $x = \gamma/(\alpha + \beta + \gamma)$ should be fairly close to 1, so you should play believe with high probability. For example, if $\alpha = 10$, $\beta = 5, \gamma = 100$, then $x = 0.87$.

From God's point of view, if γ is a very large number and this is a zero sum game, God would then play doesn't exist with high probability!! It may not make much sense to think of this as a zero sum game, at least not theologically. Maybe we should just look at this like a best response for you, rather than as a zero sum game.

So, let's suppose that God plays the strategy $Y^0 = (\frac{1}{2}, \frac{1}{2})$. What this really means is that you think that God's existence is as likely as not. What is your best response strategy? For that, we calculate $f(x) = E(X, Y^0)$, where $X = (x, 1-x), 0 \leq x \leq 1$. We get

$$f(x) = x\,\frac{\alpha + \gamma - \beta}{2} - \frac{\gamma}{2}.$$

The maximum of $f(x)$ over $x \in [0,1]$ is

$$f(x^*) = \begin{cases} \dfrac{\alpha - \beta}{2}, & \text{at } x^* = 1 \text{ if } \alpha + \gamma > \beta; \\[2ex] -\dfrac{\gamma}{2}, & \text{at } x^* = 0 \text{ if } \alpha + \gamma < \beta; \\[2ex] -\dfrac{\gamma}{2}, & \text{at any } 0 \leq x \leq 1 \text{ if } \alpha + \gamma = \beta. \end{cases}$$

Pascal argued that γ would be a very large number (and so would α) compared to β. Consequently, the best response strategy to Y^0 would be $X^* = (1, 0)$. Any rational person who thinks that God's existence is as likely as not would choose to play believe.

Are we really in a zero sum game with God? Probably not because it is unlikely that God loses if you choose to believe. A more likely opponent in a zero sum game like this might be Satan. Maybe the theological game is nonzero sum so that both players can win (and both can lose). In any case, a nonzero sum model of belief will be considered later.

1.4.4 Dominated Strategies

The goal is to find the best strategies. One way to approach this is to get rid of obviously bad strategies. Here is one way to determine strategies that can be eliminated: if one row k is worse than another row i no matter what player II does, then player I should never play the strategy k, as i would definitely be a better choice. And there is a similar argument to remove a column if there is another column that is always better for player II. Let's make this a definition.

Definition 1.4.11

1. *Row i weakly dominates row k if $a_{ij} \geq a_{kj}$ for all $j = 1, 2, \ldots, m$. We say row i strictly dominates row k if all of these inequalities are strict.*
2. *Column j weakly dominates column k if $a_{ij} \leq a_{ik}, i = 1, 2, \ldots, n$. We say column j strictly dominates row k if all of these inequalities are strict.*
3. *If a row or column is strictly dominated, then it can be dropped from the matrix.*

Remarks

1. We may also drop rows or columns that are **non-strictly dominated** but the resulting matrix may not result in all the saddle points for the original matrix. Finding all of the saddle points is not what we are concerned with so we will use non-strict dominance in what follows.

2. A row that is dropped because it is **strictly dominated** is played in a mixed strategy with probability 0. But a row that is dropped because it is equal to another row may not have probability 0 of being played. For example, suppose that we have a matrix with three rows and row 2 is the same as row 3. If we drop row 3, we now have two rows and the resulting optimal strategy will look like $X^* = (x_1, x_2)$ for the reduced game. Then for the original game the optimal strategy could be $X^* = (x_1, x_2, 0)$ or $X^* = (x_1, \frac{x_2}{2}, \frac{x_2}{2})$, or in fact $X^* = (x_1, \lambda x_2, (1 - \lambda)x_2)$ for any $0 \leq \lambda \leq 1$. The set of all optimal strategies for player I would consist of all $X^* = (x_1, \lambda x_2, (1 - \lambda)x_2)$ for any $0 \leq \lambda \leq 1$, and this is the most general description. A duplicate row is a redundant row and may be dropped to reduce the size of the matrix. But you must account for redundant strategies.

Another way to reduce the size of a matrix, which is more subtle, is to drop rows or columns by dominance through a convex combination of other rows or columns. If a row (or column) is (strictly) dominated by a convex combination of other rows (or columns), then this row (column) can be dropped from the matrix. If, for example, row k is dominated by a convex combination of two other rows, say, p and q, then we can drop row k. This means that if there is a constant $\lambda \in [0, 1]$ so that

$$a_{kj} \leq \lambda a_{pj} + (1 - \lambda)a_{qj}, \quad j = 1, \ldots, m,$$

then row k is dominated and can be dropped. Of course, if the constant $\lambda = 1$, then row p dominates row k and we can drop row k. If $\lambda = 0$ then row q dominates row k. More than two rows can be involved in the combination.

For columns, the column player wants small numbers, so column k is dominated by a convex combination of columns p and q if

$$a_{ik} \geq \lambda a_{ip} + (1 - \lambda)a_{iq}, \quad i = 1, \ldots, n.$$

The reason convex combination dominance works is because when we use mixed strategies dominance can be established because the probabilities of playing two or more dominating rows would make the probability of playing the dominated row zero.

It may be hard to spot a combination of rows or columns that dominate, but if there are suspects, the next example shows how to verify it.

Example 1.9 Consider the 3×4 game

$$A = \begin{bmatrix} 10 & 0 & 7 & 4 \\ 2 & 6 & 4 & 7 \\ 5 & 2 & 3 & 8 \end{bmatrix}.$$

It seems that we may drop column 4 right away because every number in that column is larger than each corresponding number in column 2. So now we have

$$\begin{bmatrix} 10 & 0 & 7 \\ 2 & 6 & 4 \\ 5 & 2 & 3 \end{bmatrix}.$$

There is no obvious dominance of one row by another or one column by another. However, we suspect that row 3 is dominated by a convex combination of rows 1 and 2. If that is true, we must have, for some $0 \le \lambda \le 1$, the inequalities

$$5 \le \lambda(10) + (1 - \lambda)(2), \; 2 \le 0(\lambda) + 6(1 - \lambda), \; 3 \le 7(\lambda) + 4(1 - \lambda).$$

Simplifying, $5 \le 8\lambda + 2, 2 \le 6 - 6\lambda, 3 \le 3\lambda + 4$. But this says any $\frac{3}{8} \le \lambda \le \frac{2}{3}$ will work. So, there is a λ that works to cause row 3 to be dominated by a convex combination of rows 1 and 2, and row 3 may be dropped from the matrix (i.e., an optimal mixed strategy will play row 3 with probability 0). Remember, to ensure dominance by a convex combination, all we have to show is that there are λ's that satisfy all the inequalities. We don't actually have to find them. So now the new matrix is

$$\begin{bmatrix} 10 & 0 & 7 \\ 2 & 6 & 4 \end{bmatrix}.$$

Again there is no obvious dominance, but it is a reasonable guess that column 3 is a bad column for player II and that it might be dominated by a combination of columns 1 and 2. To check, we need to have

$$7 \ge 10\lambda + 0(1 - \lambda) = 10\lambda, \; \text{and} \; 4 \ge 2\lambda + 6(1 - \lambda) = -4\lambda + 6.$$

These inequalities require that $\frac{1}{2} \le \lambda \le \frac{7}{10}$, which is okay. So there are λ's that work, and column 3 may be dropped. Finally, we are down to a 2×2 matrix

$$\begin{bmatrix} 10 & 0 \\ 2 & 6 \end{bmatrix}.$$

This game may be solved by assuming that each row and column will be used with positive probability and then solving the system of equations (see Theorem 1.9). The answer is that the value of the game is $v(A) = \frac{30}{7}$ and the optimal strategies for the original game are $X^* = (\frac{2}{7}, \frac{5}{7}, 0)$ and $Y^* = (\frac{3}{7}, \frac{4}{7}, 0, 0)$.

Example 1.10 This example shows that me may lose solutions when we reduce by non-strict dominance. Consider the game with matrix

$$A = \begin{bmatrix} 1 & 1 & 1 \\ 1 & 2 & 0 \\ 1 & 0 & 2 \end{bmatrix}.$$

Again there is no obvious dominance but it's easy to see that row 1 is dominated (non-strictly) by a convex combination of rows 2 and 3. In fact $a_{1j} = \frac{1}{2}a_{2j} + \frac{1}{2}a_{3j}, j = 1, 2, 3$. So if we drop row 1 we next see that column 1 is dominated by a convex combination of columns 2 and 3 and may be dropped. This leaves us with the reduced matrix $\begin{bmatrix} 2 & 0 \\ 0 & 2 \end{bmatrix}$. The solution for this reduced game is $v = 1, X^* = (\frac{1}{2}, \frac{1}{2}) = Y^*$ and consequently a solution of the original game is $v(A) = 1, X^* = (0, \frac{1}{2}, \frac{1}{2}) = Y^*$. However, it is easy to verify that there is in fact a pure saddle point for this game given by $X^* = (1, 0, 0), Y^* = (1, 0, 0)$ and this is missed by using non-strict domination. Dropping rows or columns that are strictly dominated, however, means that dropped row or column is never played.

Problems

1.13 Suppose A is a 2×3 matrix and A has a saddle point in pure strategies. Show that it must be true that either one column dominates another, or one row dominates the other, or both. Then find a matrix A which is 3×3 and has a saddle point in pure strategies, but no row dominates another and no column dominates another.

1.14 Consider the game with payoff matrix $\begin{bmatrix} 1 & 3 \\ 4 & 2 \end{bmatrix}$.
   (a) Find the upper and lower value, v^+ and v^-.
   (b) Show that there is no saddle point in pure strategies.
   (c) Find a saddle point in mixed strategies (you can use the graphing method or the principle of indifference… which in the end is the same calculation).
   (d) Find the value of the game allowing mixed strategies.

1.15 Solve the game with payoff matrix

$$\begin{bmatrix} 10 & 3 \\ 1 & 9 \\ 7 & 4 \end{bmatrix}.$$

That is, you need to find a pair of optimal strategies, and calculate the expected payoff to player I if both players play optimally.

1.16 Solve the game with the payoff matrix below. That is, find a saddle point in mixed strategies, and the expected payout to PI if the players both play optimally.

$$\begin{bmatrix} 4 & 6 & 5 & 6 \\ 3 & 7 & 8 & 5 \\ 7 & 4 & 7 & 3 \\ 4 & 7 & 6 & 5 \end{bmatrix}.$$

Hint: First eliminate (weakly) dominated strategies as much as possible!

1.17 Consider the following gambling game: There are two players, PI and PII. They each secretly put either $1 or $2 in their hands. They simultaneously open their hands, and let the money fall in the pot. If the total amount of money is even (so $2 or $4) PI wins and takes it. If it is odd (so, $3) PII wins and gets the pot.
   (a) Describe the possible pure strategies.
   (b) Set up a payoff matrix for PI. Hint: there will be negative numbers, since PI sometimes loses money.
   (c) Solve the game (i.e., find a saddle point, which could involve mixed strategies).
   (d) Is this game fair? That is, does either player have an advantage? If yes, who? Explain.

1.18 An entrepreneur, named Victor, outside Laguna beach can sell 500 umbrellas when it rains and 100 when the sun is out along with 1000 pairs of sunglasses. Umbrellas cost him $5 each and sell for $10. Sunglasses wholesale for 2$ and sell for $5. The vendor has 2500 dollars to buy the goods. Whatever he doesn't sell is lost as worthless at the end of the day.

(a) Assume Victor's opponent is the weather set up a payoff matrix with the elements of the matrix representing his net profit.

(b) Suppose Victor hears the weather forecast and there is a 30% chance of rain. What should he do?

1.19 You're in a bar and a stranger comes to you with a new pickup strategy.[1] The stranger proposes that you each call Heads or Tails. If you both call Heads, the stranger pays you $3. If both call Tails, the stranger pays you $1. If the calls aren't a match, then you pay the stranger $2.

(a) Formulate this as a two person game and solve it.

(b) Suppose the stranger decides to play the strategy $\tilde{Y} = \left(\frac{1}{3}, \frac{2}{3}\right)$. Find a best response and the expected payoff.

1.20 Suppose that the batter in the baseball game Problem 1.11 hasn't done his homework to learn the percentages in the game matrix. So, he uses the strategy $X^* = \left(\frac{1}{3}, \frac{1}{3}, \frac{1}{3}\right)$. What is the pitcher's best response strategy?

1.21 In general, if we have two payoff functions $f(x, y)$ for player I and $g(x, y)$ for player II, suppose that both players want to maximize their own payoff functions with the variables that they control. Then $y^* = y^*(x)$ is a best response of player II to x if

$$g(x, y^*(x)) = \max_y g(x, y), \quad y^* \in \arg\max_y g(x, y)$$

and $x^* = x^*(y)$ is a best response of player I to y if

$$f(x^*(y), y) = \max_x f(x, y), \quad x^* \in \arg\max_x f(x, y)$$

Recall that, for example, $\arg\max_x f(x, y)$ is the set of all $x's$ where the maximum of $f(x, y)$ as a function of x is achieved.

(a) Find the best responses if $f(x, y) = (C - x - y)x$ and $g(x, y) = (D - x - y)y$, where C and D are constants.

(b) Solve the best responses and show that the solution x^*, y^* satisfies $f(x^*, y^*) \geq f(x, y^*)$ for all x, and $g(x^*, y^*) \geq g(x^*, y)$ for all y.

1.22 Consider the following matrix game:

$$\begin{bmatrix} 2 & 4 & 1 & 5 \\ 1 & 0 & 7 & 4 \\ 0 & 9 & 2 & 1 \\ 5 & 6 & 3 & 4 \end{bmatrix}.$$

Five students claim to have solved the game. They give the following optimal strategies:

1. $X = (0, \frac{1}{4}, 0, \frac{3}{4}), Y = (\frac{1}{2}, 0, \frac{1}{2}, 0)$.
2. $X = (0, \frac{1}{2}, 0, \frac{1}{2}), Y = (\frac{1}{2}, 0, \frac{1}{2}, 0)$.
3. $X = (1, 0, 0, 0), Y = (0, 1, 0, 0)$.
4. $X = (0, \frac{1}{4}, 0, \frac{3}{4}), Y = (0, 0, 0, 1)$.
5. $X = (0, \frac{1}{4}, 0, \frac{3}{4}), Y = (\frac{1}{3}, 0, \frac{1}{3}, \frac{1}{3})$.

Which, if any, of them are correct?

---

1 This question appeared in a column by Marilyn Vos Savant in Parade Magazine on March 31, 2002.

1.5 The Indifference Principle and Completely Mixed Games

The Equilibrium Theorem 1.4.8, also known as the indifference principle, is a key way to find optimal strategies when we know that two or more rows or columns will be used. An important class of games is when all the rows or columns of a game will be played. Such games are called **completely mixed**.

Precisely, a game is called completely mixed if there is a mixed saddle point where both players are using all strategies with non-zero probability. In that case the indifference principle gives a way to find the saddle point. We will illustrate this with some examples.

Example 1.11   Hide and Seek: We consider a game where PII chooses a place to hide, PI chooses a place to look for PII, and PI wins if she finds PII—but the winning amount can vary. For instance, finding PII under the bed may be worth 50 points, in the closet worth 30, and behind the curtain worth 20. This would become the matrix game

$$A = \begin{bmatrix} 50 & 0 & 0 \\ 0 & 30 & 0 \\ 0 & 0 & 20 \end{bmatrix}.$$

We claim that, in this case, both players should use all strategies with some positive probability. First notice that, for any completely mixed strategy X for PI, and any response Y by PII, the expected outcome $E(X, Y) > 0$ is positive. This shows that the game has positive value $v(A) > 0$. But for any non-completely mixed strategy for PI, PII can just hide in the place PI is never looking, and get outcome 0. Next notice that if PII is not playing a totally mixed strategy neither should PI—they should not look where PII never hides! This means it is impossible for a non-completely mixed PII strategy to be part of a saddle point as well.

A formal way to see that the optimal strategies must be completely mixed is to use Remark 1.4.9. If $v(A)$ is the value of the game and Y is optimal for PII, then

$$v(A) \geq \max_{i=1,2,3} E(i, Y) = \max\{50y_1, 30y_2, 20y_3\}.$$

For any Y, optimal or not, this says $v(A) > 0$. Similarly, by Remark 1.5.9, for any optimal X for PI, $v(A) \leq \min\{50x_1, 30x_2, 20x_3\}$. If any component of X, say x_2 for instance, has $x_2 = 0$, then the minimum is 0. That would imply $v(A) \leq 0$, which we already know is not true. So any optimal strategy for PI must be completely mixed. Next, if $X^* = (x_1, x_2, x_3)$ is completely mixed and optimal for PI and $Y^* = (y_1, y_2, y_3)$ is optimal for PII with, say $y_2 = 0$, then, assuming $y_1 > 0, y_3 > 0$,

$$v(A) = E(X^*, Y^*) = 50x_1y_1 + 20x_3y_3 < 50y_1(x_1 + \tfrac{x_2}{2}) + 20y_3(x_3 + \tfrac{x_2}{2}),$$

which would say that there is a better strategy for PI than X^*, namely, $(x_1 + \tfrac{x_2}{2}, 0, x_3 + \tfrac{x_2}{2})$. This contradiction shows that no component of Y^* can be zero.

Now we know that both players use all strategies, and so we can write down a system of equations by the indifference principle to solve to find optimal strategy $X^* = (x_1, x_2, 1 - x_1 - x_2), Y^* = (y_1, y_2, 1 - y_1 - y_2)$.

Using PI's indifference

$$50y_1 = 30y_2 = 20(1 - y_1 - y_2).$$

This is really the system of two equations

$$50y_1 = 30y_2. \quad \text{and} \quad 30y_3 = 20(1 - y_1 - y_2).$$

Solving, $y_1 = \frac{6}{31}, y_2 = \frac{10}{31}$. This gives $1 - y_1 - y_2 = \frac{15}{31}$, so the optimal strategy is

$$Y^* = \left(\frac{6}{31}, \frac{10}{31}, \frac{15}{31}\right) \approx (0.194, 0.323, 0.484).$$

A similar calculation gives

$$X^* = \left(\frac{6}{31}, \frac{10}{31}, \frac{15}{31}\right).$$

In fact, in this type of hide and seek game it is always true that the seeker should look in each location with the same probability that the hider hides there. Notice that PI seeks with the highest probability the location with the lowest payoff for PI. The expected payoff to PI is

$$E_I(X^*, Y^*) = X^* A \ Y^{*T} = \frac{300}{31} = 9.6774.$$

Example 1.12 Now let's make some slight changes to the outcomes for the previous example. Say the matrix is

$$A = \begin{bmatrix} 50 & 1 & -1 \\ -1 & 30 & 2 \\ 3 & 1 & 20 \end{bmatrix}.$$

Since the game was totally mixed before, it seems like a reasonable guess that it is totally mixed now. Using that guess, the indifference equations give:

$$50y_1 + y_2 - (1 - y_1 - y_2) = -y_1 + 30y_2 + 2(1 - y_1 - y_2) = 3y_1 + y_2 + 20(1 - y_1 - y_2)$$

and

$$50x_1 - x_2 + 3(1 - x_1 - x_2) = x_1 + 30x_2 + (1 - x_1 - x_2) = -x_1 + 2x_2 + 20(1 - x_1 - x_2).$$

Solving we get

$$x_1 \simeq 0.18, x_2 \simeq 0.32, y_1 \simeq 0.21, y_2 \simeq 0.32,$$

so the solution (to two significant digits) is

$$X^* = (0.18, 0.32, 0.50), \qquad Y^* = (0.21, 0.32, 0.47).$$

These strategies are similar to but not identical to those in the previous example, which seems reasonable. The expected payoff to player I is $E(X^*, Y^*) = 10.34$.

We made a guess here, so we should now verify that this is a solution by checking that it is in fact a saddle point. But, since the strategies we found are valid, the indifference equations we solved show that any response to X^* gives the same payoff to PII (and similarly any response to Y^* gives the same payoff to PI). Thus we have in fact shown that this is a saddle point, and that the game is totally mixed. You might ask what would happen if the game was not totally mixed—we will see in the next example.

Example 1.13 Consider now the game with payout matrix

$$A = \begin{bmatrix} 3 & -1 & 0 \\ -2 & 1 & -1 \\ 2 & 0 & 2 \end{bmatrix}.$$

We don't see any obvious reason either player should avoid any of their three strategies, so we could again guess that the game is completely mixed and set up indifference equations. This time the solution is $x_1 = \frac{1}{4}, x_2 = \frac{1}{2}, y_1 = \frac{1}{2}, y_2 = \frac{3}{2}$, so

$$X^* = \left(\frac{1}{4}, \frac{1}{2}, \frac{1}{4}\right) \qquad Y^* = \left(\frac{1}{2}, \frac{3}{2}, -1\right).$$

But this is impossible: Y^* is supposed to be a probability vector, it cannot have a negative entry! So, this is not a valid saddle point, and the game is not totally mixed.

One thing to note: $X^* = (\frac{1}{4}, \frac{1}{2}, \frac{1}{4})$ looks like a valid strategy for PI, but it is still not optimal. The only saddle point in this game is in fact

$$X^* = \left(0, \frac{2}{5}, \frac{3}{5}\right) \qquad Y^* = \left(\frac{1}{5}, \frac{4}{5}, 0\right).$$

Even if you are only interested in the optimal strategy for one player, it is not enough for the equations to work out for them. You really need to solve for both players before you are sure the solution works.

Example 1.14 It can happen that games have many saddle points, some mixed and some not. Consider the game with matrix

$$A = \begin{bmatrix} -2 & 2 & -1 \\ 1 & 1 & 1 \\ 3 & 0 & 1 \end{bmatrix}.$$

You can directly check that $X^* = (0, 1, 0), Y^* = (0, 0, 1)$ is a pure saddle point, and $v(A) = 1$ because the entry in the second row and third column is simultaneously the highest in the column and the lowest in the row.

Now that we know the value, we can write down a system of inequalities to describe the possible optimal strategies for each player. For PI, they must use a strategy (x_1, x_2, x_3) where each of PII's responses gives at least the outcome 1. So,

$$-2x_1 + x_2 + 3x_3 = E(X, 1) \geq 1$$
$$2x_1 + x_2 = E(X, 2) \geq 1$$
$$-x_1 + x_2 + x_3 = E(X, 3) \geq 1$$

and of course $x_1 + x_2 + x_3 = 1$. The only solution to this system is $(x_1, x_2, x_3) = (0, 1, 0)$—you can check this by computer, or by subbing $x_3 = 1 - x_1 - x_2$ and graphing the inequalities as we do for PII's strategies below.

By the same logic, PII must play a strategy (y_1, y_2, y_3) with $y_1 + y_2 + y_3$ and

$$y_1 + y_2 + y_3 \leq 1, -2y_1 + 2y_2 - y_3 \leq 1, 3y_1 + y_3 \leq 1.$$

We replace $y_3 = 1 - y_1 - y_2$ and then get a graph of the region of points satisfying all the inequalities in (y_1, y_2) space in Figure 1.5.

Now lots of points work! In particular, $Y = (0.15, 0.5, 0.35)$ will give an optimal strategy for player II. So for this example, PI's only optimal strategy is pure, but PII has many optimal strategies, including completely mixed strategies.

Note that this game is not considered completely mixed because that requires a completely mixed optimal strategy for both players.

Remark

Question: When an optimal strategy for a player has a zero component, can you drop that row or column and then solve the reduced matrix?

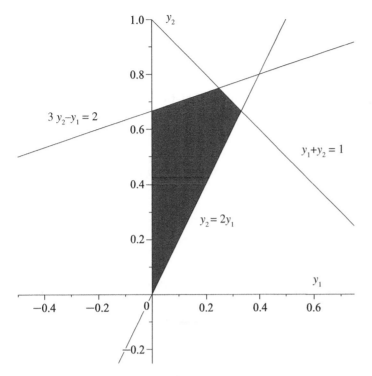

Figure 1.5 Optimal strategy set for Y.

Answer: No, in general.

Consider the game with matrix

$$A = \begin{bmatrix} 1 & 0 & -1 \\ 0 & 1 & 0 \\ -1 & -1 & 1 \end{bmatrix}.$$

Then $X^* = (0, 1, 0)$, $Y^* = (\frac{1}{2}, 0, \frac{1}{2})$ is a saddle point and $v(A) = 0$. Since player I never used strategy 1, you might think we could drop row 1 from the game, If we drop row 1 (because the optimal strategy for player I has $x_1 = 0$) to get the matrix

$$B = \begin{bmatrix} 0 & 1 & 0 \\ -1 & -1 & 1 \end{bmatrix}.$$

But then $\overline{X} = (1, 0)$, $\overline{Y} = (1, 0, 0)$ is a saddle point for B, but $\overline{Y}$ is **NOT** part of any saddle point for the original game. The original game has multiple equilibria, such as

$$X^* = \left(\tfrac{1}{3}, \tfrac{1}{3}, \tfrac{1}{3}\right), \quad Y^* = \left(\tfrac{1}{2}, 0, \tfrac{1}{2}\right).$$

1.5.1 2 × 2 Games

In this easily solved case, each player has exactly two strategies, so the matrix and strategies look like

$$A = \begin{bmatrix} a_{11} & a_{12} \\ a_{21} & a_{22} \end{bmatrix} \quad \text{player I: } X = (x, 1 - x); \quad \text{player II: } Y = (y, 1 - y).$$

For any mixed strategies we have $E(X, Y) = X A Y^T$, which, written out, is

$$E(X, Y) = xy(a_{11} - a_{12} - a_{21} + a_{22}) + x(a_{12} - a_{22}) + y(a_{21} - a_{22}) + a_{22}.$$

Now here is the theorem giving the solution of this game.

Theorem 1.5.1 *In the 2×2 game with matrix A,* **assume that there are no pure optimal strategies.** *If we set*

$$x^* = \frac{a_{22} - a_{21}}{a_{11} - a_{12} - a_{21} + a_{22}}, \qquad y^* = \frac{a_{22} - a_{12}}{a_{11} - a_{12} - a_{21} + a_{22}},$$

then $X^* = (x^*, 1 - x^*), Y^* = (y^*, 1 - y^*)$ are optimal mixed strategies for players I and II, respectively. The value of the game is

$$v(A) = E(X^*, Y^*) = \frac{a_{11}a_{22} - a_{12}a_{21}}{a_{11} - a_{12} - a_{21} + a_{22}}.$$

Remarks

1. The main assumption you need before you can use the formulas is that the game does **not** have a pure saddle point. If it does, you find it by checking $v^+ = v^-$, and then finding it directly. You don't need to use any formulas. Also, when we write down these formulas, it had better be true that $a_{11} - a_{12} - a_{21} + a_{22} \neq 0$, but if we assume that there is no pure optimal strategy, then this must be true. In other words, it isn't difficult to check that when $a_{11} - a_{12} - a_{21} + a_{22} = 0$, then $v^+ = v^-$ and that violates the assumption of the theorem.

2. For a preview of the next section where the game matrix has an inverse, we may write the formulas as

$$X^* = \frac{(1 \ 1)A^*}{(1 \ 1)A^* \begin{bmatrix} 1 \\ 1 \end{bmatrix}} \quad \text{and} \quad Y^{*T} = \frac{A^* \begin{bmatrix} 1 \\ 1 \end{bmatrix}}{(1 \ 1)A^* \begin{bmatrix} 1 \\ 1 \end{bmatrix}} \quad value(A) = \frac{\det(A)}{(1 \ 1)A^* \begin{bmatrix} 1 \\ 1 \end{bmatrix}},$$

   where

$$A^* = \begin{bmatrix} a_{22} & -a_{12} \\ -a_{21} & a_{11} \end{bmatrix} \quad \text{and} \quad \det(A) = a_{11}a_{22} - a_{12}a_{21}.$$

Recall that the inverse of a 2×2 matrix is found by swapping the main diagonal numbers, and putting a minus sign in front of the other diagonal numbers, and finally dividing by the determinant of the matrix. A^* is exactly the first two steps, but we don't divide by the determinant. The matrix we get is defined even if the matrix A doesn't have an inverse. Remember, however, that we need to make sure that the matrix doesn't have optimal pure strategies first.

Notice too, that if $\det(A) = 0$, the value of the game is zero.

Where do the formulas come from? By Theorem 1.4.6 since there is a saddle point which is completely mixed, we know that $E(X^*, 1) = v = E(X^*, 2)$ and $E(1, Y^*) = v = E(2, Y^*)$, when (X^*, Y^*) is a saddle point. This gives the equations

$$a_{11}x_1 + a_{21}(1 - x_1) = a_{12}x_1 + a_{22}(1 - x_1) \text{ and } a_{11}y_1 + a_{12}(1 - y_1) = a_{21}y_1 + a_{22}(1 - y_1).$$

Solving these equations results in the formulas in the statement.

Example 1.15 In the game with $A = \begin{bmatrix} -2 & 5 \\ 2 & 1 \end{bmatrix}$ we see that $v^- = 1 < v^+ = 2$, so there is no pure

saddle for the game. If we apply the formulas, we get the mixed strategies $X^* = (\frac{1}{8}, \frac{7}{8})$ and $Y^* = (\frac{1}{2}, \frac{1}{2})$ and the value of the game is $v(A) = \frac{3}{2}$. Notice here that

$$A^* = \begin{bmatrix} 1 & -5 \\ -2 & -2 \end{bmatrix}$$

and $\det(A) = -12$. The matrix formula for player I gives

$$X^* = \frac{(1\ 1)A^*}{(1\ 1)A^*(1\ 1)^T} = \left(\frac{1}{8}, \frac{7}{8}\right).$$

1.5.2 Completely Mixed Games and Invertible Matrix Games

In this section[1] we will solve the class of games in which the matrix A is square, say, $n \times n$ and invertible, so $\det(A) \neq 0$ and A^{-1} exists and satisfies $A^{-1}A = AA^{-1} = I_{n \times n}$. The matrix $I_{n \times n}$ is the $n \times n$ identity matrix consisting of all zeros except for ones along the diagonal. For the present let us suppose that

> Player I has an optimal strategy that is **completely mixed**, specifically, $X = (x_1, \ldots, x_n)$, and $x_i > 0, i = 1, 2, \ldots, n$. So player I plays every row with positive probability.

By the Equilibrium Theorem 1.4.8 we know that this implies that every optimal Y strategy for player II, must satisfy

$$E(i, Y) = {}_iAY^T = value(A), \text{ for every row } i = 1, 2, \ldots, n.$$

Y played against any row will give the value of the game. If we write $J_n = (1\ 1\ \ldots 1)$ for the row vector consisting of all 1s, we can write

$$AY^T = v(A)J_n^T = \begin{bmatrix} v(A) \\ \vdots \\ v(A) \end{bmatrix}. \tag{1.5.1}$$

Now, if $v(A) = 0$, then $AY^T = 0J_n^T = 0$, and this is a system of n equations in the n unknowns $Y = (y_1, \ldots, y_n)$. It is a homogeneous linear system. Since A is invertible this would have the one and only solution $Y^T = A^{-1}0 = 0$. But that is impossible if Y is a strategy (the components must add to 1). So, if this is going to work, the value of the game cannot be zero, and we get, by multiplying both sides of (1.5.1) by A^{-1}, the following equation:

$$A^{-1}A\ Y^T = Y^T = v(A)A^{-1}J_n^T.$$

This gives us Y if we knew $v(A)$. How do we get that? The extra piece of information we have is that the components of Y add to 1 (i.e., $\sum_{j=1}^n y_j = J_nY^T = 1$). So

$$J_nY^T = 1 = v(A)J_nA^{-1}J_n^T,$$

and therefore

$$v(A) = \frac{1}{J_nA^{-1}J_n^T} \text{ and then } Y^T = \frac{A^{-1}J_n^T}{J_nA^{-1}J_n^T}.$$

---

1 This involves some knowledge of basic linear algebra—see the appendix for the essentials of what we use.

We have found the only candidate for the optimal strategy for player II assuming that every component of X is greater than 0. However, if it turns out that this formula for Y has at least one $y_j < 0$, something would have to be wrong; specifically, it would **not** be true that X was completely mixed because that was our hypothesis. But, if it turns out that $y_j \geq 0$ for every component, we could try to find an optimal X for player I by the exact same method. This would give us

$$X = \frac{J_n A^{-1}}{J_n A^{-1} J_n^T}.$$

This method will work if the formulas we get for X and Y end up satisfying the condition that they are strategies. If either X or Y has a negative component, then it fails. But notice that the strategies do not have to be completely mixed as we assumed from the beginning, only bona fide strategies.

Here is a summary of what we know.

Theorem 1.5.2 *Assume that*

1. $A_{n \times n}$ *has an inverse* A^{-1}.
2. $J_n A^{-1} J_n^T \neq 0$.
3. $v(A) \neq 0$.

Set $X = (x_1, \ldots, x_n), Y = (y_1, \ldots, y_m)$, *and*

$$v \equiv \frac{1}{J_n A^{-1} J_n^T}, \qquad Y^T = \frac{A^{-1} J_n^T}{J_n A^{-1} J_n^T}, \qquad X = \frac{J_n A^{-1}}{J_n A^{-1} J_n^T}.$$

If $x_i \geq 0, i = 1, \ldots, n$ *and* $y_j \geq 0, j = 1, \ldots, n$, *we have that* $v = v(A)$ *is the value of the game with matrix A and (X, Y) is a saddle point in mixed strategies.*

Now the point is that when we have an invertible game matrix, we can always use the formulas in the theorem to calculate the number v and the vectors X and Y. If the result gives vectors with nonnegative components, then, by the Proposition 1.4.8, v must be the value, and (X, Y) is a saddle point. Notice that, directly from the formulas, $J_n Y^T = 1$ and $X J_n^T = 1$, so the components given in the formulas will automatically sum to 1; it is only the sign of the components that must be checked.

Here is a simple direct verification that (X, Y) is a saddle and v is the value assuming $x_i \geq 0, y_j \geq 0$. Let $Y' \in S_n$ be any mixed strategy and let X be given by the formula $X = \frac{J_n A^{-1}}{J_n A^{-1} J_n^T}$. Then, since $J_n^T Y' = 1$, we have

$$E(X, Y') = X A Y'^T = \frac{1}{J_n A^{-1} J_n^T} J_n A^{-1} A Y'^T$$

$$= \frac{1}{J_n A^{-1} J_n^T} J_n Y'^T$$

$$= \frac{1}{J_n A^{-1} J_n^T} = v.$$

Similarly, for any $X' \in S_n$, $E(X', Y) = v$, and so (X, Y) is a saddle and v is the value of the game.

Incidentally, these formulas match the formulas when we have a 2×2 game with an invertible matrix because then $A^{-1} = (1/\det(A)) A^*$.

In order to guarantee that the value of a game is not zero, we may add a constant to every element of A that is large enough to make all the numbers of the matrix positive. In this case the value of the new game could not be zero. Since $v(A + b) = v(A) + b$, where b is the constant added to every element, we can find the original $v(A)$ by subtracting b. Adding the constant to all the elements of A

will not change the probabilities of using any particular row or column; that is, the optimal mixed strategies are not affected by doing that.

Even if our original matrix A does not have an inverse, if we add a constant to all the elements of A, we get a new matrix $A + b$ and the new matrix **may** have an inverse (of course, it may not as well). We may have to try different values of b. Here is an example.

Example 1.16   Consider the matrix

$$A = \begin{bmatrix} 0 & 1 & -2 \\ 1 & -2 & 3 \\ -2 & 3 & -4 \end{bmatrix}.$$

This matrix has negative and positive entries, so it's possible that the value of the game is zero. The matrix does not have an inverse because the determinant of A is 0. So, let's try to add a constant to all the entries to see if we can make the new matrix invertible. Since the largest negative entry is -4, let's add 5 to everything to get

$$A + 5 = \begin{bmatrix} 5 & 6 & 3 \\ 6 & 3 & 8 \\ 3 & 8 & 1 \end{bmatrix}.$$

This matrix does have an inverse given by

$$B = \frac{1}{80} \begin{bmatrix} 61 & -18 & -39 \\ -18 & 4 & 22 \\ -39 & 22 & 21 \end{bmatrix}.$$

Next we calculate using the formulas $v = 1/(J_3 B J_3^T) = 5$, and

$$X = v(J_3 B) = \left(\frac{1}{4}, \frac{1}{2}, \frac{1}{4}\right) \quad \text{and} \quad Y = \left(\frac{1}{4}, \frac{1}{2}, \frac{1}{4}\right).$$

Since both X and Y are strategies (they have nonnegative components), the theorem tells us that they are optimal and the value of our original game is $v(A) = v - 5 = 0$.

The next example shows what can go wrong.

Example 1.17   Let

$$A = \begin{bmatrix} 4 & 1 & 2 \\ 7 & 2 & 2 \\ 5 & 2 & 8 \end{bmatrix}.$$

Then it is immediate that $v^- = v^+ = 2$ and there is a pure saddle $X^* = (0, 0, 1), Y^* = (0, 1, 0)$. If we try to use Theorem 1.5.2, we have $\det(A) = 10$, so A^{-1} exists and is given by

$$A^{-1} = \frac{1}{5} \begin{bmatrix} 6 & -2 & -1 \\ -23 & 11 & 3 \\ 2 & -\frac{3}{2} & \frac{1}{2} \end{bmatrix}.$$

If we use the formulas of the theorem, we get

$$v = -1, \; X = v(A)(J_3 A^{-1}) = \left(3, -\frac{3}{2}, -\frac{1}{2}\right),$$

and

$$Y = v(A^{-1} J_3^T)^T = \left(-\frac{3}{5}, \frac{9}{5}, -\frac{1}{5}\right).$$

Obviously these are completely messed up (i.e., wrong). The problem is that the components of X and Y are not nonnegative even though they do sum to 1.

Example 1.18 **Hide and Seek Revisited**: Suppose that we have $a_1 > a_2 > \cdots > a_n > 0$. The game matrix is

$$A = \begin{bmatrix} a_1 & 0 & 0 & \cdots & 0 \\ 0 & a_2 & 0 & \cdots & 0 \\ \vdots & \vdots & \vdots & \vdots & \vdots \\ 0 & 0 & 0 & \cdots & a_n \end{bmatrix}.$$

Because $a_i > 0$ for every $i = 1, 2, \ldots, n$, we know $\det(A) = a_1 a_2 \ldots a_n > 0$, so A^{-1} exists. It seems pretty clear that a mixed strategy for both players should use each row or column with positive probability; that is, we think that the game is completely mixed. In addition, since $v^- = 0$ and $v^+ = a_n$, the value of the game satisfies $0 \leq v(A) \leq a_n$. Choosing $X = (\frac{1}{n}, \ldots, \frac{1}{n})$ we see that

$$\min_Y X A Y^T = \min_Y \left(y_1 \frac{a_1}{n} + \cdots + y_n \frac{a_n}{n} \right) = \frac{a_n}{n} > 0,$$

so that $v(A) = \max_X \min_Y X A Y^T > 0$. This isn't a fair game for player II. It is also easy to see that

$$A^{-1} = \begin{bmatrix} \frac{1}{a_1} & 0 & 0 & \cdots & 0 \\ 0 & \frac{1}{a_2} & 0 & \cdots & 0 \\ \vdots & \vdots & \vdots & \vdots & \vdots \\ 0 & 0 & 0 & \cdots & \frac{1}{a_n} \end{bmatrix}.$$

Then, we may calculate from Theorem 1.5.2 that

$$v(A) = \frac{1}{J_n A^{-1} J_n^T} = \frac{1}{\frac{1}{a_1} + \frac{1}{a_2} + \cdots \frac{1}{a_n}},$$

$$X^* = v(A) \left(\frac{1}{a_1}, \frac{1}{a_2}, \ldots, \frac{1}{a_n} \right) = Y^*.$$

The strategies X^*, Y^* are legitimate strategies and they are optimal. Notice that for any $i = 1, 2, \ldots, n$, we obtain

$$1 < a_i \left(\frac{1}{a_1} + \frac{1}{a_2} + \cdots + \frac{1}{a_n} \right),$$

so that $v(A) < \min(a_1, a_2, \ldots, a_n) = a_n$ and we have verified that $0 < v(A) < a_n$.

1.5.3 An Application: Optimal Target Choice and Defense

Optimal target choice and defense obviously has a clear military application and is the reason why this problem was first studied. Maybe a not so obvious application is when a predator company seeks to takeover one of n prey companies.

Suppose a company, player I, is seeking to takeover one of n companies which have decreasing values to player I, namely, $a_1 > a_2 > \cdots > a_n > 0$. These companies are managed by a private equity firm which can defend exactly one of the companies from takeover. Suppose that an attack made on company i has probability $1 - p$ of being taken over if it is defended, and will definitely be taken over if it is attacked by not being defended. Player I can attack exactly one of the companies and player II can choose to defend exactly one of them. If an attack is made on an undefended company, the payoff to I is a_i. If an attack is made on a defended company, I's payoff is $(1 - p)a_i$. The payoff matrix is

I/II	1	2	3	$\cdots$	n
1	$(1 - p)a_1$	a_1	a_1	$\cdots$	a_1
2	a_2	$(1 - p)a_2$	a_2	$\cdots$	$\vdots$
3	a_3	a_3	$(1 - p)a_3$	$\cdots$	$\vdots$
$\vdots$	$\vdots$	$\vdots$	$\vdots$	$\vdots$	$\vdots$
n	a_n	a_n	a_n	$\cdots$	$(1 - p)a_n$

Let's start with $n = 3$. We want to find the value and optimal strategies.

Since $\det(A) = a_1 a_2 a_3 p^2 (3 - p) > 0$ the invertible matrix theorem can be used if $J_3 A^{-1} J^T \neq 0$ and the resulting strategies are legit. We have, after dividing by $a_1 a_2 a_3$,

$$v(A) = \frac{1}{J_3 A^{-1} J_3^T} = \frac{(3 - p)\, a_1 a_2 a_3}{a_1 a_2 + a_1 a_3 + a_2 a_3} = \frac{(3 - p)}{\frac{1}{a_3} + \frac{1}{a_2} + \frac{1}{a_1}}$$

and $X^* = v(A)\, J_3 A^{-1}$, after simplifying resulting in

$$X^* = \left(\frac{\frac{1}{a_1}}{\frac{1}{a_1} + \frac{1}{a_2} + \frac{1}{a_3}}, \frac{\frac{1}{a_2}}{\frac{1}{a_1} + \frac{1}{a_2} + \frac{1}{a_3}}, \frac{\frac{1}{a_3}}{\frac{1}{a_1} + \frac{1}{a_2} + \frac{1}{a_3}} \right).$$

The formula for $Y^{*T} = v(A) A^{-1} J_3^T$ gives

$$Y^* = \left(\frac{a_3 + a_2 + (p - 2)\,a_1}{p\,(a_1 + a_2 + a_3)}, \frac{a_3 + a_1 + (p - 2)\,a_2}{p\,(a_1 + a_2 + a_3)}, \frac{a_1 + a_2 + (p - 2)\,a_3}{p\,(a_1 + a_2 + a_3)} \right).$$

Since $p - 2 < 0$, Y^* may not be a legitimate strategy for player II. Nevertheless, player I's strategy is legitimate and is easily verified to be optimal by showing that $E(X^*, j) \geq v(A), j = 1, 2, 3$.

For example if we take $p = 0.1, a_i = 4 - i, i = 1, 2, 3$, the matrix is

$$A = \begin{bmatrix} 0.27 & 3 & 3 \\ 2 & 0.18 & 2 \\ 1 & 1 & 0.9 \end{bmatrix}$$

and we calculate from the formulas above,

$$v(A) = 1.58, \quad X^* = (0.18, 0.27, 0.54), \quad Y^* = (4.72, 2.09, -5.81),$$

which is obviously not a strategy for player II. The invertible theorem gives us the right thing for v and X^* but not for Y^*. Interestingly enough, PI attacks the highest value company with the lowest probability.

Another approach is needed because we can't seem to use this to find Y^*. The idea is still to use part of the invertible theorem for $v(A)$ and X^* but then to do a direct derivation of Y^*. We will follow an argument due to Dresher [5].[1]

Because of the structure of the payoffs it seems reasonable that an optimal strategy for player II (and I as well) will be of the form $Y = (y_1, \ldots, y_k, 0, 0, 0)$, that is, II will not defend and therefore use the columns $k + 1, \ldots, n$, for some $k > 1$. We need to find the nonzero $y_j's$ and the value of k.

---

1 A Rand corporation mathematician and game theorist (1911–1992).

We have the optimal strategy X^* for player I and we know that $x_1 > 0, x_2 > 0$. By the indifference principle and Remark 1.5.9 we then know that $E(1, Y^*) = v, E(2, Y^*) = v$. Written out, this is

$$E(1, Y^*) = (1 - p) a_1 y_1 + a_1 y_2 + \cdots + a_1 y_k = v,$$
$$E(2, Y^*) = a_2 y_1 + (1 - p) a_2 y_2 + \cdots + a_2 y_k = v.$$

Using the fact that $\sum_{j=1}^{k} y_j = 1$, we have

$$a_1 (1 - p) y_1 + a_1 \left(\sum_{j=2}^{k} y_j \right) = a_1 \left(1 - p y_1 \right),$$

$$a_2 y_1 + a_2 (1 - p) y_2 + a_2 \left(\sum_{j=3}^{k} y_j \right) = a_2 \left(1 - p y_2 \right),$$

and $a_2(1 - p y_2) = a_1(1 - p y_1)$. Solving for y_2 we get

$$y_2 = \frac{1}{p} \left[1 - \frac{a_1}{a_2} \left(1 - p y_1 \right) \right].$$

In general, the same argument using the fact that $x_1 > 0$ and $x_i > 0, i = 2, 3, \ldots, k$ results in all the $y_j's, j = 2, 3, \ldots, k$ in terms of y_1:

$$y_j = \frac{1}{p} \left[1 - \frac{a_1}{a_j} \left(1 - p y_1 \right) \right], \quad j = 2, 3, \ldots, k.$$

The index $k > 1$ is yet to be determined and remember we are setting $y_j = 0, j = k + 1, \ldots, n$. Again using the fact that $\sum_{j=1}^{k} y_j = 1$, we obtain after some algebra the formula

$$y_j = \begin{cases} \dfrac{1}{p} \left[1 - \dfrac{k - p}{a_j G_k} \right], & \text{if } j = 1, 2, \ldots, k; \\ 0, & \text{otherwise.} \end{cases}$$

Here, $G_k = \sum_{i=1}^{k} \frac{1}{a_i}$.

Now note that if we use the results from the invertible matrix theorem for v and X^*, we may define

$$v(A) = \frac{(k - p)}{G_k},$$

and

$$X^* = \left(\frac{\frac{1}{a_1}}{G_k}, \frac{\frac{1}{a_2}}{G_k}, \ldots, \frac{\frac{1}{a_k}}{G_k}, 0, 0, \ldots, 0 \right),$$

and X^* is still a legitimate strategy for player I and $v(A)$ will still be the value of the game for any value of $k \leq n$. You can easily verify that $E(X^*, j) = v(A)$ for each column j. Now we have the candidates in hand we may verify that they do indeed solve the game.

Theorem 1.5.3 *Set $G_k = \sum_{i=1}^{k} \frac{1}{a_i}$. Notice that $G_{k+1} \geq G_k$. Let k^* be the value of i giving $\max_{1 \leq i \leq n} \frac{i - p}{G_i}$. Then,*

$$X^* = \left(x_1, \ldots, x_n \right), \quad \text{where } x_i = \begin{cases} \dfrac{1}{a_i G_{k^*}}, & \text{if } i = 1, 2 \ldots, k^* \\ 0, & \text{if } i = k^* + 1, \ldots, n \end{cases}$$

is optimal for player I, and $Y^* = (y_1, y_2, \ldots, y_n)$,

$$
y_j = \begin{cases} \dfrac{1}{p}\left(1 - \dfrac{k^* - p}{a_j G_{k^*}}\right), & \text{if } j = 1, 2, \ldots, k^* \\ 0, & \text{otherwise,} \end{cases}
$$

is optimal for player II. In addition, $v(A) = \frac{k^* - p}{G_{k^*}}$ *is the value of the game.*

One point of the theorem is that PI should only attack the companies $1, 2, \ldots, k^*$ and PII should only defend those companies. It is, of course true that it is possible $k^* = n$ in which case both optimal strategies are completely mixed and all companies should be attacked and defended in some proportion.

Proof: The only thing we need to verify is that Y^* is a legitimate strategy for player II and is optimal for that player.

First, let's consider the function $f(k) = k - a_k G_k$. We have

$$
f(k+1) - f(k) = k + 1 - a_{k+1}G_{k+1} - k + a_k G_k = 1 - a_{k+1}G_{k+1} + a_k G_k.
$$

Since $a_k > a_{k+1}$,

$$
f(k+1) - f(k) = 1 - a_{k+1}G_{k+1} + a_k G_k > 1 - a_{k+1}\left(G_k - G_{k+1}\right) > 1.
$$

We conclude that $f(k+1) > f(k)$, that is, f is a strictly increasing function of $k = 1, 2, \ldots$. Also $f(1) = 1 - a_1 G_1 = 0$ and we may set $f(n+1) =$ any big positive integer. We can then find exactly one index k^* so that $f(k^*) \le p < f(k^* + 1)$ and this will be the k^* in the statement of the theorem. We will show later that it is characterized as stated in the theorem.

Now for this k^*, we have

$$
k^* - a_{k^*}G_{k^*} \le p < k^* + 1 - a_{k^*+1}G_{k^*+1} \Rightarrow a_{k^*+1} < \frac{k^* - p}{G_{k^*}} \le a_{k^*}.
$$

Since the $a_i's$ decrease, we get

$$
\frac{k^* - p}{G_{k^*}} \le a_i, \quad i = 1, 2 \ldots, k^*,
$$
$$
\frac{k^* - p}{G_{k^*}} > a_i, \quad i = k^* + 1, k^* + 2 \ldots, n.
$$

Since

$$
E(i, Y^*) = \begin{cases} \frac{k^* - p}{G_{k^*}} = v(A), & \text{if } i = 1, 2, \ldots, k^*, \\ a_i < v(A), & \text{if } i = k^* + 1, \ldots, n, \end{cases}
$$

this proves by Theorem 1.4.8 and Remark 1.5.9 that Y^* is optimal for player II. Notice also that $y_j \ge 0$ for all $j = 1, 2, \ldots, n$ since by definition

$$
y_j = \begin{cases} \dfrac{1}{p}\left(1 - \dfrac{k^* - p}{a_j G_{k^*}}\right), & \text{if } j = 1, 2, \ldots, k^*, \\ 0, & \text{otherwise,} \end{cases}
$$

and we have shown that

$$
\frac{k^* - p}{G_{k^*}} \le a_j, j = 1, 2, \ldots, k^*.
$$

Hence, Y^* is a legitimate strategy.

Finally, we show that k^* can also be characterized as follows: k^* is the value of i achieving $\max_{1 \le i \le n} \frac{i-p}{G_i} = \frac{k^*-p}{G_{k^*}}$. To see why, the function $\varphi(k) = \frac{k-p}{G_k}$ has a maximum value at some value $k = \ell$. That means $\varphi(\ell+1) \le \varphi(\ell)$ and $\varphi(\ell-1) \le \varphi(\ell)$. Using the definition of φ, we get

$$\ell - a_\ell G_\ell \le p \le \ell + 1 - a_{\ell+1} G_{\ell+1},$$

which is what k^* must satisfy. Consequently, $\ell = k^*$ and we are done. ∎

In the problems you will calculate X^*, Y^*, and v for specific values of p, a_i.

Example 1.19 Consider the optimal target choice and defense problem with $n = 3$, $p = \frac{1}{2}$, and $a_1 = 7, a_2 = 5, a_3 = 3$. The matrix becomes

$$A = \begin{bmatrix} \frac{7}{2} & 7 & 7 \\ 5 & \frac{5}{2} & 5 \\ 3 & 3 & \frac{3}{2} \end{bmatrix}.$$

Then, we should calculate the values $\varphi(k) = \frac{k-p}{G_k}$, which decides companies that are worth attacking. First, $G_1 = \frac{1}{7}, G_2 = \frac{1}{7} + \frac{1}{5} = 0.3429, G_3 = \frac{1}{7} + \frac{1}{5} + \frac{1}{3} = 0.6762$. Next,

$$\varphi(1) = \frac{1 - \frac{1}{2}}{\frac{1}{7}} = 3.5, \quad \varphi(2) = \frac{2 - \frac{1}{2}}{\frac{1}{7} + \frac{1}{5}} = 4.375, \quad \varphi(3) = \frac{3 - \frac{1}{2}}{\frac{1}{7} + \frac{1}{5} + \frac{1}{3}} \simeq 3.697.$$

This is largest for $k^* = 2$, so PI should only attack the first two companies, and should do so in proportion to $\frac{1}{a_i}$. So, $X^* = \frac{1}{G_{k^*}}(\frac{1}{7}, \frac{1}{5}, 0)$. This results in

$$X^* = \left(\frac{5}{12}, \frac{7}{12}, 0 \right).$$

Now for the defender, they should never defend company 3. Using the formula in the theorem we have

$$y_1 = 2 \left(1 - \frac{2 - \frac{1}{2}}{7\frac{12}{35}} \right) = \frac{3}{4}, \quad y_2 = 2 \left(1 - \frac{2 - \frac{1}{2}}{5\frac{12}{35}} \right) = \frac{1}{4}.$$

Therefore $Y^* = (\frac{3}{4}, \frac{1}{4}, 0)$. Furthermore, $v(A) = \frac{35}{8}$.

Another way to find Y^* is to look at the 2×2 game which remains after company 3 is ignored by both players. This has matrix

$$A = \begin{bmatrix} \frac{7}{2} & 7 \\ 5 & \frac{5}{2} \end{bmatrix}.$$

Solving gives $Y^* = \left(\frac{3}{4}, \frac{1}{4}, 0 \right)$.

Notice that, perhaps unexpectedly, PI is most likely to attack the company of middle value. This is actually the typical situation: the attacking company never attacks targets below some cutoff, is most likely to attack the targets of moderate value, and attacks the highest value targets less often. The defender's strategy is more obvious: she defends the highest value targets the most often.

Problems

1.23 In a simplified analysis of a football game suppose that the offense can only choose a pass or run, and the defense can choose only to defend a pass or run. Here is the matrix in which the payoffs are the average yards gained:

	Defense	
Offense	Run	Pass
Run	1	8
Pass	10	0

The offense's goal is to maximize the average yards gained per play. Find $v(A)$ and the optimal strategies using the explicit formulas. Check your answers by solving graphically as well.

1.24 Suppose that an offensive pass against a run defense now gives 12 yards per play on average to the offense (so the 10 in the previous matrix changes to a 12). Believe it or not, the offense should pass less, not more. Verify that and give a game theory (not math) explanation of why this is so.

1.25 Does the same phenomenon occur for the defense? To answer this, compare the original game in Problem 1.23 to the new game in which the defense reduces the number of yards per run to 6 instead of 8 when defending against the pass. What happens to the optimal strategies?

1.26 If the matrix of the game has an inverse, the formulas can be simplified to

$$X^* = \frac{(1\ 1)A^{-1}}{(1\ 1)A^{-1}\begin{bmatrix}1\\1\end{bmatrix}}, \quad Y^{*T} = \frac{A^{-1}\begin{bmatrix}1\\1\end{bmatrix}}{(1\ 1)A^{-1}\begin{bmatrix}1\\1\end{bmatrix}}, \quad v(A) = \frac{1}{(1\ 1)A^{-1}\begin{bmatrix}1\\1\end{bmatrix}}$$

where $A^{-1} = \frac{1}{\det(A)}\begin{bmatrix} a_{22} & -a_{12} \\ -a_{21} & a_{11} \end{bmatrix}$. This requires that $\det(A) \neq 0$. Construct an example to show that even if A^{-1} does not exist, the original formulas with A^{-1} replaced by A^* still hold but now $v(A) = 0$.

1.27 Solve the 2×2 games using the formulas:

(a) $\begin{bmatrix} 4 & -3 \\ -9 & 6 \end{bmatrix}$; (b) $\begin{bmatrix} 8 & 99 \\ 29 & 6 \end{bmatrix}$; (c) $\begin{bmatrix} -32 & 4 \\ 74 & -27 \end{bmatrix}$.

1.28 Give an example to show that when optimal pure strategies exist, the formulas in the theorem won't work.

1.29 Let $A = (a_{ij}), i, j = 1, 2$. Show that if $a_{11} + a_{22} = a_{12} + a_{21}$, then $v^+ = v^-$ or, equivalently, there are optimal pure strategies for both players. This means that if you end up with a zero denominator in the formula for $v(A)$, it turns out that there had to be a saddle in pure strategies for the game and hence the formulas don't apply from the outset.

1.30 There are two presidential candidates, Harry and Tom, who will choose which states they will visit to garner votes. The states are of different values: Let's say State 1 is worth 10, State 2 worth 12, and State 3 worth 14. If they go to different states, the each gain the benefit for that state. So, for instance, if tom goes to state 1 and Harry to state 3, the outcome for tom is $+6 - 10 = -4$. But, Tom is a better debater than Harry: if they go to the same state, Tom gets half of the benefit from visiting that state, and Harry gets nothing.

(a) Set this up as a matrix game.

(b) Show that, Harry plays the mixed strategy $Y = \left(\frac{1}{3}, \frac{1}{3}, \frac{1}{3}\right)$, then whatever Tom does their expected outcome is at most $\frac{13}{3}$. Conclude that the value of this game is at most $\frac{13}{3}$.

(c) If Harry plays a strategy which is not completely mixed, then Tom has a response giving them an outcome of at least 5—explain! Using this and the previous part, conclude that Harry's optimal strategy is completely mixed.

(d) Using the fact that Harry's optimal strategy is completely mixed, set up and solve a system of equations giving Tom's optimal strategy X^*.

(e) In the previous part you should have found that Tom's optimal strategy is also completely mixed. Using this, set up and solve a system of equations giving Harry's optimal strategy Y^*.

(f) Calculate the expected outcome $E(X^*, 2)$.

(g) Calculate the expected outcome $E(1, Y^*)$.

(h) Calculate value of this game. Or, if you can see how, explain why you already know the value and just write it down.

1.31 Consider Theorem 1.5.3 and let $p = 0.5, a_1 = 9, a_2 = 7, a_3 = 6, a_4 = 1$ Find $v(A), X^*, Y^*$, solving the game. Find the value of k^* first.

1.32 Consider the matrix game

$$A = \begin{bmatrix} 3 & 5 & 3 \\ 4 & -3 & 2 \\ 3 & 2 & 3 \end{bmatrix}.$$

Show that there is a saddle in pure strategies at $(1, 3)$ and find the value. Verify that $X^* = (\frac{1}{3}, 0, \frac{2}{3}), Y^* = (\frac{1}{2}, 0, \frac{1}{2})$ is also an optimal saddle point. Does A have an inverse? Find it and use the formulas in the theorem to find the optimal strategies and value.

1.33 Solve the following games:

$$(a) \begin{bmatrix} 2 & 2 & 3 \\ 2 & 2 & 1 \\ 3 & 1 & 4 \end{bmatrix}; \quad (b) \begin{bmatrix} 5 & 4 & 2 \\ 1 & 5 & 3 \\ 2 & 3 & 5 \end{bmatrix}; \quad (c) \begin{bmatrix} 4 & 2 & -1 \\ -4 & 1 & 4 \\ 0 & -1 & 5 \end{bmatrix}.$$

(d) Use dominance to solve the following game even though it has no inverse:

$$A = \begin{bmatrix} -4 & 2 & -1 \\ -4 & 1 & 4 \\ 0 & -1 & 5 \end{bmatrix}.$$

1.34 To underscore that the formulas can be used only if you end up with legitimate strategies, consider the game with matrix

$$A = \begin{bmatrix} 1 & 5 & 2 \\ 4 & 4 & 4 \\ 6 & 3 & 4 \end{bmatrix}.$$

(a) Does this matrix have a saddle in pure strategies? If so find it.
(b) Find A^{-1}.
(c) Without using the formulas, find an optimal **mixed** strategy for player II with two positive components.
(d) Use the formula to calculate Y^*. Why isn't this optimal? What's wrong?

1.35 Show that $value(A + b) = value(A) + b$ for any constant b, where, by $A + b = (a_{ij} + b)$, is meant A plus the matrix with all entries $= b$. Show also that (X, Y) is a saddle for the matrix $A + b$ if and only if it is a saddle for A.

1.36 Derive the formula for $X = \frac{J_n A^{-1}}{J_n A^{-1} J_n^T}$, assuming the game matrix has an inverse. Follow the same procedure as that in obtaining the formula for Y.

1.37 A magic square game has a matrix in which each row has a row sum that is the same as each of the column sums. For instance, consider the matrix

$$A = \begin{bmatrix} 11 & 24 & 7 & 20 & 3 \\ 4 & 12 & 25 & 8 & 16 \\ 17 & 5 & 13 & 21 & 9 \\ 10 & 18 & 1 & 14 & 22 \\ 23 & 6 & 19 & 2 & 15 \end{bmatrix}.$$

This is a magic square of order 5 and sum 65. Find the value and optimal strategies of this game and show how to solve any magic square game.

1.38 Why is the hide-and-seek game in Example 1.18 called that? Determine what happens in the hide-and-seek game if there is at least one $a_k = 0$.

1.39 Solve the hide and seek game in Example 1.18 with matrix $A_{n \times n} = (a_{ij}), a_{ii} = \frac{i}{i+1}, i = 1, 2, \ldots, n$, and $a_{ij} = 0, i \neq j$. Find the general solution and give the exact solution when $n = 5$.

1.40 For the game with matrix

$$\begin{bmatrix} -1 & 0 & 3 & 3 \\ 1 & 1 & 0 & 2 \\ 2 & -2 & 0 & 1 \\ 2 & 3 & 3 & 0 \end{bmatrix},$$

we determine that the optimal strategy for player II is $Y = (\frac{3}{7}, 0, \frac{1}{7}, \frac{3}{7})$. We are also told that player I has an optimal strategy X which is completely mixed. Given that the value of the game is $\frac{9}{7}$, find X.

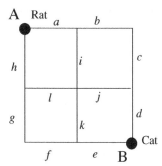

Figure 1.6 Maze for Cat versus Rat.

1.41 A triangular game is of the form

$$A = \begin{bmatrix} a_{11} & a_{12} & \cdots & a_{1n} \\ 0 & a_{22} & \cdots & a_{2n} \\ \vdots & \vdots & \vdots & \vdots \\ 0 & 0 & \cdots & a_{nn} \end{bmatrix}.$$

(a) Find conditions under which this matrix has an inverse.

(b) Now consider

$$A = \begin{bmatrix} 1 & -3 & -5 & 1 \\ 0 & 4 & 4 & -2 \\ 0 & 0 & 8 & 3 \\ 0 & 0 & 0 & 50 \end{bmatrix}.$$

Solve the game by finding $v(A)$ and the optimal strategies.

1.42 Suppose an evader (called Rat) is forced to run a maze entering at point A (see Figure 1.6). A pursuer (called Cat) will also enter the maze at point B. Rat and Cat will run exactly four segments of the maze and the game ends. If Cat and Rat ever meet at an intersection point of segments at the same time, Cat wins +1 (Rat) and the Rat loses −1 because it is zero sum, while if they never meet during the run, both Cat and the Rat win 0. In other words, if Cat finds Rat, Cat gets +1 and otherwise, Cat gets 0. We are looking at the payoffs from Cat's point of view, who wants to maximize the payoffs, while Rat wants to minimize them. Figure 1.6 shows the setup.

The strategies for Rat consist of all the choices of paths with four segments that Rat can run. Similarly, the strategies for Cat will be the possible paths it can take. With four segments it will turn out to be a 16 × 16 matrix. Using dominance to eliminate bad strategies the game reduces to the 3 × 3 game

Cat/Rat	abcj	aike	hlia
dcbi	1	0	1
djke	0	1	1
ekjd	1	1	0

Now solve the game.

1.43 In tennis two players can choose to hit a ball left, center, or right of where the opposing player is standing. Name the two players I and II and suppose that I hits the ball, while II anticipates where the ball will be hit. Suppose that II can return a ball hit right 90% of the time, a ball hit left 60% of the time, and a ball hit center 70% of the time. If II anticipates incorrectly, she can return the ball only 20% of the time. Score a return as $+1$ and a ball not returned as -1. Find the game matrix and the optimal strategies.

1.6 Finding Saddle Points in General

In this section we will see that we can completely solve all matrix games using several methods. We begin by generalizing the graphical method we introduced in the Baseball example from Section 1.2.2 to any $2 \times n$ or $m \times 2$ game. Then we use a powerful method called linear programming to solve any matrix game but which mostly requires a computer. First semester Calculus can also be used to find saddle points and we illustrate this method as well. Finally, the special case of symmetric games is considered.

1.6.1 Graphical Methods

Specifically, we will consider all $2 \times m$ and $n \times 2$ matrix games, so all games where one of the players has only 2 possible strategies. We begin by doing a 2×2 example, following the method we used in the earlier baseball example, but now with somewhat more formal language.

Example 1.20

$$A = \begin{bmatrix} 1 & 4 \\ 3 & 2 \end{bmatrix}.$$

The first step is that we must check whether there are pure optimal strategies because if there are, then we can't use the graphical method. It only finds the strategies that are mixed, but we don't need any fancy methods to find the pure ones. Since $v^- = 2$ and $v^+ = 3$, we know the optimal strategies must be mixed. Now we use Theorem 1.4.6 to find the optimal strategy $X = (x, 1 - x), 0 < x < 1$, for player I, and the value of the game $v(A)$. Here is how it goes.

Playing X against each column for player II, we get

$$E(X, 1) = XA_1 = x + 3(1 - x) \quad \text{and} \quad E(X, 2) = XA_2 = 4x + 2(1 - x).$$

We now plot each of these functions of x on the same graph in Figure 1.7. Each plot will be a straight line with $0 \leq x \leq 1$.

Now, here is how to analyze this graph from player I's perspective. First, the point at which the two lines intersect is $(x^* = \frac{1}{4}, v = \frac{10}{4})$. If player I chooses an $x < x^*$, then the best I can receive is on the highest line when x is on the left of x^*. In this case the line is $E(X, 1) = x + 3(1 - x) > \frac{10}{4}$. Player I will receive this higher payoff only if player II decides to play column 1. But player II will also have this graph and II will see that if I uses $x < x^*$ then II should definitely **not** use column 1 but should use column 2 so that I would receive a payoff on the **lower** line $E(X, 2) < \frac{10}{4}$. In other words, I will see that II would switch to using column 2 if I chose to use any mixed strategy with $x < \frac{1}{4}$. What if I uses an $x > \frac{1}{4}$? The reasoning is similar; the best I could get would happen if player II chose to use column 2 and then I gets $E(X, 2) > \frac{10}{4}$. But I cannot assume that II will play stupidly. Player II will see that if player I chooses to play an $x > \frac{1}{4}$, II will choose to play column 1 so that I receives some payoff on the line $E(X, 1) < \frac{10}{4}$.

Player I's expected
payoff

Figure 1.7 *X* against each column for player II.

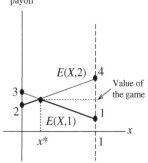

Figure 1.8 Mixed for player I versus player II's columns.

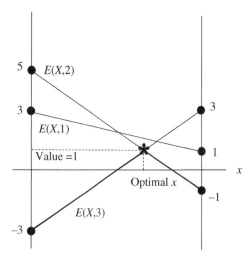

Conclusion: player I, assuming that player II will be doing her best, will choose to play $X = (x^*, 1 - x^*) = \left(\frac{1}{4}, \frac{3}{4}\right)$ and then receive exactly the payoff $= v(A) = \frac{10}{4}$. Player I will rationally choose the maximum minimum. The minimums are the bold lines and the maximum minimum is at the intersection, which is the highest point of the bold lines. Another way to put it is that player I will choose a mixed strategy so that she will get $\frac{10}{4}$ no matter what player II does, and if II does not play optimally, player I can get more than $\frac{10}{4}$.

What should player II do? Before you read the examples, try to figure it out.

Example 1.21 This method actually works just as well in the $2 \times m$ case. Consider the matrix

$$A = \begin{bmatrix} 1 & -1 & 3 \\ 3 & 5 & -3 \end{bmatrix}.$$

Consider the graph for player I first.

Looking at Figure 1.8, we see that the optimal strategy for I is the x value where the two lower lines intersect and yields $X^* = \left(\frac{2}{3}, \frac{1}{3}\right)$. Also, $v(A) = E(X^*, 3) = E(X^*, 2) = 1$. To find the optimal strategy for player II, we see that II will never use the first column, since they can figure out Player 1 will play $\left(\frac{2}{3}, \frac{1}{3}\right)$, and this is not the best response. In general, we may drop the columns (or rows in some cases) not used to get the optimal intersection point—Often that is true because the unused rows are dominated or convex-dominated (as column 1 is here) but not always. Both the second and

Figure 1.9 Mixed for player II versus 4 rows for player I.

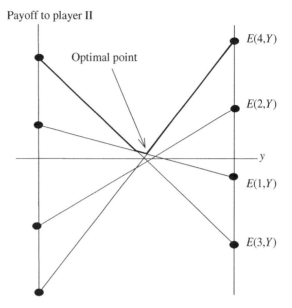

third columns are best responses, so Player 2 should play some combination of these. This means we have reduce to the subgame with the first column removed:

$$\begin{bmatrix} -1 & 3 \\ 5 & -3 \end{bmatrix}.$$

Now this has no pure equilibrium, so we can solve for player II's strategy using the indifference principle from Theorem 1.4.8. We end up solving

$$-q + 3(1 - q) = 5q - 3(1 - q)$$

giving $q = \frac{1}{2}$. Thus player II's optimal strategy for this subgame is $\left(\frac{1}{3}, \frac{2}{3}\right)$. But this is not Y^*, since the original game has 3 strategies for II! We need to record that II never plays the first column. This means $Y^* = \left(0, \frac{1}{2}, \frac{1}{2}\right)$.

Example 1.22 The $n \times 2$ case is only a little different, but we do have to be careful because we must work from the other player's point of view, which changes some calculations. Let's consider

$$A = \begin{bmatrix} -1 & 2 \\ 3 & -4 \\ -5 & 6 \\ 7 & -8 \end{bmatrix}$$

We first check that this has no pure saddle points, so we know we must look for a mixed saddle point. Let Player II use a strategy $Y = (y, 1 - y)$, and graph the payoffs $E(i, Y), i = 1, 2, 3, 4,$: Figure 1.9.

You can see the difficulty with solving games graphically; you have to be very accurate with your graphs. Carefully reading the information, it appears that the optimal strategy for Y will be determined at the intersection point of $E(4, Y) = 7y - 8(1 - y)$ and $E(1, Y) = -y + 2(1 - y)$. This occurs at the point $y^* = \frac{5}{9}$ and the corresponding value of the game will be $v(A) = \frac{1}{3}$. The optimal strategy for player II is $Y^* = \left(\frac{5}{9}, \frac{4}{9}\right)$.

Since this uses only rows 1 and 4, we may now drop rows 2 and 3 to find the optimal strategy for player I. Considering the matrix using only rows 1 and 4, we now calculate $E(X, 1) = -x + 7(1 - x)$ and $E(X, 2) = 2x - 8(1 - x)$ which intersect at $\left(x = \frac{5}{6}, v = \frac{1}{3}\right)$. We obtain that row 1 should be used with probability $\frac{5}{6}$ and row 4 should be used with probability $\frac{1}{6}$, so $X^* = \left(\frac{5}{6}, 0, 0, \frac{1}{6}\right)$.

Just to be sure, let's verify that these are indeed optimal using Theorem 1.4.6. We check that $E(i, Y^*) \leq v(A) \leq E(X^*, j)$ for all rows and columns. This gives

$$\begin{bmatrix} \frac{5}{6} & 0 & 0 & \frac{1}{6} \end{bmatrix} \begin{bmatrix} -1 & 2 \\ 3 & -4 \\ -5 & 6 \\ 7 & -8 \end{bmatrix} = \begin{bmatrix} \frac{1}{3} & \frac{1}{3} \end{bmatrix} \quad \text{and} \quad \begin{bmatrix} -1 & 2 \\ 3 & -4 \\ -5 & 6 \\ 7 & -8 \end{bmatrix} \begin{bmatrix} \frac{5}{9} \\ \frac{4}{9} \end{bmatrix} = \begin{bmatrix} \frac{1}{3} \\ -\frac{1}{9} \\ -\frac{1}{9} \\ \frac{1}{3} \end{bmatrix}.$$

Everything checks.

1.6.2 The $n \times m$ Case and Linear Programming

The graphical method is really just a visual way to illustrate the equivalence of payoffs and to see that player I seeks the maximum minimum while player II seeks the minimum maximum graphically. For a general $n \times m$ matrix A we may not be able to use a graph but we can use the same idea to set up an optimization problem. We have seen that

$$v^- = \max_{X \in S_n} \min_j E(X, j) = \max_{X \in S_n} \{v : v = \min_j E(X, j)\}.$$

Let's make a couple of changes to this equation. First, we replace $v = \min_j E(X, j)$ with $v \leq E(X, j)$ for all j. Second, we move the definition of S_m into the equation.

$$v^- = \max \left\{ v : \begin{array}{l} \text{for some } X = (x_1, \ldots, x_n), v \leq E(X, j) \text{ for all } 1 \leq j \leq m \\ X \text{ satisfies } x_i \geq 0, x_1 + \cdots + x_n = 1 \end{array} \right\}.$$

This problem is looking for $n + 1$ numbers $v, x_1, \ldots, x_n$ such that a system of linear inequalities is satisfied, and v is as large as possible subject to those constraints. We are looking to solve the optimization problem with linear constraints:

Maximize v

subject to

$v \leq E(X, j) = X A_j, \quad j = 1, 2, \ldots, m$

$x_1 + x_2 + \cdots + x_n = 1, \quad x_i \geq 0.$

There are well-known algorithms for solving a problem like this, such as the **simplex method**. This type of problem can also be solved by any computer algebra system, such as Mathematica. This will return the lower value v^- as well as the X where this is achieved, so player I's optimal strategy X^* emerges.

To find Y^*, you set up the same problem but for v^+:

$$v^+ = \min \left\{ v : \begin{array}{l} \text{for some } Y = (y_1, \ldots, y_m), v \geq E(i, Y) \text{ for all } 1 \leq i \leq n \\ Y \text{ satisfies } y_i \geq 0, y_1 + \cdots + y_m = 1 \end{array} \right\}.$$

Player II's problem is to solve

Minimize v

subject to

$$v \geq E(i, Y) = {}_i AY^T, \quad i = 1, 2, \ldots, n$$
$$y_1 + y_2 + \cdots + y_m = 1, \quad y_j \geq 0.$$

Matrix Games and Linear Programming

Linear programming is an area of optimization theory developed since World War II that is used to find the minimum (or maximum) of a linear function of many variables subject to a collection of linear constraints on the variables. It is extremely important to any modern economy to be able to solve such problems that are used to model many fundamental problems that a company may encounter. For example, the best routing of oil tankers from all of the various terminals around the world to the unloading points is a linear programming problem in which the oil company wants to minimize the total cost of transportation subject to the consumption constraints at each unloading point. But there are millions of applications of linear programming, which can range from the problems just mentioned to modeling the entire US economy. One can imagine the importance of having a very efficient way to find the optimal variables involved. Fortunately, George Dantzig,[1] in the 1940s, because of the necessities of the war effort, developed such an algorithm, called the **simplex method** that will quickly solve very large problems formulated as linear programs.

Mathematicians and economists working on game theory (including von Neumann), once they became aware of the simplex algorithm, recognized the connection.[2] After all, a game consists in minimizing and maximizing linear functions with linear things all over the place. So a method was developed to formulate a matrix game as a linear program so that the simplex algorithm could be applied.

This means that using linear programming, we can find the value and optimal strategies for a matrix game of any size without any special theorems or techniques. In many respects, this approach makes it unnecessary to know any other computational approach. The downside is that in general one needs a computer capable of running the simplex algorithm to solve a game by the method of linear programming. We will show how to set up the Mathematica commands to solve the problem.

A Little Aside on Linear Programming

A linear programming problem is a problem of the **standard form** (called the **primal program**):

> Minimize $z = \mathbf{c} \cdot \mathbf{x}$
>
> subject to $\mathbf{x} A \geq \mathbf{b}, \mathbf{x} \geq 0$,

where $\mathbf{c} = (c_1, \ldots, c_n)$, $\mathbf{x} = (x_1, \ldots, x_n)$, $A_{n \times m}$ is an $n \times m$ matrix, and $\mathbf{b} = (b_1, \ldots, b_m)$.

The primal problem seeks to minimize a **linear objective function**, $z(\mathbf{x}) = \mathbf{c} \cdot \mathbf{x}$, over a set of constraints (viz., $\mathbf{x} \cdot A \geq \mathbf{b}$) that are also linear. You can visualize what happens if you try to minimize or maximize a linear function of one variable over a closed interval on the real line.

---

1 George Bernard Dantzig was born on November 8, 1914, and died on May 13, 2005. He is considered the "father of linear programming." He was the recipient of many awards, including the National Medal of Science in 1975, and the John von Neumann Theory Prize in 1974.

2 In 1947 Dantzig met with von Neumann and began to explain the linear programming model "as I would to an ordinary mortal." After a while von Neumann ordered Dantzig to "get to the point." Dantzig relates that in "less than a minute I slapped the geometric and algebraic versions …on the blackboard." Von Neumann stood up and said "Oh, that." Dantzig relates that von Neumann then "proceeded to give me a lecture on the mathematical theory of linear programs." This story is related in the excellent article by Cottle, Johnson, and Wets in the Notices of the A.M.S., March 2007.

The minimum and maximum must occur at an endpoint. In more than one dimension this idea says that the minimum and maximum of a linear function over a variable that is in a convex set must occur on the boundary of the convex set. If the set is created by linear inequalities, even more can be said, namely, that the minimum or maximum must occur at an extreme point, or corner point, of the constraint set. The method for solving a linear program is to efficiently go through the extreme points to find the best one. That is essentially the simplex method.

For any primal there is a related linear program called the **dual program**:

Maximize $w = \mathbf{y} \, \mathbf{b}^T$

subject to $A \, \mathbf{y}^T \leq \mathbf{c}^T, \mathbf{y} \geq 0.$

A very important result of linear programming, which is called the **duality theorem**, states that if we solve the primal problem and obtain the optimal objective $z = z^*$, and solve the dual obtaining the optimal $w = w^*$, then $z^* = w^*$. In our game theory formulation to be given next, this theorem will tell us that the two objectives in the primal and the dual will give us the value of the game. Incidentally, the proof of the duality theorem would provide a proof of the theorem that says if each player uses mixed strategies in a matrix game, there is always a value and a saddle point for the game.

Returning to linear programming and matrix games let's work out an example.

Example 1.23 Consider

$$A = \begin{bmatrix} 2 & -2 & 0 \\ -4 & 4 & 4 \\ 1 & 0 & 4 \end{bmatrix}.$$

Then, we get the problem to determine Y^*.

Minimize v

subject to

$v \geq 2y_1 - 2y_2, v \geq -4y_1 + 4y_2 + 4y_3, v \geq y_1 + 4y_2, y_1, y_2, y_3 \geq 0,$

$y_1 + y_2 + y_3 = 1, y_1 \geq 0, y_2 \geq 0, y_3 \geq 0.$

Using Mathematica we apply the command

```
Minimize[{v, v>=2y1-2y2,v>=-4y1+4y2+4y3,v>=y1+4y2,
                  y1+y2+y3==1,y1>=0,y2>=0,y3>=0},{y1,y2,y3}]
```

This results in $v = \frac{4}{9}, Y^* = \left(\frac{4}{9}, \frac{5}{9}, 0\right)$. Similarly, the problem to determine X^* is

Maximize v

subject to

$v \leq 2x_1 - 4x_2 + x_3, v \leq -2x_1 + 4x_2, v \leq 4x_2 + 4x_3,$

$x_1 + x_2 + x_3 = 1, x_1 \geq 0, x_2 \geq 0, x_3 \geq 0.$

We get $v = \frac{4}{9}, x_1 = 0, x_2 = \frac{1}{9}, x_3 = \frac{8}{9}.$

Here's an example using the theory of duels.

Example 1.24   A Nonsymmetric Noisy Duel. Two players engaged in a duel with pistols begin 10 paces apart. They may shoot at paces (10, 6, 2) with accuracies (0.2, 0.4, 1.0) each. We have the following payoffs to player I at the end of the duel:

- If only player I survives, then player I receives payoff a.
- If only player II survives, player I gets payoff $b < a$. This assumes that the survival of player II is less important than the survival of player I.
- If both players survive, they each receive payoff zero.
- If neither player survives, player I receives payoff g.

We will take $a = 1, b = \frac{1}{2}, g = 0$. Then here is the expected payoff matrix for player I:

I/II	10	6	2
10	0.24	0.6	0.6
6	0.9	0.36	0.70
2	0.9	0.8	0

The pure strategies are (x, y), where $x = 10, 6, 2$ and $y = 10, 6, 2$. The elements of the matrix are obtained from the general formula

$$E(x, y) = \begin{cases} ax + b(1 - x), & \text{if } x < y; \\ ax + bx + (g - a - b)x^2, & \text{if } x = y; \\ a(1 - y) + by, & \text{if } x > y. \end{cases}$$

For example, if we look at

$$E(6, 10) = aProb(\text{II misses at } 10) + bProb(\text{II kills I at } 10)$$

$$= a(1 - 0.2) + b(0.2)$$

because if player II kills I at 10, then he survives and the payoff to I is b. If player II shoots, but misses player I at 10, then player I is certain to kill player II later (this is where the noisy part enters) and will receive a payoff of a. Similarly,

$$E(x, x) = aProb(\text{II misses}) Prob(\text{I hits})$$

$$+ bProb(\text{I misses}) Prob(\text{II hits}) + gProb(\text{I hits}) Prob(\text{II hits})$$

$$= a(1 - x)x + b(1 - x)x + g(x \cdot x).$$

The Mathematica command to find the solution of the game for player I is

```
Maximize[{v, v <= 0.24 x1 + 0.9 x2 + 0.9 x3,
  v <= 0.6 x1 + 0.36 x2 + 0.8 x3, v <= 0.6 x1 + 0.7 x2,
  x1 + x2 + x3 == 1, x1 >= 0, x2 >= 0, x3 >= 0}, {x1, x2, x3}]
```

This will give us that the value of the game is $v(A) = 0.549$, and the optimal strategy for player I is $X^* = (0.532, 0.329, 0.140)$. By a similar analysis of player II's linear programming problem, we get $Y^* = (0.141, 0.527, 0.331)$. So player I should fire at 10 paces more than half the time, even though they have the same accuracy functions, and a miss is certain death. Player II should fire at 10 paces only about 14% of the time.

Now let's look at an example in military science.

Example 1.25   Colonel Blotto Games: This is a simplified form of a military game in which the leaders must decide how many regiments to send to attack or defend two or more targets. It is an optimal allocation of forces game. In one formulation from reference [20], suppose that there are two opponents (players), which we call Red and Blue. Blue controls four regiments, and Red controls three. There are two targets of interest, say, A and B. The rules of the game say that the player who sends the most regiments to a target will win one point for the win and one point for every regiment captured at that target. A tie, in which Red and Blue send the same number of regiments to a target, gives a zero payoff. The possible strategies for each player consist of the number of regiments to send to A and the number of regiments to send to B, and so they are pairs of numbers. The payoff matrix to Blue is

Blue/Red	(3,0)	(0,3)	(2,1)	(1,2)
(4,0)	4	0	2	1
(0,4)	0	4	1	2
(3,1)	1	−1	3	0
(1,3)	−1	1	0	3
(2,2)	−2	−2	2	2

For example, if Blue plays (3, 1) against Red's play of (2, 1), then Blue sends three regiments to A while Red sends two. So Blue will win A, which gives +1 and then capture the two Red regiments for a payoff of +3 for target A. But Blue sends one regiment to B and Red also sends one to B, so that is considered a tie, or standoff, with a payoff to Blue of 0. So the net payoff to Blue is +3.

Solving Colonel Blotto with Mathematica: With software like Mathematica available we can easily solve complicated matrix games as we show on the Colonel Blotto matrix

$$A = \begin{bmatrix} 4 & 0 & 2 & 1 \\ 0 & 4 & 1 & 2 \\ 1 & -1 & 3 & 0 \\ -1 & 1 & 0 & 3 \\ -2 & -2 & 2 & 2 \end{bmatrix}.$$

The Mathematica commands to solve this game for player II directly are

```
Y={y1,y2,y3,y4}
Minimize[{v,A.Y<=v,y1+y2+y3+y4==1,y1>=0,y2>=0,y3>=0,y4>=0}
                  ,{v,y1,y2,y3,y4}]
gives output
{14/9,{v->14/9,y1->1/30,y2->7/90,y3->8/15,y4->16/45}}
```

We get $Y^* = (\frac{1}{30}, \frac{7}{90}, \frac{8}{15}, \frac{16}{45})$ and $v(A) = \frac{14}{9}$.

For player I,

```
Maximize[{v,X.A >=v,x1+x2+x3+x4+x5==1,x1>=0,x2>=0,x3>=0,x4>=0,
                  x5>=0},{v,x1,x2,x3,x4,x5}]
gives output
{14/9,{v->14/9,x1->4/9,x2->4/9,x3->0,x4->0,x5->1/9}}
```

We get $X^* = \left(\frac{4}{9}, \frac{4}{9}, 0, 0, \frac{1}{9}\right)$ and $v(A) = \frac{14}{9}$.

We may also use the Mathematica function **LinearProgramming[c,m,b]** in matrix form to solve this game.

The Mathematica command Mathematica function **LinearProgramming[c,m,b]** solves the following program: Minimize $c \cdot x$ subject to $m \cdot x \geq b, x \geq 0$. To use this command, c is the vector of the coefficients of the variables in the objective. For player II, our coefficient variables are

$c = (0, 0, 0, 0, 1)$ for the objective function $z_{II} = 0y_1 + 0y_2 + 0y_3 + 0y_4 + 1v$. The matrix m is

$$m = \begin{bmatrix} 4 & 0 & 2 & 1 & -1 \\ 0 & 4 & 1 & 2 & -1 \\ 1 & -1 & 3 & 0 & -1 \\ -1 & 1 & 0 & 3 & -1 \\ -2 & -2 & 2 & 2 & -1 \\ 1 & 1 & 1 & 1 & 0 \end{bmatrix}.$$

The last column is added for the variable v. The last row is added for the constraint $y_1 + y_2 + y_3 + y_4 = 1$. Finally, the vector b becomes

$$b = \{\{0, -1\}, \{0, -1\}, \{0, -1\}, \{0, -1\}, \{0, -1\}, \{1, 0\}\}.$$

The first number, 0, is the actual constraint, and the second number makes the inequality $\leq$ instead of the default $\geq$. The pair $\{1, 0\}$ in b comes from the equality constraint $y_1 + y_2 + y_3 + y_4 = 1$ and the second zero converts the inequality to equality. So here it is:

```
m = {{4, 0, 2, 1, -1}, {0, 4, 1, 2, -1}, {1, -1, 3, 0, -1},
     {-1, 1, 0, 3, -1}, {-2, -2, 2, 2, -1}, {1, 1, 1, 1, 0}}
c = {0, 0, 0, 0, 1}
b = {{0, -1}, {0, -1}, {0, -1}, {0, -1}, {0, -1}, {1, 0}}
LinearProgramming[c, m, b]
```

Mathematica gives the output

$$\left\{ y_1 = \frac{7}{90}, y_2 = \frac{1}{30}, y_3 = \frac{16}{45}, y_4 = \frac{8}{15}, v = \frac{14}{9} \right\}.$$

The solution for player I is found similarly, but we have to be careful because we have to put it into the form for the **LinearProgramming** in Mathematica, which always minimizes and the constraints in Mathematica form are $\geq$. Player I's problem is a maximization problem so the cost vector is

$$c = (0, 0, 0, 0, 0, -1)$$

corresponding to the variables $x_1, x_2, \ldots, x_5$, and $-v$. The constraint matrix is the transpose of the Blotto matrix but with a row added for the constraint $x_1 + \cdots + x_5 = 1$. It becomes

$$q = \begin{bmatrix} 4 & 0 & 1 & -1 & -2 & -1 \\ 0 & 4 & -1 & 1 & -2 & -1 \\ 2 & 1 & 3 & 0 & 2 & -1 \\ 1 & 2 & 0 & 3 & 2 & -1 \\ 1 & 1 & 1 & 1 & 1 & 0 \end{bmatrix}.$$

Finally, the b vector becomes

$$b = \{\{0, 1\}, \{0, 1\}, \{0, 1\}, \{0, 1\}, \{1, 0\}\}.$$

In each pair, the second number 1 means that the inequalities are of the form $\geq$. In the last pair, the second 0 means the inequality is actually $=$. For instance, the first and last constraints are, respectively

$$4x_1 + 0x_2 + x_3 - x_4 - 2x_5 - v \geq 0 \quad \text{and} \quad x_1 + x_2 + x_3 + x_4 + x_5 = 1.$$

With this setup the Mathematica command to solve is simply

LinearProgramming[c, q, b]
with output

$$\left\{ x_1 = \frac{4}{9}, x_2 = \frac{4}{9}, x_3 = 0, x_4 = 0, x_5 = \frac{1}{9}, v = \frac{14}{9} \right\}.$$

Naturally, Blue, having more regiments, will come out ahead, and a rational opponent (Red) would capitulate before the game even began. Observe also that with two equally valued targets, it is optimal for the superior force (Blue) to not divide its regiments, but for the inferior force to split its regiments, except for a small probability of doing the opposite.

1.6.3 Using Calculus

Calculus introduces one of the main topics in optimization, namely how to find minima, maxima, and saddle points of functions and we will see how it works to find completely mixed saddle points in games. For simplicity we will only consider 2×2 games because calculations get complicated for bigger games and there are better methods anyway.

Write $f(x, y) = E(X, Y)$, where $X = (x, 1 - x), Y = (y, 1 - y), 0 \leq x, y \leq 1$. Then

$$f(x, y) = (x, 1 - x) \begin{bmatrix} a_{11} & a_{12} \\ a_{21} & a_{22} \end{bmatrix} \begin{bmatrix} y \\ 1 - y \end{bmatrix} \tag{1.6.1}$$

$$= x[y(a_{11} - a_{21}) + (1 - y)(a_{12} - a_{22})] + y(a_{21} - a_{22}) + a_{22}.$$

By assumption there are no optimal pure strategies, and so the extreme points of f will be found inside the intervals $0 < x, y < 1$. But that means that we may take the partial derivatives of f with respect to x and y and set them equal to zero to find all the possible critical points. The function f has to look like the function depicted in Figure 1.10.

Figure 1.10 is the graph of $f(x, y) = X A Y^T$ taking

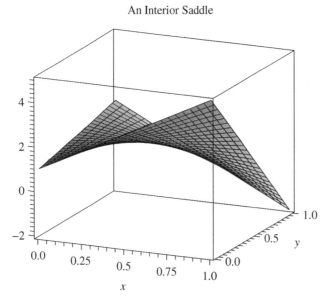

An Interior Saddle

Figure 1.10 Concave in x, convex in y, saddle at $(\frac{1}{8}, \frac{1}{2})$.

$A = \begin{bmatrix} -2 & 5 \\ 2 & 1 \end{bmatrix}$. It is a concave–convex function which has a saddle point at $x = \frac{1}{8}, y = \frac{1}{2}$. You can see now why it is called a **saddle**.

Returning to our general function $f(x, y)$ in (1.6.1), take the partial derivatives and set to zero:

$$\frac{\partial f}{\partial x} = y\alpha + \beta = 0 \quad \text{and} \quad \frac{\partial f}{\partial y} = x\alpha + \gamma = 0,$$

where we have set

$$\alpha = (a_{11} - a_{12} - a_{21} + a_{22}), \quad \beta = (a_{12} - a_{22}), \quad \gamma = (a_{21} - a_{22}).$$

Notice that if $\alpha = 0$, the partials are never zero (assuming $\beta, \gamma \neq 0$), and that would imply that there are pure optimal strategies (in other words, the min and max must be on the boundary). The existence of a pure saddle is ruled out by assumption. We solve where the partial derivatives are zero to get

$$x^* = -\frac{\gamma}{\alpha} = \frac{a_{22} - a_{21}}{a_{11} - a_{12} - a_{21} + a_{22}} \quad \text{and} \quad y^* = -\frac{\beta}{\alpha} = \frac{a_{22} - a_{12}}{a_{11} - a_{12} - a_{21} + a_{22}},$$

which is the same as our previous result on 2×2 games. How do we know this is a saddle and not a min or max of f? The reason is because if we take the second derivatives, we get the matrix of second partials (called the **Hessian**):

$$H = \begin{bmatrix} f_{xx} & f_{xy} \\ f_{yx} & f_{yy} \end{bmatrix} = \begin{bmatrix} 0 & \alpha \\ \alpha & 0 \end{bmatrix}$$

Since $\det(H) = -\alpha^2 < 0$ (unless $\alpha = 0$, which is ruled out) a theorem in elementary calculus says that an interior critical point with this condition must be a saddle.[1]

So the procedure is as follows. Look for pure strategy solutions first (calculate v^- and v^+ and see if they're equal), but if there aren't any, use the formulas to find the mixed strategy solutions.

Example 1.26 In the game with $A = \begin{bmatrix} -2 & 5 \\ 2 & 1 \end{bmatrix}$ we see that $v^- = 1 < v^+ = 2$, so there is no pure saddle for the game. If we apply the formulas, we get the mixed strategies $X^* = \left(\frac{1}{8}, \frac{7}{8} \right)$ and $Y^* = \left(\frac{1}{2}, \frac{1}{2} \right)$ and the value of the game is $v(A) = \frac{3}{2}$. Notice here that

$$A^* = \begin{bmatrix} 1 & -5 \\ -2 & -2 \end{bmatrix}$$

and $\det(A) = -12$. The matrix formula for player I gives

$$X^* = \frac{(1\ 1)A^*}{(1\ 1)A^*(1\ 1)^T} = \left(\frac{1}{8}, \frac{7}{8} \right).$$

1.6.4 Symmetric Games

Symmetric games are important classes of two-person games in which the players can use the exact same set of strategies and any payoff that player I can obtain using strategy X can be obtained by player II using the same strategy $Y = X$. The two players can switch roles, and the game doesn't

---

1 The calculus definition of a saddle point of a function $f(x, y)$ is a point so that in every neighborhood of the point there are x and y values that make f bigger and smaller than f at the candidate saddle point.

change. Such games can be quickly identified by the rule that $A = -A^T$. Any matrix satisfying this is said to be **skew symmetric**. If we want the roles of the players to be symmetric, then we need the matrix to be skew symmetric.

Why is skew symmetry the correct thing? Well, if A is the payoff matrix to player I, then the entries represent the payoffs to player I and the negative of the entries, or $-A$ represent the payoffs to player II. So player II wants to maximize the column entries in $-A$. This means that from player II's perspective, the game matrix must be $(-A)^T$ because it is always the row player by convention who is the maximizer; that is, A is the payoff matrix to player I and $-A^T$ is the payoff to player II. If we want the payoffs to player II to be the same as the payoffs to player I, then we must have the same game matrices for each player and so $A = -A^T$. If this is the case, the matrix must be square, $a_{ij} = -a_{ji}$, and the diagonal elements of A, namely, a_{ii}, must be 0. We can say more. In what follows it is helpful to keep in mind that for any appropriate size matrices $(AB)^T = B^T A^T$. Furthermore, since XAY^T is a scalar $(XAY^T)^T = XAY^T = YA^T X^T$.

Since the players may change roles without affecting the game, the following theorem seems reasonable.

Theorem 1.6.1 *For any skew symmetric game $v(A) = 0$ and if X^* is optimal for player I, then it is also optimal for player II. In particular, there are saddle points where the players are using the same strategy. These are called* **symmetric saddle points.**

Proof: Suppose X and Y are strategies. Then

$$E(X, Y) = XAY^T = X(-A^T)Y^T = -YAX^T = -E(Y, X).$$

If (X^*, Y^*) is a saddle point $v(A) = E(X^*, Y^*) = -E(Y^*, X^*)$ and for all X, Y,

$$E(X, Y^*) \leq E(X^*, Y^*) \leq E(X^*, Y) \implies -E(Y^*, X) \leq -E(Y^*, X^*) \leq -E(Y, X^*)$$

so that $E(Y^*, X) \geq E(Y^*, X^*) \geq E(Y, X^*)$. This says $v(A) = E(Y^*, X^*)$ and (Y^*, X^*) is also a saddle point. We also know $-E(Y^*, X^*) = v(A) = E(Y^*, X^*)$ which makes $v(A) = 0$. ∎

Example 1.27 A ubiquitous game is rock–paper–scissors. Each player chooses one of rock, paper, or scissors. The rules are Rock beats Scissors; Scissors beats Paper; Paper beats Rock. We'll use the following payoff matrix:

I/II	Rock	Paper	Scissors
Rock	0	$-a$	b
Paper	a	0	$-c$
Scissors	$-b$	c	0

Assume that $a, b, c > 0$. Then $v^- = \max\{-a, -c, -b\} < 0$ and $v^+ = \min\{a, c, b\} > 0$ so this game does not have a pure saddle point. Since the matrix is skew symmetric we know that $v(A) = 0$ and $X^* = Y^*$ is a saddle point, so we need only find $X^* = (x_1, x_2, x_3)$. We may use several methods to solve this and we will use the results in Remark 1.4.9. We have

$$x_2 a - x_3 b \geq 0, \quad -ax_1 + cx_3 \geq 0, \quad x_1 b - x_2 c \geq 0 \implies x_2 \frac{a}{b} \geq x_3 \geq \frac{a}{c}\frac{c}{b} x_2.$$

Therefore, we have equality throughout, and

$$x_3 \left(\frac{c}{a} + \frac{b}{a} + 1 \right) = 1 \implies x_3 = \frac{a}{a+b+c}, \quad x_1 = \frac{c}{a+b+c}, \quad x_2 = \frac{b}{a+b+c}.$$

In the standard game $a = b = c = 1$ and the optimal strategy for each player is $X^* = (\frac{1}{3}, \frac{1}{3}, \frac{1}{3})$, which is what you would expect when the payoffs are all the same. But if the payoffs are not the same, say $a = 1, b = 2, c = 3$ then the optimal strategy for each player is $X^* = (\frac{3}{6}, \frac{2}{6}, \frac{1}{6})$.

Example 1.28 Alexander Hamilton was challenged to a duel with pistols by Aaron Burr after Hamilton wrote a defamatory article about Burr in New York. We will analyze one version of such a duel by the two players as a game with the object of trying to find the optimal point at which to shoot. First, here are the rules.

Each pistol has exactly one bullet. They will face each other starting at 10 paces apart and walk toward each other, each deciding when to shoot. In a silent duel a player does not know whether the opponent has taken the shot. In a noisy duel, the players know when a shot is taken. This is important because if a player shoots and misses, he is certain to be shot by the person he missed. The Hamilton–Burr duel will be considered as a silent duel because it is more interesting. We leave as an exercise the problem of the noisy duel.

We assume that each player's accuracy will increase the closer the players are. In a simplified version, suppose that they can choose to fire at 10 paces, 6 paces, or 2 paces. Suppose also that the probability that a shot hits and kills the opponent is 0.2 at 10 paces, 0.4 at 6 paces, and 1.0 at 2 paces. An opponent who is hit is assumed killed.

If we look at this as a zero sum two-person game, the player strategies consist of the pace distance at which to take the shot. The row player is Burr (B), and the column player is Hamilton (H). Incidentally, it is worth pointing out that the game analysis should be done before the duel is actually carried out.

We assume that the payoff to both players is +1 if they kill their opponent, −1 if they are killed, and 0 if they both survive.

So here is the matrix setup for this game.

B/H	10	6	2
10	0	−0.12	−0.6
6	0.12	0	−0.2
2	0.6	0.2	0

To see where these numbers come from, let's consider the pure strategy $(6, 10)$, so Burr waits until 6 to shoot (assuming that he survived the shot by Hamilton at 10) and Hamilton chooses to shoot at 10 paces. Then the expected payoff to Burr is

$$(+1) \cdot Prob(\text{H misses at 10}) \cdot Prob(\text{Kill H at 6})$$

$$+ (-1) \cdot Prob(\text{Killed by H at 10}) = 0.8 \cdot 0.4 - 0.2 = 0.12.$$

The rest of the entries are derived in the same way.

This is a symmetric game with skew symmetric matrix so the value is zero and the optimal strategies are the same for both Burr and Hamilton, as we would expect since they have the same accuracy functions. In this example, there is a pure saddle at position $(3, 3)$ in the matrix, so that $X^* = (0, 0, 1)$ and $Y^* = (0, 0, 1)$. Both players should wait until the probability of a kill is certain.

To make this game a little more interesting, suppose that the players will be penalized if they wait until 2 paces to shoot. In this case we may use the matrix

B/H	10	6	2
10	0	−0.12	1
6	0.12	0	−0.2
2	−1	0.2	0

This is still a symmetric game with skew symmetric matrix, so the value is still zero and the optimal strategies are the same for both Burr and Hamilton. To find the optimal strategy for Burr, we can remember the formulas or, even better, the procedure. So here is what we get knowing that $E(X^*, j) \geq 0, j = 1, 2, 3$:

$$E(X^*, 1) = 0.12 \, x_2 - 1 \cdot x_3 \geq 0,$$

$$E(X^*, 2) = -0.12 \, x_1 + 0.2 \, x_3 \geq 0 \quad \text{and} \quad E(X^*, 3) = x_1 - 0.2 \, x_2 \geq 0.$$

These give us

$$x_2 \geq \frac{x_3}{0.12}, \quad \frac{x_3}{0.12} \geq \frac{x_1}{0.2}, \quad \text{and} \quad \frac{x_1}{0.2} \geq x_2,$$

which means equality all the way. Consequently

$$x_1 = \frac{0.2}{0.12 + 1 + 0.2} = \frac{0.2}{1.32}, \quad x_2 = \frac{1}{1.32}, \quad x_3 = \frac{0.12}{1.32}$$

or, $x_1 = 0.15, x_2 = 0.76, x_3 = 0.09$, so each player will shoot, with probability 0.76 at 6 paces.

In the real duel, that took place on July 11, 1804, Alexander Hamilton, who was the first US Secretary of the Treasury and widely considered to be a future president and a genius, was shot by Aaron Burr, who was Thomas Jefferson's vice president of the United States and who also wanted to be president. In fact, Hamilton took the first shot, shattering the branch above Burr's head. Witnesses report that Burr took his shot some 3 or 4 seconds after Hamilton. Whether or not Hamilton deliberately missed his shot is disputed to this day. Nevertheless, Burr's shot hit Hamilton in the torso and lodged in his spine, paralyzing him. Hamilton died of his wounds the next day. Aaron Burr was charged with murder but was later either acquitted or the charge was dropped (dueling was in the process of being outlawed). The duel was the end of the ambitious Burr's political career, and he died an ignominious death in exile. Interestingly, both Burr and Hamilton had been involved in numerous duels in the past. In a tragic historical twist, Alexander Hamilton's son, Phillip, was earlier killed in a duel on November 23, 1801.

Problems

1.44 Consider the game with matrix

$$\begin{bmatrix} 3 & -2 & 4 & 7 \\ -2 & 8 & 4 & 0 \end{bmatrix}.$$

(a) Solve the game using the graphical method.

(b) Find the best response for player I to the strategy $Y = \left(\frac{1}{4}, \frac{1}{2}, \frac{1}{8}, \frac{1}{8} \right)$.

(c) What is II's best response to I's best response?

1.45 Find the matrix for a noisy Hamilton–Burr duel and solve the game.

1.46 Each player displays either one or two fingers and simultaneously guesses how many fingers the opposing player will show. If both players guess correctly or both incorrectly, the game is a draw. If only one guesses correctly, that player wins an amount equal to the total number of fingers shown by both players. Each pure strategy has two components: the number of fingers to show and the number of fingers to guess. Find the game matrix and the optimal strategies.

1.47 This exercise shows that symmetric games are more general than they seem at first and in fact this is the main reason they are important. Assuming that $A_{n \times m}$ is **any** payoff matrix with $value(A) > 0$, define the matrix B that will be of size $(n + m + 1) \times (n + m + 1)$, by

$$B = \begin{bmatrix} 0 & A & -\vec{1} \\ -A^T & 0 & \vec{1} \\ \vec{1} & -\vec{1} & 0 \end{bmatrix}.$$

The notation $\vec{1}$, for example in the third row and first column, is the $1 \times n$ matrix consisting of all 1s. B is a skew symmetric matrix and it can be shown that if $P = (p_1, \ldots, p_n, q_1, \ldots, q_m, \gamma)$ is an optimal strategy for matrix B, then, setting

$$b = \sum_{i=1}^{n} p_i = \sum_{j=1}^{m} q_j > 0, \quad x_i = \frac{p_i}{b}, \quad y_j = \frac{q_j}{b},$$

we have $X = (x_1, \ldots, x_n), Y = (y_1, \ldots, y_m)$ as a saddle point for the game with matrix A. In addition, $value(A) = \frac{\gamma}{b}$. The converse is also true. Verify all these points with the matrix

$$A = \begin{bmatrix} 5 & 2 & 6 \\ 1 & \frac{7}{2} & 2 \end{bmatrix}.$$

1.48 Following the same procedure as that for player I, look at $E(i, Y), i = 1, 2$. with $Y = (y, 1 - y)$. Graph the lines $E(1, Y) = y + 4(1 - y)$ and $E(2, Y) = 3y + 2(1 - y), 0 \leq y \leq 1$. Now, how does player II analyze the graph to find Y^*?

1.49 Find the value and optimal X^* for the games with matrices

(a) $\begin{bmatrix} 1 & 0 \\ -1 & 2 \end{bmatrix}$ (b) $\begin{bmatrix} 3 & 1 \\ 5 & 7 \end{bmatrix}$

What, if anything, goes wrong with (b) if you use the graphical method?

1.50 Solve by the linear programming method the game with the matrix

$$A = \begin{bmatrix} 0 & -1 & 1 \\ 1 & 0 & -1 \\ -1 & 1 & 0 \end{bmatrix}$$

1.51 Curly has two safes, one at home and one at the office. The safe at home is a piece of cake to crack and any thief can get into it. The safe at the office is hard to crack and a thief has only a 15% chance of doing it. Curly has to decide where to place his gold bar (worth 1). On the other hand, if the thief hits the wrong place he gets caught (worth -1 to the thief and $+1$ to Curly). Formulate this as a two person zero sum matrix game and solve it using the graphical method.

1.52 Let z be an unknown number and consider the matrices

$$A = \begin{bmatrix} 0 & z \\ 1 & 2 \end{bmatrix} \quad \text{and} \quad B = \begin{bmatrix} 2 & 1 \\ z & 0 \end{bmatrix}$$

(a) Find $v(A)$ and $v(B)$ for any z.

(b) Now consider the game with matrix $A + B$. Find a value of z so that $v(A + B) < v(A) + v(B)$ and a value of z so that $v(A + B) > v(A) + v(B)$. Find the values of $A + B$ using the graphical method. This problem shows that the value is not a linear function of the matrix.

1.53 Suppose that we have the game matrix

$$A = \begin{bmatrix} 13 & 29 & 8 \\ 18 & 22 & 31 \\ 23 & 22 & 19 \end{bmatrix}.$$

Why can this be reduced to $B = \begin{bmatrix} 18 & 31 \\ 23 & 19 \end{bmatrix}$? Now solve the game graphically.

1.54 Two brothers, Curly and Shemp, inherit a car worth 8000 dollars. Since only one of them can actually have the car, they agree they will present sealed bids to buy the car from the other brother. The brother that puts in the highest sealed bid gets the car. They must bid in 1000 dollar units. If the bids happen to be the same, then they flip a coin to determine ownership and no money changes hands. Curly can bid only up to 5000 while Shemp can bid up to 8000.

Find the payoff matrix with Curly as the row player and the payoffs the expected net gain (since the car is worth 8000). Find v^-, v^+ and use dominance to solve the game.

1.55 In the 2×2 Nim game we saw that $v^+ = v^- = -1$. Reduce the game matrix using dominance.

1.56 Consider the matrix game

$$A = \begin{bmatrix} 2 & 0 \\ 0 & 2 \end{bmatrix}.$$

(a) Find $v(A)$ and the optimal strategies.
(b) Show that $X^* = (\frac{1}{2}, \frac{1}{2})$, $Y^* = (1, 0)$ is not a saddle point for the game even though it does happen that $E(X^*, Y^*) = v(A)$.

1.57 Use the methods of this section to solve the games

$$(a) \begin{bmatrix} 4 & -3 \\ -9 & 6 \end{bmatrix}, \quad (b) \begin{bmatrix} 4 & 9 \\ 6 & 2 \end{bmatrix}, \quad (c) \begin{bmatrix} -3 & -4 \\ -7 & 2 \end{bmatrix}.$$

1.58 Use (convex) dominance and the graphical method to solve the game with matrix

$$A = \begin{bmatrix} 0 & 5 \\ 1 & 4 \\ 3 & 0 \\ 2 & 2 \end{bmatrix}.$$

1.59 The third column of the matrix

$$A = \begin{bmatrix} 0 & 8 & 5 \\ 8 & 4 & 6 \\ 12 & -4 & 3 \end{bmatrix}$$

is dominated by a convex combination. Reduce the matrix and solve the game.

1.60 Four army divisions attack a town along two possible roads. The town has three divisions defending it. A defending division is dug in and hence equivalent to two attacking divisions. Even one division attacking an undefended road captures the town. Each commander must decide how many divisions to attack or defend each road. If the attacking commander captures a road to the town, the town falls. Score 1 to the attacker if the town falls and −1 if it doesn't.
(a) Find the payoff matrix with payoff the attacker's probability of winning the town.
(b) Find the value of the game and the optimal saddle point.

1.61 Consider the matrix game $A = \begin{bmatrix} a_4 & a_3 & a_3 \\ a_1 & a_6 & a_5 \\ a_2 & a_4 & a_3 \end{bmatrix}$, where $a_1 < a_2 < \cdots < a_5 < a_6$. Use dominance to solve the game.

1.62 Aggie and Baggie are fighting a duel each with one lemon meringue pie starting at 20 paces. They can each choose to throw the pie at either 20 paces, 10 paces, or 0 paces. The probability either player hits the other at 20 paces is $\frac{1}{3}$; at 10 paces it is $\frac{2}{3}$ and at 0 paces it is 1. If they both hit or both miss at the same number of paces, the game is a draw. If a player gets a pie in the face, the score is −1, the player throwing the pie gets +1.
Set this up as a matrix game and solve it.

1.63 Consider the game with matrix
$$A = \begin{bmatrix} -2 & 3 & 5 & -2 \\ 3 & -4 & 1 & -6 \\ -5 & 3 & 2 & -1 \\ -1 & -3 & 2 & 2 \end{bmatrix}.$$
Someone claims that the strategies $X^* = \left(\frac{1}{9}, 0, \frac{8}{9}, 0\right)$ and $Y^* = \left(0, \frac{7}{9}, \frac{2}{9}, 0\right)$ are optimal.
(a) Is that correct? Why or why not?
(b) If $X^* = \left(\frac{13}{33}, \frac{5}{33}, 0, \frac{15}{33}\right)$ is optimal and $v(A) = -\frac{26}{33}$, find Y^*.

1.64 In the baseball game Problem 1.11 it turns out that an optimal strategy for player I, the batter, is given by $X^* = (x_1, x_2, x_3) = \left(\frac{2}{7}, 0, \frac{5}{7}\right)$ and the value of the game is $v = \frac{2}{7}$. It is amazing that the batter should never expect a curveball with these payoffs under this optimal strategy. What is the pitcher's optimal strategy Y^*?

1.65 In a football game we use the matrix $A = \begin{bmatrix} 3 & 6 \\ 8 & 0 \end{bmatrix}$. The offense is the row player. The first row is Run, the second is Pass. The first column is Defend against Run, the second is Defend against the Pass.
(a) Use the graphical method to solve this game.
(b) Now suppose the offense gets a better quarterback so the matrix becomes $A = \begin{bmatrix} 3 & 6 \\ 12 & 0 \end{bmatrix}$. What happens?

1.66 Two players Reinhard and Carla play a number game. Carla writes down a number either 1, 2 or 3. Reinhard chooses a number (again 1, 2 or 3) and guesses that Carla has written down that number. If Reinhard guesses right he wins $1 from Carla; if he guesses wrong, Carla tells him if his number is higher or lower and he gets to guess again. If he is right, no money changes hands but if he guesses wrong he pays Carla $1.

(a) Find the game matrix with Reinhard as the row player and find the upper and lower values. A strategy for Reinhard is of the form

[first guess, guess if low, guess if high]

(b) Find the value of the game by first noticing that Carla's strategy 1 and 3 are symmetric as are $[1, -, 2]$, $[3, 2, -]$, and $[1, -, 3]$, $[3, 1-]$ for Reinhard. Then modify the graphical method slightly to solve.

1.67 We have an infinite sequence of numbers $0 < a_1 \leq a_2 \leq a_3 \leq \cdots$. Each of 2 players chooses an integer independent of the other player. If they both happen to choose the same number k, then player I receives a_k dollars from player II. Otherwise, no money changes hand. Assume that $\sum_{k=1}^{\infty} \frac{1}{a_k} < \infty$.

(a) Find the game matrix (it will be infinite). Find v^+, v^-.

(b) Find the value of the game if mixed strategies are allowed and find the saddle point in mixed strategies. Use Theorem 1.4.6.

(c) Assume next that $\sum_{k=1}^{\infty} \frac{1}{a_k} = \infty$. Show that the value of the game is $v = 0$ and every mixed strategy for player I is optimal but there is no optimal strategy for player II.

1.68 **Calculus to Find the Interior Saddle Points:** Consider

$$A = \begin{bmatrix} 4 & -3 & -2 \\ -3 & 4 & -2 \\ 0 & 0 & 1 \end{bmatrix}.$$

A strategy for each player is of the form $X = (x_1, x_2, 1 - x_1 - x_2)$, $Y = (y_1, y_2, 1 - y_1 - y_2)$, so we consider the function $f(x_1, x_2, y_1, y_2) = XAY^T$. Now solve the system of equations

$$f_{x_1} = f_{x_2} = f_{y_1} = f_{y_2} = 0$$

to get X^* and Y^*. If these are legitimate completely mixed strategies, then you can verify that they are optimal and then find $v(A)$. Carry out these calculations for A and verify that they give optimal strategies.

1.69 Show that for any strategy $X = (x_1, \ldots, x_m) \in S_m$ and any numbers $b_1, \ldots, b_m$, it must be that

$$\max_{X \in S_m} \sum_{i=1}^{m} x_i b_i = \max_{1 \leq i \leq m} b_i \quad \text{and} \quad \min_{X \in S_m} \sum_{i=1}^{m} x_i b_i = \min_{1 \leq i \leq m} b_i.$$

1.70 The properties of optimal strategies (Remark 1.4.9) show that $X^* \in S_n$ and $Y^* \in S_m$ are optimal if and only if $\min_j E(X^*, j) = \max_i E(i, Y^*)$. The common value will be the value of the game. Verify this.

1.71 Show that if (X^*, Y^*) and (X^0, Y^0) are both saddle points for the game with matrix A, then so is (X^*, Y^0) and (X^0, Y^*). In fact, show that (X_λ, Y_β) where $X_\lambda = \lambda X^* + (1 - \lambda)X^0$, $Y_\beta = \beta Y^* + (1 - \beta)Y^0$ and λ, β any numbers in $[0, 1]$, is also a saddle point. Thus if there are two saddle points, there are an infinite number.

1.7 Existence of Saddle Points: The Von Neumann Minimax Theorem

So far we have understood what saddle points are, why there are a reasonable definition of optimal strategies, and a number of ways to look for them. We have also asserted that, if we allow mixed strategies, two player matrix games will always have saddle points if we allow mixed strategies. How do we know that for a fact? It will follow from the celebrated von Neumann inimax Theorem, due to John von Neumann.

Remark Before stating the theorem, it is helpful to realize that not every game will have a saddle point. For instance, here is a silly game that does not: two players each pick a number. Whoever picks the highest number wins $1. There cannot possibly be a saddle point because, as soon as you pick a number, you will wish you picked a bigger one. There will have to be some assumptions on games to know that saddle points exist.

1.7.1 Statement of the Minimax Theorem

We start by considering general functions of two variables $f = f(x, y)$, and give the definition of a saddle point for an arbitrary of function f.

Definition 1.7.1 *Let C and D be sets. A function $f : C \times D \to \mathbb{R}$ has at least one saddle point (x^*, y^*) with $x^* \in C$ and $y^* \in D$ if*

$$f(x, y^*) \leq f(x^*, y^*) \leq f(x^*, y) \quad \text{for all} \quad x \in C, y \in D.$$

We can think of this function as a game: Player I gets to choose a point $x \in C$ and player II gets to choose a point in $y \in D$. The payoff is $f(x, y)$ to Player I. This type of game is called a **continuous game**. We can define upper and lower value by

$$v^+ = \min_{y \in D} \max_{x \in C} f(x, y), \quad \text{and} \quad v^- = \max_{x \in C} \min_{y \in D} f(x, y).$$

You can check as before that $v^- \leq v^+$. If $v^+ = v^-$ we say, as usual, that the **game has a value** $v = v^+ = v^-$. The **von Neumann Minimax Theorem** gives conditions on $f, C,$ and D so that the associated game has a value $v = v^+ = v^-$. It will be used to determine what we need to do in matrix games in order to get a value.

Theorem 1.7.2 *Let $f : C \times D \to \mathbb{R}$ be a continuous function. Let $C \subset \mathbb{R}^n$ and $D \subset \mathbb{R}^m$ be convex, closed, and bounded.[1] Suppose that as a function of $x, f(x, y)$ is concave and as a function of $y, f(x, y)$ is convex. Then*

$$v^+ = \min_{y \in D} \max_{x \in C} f(x, y) = \max_{x \in C} \min_{y \in D} f(x, y) = v^-.$$

Von Neumann's theorem tells us what we need in order to guarantee that our game has a value. It is critical that we are dealing with a concave–convex function, and that the strategy sets be convex and bounded. The bounded assumption eliminates the silly game of choosing the largest number.

---

1 Convex means the line segment connecting any two points in the set is also in the set. Closed means the set contains all its limit points, and Bounded means the set can be jammed inside a large enough ball.

If you're interested in seeing a proof of this major theorem refer to the appendix Section 1.9 of this chapter. There is also a really cool proof of the theorem for matrix games that uses evolutionary game theory at the end of the book.

One observation: Recall the common calculus test for twice differentiable functions of one variable. If $g = g(x)$ is a function of one variable and has at least two derivatives, then g is convex if $g'' \geq 0$ and g is concave if $g'' \leq 0$.

Example 1.29 For an example of the use of von Neumann's theorem, suppose we look at

$$f(x, y) = 4xy - 2x - 2y + 1 \quad \text{on} \quad 0 \leq x, y \leq 1.$$

This function has $f_{xx} = 0 \geq 0, f_{yy} = 0 \leq 0$, so it is convex in y for each x and concave in x for each y. Since $(x, y) \in [0, 1] \times [0, 1]$, and the square is closed and bounded, von Neumann's theorem guarantees the existence of a saddle point for this function. To find it, solve $f_x = f_y = 0$ to get $x = y = \frac{1}{2}$. The Hessian for f, which is the matrix of second partial derivatives, is given by

$$H(f, [x, y]) = \begin{bmatrix} f_{xx} & f_{xy} \\ f_{yx} & f_{yy} \end{bmatrix} = \begin{bmatrix} 0 & 4 \\ 4 & 0 \end{bmatrix}.$$

Since $\det(H) = -16 < 0$ we are guaranteed by elementary calculus that $(x = \frac{1}{2}, y = \frac{1}{2})$ is an interior saddle for f. Here is a picture of f:

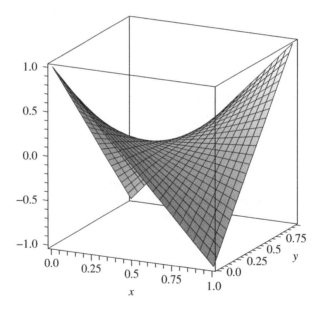

Incidentally, another way to write our example function would be

$$f(x, y) = (x, 1 - x)A(y, 1 - y)^T, \quad \text{where} \quad A = \begin{bmatrix} 1 & -1 \\ -1 & 1 \end{bmatrix}.$$

Obviously, not all functions will have saddle points. For instance, you will show in an exercise that $g(x, y) = (x - y)^2$ is not concave–convex and in fact does not have a saddle point in $[0, 1] \times [0, 1]$.

1.7.2 Von Neumann's Theorem Guarantees Matrix Games Have Saddle Points

Does a game with matrix A have a saddle point in mixed strategies? von Neumann's minimax theorem tells us the answer is **Yes**.

The minimax theorem basically says that if we have a continuous function $f(x, y)$ which is concave in x and convex in y (called a saddle function, and the variables x and y are members of convex sets, then $\min_y \max_x f(x, y) = \max_y \min_y f(x, y)$. To apply this to a matrix game all we need to do is define the function

$$f(X, Y) \equiv E(X, Y) = XAY^T$$

and the sets S_n for X, and S_m for Y. For any $n \times m$ matrix A, this function is concave in X and convex in Y. Actually, it is even **linear** in each variable when the other variable is fixed. Recall that any linear function is both concave and convex, so our function f is concave–convex and certainly continuous. The second requirement of von Neumann's theorem is that the sets S_n and S_m be convex sets. This is very easy to check and we leave that as an exercise for the reader. These sets S_n and S_m are also closed and bounded. Consequently, we may apply the general minimax theorem to conclude the following.

Theorem 1.7.3   *For any $n \times m$ matrix A, we have*

$$v^+ = \min_{Y \in S_m} \max_{X \in S_n} XAY^T = \max_{X \in S_n} \min_{Y \in S_m} XAY^T = v^-.$$

The common value is denoted $v(A)$, or value(A), and that is the **value of the game**. *In addition, there is at least one saddle point $X^* \in S_n, Y^* \in S_m$ so that*

$$E(X, Y^*) \leq E(X^*, Y^*) = v(A) \leq E(X^*, Y), \quad \text{forall} \quad X \in S_n, Y \in S_m.$$

Remark The theorem tells us that a saddle point for a matrix game always exists if mixed strategies are used. It does not give us a way to find it. Also, there could be more than one saddle point.

Problems

1.72   Use any method of your choice to solve the games with the following matrices.

(a) $\begin{bmatrix} 0 & 3 & 3 & 2 & 4 \\ 4 & 4 & 3 & 1 & 4 \end{bmatrix}$;   (b) $\begin{bmatrix} 4 & 4 & -4 & -1 \\ -4 & -2 & 4 & 4 \\ 2 & -4 & -1 & -5 \\ -3 & 1 & 0 & -4 \end{bmatrix}$;   (c) $\begin{bmatrix} 2 & -5 & 3 & 0 \\ -4 & -5 & -5 & -6 \\ 3 & -4 & -1 & -2 \\ 0 & 4 & 1 & 3 \end{bmatrix}$

1.73   Let $f(x, y) = x^2 + y^2, C = D = [-1, 1]$. Find $v^+ = \min_{y \in D} \max_{x \in C} f(x, y)$ and $v^- = \max_{x \in C} \min_{y \in D} f(x, y)$.

1.74   Let $f(x, y) = y^2 - x^2, C = D = [-1, 1]$.
(a) Find $v^+ = \min_{y \in D} \max_{x \in C} f(x, y)$ and $v^- = \max_{x \in C} \min_{y \in D} f(x, y)$.
(b) Show that $(0, 0)$ is a pure saddle point for $f(x, y)$.

1.75   Let $f(x, y) = (x - y)^2, C = D = [-1, 1]$. Find $v^+ = \min_{y \in D} \max_{x \in C} f(x, y)$ and $v^- = \max_{x \in C} \min_{y \in D} f(x, y)$.

1.76 Solve this game by setting it up as a linear program:

$$\begin{bmatrix} -2 & 3 & 3 & 4 & 1 \\ 3 & -2 & -5 & 2 & 4 \\ 4 & -5 & -1 & 4 & -1 \\ 2 & -4 & 3 & 4 & -3 \end{bmatrix}.$$

In each of the following problems, set up the payoff matrices and solve the games using the linear programming method with both formulations, that is, both with and without transforming variables. You may use Mathematica or any software you want to solve these linear programs.

1.77 Consider the Submarine versus Bomber game. The board is a 3×3 grid in Figure 1.11.

1	2	3
4	5	6
7	8	9

Figure 1.11 3×3 Submarine–Bomber game.

A submarine (which occupies two squares) is trying to hide from a plane which can deliver torpedoes. The bomber can fire one torpedo at a square in the grid. If it is occupied by a part of the submarine, the sub is destroyed (score 1 for the bomber). If the bomber fires at an unoccupied square, the sub escapes (score 0 for the bomber). The bomber should be the row player.
  (a) Formulate this game as a matrix and solve it. (Hint: there are 12 pure strategies for the sub and 9 for the bomber.)
  (b) Using symmetry the game can be reduced to a 3×2 game. Find the reduction and then solve the resulting game.

1.78 There are two presidential candidates, John and Dick, who will choose which states they will visit to garner votes. Suppose that there are 4 states that are in play, but candidate John only has the money to visit 3 states. If a candidate visits a state, the gain (or loss) in poll numbers is indicated as the payoff in the matrix

John/Dick	State 1	State 2	State 3	State 4
State 1	1	−4	6	−2
State 2	−8	7	2	1
State 3	11	−1	3	−3

Solve the game.

1.79 Assume that we have a silent duel but the choice of a shot may be taken at 10,8,6,4, or 2 paces. The accuracy, or probability of a kill is 0.2, 0.4, 0.6, 0.8, and 1, respectively, at the paces. Set up and solve the game.

1.80 Consider the Cat vs Rat game in Problem 1.42. Suppose each player moves exactly 3 segments. The payoff to the cat is 1 if they meet and 0 otherwise. Find the matrix and solve the game.

1.81 Drug runners can use three possible methods for running drugs through Florida: small plane, main highway, or back roads. The cops know this, but they can only patrol one of these methods at a time. The street value of the drugs is $100,000 if the drugs reach New York using the main highway. If they use the back roads, they have to use smaller-capacity cars so the street value drops to $80,000. If they fly, the street value is $150,000. If they get stopped, the drugs and the vehicles are impounded, they get fined, and they go to prison. This represents a loss to the drug kingpins of $90,000, by highway, $70,000 by back roads, and $130,000 if by small plane. On the main highway they have a 40% chance of getting caught if the highway is being patrolled, a 30% chance on the back roads, and a 60% chance if they fly a small plane (all assuming that the cops are patrolling that method). Set this up as a zero sum game assuming that the cops want to minimize the drug kingpins gains, and solve the game to find the best strategies the cops and drug lords should use.

1.82 LUC (Loyola University Chicago) is about to play UIC (University of Illinois Chicago) for the state tennis championship. LUC has two players: A and B, and UIC has three players: X,Y,Z. The following facts are known about the relative abilities of the players: X always beats B; Y always beats A; A always beats Z. In any other match each player has a probability $\frac{1}{2}$ of winning. Before the matches, the coaches must decide on who plays who. Assume that each coach wants to maximize the expected number of matches won (these are singles matches and there are 2 games so each coach must pick who will play game 1 and who will play game 2. The players do not have to be different in each game).
Set this up as a matrix game. Find the value of the game and the optimal strategies.

1.83 Player II chooses a number $j \in \{1, 2, \dots, n\}, n \geq 2$, while I tries to guess what it is by guessing an $i \in \{1, 2, \dots, n\}$. If the guess is correct (i.e., $i = j$) then II pays +1 to player I. If $i > j$ player II pays player I the amount b^{i-j} where $0 < b < 1$ is fixed. If $i < j$, player I wins nothing.
(a) Find the game matrix.
(b) Solve the game.

1.84 Consider the asymmetric silent duel in which the two players may shoot at paces $(10, 6, 2)$ with accuracies $(0.2, 0.4, 1.0)$ for player I, but, since II is a better shot, $(0.6, 0.8, 1.0)$ for player II. Given that the payoffs are +1 to a survivor and -1 otherwise (but 0 if a draw), set up the matrix game and solve it.

1.85 Suppose that there are four towns connected by a highway as in the following diagram:

Assume that 15% of the total populations of the four towns are nearest to town 1, 30% nearest to town two, 20% nearest to town three, and 35% nearest to town four. There are two superstores, say, I and II, thinking about building a store to serve these four towns. If both stores are in the same town or in two different towns but with the same distance to a town, then I will get a 65% market share of business. Each store gets 90% of the business of the town in which they put the store if they are in two different towns. If store I is closer to a town than II is, then I gets 90% of the business of that town. If store I is farther than store II from a town, store I still gets 40% of that town's business, except for the town II is in. Find the payoff matrix and solve the game.

1.86 Two farmers are having a fight over a disputed six-yard-wide strip of land between their farms. Both farmers think that the strip is theirs. A lawsuit is filed and the judge orders them to submit a confidential settlement offer to settle the case fairly. The judge has decided to accept the settlement offer that concedes the most land to the other farmer. In case both farmers make no concession or they concede equal amounts, the judge will favor farmer II and award her all 6 yards. Formulate this as a constant sum game assuming that both farmers' pure strategies must be the yards that they concede: 0, 1, 2, 3, 4, 5, 6. Solve the game. What if the judge awards three yards if equal concessions?

1.87 Two football teams, B and P, meet for the Superbowl. Each offense can play run right (RR), run left (RL), short pass (SP), deep pass (DP), or screen pass (ScP). Each defense can choose to blitz (BL), defend a short pass (DSP), or defend a long pass (DLP), or defend a run (DR). Suppose that team B does a statistical analysis and determines the following payoffs when they are on defense:

B/P	RR	RL	SP	DP	ScP
BL	−5	−7	−7	5	4
DSP	−6	−5	8	6	3
DLP	−2	−3	−8	6	−5
DR	3	3	−5	−15	−7

A negative payoff represents the number of yards gained by the offense, so a positive number is the number of yards lost by the offense on that play of the game. Solve this game and find the optimal mix of plays by the defense and offense. (**Caution**: This is a matrix in which you might want to add a constant to ensure $v(A) > 0$. Then subtract the constant at the end to get the real value. You do not need to do that with the strategies.)

1.88 Let $a > 0$. Use the graphical method to solve the game in which player II has an infinite number of strategies with matrix

$$\begin{bmatrix} a & 2a & \frac{1}{2} & 2a & \frac{1}{4} & 2a & \frac{1}{6} & \cdots \\ a & 1 & 2a & \frac{1}{3} & 2a & \frac{1}{5} & 2a & \cdots \end{bmatrix}.$$

Pick a value for a and solve the game with a finite number of columns to see what's going on.

1.89 A Latin square game is a square game in which the matrix A is a Latin square. A Latin square of size n has every integer from 1 to n in each row and column.

(a) Solve the game of size 5

$$A = \begin{bmatrix} 1 & 2 & 3 & 4 & 5 \\ 2 & 4 & 1 & 5 & 3 \\ 3 & 5 & 4 & 2 & 1 \\ 4 & 1 & 5 & 3 & 2 \\ 5 & 3 & 2 & 1 & 4 \end{bmatrix}.$$

(b) Prove that a Latin square game of size n has $v(A) = \frac{(n+1)}{2}$. You may need the fact that $1 + 2 + \cdots + n = \frac{n(n+1)}{2}$.

1.90 Consider the game with matrix $A = \begin{bmatrix} -3 & -3 & 2 \\ -1 & 3 & -2 \\ 3 & -1 & -2 \\ 2 & 2 & 3 \end{bmatrix}$.

(a) Show that

$$X^* = \left(\frac{3}{8}\left(1 + \frac{\lambda}{3}\right), \frac{5}{16}(1 - \lambda), \frac{5}{16}(1 - \lambda), \frac{\lambda}{2} \right), \quad Y^* = \left(\frac{1}{4}, \frac{1}{4}, \frac{1}{2} \right)$$

is not a saddle point of the game if $0 \leq \lambda \leq 1$.
(b) Solve the game.

1.91 Two players, Curly and Shemp, are betting on the toss of a fair coin. Shemp tosses the coin, hiding the result from Curly. Shemp looks at the coin. If the coin is heads, Shemp says that he has heads and demands $1 from Curly. If the coin is tails, then Shemp may tell the truth and pay Curly $1, or he may lie and say that he got a head and demands $1 from Curly. Curly can challenge Shemp whenever Shemp demands $1 to see whether Shemp is telling the truth, but it will cost him $2 if it turns out that Shemp was telling the truth. If Curly challenges the call and it turns out that Shemp was lying, then Shemp must pay Curly $2. Find the matrix and solve the game.

1.92 Wiley Coyote[1] is waiting to nab Speedy who must emerge from a tunnel with three exits, A, B, and C. B and C are relatively close together, but far away from A. Wiley can lie in wait for Speedy near an exit, but then he will catch Speedy only if Speedy uses this exit. But Wiley has two other options. He can wait between B and C, but then if Speedy comes out of A he escapes while if he comes out of B or C, Wiley can only get him with probability p. Wiley's last option is to wait at a spot overlooking all three exits, but then he catches Speedy with probability q no matter which exit Speedy uses.
(a) Find the matrix for this game for arbitrary p, q.
(b) Show that using convex dominance, Wiley's options of waiting directly at an exit dominate one or the other or both of his other two options if $p < \frac{1}{2}$ or if $q < \frac{1}{3}$.
(c) Let $p = 3/4, q = 2/5$. Solve the game.

1.93 Left and Right bet either $1 or $2. If the total amount wagered is even, then Left takes the entire pot; if it is odd, then Right takes the pot.
(a) Set up the game matrix and analyze this game.
(b) Suppose now that the amount bet is not necessarily 1 or 2: Left and Right choose their bet from a set of positive integers. Suppose the sets of options are the following
  1. Left and Right can bet any number in $L = R = \{1, 2, 3, 4, 5, 6\}$;
  2. Left and Right can bet any number in $L = R = \{2, 4, 6, 8, 9, 13\}$.
  Analyze each of these cases.

---

1 Versions of this and the remaining problems appeared on exams at the London School of Economics.

(c) Suppose Left may bet any amount in $L = \{1, 2, 31, 32\}$ and Right any amount in $R = \{2, 5, 16, 17\}$. The payoffs are as follows:

1. If Left $\in \{1, 2\}$ and Right $\in \{2, 5\}$ and Left + Right even $\implies$ Right pays Left the sum of the amounts each bet.
2. If Left $\in \{1, 2\}$ and Right $\in \{2, 5\}$ and Left + Right odd $\implies$ Left pays Right the sum of the amounts each bet.
3. If Left $\in \{31, 32\}$ **or** Right $\in \{16, 17\}$, and Left + Right odd $\implies$ Right pays Left the sum of the amounts each bet.
4. If Left $\in \{31, 32\}$ **or** Right $\in \{16, 17\}$, and Left + Right even $\implies$ Left pays Right the sum of the amounts each bet.

1.94 The evil Don Barzini has imprisoned Tessio and three of his underlings in the Corleone family (total value 4) somewhere in Brooklyn, and Clemenza and his two assistants somewhere in Queens (total value 3). Sonny sets out to rescue either Tessio or Clemenza and their associates; Barzini sets out to prevent the rescue but he doesn't know where Sonny is headed. If they set out to the same place, Barzini has an even chance of arriving first, in which case Sonny returns without rescuing anybody. If Sonny arrives first he rescues the group. The payoff to Sonny is determined as the number of Corleone family members rescued. Describe this scenario as a matrix game and determine the optimal strategies for each player.

1.95 You own two companies Uno and Due. At the end of a fiscal year Uno should have a tax bill of $3 million, and Due should have a tax bill of $1 million. By cooking the books you can file tax returns making it look as though you should not have to pay any tax at all for either Uno or Due, or both. Due to limited resources, the IRS (internal revenue service) can only audit one company. If a company with cooked books is audited, the fraud is revealed and all unpaid taxes will be levied plus a penalty of $p\%$ of the tax due.

(a) Set this up as a matrix game. Your payoffs will depend on p.
(b) In the case the fine is $p = 50\%$, what strategy should you adopt to minimize your expected tax bill?
(c) For what values of p is complete honesty optimal?

1.96 **A Nonsymmetric Noisy Duel.** We consider a nonsymmetric duel at which the two players may shoot at paces $(10, 6, 2)$ with accuracies $(0.2, 0.4, 1.0)$ each.

- If only player I survives, then player I receives payoff a.
- If only player II survives, player I gets payoff $b < a$. This assumes that the survival of player II is less important than the survival of player I.
- If both players survive, they each receive payoff zero.
- If neither player survives, player I receives payoff g.

We will take $a = 1, b = \frac{1}{2}, g = 0$.
(a) Find the expected payoff matrix for player I.
(b) Solve the game.

1.8 Review Problems

Problems

Give the precise definition, complete the statement, or answer True or False. If False, give the correct statement.

1.97 Complete the statement: In order for (X^*, Y^*) to be a saddle point and v to be the value, it is necessary and sufficient that $E(X^*, j)$ _____, for all _____ and $E(i, Y^*)$ _____ for all _____ .

1.98 State the Von Neumann Minimax theorem for two person zero sum games with matrix A.

1.99 Suppose (X^*, Y^*) is a saddle point, $v(A)$ is the value of the game and $x_k > 0$ for the kth component of X^*. What is $E(k, Y^*)$? What is the most you can say if $x_k = 0$ for $E(k, Y^*)$?

1.100 Suppose A is a game matrix and Y^0 is a given strategy for player II. Define what it means for X^* to be a best response strategy to Y^0 for player I.

1.101 Suppose $A_{n \times n}$ has an inverse A^{-1}. What three assumptions do you need to be able to use the formulas
$$v := 1/J_n A^{-1} J_n^T, \quad X^* = v J_n A^{-1}, \quad Y^* = v A^{-1} J_n^T$$
to conclude that $v = v(A)$ and (X^*, Y^*) are optimal mixed strategies?

1.102 $v(A)$ is the value and (X^*, Y^*) is a Saddle Point of the game with matrix A if and only if....

1.103 $v(A)$ is the value and (i^*, j^*) is a pure saddle point of the game with matrix A if and only if ....

1.104 For any game $v^+ \leq v(A) \leq v^-$.

1.105 For any matrix game $E(X, Y) = XAY^T$ and the value satisfies
$v(A) = \max_X \min_Y E(X, Y) = \min_Y \max_X E(X, Y)$.

1.106 If (X^*, Y^*) is a saddle point for a matrix game, then X^* is a best response strategy to Y^* for player I, but the reverse is not necessarily true.

1.107 If Y is an optimal strategy for player II in a zero sum game and $E(i, Y) < value(A)$ for some row i, then for any optimal strategy X for player I, we must have .... .

1.108 The graphical method for solving a $2 \times m$ game won't work if .... .

1.109 (X^*, Y^*) is a saddle point if and only if $\min_{j=1,2,\dots,m} E(X^*, j) = \max_{i=1,2,\dots,n} E(i, Y^*)$.

1.110 If (Y^0, X^0) is a saddle point for the game A, then $E(Y^0, X) \geq E(Y^0, X^0) \geq E(Y, X^0)$, for all strategies X for player II and Y for player I.

1.111 If $X^* = (x_1^*, \dots, x_n^*)$ is an optimal strategy for player I and $x_1^* = 0$, then we may drop row 1 when we look for the optimal strategy for player II.

1.112 For a game matrix $A_{n \times m}$ and given strategies $(X^* = (x_1, \dots, x_n), Y^* = (y_1, \dots, y_m))$, then $E(X^*, Y^*) = \sum_{i=1}^{n} x_i \, E(i, Y^*) = \sum_{j=1}^{m} y_j \, E(X^*, j)$

1.113 If $i^* = 2, j^* = 3$, is a pure saddle point in the game with matrix $A_{3 \times 4}$, then using mixed strategies, $X^* =$ _____, $Y^* =$ _____ is a mixed saddle point for A.

1.114 For a 2×2 game with matrix $A = \begin{bmatrix} a_{11} & a_{12} \\ a_{21} & a_{22} \end{bmatrix}$, the value of the game is $\frac{\det(A)}{J_2 A^* J_2^T}$, where

$$A^* = \begin{bmatrix} \text{\_\_\_\_} & \text{\_\_\_\_} \\ \text{\_\_\_\_} & \text{\_\_\_\_} \end{bmatrix}$$

1.115 The invertible matrix theorem says that if A^{-1} exists, then $v(A) = \frac{1}{J_n A^{-1} J_n^T}$ is the value of the game.

1.116 If $f(x, y) = (x, 1-x) \begin{bmatrix} a_{11} & a_{12} \\ a_{21} & a_{22} \end{bmatrix} \begin{bmatrix} y \\ 1-y \end{bmatrix}$, then a mixed saddle point for the game $X^* = (x^*, 1-x^*), Y^* = (y^*, 1-y^*), 0 < x^*, y^* < 1$, satisfies

$$\frac{\partial f}{\partial x}(x^*, y^*) = \frac{\partial f}{\partial y}(x^*, y^*) = 0.$$

1.117 If A is a skew symmetric game with $v(A) > 0$, then there is always a saddle point in which $X^* = -Y^*$.

1.118 If A is skew symmetric then for any strategy X for player I, $E(X, X) = 0$.

1.9 Appendix: A Proof of the von Neumann Minimax Theorem

Once von Neumann proved the minimax theorem many different proofs were constructed. The proof we present here uses only elementary properties of convex functions and some advanced calculus (seeDevinatz [4]). At the end of the last chapter is a newer proof using evolutionary game theory.

Theorem 1.9.1 *Let $f : C \times D \to \mathbb{R}$ be a continuous function. Let $C \subset \mathbb{R}^n$ and $D \subset \mathbb{R}^m$ be convex, closed, and bounded.*[1] *Suppose that as a function of $x, f(x, y)$ is concave and as a function of $y, f(x, y)$ is convex. Then*

$$v^+ = \min_{y \in D} \max_{x \in C} f(x, y) = \max_{x \in C} \min_{y \in D} f(x, y) = v^-.$$

---

1 Convex means the line segment connecting any two points in the set is also in the set. Closed means the set contains all its limit points, and Bounded means the set can be jammed inside a large enough ball.

Proof:

1. Assume first that f is *strictly* concave–convex, meaning that

$$f(\lambda x + (1 - \lambda)z, y) > \lambda f(x, y) + (1 - \lambda)f(z, y), \quad 0 < \lambda < 1,$$
$$f(x, \mu y + (1 - \mu)w) < \mu f(x, y) + (1 - \mu)f(x, w), \quad 0 < \mu < 1.$$

The advantage of doing this is that for each $x \in C$ there is one and only one $y = y(x) \in D$ (y depends on the choice of x) so that

$$f(x, y(x)) = \min_{y \in D} f(x, y) := g(x).$$

This defines a function $g: C \to \mathbb{R}$ that is continuous (since f is continuous on the closed bounded sets $C \times D$ and thus is uniformly continuous). Furthermore, $g(x)$ is concave since

$$g(\lambda x + (1 - \lambda)z) \geq \min_{y \in D} (\lambda f(x, y) + (1 - \lambda)f(z, y)) \geq \lambda g(x) + (1 - \lambda)g(z).$$

So, there is a point $x^* \in C$ at which g achieves its maximum:

$$g(x^*) = f(x^*, y(x^*)) = \max_{x \in C} \min_{y \in D} f(x, y).$$

2. Let $x \in C$ and $y \in D$ be arbitrary. Then, for any $0 < \lambda < 1$, we obtain

$$f(\lambda x + (1 - \lambda)x^*, y) > \lambda f(x, y) + (1 - \lambda)f(x^*, y)$$
$$\geq \lambda f(x, y) + (1 - \lambda)f(x^*, y(x^*))$$
$$= \lambda f(x, y) + (1 - \lambda)g(x^*).$$

Now take $y = y(\lambda x + (1 - \lambda)x^*) \in D$ to get

$$g(x^*) \geq f(\lambda x + (1 - \lambda)x^*, y(\lambda x + (1 - \lambda)x^*)) = g(\lambda x + (1 - \lambda)x^*)$$
$$\geq g(x^*)(1 - \lambda) + \lambda f(x, y(\lambda x + (1 - \lambda)x^*)),$$

where the first inequality follows from the fact that $g(x^*) \geq g(x)$, $\forall x \in C$. As a result, we have

$$g(x^*)[1 - (1 - \lambda)] = g(x^*)\lambda \geq \lambda f(x, y(\lambda x + (1 - \lambda)x^*)),$$

or

$$f(x^*, y(x^*)) = g(x^*) \geq f(x, y(\lambda x + (1 - \lambda)x^*)) \quad \text{for all } x \in C.$$

3. Sending $\lambda \to 0$, we see that $\lambda x + (1 - \lambda)x^* \to x^*$ and $y(\lambda x + (1 - \lambda)x^*) \to y(x^*)$. We obtain

$$f(x, y(x^*)) \leq f(x^*, y(x^*)) := v, \quad \text{for any } x \in C.$$

Consequently, with $y^* = y(x^*)$

$$f(x, y^*) \leq f(x^*, y^*) = v, \quad \forall x \in C.$$

In addition, since $f(x^*, y^*) = \min_y f(x^*, y) \leq f(x^*, y)$ for all $y \in D$, we get

$$f(x, y^*) \leq f(x^*, y^*) = v \leq f(x^*, y), \quad \forall x \in C, \quad y \in D.$$

This says that (x^*, y^*) is a saddle point and the minimax theorem holds, since

$$\min_y \max_x f(x, y) \leq \max_x f(x, y^*) \leq v \leq \min_y f(x^*, y) \leq \max_x \min_y f(x, y),$$

and so we have equality throughout because the right side is always less than the left side.

4. The last step would be to get rid of the assumption of strict concavity and convexity. Here is how it goes. For $\varepsilon >$, set

$$f_\varepsilon(x,y) \equiv f(x,y) - \varepsilon|x|^2 + \varepsilon|y|^2, \quad |x|^2 = \sum_{i=1}^{n} x_i^2, \quad |y|^2 = \sum_{j=1}^{m} y_j^2.$$

This function will be strictly concave–convex, so the previous steps apply to f_ε. Therefore, we get a point $(x_\varepsilon, y_\varepsilon) \in C \times D$ so that $v_\varepsilon = f_\varepsilon(x_\varepsilon, y_\varepsilon)$ and

$$f_\varepsilon(x, y_\varepsilon) \le v_\varepsilon = f_\varepsilon(x_\varepsilon, y_\varepsilon) \le f_\varepsilon(x_\varepsilon, y), \quad \forall x \in C, y \in D.$$

Since $f_\varepsilon(x, y_\varepsilon) \ge f(x, y_\varepsilon) - \varepsilon|x|^2$ and $f_\varepsilon(x_\varepsilon, y) \le f(x_\varepsilon, y) + \varepsilon|y|^2$, we get

$$f(x, y_\varepsilon) - \varepsilon|x|^2 \le v_\varepsilon \le f(x_\varepsilon, y) + \varepsilon|y|^2, \quad \forall (x, y) \in C \times D.$$

Since the sets C, D are closed and bounded, we take a sequence $\varepsilon \to 0$, $x_\varepsilon \to x^* \in C$, $y_\varepsilon \to y^*$ $\in D$ and also $v_\varepsilon \to v \in \mathbb{R}$. Sending $\varepsilon \to 0$, we get

$$f(x, y^*) \le v \le f(x^*, y), \quad \forall (x, y) \in C \times D.$$

This says that $v^+ = v^- = v$ and (x^*, y^*) is a saddle point. ∎

Bibliographic Notes

The classic reference for the material in this chapter is the wonderful two-volume set of books by S. Karlin[1] [20], one of the pioneers of game theory when he was a member of the Rand Institute. The proof of the von Neumann minimax theorem, which uses only advanced calculus, is original to Karlin. On the other hand, the birth of game theory can be dated to the appearance of the seminal book by von Neumann and Morgenstern, reference[31].

The idea of using mixed strategies in order to establish the existence of a saddle point in this wider class is called **relaxation**. This idea is due to von Neumann and extends to games with continuous strategies, as well as the modern theory of optimal control, differential games, and the calculus of variations. It has turned out to be one of the most far-reaching ideas of the twentieth century.

The graphical solution of matrix games is extremely instructive as to the objectives of each player. Apparently, this method is present at the founding of game theory, but its origin is unknown to the author of this book.

Poker models have been considered from the birth of game theory. Any gambler interested in poker should consult the book by Karlin [20] for much more information and results on poker models. We will also study poker models when we look at continuous games.

The ideas of a best response strategy and viewing a saddle point as a pair of strategies, each of which is a best response to the other, leads naturally in later chapters to a Nash equilibrium for more general games.

Invertible matrices are discussed by Owen [33] and Ferguson [6] and in many other references. Symmetric games were studied early in the development of the theory because it was shown (see Problem 1.47, from Karlin [20]) that any matrix game could be symmetrized so that techniques developed for symmetric games could be used to yield information about general games. Before the advent of high-speed computers, this was mostly of theoretical interest because the symmetrization made the game matrix much larger. Today, this is not much of a problem.

---

1 June 8, 1924–December 18, 2007. He was was Professor of Mathematics, Stanford University.

The introduction of the Burr–Hamilton duel is an example of a game of timing that is properly considered in the continuous-time setting discussed in later chapters. For the full generality see Karlin's book [20].

The fundamental idea of using linear programming to solve games, and the use of the duality theorem to even prove the existence of value of a matrix game, might have been known to several mathematicians at the same time. The exact originator of this idea is unknown to the present author but possibly may be due to Dantzig, von Neumann, Karlin, or others.

Problem 1.85 is a classic example of the use of game theory in a business problem. Here it is used to determine the location of a business depending on marketing conditions and the location of a competitor's store. Versions of this problem appear throughout the literature, but this version is due to R. Bronson. Problem 1.86 is an example of a dispute resolution problem (refer to Winston's book [44]) presaging the idea of cooperative games considered in Chapter 6. Problem 1.88 is an instructive problem on the use of the graphical method and appears in at least the books by [20] and [16]. The Colonel Blotto game is an early application of game theory to military theory. There is an entire class of games known as Blotto games.

An AI Poem on Gambling

This poem, which is apt for many of the games in this book, was unbelievably generated by Chat-GPT in under one second:

In a hazy room, where fortunes loom,
There's a dance with fate, in the
gambler's tomb.
With cards and dice, and a spinning wheel,
Hope and despair intertwine, a surreal ordeal.

The roulette wheel spins, a whirlwind of
chance,
Red or black, the outcome to enhance.
Heartbeats quicken, as the ball takes flight,
Destiny twirls, in the depths of the night.

Blackjack tables beckon, with cards concealed,
Players' hopes congealed, as strategies are
revealed.
Aces and faces, kings and queens collide,
Fortunes shift swiftly, with every turn of
the tide.

Eyes ablaze with a glint of greed,
Players gather, fueled by their need.
Chips clink and jingle, as bets are laid,
Each soul entangled, by the game's charade.

Poker faces hide tales of deceit,
Bluffing and raising, opponents they meet.
Chips rise and fall, with every hand dealt,
As fortunes fluctuate, like cards in a welt.

But amidst the thrill and the alluring spell,
Lies a cautionary tale that history can tell.
For in this dance with luck and desire,
Many find themselves consumed by the fire.

So tread with care, in this realm of chance,
For gambling's allure can swiftly enhance.
Enjoy the game, but know the stakes,
Lest your life and fortune, it overtakes.

2

Two-Person Nonzero Sum Games

But war's a game, which, were their subjects wise, Kings would not play at.
–William Cowper, The Winter Morning Walk

I made a game effort to argue but two things were against me: the umpires and the rules.
–Leo Durocher

All things are subject to interpretation. Whichever interpretation prevails at a given time is a function of power and not truth.
–Friedrich Nietzsche

If past history was all there was to the game, the richest people would be librarians.
–Warren Buffett

2.1 The Basics

The previous chapter considered two-person games in which whatever one player gains, the other loses. This is far too restrictive for many games, especially games in business, economics or politics, where both players can win something or both players can lose something. We no longer assume that the game is zero-sum or even constant-sum. All players will have their own individual payoff matrix and the goal of maximizing their own individual payoff. We will have to reconsider what we mean by a solution, how to get optimal strategies, and exactly what a strategy is.

In a two-person nonzero sum game, we simply assume that each player has her or his own payoff matrix. Suppose that the payoff matrices are

$$
A = \begin{bmatrix} a_{11} & a_{12} & \cdots & a_{1m} \\ \vdots & \vdots & \vdots & \vdots \\ a_{n1} & a_{n2} & \cdots & a_{nm} \end{bmatrix}, \qquad B = \begin{bmatrix} b_{11} & b_{12} & \cdots & b_{1m} \\ \vdots & \vdots & \vdots & \vdots \\ b_{n1} & b_{n2} & \cdots & b_{nm} \end{bmatrix}
$$

For example, if player I plays row 1 and player II plays column 2, then the payoff to player I is a_{12} and the payoff to player II is b_{12}. In a zero-sum game we always had $a_{ij} + b_{ij} = 0$, or more generally $a_{ij} + b_{ij} = k$, where k is a fixed constant. In a nonzero sum game we do not assume that. Instead, the payoff when player I plays row i and player II plays column j is now a pair of numbers (a_{ij}, b_{ij}) where the first component is the payoff to player I and the second number is the payoff to player II. The individual rows and columns are called **pure strategies** for the players. Finally, every zero-sum

Game Theory: An Introduction, Third Edition. E. N. Barron.
© 2024 John Wiley & Sons, Inc. Published 2024 by John Wiley & Sons, Inc.

game can be put into the bimatrix framework by taking $B = -A$, so this is a true generalization of the theory in the first chapter. Let's start with a simple example.

Example 2.1 Two students have an exam tomorrow. They can choose to study or go to a party. The payoff matrices, written together as a bimatrix, are given by

I/II	Study	Party
Study	$(2,2)$	$(3,1)$
Party	$(1,3)$	$(4,-1)$

If they both study, they each receive a payoff of 2, perhaps in grade point average (GPA) points. If player I studies and player II parties, then player I receives a better grade because the curve is lower. But player II also receives a payoff from going to the party (in good time units). If they both go to the party, player I has a really good time, but player II flunks the exam the next day, and his girlfriend is stolen by player I, so his payoff is -1. What should they do?

Players can choose to play pure strategies, but we know that greatly limits their options. If we expect optimal strategies to exist Chapter 1 has shown we must allow mixed strategies. A mixed strategy for player I is again a vector (or matrix) $X = (x_1, \ldots, x_n) \in S_n$ with $x_i \geq 0$ representing the probability that player I uses row i, and $x_1 + x_2 + \cdots + x_n = 1$. Similarly, a mixed strategy for player II is $Y = (y_1, \ldots, y_m) \in S_m$, with $y_j \geq 0$ and $y_1 + \cdots + y_m = 1$. Now given the player's choice of mixed strategies, each player will have their own expected payoffs given by

$$E_I(X, Y) = XAY^T \text{ for player I,}$$

$$E_{II}(X, Y) = XBY^T \text{ for player II.}$$

We need to define a concept of optimal play that should reduce to a saddle point in mixed strategies in the case $B = -A$. It is a fundamental and far-reaching definition due to another genius of mathematics who turned his attention to game theory in the middle twentieth century, John Nash[1].

Definition 2.1.1 *A pair of mixed strategies $(X^* \in S_n, Y^* \in S_m)$ is a Nash equilibrium if $E_I(X, Y^*) \leq E_I(X^*, Y^*)$ for every mixed $X \in S_n$ and $E_{II}(X^*, Y) \leq E_{II}(X^*, Y^*)$ for every mixed $Y \in S_m$. If (X^*, Y^*) is a Nash equilibrium, we denote by $v_A = E_I(X^*, Y^*)$ and $v_B = E_{II}(X^*, Y^*)$ as the optimal payoff to each player. Written out with the matrices, (X^*, Y^*) is a Nash equilibrium if*

$$E_I(X^*, Y^*) = X^*AY^{*T} \geq XAY^{*T} = E_I(X, Y^*), \quad \text{for every } X \in S_n,$$

$$E_{II}(X^*, Y^*) = X^*BY^{*T} \geq X^*BY^T = E_{II}(X^*, Y), \quad \text{for every } Y \in S_m.$$

We will frequently abbreviate Nash equilibrium as NE.

This says that neither player can gain any expected payoff if either one chooses to deviate from playing the Nash equilibrium, **assuming that the other player is implementing his or her piece of the Nash equilibrium.** On the other hand, if it is known that one player will not be using his piece of the Nash equilibrium, then the other player may be able to increase her payoff by using some strategy other than that in the Nash equilibrium. The player then uses a **best-response strategy**. In fact, the definition of a Nash equilibrium says that each strategy in a Nash equilibrium is a best response strategy against the opponent's Nash strategy. Here is a precise definition for two players.

---

1 See Appendix D for a short biography.

Definition 2.1.2 *A strategy $X^0 \in S_n$ is a **best response strategy** to a given strategy $Y^0 \in S_m$ for player II, if*

$$E_I(X^0, Y^0) = \max_{X \in S_n} E_I(X, Y^0).$$

*Similarly, a strategy $Y^0 \in S_m$ is a **best response strategy** to a given strategy $X^0 \in S_n$ for player I, if*

$$E_{II}(X^0, Y^0) = \max_{Y \in S_m} E_{II}(X^0, Y).$$

In particular, another way to define a Nash equilibrium (X^*, Y^*) is that X^* maximizes $E_I(X, Y^*)$ over all $X \in S_n$ and Y^* maximizes $E_{II}(X^*, Y)$ over all $Y \in S_m$. X^* is a best response to Y^* and Y^* is a best response to X^*.

Remark It is important to notice what a Nash equilibrium is not. A Nash equilibrium does not maximize the payoff to each player. That would make the problem trivial because it would simply say that the Nash equilibrium is found by maximizing a function (the first player's payoff) of two variables over both variables and then requiring that it also maximize the second player's payoff at the *same* point. That would be a rare occurrence but trivial from a mathematical point of view to find (if one existed).

If $B = -A$, a bimatrix game is a zero-sum two-person game and a Nash equilibrium is the same as a saddle point in mixed strategies. It is easy to check that from the definitions because $E_I(X, Y) = XAY^T = -E_{II}(X, Y)$.

Note that a Nash equilibrium in pure strategies will be a row i^* and column j^* satisfying

$$a_{ij^*} \leq a_{i^*j^*} \text{ and } b_{i^*j} \leq b_{i^*j^*}, i = 1, \ldots, n, j = 1, \ldots, m.$$

So $a_{i^*j^*}$ is the largest number in column j^* and $b_{i^*j^*}$ is the largest number in row i^*. In the bimatrix game a Nash equilibrium in pure strategies must be the **pair that is, at the same time, the largest first component in the column and the largest second component in the row**.

Just as in the zero-sum case, a pure strategy can always be considered as a mixed strategy by concentrating all the probability at the row or column, which should always be played.

As in Chapter 1 we will use the notation that if player I uses the pure strategy row i, and player II uses mixed strategy $Y \in S_m$, then the expected payoffs to each player are

$$E_I(i, Y) = {}_iAY^T \text{ and } E_{II}(i, Y) = {}_iBY^T.$$

Similarly, if player II uses column j and player I uses the mixed strategy X, then

$$E_I(X, j) = XA_j \text{ and } E_{II}(X, j) = XB_j.$$

The questions we ask for a given bimatrix game are

1. Is there a Nash equilibrium using pure strategies?
2. Is there a Nash equilibrium using mixed strategies? Maybe more than one?
3. How do we compute these?

To start, we consider the classic example.

2.1.1 Prisoner's Dilemma

Two criminals have just been caught after committing a crime. The police interrogate the prisoners by placing them in separate rooms so that they cannot communicate and coordinate their stories.

The goal of the police is to try to get one or both of them to confess to having committed the crime. We consider the two prisoners as the players in a game in which they have two pure strategies: confess or don't confess. Their prison sentences, if any, will depend on whether they confess and agree to testify against each other. But if they both confess, no benefit will be gained by testimony that is no longer needed. If neither confesses, there may not be enough evidence to convict either of them of the crime. The following matrices represent the possible payoffs remembering that they are set up to maximize the payoff.

Prisoner I/II	Confess	Don't confess
Confess	$(-5, -5)$	$(0, -20)$
Don't confess	$(-20, 0)$	$(-1, -1)$

The individual matrices for the two prisoners are

$$A = \begin{bmatrix} -5 & 0 \\ -20 & -1 \end{bmatrix}, \qquad B = \begin{bmatrix} -5 & -20 \\ 0 & -1 \end{bmatrix}$$

The numbers are negative because they represent the number of years in a prison sentence and each player wants to maximize the payoff, that is, minimize their own sentences.

To see whether there is a Nash equilibrium in pure strategies, we are looking for a payoff pair (a, b) in which a is the largest number in a column and b is the largest number in a row simultaneously. There may be more than one such pair. Looking at the bimatrix is the easiest way to find them. The **systematic way** is to put a bar over the first number that is the **largest in each column and put a bar over the second number that is the largest in each row.** Any pair of numbers that both have bars is a Nash equilibrium in pure strategies. In the prisoner's dilemma problem we have

Prisoner I/II	Confess	Don't confess
Confess	$\boxed{(-\bar{5}, -\bar{5})}$	$(\bar{0}, -20)$
Don't confess	$(-20, \bar{0})$	$(-1, -1)$

We see that there is exactly one pure Nash equilibrium at (confess, confess), they should both confess and settle for five years in prison each. If either player deviates from *confess*, while the other player still plays *confess*, then the payoff to the player who deviates goes from -5 to -20.

Wait a minute: clearly both players can do better if they both choose *don't confess* because then there is not enough evidence to put them in jail for more than one year. But there is an incentive for each player to **not** play *don't confess*. If one player chooses *don't confess* and the other chooses *confess*, the payoff to the confessing player is 0–he won't go to jail at all! The players are rewarded for a **betrayal of the other prisoner**, and so that is exactly what will happen. This reveals a major reason why conspiracies almost always fail. As soon as one member of the conspiracy senses an advantage by confessing, that is exactly what they will do, and then the game is over.

The payoff pair $(-1, -1)$ is **unstable** in the sense that a player can do better by deviating, assuming that the other player does not, whereas the payoff pair $(-5, -5)$ is **stable** because neither player can improve their own individual payoff if they both play it. Even if they agree before they are caught to not confess, it would take extraordinary will power for both players to stick with that agreement in the face of the numbers. In this sense, the Nash equilibrium is self-enforcing.

Any bimatrix game in which the Nash equilibrium gives both players a lower payoff than if they cooperated is a prisoner's dilemma game. In a prisoner's dilemma game, the players will choose to not cooperate. That is pretty pessimistic and could lead to lots of bad things which could, but don't usually happen in the real world. Why not? One explanation is that many such games are not played just once but repeatedly over long periods of time (like arms control, for example). Then the players need to take into account the costs of not cooperating. The conclusion is that a game which is repeated many times may not exhibit the same features as a game played once. We will see in Chapter 5 that cooperation can be a Nash Equilibrium in a game which is repeated an indefinite number of times.

Notice that in matrix A in the prisoner's dilemma, the first row is always better than the second row for player I. This says that for player I, row 1 (i.e., confess) **strictly dominates** row 2 and hence row 2 can be eliminated from consideration by player I, no matter what player II does because player I will never play row 2. Similarly, in matrix B for the other player, column 1 strictly dominates column 2, so player II, who chooses the columns, would never play column 2. Once again, player II would always confess. This problem can be solved by domination.

Example 2.2 This example explains what is going on when you are looking for the pure Nash equilibria. Let's look at the game with matrix

I/II	A	B	C
a	$(1, 0)$	$(1, 3)$	$(3, 0)$
b	$(0, 2)$	$(0, 1)$	$(3, 0)$
c	$(0, 2)$	$(2, 4)$	$(5, 3)$

Consider player II.

1. If II plays A, the best response for player I is strategy a.
2. If II plays B, I's best response is c.
3. If II plays C, I's best response is c.

Now consider player I.

1. If I plays a, the best response for player II is strategy B.
2. If I plays b, II's best response is A.
3. If I plays c, II's best response is B.

The only pair which is a best response to a best response is *If II plays B, then I plays c; If I plays c, then II plays B*. It is the only pair $(x, y) = (2, 4)$ with x the largest first number in all the rows and y is the largest second number of all the columns, simultaneously.

Example 2.3 Can a bimatrix game have more than one Nash equilibrium? Absolutely. If we go back to the study–party game and change one number, we will see that it has two Nash equilibria:

I/II	Study	Party
Study	$\boxed{(2, 2)}$	$(3, 1)$
Party	$(1, 3)$	$\boxed{(4, 4)}$

There is a Nash equilibrium at payoff $(2, 2)$ and at $(4, 4)$. Which one is better? In this example, since both students get a payoff of 4 by going to the party, that Nash point is clearly better for both. But if player I decides to study instead, then what is best for player II?

In the next example we get an insight into why countries want to have nuclear bombs.

Example 2.4 The Arms Race: Suppose that two countries have the choice of developing or not developing nuclear weapons. There is a cost of the development of the weapons in the price that the country might have to pay in sanctions, etc. But there is also a benefit to having nuclear weapons in prestige, defense, deterrence, and so on. Of course, the benefits of nuclear weapons disappear if a country is actually going to use the weapons. If one considers the outcome of attacking an enemy country with nuclear weapons and the risk of having your own country vaporized in retaliation, a rational person would certainly consider the Cold War acronym MAD, Mutually Assured Destruction, as completely accurate.

Suppose that we quantify the game using a bimatrix in which each player wants to maximize the payoff.

Country I/II	Nuclear	Conventional
Nuclear	$(1,1)$	$(10,-5)$
Conventional	$(-5,10)$	$(1,1)$

To explain these numbers, if country I goes nuclear and country II does not, then country I can dictate terms to country II to some extent because a war with I, who holds nuclear weapons and will credibly use them, would result in destruction of country II, which has only conventional weapons. The result is a representative payoff of 10 to country I and -5 to country II, as I's lackey now. On the other hand, if both countries go nuclear, neither country has an advantage or can dictate terms to the other because war would result in mutual destruction, assuming that the weapons are actually used. Consequently, there is only a minor benefit to both countries going nuclear, represented by a payoff of 1 to each. That is the same as if they remain conventional because then they do not have to spend money to develop the bomb, dispose of nuclear waste, and so on.

We see from the bimatrix that we have a Nash equilibrium at the pair $(1,1)$ corresponding to the strategy (nuclear, nuclear). The pair $(1,1)$ when both countries maintain conventional weapons is **not** a Nash equilibrium because each player can improve its own payoff by unilaterally deviating from this. Observe, too, that if one country decides to go nuclear, the other country clearly has no choice but to do likewise. The only way that the situation could change would be to reduce the benefits of going nuclear, perhaps by third-party sanctions or in other ways.

This simple matrix game captures the theoretical basis of the MAD policy of the United States and the former Soviet Union during the cold war. Once the United States possessed nuclear weapons, the payoff matrix showed the Soviet Union that it was in their best interest to also own them and to match the US nuclear arsenal to maintain the MAD option.

Since World War II the nuclear nonproliferation treaty has attempted to have the parties reach an agreement that they would not obtain nuclear weapons, in other words to agree to play (*conventional, conventional*). Countries which have instead played *Nuclear* include India, Pakistan, North Korea, and Israel[1]. Iran is playing *Nuclear* covertly, and Libya, which once played *Nuclear*, switched to *conventional* because it's payoff changed.

The lesson to learn here is that once one government obtains nuclear weapons, it is a Nash equilibrium–and self-enforcing equilibrium–for opposing countries to also obtain the weapons.

Example 2.5 Do all bimatrix games have Nash equilibrium points in pure strategies? Game theory would be pretty boring if that were true. For example, if we look at the game

$$\begin{bmatrix} (2,0) & (1,3) \\ (0,1) & (3,0) \end{bmatrix},$$

---

1 Israel has never admitted to having nuclear weapons, but it is widely accepted as true and reported as a fact. The secrecy on the part of the Israelis also indicates an implicit understanding of the bimatrix game here.

there is no pair (a, b) in which a is the largest in the column and b is the largest in the row. In a case like this it seems reasonable that we use mixed strategies. In addition, even though a game might have pure strategy Nash equilibria, it could also have a mixed strategy Nash equilibrium.

In the next section we will see how to solve such games and find the mixed Nash equilibria.

Finally, we end this section with a concept that is a starting point for solving bimatrix games and that we will use extensively when we discuss cooperation. Each player asks, what is the worst that can happen to me in this game?

The amount that player I can be guaranteed to receive is obtained by assuming that player II is actually trying to minimize player I's payoff. In the bimatrix game with two players with matrices $(A_{n \times m}, B_{n \times m})$, we consider separately the two games arising from each matrix. Matrix A is considered as the matrix for a **zero-sum** game with player I against player II (player I is the row player = maximizer and player II is the column player = minimizer). The value of the game with matrix A is the guaranteed amount for player I. Similarly, the amount that player II can be guaranteed to receive is obtained by assuming player I is actively trying to minimize the amount that II gets. For player II, the zero sum game is B^T because the row player is always the maximizer. Consequently, player II can guarantee that she will receive the value of the game with matrix B^T. The formal definition is summarized below.

Definition 2.1.3 *Consider the bimatrix game with matrices (A, B). The* **safety value** *for player I is* value(A). *The* **safety value** *for player II in the bimatrix game is* value(B^T).

If A has the saddle point (X^A, Y^A), then X^A is called the **maxmin strategy for player I.**
If B^T has saddle point (X^{B^T}, Y^{B^T}), then X^{B^T} is the **maxmin strategy for player II.**

In the prisoner's dilemma game, the safety values are both -5 to each player, as you can quickly verify.

In the game with matrix

$$\begin{bmatrix} (2,0) & (1,3) \\ (0,1) & (3,0) \end{bmatrix},$$

we have

$$A = \begin{bmatrix} 2 & 1 \\ 0 & 3 \end{bmatrix}, B^T = \begin{bmatrix} 0 & 1 \\ 3 & 0 \end{bmatrix}.$$

Then $v(A) = \frac{3}{2}$ is the safety value for player I and $v(B^T) = \frac{3}{4}$ is the safety value for player II.

The maxmin strategy for player I is $X = (\frac{3}{4}, \frac{1}{4})$, and the implementation of this strategy guarantees that player I can get at least her safety level. In other words, if I uses $X = (\frac{3}{4}, \frac{1}{4})$, then $E_I(X, Y) \geq v(A) = \frac{3}{2}$ no matter what Y strategy is used by II. In fact

$$E_I\left(\left(\frac{3}{4}, \frac{1}{4}\right), Y\right) = \frac{3}{2}(y_1 + y_2) = \frac{3}{2}, \text{ for any strategy } Y = (y_1, y_2).$$

The maxmin strategy for player II is $Y = X^{B^T} = (\frac{3}{4}, \frac{1}{4})$, which she can use to get at least her safety value of $\frac{3}{4}$.

Is there a connection between the safety levels and a Nash equilibrium? The safety levels are the guaranteed amounts each player can get by using their own individual maxmin strategies, so any rational player must get at least the safety level in a bimatrix game. In other words, it has to be true that if (X^*, Y^*) is a Nash equilibrium for the bimatrix game (A, B), then

$$E_I(X^*, Y^*) = X^* A Y^{*T} \geq value(A) \text{ and } E_{II}(X^*, Y^*) = X^* B Y^{*T} \geq value(B^T).$$

This would say that in the bimatrix game, if players use their Nash points, they get at least their safety levels. That's what it means to be **individually rational**. Precisely, two strategies X, Y are individually rational if $E_I(X, Y) \geq v(A)$ and $E_{II}(X, Y) \geq v(B^T)$.

Here's why any Nash equilibrium is individually rational.

Proof: It's really just from the definitions. The definition of Nash equilibrium says

$$E_I(X^*, Y^*) = X^* AY^{*T} \geq E_I(X, Y^*) = X\, AY^{*T}, \text{ for all } X \in S_n.$$

But if that is true for all mixed X, then

$$E_I(X^*, Y^*) \geq \max_{X \in S_n} X\, AY^{*T} \geq \min_{Y \in S_m} \max_{X \in S_n} X\, AY^T = value(A).$$

The other part of a Nash definition gives us

$$E_{II}(X^*, Y^*) = X^* BY^{*T} \geq \max_{Y \in S_m} X^* BY^T$$

$$= \max_{Y \in S_m} YB^T X^{*T} \qquad (\text{since } X^* BY^T = Y\, B^T X^{*T})$$

$$\geq \min_{X \in S_n} \max_{Y \in S_m} YB^T X^T = value(B^T).$$

Each player does at least as well as assuming the worst. ∎

Problems

2.1 Show that (X^*, Y^*) is a saddle point of the game with matrix A if and only if (X^*, Y^*) is a Nash equilibrium of the bimatrix game $(A, -A)$.

2.2 Suppose that a married couple, both of whom have just finished medical school, now have choices regarding their residencies. One of the new doctors has three choices of programs, while the other has two choices. They value their prospects numerically on the basis of the program itself, the city, staying together, and other factors, and arrive at the bimatrix

$$\begin{bmatrix} (5.2, 5.0) & (4.4, 4.4) & (4.4, 4.1) \\ (4.2, 4.2) & (4.6, 4.9) & (3.9, 4.3) \end{bmatrix}$$

(a) Find all the pure Nash equilibria. Which one should be played?

(b) Find the safety levels for each player.

2.3 Consider the bimatrix game that models the game of chicken:

I/II	Turn	Straight
Turn	$(19, 19)$	$(-42, 68)$
Straight	$(68, -42)$	$(-45, -45)$

Two cars are headed toward each other at a high rate of speed. Each player has two options: turn off, or continue straight ahead. This game is a macho game for reputation, but leads to mutual destruction if both play straight ahead.

(a) There are two pure Nash equilibria. Find them.

(b) Verify by using the definition of mixed Nash equilibrium that the mixed strategy pair $X^* = (\frac{3}{52}, \frac{49}{52})$, $Y^* = (\frac{3}{52}, \frac{49}{52})$ is a Nash equilibrium, and find the expected payoffs to each player.

(c) Find the safety levels for each player.

2.4 We may eliminate a row or a column by dominance. If $a_{ij} \geq a_{i'j}$ for *every* column j, and we have strict inequality for *some* column j, then we may eliminate row i'. If player I drops row i', then the entire pair of numbers in that row are dropped. Similarly, if $b_{ij} \geq b_{ij'}$ for every row i, and we have strict inequality for some row i, then we may drop column j', and all the pairs of payoffs in that column. If the inequalities are all strict, this is called **strict dominance**, otherwise it is **weak dominance**.

(a) By this method, solve the game

$$\begin{bmatrix} (3,-1) & (2,1) \\ (-1,7) & (1,3) \\ (4,-3) & (-2,9) \end{bmatrix}.$$

(b) Find the safety levels for each player.

2.5 When we have weak dominance in a game, the order of removal of the dominated row or column makes a difference. Consider the matrix

$$\begin{bmatrix} (10,0) & (5,1) & (4,-2) \\ (10,1) & (5,0) & (1,-1) \end{bmatrix}$$

(a) Suppose player I moves first. Find the result.
(b) What happens if player II moves first?

2.6 Suppose two merchants have to choose a location along the straight road. They may choose any point in $\{1, 2, \ldots, n\}$. Assume there is exactly one customer at each of these points and a customer will always go to the nearest merchant. If the two merchants are equidistant to a customer then they share that customer, i.e., $\frac{1}{2}$ the customer goes to each store. For example, if $n = 11$ and if player I chooses location 3 and player II chooses location 8, then the payoff to I is 5 and the payoff to II is 6.

(a) Suppose $n = 5$. Find the bimatrix and find the Nash equilibrium.
(b) What do you think is the Nash equilibrium in general if $n = 2k + 1$, i.e., n is an odd integer.

2.7 Two airlines serve the route O'Hare airport (ORD) to Los Angeles airport (LAX). Naturally they are in competition for passengers who make their decision based on airfares alone. Lower fares attract more passengers and increases the load factor (the number of bodies in seats). Suppose the bimatrix is given as follows where each airline can choose to set the fare at Low=$250 or High=$700

	Low	High
Low	$(-50, -10)$	$(175, -20)$
High	$(-100, 200)$	$(100, 100)$

The numbers are in millions. Find the pure Nash equilibria.

2.8 Consider the game with bimatrix

	A	B	C
a	$(1, 1)$	$(3, x)$	$(2, 0)$
b	$(2x, 3)$	$(2, 2)$	$(3, 1)$
c	$(2, 1)$	$(1, x)$	$(x^2, 4)$

(a) Find x so that the game has no pure Nash equilibria.
(b) Find x so that the game has exactly two pure Nash equilibria.
(c) Find x so that the game has exactly three pure Nash equilibria. Is there an x so that there are four pure Nash equilibria?

2.2 2×2 Bimatrix Games, Best Response, Equality of Payoffs

Now we will analyze all two-person 2×2 nonzero sum games. We are after a method to find all Nash equilibria for a bimatrix game, mixed and pure. Let $X = (x, 1 - x), Y = (y, 1 - y), 0 \le x \le 1, 0 \le y \le 1$ be mixed strategies for players I and II, respectively. As in the zero-sum case, X represents the mixture of the rows that player I has to play; specifically, player I plays row 1 with probability x and row 2 with probability $1 - x$. Similarly for player II and the mixed strategy Y. Now we may calculate the expected payoff to each player. As usual,

$$E_I(X, Y) = X\,AY^T = (x, \ 1 - x) \begin{bmatrix} a_{11} & a_{12} \\ a_{21} & a_{22} \end{bmatrix} \begin{bmatrix} y \\ 1 - y \end{bmatrix},$$

$$E_{II}(X, Y) = X\,BY^T = (x, \ 1 - x) \begin{bmatrix} b_{11} & b_{12} \\ b_{21} & b_{22} \end{bmatrix} \begin{bmatrix} y \\ 1 - y \end{bmatrix},$$

are the expected payoffs to I and II, respectively. It is the goal of each player to maximize her own expected payoff **assuming that the other player is doing her best to maximize her own payoff** with the strategies she controls.

Let's proceed to figure out how to find mixed Nash equilibria. First, we need a definition of a certain set.

Definition 2.2.1 *Let $X = (x, 1 - x), Y = (y, 1 - y)$ be strategies, and set $f(x, y) = E_I(X, Y)$, and $g(x, y) = E_{II}(X, Y)$. The **rational reaction set** for player I is the set of points*

$$R_I = \{(x, y) \mid 0 \le x, y \le 1, \max_{0 \le z \le 1} f(z, y) = f(x, y)\},$$

and the rational reaction set for player II is the set

$$R_{II} = \{(x, y) \mid 0 \le x, y \le 1, \max_{0 \le w \le 1} g(x, w) = g(x, y)\}.$$

A point $(x^*, y) \in R_I$ means that x^* is the point in $[0, 1]$ where $x \mapsto f(x, y)$ is maximized for y fixed. Then $X^* = (x^*, 1 - x^*)$ is a best response to $Y = (y, 1 - y)$. Similarly, if $(x, y^*) \in R_{II}$, then $y^* \in [0, 1]$ is a point where $y \mapsto g(x, y)$ is maximized for x fixed. The strategy $Y^* = (y^*, 1 - y^*)$ is a best response to $X = (x, 1 - x)$. Consequently, a point (x^*, y^*) in both R_I and R_{II} says that $X^* = (x^*, 1 - x^*)$ and $Y^* = (y^*, 1 - y^*)$, as best responses to each other, is a Nash equilibrium.

Calculation of the rational reaction sets for 2×2 games. First we will show how to find the rational reaction sets for the bimatrix game with matrices (A, B). Let $X = (x, 1 - x)$, $Y = (y, 1 - y)$ be any strategies and define the functions

$$f(x, y) = E_I(X, Y) \text{ and } g(x, y) = E_{II}(X, Y).$$

The idea is to find for a fixed $0 \le y \le 1$, the best response of player I to y; that is, find $x \in [0, 1]$ which will give the largest value of player I's payoff $f(x, y)$ for a given y. Accordingly, we seek

$$\max_{0 \le x \le 1} f(x, y) = \max_{0 \le x \le 1} x E_I(1, Y) + (1 - x) E_I(2, Y)$$

$$= \max_{0 \le x \le 1} x[E_{\mathrm{I}}(1, Y) - E_{\mathrm{I}}(2, Y)] + E_{\mathrm{I}}(2, Y)$$

$$= \begin{cases} E_{\mathrm{I}}(2, Y) & \text{at } x = 0 \text{ if } E_{\mathrm{I}}(1, Y) < E_{\mathrm{I}}(2, Y); \\ E_{\mathrm{I}}(1, Y) & \text{at } x = 1 \text{ if } E_{\mathrm{I}}(1, Y) > E_{\mathrm{I}}(2, Y); \\ E_{\mathrm{I}}(2, Y) & \text{at any } 0 < x < 1 \text{ if } E_{\mathrm{I}}(1, Y) = E_{\mathrm{I}}(2, Y). \end{cases}$$

If we use the elements of the matrices we can be more explicit. We know that

$$f(x, y) = (x \ \ 1 - x) \begin{bmatrix} a_{11} & a_{12} \\ a_{21} & a_{22} \end{bmatrix} \begin{bmatrix} y \\ 1 - y \end{bmatrix}$$

$$= x[(a_{11} - a_{12} - a_{21} + a_{22})y + a_{12} - a_{22}] + (a_{21} - a_{22})y + a_{22}.$$

Now we see that in order to make this as large as possible we need only look at the term $M = [(a_{11} - a_{12} - a_{21} + a_{22})y + a_{12} - a_{22}]$. The graph of f for a fixed y is simply a straight line with slope M.

- If $M > 0$ the maximum of the line will occur at $x^* = 1$.
- If $M < 0$ the maximum of the line will occur at $x^* = 0$.
- If $M = 0$ it won't matter where x is because the line will be horizontal.

Now

$$M = 0 \implies (a_{11} - a_{12} - a_{21} + a_{22})y + a_{12} - a_{22} = 0 \implies y = \frac{a_{22} - a_{12}}{a_{11} - a_{12} - a_{21} + a_{22}}$$

and for this given y, the best response is any $0 \le x \le 1$.

If $M > 0$, the best response to a given y is $x^* = 1$, while if $M < 0$ the best response is $x^* = 0$. But we know that

$$M > 0 \Leftrightarrow y > \frac{a_{22} - a_{12}}{a_{11} - a_{12} - a_{21} + a_{22}}$$

and

$$M < 0 \Leftrightarrow y < \frac{a_{22} - a_{12}}{a_{11} - a_{12} - a_{21} + a_{22}}.$$

What we have found is a (set valued) function, called the **best response function**:

$$x^* = BR_{\mathrm{I}}(y) = \begin{cases} 1, & \text{if } y > \dfrac{a_{22} - a_{12}}{a_{11} - a_{12} - a_{21} + a_{22}}; \\ 0, & \text{if } y < \dfrac{a_{22} - a_{12}}{a_{11} - a_{12} - a_{21} + a_{22}}; \\ [0, 1], & \text{if } y = \dfrac{a_{22} - a_{12}}{a_{11} - a_{12} - a_{21} + a_{22}}. \end{cases}$$

The best response function comes from maximizing a linear function over the interval $0 \le x \le 1$.

The rational reaction set for player I is then the graph of the best response function for player I. In symbols,

$$R_{\mathrm{I}} = \{(x, y) \in [0, 1] \times [0, 1] \mid x = BR_{\mathrm{I}}(y)\}.$$

In a similar way we may consider the problem for player II. For player II we seek

$$\max_{0 \le y \le 1} g(x, y) = \max_{0 \le y \le 1} (x \ \ 1 - x) \begin{bmatrix} b_{11} & b_{12} \\ b_{21} & b_{22} \end{bmatrix} \begin{bmatrix} y \\ 1 - y \end{bmatrix}$$

$$. = \max_{0 \le y \le 1} y[(b_{11} - b_{12} - b_{21} + b_{22})x + b_{21} - b_{22}] + (b_{12} - b_{22})x + b_{22}.$$

Now we see that in order to make this as large as possible we need only look at the term $R = [(b_{11} - b_{12} - b_{21} + b_{22})x + b_{21} - b_{22}]$. The graph of g for a fixed x is a straight line with slope R.

- If $R > 0$ the maximum of the line will occur at $y^* = 1$.
- If $R < 0$ the maximum of the line will occur at $y^* = 0$.
- If $R = 0$ it won't matter where y is because the line will be horizontal.

Just as before we derive the **best response function** for player II:

$$y^* = BR_{II}(x) = \begin{cases} 1, & \text{if } x > \dfrac{b_{22} - b_{21}}{b_{11} - b_{12} - b_{21} + b_{22}}; \\ 0, & \text{if } x < \dfrac{b_{22} - b_{21}}{b_{11} - b_{12} - b_{21} + b_{22}}; \\ [0, 1], & \text{if } x = \dfrac{b_{22} - b_{21}}{b_{11} - b_{12} - b_{21} + b_{22}}. \end{cases}$$

The rational reaction set for player II is the graph of the best response function for player II. That is,

$$R_{II} = \{(x, y) \in [0, 1] \times [0, 1] \mid y = BR_{II}(x)\}.$$

The points of $R_I \cap R_{II}$ are exactly the (x, y) for which $X = (x, 1 - x)$, $Y = (y, 1 - y)$ are Nash equilibria for the game. Here is a specific example.

Example 2.6 The bimatrix game with the two matrices

$$A = \begin{bmatrix} 2 & -1 \\ -1 & 1 \end{bmatrix}, \qquad B = \begin{bmatrix} 1 & -1 \\ -1 & 2 \end{bmatrix},$$

will have multiple Nash equilibria. Two of them are obvious; the pair $(2, 1)$ and the pair $(1, 2)$ in (A, B) have the property that the first number is the largest in the first column and at the same time the second number is the largest in the first row. So we know that $X^* = (1, 0), Y^* = (1, 0)$ is a Nash point (with payoff 2 for player I and 1 for player II) as is $X^* = (0, 1), Y^* = (0, 1)$, (with payoff 1 for player I and 2 for player II).

For these matrices

$$f(x, y) = X A Y^T = x(5y - 2) - 2y + 1$$

and the maximum occurs at x^* which is the best response to a given $y \in [0, 1]$, namely,

$$x^* = BR_I(y) = \begin{cases} 1, & \text{if } y > \frac{2}{5}; \\ 0, & \text{if } y < \frac{2}{5}; \\ [0, 1], & \text{if } y = \frac{2}{5}. \end{cases}$$

Here is what this looks like (Figure 2.1).

Similarly,

$$g(x, y) = X B Y^T = y(5x - 3) - 3x + 2$$

and and the maximum for g when x is given occurs at y^*,

$$y^* = BR_{II}(x) = \begin{cases} 1, & \text{if } x > \frac{3}{5}; \\ 0, & \text{if } x < \frac{3}{5}; \\ [0, 1], & \text{if } x = \frac{3}{5}. \end{cases}$$

We graph this as the dotted line on the same graph as before.

Figure 2.1 Best response function for player I.

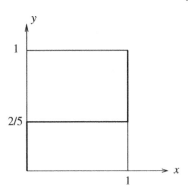

As you can see the best response functions cross at $(x, y) = (0, 0)$, and $(1, 1)$ on the boundary of the square, but also at $(x, y) = (\frac{3}{5}, \frac{2}{5})$ in the interior, corresponding to the unique mixed Nash equilibrium $X^* = (\frac{3}{5}, \frac{2}{5})$, $Y^* = (\frac{2}{5}, \frac{3}{5})$. Notice that the point of intersection is **not** a Nash equilibrium but it gives you the first component of the two strategies, which do give the equilibrium. The expected payoffs are

$$E_I(X^*, Y^*) = \frac{1}{5} \text{ and } E_{II}(X^*, Y^*) = \frac{1}{5}.$$

This is curious because the expected payoffs to each player are **much less** than they could get at the other Nash points.

We will see pictures like Figure 2.2 again in Section 2.3 when we consider an easier way to get Nash equilibria .

Our next proposition shows that in order to check if strategies (X^*, Y^*) is a Nash equilibrium, we need only check inequalities in the definition against only pure strategies. This is similar to the result in zero sum games that (X^*, Y^*) is a saddle and v is the value if and only if $E(X^*, j) \geq v$ and $E(i, Y^*) \leq v$, for all rows and columns.

Proposition 2.2.2 *(X^*, Y^*) is a Nash equilibrium if and only if*

$$E_I(i, Y^*) = {}_iAY^{*T} \leq X^*AY^{*T} = E_I(X^*, Y^*), \quad i = 1, \ldots, n, \tag{2.2.1}$$

$$E_{II}(X^*, j) = X^*B_j \leq X^*BY^{*T} = E_{II}(X^*, Y^*), \quad j = 1, \ldots, m. \tag{2.2.2}$$

Figure 2.2 Best response function for players I and II.

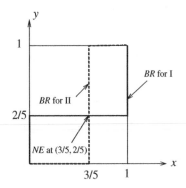

Proof: To see why this is true, we first note that if (X^*, Y^*) is a Nash equilibrium, then the inequalities must hold by definition (simply choose pure strategies for comparison). We need only show that the inequalities are sufficient.

Suppose that the inequalities hold for (X^*, Y^*). Let $X = (x_1, x_2, \ldots, x_n)$ and $Y = (y_1, y_2, \ldots, y_m)$ be any mixed strategies. Successively multiply (2.2.1) by $x_i \geq 0$ and sum over i to get

$$E(X, Y^*) = \sum_{i=1}^{n} x_i E_I(i, Y^*) = \sum_i x_i \, {}_iAY^{*T} = X \, AY^{*T} \leq E_I(X^*, Y^*).$$

Similarly,

$$E(X^*, Y) = \sum_{j=1}^{m} y_j E_{II}(X^*, j) = \sum_j X^* B_j y_j = X^* BY^T \leq E_{II}(X^*, Y^*). \qquad \blacksquare$$

One important use of this result is as a check to make sure that we actually have a Nash equilibrium. The next example illustrates that.

Example 2.7 Someone says that the bimatrix game

$$\begin{bmatrix} (2,1) & (-1,-1) \\ (-1,-1) & (1,2) \end{bmatrix},$$

has a Nash equilibrium at $X^* = (\frac{3}{5}, \frac{2}{5})$, $Y^* = (\frac{2}{5}, \frac{3}{5})$. To check that, first compute $E_I(X^*, Y^*) = E_{II}(X^*, Y^*) = \frac{1}{5}$. Now check to make sure that this number is at least as good as what could be gained if the other player plays a pure strategy. You can readily verify that, in fact, $E_I(1, Y^*) = \frac{1}{5} = E_I(2, Y^*)$ and also $E_{II}(X^*, 1) = E_{II}(X^*, 2) = \frac{1}{5}$, so we do indeed have a Nash point.

The most important theorem of this section gives us a way to find a mixed Nash equilibrium of any two-person bimatrix game. It is a necessary condition that may be used for computation and is very similar to the equilibrium theorem for zero-sum games. If you recall in Theorem 1.4.8, we had the result that in a zero-sum game whenever a row, say, row k, is played with positive probability against Y^*, then the expected payoff $E(k, Y^*)$ must give the value of the game. The next result says the same thing, but now for each of the two players. This is called the **equality of payoffs theorem.**

Theorem 2.2.3 *(Equality of Payoffs Theorem)* *Suppose that*

$$X^* = (x_1, x_2, \ldots, x_n), \quad Y^* = (y_1, y_2, \ldots, y_m)$$

is a Nash equilibrium for the bimatrix game (A, B).

For any row k that has a positive probability of being used, $x_k > 0$, we have $E_I(k, Y^*) = E_I(X^*, Y^*) \equiv v_I$.

For any column j that has a positive probability of being used, $y_j > 0$, we have $E_{II}(X^*, j) = E_{II}(X^*, Y^*) \equiv v_{II}$. That is,

$$x_k > 0 \implies E_I(k, Y^*) = v_I$$
$$y_j > 0 \implies E_{II}(X^*, j) = v_{II}.$$

Proof: We know from Proposition 2.2.2 that since we have a Nash point, $E_I(X^*, Y^*) = v_I \geq E_I(i, Y^*)$ for any row i. Now, suppose that row k has positive probability of being played against Y^* and that it gives player I a strictly smaller expected payoff $v_I > E_I(k, Y^*)$. Then $v_I \geq E_I(i, Y^*)$ for all the rows $i = 1, 2, \ldots, n, i \neq k$, and $v_I > E_I(k, Y^*)$ together imply that

$$x_i v_I \geq x_i E_I(i, Y^*), i \neq k, \text{ and } x_k v_I > x_k E_I(k, Y^*).$$

Adding up all these inequalities, we get

$$\sum_{i=1}^{n} x_i v_I = v_I > \sum_{i=1}^{n} x_i E_I(i, Y^*) = E_I(X^*, Y^*) = v_I.$$

This contradiction says it must be true that $v_I = E_I(k, Y^*)$. The only thing that could have gone wrong with this argument is $x_k = 0$. (Why?) Now you can argue in the same way for the assertion about player II. Observe, too, that the argument we just gave is basically the identical one we gave for zero sum games. ∎

Remarks

1. It is very important to note that knowing player I will play a row with positive probability gives an equation for player II's strategy, not player I's. Similarly, knowing player I uses a column with positive probability allows us to find the strategy for player I.

2. If two or more rows are used with positive probability, then player II's strategy in a Nash equilibrium must satisfy the property that it is chosen to make player I **indifferent** as to which of the rows are played. For this reason, it is also called the **indifference principle**.

The idea now is that we can find the (completely) mixed Nash equilibria by solving a system of equations rather than inequalities for player II:

$$_k AY^{*T} = E_I(k, Y^*) = E_I(s, Y^*) = {}_s AY^{*T}, \text{ assuming that } x_k > 0, x_s > 0,$$

and

$$X^* B_j = E_{II}(X^*, j) = E_{II}(X^*, r) = X^* B_r, \text{ assuming that } y_j > 0, y_r > 0.$$

This won't be enough to solve the equations, however. You need the additional condition that the components of the strategies must sum to one:

$$x_1 + x_2 + \cdots + x_n = 1, \quad y_1 + y_2 + \cdots + y_m = 1.$$

Example 2.8 As a simple example of this theorem, suppose that we take the matrices

$$A = \begin{bmatrix} -4 & 2 \\ 2 & 1 \end{bmatrix}, \qquad B = \begin{bmatrix} 1 & 0 \\ 2 & 3 \end{bmatrix}.$$

Suppose that $X = (x_1, x_2), Y = (y_1, y_2)$ is a mixed Nash equilibrium. If $0 < x_1 < 1$, then, since both rows are played with positive probability, by the equality of payoffs Theorem 2.2.3, $v_I = E_I(1, Y) = E_I(2, Y)$. So the equations

$$2y_1 + y_2 = -4y_1 + 2y_2, \text{ and } y_1 + y_2 = 1,$$

have solution $y_1 = 0.143$, $y_2 = 0.857$, and $v_I = 1.143$. Similarly, $E_{II}(X, 1) = E_{II}(X, 2)$ gives

$$x_1 + 2x_2 = 3x_2 \text{ and } x_1 + x_2 = 1 \implies x_1 = x_2 = 0.5, v_{II} = 1.5.$$

Notice that we can find the Nash point without actually knowing v_I or v_{II}. Also, assuming that $E_I(1, Y) = v_I = E_I(2, Y)$ gives us the optimal Nash point for player II, and assuming $E_{II}(X, 1) = E_{II}(X, 2)$ gives us the Nash point for player I. In other words, the Nash point for II is found from the payoff function for player I and vice versa.

Problems

2.9 Suppose we have a two-person matrix game, which results in the following best response functions. Find the Nash equilibria if they exist.

(a)

$$BR_1(y) = \begin{cases} 0, & \text{if } y \in [0, \frac{1}{4}); \\ [0,1], & \text{if } y = \frac{1}{4}; \\ 1, & \text{if } y \in (\frac{1}{4}, 1]. \end{cases} \qquad BR_2(x) = \begin{cases} 1, & \text{if } x \in [0, \frac{1}{2}); \\ [0,1], & \text{if } x = \frac{1}{2}; \\ 0, & \text{if } x \in (\frac{1}{2}, 1]. \end{cases}$$

(b)

$$BR_1(y) = \{x \in [0,1] \mid 1 - 3y \le x \le 1 - 2y\}$$

$$BR_2(x) = \{y \in [0,1] \mid x \le y \le \frac{2}{3}x + \frac{1}{3}\}.$$

2.10 Verify that the inequalities in Proposition 2.2.2 are sufficient for a Nash equilibrium in a 2×2 game.

2.11 Apply the method of this section to analyze the modified study–party game:

I/II	Study	Party
Study	$(2, 2)$	$(3, 1)$.
Party	$(1, 3)$	$(4, 4)$

Find all Nash equilibria and graph the rational reaction sets.

2.12 Consider the Stop-Go game. Two drivers meet at an intersection at the same time. They have the options of stopping and waiting for the other driver to continue, or going.

Here's the payoff matrix in which the player who stops while the other goes loses a bit less than if they both stopped.

I/II	Stop	Go
Stop	$(1, 1)$	$(1 - \varepsilon, 2)$
Go	$(2, 1 - \varepsilon)$	$(0, 0)$

Assume that $0 < \varepsilon < 1$. Find all Nash equilibria.

2.13 Determine all Nash equilibria and graph the rational reaction sets for the game

$$\begin{bmatrix} (-10, 5) & (2, -2) \\ (1, -1) & (-1, 1) \end{bmatrix}$$

2.14 There are two companies each with exactly one job opening. Suppose firm 1 offers the pay p_1 and firm 2 offers pay p_2 with $p_1 < p_2 < 2p_1$. There are two prospective applicants each of whom can apply to only one of the two firms. They make simultaneous and independent decisions. If exactly one applicant applies to a company, that applicant gets the job. If both apply to the same company, the firm flips a fair coin to decide who is hired and the other is unemployed (payoff zero).

(a) Find the game matrix.

(b) Find all Nash equilibria.

2.15 This problem looks at the general 2×2 game to compute a mixed Nash equilibrium:

$$\begin{bmatrix} (A,a) & (B,b) \\ (C,c) & (D,d) \end{bmatrix}$$

(a) Assume $a - b + d - c \neq 0, d \neq c$. Use equality of payoffs to find $X^* = (x, 1 - x)$.

(b) What do you have to assume about A, B, C, D to find $Y^* = (y, 1 - y), 0 < y < 1$? Now find y.

2.16 This game is the nonzero sum version of Pascal's wager (see Example 1.8).

You/God	God reveals	God hidden
Believe	(α, A)	$(-\beta, B)$
Don't believe	$(-\gamma, -\Gamma)$	$(0, -\Delta)$

Since God choosing to not exist is paradoxical, we change God's strategies to Reveals, or is Hidden. Your payoffs are explained as in Example 1.8. If you choose Believe, God obtains the positive payoffs $A > 0, B > 0$. If you choose to Not Believe, and God chooses Reveal, then you receive $-\gamma$ but God also receives $-\Gamma$. If God chooses to remain Hidden, then He receives $-\Delta$ if you choose to Not Believe. Assume $A, B, \Gamma, \Delta > 0$ and $\alpha, \beta, \gamma > 0$.

(a) Determine when there are only pure Nash equilibria and find them under those conditions.

(b) Find conditions under which a single mixed Nash equilibrium exists and determine what it is.

(c) Find player I's best response to the strategy $Y^0 = (\frac{1}{2}, \frac{1}{2})$.

2.17 Verify by checking against pure strategies that the mixed strategies $X^* = (\frac{3}{4}, 0, \frac{1}{4})$ and $Y^* = (0, \frac{1}{3}, \frac{2}{3})$ is a Nash equilibrium for the game with matrix

$$\begin{bmatrix} (x, 2) & (3, 3) & (1, 1) \\ (y, y) & (0, z) & (2, w) \\ (a, 4) & (5, 1) & (0, 7) \end{bmatrix},$$

where x, y, z, w, a are arbitrary.

2.18 Two radio stations (WSUP and WHAP) have to choose formats for their broadcasts. There are three possible formats: Rhythm and Blues (RB), Elevator Music (EM) or all talk (AT). The audiences for the three formats are 50%, 30%, and 20%, respectively. If they choose the same formats they will split the audience for that format equally, while if they choose different formats, each will get the total audience for that format.

(a) Model this as a nonzero sum game.

(b) Find all the Nash equilibria.

2.19 In a modified story of the prodigal son, a man had two sons, the prodigal and the one who stayed home. The man gave the prodigal son his share of the estate, which he squandered, and told the son who stayed home that all that he (the father) has is his (the son's). When the man died, the prodigal son again wanted his share of the estate. They each tell the judge (it ends up in court) a share amount they would be willing to take, either $\frac{1}{4}, \frac{1}{2}$, or $\frac{3}{4}$. Call the shares for each player $I_i, II_i, i = 1, 2, 3$. If $I_i + II_j > 1$, all the money goes to the game theory society. If $I_i + II_j \leq 1$, then each gets the share they asked for and the rest goes to an antismoking group.

(a) Find the game matrix and find all pure Nash equilibria.

(b) Find at least two distinct mixed Nash equilibria using the equality of payoffs Theorem 2.2.3.

2.20 Willard is a salesman with an expense account for travel. He can steal (S) from the account by claiming false expenses or be honest (H) and accurately claim the expenses incurred. Willard's boss is Fillmore. Fillmore can check (C) into the expenses claimed or trust (T) that Willard is honest. If Willard cheats on his expenses he benefits by the amount b assuming he isn't caught by Fillmore. If Fillmore doesn't check, then Willard gets away with cheating. If Willard is caught cheating he incurs costs p which may include getting fired and paying back the amount he stole. Since Willard is a clever thief, we let $0 < \alpha < 1$ be the probability that Willard is caught if Fillmore investigates.

If Fillmore investigates he incurs cost c no matter what. Finally, let λ be Fillmore's loss if Willard cheats on his expenses but is not caught. Assume all of $b, p, \alpha, \lambda > 0$.

(a) Find the game bimatrix.

(b) Determine conditions on the parameters so that there is one mixed Nash equilibrium and find it.

2.21 Use the equality of payoffs Theorem 2.2.3 to solve the welfare game. In the welfare game the state, or government, wants to aid a pauper if he is looking for work but not otherwise. The pauper looks for work only if he cannot depend on welfare, but he may not be able to find a job even if he looks. The game matrix is

G/P	Look for work	Be a bum
Welfare	$(3, 2)$	$(-1, 3)$
No welfare	$(-1, 1)$	$(0, 0)$

Find all Nash equilibria and graph the rational reaction sets.

2.3 Interior Mixed Nash Points by Calculus

Whenever we are faced with a problem of maximizing or minimizing a function, we are taught in calculus that we can find them by finding critical points and then trying to verify that they are minima or maxima or saddle points. Of course, a critical point doesn't have to be any of these special points. When we look for Nash equilibria that simply supply the maximum expected payoff, assuming that the other players are doing the same for their payoffs, why not apply calculus? That's exactly what we can do, and it will give us all the interior, that is, completely mixed Nash points. The reason it works here is because of the nature of functions like $f(x, y) = XAY^T$. Calculus cannot give us the pure Nash equilibria because those are achieved on the boundary of the strategy region.

The easy part in applying calculus is to find the partial derivatives, set them equal to zero, and see what happens. Here is the procedure.

2.3.1 Calculus Method for Interior Nash

1. The payoff matrices are $A_{n \times m}$ for player I and $B_{n \times m}$ for player II. The expected payoff to I is $E_I(X, Y) = X \, AY^T$, and the expected payoff to II is $E_{II}(X, Y) = X \, BY^T$.

2. Let $x_n = 1 - (x_1 + \cdots + x_{n-1}) = 1 - \sum_{i=1}^{n-1} x_i$, $y_m = 1 - \sum_{j=1}^{m-1} y_j$ so each expected payoff is a function only of $x_1, \ldots, x_{n-1}, y_1, \ldots, y_{m-1}$. We can write

$$E_1(x_1, \ldots, x_{n-1}, y_1, \ldots, y_{m-1}) = E_I(X, Y),$$
$$E_2(x_1, \ldots, x_{n-1}, y_1, \ldots, y_{m-1}) = E_{II}(X, Y).$$

3. Take the partial derivatives and solve the system of equations $\partial E_1/\partial x_i = 0$, $\partial E_2/\partial y_j = 0$, $i = 1, \ldots, n-1$, $j = 1, \ldots, m-1$.

4. If there is a solution of this system of equations which satisfies the constraints $x_i \geq 0$, $y_j \geq 0$ and $\sum_{i=1}^{n-1} x_i \leq 1$, $\sum_{j=1}^{m-1} y_j \leq 1$, then this is the mixed strategy Nash equilibrium.

It is important to observe that we do **not** maximize $E_1(x_1, \ldots, x_{n-1}, y_1, \ldots, y_{m-1})$ over **all** variables x and y, but only over the x variables. Similarly, we do **not** maximize $E_2(x_1, \ldots, x_{n-1}, y_1, \ldots, y_{m-1})$ over **all** variables x and y, but only over the y variables. A Nash equilibrium for a player is a maximum of the player's payoff over those variables that player controls, assuming that the other player's variables are held fixed.

Example 2.9 Let's consider again the bimatrix game with the two matrices

$$A = \begin{bmatrix} 2 & -1 \\ -1 & 1 \end{bmatrix}, \qquad B = \begin{bmatrix} 1 & -1 \\ -1 & 2 \end{bmatrix}.$$

What happens if we use the calculus method on this game? First, set up the functions (using $X = (x, 1-x)$, $Y = (y, 1-y)$)

$$E_1(x, y) = [2x - (1-x)]y + [-x + (1-x)](1-y),$$
$$E_2(x, y) = [x - (1-x)]y + [-x + 2(1-x)](1-y).$$

Player I wants to maximize E_1 for each fixed y, so we take

$$\frac{\partial E_1(x, y)}{\partial x} = 3y + 2y - 2 = 5y - 2 = 0 \implies y = \frac{2}{5}.$$

Similarly, player II wants to maximize $E_2(x, y)$ for each fixed x, so

$$\frac{\partial E_2(x, y)}{\partial y} = 5x - 3 = 0 \implies x = \frac{3}{5}.$$

Everything works to give us $X^* = (\frac{3}{5}, \frac{2}{5})$ and $Y^* = (\frac{2}{5}, \frac{3}{5})$ is a Nash equilibrium for the game, just as we had before. We do not get the pure Nash points for this problem. But those are easy to get by determining the pairs of payoffs that are simultaneously the largest in the column and the largest in the row, just as we did before. We don't need calculus for that.

Example 2.10 Two partners have two choices for where to invest their money, say, O_1, O_2 where the letter O stands for opportunity, but they have to come to an agreement. If they do not agree on joint investment, there is no benefit to either player. We model this using the bimatrix

	O_1	O_2
O_1	(1, 2)	(0, 0)
O_2	(0, 0)	(2, 1)

If player I chooses O_1 and II chooses O_1 the payoff to I is 1 and the payoff to II is 2 units because II prefers to invest the money into O_1 more than into O_2. If the players do not agree on how to invest, then each receives 0.

To solve this game, first notice that there are two pure Nash points at (O_1, O_1) and (O_2, O_2), so total cooperation will be a Nash equilibrium. We want to know if there are any mixed Nash points. We will start the analysis from the beginning rather than using the formulas from Section 2.3. The reader should verify that our result will match if we do use the formulas. We will derive the rational reaction sets for each player directly.

Set

$$A = \begin{bmatrix} 1 & 0 \\ 0 & 2 \end{bmatrix}, B = \begin{bmatrix} 2 & 0 \\ 0 & 1 \end{bmatrix}.$$

Then player I's expected payoff (using $X = (x, 1-x), Y = (y, 1-y)$) is

$$E_1(x, y) = (x, 2(1-x)) \cdot (y, 1-y)^T$$
$$= xy + 2(1-x)(1-y) = 3xy - 2x - 2y + 2.$$

Keep in mind that player I wants to make this as large as possible for any fixed $0 \le y \le 1$. We need to find the maximum of E_1 as a function of $x \in [0, 1]$ for each fixed $y \in [0, 1]$.

Write E_1 as

$$E_1(x, y) = x(3y - 2) - 2y + 2.$$

If $3y - 2 > 0$, then $E_1(x, y)$ is maximized as a function of x at $x = 1$. If $3y - 2 < 0$, then the maximum of $E_1(x, y)$ will occur at $x = 0$. If $3y - 2 = 0$, then $y = \frac{2}{3}$ and $E_1(x, \frac{2}{3}) = \frac{2}{3}$; that is, we have shown that

$$\max_{0 \le x \le 1} E_1(x, y) = \begin{cases} y & \text{if } 3y - 2 > 0 \implies y > \frac{2}{3}, \text{ achieved at } x = 1; \\ \frac{2}{3} & \text{if } y = \frac{2}{3}, \text{ achieved at any } x \in [0, 1]; \\ -2y + 2 & \text{if } 3y - 2 < 0 \implies y < \frac{2}{3}, \text{ achieved at } x = 0. \end{cases}$$

Recall that the set of points where the maximum is achieved by player I for each fixed y for player II is the rational reaction set for player I:

$$R_{\mathrm{I}} = \{(x^*, y) \in [0, 1] \times [0, 1] \mid \max_{0 \le x \le 1} E_1(x, y) = E_1(x^*, y)\}.$$

In this example we have shown that

$$R_{\mathrm{I}} = \left\{ (1, y), \frac{2}{3} < y \le 1 \right\} \cup \left\{ \left(x, \frac{2}{3} \right), 0 \le x \le 1 \right\} \cup \left\{ (0, y), 0 \le y < \frac{2}{3} \right\}.$$

This is the rational reaction set for player I because no matter what II plays, player I should use an $(x, y) \in R_{\mathrm{I}}$. For example, if $y = \frac{1}{2}$, then player I should use $x = 0$ and I receives payoff $E(0, \frac{1}{2}) = 1$; if $y = \frac{15}{16}$, then I should choose $x = 1$ and I receives $E_1(1, \frac{15}{16}) = \frac{15}{16}$; and when $y = \frac{2}{3}$, it doesn't matter what player I chooses because the payoff to I will be exactly $\frac{2}{3}$ for any $x \in [0, 1]$.

Next, we consider $E_2(x, y)$ in a similar way. Write

$$E_2(x, y) = 2xy + (1-x)(1-y) = 3xy - x - y + 1 = y(3x - 1) - x + 1.$$

Player II wants to choose $y \in [0, 1]$ to maximize this, and that will depend on the coefficient of y, namely, $3x - 1$. We see as before that

$$\max_{y \in [0,1]} E_2(x, y) = \begin{cases} -x + 1 & \text{if } 0 \le x < \frac{1}{3} \text{ achieved at } y = 0; \\ \frac{2}{3} & \text{if } x = \frac{1}{3} \text{ achieved at any } y \in [0, 1] \\ 2x & \text{if } \frac{1}{3} < x \le 1 \text{ achieved at } y = 1. \end{cases}$$

The rational reaction set for player II is the set of points where the maximum is achieved by player II for each fixed x for player II:

$$R_{\text{II}} = \{(x, y^*) \in [0, 1] \times [0, 1] \mid \max_{0 \le y \le 1} E_2(x, y) = E_2(x, y^*)\}.$$

We see that in this example

$$R_{\text{II}} = \left\{ (x, 0), 0 \le x < \frac{1}{3} \right\} \cup \left\{ \left(\frac{1}{3}, y \right), 0 \le y \le 1 \right\} \cup \left\{ (x, 1), \frac{1}{3} < x \le 1 \right\}.$$

Here is the graph of R_{I} and R_{II} on the same set of axes:

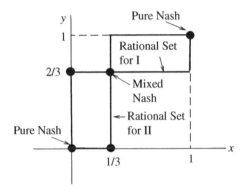

The zigzag lines form the rational reaction sets of the players. For example, if player I decides for some reason to play O_1 with probability $x = \frac{1}{2}$, then player II would rationally play $y = 1$. Where the zigzag lines cross (which is the set of points $R_{\text{I}} \cap R_{\text{II}}$) are all the Nash points; that is, the Nash points are at $(x, y) = (0, 0), (1, 1)$ and $(\frac{1}{3}, \frac{2}{3})$. So either they should always cooperate to the advantage of one player or the other, or player I should play O_1 one-third of the time and player II should play O_1 two-thirds of the time. The associated expected payoffs are

$$E_1(0, 0) = 2, \quad E_2(0, 0) = 1,$$

$$E_1(1, 1) = 1, \quad E_2(1, 1) = 2,$$

and

$$E_1\left(\frac{1}{3}, \frac{2}{3}\right) = \frac{2}{3} = E_2\left(\frac{1}{3}, \frac{2}{3}\right).$$

Only the mixed strategy Nash point $(X^*, Y^*) = ((\frac{1}{3}, \frac{2}{3}), (\frac{2}{3}, \frac{1}{3}))$ gives the same expected payoffs to the two players. This seems to be a problem. Only the mixed strategies gives the same payoff to each player, but it will result in **less** for each player than they could get if they play the pure Nash points. Permitting the other player the advantage results in a **bigger** payoff to both players! If one player decides to cave, they both can do better, but if both players insist that the outcome be **fair**, whatever that means, then they both do worse.

Calculus will give us the interior mixed Nash very easily:

$$\frac{\partial E_1(x, y)}{\partial x} = 3y - 2 = 0 \implies y = \frac{2}{3}, \text{ and}$$

$$\frac{\partial E_2(x, y)}{\partial y} = 3x - 1 = 0 \implies x = \frac{1}{3}.$$

The method of solution we used in this example gives us the entire rational reaction set for each player. It is essentially the way Proposition 2.2.2 was proved .

Another point to notice is that the rational reaction sets and their graphs do **not** indicate what the **payoffs** are to the individual players, only what their **strategies** should be.

Finally, we record here the definition of the rational reaction sets in the general case with arbitrary size matrices.

Definition 2.3.1 *The rational reaction set for each player are defined as follows:*

$$R_{\mathrm{I}} = \{(X, Y) \in S_n \times S_m \mid E_{\mathrm{I}}(X, Y) = \max_{p \in S_n} E_{\mathrm{I}}(p, Y)\},$$

$$R_{\mathrm{II}} = \{(X, Y) \in S_n \times S_m \mid E_{\mathrm{II}}(X, Y) = \max_{t \in S_m} E_{\mathrm{II}}(X, t)\}.$$

The set of all Nash equilibria is then the set of all common points $R_{\mathrm{I}} \cap R_{\mathrm{II}}$.

We can write down the system of equations that we get using calculus in the general case, after taking the partial derivatives and setting to zero. We start with

$$E_{\mathrm{I}}(X, Y) = X \, AY^T = \sum_{j=1}^{m} \sum_{i=1}^{n} x_i a_{ij} y_j.$$

Following the calculus method (2.3.1), we replace[1] $x_n = 1 - \sum_{k=1}^{n-1} x_k$ and do some algebra:

$$
\begin{aligned}
X \, AY^T &= \sum_{j=1}^{m} \sum_{i=1}^{n} x_i a_{ij} y_j \\
&= \sum_{j=1}^{m} \left(\sum_{i=1}^{n-1} x_i a_{ij} y_j + \left(1 - \sum_{k=1}^{n-1} x_k \right) a_{nj} y_j \right) \\
&= \sum_{j=1}^{m} \left(a_{nj} y_j + \sum_{i=1}^{n-1} x_i a_{ij} y_j - \sum_{k=1}^{n-1} x_k (a_{nj} y_j) \right) \\
&= \sum_{j=1}^{m} \left(a_{nj} y_j + \sum_{i=1}^{n-1} x_i [a_{ij} - a_{nj}] y_j \right) \\
&= E_{\mathrm{I}}(x_1, \ldots, x_{n-1}, y_1, \ldots, y_m).
\end{aligned}
$$

But then, for each $k = 1, 2, \ldots, n-1$, we obtain

$$\frac{\partial E_{\mathrm{I}}(x_1, \ldots, x_{n-1}, y_1, \ldots, y_m)}{\partial x_k} = \sum_{j=1}^{m} y_j [a_{kj} - a_{nj}].$$

Similarly, for each $s = 1, 2, \ldots, m-1$, we get the partials

$$\frac{\partial E_2(x_1, \ldots, x_n, y_1, \ldots, y_{m-1})}{\partial y_s} = \sum_{i=1}^{n} x_i [b_{is} - b_{im}].$$

---

1 Alternatively we may take the partial derivative of the function with a Lagrange multiplier $E_1(\vec{x}, \vec{y}) - \lambda(\sum_i x_i - 1)$. Taking a partial with respect to x_k shows that $\partial E_1 / \partial x_k = \partial E_1 / \partial x_n$ for all $k = 1, 2, \ldots, n-1$. This gives us the same system as (2.3.1).

So, the system of equations we need to solve to get an interior Nash equilibrium is

$$
\left.\begin{aligned}
\sum_{j=1}^{m} y_j [a_{kj} - a_{nj}] &= 0, & k = 1, 2, \dots, n-1, \\
\sum_{i=1}^{n} x_i [b_{is} - b_{im}] &= 0, & s = 1, 2, \dots, m-1, \\
x_n = 1 - \sum_{i=1}^{n-1} x_i, \quad y_m &= 1 - \sum_{j=1}^{m-1} y_j.
\end{aligned}\right\}
\tag{2.3.1}
$$

Once these are solved, we check that $x_i \geq 0, y_j \geq 0$. If these all check out, we have found a Nash equilibrium $X^* = (x_1, \dots, x_n)$ and $Y^* = (y_1, \dots, y_m)$. Notice also that the equations are really two separate systems of linear equations and can be solved separately because the variables x_i and y_j appear only in their own system. Also notice that these equations are really nothing more than the equality of payoffs Theorem 2.2.3, because, for example

$$
\sum_{j=1}^{m} y_j [a_{kj} - a_{nj}] = 0 \implies \sum_{j=1}^{m} y_j a_{kj} = \sum_{j=1}^{m} y_j a_{nj},
$$

which is the same as saying that for $k = 1, 2, \dots, n-1$, we have

$$
E_I(k, Y^*) = \sum_{j=1}^{m} y_j a_{kj} = \sum_{j=1}^{m} y_j a_{nj} = E_I(n, Y^*).
$$

All the payoffs are equal. This, of course, assumes that each row of A is used by player I with positive probability in a Nash equilibrium, but that is our assumption about an **interior** Nash equilibrium. These equations won't necessarily work for the pure Nash or the ones with zero components.

Example 2.11 We are going to use Eq. (2.3.1) to find interior Nash points for the following bimatrix game:

$$
A = \begin{bmatrix} -2 & 5 & 1 \\ -3 & 2 & 3 \\ 2 & 1 & 3 \end{bmatrix}, \quad B = \begin{bmatrix} -4 & -2 & 4 \\ -3 & 1 & 4 \\ 3 & 1 & -1 \end{bmatrix}.
$$

The system of Eq. (2.3.1) for an interior Nash point become

$$
-2y_1 + 6y_2 - 2 = 0, \quad -5y_1 + y_2 = 0
$$

and

$$
-12x_1 - 11x_2 + 4 = 0, \quad -8x_1 - 5x_2 + 2 = 0.
$$

There is one and only one solution given by $y_1 = \frac{1}{14}, y_2 = \frac{5}{14}$ and $x_1 = \frac{1}{14}, x_2 = \frac{4}{14}$, so we have $y_3 = \frac{8}{14}, x_3 = \frac{9}{14}$, and our interior Nash point is

$$
X^* = \left(\frac{1}{14}, \frac{4}{14}, \frac{9}{14} \right) \text{ and } Y^* = \left(\frac{1}{14}, \frac{5}{14}, \frac{8}{14} \right).
$$

The expected payoffs to each player are $E_I(X^*, Y^*) = X^* A Y^{*T} = \frac{31}{14}$ and $E_{II}(X^*, Y^*) = X^* B Y^{*T} = \frac{11}{14}$. It appears that player I does a lot better in this game with this Nash.

We have found the interior, or mixed Nash point. There are also two pure Nash points and they are $X^* = (0, 0, 1), Y^* = (1, 0, 0)$ with payoffs $(2, 3)$ and $X^* = (0, 1, 0), Y^* = (0, 0, 1)$ with payoffs $(3, 4)$. With multiple Nash points, the game can take on one of many forms.

Example 2.12 Here is a last example in which the equations do not work (see Problem 2.22) because it turns out that one of the columns should never be played by player II. That means that the mixed Nash is not in the interior, but on the boundary of $S_n \times S_m$.

Let's consider the game with payoff matrices

$$A = \begin{bmatrix} 2 & 3 & 4 \\ 0 & 4 & 3 \end{bmatrix}, \qquad B = \begin{bmatrix} 0 & 1 & -1 \\ 2 & 0 & 1 \end{bmatrix}.$$

You can calculate that the safety levels are $value(A) = 2$, with pure saddle $X_A = (1,0), Y_A = (1,0,0)$, and $value(B^T) = \frac{2}{3}$, with saddle $X_B = (\frac{1}{3}, \frac{2}{3}, 0), Y_B = (\frac{2}{3}, \frac{1}{3})$. These are the amounts that each player can get assuming that both are playing in a zero-sum game with the two matrices.

Now, let $X = (x, 1-x), Y = (y_1, y_2, 1 - y_1 - y_2)$ be a Nash point for the bimatrix game. Calculate

$$E_1(x, y_1, y_2) = XAY^T = x[y_1 - 2y_2 + 1] + y_2 - 3y_1 + 3.$$

Player I wants to maximize $E_1(x, y_1, y_2)$ for given fixed $y_1, y_2 \in [0,1], y_1 + y_2 \le 1$, using $x \in [0,1]$. So we look for $\max_x E_1(x, y_1, y_2)$. For fixed y_1, y_2, we see that $E_1(x, y_1, y_2)$ is a straight line with slope $y_1 - 2y_2 + 1$. The maximum of that line will occur at an endpoint depending on the sign of the slope. Here is what we get:

$$\max_{0 \le x \le 1} x[y_1 - 2y_2 + 1] + y_2 - 3y_1 + 3$$

$$= \begin{cases} -2y_1 - y_2 + 4 & \text{if } y_1 > 2y_2 - 1; \\ y_2 - 3y_1 + 3 & \text{if } y_1 = 2y_2 - 1; \\ y_2 - 3y_1 + 3 & \text{if } y_1 < 2y_2 - 1. \end{cases} = \begin{cases} E_1(1, y_1, y_2) & \text{if } y_1 > 2y_2 - 1; \\ E_1(x, 2y_2 - 1, y_2) & \text{if } y_1 = 2y_2 - 1; \\ E_1(0, y_1, y_2) & \text{if } y_1 < 2y_2 - 1. \end{cases}$$

Along any point of the straight line $y_1 = 2y_2 - 1$ the maximum of $E_1(x, y_1, y_2)$ is achieved at any point $0 \le x \le 1$. We end up with the following set of maximums for $E_1(x, y_1, y_2)$:

$$R_1 = \{(x, y_1, y_2) \mid [(1, y_1, y_2), y_1 > 2y_2 - 1], \text{ or}$$

$$[(x, 2y_2 - 1, y_2), 0 \le x \le 1], \text{ or } [(0, y_1, y_2), y_1 < 2y_2 - 1]\},$$

which is exactly the rational reaction set for player I. This is a set in three dimensions.

Now we go through the same procedure for player II, for whom we calculate,

$$E_2(x, y_1, y_2) = X BY^T = y_1 + y_2(3x - 1) + (-2x + 1).$$

Player II wants to maximize $E_2(x, y_1, y_2)$ for given fixed $x \in [0,1]$ using $y_1, y_2 \in [0,1], y_1 + y_2 \le 1$. We look for $\max_{y_1, y_2} E_2(x, y_1, y_2)$.

Here is what we get:

$$\max_{\substack{y_1 + y_2 \le 1, \\ y_1, y_2 \ge 0}} y_1 + y_2(3x - 1) + (-2x + 1)$$

$$= \begin{cases} -2x + 2 & \text{if } 3x - 1 < 1; \\ \frac{2}{3} & \text{if } 3x - 1 = 1; \\ x & \text{if } 3x - 1 > 1. \end{cases}$$

$$= \begin{cases} E_2(x, 1, 0) & \text{if } 0 \le x < \frac{2}{3}; \\ E_2\left(\frac{2}{3}, y_1, y_2\right) = \frac{2}{3} & \text{if } x = \frac{2}{3}, y_1 + y_2 = 1; \\ E_2(x, 0, 1) & \text{if } \frac{2}{3} < x \le 1. \end{cases}$$

To explain where this came from, let's consider the case $3x - 1 < 1$. In this case, the coefficient of y_2 is less than the coefficient of y_1 (which is 1), so the maximum will be achieved by taking $y_2 = 0$ and $y_1 = 1$ because that gives the biggest weight to the largest coefficient. Then plugging in $y_1 = 1, y_2 = 0$ gives payoff $-2x + 2$. If $3x - 1 = 1$ the coefficients of y_1 and y_2 are the same, then we can take any y_1 and y_2 as long as $y_1 + y_2 = 1$. Then the payoff becomes $(y_1 + y_2)(3x - 1) + (-2x + 1) = x$, but $3x - 1 = 1$ requires that $x = \frac{2}{3}$, so the payoff is $\frac{2}{3}$. The case $3x - 1 > 1$ is similar.

We end up with the following set of maximums for $E_2(x, y_1, y_2)$:

$$R_{\mathrm{II}} = \left\{ (x, y_1, y_2) \mid \left[(x, 1, 0), 0 \leq x < \frac{2}{3}\right], \text{ or } \left[\left(\frac{2}{3}, y_1, y_2\right), y_1 + y_2 = 1\right], \text{ or } \left[(x, 0, 1), \frac{2}{3} < x \leq 1\right]\right\},$$

which is the rational reaction set for player II.

The graph of R_{I} and R_{II} on the same graph (in three dimensions) will intersect at the mixed Nash equilibrium points. In this example, the Nash equilibrium is given by

$$X^* = \left(\frac{2}{3}, \frac{1}{3}\right), \quad Y^* = \left(\frac{1}{3}, \frac{2}{3}, 0\right).$$

Then, $E_1(\frac{2}{3}, \frac{1}{3}, \frac{2}{3}) = \frac{8}{3}$ and $E_2(\frac{2}{3}, \frac{1}{3}, \frac{2}{3}) = \frac{2}{3}$ are the payoffs to each player. We could have simplified the calculations significantly if we had noticed from the beginning that we could have eliminated the third column from the bimatrix because column 3 for player II is dominated by column 1, and so may be dropped.

Problems

2.22 Write down Eq. (2.3.1) for the game

$$\begin{bmatrix} (2, 0) & (3, 2) & (4, 1) \\ (0, 2) & (4, 0) & (3, 1) \end{bmatrix}.$$

Try to solve the equations. What, if anything, goes wrong?

2.23 The game matrix in the welfare problem is

G/P	Look for work	Be a bum
Welfare	$(3, 2)$	$(-1, 3)$
No welfare	$(-1, 1)$	$(0, 0)$

Write the system of equations for an interior Nash and solve them.

2.24 Consider the game with bimatrix

$$\begin{bmatrix} (5, 4) & (3, 6) \\ (6, 3) & (1, 1) \end{bmatrix}.$$

(a) Find the safety levels and maxmin strategies for each player.
(b) Find all Nash equilibria using the equality of payoffs theorem for the mixed strategy.
(c) Verify that the mixed Nash equilibrium is individually rational.
(d) Verify that $X^* = (\frac{1}{4}, \frac{3}{4})$, $Y^* = (\frac{5}{8}, \frac{3}{8})$ is not a Nash equilibrium.

2.25 In this problem we consider the analog of the invertible matrix Theorem 1.5.2 for zero-sum games. Consider the nonzero sum game $(A_{n \times n}, B_{n \times n})$ and suppose that A^{-1} and B^{-1} exist.

(a) Show that if $J_n A^{-1} J_n^T \neq 0, J_n B^{-1} J_n^T \neq 0$ and

$$X^* = \frac{J_n B^{-1}}{J_n B^{-1} J_n^T}, \quad Y^{*T} = \frac{A^{-1} J_n^T}{J_n A^{-1} J_n^T}$$

are strategies, then (X^*, Y^*) is a Nash equilibrium, and

$$v_I = \frac{1}{J_n A^{-1} J_n^T} = E_I(X^*, Y^*), \quad v_{II} = \frac{1}{J_n B^{-1} J_n^T} = E_{II}(X^*, Y^*).$$

(b) Use the previous result to solve the game

$$\begin{bmatrix} (0,9) & (8,2) & (7,3) \\ (4,7) & (2,8) & (9,1) \\ (10,-1) & (5,5) & (1,9) \end{bmatrix}.$$

(c) Suppose the game is 2×2 with matrices $A_{2 \times 2}, B_{2 \times 2}$. What is the analog of the formulas for a saddle point when the matrices do not have an inverse? Use the formulas you obtain to solve the games

$$(1) \begin{bmatrix} (1,3) & (-1,-2) \\ (-2,-1) & (4,0) \end{bmatrix} \quad \text{and} \quad (2) \begin{bmatrix} (1,3) & (-2,-2) \\ (-2,-1) & (4,0) \end{bmatrix}$$

2.26 **Chicken game** In this version of the game the matrix is

I/II	Straight	Swerve
Straight	$(-a, -a)$	$(2, 0)$
Swerve	$(0, 2)$	$(1, 1)$

Find the mixed Nash equilibrium and show that as $a > 0$ increases, so does the expected payoff to each player under the mixed Nash.

2.27 Find all possible Nash equilibria for the game and the rational reaction sets:

$$\begin{bmatrix} (a,a) & (0,0) \\ (0,0) & (b,b) \end{bmatrix}.$$

Consider all cases $(a > 0, b > 0)$, $(a > 0, b < 0)$, and so on.

2.28 Show that a 2×2 symmetric game

$$A = \begin{bmatrix} a_{11} & a_{12} \\ a_{21} & a_{22,} \end{bmatrix}, \quad B = A^T,$$

has exactly the same set of Nash equilibria as does the symmetric game with matrix

$$A' = \begin{bmatrix} a_{11} - a & a_{12} - b \\ a_{21} - a & a_{22} - b, \end{bmatrix}, \quad B = A'^T,$$

for any a, b.

2.29 Consider the game with matrix

$$\begin{bmatrix} (2,0) & (3,1) & (4,-1) \\ (0,2) & (4,0) & (3,1) \end{bmatrix}.$$

(a) Find the best response sets for each player.
(b) Find the Nash equilibria and verify that they are in the intersection of the best response sets.

2.30 A game called the **Battle of the Sexes** is a game between a husband and wife trying to decide about cooperation. On a given evening, the husband wants to see wrestling at the stadium, while the wife wants to attend a concert at orchestra hall. Neither the husband nor the wife wants to go to what the other has chosen, but neither do they want to go alone to their preferred choice. They view this as a two-person nonzero sum game with matrix

H/W	Wr	Co
Wr	$(2, 1)$	$(-1, -1)$
Co	$(-1, -1)$	$(1, 2)$

If they decide to cooperate and both go to wrestling, the husband receives 2 and the wife receives 1, because the husband gets what he wants and the wife partially gets what she wants. The rest are explained similarly. This is a model of compromise and cooperation. Use the method of this section to find all Nash equilibria. Graph the rational reaction sets.

2.31 (**Hawk–Dove game**) Two companies both want to take over a sales territory. They have the choices of defending the territory and fighting if necessary, or act as if willing to fight but if the opponent fights (F), then backing off (Bo). They look at this as a two-person nonzero sum game with matrix

I/II	F	Bo
F	$(-1, -1)$	$(2, 0)$
Bo	$(0, 2)$	$(0, 0)$

Solve this game and graph the rational reaction sets.

2.32 (**Stag–Hare game**) Two hunters are pursuing a stag. Each hunter has the choice of going after the stag (S), which will be caught if they both go after it and it will be shared equally, or peeling off and going after a rabbit (R). Only one hunter is needed to catch the rabbit and it will not be shared. Look at this as a nonzero sum two-person game with matrix

I/II	S	R
S	$(2, 2)$	$(0, 1)$
R	$(1, 0)$	$(1, 1)$

This assumes that they each prefer stag meat to rabbit meat, and they will each catch a rabbit if they decide to peel off. Solve this game and graph the rational reaction sets.

2.33 We are given the following game matrix

$$\begin{bmatrix} (0, 0) & (50, 40) & (40, 50) \\ (40, 50) & (0, 0) & (50, 40) \\ (50, 40) & (40, 50) & (0, 0) \end{bmatrix}.$$

(a) There is a unique mixed Nash equilibrium for this game. Find it.

(b) Suppose either player deviates from using her Nash equilibrium. Find the best response of the other player and show that it results in a higher payoff.

2.3.2 Existence of a Nash Equilibrium for Bimatrix Games

The set of all Nash equilibria is determined by the set of all common points $R_I \cap R_{II}$. How do we know whether this intersection has any points at all? It might have occurred to you that we have no guarantee that looking for a Nash equilibrium in a bimatrix game with matrices (A, B) would ever be successful. So maybe we need a guarantee that what we are looking for actually exists. That is what Nash's theorem gives us. We will not prove the theorem.

Theorem 2.3.2 *There exists $X^* \in S_n$ and $Y^* \in S_m$ so that*

$$E_I(X^*, Y^*) = X^*AY^{*T} \geq E_I(X, Y^*),$$
$$E_{II}(X^*, Y^*) = X^*BY^{*T} \geq E_{II}(X^*, Y),$$

for any other mixed strategies $X \in S_n, Y \in S_m$.

The theorem guarantees at least one Nash equilibrium if we are willing to use mixed strategies. In the zero sum case, this theorem reduces to von Neumann's minimax theorem (show that). The existence of a Nash equilibrium depends on showing that there is always a best response strategy to a best response strategy. Here's what we mean by a best response set of strategies.

Definition 2.3.3 *The **best response sets** for each player are defined as*

$$BR_I(Y) = \{X \in S_n \mid E_I(X, Y) = \max_{p \in S_n} E_I(p, Y)\},$$
$$BR_{II}(X) = \{Y \in S_m \mid E_{II}(X, Y) = \max_{q \in S_m} E_I(X, q)\}.$$

The difference between the best response set and the rational reaction set is that the rational reaction set R_I consists of the **pairs of strategies** (X, Y) for which $E_I(X, Y) = \max_p E_I(p, Y)$, whereas the set $BR_I(Y)$ consists of the strategy (or collection of strategies) X for player I that is the best response to a **fixed** Y.

It seems natural that our Nash equilibrium should be among the best response strategies to the opponent. Written out, this means $X^* \in BR_I(Y^*)$ so that

$$E_I(X^*, Y^*) = \max_{p \in S_n} E_I(p, Y^*) \geq E_I(X, Y^*) \text{ for all } X \in S_n$$

and $Y^* \in BR_{II}(X^*)$ so that

$$E_{II}(X^*, Y^*) = \max_{t \in S_m} E_{II}(X^*, t) \geq E_{II}(X^*, Y) \text{ for all } Y \in S_m.$$

That's what a Nash equilibrium is. The trick is to show that $X^* \in BR_I(Y^*)$ and $Y^* \in BR_{II}(X^*)$. That's what John Nash did.

Finally, it is important to understand the difficulty in obtaining the existence of a Nash equilibrium. If our problem was

$$\max_{X \in S_n, Y \in S_m} E_I(X, Y) \text{ and } \max_{X \in S_n, Y \in S_m} E_{II}(X, Y),$$

then the existence of an (X_I, Y_I) providing the maximum of E_I is immediate from the fact that $E_I(X, Y)$ is a continuous function over a closed and bounded set. The same is true for the existence of an (X_{II}, Y_{II}) providing the maximum of $E_{II}(X, Y)$. The problem is that we don't know whether $X_I = X_{II}$ and $Y_I = Y_{II}$. In fact, that is very unlikely and it is a stronger condition than what is required of a Nash equilibrium. Moreover, a player cannot control the actions of the other player but only her own.

2.4 Nonlinear Programming Method for Nonzero Sum Two-Person Games

We have shown how to calculate the pure Nash equilibria in the 2×2 case, and the system of equations that will give a mixed Nash equilibrium in more general cases. In this section we present a method (introduced by Lemke and Howson [21] in the 1960s) of finding all Nash equilibria for arbitrary two-person nonzero sum games with any number of strategies. We will formulate the problem of finding a Nash equilibrium as a **nonlinear program**, in contrast to the formulation of solution of a zero sum game as a **linear program**.

In general, a nonlinear program is simply an optimization problem involving nonlinear functions and nonlinear constraints. For example, if we have an objective function f and constraint functions $h_1, \ldots, h_k$, the problem

$$\text{Minimize } f(x_1, \ldots, x_n) \text{ subject to } h_j(x_1, \ldots, x_n) \leq 0, \; j = 1, \ldots, k,$$

is a fairly general formulation of a nonlinear programming problem. In general, the functions f, h_j are not linear, but they could be. In that case, of course, we have a linear program, which is solved by the simplex algorithm. If the function f is quadratic and the constraint functions are linear, then this is called a **quadratic programming problem** and special techniques are available for those. Nonlinear programs are more general and the techniques to solve them are more involved, but fortunately there are several methods available, both theoretical and numerical. Nonlinear programming is a major branch of operations research and is under very active development. In our case, once we formulate the game as a nonlinear program, we will use the packages developed in Mathematica to solve them numerically. The first step is to set it up.

Theorem 2.4.1 *Consider the two-person game with matrices (A, B) for players I and II. Then, $(X^* \in S_n, Y^* \in S_m)$ is a Nash equilibrium if and only if they satisfy, along with scalars p^*, q^* the nonlinear program*

$$\max_{X,Y,p,q} XAY^T + XBY^T - p - q$$

subject to

$$AY^T \leq pJ_n^T$$
$$B^T X^T \leq qJ_m^T \qquad \text{(equivalently } XB \leq qJ_m\text{)}$$
$$x_i \geq 0, y_j \geq 0, \quad XJ_n = 1 = Y \, J_m^T$$

where $J_k = (1 \; 1 \; 1 \; \cdots 1)$ is the $1 \times k$ row vector consisting of all 1s. In addition, $p^* = E_I(X^*, Y^*)$, and $q^* = E_{II}(X^*, Y^*)$.

Remark Expanded, this program reads as

$$\max_{X,Y,p,q} \sum_{i=1}^{n} \sum_{j=1}^{m} x_i a_{ij} y_j + \sum_{i=1}^{n} \sum_{j=1}^{m} x_i b_{ij} y_j - p - q$$

subject to

$$\sum_{j=1}^{m} a_{ij} y_j \leq p, \; i = 1, 2, \ldots, n,$$

$$\sum_{i=1}^{n} x_i b_{ij} \leq q, \; j = 1, 2, \ldots, m,$$

$$x_i \geq 0, y_j \geq 0, \quad \sum_{i=1}^{n} x_i = \sum_{j=1}^{m} y_j = 1.$$

You can see that this is a nonlinear program because of the presence of the terms $x_i y_j$. That is why we need a nonlinear programming method, or a quadratic programming method because it falls into that category.

We will skip the proof of this theorem and just note that this theorem does provide a way to reduce finding Nash equilibria to maximizing a function using all the variables subject to constraints.

Remark It is **not** necessarily true that $E_I(X_1, Y_1) = p_1 = p^* = E_I(X^*, Y^*)$ and $E_{II}(X_1, Y_1) = q_1 = q^* = E_{II}(X^*, Y^*)$. Different Nash points can, and usually do, give different expected payoffs, as we have seen many times.

Using this theorem and some nonlinear programming implemented in Mathematica (see the commands for this example in the appendix), we can numerically solve any two-person nonzero sum game. For an example, suppose that we have the matrices

$$A = \begin{bmatrix} -1 & 0 & 0 \\ 2 & 1 & 0 \\ 0 & 1 & 1 \end{bmatrix}, \quad B = \begin{bmatrix} 1 & 2 & 2 \\ 1 & -1 & 0 \\ 0 & 1 & 2 \end{bmatrix}.$$

Now, before you get started looking for mixed Nash points, you should first find the pure Nash points. For this game we have two Nash points at $(X_1 = (0, 1, 0), Y_1 = (1, 0, 0))$ with expected payoff $E_I^1 = 2, E_{II}^1 = 1$, and $(X_2 = (0, 0, 1) = Y_2)$ with expected payoffs $E_I^2 = 1, E_{II}^2 = 2$.

On the other hand, we may solve the nonlinear programming problem using Mathematica (see the appendix) with these matrices and obtain a mixed Nash point

$$X_3 = \left(0, \frac{2}{3}, \frac{1}{3}\right), \quad Y_3 = \left(\frac{1}{3}, 0, \frac{2}{3}\right), \quad p = E_I(X_3, Y_3) = \frac{2}{3}, \quad q = E_{II}(X_3, Y_3) = \frac{2}{3}.$$

This gives us the result $p = 0.66, q = 0.66$. The commands also tell us that the value of the objective function at the optimal points is zero, which is what the theorem guarantees, viz., $\max f(X, Y, p, q) = 0$.

Example 2.13 Suppose that two countries are involved in an arms control negotiation. Each country can decide to either cooperate or not cooperate (don't). For this game, one possible bimatrix payoff situation may be

	Cooperate	Don't
Cooperate	$(1, 1)$	$(0, 3)$
Don't	$(3, 0)$	$(2, 2)$

This game has a pure Nash equilibrium at $(2, 2)$, so these countries will not actually negotiate in good faith. This would lead to what we might call **deadlock** because the two players will decide not to cooperate. If a third party managed to intervene to change the payoffs, you might get the following payoff matrix:

	Cooperate	Don't
Cooperate	$(3, 3)$	$(-1, -3)$ ·
Don't	$(3, -1)$	$(1, 1)$

What's happened is that both countries will receive a greater reward if they cooperate and the benefits of not cooperating on the part of both of them have decreased. In addition, we have lost symmetry. If the row player chooses to cooperate but the column player doesn't, they both lose, but player I will lose less than player II. On the other hand, if player I doesn't cooperate but player II does cooperate, then player I will gain and player II will lose, although not as much. Will that guarantee that they both cooperate? Not necessarily. We now have pure Nash equilibria at **both** $(3, 3)$ and $(1, 1)$. Is there also a mixed Nash equilibrium? If we apply the Mathematica commands, or use the calculus method, or the formulas, we obtain

$$X_1 = \left(\frac{1}{4}, \frac{3}{4}\right), Y_1 = (1,0), p_1 = 3, q_1 = 0,$$
$$X_2 = (0,1), Y_2 = (0,1), p_2 = q_2 = 1,$$
$$X_3 = (1,0) = Y_3, p_3 = q_3 = 3.$$

Here are the Mathematica commands for this problem

```
A={{3,-1},{3,1}}
B={{3,-3},{-1,1}}
f[X_]=X.B
g[Y_]=A.Y
FindMaximum[{f[{x,y}].{a,b}+{x,y}.g[{a,b}]-p-q,
f[{x,y}][[1]]-q<=0,f[{x,y}][[2]]-q<=0,
g[{a,b}][[1]]-p<=0,g[{a,b}][[2]]-p<=0,
x+y==1,x>=0,y>=0,a>=0,b>=0,a+b==1},{x,y,a,b,p,q}]
Out= {-3.23669*10^-7,
        {x->0.579182,y->0.420818,a->1.,b->2.50898*10^-8,p->3.,q->1.31673}}
```

This finds another Nash equilibrium $X^* = (0.58, 0.42)$, $Y^* = (1,0)$ with payoffs $E_I = 3, E_{II} = 1.32$. Actually, by graphing the rational reaction sets we see that any mixed strategy $X = (x, 1-x)$, $Y = (1,0), \frac{1}{4} \le x \le 1$, is a Nash point.

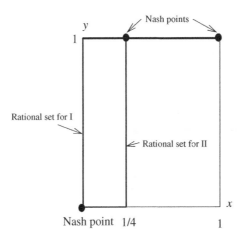

Each player receives the most if both cooperate, but how does one move them from *Don't Cooperate* to *Cooperate*? If the game is played an indefinite number of times and the chances the game is played again and again is large enough, cooperation can result. That is the subject of Chapter 5 on repeated games.

Summary of Methods for Finding Mixed Nash Equilibria

The methods we have for finding the mixed strategies for nonzero sum games are recapped here.

1. Equality of payoffs. Suppose that we have mixed strategies $X^* = (x_1, \ldots, x_n)$ and $Y^* = (y_1, \ldots, y_m)$. For any rows $k_1, k_2, \ldots$ that have a positive probability of being used, the expected payoffs to player I for using any of those rows must be equal: $E_I(k_r, Y^*) = E_I(k_s, Y^*) = E_I(X^*, Y^*)$.

You can find Y^* from these equations. Similarly, for any columns j that have a positive probability of being used, we have $E_{\text{II}}(X^*, j_r) = E_{\text{II}}(X^*, j_s) = E_{\text{II}}(X^*, Y^*)$. You can find X^* from these equations.

2. You can use the calculus method directly by computing

$$f(x_1, \ldots, x_{n-1}, y_1, \ldots, y_{m-1}) = \left(x_1, \ldots, x_{n-1}, 1 - \sum_{i=1}^{n-1} x_i \right) A \begin{bmatrix} y_1 \\ y_2 \\ \vdots \\ 1 - \sum_{j=1}^{m-1} y_j \end{bmatrix}$$

and then

$$\frac{\partial f}{\partial x_i} = 0, \quad i = 1, 2, \ldots, n - 1.$$

This will let you find Y^*. Next, compute

$$g(x_1, \ldots, x_{n-1}, y_1, \ldots, y_{m-1}) = \left(x_1, \ldots, x_{n-1}, 1 - \sum_{i=1}^{n-1} x_i \right) B \begin{bmatrix} y_1 \\ y_2 \\ \vdots \\ 1 - \sum_{j=1}^{m-1} y_j \end{bmatrix}$$

and then

$$\frac{\partial g}{\partial y_j} = 0, \quad j = 1, 2, \ldots, m - 1.$$

From these you will find X^*.

3. You can use the system of equations to find interior Nash points given by

$$\sum_{j=1}^{m} y_j [a_{kj} - a_{nj}] = 0, \qquad k = 1, 2, \ldots, n - 1$$

$$\sum_{i=1}^{n} x_i [b_{is} - b_{im}] = 0, \qquad s = 1, 2, \ldots, m - 1.$$

$$x_n = 1 - \sum_{i=1}^{n-1} x_i, \quad y_m = 1 - \sum_{j=1}^{m-1} y_j.$$

4. If you are in the 2×2 case, or if you have square invertible $n \times n$ matrices you may use the formulas derived in Problem (2.25).

5. In the 2×2 case you can find the rational reaction sets for each player and see where they intersect. This gives all the Nash equilibria including the pure ones.

6. Use the nonlinear programming method: set up the objective, the constraints, and solve.

Problems

2.34 Consider the following bimatrix for a version of the game of chicken (see Problem 2.3):

I/II	Avoid	Straight
Avoid	$(1, 1)$	$(-1, 2)$
Straight	$(2, -1)$	$(-3, -3)$

(a) What is the explicit objective function for use in the Lemke–Howson algorithm?

(b) What are the explicit constraints?

(c) Solve the game.

2.35 Suppose you are told that the following is the nonlinear program for solving a game with matrices (A, B):

$$\text{obj} = (90 - 40x)y + (60x + 20)(1 - y)$$
$$+ (40x + 10)y + (80 - 60x)(1 - y) - p - q$$
$$\text{cnst1} = 80 - 30y \le p$$
$$\text{cnst2} = 20 + 70y \le p$$
$$\text{cnst3} = 10 + 40x \le q$$
$$\text{cnst4} = 80 - 60x \le q$$
$$0 \le x \le 1, 0 \le y \le 1.$$

Find the associated matrices A and B and then solve the problem to find the Nash equilibrium.

2.36 Suppose that the wife in a battle of the sexes game has an additional strategy she can use to try to get the husband to go along with her to the concert, instead of wrestling. Call it strategy Z. The payoff matrix then becomes

H/W	Wr	Co	Z
Wr	$(2,1)$	$(0,0)$	$(-1,1)$
Co	$(0,0)$	$(1,2)$	$(1,3)$

Find all Nash equilibria.

2.37 Since every two-person **zero-sum** game can be formulated as a bimatrix game, show how to modify the Lemke–Howson algorithm to be able to calculate saddle points of zero sum two-person games. Then use that to find the value and saddle point for the game with matrix

$$A = \begin{bmatrix} 2 & -1 & 2 \\ -1 & 2 & -3 \\ 4 & -3 & 5 \\ -3 & \frac{1}{2} & -9 \end{bmatrix}.$$

Check your answer by using the linear programming method to solve this game.

2.38 Use nonlinear programming to find all Nash equilibria for the game and expected payoffs for the game with bimatrix

$$\begin{bmatrix} (1,2) & (0,0) & (2,0) \\ (0,0) & (2,3) & (0,0) \\ (2,0) & (0,0) & (-1,6) \end{bmatrix}.$$

2.39 Consider the game with bimatrix

$$\begin{bmatrix} (-3,-4) & (2,-1) & (0,6) & (1,1) \\ (2,0) & (2,2) & (-3,0) & (1,-2) \\ (2,-3) & (-5,1) & (-1,-1) & (1,-3) \\ (-4,3) & (2,-5) & (1,2) & (-3,1) \end{bmatrix}.$$

Find as many Nash equilibria as you can by adjusting the starting point in the Mathematica commands for Lemke–Howson.

2.5 Correlated Equilibria

2.5.1 Motivating Example

The game we consider is the traffic light problem with matrix

	Stop	Go
Stop	(0,0)	(0,1)
Go	(1,0)	(-100,-100)

There are two pure strategy NEs at (Stop, Go) and (Go, Stop). In addition, there is a mixed strategy NE $X^* = \left(\frac{100}{101}, \frac{1}{101} \right) = Y^*$ so both players play Stop with probability $\frac{100}{101}$ and Go with probability $\frac{1}{101}$ with expected payoff 0 to both players because

$$E_I(X^*, Y^*) = E_{II}(X^*, Y^*) = \left(\frac{100}{101}, \frac{1}{101} \right) \begin{bmatrix} 0 & 0 \\ 1 & -100 \end{bmatrix} \left(\frac{100}{101}, \frac{1}{101} \right)^T = 0.$$

The pure NEs are bad because there is always a player who gets nothing; the mixed strategy is even worse because neither player gets anything.

Set P to be a probability distribution matrix with entries $p_{ij} = x_i y_j$ where $x_1 = \frac{100}{101} = y_1, x_2 = 1 - x_1$, $y_2 = 1 - y_1$. This represents the probability I plays row i and II plays column j assuming independence. The P matrix is then

		Stop	Go
$P =$	Stop	0.98	0.0098
	Go	0.0098	0.98

This is the distribution matrix of the choices under the mixed NE assuming they choose independently. We have already shown that this joint choice of rows and columns results in expected payoff to both players of 0. This gives a total payoff to both players, called the **social welfare**, a value of zero.

Instead let's consider the following distribution for the choices of rows and columns:

	Stop	Go
Stop	0	0.5
Go	0.5	0

In words, the probability they both choose Stop is 0, they both choose Go is 0, and the probability one chooses Stop and one chooses Go is $\frac{1}{2}$. Denote by $P = (p_{ij})$ the matrix for this distribution of choices. If we calculate the expected payoff to player I and II we get

$$E_I(P) = \sum_i \sum_j a_{ij} p_{ij} = p_{21}(1) = 0.5, \text{ and } E_{II}(P) = \sum_i \sum_j b_{ij} p_{ij} = p_{12}(1) = 0.5.$$

The social welfare, $E_I(P) + E_{II}(P)$, with this distribution is 1 and clearly, both players do better than using the NEs. Moreover, the distribution P cannot be obtained using **any** mixed strategies for the players. Indeed, if $X = (x_1, x_2)$, $Y = (y_1, y_2)$ we would need

$$x_1 y_1 = 0, x_1 y_2 = 0.5, x_2 y_1 = 0.5, \text{ and } x_2 y_2 = 0.$$

Consider the first two equalities. If $x_1 = 0$ the second equality cannot hold. Therefore it must be $y_1 = 0$ which contradicts $x_2 y_1 = 0.5$. These equalities are inconsistent. No way can any mixed strategies be found which will work. What this says is that a distribution of choices which are not chosen independently can be better than any Nash equilibrium and that is our motivation for studying **correlated**, or dependent distributions. In the traffic light problem the way to make their choices dependent is to use a traffic light.

2.5.2 Definition of Correlated Equilibrium and Social Welfare

A Nash equilibrium assumes that the players choose their strategies independently of one another. This seems to be an unnecessary and often unrealistic restriction on how to play the game. For example, in a Battle of the Sexes game it is conceivable that a husband and wife will make their decisions based on the outcome of some external event like the toss of a coin or the weather. In many situations, players choose their strategies on the basis of some random external event. Their choices of strategies would then be correlated, or, in classical probability terms, their choices are dependent. Then, how do the players choose their strategies? How did we choose the distribution matrix in the traffic light problem? That question will be answered in this section.

Given the game with matrices (A, B) let

$$p_{ij} = Prob(\text{I plays row i of A and II plays column j of B}).$$

Let P be the matrix with components $p_{ij}, i = 1, 2, \ldots, n, j = 1, 2, \ldots, m$. Denote the n rows of P as $_iP$ and the m columns of P as P_j. The matrix P is called a **distribution on the pure strategies of both players**. If we have mixed strategies $X = (x_1, \ldots, x_n)$ for player I and $Y = (y_1, \ldots, y_m)$ for player II, then if the players choose independently the joint distribution is $p_{ij} = x_i y_j$.

Denote by I_i and II_j the situation in which I plays row i and II plays column j, respectively. Then $p_{ij} = Prob(I_i, II_j)$. As just mentioned, we assumed independence of the choices which would give us $p_{ij} = Prob(I_i) \times Prob(II_j)$. This assumption is now dropped.

Suppose player I plays row i. Then, by Bayes' rule and the law of total probability, the probability player II plays II_j given player I plays I_i is

$$Prob(II_j \mid I_i) = \frac{Prob(I_i, II_j)}{Prob(I_i)} = \frac{Prob(I_i, II_j)}{\sum_{k=1}^{m} Prob(I_i, II_k)} = \frac{p_{ij}}{\sum_{k=1}^{m} p_{ik}}.$$

Now we calculate the expected payoff to player I given that she plays row I_i :

$$E_P(I \mid I_i) = \text{Conditional expected payoff to I given } I_i$$

$$= \sum_{j=1}^{m} a_{ij} Prob(II_j \mid I_i)$$

$$= \sum_{j=1}^{m} a_{ij} \left(\frac{p_{ij}}{\sum_{k=1}^{m} p_{ik}} \right)$$

$$= \frac{\sum_{j=1}^{m} a_{ij} p_{ij}}{\sum_{k=1}^{m} p_{ik}}.$$

The idea now is that player I will play row i if it is a best response. In order for the distribution $P = (p_{ij})$ to be good for player I, the expected conditional payoff of player I given that I plays I_i should be at least as good if the payoffs for player I are switched to another row q. That is, the payoff to I, given she plays row i should be at least as good as her payoff for playing some other row q using the same information about her playing row i. In symbols,

$$E_P(I \mid I_i) = \frac{\sum_j a_{ij} p_{ij}}{\sum_k p_{ik}} \geq \frac{\sum_j a_{qj} p_{ij}}{\sum_k p_{ik}}.$$

This holds because the information given, I uses I_i, does not change, but the payoffs do. This is *not* saying $E_P(I \mid I_i) \geq E_P(I \mid I_q)$ because the information is that I uses I_i but I chooses to use I_q instead. Now we may cancel the denominators to get the requirement,

$$\sum_{j=1}^{m} a_{ij} p_{ij} \geq \sum_{j=1}^{m} a_{qj} p_{ij}.$$

This should be true for all rows $i = 1, 2, \ldots, n, q = 1, 2, \ldots, n$.

Similarly, for player II starting from the requirement that for all columns $j = 1, 2, \ldots, m$

$$E_P(II \mid II_j) = \text{Conditional expected payoff to II given II plays } II_j$$

$$= \sum_{i=1}^{n} b_{ij} Prob(I_i \mid II_j)$$

$$\geq \sum_{i=1}^{n} b_{ir} Prob(I_i \mid II_j),$$

for any columns $r = 1, 2, \ldots, m, j = 1, 2, \ldots, m$. Simplifying the conditional probabilities and canceling the denominator gives

$$\sum_{i=1}^{n} b_{ij} p_{ij} \geq \sum_{i=1}^{n} b_{ir} p_{ij},$$

for all columns $j = 1, 2, \ldots, m, r = 1, 2, \ldots, m$.

In the 2×2 case we have for the correlated equilibrium matrix P

$$P = \begin{array}{c|cc} & 1 & 2 \\ \hline 1 & p_{11} & p_{12} \\ 2 & p_{21} & p_{22} \end{array}$$

where p_{ij} is the probability player I plays row i **and** player II plays column j. The probability player I plays row i is $p_{i1} + p_{i2}$. Player II will play column 1 given that player I plays row 1 with probability $\frac{p_{11}}{p_{11}+p_{12}}$. If player I plays row 1 under a correlated equilibrium, her expected payoff is

$$\frac{p_{11}}{p_{11} + p_{12}} a_{11} + \frac{p_{12}}{p_{11} + p_{12}} a_{12},$$

while if she deviates from playing row 1 and instead plays row 2, her payoff will be

$$\frac{p_{11}}{p_{11} + p_{12}} a_{21} + \frac{p_{12}}{p_{11} + p_{12}} a_{22}.$$

Under a correlated equilibrium we must have

$$\frac{p_{11}}{p_{11} + p_{12}} a_{11} + \frac{p_{12}}{p_{11} + p_{12}} a_{12} \geq \frac{p_{11}}{p_{11} + p_{12}} a_{21} + \frac{p_{12}}{p_{11} + p_{12}} a_{22}.$$

Multiplying through by $p_{11} + p_{12}$ we have the requirement

$$p_{11}a_{11} + p_{12}a_{12} \geq p_{11}a_{21} + p_{12}a_{22}.$$

A similar inequality must hold for the matrix B for player II. That's the way it works.

Now we can state the precise definition which is due to Robert Aumann[1].

Definition 2.5.1 *A distribution $P = (p_{ij})$ is a **correlated equilibrium** if*

$$\sum_{j=1}^{m} a_{ij} p_{ij} \geq \sum_{j=1}^{m} a_{qj} p_{ij},$$

for all rows $i = 1, 2, \ldots, n, q = 1, 2, \ldots, n$, and

$$\sum_{i=1}^{n} b_{ij} p_{ij} \geq \sum_{i=1}^{n} b_{ir} p_{ij},$$

for all columns $j = 1, 2, \ldots, m, r = 1, 2, \ldots, m$.

The definition of correlated equilibrium essentially states that player I's expected payoff when the referee recommends row i is maximized when the player actually plays row i instead of any other row q, and this has to hold for all rows i and q. A similar statement is made for player II.

Before we see how to calculate a correlated equilibrium we look at some examples.

Example 2.14 The classic example is the Game of Chicken. Consider the specific game

I/II	Avoid	Straight
Avoid	$(3,3)$	$(1,4)$
Straight	$(4,1)$	$(0,0)$

First consider the Nash equilibria. There are two pure Nash equilibria at $(4,1)$ and $(1,4)$. There is also a single mixed Nash equilibrium $X^* = Y^* = (\frac{1}{2}, \frac{1}{2})$ which gives payoffs $(2,2)$ to each player.

I claim that

$$P = \begin{bmatrix} \frac{1}{3} & \frac{1}{3} \\ \frac{1}{3} & 0 \end{bmatrix},$$

is a correlated equilibrium distribution P. This means that all combinations of Avoid and Straight will be played with equal probabilities and (*Straight, Straight*) will never be played. By definition, we have to check if

$$a_{11}p_{11} + a_{12}p_{12} \geq a_{21}p_{11} + a_{22}p_{12}$$
$$a_{21}p_{21} + a_{22}p_{22} \geq a_{11}p_{21} + a_{12}p_{22}$$

and

$$b_{11}p_{11} + b_{21}p_{21} \geq b_{12}p_{11} + b_{22}p_{21}$$
$$b_{12}p_{12} + b_{22}p_{22} \geq b_{11}p_{12} + b_{21}p_{22}.$$

---

1 Born on June 8, 1930, he is a Professor at the Center for the Study of Rationality in the Hebrew University of Jerusalem in Israel. He also holds a visiting position at Stony Brook University, and is one of the founding members of the Stony Brook Center for Game Theory. Aumann was awarded the Nobel Prize in Economics (along with Schelling) in 2005.

These are the general conditions for any 2×2 game. A compact way of remembering this is

$$({}_1A) \cdot ({}_1P) \geq ({}_2A) \cdot ({}_1P) \quad \text{and} \quad ({}_2A) \cdot ({}_2P) \geq ({}_1A) \cdot ({}_2P)$$
$$(B_1) \cdot (P_1) \geq (B_2) \cdot (P_1) \quad \text{and} \quad (B_2) \cdot (P_2) \geq (B_1) \cdot (P_2),$$

where we recall that ${}_iA$ is the *ith* row of A and B_j is the *jth* column of B. For our chicken problem we have

$$ {}_1A \cdot {}_1P = \frac{4}{3} \geq {}_2A \cdot {}_1P = \frac{4}{3}$$
$$ {}_2A \cdot {}_2P = \frac{4}{3} \geq {}_1A \cdot {}_2P = 1$$

and

$$ {}_1B \cdot {}_1P = \frac{4}{3} \geq {}_2B \cdot {}_1P = \frac{4}{3}$$
$$ {}_2B \cdot {}_2P = \frac{4}{3} \geq {}_1B \cdot {}_2P = 1.$$

We conclude that P is indeed a correlated equilibrium. Unfortunately, it is not the only correlated equilibrium. For example, both

$$P = \begin{bmatrix} \frac{1}{4} & \frac{1}{4} \\ \frac{1}{4} & \frac{1}{4} \end{bmatrix}, \quad \text{and} \quad P = \begin{bmatrix} \frac{1}{4} & \frac{1}{2} \\ \frac{1}{4} & 0 \end{bmatrix}$$

are also correlated equilibria, as you can readily check. The question is how do we find a correlated equilibrium and how do we pick out a good one.

We will set our goal as trying to make the **social welfare** as large as possible. Here's what that means.

Definition 2.5.2 *The social welfare payoff of a game is the maximum sum of each individual payoff. That is*

$$\text{Maximum } a_{ij} + b_{ij}, \quad i = 1, 2, \ldots, n, \quad j = 1, 2, \ldots, m.$$

*The social welfare of a **pure strategy** (i^*, j^*) is $a_{i^*j^*} + b_{i^*j^*}$.*
*The expected social welfare of a **mixed pair** (X, Y) is*

$$\sum_{i=1}^{n} \sum_{j=1}^{m} (a_{ij} + b_{ij}) x_i y_j = E_I(X, Y) + E_{II}(X, Y).$$

*The expected social welfare of a **distribution** P is*

$$\sum_{i=1}^{n} \sum_{j=1}^{m} (a_{ij} + b_{ij}) p_{ij} = E_P(I) + E_P(II).$$

Example 2.15 Chicken Game-Continued

Each pure Nash equilibrium of the chicken game has social welfare of 5, while the mixed one gets an expected social welfare of 4, as you can easily check.

The maximum social welfare is 6 and it occurs at (Avoid,Avoid) which is not a Nash equilibrium.

If we use the correlated equilibrium $P = \begin{bmatrix} \frac{1}{3} & \frac{1}{3} \\ \frac{1}{3} & 0 \end{bmatrix}$ we get

$$\sum_{i=1}^{2}\sum_{j=1}^{2}(a_{ij} + b_{ij})p_{ij} = E_P(I) + E_P(II) = \frac{21}{4}.$$

Notice that the expected sum of payoffs in this correlated equilibrium is bigger than that of any Nash equilibrium. This is not necessarily the case. A correlated equilibria may give a higher or lower social welfare than any given Nash equilibrium. The question we answer is how do we find the correlated equilibrium which gives the greatest social welfare?

Remark: Existence of Correlated Equilibrium. How do we know a game has a correlated equilibrium? That's actually easy because we know that a game always has a Nash equilibrium, say (X^*, Y^*). To turn this into a correlated equilibrium do the following. If $X^* = (x_1, \ldots, x_n), Y^* = (y_1, \ldots, y_m)$, set $p_{ij} = x_i y_j$ and then P is the $n \times m$ matrix with elements p_{ij}. In matrix notation $P = (X^*)^T Y^*$.

You now need to verify that P satisfies the definition of a correlated equilibrium. That is directly from the definitions and you will verify that in the problems. Observe that in our motivating example at the start of this section we showed by example that a given correlated distribution need not be given as product of any mixed strategies, i.e., a joint distribution does not have to come from two independent distributions, but it is always true that two independent distributions gives a joint distribution. If we start with a Nash equilibrium (X^*, Y^*), then $P = (X^*)^T Y^*$ is definitely a correlated equilibrium, but it may not be a good one.

Before we consider the problem of determining the correlated equilibrium giving the greatest social welfare let's consider another example.

Example 2.16   Battle of the Sexes

The game between a husband and wife is given by

I/II	Home	Wrestle
Home	$(10, 10)$	$(5, 13)$
Wrestle	$(13, 5)$	$(0, 0)$

One of the spouses must stay home since they are new parents and neither trusts a baby sitter. Both would prefer to go to Wrestling if the other stays home. Each spouse must choose whether to stay Home or go Wrestling without communicating with the other spouse, i.e., completely independently.

First consider the following Nash equilibria:

X	Y	Payoffs	Social welfare
$(1, 0)$	$(0, 1)$	$(5, 13)$	18
$(0, 1)$	$(1, 0)$	$(13, 5)$	18
$(\frac{5}{8}, \frac{3}{8})$	$(\frac{5}{8}, \frac{3}{8})$	$(\frac{65}{8}, \frac{65}{8})$	16.25

There are two equilibria in pure strategies, and a mixed equilibrium where each player goes Wrestling with probability $\frac{3}{8}$. The expected payoff of the mixed Nash is $\frac{65}{8}$ for each spouse. Each pure Nash gives a higher social welfare than the mixed Nash.

Suppose the two spouses agree to flip a single fair coin to decide for them. In other words, they will agree to play the distribution

$$P = \begin{bmatrix} \frac{1}{4} & \frac{1}{4} \\ \frac{1}{4} & \frac{1}{4} \end{bmatrix}.$$

If they do that, then

$$E_I(P) = \sum_{i=1}^{2}\sum_{j=1}^{2} a_{ij}p_{ij} = \frac{1}{4}(28) = 7 = E_{II}(P),$$

and the social welfare under P is 14.

If they were to use a biased coin that would correspond to each player using their mixed strategy, we get $P = (x_i y_j)$ giving the correlated equilibrium

$$\begin{bmatrix} \frac{25}{64} & \frac{15}{64} \\ \frac{15}{64} & \frac{9}{64} \end{bmatrix}.$$

Under this distribution the social welfare is 16.25, as we expect.

Can they do better? Of course each pure Nash, if they agreed to play those, would do better. It looks like what we want is to maximize the social welfare over all possible correlated equilibria. That problem is

$$\text{Maximize } \sum_{i=1}^{2}\sum_{j=1}^{2}(a_{ij} + b_{ij})p_{ij} = 20p_{11} + 18p_{21} + 18p_{12}$$

Subject to P must be a correlated equilibrium

The constraints in this problem are

$$p_{11} + p_{21} + p_{12} + p_{22} = 1,$$
$$p_{ij} \geq 0$$
$$13p_{11} \leq 10p_{11} + 5p_{12}$$
$$13p_{21} \leq 10p_{21} + 5p_{22}$$
$$10p_{11} + 5p_{21} \leq 13p_{11}$$
$$10p_{12} + 5p_{22} \leq 13p_{12}.$$

This is a linear programming problem for the unknown P. The solution of this program gives that the maximum social welfare is achieved with the following distribution:

$$P = \begin{bmatrix} \frac{5}{11} & \frac{3}{11} \\ \frac{3}{11} & 0 \end{bmatrix}.$$

This is a correlated equilibrium giving a payoff of 9.45 to each spouse, which is better than any of the Nash equilibria and can be calculated and achieved without a prior agreement between the two players.

How do we find a correlated equilibrium in general? It turns out that finding a correlated equilibrium is actually computationally (on a computer) easier than finding a Nash equilibrium. The algorithm for a correlated equilibrium, which maximizes the social welfare is very simple. It is the following linear programming problem:

LP Problem for a Correlated Equilibrium

Maximize $\sum_{i,j} p_{ij}(a_{i,j} + b_{i,j})$

over variables $P = (p_{ij})$ subject to

$$\sum_j a_{ij} p_{ij} \geq \sum_j a_{qj} p_{ij}$$

for all rows $i = 1, 2, \ldots, n, q = 1, 2, \ldots, n$, and

$$\sum_{i=1}^n b_{ij} p_{ij} \geq \sum_{i=1}^n b_{ir} p_{ij}$$

for all columns $j = 1, 2, \ldots, m, r = 1, 2, \ldots, m$.

$$\sum_{i,j} p_{ij} = 1$$

$$p_{i,j} \geq 0.$$

Our objective function here is chosen to be the expected social welfare of the two players, namely $E_I(P) + E_{II}(P)$. However, this is not the only possible objective. In fact we may take any linear function $f(P)$, which depends on the payoffs for each player to find a correlated equilibrium. All that is necessary for P to be a correlated equilibrium is that the constraints are satisfied. This means that in general there are many correlated equilibria of a game but we choose the one which gives both players the maximum possible payoffs.

Here are the commands using Mathematica for our Battle of the Sexes problem:

```
A={{10,5},{13,0}}
B={{10,13},{5,0}}
P = {{p11, p12}, {p21, p22}}
Maximize[
{Sum[Sum[(A[[i, j]] + B[[i, j]]) P[[i, j]], {j, 1, 2}], {i, 1,
2}],
A[[1]].P[[1]] >= A[[2]].P[[1]],
A[[2]].P[[2]] >= A[[1]].P[[2]],
B[[All, 1]].P[[All, 1]] >=B[[All, 2]].P[[All, 1]],
B[[All, 2]].P[[All, 2]] >= B[[All, 1]].P[[All, 2]],
p11 + p12 + p21 + p22 == 1,
p11 >= 0, p22 >= 0, p12 >= 0, p21 >= 0},
{p11, p12, p21, p22}]
```

Mathematica gives us that the maximum is $\frac{208}{11}$ achieved at

$$p_{11} = \frac{5}{11}, \ p_{12} = \frac{3}{11}, \ p_{21} = \frac{3}{11}, \ p_{22} = 0.$$

You can replace the Maximize command with FindMaximum. They both work.

Problems

2.40 Let A, B be 2×2 games and $P_{2\times2} = (p_{ij})$ a probability distribution. Show that P is a correlated equilibrium if and only if

$$p_{11}(a_{11} - a_{21}) \geq p_{12}(a_{22} - a_{12})$$
$$p_{22}(a_{22} - a_{12}) \geq p_{21}(a_{11} - a_{21})$$
$$p_{11}(b_{11} - b_{12}) \geq p_{21}(b_{22} - b_{21})$$
$$p_{22}(b_{22} - b_{21}) \geq p_{12}(b_{11} - b_{12}).$$

The payoffs in terms of expected social welfare are

$$u_A = p_{11}a_{11} + p_{22}a_{22} + p_{21}a_{21} + p_{12}a_{12}$$
$$u_B = p_{11}b_{11} + p_{22}b_{22} + p_{21}b_{21} + p_{12}b_{12}.$$

2.41 Verify that if (X^*, Y^*) is a Nash equilibrium for the game (A, B), then $P = (X^*)^T Y^* = (x_i y_j)$ is a correlated equilibrium.

2.42 Consider the following game of chicken:

I/II	Avoid	Straight
Avoid	$(4, 4)$	$(1, 5)$
Straight	$(5, 1)$	$(0, 0)$

Show that the following are all correlated equilibria of this game:

$$P_1 = \begin{bmatrix} 0 & 1 \\ 0 & 0 \end{bmatrix}, \quad P_2 = \begin{bmatrix} 0 & 0 \\ 1 & 0 \end{bmatrix},$$

$$P_3 = \begin{bmatrix} \frac{1}{4} & \frac{1}{4} \\ \frac{1}{4} & \frac{1}{4} \end{bmatrix}, \quad P_4 = \begin{bmatrix} 0 & \frac{1}{2} \\ \frac{1}{2} & 0 \end{bmatrix}, \quad P_5 = \begin{bmatrix} \frac{1}{3} & \frac{1}{3} \\ \frac{1}{3} & 0 \end{bmatrix}.$$

Show also that P_4 gives a bigger social welfare than P_1, P_2, or P_3.

2.43 Consider the game in which each player has two strategies Wait and Go. The game matrix is

I/II	Wait	Go
Wait	$(1, 1)$	$(1 - \varepsilon, 2)$
Go	$(2, 1 - \varepsilon)$	$(0, 0)$

(a) Find the correlated equilibria corresponding to the Nash equilibria.

(b) Find the correlated equilibrium corresponding to maximizing the social welfare for any $0 < \varepsilon < 1$.

2.44 Consider the game with payoff matrix

$$\begin{bmatrix} (6, 4) & (1, 6) \\ (7, 1) & (0, 0) \end{bmatrix}$$

1. Show that $p_{11} = \frac{2}{9}, p_{21} = \frac{4}{9}, p_{12} = \frac{1}{3}$ is a correlated equilibrium by directly checking that neither player has an incentive to change, whatever they are told to do.

2. Show that $p_{11} = \frac{2}{9}, p_{21} = \frac{1}{3}, p_{12} = \frac{4}{9}$ is not a correlated equilibrium.

3. Write down the linear program that finds the correlated equilibrium that maximizes social welfare (i.e. what needs to be maximized, and how).
4. Solve to find the socially optimal correlated equilibrium.
5. Now assume you care twice as much about player I as player II. Modify your program to find the correlated equilibrium that maximizes $2E_I + E_{II}$. Show that the solution is just $p_{21} = 1$ (which is really a pure Nash equilibrium)!
6. How much more weight do you need to give to PI before the correlated equilibrium and the pure Nash equilibrium become the same?

2.45 Consider the bimatrix game (which is a form of nonsymmetric chicken) given by

$$\begin{bmatrix} (4,6) & (1,7) \\ (8,1) & (0,0) \end{bmatrix}.$$

1. Which of the following probability matrices are correlated equilibria?

$(i) \begin{bmatrix} \frac{1}{9} & \frac{6}{9} \\ \frac{2}{9} & 0 \end{bmatrix}, \quad (ii) \begin{bmatrix} \frac{1}{9} & \frac{2}{9} \\ \frac{6}{9} & 0 \end{bmatrix}$

$(iii) \begin{bmatrix} \frac{1}{10} & \frac{4}{10} \\ \frac{1}{10} & \frac{4}{10} \end{bmatrix}, \quad (iv) \begin{bmatrix} 0 & \frac{2}{3} \\ \frac{1}{3} & 0 \end{bmatrix}$

2. Set up and solve a linear system to find the socially optimal correlated equilibrium.
3. Find the Nash equilibria of this game. Verify that, for each Nash equilibrium (X^*, Y^*), the matrix $P = (X^*)^T Y^*$ is a correlated equilibria.
4. It is a fact that, if probability matrices P_1 and P_2 are both correlated equilibria, so is the average $\frac{1}{2}(P_1 + P_2)$. Why?
5. Using the previous part, explain how you could have known right away that (iii) and (iv) were correlated equilibria.
6. Also, using the previous parts, write down at least one other correlated equilibrium.
7. What is the worst possible correlated equilibrium?

2.6 Choosing Among Several Nash Equilibria (Optional)

The concept of a Nash equilibrium is the most widely used idea of equilibrium in most fields, certainly in economics. But there is obviously a problem we have to deal with. How do we choose among the Nash equilibria in games where there are more than one? This must be addressed if we ever want to be able to predict the outcome of a game, which is, after all, the reason why we are studying this to begin with. But we have to warn the reader that many different criteria are used to choose a Nash equilibrium and there is no definitive way to make a choice.

The first idea is to use some sort of stability. Intuitively that means that we start with any strategy, say, for player II. Then, we calculate the best response strategy for player I to this first strategy, then we calculate the best response strategy for player II to the best response for player I, and so on. There are many different things that could happen with this procedure. Here is our example.

Example 2.17 Let's carry out the repeated best response idea for the two-person zero-sum game

$$A = \begin{bmatrix} 2 & 1 & 3 \\ 3 & 0 & -2 \\ 0 & -1 & 4 \end{bmatrix}.$$

Notice that $v^+ = v^- = 1$ and we have a saddle point at $X^* = (1, 0, 0)$, $Y^* = (0, 1, 0)$. Suppose that we start with a strategy for player II, $Y^0 = (0, 0, 1)$, so player II starts by playing column 3. The table summarizes the sequence of best responses:

	Best response strategy for II	Best response strategy for I	Payoff to I
Step 0	Column 3(Start)	Row 3	4
Step 1	Column 2	Row 1	1
Step 2	Column 2	Row 1	1
⋮	⋮	⋮	⋮

The best response to player II by player I is to play row 3 (and receive 4). The best response to that by player II is to play column 2 (and II receives −1). The best response to II's choice of column 2 is now to play row 1 (and I receives 1). The best response of player II to player I's choice of row 1 is to play column 2, and that is the end. We have arrived at the one and only saddle point of the matrix, namely, I plays row 1 and II plays column 2.

Similarly, you can check that this convergence to the saddle point will happen no matter where we start with a strategy, and no matter who chooses first. This is a really stable saddle point. Maybe it is because it is the only saddle point?

Nope. That isn't the thing going on, because here is a matrix with only one saddle but the best response sequence doesn't converge to it:

$$A = \begin{bmatrix} 6 & 3 & 2 \\ 5 & 4 & 5 \\ 2 & 3 & 6 \end{bmatrix}.$$

The value of this game is $v(A) = 4$, and there is a unique saddle at row 2 column 2. Now suppose that the players play as in the following table:

	Best response strategy for II	Best response strategy for I
Step 0	Column 1	Row 1
Step 1	Column 3	Row 3
Step 2	Column 1	Row 1
⋮	⋮	⋮

This just cycles from corner payoff to corner payoff and doesn't get to the saddle. Only starting with row 2 or column 2 would bring us to the saddle, and that after only one step.

Stability seems to be a desirable characteristic for saddles and also for Nash points, especially with games that will be played many times. The next example considers a sort of stability to determine the choice of a Nash equilibrium among several candidates.

Example 2.18 Let's consider the bimatrix game

$$\begin{bmatrix} (-a, -a) & (0, -b) \\ (-b, 0) & (-c, -c) \end{bmatrix}.$$

Assume $a > b, c > 0$. Because the game is symmetric with respect to payoffs, it doesn't matter whether we talk about the row or column player. Then we have two pure Nash equilibria at $(-b, 0)$ and at $(0, -b)$. By calculating

$$E_1(x, y) = (x \quad 1 - x) \begin{bmatrix} -a & 0 \\ -b & -c \end{bmatrix} \begin{bmatrix} y \\ 1 - y \end{bmatrix} = -ayx - (1 - x)[(b - c)y + c],$$

we see that setting

$$\frac{\partial E_1(x,y)}{\partial x} = -(a - b + c)y + c = 0 \implies y = \frac{c}{a - b + c}.$$

Similarly, $x = \frac{c}{a - b + c}$. Defining $h = \frac{c}{a - b + c}$, we see that we also have a mixed Nash equilibrium at $X = (h, 1 - h) = Y$. Here is the table of payoffs for each of the three equilibria:

	$E_1(x,y)$	$E_2(x,y)$
$x = 0, y = 1$	$-b$	0
$x = 1, y = 0$	0	$-b$
$x = h, y = h$	z	z

where $z = E_1(h, h) = -h^2(a - b + c) - h(b - 2c) - c$. To be concrete, suppose $a = 3, b = 1, c = 2$. Then $h = \frac{1}{2}$ gives the mixed Nash equilibrium $X = Y = (\frac{1}{2}, \frac{1}{2})$. These are the payoffs:

Equilibrium	Payoff to I	Payoff to II
$x = y = \frac{1}{2}$	$-\frac{3}{2}$	$-\frac{3}{2}$
$x = 1, y = 0$	0	-1
$x = 0, y = 1$	-1	0

Now suppose that we go through a thought experiment. Both players are trying to maximize their own payoffs without knowing what the opponent will do. Player I sees that choosing $x = 1$ will do that, but only if II chooses $y = 0$. Player II sees that choosing $y = 1$ will maximize her payoff, but only if I chooses $x = 0$. Without knowing the opponent's choice, they will end up playing $(x = 1, y = 1)$, resulting in the nonoptimal payoff $E_1(1, 1) = -3 = E_2(1, 1)$.

If there are **many players playing this game whenever two players encounter each other** and they each play nonoptimally, namely, $x = 1, y = 1$, they will all receive less than they could otherwise get. Eventually, one of the players will realize that everyone else is playing nonoptimally $(x = 1)$ and decide to switch to playing $x = 0$ for player I, or $y = 0$ for player II. Knowing, or believing, that in the next game a player will be using $y = 1$, then player I can switch and use $x = 0$, receiving -1, instead of -3.

Now, we are at an equilibrium $x = 0, y = 1$. But there is nothing to prevent other players from reaching this conclusion as well. Consequently others also start playing $x = 0$ or $y = 0$, and now we move again to the nonoptimal play $x = 0, y = 0$, resulting in payoffs -2 to each player.

If this reasoning is correct, then we could cycle forever between $(x = 1, y = 1)$ and $(x = 0, y = 0)$, until someone stumbles on trying $x = h = \frac{1}{2}$. If player I uses $x = \frac{1}{2}$ and everyone else is playing $y = 0$, then player I gets $E_1(\frac{1}{2}, 0) = -1$ and player II gets $E_2(\frac{1}{2}, 0) = -\frac{3}{2}$. Eventually, everyone will see that $\frac{1}{2}$ is a better response to 0 and everyone will switch to $h = \frac{1}{2}$, that is, half the time playing the first row (or column) and half the time playing the second row (or column), When that happens, everyone receives $-\frac{3}{2}$.

In addition, notice that since $E_1(x, \frac{1}{2}) = E_2(\frac{1}{2}, y) = -\frac{3}{2}$ for any x, y, **no strategy chosen by either player can get a higher payoff if the opposing player chooses $h = \frac{1}{2}$.** That means that once a player hits on using $x = \frac{1}{2}$ or $y = \frac{1}{2}$ the cycling is over. No one can do better with a unilateral switch to something else, and there is no incentive to move to another equilibrium.

This Nash equilibrium $x = y = \frac{1}{2}$ is the only one that allows the players to choose without knowing the other's choice and then have no incentive to do something else. It is stable in that sense.

This strategy is called **uninvadable**, or an **evolutionary stable strategy**, and shows us, sometimes, one way to pick the **right** Nash equilibrium when there are more than one. We will discuss this in much more depth when we discuss evolutionary stable strategies and population games in Chapter 8.

Stability is one criterion for picking a good Nash equilibrium, but there are others. Another criterion taking into account the idea of risk is discussed with an example.

Example 2.19 **Entry Deterrence:** There are two players producing gadgets. Firm (player) A is already producing and selling the gadgets, while firm (player) B is thinking of producing and selling the gadgets and competing with firm A. Firm A has two strategies: (1) join with firm B to control the total market (perhaps by setting prices), or (2) resist firm B and make it less profitable or unprofitable for firm B to enter the market (perhaps by lowering prices unilaterally). Firm B has the two strategies to (1) enter the market and compete with firm A or (2) move on to something else. Here is the bimatrix:

A/B	Enter	Move on
Resist	$(0, -1)$	$(10, 0)$
Join	$(5, 4)$	$(10, 0)$

The idea is that a war hurts both sides, but sharing the market means less for firm A. There are two pure Nash equilibria:

$$X_1 = (1, 0), Y_1 = (0, 1),$$
$$X_2 = (0, 1), Y_2 = (1, 0).$$

The associated payoffs for these Nash equilibria are

$$(A_1, B_1) = (10, 0), \quad (A_2, B_2) = (5, 4).$$

Assuming that firm B actually will make the first move, if firm B decides to enter the market, firm A will get 0 if A resists, but 5 if firm A joins with firm B. If firm B decides to not enter the market, then firm A gets 10. So, we consider the two Nash points (resist, move on) and (join, enter) in which the second Nash point gives firm A significantly less (5 versus 10). So long as firm B does not enter the market, firm A can use either resist or join, but with any chance that firm B enters the market, firm A definitely prefers to join.

If firm B moves first, the best choice is Y_2, which will yield it B_2 (if A uses X_2). Since B doesn't know what A will do, suppose that B just chooses Y_2 in the hope that A will use X_2. Now, firm A looks over the payoffs and sees that X_1 gives A a payoff of 10 (if B plays Y_1). So the best for firm A is X_1, which is resist. So, if each player plays the best for themselves without regard to what the other player will do, they will play $X = (1, 0), Y = (1, 0)$ with the result that A gets 0 and B gets -1. This is the worst outcome for both players. Now, what?

In the previous example, we did not account for the fact that if there is any positive probability that player B will enter the market, then firm A must take this into account in order to reduce the risk. From this perspective, firm A would definitely not play resist because if $Y^* = (\varepsilon, 1 - \varepsilon)$, with $\varepsilon > 0$ a very small number, then firm A is better off playing join. Economists say that equilibrium X_2, Y_2 **risk dominates** the other equilibrium and so that is the **correct one**. A risk-dominant Nash equilibrium will be **correct** the more uncertainty exists on the part of the players as to which strategy an opponent will choose; that is, the more risk and uncertainty, the more likely the risk-dominant Nash equilibrium will be played.

Finally, we end this section with a definition and discussion of **Pareto-optimality**. Pareto-optimality is yet another criterion used to choose among several Nash equilibria. Here is the definition and we will use this again later in the book.

Definition 2.6.1   *Given a collection of payoff functions*

$$(u_1(q_1, \ldots, q_n), \ldots, u_n(q_1, \ldots, q_n)),$$

for an n-person nonzero sum game, where the q_i is a pure or mixed strategy for player $i = 1, 2, \ldots, n$, we say that $(q_1^*, \ldots, q_n^*)$ is Pareto-optimal if there does not exist any other strategy for any of the players that increases her payoff without making other players worse off, namely, decreasing at least one other player's payoff.

From this perspective, it is clear that $(5, 4)$ is the Pareto-optimal payoff point for the firms in entry deterrence because if either player deviates from using X_2, Y_2, then at least one of the two players does worse. On the other hand, if we look back at the prisoner's dilemma problem at the beginning of this chapter we showed that $(-5, -5)$ is a Nash equilibrium, but it is **not** Pareto-optimal because $(-1, -1)$ simultaneously improves both their payoffs. Unfortunately, $(-1, -1)$ is not a Nash equilibrium in that prisoner's dilemma game.

Closely related to Pareto-optimality is the concept of a **payoff-dominant Nash equilibrium**, which was introduced by Nobel Prize winners Harsanyi and Selten.

Definition 2.6.2   *A Nash equilibrium is* **payoff-dominant** *if it is Pareto-optimal compared to all other Nash equilibria in the game.*

Naturally, **risk-dominant** and **payoff-dominant** are two different things. Here is an example, commonly known as the **stag hunt game**:

	Hunt	Gather
Hunt	$(5, 5)$	$(0, 4)$
Gather	$(4, 0)$	$(2, 2)$

This is an example of a **coordination** game. The idea is that if the players can coordinate their actions and hunt, then they can both do better. Gathering alone is preferred to gathering together, but hunting alone is much worse than gathering alone.

The pure Nash equilibria are (hunt,hunt) and (gather, gather). There is also a mixed Nash equilibrium at $X_1 = (\frac{2}{3}, \frac{1}{3}), Y_1 = (\frac{2}{3}, \frac{1}{3})$. The following table summarizes the Nash points and their payoffs:

$X_1 = (\frac{2}{3}, \frac{1}{3}) = Y_1$	$E_I = \frac{10}{3}$ $E_{II} = \frac{10}{3}$
$X_2 = (0, 1) = Y_2$	$E_I = 2$ $E_{II} = 2$
$X_3 = (1, 0) = Y_3$	$E_I = 5$ $E_{II} = 5$

The Nash equilibrium X_3, Y_3 is payoff-dominant because no player can do better no matter what. But the Nash equilibrium X_2, Y_2 risk dominates X_3, Y_3 (i.e., (gather,gather) risk dominates (hunt,hunt), The intuitive reasoning is that if either player is not absolutely certain that the other player will join the hunt, then the player who was going to hunt sees that she can do better by gathering. The result is that both players end up playing gather in order to minimize the risk of getting zero. Even though they both do worse, (gather, gather) is risk-dominant. Notice the resemblance to the prisoner's dilemma game in which both players choose to confess because that is the risk-dominant strategy.

Problems

2.46 In the following games solved earlier, determine which, if any, Nash equilibrium is payoff dominant, risk-dominant, and Pareto-optimal.

(a) (**The Game of Chicken**) :

I/II	Avoid	Straight
Avoid	$(1, 1)$	$(-1, 2)$
Straight	$(2, -1)$	$(-3, -3)$

(b) (**Arms Control**)

	Cooperate	Don'
Cooperate	$(3, 3)$	$(-1, -3)$
Don'	$(3, -1)$	$(1, 1)$

(c)

$$\begin{bmatrix} (1, 2) & (0, 0) & (2, 0) \\ (0, 0) & (2, 3) & (0, 0) \\ (2, 0) & (0, 0) & (-1, 6) \end{bmatrix}$$

Bibliographic Notes

The example games considered in this chapter (prisoner's dilemma, chicken, battle of sexes, hawk–dove, stag–hare, etc.) are representative examples of the general class of games of pure competition, or cooperation, and so on. Rational reaction is a standard concept appearing in all game theory texts (see References [41], [27], [6], for example).

John Nash was the first mathematician to introduce and prove the existence of a Nash equilibrium. Other concepts of equilibrium also exist and our example of Kantian equilibrium in Chapter 4 is only a small taste. The nonlinear programming approach to the calculation of a Nash equilibrium originates with Lemke and Howson [21]. The main advantage of their approach is the fact that it combines two separate optimization problems over two sets of variables into one optimization problem over one set of variables; that is, instead of maximizing $E_I(X, Y)$ over X for fixed Y and $E_{II}(X, Y)$ over Y for fixed X, we maximize $E_I(X, Y) + E_{II}(X, Y) - p - q$ over all variables (X, Y, p, q) at the same time. Computationally, this is much easier to carry out on a computer, which is the main advantage of the Lemke–Howson algorithm.

All the games of this chapter can be solved using the software package **Gambit** as a blackbox in which the matrices are entered and all the Nash equilibria as well as the expected payoffs are calculated. We will introduce and use Gambit extensively in the next chapter.

3

Games in Extensive Form: Sequential Decision Making

Competition exists to choose who wins the prize when the prize can't be shared.

–Andrew Harvey

History is an account, mostly false, of events, mostly unimportant, which are brought about by rulers, mostly knaves, and soldiers, mostly fools.

–Ambrose Bierce

I hate competition. I hate it so much that I decided to become the best. I'm not saying I'm the best, but I'm not saying I'm not either.

–Michael Jordan

Not the power to remember, but its very opposite, the power to forget, is a necessary condition for our existence.

–Saint Basil

Games often involve players making moves in sequence, sometimes with knowledge of the opponent's move and sometimes not. These games can be modeled using a game tree illustrating a player's moves and then the opponent's choices at each stage of the game. Even simultaneous move games can be modeled with such trees using so-called information sets. Thus, every game we have studied so far and many more can be modeled using a game tree. Such games are a major generalization of matrix games although these sequential games and matrix games turn out to be equivalent. We have already seen examples of sequential games with the examples of 2×2 Nim and poker. This chapter introduces the study of extensive form games.

3.1 Introduction to Game Trees/Extensive form of Games

3.1.1 Gambit

In this chapter, we will consider games in which player decisions are made simultaneously (as in a matrix game) or sequentially (one after the other), or both. These are interchangeably called **sequential games, extensive form games, or dynamic games.** When moves are made sequentially various questions regarding the information available to the next mover must be answered. Every game in which the players choose simultaneously can be modeled as a sequential game with no information. Consequently, every matrix game can be viewed as a sequential game. Conversely, every sequential game can be reduced to a matrix game if strategies are defined correctly.

Every sequential game is represented by a game tree. The tree has a starting **node** and branches to **decision nodes**, which is when a decision needs to be made by a player. When the end of the decision-making process is complete, the payoffs to each player are computed. Thus, decisions need not be made simultaneously or without knowledge of the other players decisions. In addition extensive form games can also be used to describe games with simultaneous moves if so-called information sets are used correctly.

Game trees can be used to incorporate information passed between the players. In a **complete (or perfect) information** game, players know all there is to know–the moves and payoffs of all players. In a perfect information game, the players know the moves and payoffs of the players at any given moment in time. In other words, each player sees the board and all the moves the opponent has made. In a game with **imperfect information**, one or both players must make a decision at certain times not knowing the previous move of the other player or the outcome of some random event (like a coin toss). Each player at each decision node will have an **information set** associated with that node.

- **If there is more than one node in an information set, then it is a game of incomplete information.**
- **If there is only one node in each information set, it is a game of perfect information.**

This chapter will present examples to illustrate the basic concepts, without going into the technical definitions. We will rely to a great extent on the open-source software project **GAMBIT**[1] to model and solve the games as they get more complicated.

We start with a simple example which illustrates the difference between **sequential and simultaneous moves**.

Example 3.1   Two players select a direction. Player I may choose either Up or Down, while player II may choose either Left or Right. Now we look at the two cases when the players choose sequentially, and simultaneously.

Sequential Moves. Player I goes first, choosing either Up or Down, then player II, knowing player I's choice, chooses either Left or Right. At the end of each branch, the payoffs are computed.

As in Chapter 1, a strategy is a plan for each player to carry out at whichever node one finds oneself. A strategy for player I, who goes first in this game and makes only one move, is simply Up or Down. The strategies for player II are more complicated because she has to account for player I's choices.

LL	If I chooses U, then Left; If I chooses Down, then Left
LR	If I chooses U, then Left; If I chooses Down, then Right
RL	If I chooses U, then Right; If I chooses Down, then Left
RR	If I chooses U, then Right; If I chooses Down, then Right

Gambit automatically numbers the decision nodes for each player. In the diagram Figure 3.1, 2 : 1 means that this is node 1 and information set 1 for player 2. Similarly, 2 : 2 signifies information set

---

1 McKelvey, Richard D., McLennan, Andrew M., and Turocy, Theodore L. (2016). Gambit: Software Tools for Game Theory, Version 16.0.1. http://www.gambit-project.org.

Figure 3.1 Sequential moves.

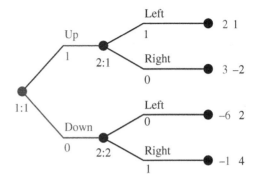

2 for player 2. When each player has a new information set at each node where they have to make a decision, it is a **perfect information game**.

Once the tree is constructed and the payoffs calculated, this game can be summarized in the matrix, or strategic form of the game,

I/II	LL	LR	RL	RR
Up	(2, 1)	(2, 1)	(3, −2)	(3, −2)
Down	(−6, 2)	(−1, 4)	(−6, 2)	(−1, 4)

Now we can analyze the bimatrix game as before. When we use Gambit to analyze the game, it produces Nash equilibria and the corresponding payoffs to each player. In the diagram in Figure 3.1, the numbers below the branches represent the probability, calculated by Gambit, the player should play that branch. In this case, player 1 should play Up all the time; player 2 should play LR all the time. You can see from the matrix that it is a pure Nash equilibrium.

One last remark about this is that when we view the extensive game in strategic form, on Gambit it produces the matrix

I/II	11	12	21	22
1	(2, 1)	(2, 1)	(3, −2)	(3, −2)
2	(−6, 2)	(−1, 4)	(−6, 2)	(−1, 4)

Gambit labels the strategies for a player according to the following scheme:

> **ij = the ith action is taken at information set 1, the jth action is taken at info set 2**

A Branch at a Node Is Also Called an Action. Translated, 21 is the player II strategy to take action 2 (Right) at information set 2 : 1 (I chose Up), and take action 1 (Left) at information set 2 : 2 (I chose Down). In other words, it is strategy RL.

Simultaneous Moves. The idea now is that players I and II will choose their move at the same time, i.e., without the knowledge of the opponent's move. Each player makes a choice and they calculate the payoff. That is typically the set up we have studied in earlier chapters when we set up a matrix game. This is considering the reverse situation now–modeling simultaneous moves using a game tree. The way to model this in Gambit is illustrated in the Figure 3.2.

It looks almost identical to the Figure 3.1 with sequential moves but with a crucial difference, namely the dotted lines connecting the end nodes of player I's choice. This dotted line groups the two nodes into an **information set for player II**. What that means is that player II knows only that player I has made a move landing her in information set 2 : 1, but not exactly what player I's

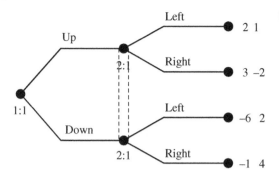

Figure 3.2 Simultaneous moves.

choice has been. Player II does not know which action U or D has been chosen by I. In fact, in general, the opposing player never knows which precise node has been reached in the information set, only that it is one of the nodes in the set.

The strategies for the simultaneous move game are really simple because neither player knows what the other player has chosen and therefore cannot base her decision on what the other player has done. That means that *L* is a strategy for player II and *R* is a strategy for player II but *L if Up and R if Down* is not because player II cannot know whether *Up* or *Down* has been chosen by player I.

The game matrix becomes

I/II	L	R
Up	$(2, 1)$	$(3, -2)$
Down	$(-6, 2)$	$(-1, 4)$

This game has a unique Nash equilibrium at (Up,L).

Given a dynamic game we may convert to a regular matrix game by listing all the strategies for each player and calculating the outcomes when the strategies are played against each other. Conversely given the matrix form of a game we can derive the equivalent game tree.

One of the primary advantages of studying games in extensive form is that chance moves can easily be incorporated. We have already seen this in the poker example in Chapter 1. Here's another example.

Example 3.2 Let's consider a Russian roulette game. In this version of the game, player I is handed a gun which is a six-shooter with one bullet in one of the 6 chambers. Player I goes first and decides whether to pass the gun or spin the chamber and fire. If player I shoots, the gun has a $\frac{1}{6}$ chance of firing (after spinning the chambers). If player I has chosen to pass, or if player I has chosen to spin/fire and lives, then the gun is passed to player II, who then has the same choices with consequent payoffs. The new feature of the game tree is that a third player, Nature, determines whether or not the one bullet is properly set up as the outcome of the spinning. Nature provides the chance to move and earns no payoff. This is an example of a perfect information game with chance moves (which we have also seen in the poker example of Chapter 1). It's perfect information in the sense that both players can see all of the opponent's choices, even though the outcome of the chance move is unknown before the spinning actually takes place. The game tree in Figure 3.3 is from Gambit.

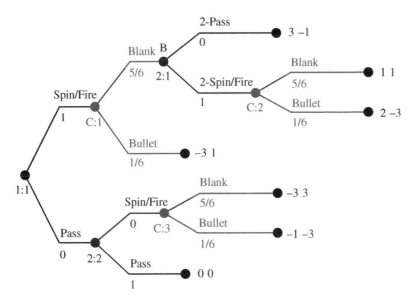

Figure 3.3 Russian roulette.

The strategies for player I are I1: Spin and Fire, and I2: Pass. There's nothing else player I has to do. For player II, the strategies are

II1	If I Spins, Pass; If I Passes, Spin
II2	If I Spins, Pass; If I Passes, Pass
II3	If I Spins, Spin; If I Passes, Spin
II4	If I Spins, Spin; If I Passes, Pass

By Spin, we mean Spin and Fire. The game matrix is then,

I/II	II1	II2	II3	II4
I1	$(2, -\frac{2}{3})$	$(2, -\frac{2}{3})$	$(\frac{17}{36}, \frac{4}{9})$	$(\frac{17}{36}, \frac{4}{9})$
I2	$(-\frac{8}{3}, 2)$	$(0, 0)$	$(-\frac{8}{3}, 2)$	$(0, 0)$

To see how the payoffs are derived, let's consider I1 versus II3. In this case, the expected payoff to I is

$$E_I = \frac{5}{6}(\frac{5}{6}(1) + \frac{1}{6}(2)) + \frac{1}{6}(-3) = \frac{17}{36}.$$

The expected payoff to player II is

$$E_{II} = \frac{5}{6}(\frac{5}{6}(1) + \frac{1}{6}(-3)) + \frac{1}{6}(1) = \frac{16}{36}.$$

The remaining entries are derived in the same way using elementary probability, the strategies, and the tree. This game has a pure Nash equilibrium given by $X^* = (1, 0), Y^* = (0, 0, 1, 0)$ or $Y^* = (0, 0, 0, 1)$. The expected payoffs to each player are $\frac{35}{36}$ and $\frac{5}{18}$. In the tree, the numbers below each branch represent the probability of taking that branch in a Nash equilibrium. The numbers below the chance moves represent the probability of that branch as an outcome. Notice also that chance nodes are indicated with a C on the tree. That is, C:1, C:2, C:3 are the information sets for Nature.

Remark

It is worth emphasizing that a **perfect information game** is a game in which **every node is a separate information set. If two or more nodes are included in the same information set, then it is a game with incomplete information.** The nodes comprising an information set are connected with dotted lines.

An interesting consequence of complete information games is the following theorem. In a certain sense this makes perfect information games less interesting since they always have a Nash equilibrium and if the players play their strategy in the Nash equilibrium, the game will always have the same outcome. Chess is such a game but what makes it an interesting game is that the number of strategies may be more than the number of electrons in the universe. Finding and implementing a Nash equilibrium for chess is essentially impossible.

Theorem 3.1.1 *Any complete information game that has a finite game tree has a pure strategy Nash Equilibrium.*

Remark

Every extensive game we consider can be reduced to strategic form once we know the strategies for each player. Then we can apply all that we know about matrix games to find Nash equilibria. How do we reduce a game tree to the matrix form, i.e., extensive to strategic? First, we must find all the pure strategies for each player. Remember that a pure strategy is a plan which tells the player exactly what move to make from the start to the finish of the game depending on the players knowledge of the opponent's moves. In many, if not most extensive games, it is not feasible to list all of the pure strategies because there are too many. In that case we must have a method of solving the game directly from the tree. If the number of pure strategies is not too large then the procedure to reduce to a strategic game is simple: play each strategy against an opposing strategy and tabulate the payoffs.

Example 3.3 Converting from a Tree to a Matrix. Suppose we are given the following game tree with chance moves depicted in extensive form in Figure 3.4.

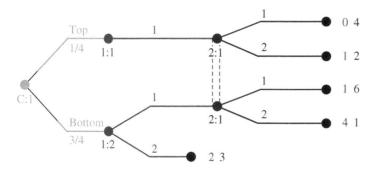

Figure 3.4 Convert tree to matrix.

This game starts with a chance move. With probability $\frac{1}{4}$ player I moves to 1 : 1; with probability $\frac{3}{4}$ player I moves to 1 : 2. From 1 : 1 player I has no choices to make and it becomes player II's turn at 2 : 1. Player II then has two choices.

From 1 : 2, player I can choose either action 1 or action 2. If action 2 is chosen, the game ends with I receiving 2 and II receiving 3. If action 1 is chosen player II then has two choices, action 1 or action 2. Now since there is an information set consisting of two decision nodes for player II at 2 : 1, the tree models the fact that when player II makes her move she does not know if player I has made a move from Top or from Bottom. On the other hand, because nodes at 1 : 1 and 1 : 2 are not connected (i.e., they are not in the same information set), player I knows whether he is at Top or Bottom. If these nodes were in the same information set, the interpretation would be that player I does not know the outcome of the chance move.

This game is not hard to convert to a matrix game. Player I has two strategies and so does player II. point we may analyze the game an

I/II	Action 1	Action 2
11-If Top, 1;If Bottom, 1	$(\frac{3}{4}, \frac{11}{2})$	$(\frac{13}{4}, \frac{5}{4})$
12-If Top 1;If Bottom, 2	$(\frac{3}{2}, \frac{13}{4})$	$(\frac{7}{4}, \frac{11}{4})$

At this point we may analyze the game and see that there is a Nash equilibrium in which player I will always play Top and player II will always play action 1.

Here's where the numbers come from in the matrix. For example, if we play 11 against 1, we have the expected payoff for player II is $\frac{1}{4}(4) + \frac{3}{4}(6) = \frac{11}{2}$. The expected payoff for player I is $\frac{1}{4}(0) + \frac{3}{4}(1) = \frac{3}{4}$. If we play 12 against 2, the expected payoff to player II is $\frac{1}{4}(2) + \frac{3}{4}(3) = \frac{11}{4}$ and the expected payoff to player I is $\frac{1}{4}(1) + \frac{3}{4}(2) = \frac{7}{4}$.

Example 3.4   Poker with Call, Raise, Fold. Our next example models and solves a classic end-game situation in poker and illustrates further the use of information sets. This example proves that bluffing is an integral part of optimal play in poker. For a reader who is unaware of the term *bluffing*, it means that a player bets while holding a losing card with the hope that the opponent will think the player has a card that will beat her. Here's the model.

Two players are in a card game. A card is dealt to player I and it is either an Ace or a Two. Player I knows the card but player II does not. Player I has the choice now of either (B)etting or (F)olding. Player II, without knowing the card I has received, may either (C)all, (R)aise, or (F)old. If player II raises, then player I may either (C)all, or (F)old. The game is then over and the winner takes the pot. Initially the players ante \$1 and each bet or raise adds another dollar to the pot. The tree is illustrated in Figure 3.5.

Every node is a separate information set except for the information set labeled at 2 : 1 for player II. This is required since player II does not know which card player I was dealt. If instead we had created a separate information set for each node (making it a perfect information game), then it would imply that player I and II both knew the card. That isn't poker.

Player II has only 3 strategies because II has one information set and 3 actions possible from that set. Player I has 16 possible strategies because I has 4 information sets with 2 actions possible from each set (so $2^4 = 16$ strategies). Although there are many strategies for this game, most of them are dominated (strictly or weakly) and Gambit reduces the matrix to a simple 2×3,

I/II	C	R	F
BCF_-	$(\frac{1}{2}, -\frac{1}{2})$	$(1, -1)$	$(0, 0)$
$BCBF$	$(0, 0)$	$(\frac{1}{2}, -\frac{1}{2})$	$(1, -1)$

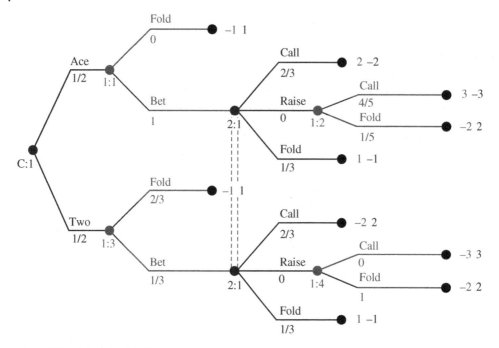

Figure 3.5 Poker with bluffing.

Player II has 3 strategies: (1) Call, (2) Raise, (3) Fold. After eliminating the dominated strategies, player I has two strategies left. Here's what they mean. First, $BCF\_$ means Bet, Call, Fold, arbitrary, at info sets $1:1, 1:2, 1:3, 1:4$, respectively. If player I Folds at $1:3$, then $1:4$ is never reached and it doesn't matter what I does there. Here's the summary.

Strategy $BCF\_$

1. take action 2 (Bet if you have an Ace) at info node $1:1$
2. take action 1 (Call if II Raises and you have an Ace) at info node $1:2$
3. take action 1 (Fold if you have a Two) at info node $1:3$

Strategy $BCBF$

1. take action 2 (Bet if you have an Ace) at info node $1:1$;
2. take action 1 (Call if II Raises and you have Ace) at info node $1:2$;
3. take action 2 (Bet if you have a Two) at info node $1:3$;
4. take action 2 (Fold if II Raises and you have Two) at info node $1:4$.

Notice that the middle column of the matrix may be dropped by dominance. Raising by player II is never part of a Nash equilibrium in this game. The matrix can be reduced to the 2×2 zero sum game with matrix $\begin{bmatrix} \frac{1}{2} & 0 \\ 0 & 1 \end{bmatrix}$.

This game has a Nash equilibrium leading to the payoff of $\frac{1}{3}$ to player I and $-\frac{1}{3}$ to player II. The equilibrium tells us that if player I gets an Ace, then always Bet; player II doesn't know if I has an Ace or a Two but II should Call with probability $\frac{2}{3}$ and Fold with probability $\frac{1}{3}$. Player II should never Raise. On the other hand, if player I gets a Two, player I should Bet with probability $\frac{1}{3}$ and Fold with probability $\frac{2}{3}$; player II (not knowing I has a Two) should Call with probability $\frac{2}{3}$ and Fold with probability $\frac{1}{3}$. Since player II should never Raise a player I bet, the response by player I to a Raise never arises.

The conclusion is that player I should bluff one-third of the time with a losing card. Player II should Call the bet two-thirds of the time. The result is that, on average player I wins $\frac{1}{3}$.

Extensive games can be used to model quite complicated situations–like the Greeks versus the Persians in the Battle of Marathon.

Example 3.5 The Persians have decided to attack a village on a Greek-defended mountain. They don't know if the Persians have decided to come by the **Road** to the village or **Climb** up the side of the mountain. The Greek force is too small to defend both possibilities. They must choose to defend either the Road to the village or the path up the side of the mountain, but not both.

But the Persians are well aware of the Greek warrior reputation for ferocity so if the Persians meet the Greeks, they will retreat. This scores one for the Greeks. If the two forces do not engage the Persians will reach a Greek arms depot and capture their heavy weapons which gives them courage to fight. If the Persians reach the arms depot the Greeks learn of it and now both sides must plan what to do as the Persians return from the arms depot. The Greeks have two choices. They can either lay an ambush on the unknown path of return or move up and attack the Persians at the arms depot. The Persians have two choices. They can either withdraw immediately by day or wait for night.

If the Persians meet an ambush by day, they will be destroyed (score one for the Greeks). If the Persians meet the ambush at night, they can make it through with a small loss (score one for the Persians). If the Greeks attack the arsenal and the Persians have already gone, the Persians score one for getting away. If the Greeks attack and find the Persians waiting for night both sides suffer heavy losses and each scores zero. Assume the Greeks, having an in depth knowledge of game theory, will assume the Persians make the first move, but this move is unknown to the Greeks.

This is a zero sum game which is most easily modeled in extensive form using Gambit.

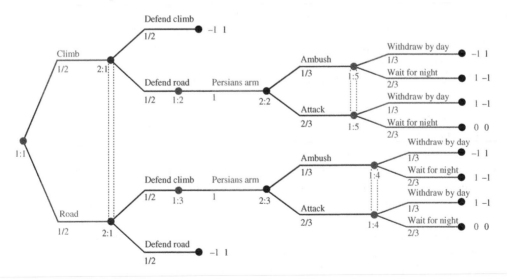

The solution is obtained by having Gambit find the equilibrium. It is illustrated on the figure by the probabilities below the branches. We describe it in words. The Persians should flip a fair coin to decide if they will climb the side of the mountain or take the road. Assuming the Greeks have not defended the chosen Persian route, the Persians reach the arms depot. Then they flip a three-sided coin so that with probability $\frac{1}{3}$ they will withdraw by day and with probability $\frac{2}{3}$, they will wait for night.

The Greeks will flip a fair coin to decide if they will defend the road or defend the climb up the side of the mountain. Assuming they defended the wrong approach and the Persians have reached the arms depot, the Greeks will then flip a three-sided coin so that with probability $\frac{1}{3}$ they will lay an ambush and with probability $\frac{2}{3}$ they will attack the depot hoping the Persians haven't fled. This is a zero sum game with value $+\frac{1}{3}$ for the Greeks. If you think the payoffs to each side would make this a non zero sum game, all you have to do is change the numbers at the terminal node and resolve the game using Gambit.

The next example shows that information sets can be a little tricky. You have to keep in mind what information is revealed at each node.

Example 3.6   A Simple Naval Warfare Game. Consider a part of the ocean as a 1×4 grid. The grid is

| 1 | 2 | 3 | 4 |

A submarine is hiding in the grid that the navy is trying to locate and destroy. Assume that the sub occupies two consecutive squares. In a search algorithm the navy must pick a square to bomb, see if it misses or hits the sub, pick another square to bomb, etc. until the sub is destroyed. The payoff to the sub is the number of squares the navy must bomb in order to hit both parts of the sub. Assume that if a square is bombed and part of the sub is hit, the navy knows it.

The strategies for the submarine are clear: Hide in 1–2 Hide in 2–3 or Hide in 3–4. The strategies for the bomber are more complicated because of the sequential nature of the choices to bomb and the knowledge of whether a bomb hit or missed.

The first player is the submarine which chooses to hide in 1–2, 2–3, or 3–4. The navy bomber is player II and he is the last player to make any decisions. In each case, he decides to shoot at 1,2,3, or 4 for the first shot. The information set 2 : 1 in Figure 3.6 indicates that the bomber does not know in which of the 3 pairs of squares the sub is hidden. Without that information set, player II would know exactly where the sub was and that is not a game.

Here's an explanation of how this is analyzed referring to Figure 3.6. Let's assume that the sub is in 1-2. The other cases are similar.

- If player II takes his first shot at 1, then it will be a hit and II knows the sub is in 1-2 and second shot will be at 2. The game is over and the payoff to I is 2.
- If player II takes the first shot at 2, it will be a hit and the rest of the sub must be in 1 or 3. If the bomber fires at 1, the payoff will be 2; if the sub fires at 3, the payoff will be 3 because the sub will miss the second shot but then conclude the sub must be in 1-2.
- If player II takes the first shot at 3, it will be a miss. Then player II knows the sub must be in 1-2 so the payoff will be 3.
- If player II takes the first shot at 4, it will be a miss. Player II then may fire at 1, 2, or 3. If the bomber fires at 1, it will be a hit and the sub must be in 1-2, giving a payoff of 3. If the bomber fires at 3, it will be a miss. The sub must then be in 1-2 and the payoff is 4. If the bomber fires at 2, it will be a hit but the sub could be in 1-2 or 2-3. This will require another shot. The resulting payoff is either 3 or 4.

Here is the extensive form of the game.

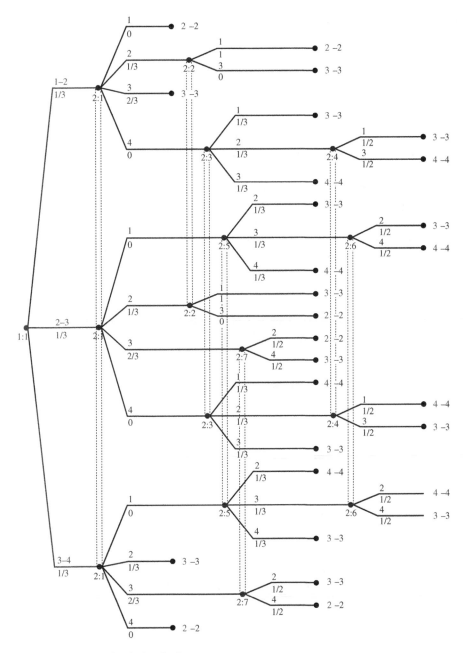

Figure 3.6 Submarine in 1×4 grid.

The information sets in this game are a bit tricky. First, player II, the bomber, at the first shot has the information set 2 : 1 because the bomber doesn't know which two squares the sub occupies. At the second shot, player II will know if the first shot hit or missed but cannot know more than that. Thus, information set 2 : 2 means the bomber's first shot at square 2 hit, but it is unknown if the sub is in $1 - 2$ or $2 - 3$. Information set 2 : 3 means the first shot at 4 missed, but it is unknown if that means the sub is in $1 - 2$ or $2 - 3$. The rest of the information sets are similar.

After eliminating dominated strategies, the game matrix reduces to

Sub/Bomber	1	2	3
12	2	3	3
23	3	2	3
34	3	3	2

We are labeling strategies 1, 2 , and 3 for the bomber.

The matrix game may be solved using the invertible matrix theorem, Mathematica, or Gambit. The value of this game is $v(A) = \frac{8}{3}$ so it takes a little over 2 bombs, on average, to find and kill the sub. The sub should hide in $1 - 2, 2 - 3$, or $3 - 4$ with equal probability, $\frac{1}{3}$. The bomber should play strategies 1, 2, or 3, with equal probability. The strategies depend on which node player II happens to be at so we will give the description in a simple to follow plan.

The bomber should fire at 2 with probability $\frac{1}{3}$ and at 3 with probability $\frac{2}{3}$[1] with the first shot and never fire at 1 or 4 with the first shot. If the first shot is at 2 and it is a hit, the next shot should be at 1. The game is then over because if it is a miss the bomber knows the sub is in $3 - 4$. If the first shot is at 3 and it is a hit, the next shot should be at 2 or 4 with probability $\frac{1}{2}$, and the game is over.

In all of these games we assume the players have **perfect recall**. That means each player remembers all of her moves and all her choices in the past. Here is what Gambit shows you when you have a game without perfect recall:

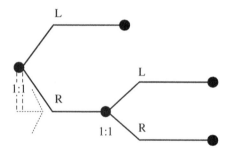

Figure 3.7 Game without perfect recall.

This models a player who chooses initially Left or Right and then subsequently also must choose Left or Right. The start node 1 : 1 in Figure 3.7 and the second node also with label 1 : 1 are in the **same information set** which means that the player has forgotten that she went Right the first time. This type of game is not considered here and you should beware of drawings like this when you solve the problems using Gambit.

Problems

3.1 Convert the following extensive form game to a strategic form game. Be sure to list all of the pure strategies for each player and then use the matrix to find the Nash equilibrium.

---

1 It seems reasonable that another Nash equilibrium by symmetry would specify that the bomber should fire at 2 with probability $\frac{2}{3}$ and 3 with probability $\frac{1}{3}$.

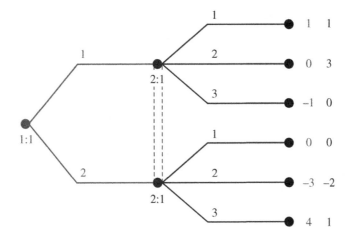

3.2 Given the game matrix draw the tree. Assume the players make their choices simultaneously. Solve the game.

I/II	a	b
A	(3, 3)	(0, 4)
B	(4, 0)	(2, −1)

3.3 Consider the three-player game in which each player can choose Up or Down. In matrix form the game is

III plays Up

I/II	Up	Down
Up	(1, 2, −1)	(−2, 1, 3)
Down	(−1, 3, 0)	(2, −1, 4)

III plays Down

I/II	Up	Down
Up	(1, −2, 4)	(1, −1, −3)
Down	(−1, 3, −1)	(2, 2, 1)

(a) Draw the extensive form of the game assuming the players choose simultaneously.

(b) Use Gambit to solve the game and show that the mixed Nash Equilibrium is $X^* = (0.98159, 0.01841), Y^* = (0.54869, 0.45131), Z^* = (0.47720, 0.5228)$.

3.4 In problem 1.94 we considered the game in which Sonny is trying to rescue family members Barzini has captured. With the same description as in that problem, draw the game tree and solve the game.

3.5 Aggie and Baggie are each armed with a single coconut cream pie. Since they are the friends of Larry and Curly, naturally, instead of eating the pies they are going to throw them at each other. Aggie goes first. At 20 paces she has the option of hurling the pie at Baggie or passing. Baggie then has the same choices at 20 paces but if Aggie hurls her pie and misses, the game is over because Aggie has no more pies. The game could go into a second round but now at 10 paces, and then a third round at 0 paces. Aggie and Baggie have the same probability of hitting the target at each stage: $\frac{1}{3}$ at 20 paces, $\frac{3}{4}$ at 10 paces, and 1 at 0 paces. If Baggie gets a pie in the face, Aggie scores +1, and if Aggie gets a pie in the face Baggie scores +1. This is a zero sum game.

(a) Draw the game tree and convert to a matrix form game.

(b) Solve the game.

3.6 One possibility not accounted for in a Battle of the Sexes game is the choice of player I to go to a bar instead of the Ballet or Wrestling. Suppose the husband, who wants to go to Wrestling, has a 15% chance of going to a bar instead. His wife, who wants to go to the Ballet also has a 25% chance of going to the same bar instead. If they both meet at Wrestling, the husband gets 4, the wife gets 2. If they both meet at the Ballet, the wife gets 4, and the husband gets 1. If they both meet at the bar, the husband gets 4 and the wife also gets 4. If they decide to go to different places they each get 2. Assume the players make their decisions without knowledge of the other's choice. Draw the game and solve using Gambit.

3.7 The tree below describes a zero-sum game. The labels at each vertex say which player gets to make the choice, and the dotted line means that P_1 cannot tell which of these nodes they are at. I have labeled the various choices P_1 can make with capital letters, and those P_2 can make with lower case letters. Payoffs are all, as usual, to P_1.

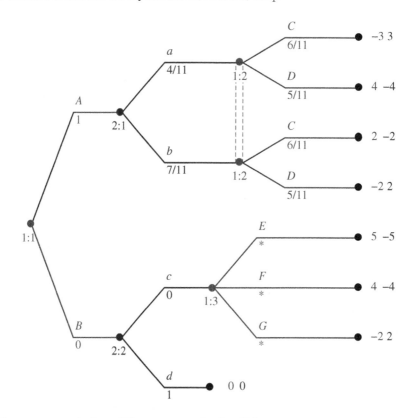

(a) Convert to matrix form. Hint: your matrix should be 5 × 4.

(b) Reduce using dominance.

(c) Solve the reduced game graphically, and from there give the solution for the full game. Make sure to answer in a way that can be understood from the tree, since this is what was given in the question.

(d) Can you see how to reduce by dominance in the original tree, without using the matrix? Explain briefly.

3.8 Consider the bimatrix game with payoffs given by

$$\begin{bmatrix} (6,5) & (3,2) \\ (4,1) & (7,9) \end{bmatrix}$$

1. Graph the tree assuming simultaneous choices by the two players.
2. Solve the game and find the payoff to each player for each equilibrium.
3. What should the players do? Although the answer is often unclear in non-zero-sum games, in this case there is a fairly clear answer.

3.9 Consider the following two-player game: Each player initially puts $1 in the pot. Player I has two cards, an ace and a king. She gets to discard one of these, of her choice. She then announces either "I kept ace" or "I kept king" to PII (you should think of this as two turns for PI: a turn where she discards a card, and a turn where she makes an announcement). Now it is PIIs turn. She can either fold, or raise $2. If she folds the game is over and PI gets the pot, but if she raises, both players must add $2 to the pot. In that case, if PI kept an ace PI gets the pot, and if PI kept a king then PII gets the pot.

Additionally, if PI lied in either case, she needs to pay a $1 to a third party. Finally, in the special case where PI kept a king, lied, and PII folded (so PII believed the lie), PI's friend will pay her (namely PI) $5 as a sort of reward.

1. Draw the game tree, indicating all necessary information sets.
2. Use Gambit to find all Nash equilibria.
3. Explain how the two players should play and give the expected payoff to each.

3.2 Backward Induction and Subgame Perfect Equilibrium

Backward induction is a simple way to solve a game with perfect information and perfect recall. It is based on the simple idea that the players will start at the end of the tree, look at the payoffs, choose the action giving them the largest payoff, and work backward to the beginning of the game. This has obvious limitations we will go over. But first let's look at an example.

Example 3.7 Figure 3.8 depicts a two-player game with perfect information and perfect recall. To find the equilibrium given by backward induction, we start at the end and look at the payoffs to the last player to choose a branch.

In this case player 1 makes the last choice. For the node at $1:2$, player 1 would clearly choose branch 2 with payoff $(7,7)$. For the node at $1:3$, player 1 would choose branch 2 with payoff $(4,1)$. Finally, for the node at $1:4$, player 1 would choose branch 3 with payoff $(5,2)$. Eliminate all the terminal branches player 1 would not choose. The result is the following tree in Figure 3.9.

Now we work backward to player 2's nodes. At $2:1$ player 2 would choose branch 1 since that gives payoff $(7,7)$. At $2:2$, player 2 chooses branch 2 with payoff $(8,8)$. Eliminating all other branches gives us the final game tree in Figure 3.10.

Working back to the start node $1:1$, we see from Figure 3.10 that player 1 would choose branch 2, resulting in payoff $(8,8)$. We have found an equilibrium by backward induction. The backward induction equilibrium is that player 1 chooses branch 2 and player 2 chooses branch 2. Gambit

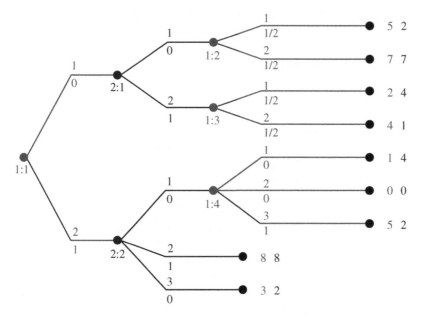

Figure 3.8 The full tree.

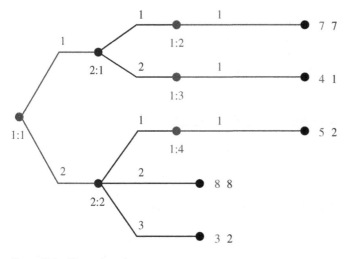

Figure 3.9 The reduced tree.

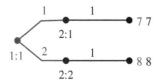

Figure 3.10 The final reduction.

solves the game and gives three Nash equilibria including the backward induction one. The other two equilibria give a lower payoff to each player, but they are equilibria.

Example 3.8   We will see in this example that backward induction picks out the reasonable Nash equilibrium just as in the last example. Suppose there are two firms A and B which produce competing widgets. Firm A is thinking of entering a market in which B is the only seller of widgets. Here is a game tree for this model. 2

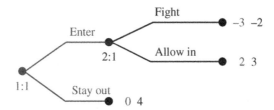

The matrix for this game is

A/B	Fight	Allow in
Enter	$(-3, -2)$	$(2, 3)$
Stay out	$(0, 4)$	$(0, 4)$

There are two pure Nash equilibria at $(0, 4)$ and $(2, 3)$ corresponding to (Stay Out, Fight) and (Enter, Allow In), respectively. Now something just doesn't make sense with the Nash equilibrium (Stay Out, Fight). Why would firm B Fight, when firm A Stays Out of the market? That Nash equilibrium is just not credible.

Now let's see which Nash equilibrium we get from backward induction. The last move chosen by B will be Allow In from node $2:1$. Then A, knowing B will play Allow In will choose Enter since that will give A a larger payoff. Thus the Nash equilibrium determined by backward induction is (Enter, Allow In) resulting in payoff $(2, 3)$. The incredible Nash equilibrium at $(0, 4)$ is not picked up by backward induction.

A more complicated example gives an application to conflicts with more options.

Example 3.9   Solomon's Decision. A biblical story has two women before King Solomon in a court case. Both women are claiming that a child is theirs. Solomon decides that he will cut the child in half and give one-half to each woman. Horrified, one of the women offers to give up her claim to the kid so that the child is not cut in two. Solomon awards the child to the woman willing to give up her claim, deciding that she is the true mother. The following game analysis (due to Herve Moulin[1]) gives another way to decide who is the true mother without threatening to cut up the child.

Solomon knows that one of the women is lying but not which one, but the women know who is lying. Solomon assumes the child is worth C_T to the real mother and worth C_F to the false mother, with $C_T \gg C_F$. In order to decide who is the true mother Solomon makes the women play the following game:

To begin, Woman 1 announces Mine or Hers. If she announces Hers, Woman 2 gets the child and the game is over. If she announces Mine, Woman 2 either Agrees or Objects. If Woman 2 Agrees,

1 George A. Peterkin Professor of Economics at Rice University.

then Woman 1 gets the child and the game is over. If Woman 2 Objects, then Woman 2 pays a penalty of \$V where $C_F < V < C_T$, but she does get the child, and Woman 1 pays a penalty of \$W>0.

We will consider two cases to analyze this game. In the analysis, we make no decision based on the actual values of C_T, C_F, which are unknown to Solomon.

Case 1: Woman 1 Is the True Mother. The tree when Woman 1 is the true mother is in Figure 3.11.

In view of the fact that $C_F - V < 0$, we see that Woman 2 will always choose to Agree, if Woman 1 chooses Mine. Eliminating the Object branch, we now compare the payoffs $(C_T, 0)$ with $(0, C_F)$, for Woman 1. Since $C_T > 0$, woman 1 will always call Mine. By backward induction, we have the equilibrium that Woman 1 should call Mine, and Woman 2 should Agree. The result is that Woman 1, the true mother, ends up with the child.

Case 2: Woman 2 Is the True Mother. The tree when Woman 2 is the true mother is in Figure 3.12.

Now we have $C_T - V > 0$, and Woman 2 will always choose to Object, if Woman 1 chooses Mine. Eliminating the Agree branch, we now compare the payoffs $(-W, C_T - V)$ with $(0, C_T)$, for Woman 1. Since $-W < 0$, Woman 1 will always call Hers. By backward induction we have the equilibrium

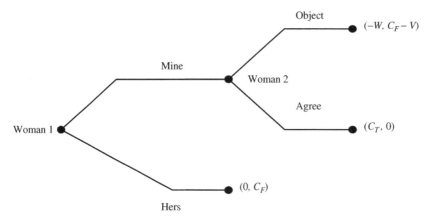

Figure 3.11 Woman 1 is the true mother.

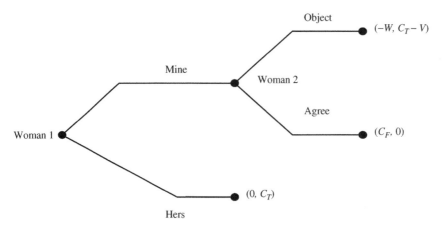

Figure 3.12 Woman 2 the true mother.

that Woman 1 should call Hers, and Woman 2 should Object if Woman 1 calls Mine. The result is that Woman 2, the true mother, ends up with the child.

Problems

3.10 Modify Solomon's decision problem so that each Woman actually loses the value she places on the child if she is not awarded the child. The value to the true mother is assumed to satisfy $C_T > 2C_F$. Assume that Solomon knows both C_T and C_F and so can set $V > 2C_F$. Apply backward induction to find the Nash equilibrium and verify that the true mother always gets the child.

3.11 Find the Nash equilibrium using backward induction for the tree

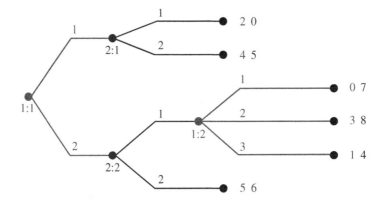

3.12 Two players ante $1. Player I is dealt an Ace or a Two with equal probability. Player I sees the card but player II does not. Player I can now decide to Raise, or Fold. If he Folds, then player II wins the pot if the card is a Two while player I wins the pot if the card is an Ace. If player I Raises, he must add either $1 or $2 to the pot. Player II now decides to Call or Fold. If she Calls and the card was a Two, then II wins the pot. If the card was an Ace, I wins the pot. If she Folds, player I takes the pot. Model this as an extensive game and solve it using Gambit.

3.13 Consider the following game show game with two contestants. Initially, there is a prize pool of $20,000. The players take turns. At each turn, they may either
- Shout "give me my money!" In that case, the prize gets split between the two contestants, with the player who said "give me my money!" getting $20,000 more than her opponent (so, if PI says give me my money on her first turn, she gets the full $20,000, but if PI passes on her first turn $10,000 is added to the pot. If PII says give me my money on her first turn, the pot is split $25,000 and $5000).
- At each turn if a player shouts "play on!", the prize goes up by $10,000, and it becomes the other player's turn.

After each player has had three turns, if no one has claimed the money, it gets split evenly between the two players (so each should get $40,000).
  (a) Draw a game tree for the game, making sure you indicate payoffs and whose turn it is appropriately.

(b) Solve by backward induction to find the unique Nash equilibrium.

(c) Comment on whether this answer makes sense, and if you think players would play this way.

(d) Redo the question if the player who says "give me my money" only gets $2000 more than their opponent. Describe how it is different.

(e) What if the player who says "give me my money" gets $10,000 more? Why is this case interesting?

3.14 Consider the following game tree in Figure 3.13.

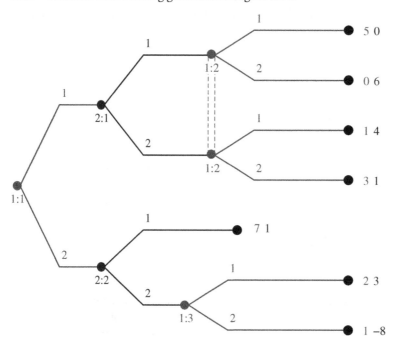

Figure 3.13 Backward induction.

Reduce as much as possible using backward induction (i.e., eliminating strategies that are obviously bad looking at the tree). **Caution: the information set stops you from reducing all the way!**

3.15 Consider the game in Figure 3.14.

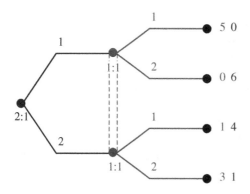

Figure 3.14 Backward induction.

Convert to a matrix game, and find the Nash equilibrium (there should only be one). **Caution:** *PII* **is now going first!**

(a) Finish solving the original game by backward induction.

(b) Convert to a matrix game, and show that there is a Nash equilibrium with payoff 7 to PI. What is going on? Which calculation is correct?

3.2.1 Subgame Perfect Equilibrium

An extensive form game may have lots of Nash equilibria and often the problem is to decide which of these, if any, are actually useful, implementable, or practical. Backward induction, which we considered in Section 3.2 was a way to pick out a good equilibrium. The concept of **subgame perfection** is a way to eliminate or pick out certain Nash equilibria which are the most implementable. The idea is that at any fixed node, a player must make a decision which is optimal from that point on. In fact it makes sense that a player will choose to play according to a Nash equilibrium no matter at which stage of the game the player finds herself, because there is nothing that can be done about the past. A good Nash equilibrium for a player is one which is a Nash equilibrium for **every** subgame. That is essentially the definition of subgame perfect equilibrium.

We assume that our game has perfect recall. Each player remembers all her past moves. In particular, this means that an information set cannot be split into separate nodes.

For instance, in the following game in Figure 3.15, there are no proper subgames. This is a game in which player I holds an Ace or a Two. Player II without knowing which card player I holds, guesses that I holds an Ace or I holds a Two. If player II is correct, II wins $1, and loses $1 if incorrect.

Now here is the precise definition of a subgame perfect Nash equilibrium.

Definition 3.2.1 *A **subgame perfect equilibrium** for an extensive form game is a Nash equilibrium whose restriction to any subgame is also a Nash equilibrium of this subgame.*

Example 3.10 We return to the Up-Down-Left-Right game given in Figure 3.16.

There are two subgames in this game but only player II moves in the subgames. The game has three pure Nash equilibria:

1. player I chooses Right, player II chooses Down no matter what–payoff (1, 3);
2. player I chooses Left and II chooses Up no matter what–payoff (3, 1);
3. player I chooses Left and II plays Up (if I plays Left), and Down (if I plays Right)–payoff (3, 1).

Figure 3.15 An extensive game with no proper subgames.

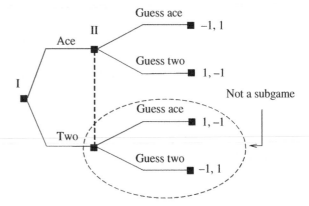

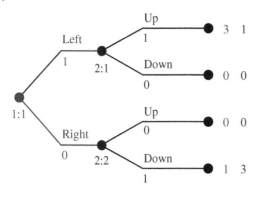

Figure 3.16 Up-down-left-right.

Which of these, if any are subgame perfect? Start at player II's moves. If I chooses Left, then II clearly must choose Up since otherwise II gets 0. If I chooses Right, then II must choose Down. This means that for each subgame, the equilibrium is that player II chooses Up in the top subgame and Down in the bottom subgame. Thus strategy (3) is subgame perfect; (1) and (2) are not.

You may have noticed the similarity in finding the subgame perfect equilibrium and using backward induction. Directly from the definition of subgame perfect, we know that any Nash equilibrium found by backward induction must be subgame perfect. The reason is because backward induction uses a Nash equilibrium from the end of the tree to the beginning. That means it has to be subgame perfect. This gives us a general rule: **To find a subgame perfect equilibrium, use backward induction.**

How do we know in general that a game with multiple Nash equilibria, one of them must be subgame perfect? The next theorem guarantees it.

Theorem 3.2.2 *Any finite tree game in extensive form has at least one subgame perfect equilibrium.*

We can say more about subgame perfect equilibria if the game has perfect information (each node has its own information set).

Theorem 3.2.3 *(1) Any finite game in extensive form with perfect information has at least one subgame perfect equilibrium in pure strategies.*

(2) Suppose all payoffs at all terminal nodes are distinct for any player. Then there is one and only one subgame perfect equilibrium.

(3) If a game has no proper subgames, then every Nash equilibrium is subgame perfect.

The proof of this theorem is by construction because a subgame perfect equilibrium in pure strategies can be found by backward induction. Just start at the end and work backward as we did earlier. If all the payoffs are distinct for each player then this will result in a unique path back through the tree.

Example 3.11 In the following tree in Figure 3.17, there are three subgames: (1) the whole tree, (2) the game from node 2 : 1, and (3) the game from node 2 : 2. The subtree from information set 1 : 2 does not constitute two more subgames because that would slice up an information set.

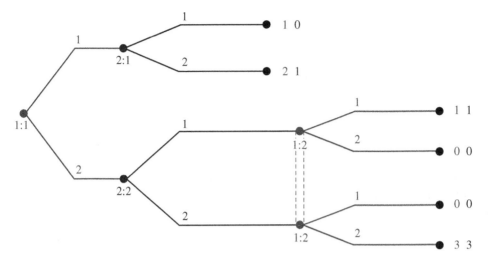

Figure 3.17 Subgame perfect equilibrium.

The equivalent strategic form of the game is

I/II	11	12	21	22
1_	(1,0)	(1,0)	**(2,1)**	(2,1)
21	**(1,1)**	(0,0)	(1,1)	(0,0)
22	(0,0)	**(3,3)**	(0,0)	**(3,3)**

This game has 4 pure Nash equilibria shown as boxed in the matrix. Now we seek the subgame perfect equilibrium.

Begin by solving the subgame starting from 2 : 2. This subgame corresponds to rows 2 and 3, and columns 2 and 3 in the matrix for the entire game. In other words, the subgame matrix is

I/II	21	22
21	(1,1)	(0,0)
22	(0,0)	(3,3)

This subgame has 3 Nash equilibria resulting in payoffs $(3,3), (\frac{3}{4}, \frac{3}{4}), (1,1)$.

Next we consider the subgame starting from 1 : 1 with the subgame from 2 : 2 replaced by the payoffs previously calculated. Think of 2 : 2 as a terminal node.

With payoff $(1, 1)$ (at node 2 : 2), we solve by backward induction. Player 2 will use action 2 from 2 : 1. Player 1 chooses action 1 from 1 : 1. The equilibrium payoffs are $(2, 1)$. Thus a subgame perfect equilibrium consists of player 1 always taking branch 1, and player 2 always using branch 2 from 2 : 1.

With payoff $(\frac{3}{4}, \frac{3}{4})$ at node 2 : 2, the subgame perfect equilibrium is the same.

With payoff $(3, 3)$ at node 2 : 2, Player 2 will choose action 2 from 2 : 1. Player 1 always chooses action 2 from 1 : 1. The equilibrium payoffs are $(3, 3)$ and this is the subgame perfect equilibrium.

This game does not have a unique subgame perfect equilibrium since the subgame perfect equilibria are the strategies giving the payoffs $(3, 3)$ and $(2, 1)$. But we have eliminated the Nash equilibrium giving $(1, 1)$.

Example 3.12 This example will indicate how to use Gambit to find subgame perfect equilibria for more complex games. Suppose we consider a two-stage game of Russian roulette depicted in Figure 3.18. Player I begins by making the choice of spinning and firing or passing. The gun is a six shooter with one bullet in the chambers. If player I passes, or survives, then player II has the same two choices. If a player doesn't make it in round 1, the survivor gets 2 and the loser gets 0.

In round 2, (if there is a round 2) if both players passed in round 1, the gun gets an additional bullet and there is no option to pass. If player I passed and player II spun in round 1, in round 2, then player I must spin and the gun now has 3 bullets. If player I survives that, then player II must spin but the gun has 1 bullet.

If player I spun in round 1 and player II passed, then in the second round player I must spin with 1 bullet. If player I is safe, then player II must spin with 3 bullets. In all cases, a survivor gets 2 and a loser gets 0.

The game tree using Gambit is shown in Figure 3.18.

Gambit gives us the Nash equilibrium that in round 1 player I should spin. If I survives, II should also spin. The expected payoff to player I is 1.667 and the expected payoff to player II is 1.722. Is this a subgame perfect equilibrium? Let's use backward induction.

Gambit has the very useful feature that if we click on a node, it will tells us what the **node value** is. The node value to each player is the payoff to the player if that node is reached. If we click on the nodes $2:2$, $2;4$, and $2:5$, we get the payoffs to each player if those nodes are reached. Then we delete the subtrees from those nodes and specify the payoffs at the node as the node values for each player. This is the resulting tree in Figure 3.19.

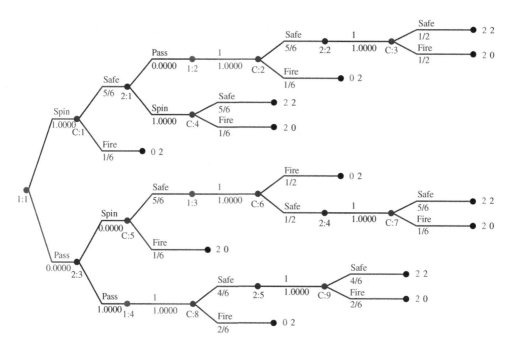

Figure 3.18 2 Stage Russian roulette.

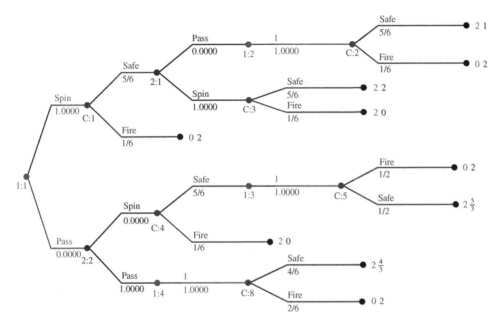

Figure 3.19 The first reduction.

In this tree player I makes the last move but he has no choices to make.

Now let's move back a level by deleting the subtrees from 1 : 2, 1 : 3, and 1 : 4. The result is depicted in Figure 3.20.

Player II should spin if player I spins and survives because his expected payoff is either $\frac{5}{6} \times 2 > \frac{7}{6}$. If player I passes, player II should also pass because $\frac{5}{6} \times \frac{11}{6} < \frac{14}{9}$.

Finally, replacing the subtrees at 2 : 1 and 2 : 2 we get Figure 3.21

In this last tree player I should clearly spin, because his expected payoff is $\frac{5}{6} \times 2 + \frac{1}{6} \times 0 = \frac{10}{6} = 1.66$ if he spins, while it is $\frac{4}{3} = 1.33$ if he passes. We have found the subgame perfect equilibrium:

1. Player I at the start of the game should spin.
2. If player I survives, player II should spin.

This is the Nash equilibrium Gambit gave us at the beginning. Since it is actually the unique Nash equilibrium and we know a game of this type must have at least one subgame perfect equilibrium, it had to be subgame perfect, as we verified.

Incidentally, if we take half of the expected payoffs to each player we will find the probability a player survives the game. Directly, we calculate

$$Prob(\text{I Lives}) = \frac{5}{6} = 0.833$$

and

$$Prob(\text{II Lives}) = Prob(\text{II Lives}|\text{I Lives})Prob(\text{I Lives})$$
$$+ Prob(\text{II Lives}|\text{I Dies})Prob(\text{I Dies})$$
$$= \frac{5}{6} \times \frac{5}{6} + 1 \times \frac{1}{6} = 0.8611.$$

Naturally, player II has a slightly better chance of surviving because player I could die with the first shot.

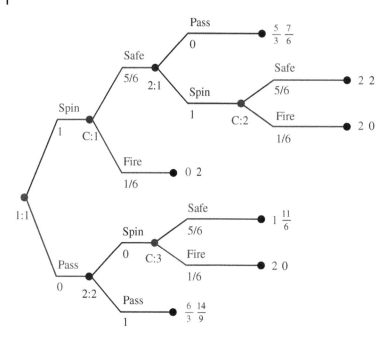

Figure 3.20 The second reduction.

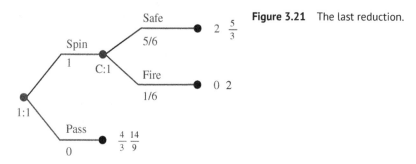

Figure 3.21 The last reduction.

In the next section we will consider some examples of modeling games and finding subgame perfect equilibria.

3.2.2 Examples of Extensive Games Using Gambit

Many extensive form games are much too complex to solve by hand. Even determining the strategic game from the game tree can be a formidable job. In this subsection we illustrate the use of Gambit to solve some interesting games.

Example 3.13 Gale's Roulette. There are two players and three wheels. The numbers on Wheel 1 are 1, 3, 9; on Wheel 2 are 0, 7, 8; and on Wheel 3 are 2, 4, 6. When spun, each of the numbers on the wheel are equally likely. Player I chooses a wheel and spins it. While the wheel player I has chosen is still spinning, player II chooses one of the two remaining wheels and spins it. The winner is the player whose wheel stops on a higher number and the winner receives $1 from the loser.

To summarize, the numbers on the wheels are

$$W1 : 1, 3, 9$$
$$W2 : 0, 7, 8$$
$$W3 : 2, 4, 6$$

We first suppose that I has chosen wheel 1 and II has chosen wheel 2. We will find the chances player I wins the game.

Let X_1 be the number that comes up from wheel 1 and Y_2 the number from wheel 2. Then,

$$Prob(X_1 > Y_2) = Prob((X_1 = 1, Y_2 = 0) \cup (X_1 = 3, Y_2 = 0) \cup (X_1 = 9))$$
$$= \frac{1}{3} \times \frac{1}{3} + \frac{1}{3} \times \frac{1}{3} + \frac{1}{3} = \frac{5}{9}.$$

Similarly, if Z_3 is the number which comes up on wheel 3, then if player II chooses wheel 3 the probability player I wins is

$$Prob(X_1 > Z_3) = Prob((X_1 = 3, Z_3 = 2) \cup (X_1 = 9)) = \frac{4}{9}$$

Finally, if player I chooses wheel 2 and player II chooses wheel 3 we have

$$Prob(X_2 > Z_3) = Prob(X_2 = 7 \cup X_2 = 8)$$
$$= \frac{1}{3} + \frac{1}{3} = \frac{2}{3}$$

Next in Figure 3.22 we may draw the extensive form of the game and use Gambit to solve it. Notice that this is a perfect information, constant sum game.

Gambit gives us 4 Nash equilibria all with the payoffs $\frac{4}{9}$ to player I and $\frac{5}{9}$ to player II. There are two subgame perfect Nash equilibria. The first equilibrium consists of player I always choosing wheel 2 and player II subsequently choosing wheel 1. The second subgame perfect equilibrium is player I chooses wheel 1 and player II subsequently choosing wheel 3. Both of these are constructed using backward induction. The expected winnings for player I is $1 \cdot \frac{4}{9} - 1 \cdot \frac{5}{9} = -\frac{1}{9}$.

Our next example has been studied by many economists and is a cornerstone of the field of behavioral economics and political science which tries to predict and explain irrational behavior.

Figure 3.22 Gale's roulette.

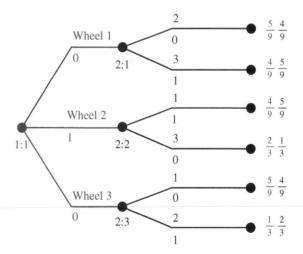

Example 3.14 This game is commonly known as the **centipede game** because of the way the tree looks. The game is a perfect information, perfect recall game. Each player has two options at each stage, Stop, or Continue. Player I chooses first so that player I chooses in odd stages $1, 3, 5, \ldots$. Player II chooses in even-numbered stages $2, 4, 6, \ldots$.

The game begins with player I having 1 and player II having 0. In each odd round, player I can either Stop the game and receive her portion of the pot, or give up 1 and Go in the hope of receiving more later. If player I chooses Go, player II receives an additional 3. In each even round player II has the same choices of Stop or Go. If II chooses Go her payoff is reduced by 1 but player I's payoff is increased by 3. The game ends at round 100 or as soon as one of the players chooses Stop.

We aren't going to go 100 rounds but we can go 6. Here's Figure 3.23 for the extensive game.

Gambit can solve this game pretty quickly but we will do so by hand. Before we begin it is useful to see the equivalent matrix form for the strategic game. Gambit gives us the matrix

I/II	1_ _	21_	221	222
1_ _	$(1,0)$	$(1,0)$	$(1,0)$	$(1,0)$
21_	$(0,3)$	$(3,2)$	$(3,2)$	$(3,2)$
221	$(0,3)$	$(2,5)$	$(5,4)$	$(5,4)$
222	$(0,3)$	$(2,5)$	$(4,7)$	$(7,6)$

In the notation for the pure strategies for player I, for example $1\_\_$ means take action 1 (stop) at node $1:1$ (after that it doesn't matter), and 221 means take action 2 (go) at $1:1$, then take action 2 (go) at $1:2$, and take action 1 (stop) at $1:3$. Immediately, we can see that we may drop column 4 since every second number in the pair is smaller (or equal) to the corresponding number in column 3. Here's how the dominance works.

1. Drop column 4.
2. Drop row 4.
3. Drop column 3.
4. Drop row 3.
5. Drop column 2.
6. Drop row 2.

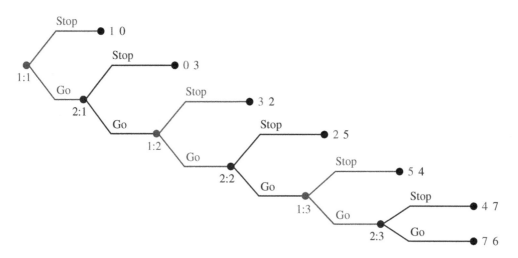

Figure 3.23 6 Stage centipede game.

We are left with a unique Nash equilibrium at row 1, column 1, which corresponds to player I stops the game at stage 1.

Now we work this out using backward induction. Starting at the end of the tree, the best response for player II is Stop. Knowing that, the best response in stage 5 for player I is Stop, and so on back to stage 1. The unique subgame perfect equilibrium is Stop as soon as each player gets a chance to play.

It's interesting that Gambit gives us this solution as well, but Gambit also tells us that if at stage 1 player I plays Go instead of Stop, then player II should Stop with probability $\frac{2}{3}$ and Go with probability $\frac{1}{3}$ at the second and each remaining even stage. Player I should play Stop with probability $\frac{1}{2}$ and Go with probability $\frac{1}{2}$ at each remaining odd stage.

Does the Nash equilibrium predict outcomes in a centipede game? The answer is No. Experiments have shown that only rarely do players follow the Nash equilibrium. It was also rare that players always played Go to get the largest possible payoff. In the six-stage game, the vast majority of players stopped after three or four rounds.

What could be the reasons for actually observing irrational behavior? One possibility is that players actually have some degree of altruism so that they actually care about the payoff of their opponent. Of course it is also possible that some selfish players anticipate that their opponents are altruists and so continue the game. These hypotheses can and are tested by behavioral economists.

3.3 Behavior Strategies in Extensive Games

We know that games possess a mixed strategy Nash Equilibrium which specifies the probability each row or column is played by each player. In an extensive game we can find the NEs by converting it to a matrix game and using the usual methods. The mixed strategy then gives the probability of playing each of the strategies for each player winding their way through the game tree. The mixed strategy is a complete plan of action for the whole game. In contrast, what if a player could decide which action to take at each node based on a probability distribution for that node? That's the basic idea behind a **behavioral strategy**.

Here is an example which should help.

Example 3.15 Consider the game snippet in Figure 3.24:

Figure 3.24 Game snippet.

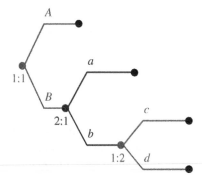

Player I chooses between A and B. If Player I chooses B, then Player II gets to choose between a and b. If Player II chooses b, then Player I must choose between c and d. Player I has the **pure**

strategies Ac, Ad, Bc, Bd. Player I's **mixed** strategies is a probability distribution over all the pure strategies player I has. We know that a mixed strategy for player I looks like (p_1, p_2, p_3, p_4), $p_1 + p_2 + p_3 + p_4 = 1$. For instance, $(0, 0, \frac{1}{2}, \frac{1}{2})$ is the mixed strategy to never play Ac or Ad and play Bc and Bd with probabilities $\frac{1}{2}$.

A behavioral strategy for Player I is as follows: At the first information set where I has to make a move, i.e., at node $1:1$, the behavioral strategy is a distribution on the two actions A and B, namely $(p, 1 - p)$. At the second information set where I has to move, i.e., at node $1:2$, the behavioral strategy is a randomization between c and d, i.e., $(q, 1 - q), 0 \leq q \leq 1$.

A particular behavioral strategy for Player I is to choose $p = \frac{1}{3}, q = \frac{1}{4}$, i.e., at the first move choose A with probability $\frac{1}{3}$ and B with probability $\frac{2}{3}$, and at the second move choose c with probability $q = \frac{1}{4}$ and d with probability $1 - q = \frac{3}{4}$.

Is there a p and a q so that the behavioral strategy gives exactly the same thing as the mixed strategy $(0, 0, \frac{1}{2}, \frac{1}{2})$? Clearly, $p = 0, q = \frac{1}{2}$ will work. Can this always be done? And can a behavioral strategy be constructed from a mixed strategy? In other words, are behavioral and mixed strategies equivalent? What does equivalence mean exactly?

Definition 3.3.1 *For a player i in an extensive game, two strategies (mixed or behavioral or mixed and behavioral) are equivalent if for each pure strategy of the other players, the probability distributions of the outcomes generated by the two strategies at the end of the game are the same.*

Equivalence means the strategies produce exactly the same outcomes at the end of the game.

Example 3.16   Consider the game in Figure 3.25

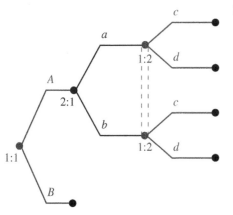

Figure 3.25   A simple game.

Player I has pure strategies (Ac, Ad, Bc, Bd). Let's look at the mixed strategy $X_1 = (\frac{1}{5}, \frac{1}{10}, \frac{2}{5}, \frac{3}{10})$. Can we construct an equivalent behavioral strategy for player I? We have to define probabilities for taking an action at each information set.

$$p_{1:1} = Prob(\text{taking action } A) = Prob(A \cap c) + P(A \cap d) = \frac{1}{5} + \frac{1}{10} = \frac{3}{10}$$

$$1 - p_{1:1} = Prob(\text{taking action } B) = \frac{7}{10}$$

$$p_{12} = Prob(\text{taking action c}|A) = \frac{Prob(c \cap A)}{Prob(A)} = \frac{\frac{1}{5}}{\frac{3}{10}} = \frac{2}{3}$$

$$1 - p_{12} = Prob(\text{taking action d}|A) = \frac{Prob(d \cap A)}{Prob(A)} = \frac{\frac{1}{10}}{\frac{3}{10}} = \frac{1}{3}$$

Thus the only equivalent behavior strategy must be $B_1 = (\frac{3}{10}, \frac{7}{10}, \frac{2}{3}, \frac{1}{3})$ where each entry is the probability for player I of taking the corresponding action at each node.

The question in general is which behavioral strategies are equivalent to a given mixed strategy for player I, say $X_1 = (x_{Ac}, x_{Ad}, x_{Bc}, x_{Bd})$.?

To answer this, in player I's first information set we have

$$Prob(\text{A is chosen}) = Prob(A \cap c) + P(A \cap d) = x_{Ac} + x_{Ad}.$$

In player I's second information set, the probability of choosing c must be consistent with earlier steps. This means

$$Prob(\text{c is chosen}) = Prob(c|A) = \frac{P(c \cap A)}{P(A)} = \frac{x_{Ac}}{x_{Ac} + x_{Ad}}.$$

These are the probabilities which must be assigned to each information set for player I in order to match with the mixed strategy X_1.

Conversely, if we are given a behavioral strategy, say (p, q), where $p = Prob(\text{choose A})$ and $q = Prob(\text{choose c})$, which mixed strategy is equivalent to that? To answer that, first the pure strategies for player I are Ac, Ad, Bc, Bd. The probability of playing strategy Ac is pq and Ad is $p(1 - q)$, etc.. We need

$$(x_{Ac}, x_{Ad}, x_{Bc}, x_{Bd}) = (pq, p(1 - q), (1 - p)q, (1 - p)(1 - q)).$$

Notice that if we are given $p = 0$, this means player I never takes action A, so certainly if, say $x_{Ac} \neq 0$, there is no way to make them equivalent. Only Bc and Bd are consistent and any mixed strategy with $X_1 = (0, 0, x_{Bc}, x_{Bd})$ is equivalent and we may solve

$$x_{Bc} = (1 - p)q, x_{Bd} = (1 - p)(1 - q).$$

The following theorem, known as **Kuhn's theorem**, gives the answer to the question of equivalence.

Theorem 3.3.2 *Suppose we have an extensive game with **perfect recall**. Let $X_i = (x_1, x_2, \ldots, x_n)$ be a mixed strategy for a player i. Then, there is a behavioral strategy for this player which is equivalent to X_i. The converse is not necessarily true.*

Behavior strategies will arise again when we calculate sequential equilibria for Bayesian games.

Problems

3.16 Consider the following game in extensive form depicted in the figure

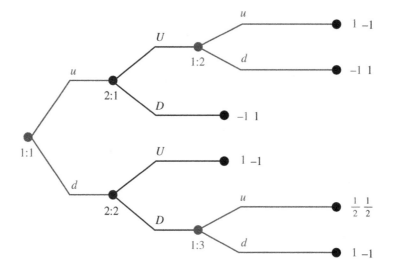

(a) Find the strategic form of the game.
(b) Find all the Nash equilibria.
(c) Find the subgame perfect equilibrium using backward induction.

3.17 BAT (British American Tobacco) is thinking of entering a market to sell cigarettes currently dominated by PM (Phillip Morris). PM knows about this and can choose to either be passive and give up market share, or be tough and try to prevent BAT from gaining any market share. BAT knows that PM can do this and then must make a decision of whether to fight back, enter the market but act passively, or just stay out entirely. Here's the tree:

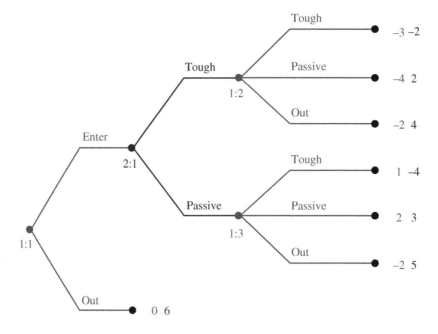

Find the subgame perfect Nash equilibrium.

3.18 In a certain card game, player 1 holds two Kings and one Ace. He discards one card, lays the other two face down on the table, and announces either 2 Kings or Ace King. Ace King is a better hand than 2 Kings. Player 2 must either Fold or Bet on player 1's announcement. The hand is then shown and the payoffs are as follows:

1. if player 1 announces the hand truthfully and player 2 Folds, player 1 wins $1 from player 2;
2. if player 1 lies that the hand is better than it is and player 2 Folds, player 1 wins $2 from player 2;
3. if player 1 lies that the hand is worse than it is and player 2 Folds, player 2 wins $2 from player 1;
4. if player 2 Bets player 1 is lying and if player 1 is actually lying the above payoffs are doubled and reversed. If player 1 is not lying, player 1 wins $2 from player 2.
    (a) Use Gambit to draw a game tree.
    (b) Give a complete list of the pure strategies for each player as given by Gambit and write down the game matrix.
    (c) Find the value of the game and the optimal strategies for each player.
    (d) Modify the game so that player 1 first chooses one of the 3 cards randomly and tosses it. Player 1 is left with 2 cards–either an A and K, or two K's. Player 1 knows the cards he has but player 2 does not and the game proceeds as before. Solve the game.
    (e) As in the previous part, assume that player 1 does not know the outcome of the random discard. That is, he does not actually know if he has 1A1K, or 2K. The payoffs remain as before replacing *lying* with *incorrect* and *truthfully* with *correct*.

3.19 Solve the Beer or Quiche game assuming Curly has probability $\frac{1}{10}$ of being weak and probability $\frac{9}{10}$ of being strong.

3.20 Curly has two safes, one at home and one at the office. The safe at home is a piece of cake to crack and any thief can get into it. The safe at the office is hard to crack and a thief has only a 15% chance of getting at the gold if it is at the office. Curly has to decide where to place his gold bar (worth 1). On the other hand, the thief has a 50% chance of getting caught if he tries to hit the office and a 20% chance if he hits the house. If he gets the gold, he wins 1; if he gets caught, he not only doesn't get the gold but he goes to the joint (worth −2). Find the Nash equilibrium and expected payoffs to each player.[1]

3.21 Jack and Jill are commodities traders who go to a bar after the market closes. They are trying to determine who will buy the next round of drinks. Jack has caught a fly, and he also has a realistic fake fly. Jill, who is always prepared has a fly swatter. They are going to play the following game. Jack places either the real or fake fly on the table covered by his hand. When Jack says 'ready,' he uncovers the fly (fake or real) and as he does so Jill can either swat the fly or pass. If Jill swats the real fly, Jack buys the drinks; if she swats the fake fly, Jill buys the drinks. If Jill thinks the real fly is the fake and she passes on the real fly, the fly flies away, the game is over, and Jack and Jill split the round of drinks. If Jill passes on the fake fly, then they will give it another go. If in the second round Jill again passes on the fake fly, then Jack buys the round and the game is over. A round of drinks costs $2.

---

1 A version of this and the next 4 problems appeared on an exam at the London School of Economics.

(a) Use Gambit to solve the game.

(b) What happens if Jill only has a 75% chance of nailing the real fly when she swats at it?

3.22 Two opposing navys (French and British) are offshore an island while the admirals are deciding whether or not to attack. Each navy is either strong or weak with equal probability. The status of their own navy is known to the admiral but not to the admiral of the opposing navy. A navy captures the island if it either attacks while the opponent does not attack, or if it attacks as strong while the opponent is weak. If the navys attack and they are of equal strength, then neither captures the island.

The island is worth 8 if captured. The cost of fighting is 3 if it is strong and 6 if it is weak. There is no cost of attacking if the opponent does not attack and there is no cost if no attack takes place. What should the admirals do?

3.23 Solve the six-stage centipede game by finding a correlated equilibrium. Since the objective function is the sum of the expected payoffs this would be the solution assuming that each player cares about the total, not the individual.

3.24 There are 3 gunfighters A, B, C. Each player has 1 bullet and they will fire in the sequence A then B then C, assuming the gunfighter whose turn comes up is still alive. The game ends after all three players have had a shot. If a player survives, that player gets 2, while if the player is killed, the payoff to that player is -1.

(a) First assume that the gunfighters all have probability $\frac{1}{2}$ of actually hitting the person they are shooting at. (Hitting the player results in a kill). Find the extensive game and as many Nash equilibria as you can using Gambit.

(b) Since the game is perfect information and perfect recall we know there is a subgame perfect equilibrium. Find it.

(c) Now assume that gunfighters have different accuracies. A's accuracy is 40%, B's accuracy is 60% and C's accuracy is 80%. Solve this game.

3.25 Moe, Larry, and Curly are in a two round truel. In round 1, Larry fires first, then Moe, then Curly. Each stooge gets one shot in round 1. Each stooge can **choose to fire at one of the other players, or to deliberately miss by firing in the air.** In round 2, all survivors are given a second shot, in the same order of play.

Suppose Larry's accuracy is 30 percent, Moe's is 80 percent, and Curly's is 100 percent. For each player, the preferred outcome is to be the only survivor, next is to be one of 2 survivors, next is the outcome of no deaths, and the worst outcome is that you get killed. Assume an accurate shot results in death of the one shot at. Who should each player shoot at and find the probabilities of survival under the Nash equilibrium.

3.26 Two players will play a Battle of the Sexes game with bimatrix

I/II	A	B
A	(3, 1)	(0, 0)
B	(0, 0)	(1, 3)

Before the game is played, player I has the option of burning $1 with the result that it would drop I's payoff by 1 in every circumstance.

(a) Draw the extensive form of this game where player II can observe whether or not player I has burned the dollar. Find the strategic form of the game.

(b) Find all the Nash equilibria of the game and determine which are subgame perfect.

(c) Solve the game when player II cannot observe whether or not player I has burned the dollar.

3.27 There are two players involved in a 2 step centipede game. Each player may be either Rational (with probability $\frac{19}{20}$) or an Altruist (with probability $\frac{1}{20}$). Neither player knows the other players type.

If both players are rational, each player may either Take the payoff or Pass. Player I moves first and if he Takes, then the payoff is 0.8 to I and 0.2 to II. If I Passes, then II may either Take or Pass. If II Takes, the payoff is 0.4 to I and 1.60 to II. If II Passes the next round begins. Player I may Take or Pass (payoffs 3.20, 0.80 if I Takes) then II may Take or Pass (payoffs 1.60, 6.40 if II Takes and 12.80, 3.20 if II Passes). The game ends.

If player I is rational and player II is altruistic, then player II will Pass at both steps. The payoffs to each player are the same as before if I Takes, or the game is at an end.

The game is symmetric if player II is rational and player I is altruistic.

If both players are altruistic then they each Pass at each stage and the payoffs are 12.80, 3.20 at the end of the second step.

Draw the extensive form of this game and find as many Nash equilibria as you can. Compare the result with the game in which both players are rational.

3.28 Consider the game tree:

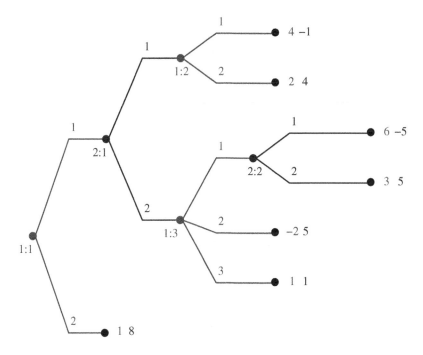

1. Use backward induction to find a subgame perfect equilibrium.
2. Convert to a matrix game. You should end up with a 7×3 bimatrix.
3. Show that there is a pure Nash equilibrium for the matrix game with payoffs (1, 8).
4. State what it is about the strategy from (3) that makes it non-subgame perfect.
5. Find a second Nash equilibrium which does have payoffs (1, 8).
6. State what it is about the strategy from (5) that makes it non-subgame perfect.

3.29 There are two companies, C1 and C2, who each need to decide to increase production, or not. C2 has a spy in C1, so C2 can see C1's decision before they make their own. That is, C1 decides first, and C2 can see their decision before deciding. But C1 is more flexible than C2, and can change their decision afterwards at a cost of a million dollars, while C2's decision is final. The expected profits of the companies are:

- 4 million each at current production levels.
- If one company increases production and the other does not, 6 million for the one who increased and 3 million of the one that did not (because she needs to reduce prices).
- If both increase, only 1 million each (prices will drop a lot).
- Also C1 pays the 1 million penalty if she changed their mind.
    1. Draw the game tree, indicating whose turn it is at each node, what the choices are, and all payoffs. It is a complete information game.
    2. Find a subgame perfect equilibrium by backward induction.
    3. Convert to a bimatrix game. Your matrix should be 4 × 4. Indicate clearly what the strategies are.
    4. Find a non-subgame perfect equilibrium which gives PI a better payoff than the equilibrium in part (3). Discuss whether this is realistic, and under what circumstances it might be played.
    5. Is the option to switch helping PI?

3.30 Consider the game tree

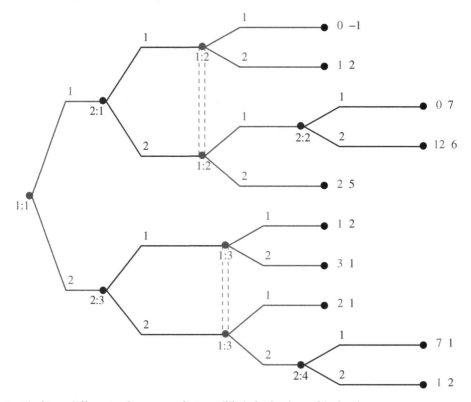

1. Find two different subgame perfect equilibria by backward induction.
2. Discuss if there is any reason to favor one of the equilibria over the other.

3.31 Consider the game with Gale's roulette. Suppose we change the payoff to the following rule: If the high number on the winning wheel is N, then the loser pays the winner $\$N$. Draw the game tree using Gambit and solve the game.

3.32 This is about the famous Monte-Hall problem. The set up is: Monte runs a game show. There are 3 doors; behind one is a car, and behind the other two are goats. The contestant is trying to pick the door with the car; if they do, they win it. The play works as follows:
- The car and goats are placed randomly.
- Contestant picks a door.
- Monte opens a different door to reveal a goat (notice this is always possible, since Monte knows where the car is).
- Contestant is given the chance to change doors or not.
- Contestant's door is opened to reveal either a goat or a car; if it is the car, they win it.
    (a) Draw the game tree, showing the possible choices due to chance (i.e., where the car actually is), of the contestant, and of Monte. It is a fairly large tree.
    (b) Calculate PI's expected payoff (i.e., probability of winning) if they do not switch doors.
    (c) Calculate PI's probability of winning if they do switch doors.

3.33 Consider Monte-Hall but where the prize is not placed randomly behind the doors. Instead, it is behind door 1 half the time, behind door 2 one-third of the time, and behind door 3 one-sixth of the time. The contestant knows this. Now what strategy should the contestant use? How likely are they to win the car?

3.34 Let's make Monte-Hall a little more interesting: Monte may or may not open a door, but if he chooses not to open a door, he must pay the contestant $\$10,000$. The player may switch doors after Monte makes his choice, regardless of whether or not Monte opens a door. Assume the car is worth $\$30,000$.
    (a) Draw the game tree, where the first branch is whether the contestant picks car or goat. Fill in the expected payout to contestant at each node.
    (b) List the possible strategies of each player and express the game as a matrix game.
    (c) Solve the game. How often does Monte open a door with an optimal strategy?

3.4 Extensive Games with Imperfect Information

We introduce the important idea of Bayesian Games with several examples. These are all games in which knowledge of one or both players' types or characteristics are unknown to one or both players. These are referred to as Bayesian Games because of the critical role Bayes rule plays in determining the belief by a player about the other players' move.

The difference between a complete information and incomplete information game is that in a complete information game all the players know all the relevant details. For example, in the classic Cournot Duopoly games all production costs and prices and demands are known. Obviously, in many, if not most games of interest this is not the case. In a duopoly, why would one company know all the information of a competitor. In war, how could an opponent know if a combatant is willing to use nuclear weapons? In an auction how could a bidder know the valuations of the other bidders? These are all incomplete information games.

In this section, we will present several examples illustrating Bayesian games.

Example 3.17   This is a classic example called the **Beer or Quiche game**. Curly[1] is at a diner having breakfast and in walks Moe. We all know Moe is a bully and he decides he wants to have some fun by bullying Curly out of his breakfast. But we also know that Moe is a coward so he plays his tricks only on those he considers wimps. Moe doesn't know whether Curly is a wimp or will fight (because if there is even a hint of limburger cheese, Curly is a beast) but he estimates that about $\frac{1}{3}$ of men eating breakfast at this diner will fight and $\frac{2}{3}$ will cave.

Moe decides to base his decision about Curly on what Curly is having for breakfast. If Curly is having a quiche, Moe thinks Curly definitely is a wimp. On the other hand, if Curly is having a beer for breakfast, then Moe assumes Curly is not a wimp and will fight. Curly knows if he is a wimp, but Moe doesn't, and both fighters and wimps could order either quiche or beer for breakfast so the signal is not foolproof. Also, wimps sometimes choose to fight and don't always cave.

Curly gets to choose what he has for breakfast and Moe gets to choose if he will fight Curly or cave. Curly will get 2 points if Moe caves and an additional 1 point for not having to eat something he doesn't really want. Moe gets 1 point for correctly guessing Curly's status. Finally, if Curly is tough, he really doesn't like quiche but he may order it to send a confusing signal.

Here's the game tree in Figure 3.26.

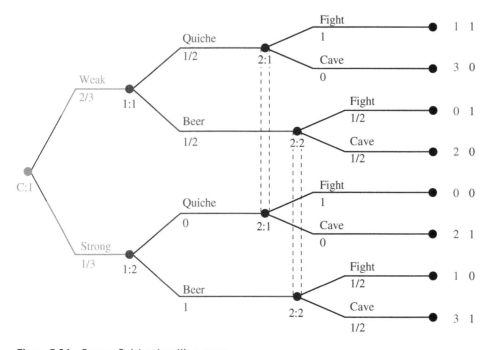

Figure 3.26   Beer or Quiche signalling game.

Notice that there are no strict subgames of this game since any subgame would cut through an information set, and hence every Nash equilibrium is subgame perfect. Let's first consider the

---

1 My favorite stooges, Moe Howard, Larry Fine, Curly Howard, and Shemp Howard.

equivalent strategic form of this game. The game matrix is

Curly/Moe	11	12	21	22
11	$\left(\frac{2}{3},\frac{2}{3}\right)$	$\left(\frac{2}{3},\frac{2}{3}\right)$	$\left(\frac{8}{3},\frac{1}{3}\right)$	$\left(\frac{8}{3},\frac{1}{3}\right)$
12	$\left(1,\frac{2}{3}\right)$	$\left(\frac{5}{3},1\right)$	$\left(\frac{7}{3},0\right)$	$\boxed{\left(3,\frac{1}{3}\right)}$
21	$\left(0,\frac{2}{3}\right)$	$\left(\frac{4}{3},0\right)$	$\left(\frac{2}{3},1\right)$	$\left(2,\frac{1}{3}\right)$
22	$\left(\frac{1}{3},\frac{2}{3}\right)$	$\left(\frac{7}{3},\frac{1}{3}\right)$	$\left(\frac{1}{3},\frac{2}{3}\right)$	$\left(\frac{7}{3},\frac{1}{3}\right)$

The labels for the strategies correspond to the labels given in the game tree. In the figure, action 1 = quiche, 2 = beer for player I and action 1 = fight, 2 = cave for player II. Then strategy 11 for player I means to take action 1 (order a quiche) if player I is weak and action 1 (quiche) if player I is strong. Strategy 12 means to play action 1 (quiche) at if at $1:1$ (weak) and action 2 (order a beer) if at $1:2$ (strong). For player II, strategy 22 means play cave (action 2) if I plays quiche (info set $2:1$) and play cave (action 2) if I plays beer (info set $2:2$).

As an example to see how the numbers arise in the matrix, let's calculate the payoff if Curly plays 11 and Moe plays 21. Curly orders Quiche whether he is strong or weak. Moe caves if he spots a Quiche but fights if he spots Curly with a Beer. The expected payoff to Curly is

$$Prob(weak) \times 3 + Prob(strong) \times 2 = \frac{2}{3} \times 3 + \frac{1}{3} \times 2 = \frac{8}{3},$$

and the payoff to Moe is

$$Prob(weak) \times 0 + Prob(strong) \times 1 = \frac{1}{3} \times 1 = \frac{1}{3}.$$

It is easy to check that row 3 and column 4 may be dropped because of dominance. The resulting 3×3 matrix can be solved using equality of payoffs. We get $X^* = (0,\frac{1}{2},0,\frac{1}{2})$, $Y^* = (\frac{1}{2},\frac{1}{2},0,0)$. Curly should play strategy 12 and strategy 22 with probability $\frac{1}{2}$, while Moe should play strategy 11 and strategy 12 with probability $\frac{1}{2}$. In words, given that Curly is weak, he should order quiche with probability $\frac{1}{2}$ and beer with probability $\frac{1}{2}$, and then Moe should fight if he sees the quiche but only fight with probability $\frac{1}{2}$ if he sees the beer. Given that Curly is strong, Curly should always order beer, and then Moe should fight with probability $\frac{1}{2}$. Of course Moe doesn't know if Curly is weak or strong but he does know there is only a $\frac{1}{3}$ chance he is strong. The payoffs are $\frac{4}{3}$ to Curly and $\frac{2}{3}$ to Moe.

Informally, a **Bayesian game** is a game in which a player has a particular characteristic, or is a particular type, with some probability, like weak or strong in the Beer-Quiche game. Practically, this means that in Gambit we have to account for the characteristic with a chance move. This is a very useful modeling technique which we have already encountered many times. The next example further illustrates this.

Example 3.18 A prisoner's dilemma game in which one of the players has a conscience can be set up as a Bayesian game if the other player doesn't know about the conscience. For example, under normal circumstances the game matrix is given by

I/II	C	D
C	(5, 5)	(0, 8)
D	(8, 0)	(1, 1)

On the other hand, if player II develops a conscience and is less rewarded for playing *D* the game matrix is

I/II	C	D
C	(5, 5)	(0, 2)
D	(8, 0)	(1, −5)

Let's call the first game the game that is played if player II is Normal, while the second game is played if II is Abnormal. Player I does not know ahead of time which game will be played when she is picked up by the cops, but let's say she believes the game is Normal with probability $0 < p < 1$ and Abnormal with probability $1 - p$. In this example we take $p = \frac{2}{3}$.

We can solve this game by using the Chance player who chooses at the start of the game which matrix will be used. In other words, the Chance player determines if player II is a type who has a conscience or not. Player II knows if they have a conscience, but player I does not. As an extensive game, we model this with an information set in which player I does not know how the probability came out. Here's the picture in Figure 3.27.

The equivalent strategic form of the game is

I/II	11	12	21	22
1	(5, 5)	$\left(\frac{10}{3}, 4\right)$	$\left(\frac{5}{3}, 7\right)$	(0, 6)
2	(8, 0)	$\left(\frac{17}{3}, -\frac{5}{3}\right)$	$\left(\frac{10}{3}, \frac{2}{3}\right)$	(1, −1)

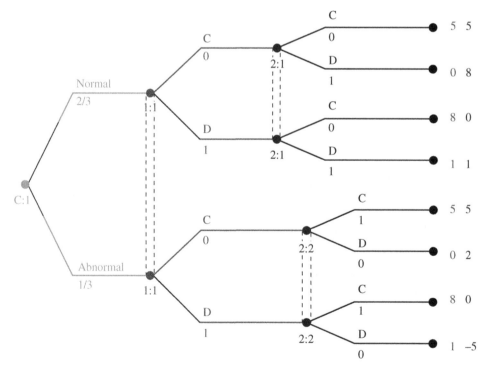

Figure 3.27 Prisoner's dilemma with a conscience.

Since player I only makes a decision once at the start of the game, player I has only the two strategies $1 = C$ and $2 = D$. For player II we have the 4 strategies:

Strategy	Description
11	If Normal, play C; If Abnormal, play C
12	If Normal, play C; If Abnormal, play D
21	If Normal, play D; If Abnormal, play C
22	If Normal, play D; If Abnormal, play D

From the strategic form for the game, we see immediately that for player I strategy 2 (i.e., D) dominates strategy 1 and player I will always play D. Once that is determined, we see that strategy 21 dominates all the others and player II will play D, if the game is normal, but play C if the game is Abnormal. That is the Nash equilibrium for this game.

Here is an example in which the observations are used to infer a players type. Even though the game is similar to the Conscience game, it has a crucial difference through the information sets.

Example 3.19 Suppose a population has two types of people, Smart and Stupid. Any person (Smart or Stupid) wanting to enter the job market has two options, get educated by going to College, or Party. A potential employer can observe a potential employee's educational level, but not if the potential employee is Smart or Stupid. Naturally, the employer has the decision to hire (H) or pass over (P) an applicant. We assume that there is a $\frac{5}{8}$ chance the student is Stupid and a $\frac{3}{8}$ chance the student is Smart. The tree is in Figure 3.28.

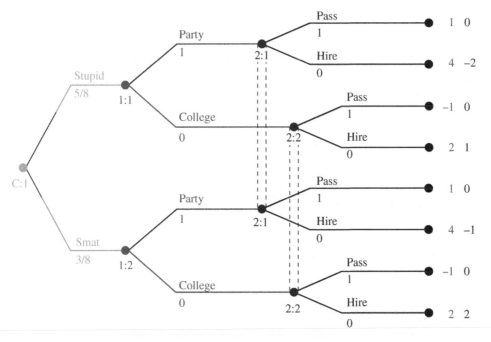

Figure 3.28 Party or college?

The strategic form of this game is given by

I/II	11	12	21	22
11	$(1,0)$	$(1,0)$	$\left(4,-\frac{13}{8}\right)$	$\left(4,-\frac{13}{8}\right)$
12	$\left(\frac{1}{4},0\right)$	$\left(\frac{11}{8},\frac{3}{4}\right)$	$\left(\frac{17}{8},-\frac{5}{4}\right)$	$\left(\frac{13}{4},-\frac{1}{2}\right)$
21	$\left(-\frac{1}{4},0\right)$	$\left(\frac{13}{8},\frac{5}{8}\right)$	$\left(\frac{7}{8},-\frac{3}{8}\right)$	$\left(\frac{11}{4},\frac{1}{4}\right)$
22	$(-1,0)$	$\boxed{\left(2,\frac{11}{8}\right)}$	$(-1,0)$	$\left(2,\frac{11}{8}\right)$

Here for player I, 1=party, 2=college; for player II, 1=pass, 2=hire. Strategies 11, 21, and 22 for player II are dominated by strategy 12; then strategy 22 for player I dominates the others. Player I, the student, should always play 22 and player II (the employer) should always play 12. Thus the pure Nash equilibrium is $X = (0,0,0,1), Y = (0,1,0,0)$. What does this translate to? Pure strategy 22 for player I (the student) means that at the first info set (i.e., at $1:1$), the student should take action 2 (college), and at the second info set (i.e., at $1:2$), the student should take action 2 (college). The Nash equilibrium 12 for player II says that at info set $2:1$ (i.e., when student chooses party), action 1 should be taken (Pass); at info set $2:2$ (student chooses college), the employer should take action 2 (Hire).

There are actually four Nash equilibria for this game:

1. $X_1 = (0,0,0,1), Y_1 = (0,1,0,0)$, I's payoff $= 2$, II's payoff $= \frac{11}{8}$
2. $X_2 = (0,0,0,1), Y_2 = (0,\frac{2}{3},0,\frac{1}{3})$, same as (1)
3. $X_3 = (1,0,0,0), Y_3 = (\frac{1}{3},\frac{2}{3},0,0)$, I's payoff $= 1$, II's payoff $= 0$,
4. $X_4 = (1,0,0,0), Y_4 = (1,0,0,0)$, same as (3).

We have seen that (X_1, Y_1) is the dominant equilibrium but there are others. For example X_4, Y_4 says that the student should always Party, whether he is Smart or Stupid, and the employer should always Pass. This leads to payoffs of 1 for the student and 0 for the employer. (This is the Nash depicted in the Figure 3.28.)

In summary, the employer will hire a college graduate and always pass on hiring a partier.

3.4.1 Bayesian Games and Bayesian Equilibria

What exactly is a Bayesian Equilibrium? **A Bayesian Equilibrium (BE) is a Nash Equilibrium (NE) which maximizes the expected payoff to each player.**

A **Perfect Bayesian Equilibrium (PBE)** is similar to subgame perfect Nash equilibrium in that it is found by working back along the game tree maximizing the **expected payoff at each information set**. The difference is that the probabilities used to calculate the expected values are the **beliefs** of the player who is faced with a decision as to which node in the information set was reached by the opponent. This belief business is a new wrinkle in the analysis of the game.

The third type of equilibrium considered is a **sequential equilibrium**. A sequential equilibrium is essentially the same as a Perfect Bayesian Equilibrium although there are some caveats which we will not go into in this book. Refer to reference [8] for a lot more about this. We will use the terms Perfect Bayesian Equilibrium and sequential equilibrium interchangeably.

In the games considered here, one interpretation of chance moves is that chance determines a player's type. Player i has type t_i and this can be determined by chance. An example will make this clear.

Example 3.20 There are two possibilities for Dictator P when he attacks a sovereign country. With probability p, P will use tactical nuclear weapons and with probability $1 - p$ he uses only conventional weapons. P of course is aware if he will use nuclear weapons but the sovereign nation is not. P has three pure strategies: Attack in force (AF), Attack weakly (AW), Do Not Attack (DNA). The sovereign nation has two strategies: Defend (D), or Capitulate (C). The game matrices in both cases in which P uses nuclear or conventional weapons are as follows.

Nuclear p	D	C
AF	$(-1, 1)$	$(3, -2)$
AW	$(-2, 3)$	$(2, -1)$
DNA	$(0, 2)$	$(2, 0)$

Conventional $(1 - p)$	D	C
AF	$(1, 1)$	$(3, 0)$
AW	$(0, 3)$	$(2, 2)$
DNA	$(0, 2)$	$(2, 0)$

We can say that player I, the row player, dictator P is one of two types labeled t_1, i.e., he will use nuclear weapons, or t_2, he will only use conventional weapons. Player II, the sovereign country does not know which type player I will be, but she does know, or has a pretty good idea about p, called the prior probability of P being type t_1.

Figure 3.29 is our model of this game. Nature chooses the type of player I. Player I knows her type but player II does not know player I's type. This results in the three information sets for player II depending on the action player I chooses. Later we will see that the choice of action by player I is a sort of signal to player II as to player I's type. Observe that this game has no proper subgames. Player I has nine pure strategies and player II has eight pure strategies. The NE, which is the Bayes Equilibrium, is for player I, if she is nuclear type t_1 should always play DNA; if she is conventional type t_2 should play AF. Player II, should always play D, no matter what. The resulting payoffs are $\frac{1}{2}$ to player I and $\frac{3}{2}$ to player II. These payoffs arise from the calculation

$$E_I = 0 \times \tfrac{1}{2} + 1 \times \tfrac{1}{2} = \tfrac{1}{2}, \quad E_{II} = 2 \times \tfrac{1}{2} + 1 \times \tfrac{1}{2} = \tfrac{3}{2}.$$

By considering the matrices separately, we may find the Nash Equilibrium. First, if the game is nuclear, we see that D dominates C for player II and then DNA dominates for player I. Next, if the game is conventional, again D dominates C for player II and then AF dominates for player I.

Even in this simple model, we have determined that P should not attack if he is willing to use nuclear weapons but, with only conventional weapons he should attack in force. Player II should always defend and not capitulate[1].

Bayesian games also involve the concept of a **belief system** for each player which we explain next.

Remarks 1. What exactly is a **belief** of a player? When there is more than one node at an information set, the player who has to choose which action to take from that information set will have a **belief as to which particular node in the info set was reached by the opponent**. This is modeled by assigning a probability quantifying the chances that each node in the info set was reached. **This probability of arriving at a particular node in an info set is called the belief at that node.**

2. Bayes' rule is used to ensure that **beliefs of a player are consistent with a strategy**. A simple way to calculate these beliefs is to look at the information set I containing nodes $i = 1, 2, \ldots, n$. **If node i in information set I is chosen with probability p_i, then the belief of the player at this**

---

1 Eerily, as of 2024, this seems to be exactly what is happening in the Ukraine war.

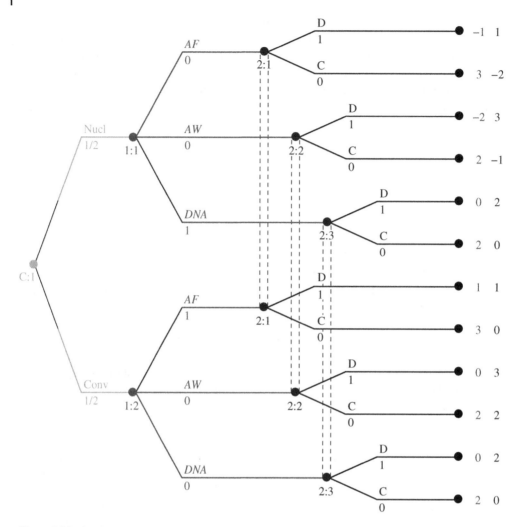

Figure 3.29 Nuclear or conventional.

node on this information set should be as follows:

$$\mu_i = \begin{cases} \dfrac{p_i}{p_1 + \cdots + p_n}, & \text{if } p_1 + \cdots + p_n \neq 0, \\ \text{arbitrary}, & \text{if } p_1 + \cdots + p_n = 0, \end{cases} \tag{3.4.1}$$

Where does this come from? The reason for this is exactly Bayes' rule since if I is reached and I consists of nodes $i = 1, 2, \ldots, n$ then $Prob(I|i) = 1$, and we have the conditional probability of being at node i within info set I is, by Bayes' rule,

$$\mu_i = Prob(i|I) = \frac{Prob(I|i)Prob(i)}{Prob(I|i)Prob(i) + \sum_{j \neq i} Prob(I|j)Prob(j)} = \frac{p_i}{p_1 + \cdots + p_n}$$

The beliefs are labeled throughout as μ_i for each node $i \in I$, or as μ_I^i to denote dependence on the infomation set. If we are at some information set, say for player II, with only two nodes we label the beliefs at that information set as μ_{21} and $1 - \mu_{21}$.

A Perfect Bayesian Equilibrium (PBE) has requirements.

1. At each info set the player with that move has a belief about which node within the info set has been reached.
2. Given the beliefs at each info set a player must choose the action which maximizes the expected payoff of the player using the beliefs as the probability weights at that info set.
3. The strategy profile is **sequentially rational given** μ, i.e., each player's strategy is optimal in the part of the game that follows each of her information sets, given the strategy profile and her belief about the history in the information set that has occurred.

The next theorem shows that a PBE exists.

Theorem 3.4.1

- *Any PBE must be a NE.*
- *In a finite dynamic game with incomplete information, a NE and therefore a PBE exists. Backward induction starting from the info sets at the end ensures perfection and one can construct a belief system supporting these strategies, so the result is a PBE.*

Several examples will illustrate the definitions and illustrate how a PBE can be used in some cases to determine the best NE to be played.

Example 3.21 Suppose two players are in a game in which a part of the payoff is determined by a probability distribution. Let's say

	L	R
T	$(X,9)$	$(3,6)$
B	$(6,0)$	$(6,9)$

where $Prob(X = 12) = \frac{2}{3}$, $Prob(X = 0) = \frac{1}{3}$.

1. Player I knows the value of X but II does not. When $X = 12$, we can say player I is of type t_1 and when $X = 0$, player I is type t_2. Player I knows her type but player II does not know player I's type. The model involves an info set for player II in which player II has no knowledge of player I's move or type.

Let's find the Bayesian equilibrium for the game in Figure 3.30. The strategic form of the game is given in the matrix

	L	R
TT	$(8,9)$	$(3,6)$
TB	$(10,6)$	$(4,7)$
BT	$(4,3)$	$(5,8)$
BB	$(6,0)$	$\boxed{(6,9)}$

As an example of how these numbers are obtained, we calculate the expected payoffs for *BT versus R*. We have

$$E_I(BT, R) = \frac{2}{3} \times 6 + \frac{1}{3} \times 3 = \frac{15}{3} = 5, \quad \text{and} \quad E_{II}(BT, R) = \frac{2}{3} \times 9 + \frac{1}{3} \times 6 = \frac{24}{3} = 8.$$

The best response to the best response gives us the unique NE and therefore Bayesian equilibrium (BB, R) with expected payoffs $(6, 9)$.

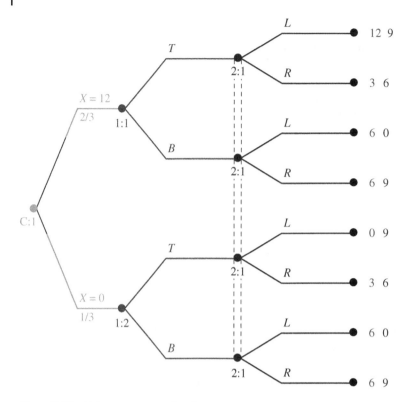

Figure 3.30 Unknown types and actions.

In fact, from the theorem we may conclude that the unique Nash equilibrium must be the perfect Bayesian equilibrium. In the exercises you will see that you can construct the PBE using sequential rationality.

2. On the other hand, if we wanted to model player II as seeing player I's choice of actions but not his type, the game looks like this.

The game in Figure 3.31 has two info sets for player II modeling if I has played T or B but II does not know what is the state of Nature. This is an example of a **signalling game**. The game matrix is then

	LL	LR	RL	RR
TT	(8, 9)	(8, 9)	(3, 6)	(3, 6)
TB	(10, 6)	(10, 9)	(4, 4)	(4, 7)
BT	(4, 3)	(4, 9)	(5, 2)	(5, 8)
BB	(6, 0)	(6, 9)	(6, 0)	(6, 9)

The Bayesian equilibrium for the game in Figure 3.31 can easily be seen to be (TB, LR) giving expected payoffs $(10, 9)$. It is calculated as

$$E_I(TB, LR) = \frac{2}{3}12 + \frac{1}{3}6 = 10, \quad \text{and} \quad E_{II}(TB, LR) = \frac{2}{3}9 + \frac{1}{3}9 = 9.$$

However, there is also a BE at (BB, RR) giving a payoff $(6, 9)$ but it is **not a PBE because it is not sequentially rational.** To see this and the fact that (TB, LR) is a PBE, we let $p =$ probability

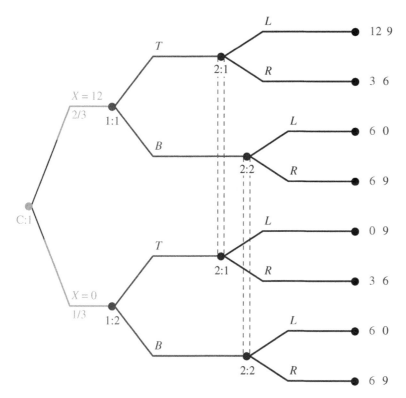

Figure 3.31 Unknown actions.

I plays T, if $X = 12$ and q = probability I plays T if $X = 0$. Also, r = probability II plays L at $2:1$ and s = probability II plays L at $2:2$. We calculate starting with player II at the end of the game and give the values of p, q, r, s which maximize the expected payoffs. The beliefs at top node of $2:1$ and $2:2$ are

$$\mu_{21} = \frac{2/3\, p}{2/3\, p + 1/3\, q} = \frac{2p}{2p+q} \quad \text{and} \quad \mu_{22} = \frac{2(1-p)}{2(1-p)+1-q}.$$

$$E_{II}(r, 2:1) = \mu_{21}(9r + 6(1-r)) + (1 - \mu_{21})(9r + 6(1-r)) = 3r + 6$$
$$\implies \text{maximized when } r = 1$$
$$E_{II}(s, 2:2) = 9 - 9s \implies s = 0$$
$$E_{I}(p, 1:1) = p(12) + (1 - p)6 = 6p + 6 \implies p = 1$$
$$E_{I}(q, 1:2) = 6 - 6q \implies q = 0$$

We conclude that the expected payoffs of each player are maximized when $p = 1, q = 0, r = 1, s = 0$. This says the sequential equilibrium is for player I to play T if $X = 12$ and B if $X = 0$ while player II should play L if I plays T and R if I plays B, i.e., (TB, LR) is a PBE. It is also true that while (BB, RR) is a NE, it is not a PBE and it gives a lower expected payoff to player I.

3. The game in Figure 3.32 models the fact that player II is unaware of whether player I has played T or B in either state of Nature, but does know the choice of Nature. If player II observes player I's type but not his actions, the game is modeled as

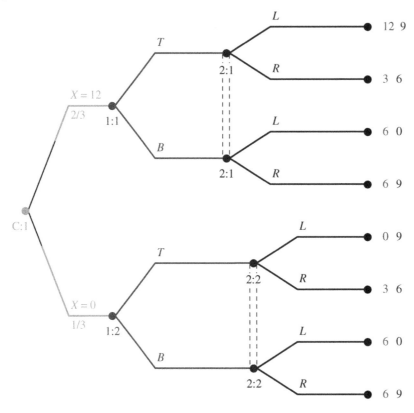

Figure 3.32 Unknown type.

Here is the matrix for this game which you should compare with the matrices of the previous games.

	LL	LR	RL	RR
TT	(8, 9)	(9, 8)	(2, 7)	(3, 6)
TB	(10, 6)	(10, 9)	(4, 4)	(4, 7)
BT	(4, 3)	(5, 2)	(4, 9)	(5, 8)
BB	(6, 0)	(6, 3)	(6, 6)	(6, 9)

The game has three Bayesian equilibria given by (BB, RR), with expected payoffs $(6, 9)$, $((\frac{3}{4}T, B), (\frac{1}{3}L, R))$, with expected payoffs $(6, \frac{15}{2})$, and (TB, LR) with expected payoffs $(10, 9)$. Let's calculate to find all PBEs. Let p =probability I plays T, if $X = 12$ and q =probability I plays T if $X = 0$. Also, r =probability II plays L at $2:1$ and s =probability II plays L at $2:2$. We calculate and give the values of p, q, r, s which maximize the expected payoffs:

$$\mu_{21} = p \implies$$

$$E_{II}(r, 2:1) = \mu_{21}(9r + 6(1 - r)) + (1 - \mu_{21})(9(1 - r))$$

$$= r(12p - 9) + 9 - 3p \implies r = \begin{cases} 1, & p > \frac{3}{4} \\ 0, & p < \frac{3}{4} \\ \text{arb.,} & p = \frac{3}{4} \end{cases}$$

$$\mu_{22} = q \implies E_{II}(s, 2:2) = s(12q - 9) + 9 - 3q \implies s = \begin{cases} 1, & q > \frac{3}{4} \\ 0, & q < \frac{3}{4} \\ \text{arb.}, & q = \frac{3}{4} \end{cases}$$

$$E_I(p, 1:1) = p(9r - 3) + 6 \implies p = \begin{cases} 1, & r > \frac{1}{3} \\ 0, & r < \frac{1}{3} \\ \text{arb.}, & r = \frac{1}{3} \end{cases}$$

$$E_I(q, 1:2) = q(-3 - 3s) + 6 \implies q = 0 \text{ always}.$$

Now we can put this all together.

If $r > \frac{1}{3}$. $\implies p = 1 \implies r = 1$. If $r < \frac{1}{3} \implies p = 0 \implies r = 0$. If $r = \frac{1}{3} \implies p = \text{arb.} \implies p = \frac{3}{4}$. And in all cases, $q = 0$ and so $s = 0$. We conclude that we have 3 PBEs:

- $p = 1, r = 1, q = 0, s = 0$ corresponding to (TB, LR).
- $p = 0, r = 0, q = 0, s = 0$ corresponding to (BB, RR).
- $p = \frac{3}{4}, r = \frac{1}{3}, q = 0, s = 0$ corresponding to $((\frac{3}{4}T, B), (\frac{1}{3}L, R))$.

Example 3.22 Consider the game in which player I may be one of two types t_1 or t_2 with equal probability. Player II is unaware of player I's type. If player I is of type t_1 or of type t_2. Players move simultaneously. The extensive form of this game in which Nature determines the type of player I is in Figure 3.33.

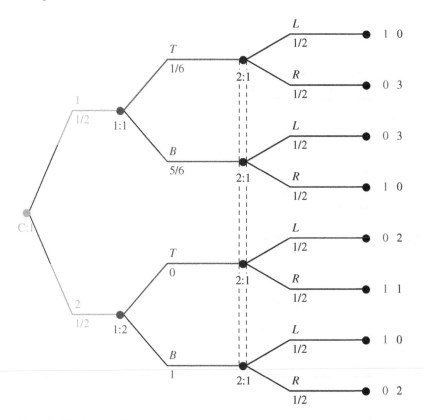

Figure 3.33 Nature chooses type.

The strategic form of the game along side the separated type games is

	L	R
TT	$\left(\frac{1}{2},1\right)$	$\left(\frac{1}{2},2\right)$
TB	$(1,0)$	$\left(0,\frac{5}{2}\right)$,
BT	$\left(0,\frac{5}{2}\right)$	$\left(1,\frac{1}{2}\right)$
BB	$\left(\frac{1}{2},\frac{3}{2}\right)$	$\left(\frac{1}{2},1\right)$

$Prob(t_1) = \frac{1}{2}$	L	R
T	$(1,0)$	$(0,3)$,
B	$(0,3)$	$(1,0)$

$Prob(t_2) = \frac{1}{2}$	L	R
T	$(0,2)$	$(1,1)$
B	$(1,0)$	$(0,2)$

To see how the matrix form of the game is obtained, let's consider *BB* versus *L*. The expected payoff to player I and player II is then, respectively,

$$\frac{1}{2}(0) + \frac{1}{2}(1) = \frac{1}{2} \quad \text{and} \quad \frac{1}{2}(3) + \frac{1}{2}(0) = \frac{3}{2}.$$

The remaining entries in the matrix are derived in the same way.

This game has a mixed NE given by $X^* = (0, \frac{1}{6}, 0, \frac{5}{6})$, $Y^* = (\frac{1}{2}, \frac{1}{2})$. This means player I plays *TB* with probability $\frac{1}{6}$ and *BB* with probability $\frac{5}{6}$ and player II plays *L* or *R* with equal probability. That is, player I will play *T* if she is type I and *B* if she is type 2 with probability $\frac{1}{6}$. With probability $\frac{5}{6}$, player I will play *B* if she is type 1 and also *B* if she is type 2. The expected payoffs to each player is $\frac{1}{2}$ to player I and $\frac{5}{4}$ to player II.

There are more NEs for player I. Player I also has the mixed strategy $X = (\frac{2}{3}, 0, \frac{1}{3}, 0)$ with resulting payoff $\frac{1}{2}$ to I and $\frac{3}{2}$ to II. In addition, $X = (0, \frac{4}{9}, \frac{5}{9}, 0)$ and $X = (\frac{1}{3}, 0, 0, \frac{2}{3})$ are also NE with payoffs $(\frac{1}{2}, \frac{25}{18})$, and $(\frac{1}{2}, \frac{4}{3})$, to players I and II, respectively. Because of convexity, any convex combination of the NEs is also an NE. Player I always gets $\frac{1}{2}$ but player II's payoff depends on the particular NE. Note that this game has no proper subgames.

Now we will find the Perfect Bayes Equilibrium and see how to get sequential rationality. This is very similar to deriving a subgame perfect NE but we have to deal with information sets and beliefs as to which node was hit in that set.

We start at the end of the tree with player II. Let p denote the probability player II plays *L*. Denote by q the probability player I plays T if her type is t_1 and r if she is type t_2. For info set $2:1$ there are 4 nodes and player II does not know at which node player I arrived but there are probabilities (beliefs) assigned to each node by player II. We may now calculate the beliefs of player II that she is at a particular node in info set $2:1$. We have at each node within $2:1$,

$$\mu_1 = \frac{\frac{1}{2}q}{\frac{1}{2}(q+(1-q)+r+(1-r))} = \frac{1}{2}q \qquad \mu_2 = \frac{1}{2}(1-q)$$

$$\mu_3 = \frac{1}{2}r \qquad \mu_4 = \frac{1}{2}(1-r)$$

Next we calculate the expected payoff to player II based on these beliefs as

$$E_{II}(p, 2:1) = \mu_1(3(1-p)) + \mu_2(3p) + \mu_3(2p+1-p) + \mu_4(2(1-p))$$
$$= p(\frac{1}{2} - 3q + \frac{3}{2}r) + 3\mu_1 + \mu_3 + 2\mu_4.$$

This is maximized when $p = \begin{cases} 1, & \frac{1}{2} - 3q + \frac{3}{2}r > 0, \\ 0, & \frac{1}{2} - 3q + \frac{3}{2}r < 0, \\ \text{arb.,} & \frac{1}{2} - 3q + \frac{3}{2}r = 0. \end{cases}$

Now we calculate the expected payoff to player I at $1:1$ and $1:2$ to get

$$E_I(q, 1:1) = qp + (1-q)(1-p) = q(2p-1) - p + 1$$
$$E_I(r, 1:2) = r(1-p) + (1-r)p = r(1-2p) + p.$$

If $p = 1$ then $E_I(q, 1:1)$ is maximized when $q = 1$ and $E_I(r, 1:2)$ is maximized when $r = 0$. But then $\frac{1}{2} - 3q + \frac{3}{2}r = -\frac{5}{2} < 0$ which implies $p = 0$ and $p \neq 1$. So $p < 1$.

If $p = 0$ then $E_I(q, 1:1)$ is maximized when $q = 0$ and $E_I(r, 1:2)$ is maximized when $r = 1$. But then $\frac{1}{2} - 3q + \frac{3}{2}r = 2 > 0$ which implies $p = 1$ and $p \neq 0$. So $0 < p < 1$.

Finally, for $0 < p < 1$ we must have $\frac{1}{2} - 3q + \frac{3}{2}r = 0$, or $q = \frac{1}{6} + \frac{1}{2}r$.

Now that we know $0 < p < 1$ suppose that $p < \frac{1}{2}$. In this case $E_I(q, 1:1)$ is maximized for $q = 0$ and $E_I(r, 1:2)$ is maximized for $r = 1$. But we must have $q = \frac{1}{6} + \frac{1}{2}r$ which is not true for $q = 0$, $r = 1$. But $p > \frac{1}{2}$ also implies that $q = 1, r = 0$ and again $q \neq \frac{1}{6} + \frac{1}{2}r$. This forces $p = \frac{1}{2}$.

We have derived that the equilibrium which is consistent with the beliefs of player II at info set $2:1$ and which maximizes the expected payoff at each info set is player II plays L or R with probability $\frac{1}{2}$ and player I plays T with probability q if type t_1 and r if type t_2 but q and r are restricted to satisfy $q = \frac{1}{6} + \frac{1}{2}r$. Since both $0 \leq q \leq 1, 0 \leq r \leq 1$, we must have $\frac{1}{6} \leq q \leq \frac{4}{6}$. Any such equilibrium is a sequential equilibrium and a Perfect Bayes equilibrium. It is an exercise to determine which, if any, of the NEs mentioned above, are not PBEs.

Example 3.23   Selten's Horse. This is a famous example due to R. Selten[1] which shows again that a NE need not be a PBE. If you redraw the tree in Figure 3.34, you will see it can be put into the outline of a horse.

The game matrix is

III plays L	c	d	III plays R	c	d
C	$(1,1,1)$	$(4,4,0)$	C	$\boxed{(1,1,1)}$	$(0,0,1)$
D	$\boxed{(3,3,3)}$	$(3,3,3)$	D	$(0,0,0)$	$(0,0,0)$

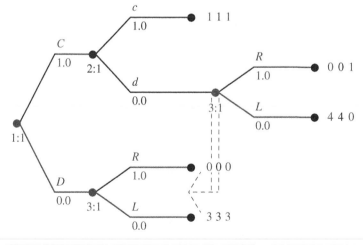

Figure 3.34   Selten's horse.

This is a three-player game. Player I moves first and chooses C or D. Player II, knowing I chose C then chooses either c or d. Player III must make a move knowing only that either player II chose d or player I chose D. Player III needs to have beliefs, labeled μ_{31} when he is at info set $3:1$ about whether he arrived there from player I playing D or player II playing d. It is clear that the tree form of this game is much easier to follow than the matrices–the extensive form is better than the strategic form.

If player I plays D with probability p and player II plays d with probability q, then the probability of reaching the top node of $3:1$ is $(1-p)q$ and the bottom node is p. According to Bayes rule the belief of player III that player I chose D is

$$\mu_{31}(\text{lower}) = Prob(\text{lower}|3:1)$$
$$= \frac{Prob(3:1|\text{lower})Prob(\text{lower})}{Prob(3:1|\text{lower})Prob(\text{lower}) + Prob(3:1|\text{upper})Prob(\text{upper})}$$
$$= \frac{p}{p+(1-p)q}.$$

Of course if $q = 0$ then $\mu_{31}(\text{lower}) = 1$. Info sets that are not reached do not have any restriction on the beliefs. Notice too that the probability of reaching the top node of $3:1$, $(1-p)q$, plus the probability of reaching the bottom node, p, do not add to 1, and they don't have to. That is the difference between these probabilities and the beliefs. In Selten's horse, if player I played D, then $\mu_{31}(\text{lower}) = 1$ since if $3:1$ is reached, player III believes it was because player I played D. There's no other way for $3:1$ (lower) to be reached. Summarizing, we have

$$\mu_{31}(\text{upper}) = \frac{(1-p)q}{(1-p)q+p}, \quad \text{and} \quad \mu_{31}(\text{lower}) = 1 - \mu_{31}(\text{upper}) = \frac{p}{(1-p)q+p}$$

There are 2 pure NEs (C,c,R) and (D,c,L). However, let's consider the NE (D,c,L). If player I plays D then $\mu_{31}(\text{lower}) = 1$ and the best play for III is L. If that happens, player II will play d and not c. In other words, player II looks at what happens if player I should choose D and anticipates that player III will play L. This leads player II to decide to play d. Therefore (D,c,L) is ruled out by sequential rationality and it is not a Bayesian Equilibrium even though it is a NE.

It is not hard for you to check that (C,c,R) is indeed a PBE and we leave that as an exercise.

Example 3.24 How Do We Distinguish Perfection? Consider the game

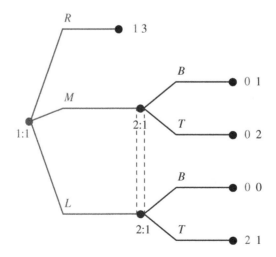

	B	T
R	(1, 3)	(1, 3)
M	(0, 1)	(0, 2)
L	(0, 0)	(2, 1)

In this game, player I gets 1 with certainty or plays the game with player II with matrix

	B	T
M	(0, 1)	(0, 2)
L	(0, 0)	(2, 1)

What would you do if you were player I? Clearly (L, T) is a strict NE so it seems that player I should always play L and player II should always play T in this game.

Let's find all Perfect Bayesian equilibria. Let's label the probability of player I playing R as p, M as q and player II playing B as a and playing T as $1 - a$. The beliefs of player II at $2:1$ are $\mu_1 = \frac{q}{1-p}$ and $\mu_2 = \frac{1-p-q}{1-p}$. Then, the expected payoff to player II at $2:1$ is

$$E_{II}(a, 2:1) = \mu_1(a \cdot 1 + (1 - a)2) + \mu_2(a \cdot 0 + (1 - a)1)$$
$$= \frac{q}{1-p}(2 - a) + \frac{1-p-q}{1-p}(1 - a)$$
$$= 1 - a + \frac{q}{1-p}.$$

This is maximized by taking $a = 0$ which means player II will always play T. However, it is undefined if $p = 1$ meaning that if player I plays R it doesn't matter what a is. If $p < 1$, we have

$$E_I(a, 2:1) = \mu_1(a \cdot 0 + (1 - a)0) + \mu_2(a \cdot 0 + (1 - a)2)$$
$$= \frac{1-p-q}{1-p}(2 - 2a) = 2 - 2a - \frac{q}{1-p}(2 - 2a).$$

If $a = 0$ then $E_I(0, 2:1) = 2\frac{1-p-q}{1-p}$. Now let's consider what happens at $1:1$.

$$E_I(p, q, 1:1) = p \cdot 1 + q(a \cdot 0 + (1 - a) \cdot 0) + (1 - p - q)(a \cdot 0 + (1 - a)2)$$
$$= p + (1 - p - q)(2 - 2a).$$

If $a = 0$ we have $E_I(p, q, 1:1) = p + (1 - p - q)2 = 2 - p - 2q$. This is maximized when $p = 0, q = 0$. We conclude that the unique sequential equilibrium is (L, T).

There are two pure NE for this game (R, B) and (L, T). There is one mixed NE $(R, (\frac{1}{2}B, \frac{1}{2}T))$. Player II will not get a move if player I chooses R. If player II does get a move she should choose T because T dominates B. We know that player II would never want to play B. Therefore, the NE (R, B) is a bad equilibrium and should somehow be eliminated even though we cannot eliminate it due to subgame perfection. This is where the idea of a Perfect Bayesian equilibrium comes in.

If we look at the beliefs consistent with the NE (R, T) we have the probability of being at any node in info set $2:1$ is 0, so the belief of player 2 is that $2:1$ will never be reached. That means that even though (R, T) is a NE, it is not a Bayes or Perfect Bayes equilibrium.

For (L, T), the beliefs for player 2 are 0 at the top node and 1 at the bottom node. This means that the beliefs are consistent with the NE and $2:1$ has a positive probability of being reached, so it is a Perfect BE.

This means that (L, T) is the unique PBE and we have eliminated the bad NE equilibrium.

3.4.1.1 Separating and Pooling PBEs

When the player types are uncertain, there are two types of PBEs:

Definition 3.4.2 *An equilibrium is separating if the types of a player result in different behaviors. An equilibrium is pooling if the types result in the same behavior.*

The next example illustrates pooling and separating equilibria.

Example 3.25 The Gift Game.
Player I is either a friend (F) (with probability p) or an enemy (E) (with probability $1 - p$). Player I can either give a gift if he is a friend (GF) or enemy (GE), or not give if he is either type (NF or NE). Player II can either accept the gift (A), or reject it (R). The payoffs are shown in Figure 3.35. Player II's payoff depends on player I's type. The model includes the fact that player II is unaware of player I's type but player II is aware if player I sent a gift. The figure chooses $p = \frac{2}{3}$ but we will work with a general $0 < p < 1$.

The strategic form of this game for general p is

	A	R
NF NE	$(0,0)$	$\boxed{(0,0)}$
NF GE	$(1-p,-1+p)$	$(-1+p,0)$
GF NE	(p,p)	$(-p,0)$
GF GE	$\boxed{(1,2p-1)}$	$(-1,0)$

Let p =probability player I is a Friend, r =probability of not giving a gift if she is a Friend and s =probability of not giving a gift if she is an enemy. Let q =probability player II will accept a gift and μ be the belief of player II at $2:1$ of the type of player I. We have, at the top node of $2:1$,

$$\mu = \mu_{21} = \frac{p(1-r)}{p(1-r)+(1-p)(1-s)}.$$

The expected payoff of player II at $2:1$ is

$$E_{II}(q, 2:1) = \mu(q) + (1-\mu)(-q) = q(2\mu - 1)$$

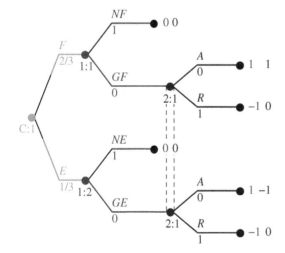

Figure 3.35 The Gift game.

which is maximized for $q = \begin{cases} 1, & \mu > \frac{1}{2} \\ 0, & \mu < \frac{1}{2} \\ \text{arb.,} & \mu = \frac{1}{2}. \end{cases}$ Next, the expected payoff to player I at $1:1$ and $1:2$ is

$$E_I(r, 1:1) = p(1-r)(2q-1), \qquad E_I(s, 1:2) = (1-p)(1-s)(2q-1).$$

If $q > \frac{1}{2}$, these are maximized for $r = 0, s = 0$, respectively, which then implies $q = 1$ and $\mu = p > \frac{1}{2}$. If $q < \frac{1}{2}$ the maximum occurs for $r = 1, s = 1$, respectively, and μ is indeterminate. If $q = \frac{1}{2}$ both r and s are arbitrary.

Considering maximizing $E_{II}(q, 2:1)$, $q > \frac{1}{2}$ iff $q = 1, \mu > \frac{1}{2}, r = 0, s = 0$, and $q < \frac{1}{2}$ iff $q = 0, \mu < \frac{1}{2}, r = 1, s = 1$. Thus there are two pure PBE, namely $((GF, GE), A)$ and $((NF, NE), R)$. Note that both of these are pooling PBEs for any value of p. The pure equilibrium $(GF, GE), A)$ will be played if the probability that player I is a Friend is greater than $\frac{1}{2}$.

In the case $q = 0, \mu < \frac{1}{2}, r = 1, s = 1$ by definition of μ we have $\mu = 0/0$ which means the beliefs are indeterminate because $2:1$ is never reached.

Finally, if $q = \frac{1}{2}$, r and s are arbitrary but $\mu = \frac{1}{2}$. Equivalently, $\frac{p}{1-p} = \frac{1-s}{1-r}$. In this case, we have a continuum of PBEs given by player II plays A or R with probability $q = \frac{1}{2}$ and player I plays NF with probability r if she is a Friend and NE with probability s if she is an enemy, with $\frac{p}{1-p} = \frac{1-s}{1-r}$. For instance, if $p = \frac{2}{3}$ then $s = 2r - 1$ and $\frac{1}{2} \le r \le 1$.

There are no separating equilibria. To see this suppose player I plays (NF, GE). In this case $r = 1, s = 0$ and so $\mu = 0$. Since GE is played by I, this results in player II choosing $q = 0$, i.e., R. But then $s = 1$ which means player I would prefer NE when her type is enemy. This contradicts $s = 0$ and this cannot be sequentially rational. The remaining cases are similar.

In summary:

- There is no separating PBE.
- There is always a pooling PBE in which no gifts are given because player II believes receiving a gift signals I is an enemy.
- If the probability I is a Friend ($p \ge \frac{1}{2}$), then there is a pooling PBE where I always gives a gift and II accepts the gift.

For a given equilibrium in a given extensive game, an information set is on the equilibrium path if it will be reached with positive probability if the game is played according to the equilibrium strategies, and is off the equilibrium path if it is certain not to be reached if the game is played according to the equilibrium strategies.

Example 3.26 The game we consider is in Figure 3.36.

We want to find all sequential equilibria. To do this, we must maximize each player's expected payoff at each node and to do that we work backward calculating the beliefs that a player is at a particular node when the info set has more than two nodes.

At $2:2$, it is clear that the best choice for player II is action 2, so $y = 0$. Next we consider info set $1:2$. The beliefs for this info set are

$$\mu_{12} = \frac{0.95pq}{0.95pq + 0.05p} = \frac{19q}{19q + 1}, \quad \text{and} \quad 1 - \mu_{12}.$$

Here $p =$ probability of playing action 1 at $1:1$ and $q =$ probability of playing action 1 at $2:1$. Then player I has expected payoff on $1:2$ as

$$E_I(\mu_{12}) = \mu_{12}(3 \cdot x + (1-x)4) + (1 - \mu_{12})(8 \cdot x + (1-x)4) = x(4 - 5\mu_{12}) + 4,$$

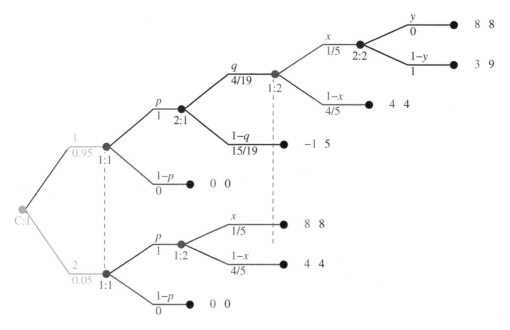

Figure 3.36 Sequential Equilibrium

where x = probability player I plays action 1 at $1:2$. This is maximized when

$$x = \begin{cases} 1, & \mu_{12} < \frac{4}{5}, \\ 0, & \mu_{12} > \frac{4}{5}, \\ \text{arb.}, & \mu_{12} = \frac{4}{5}. \end{cases}$$

Since $\mu_{12}(1) = \frac{19q}{19q+1}$ this optimal x can be expressed in terms of q. The calculation gives

$$x = \begin{cases} 1, & q < \frac{4}{19}, \\ 0, & q > \frac{4}{19}, \\ \text{arb.}, & q = \frac{4}{19}. \end{cases}$$

Next the expected payoff for player II at $2:1$ is

$$E_{II}(2:1) = q(9 \cdot x + (1-x)4) + (1-q)5 = q(5x-1) + 5,$$

which is maximized for $q = \begin{cases} 1, & x > \frac{1}{5}, \\ 0, & x < \frac{1}{5}, \\ \text{arb.}, & x = \frac{1}{5}. \end{cases}$

The choices at $2:1$ combined imply the following:

$$x < \tfrac{1}{5} \implies q = 0 \implies x = 1 \qquad \text{impossible}$$
$$x > \tfrac{1}{5} \implies q = 1 \implies x = 0 \qquad \text{impossible}$$
$$x = \tfrac{1}{5} \implies q = \tfrac{4}{19}, \mu_{12} = \tfrac{4}{5}, \qquad \text{this is ok.}$$

We conclude that we must have $x = \frac{1}{5}, q = \frac{4}{19}, \mu_{12} = \frac{4}{5}$.

The expected payoffs to players I and II are then given by

$$E(1:2, \text{lower}) = (E_I(1:2, \text{lower}), E_{II}(1:2, \text{lower})) = \tfrac{1}{5}(8,8) + \tfrac{4}{5}(4,4) = (\tfrac{24}{5}, \tfrac{24}{5})$$

$$E(1:2, \text{upper}) = (E_I(1:2, \text{upper}), E_{II}(1:2, \text{upper}))\tfrac{1}{5}(3,9) + \tfrac{4}{5}(4,4) = (\tfrac{19}{5}, 5)$$

$$E(2:1) = \tfrac{4}{19}(\tfrac{19}{5}, 5) + \tfrac{15}{19}(-1, 5) = (\tfrac{1}{5 \cdot 19}, 5)$$

The next step is to calculate the expected payoff for player I at $1:1$. We have (referring to the tree)

$$E_I(1:1) = 0.95 p E_I(2:1) + 0.05 p E_I(1:2) = 0.95 p \tfrac{1}{5 \cdot 19} + 0.05 p \tfrac{24}{5} = \tfrac{p}{4}.$$

This is maximized when $p = 1$. This means player I should always play action 1. The expected payoff is $\tfrac{1}{4}$ for player I and

$$E_{II} = 0.95 \cdot 5 + 0.05 \tfrac{24}{5} = 4.99$$

for player II. The only sequential equilibrium is therefore $p = 1, x = \tfrac{1}{5}, q = \tfrac{4}{19}, y = 0, \mu_{12} = \tfrac{4}{5}$. This is depicted in Figure 3.36.

Example 3.27 How Uncertainty Affects the Used Car Market. In a used car market there are only two types of cars: poor-quality cars (lemons) or good quality cars (peaches). If all information about the cars is known, which is really unrealistic, the cost of production is c_p for peaches and c_ℓ for lemons where $c_p > c_\ell$.

In the usual case, the quality of the car is not known by the buyer. Suppose that the probability a car is a lemon is $0 < b < 1$ and the probability it is a peach is $1 - b$. This is known by everyone. If a car costs c to produce and p to buy then the payoff to the buyer is

$$u = \begin{cases} c - p, & \text{if the deal is made} \\ 0, & \text{otherwise} \end{cases}$$

Unfortunately the cost to produce a car is unknown exactly and can only be considered as a random variable, say X. Assume $Prob(X = c_\ell) = b$ and $Prob(X = c_p) = 1 - b$. The expected payoff to the buyer is then

$$E(u) = E(X) - p = b c_\ell + (1 - b)c_p - p = b(c_\ell - c_p) - (p - c_p),$$

for a given observed price p. It is clear that $E(u) \geq 0$ if and only if $b(c_\ell - c_p) \geq p - c_p$, or $0 \leq b(c_p - c_\ell) \leq c_p - p$. Therefore $E(u) > 0$ if and only if $c_p > p$, the cost to produce a peach is greater than the price to buy a car. This means no buyer would ever pay the price for a peach and no seller would ever offer a peach for sale at less than c_p. The result is that there will not be a market for peaches and the entire market will be lemons. But we know that in real life that isn't true.

What's wrong here? The problem is the asymmetrical information available to the players. The seller knows the condition of the car while the buyer does not. Would you pay a premium for a car if there is a chance the car is a lemon? Since sellers of peaches will not sell the car below the premium price no peaches will be sold and the market will only be for lemons.This is an example of what economists call the **lemon principle** and comes up in more than just the used car market. For example in investments, the subjective value of a stock may differ between buyers and sellers. In online sales, the risk of the quality of an item or other risks unknown to the buyer, may lead to the lemon problem.

Several solutions to this problem have been proposed, including certain guarantees, or more information to the buyer. Let's consider a specific situation.

The Market for Lemons. A seller wants to sell his old car. The seller is aware of the quality of the car; the buyer is not. The car is valued as either High quality, i.e., a peach (with probability $\tfrac{4}{5}$)

or Lemon quality (probability $\frac{1}{5}$). If the car is rated as High, the buyer values it at $20,000 while the seller values it at $10,000 because the buyer needs the car and the seller wants to unload it. If the car is a lemon, both buyer and seller personally value it as 0. The seller's asking price for a car is either $5000 or $15,000. The buyer can accept or reject and there is no negotiation.

In order to make the car more appealing, the seller has the option to get the car inspected at a cost of $200. If the car is a lemon, the seller needs to pay $15,200 to have his car fixed including the inspection. In this case, the buyer's valuation for the serviced car is $20,000 and the seller's is $10,000. Now, the seller has four choices: $5000 with inspection (5i), $15,000 with inspection (15i), $5000 without inspection (5), and $15,000 without inspection (15).

The game tree for this problem to keep this all straight is in Figure 3.37

Let's see how the payoffs are obtained. If the car is High quality and player I, the seller, asks 20K and player II, the buyer Accepts the deal, player I nets 4.8K because he values the car at 15K plus the cost of the inspection leads to $20 - 15.2 = 4.8$. The payoff to player II will be 5K because she

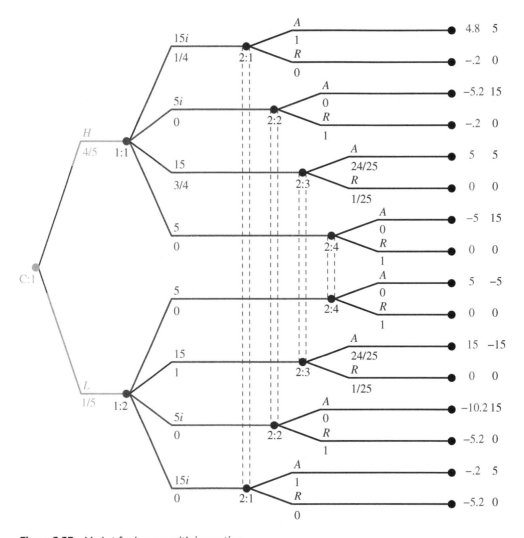

Figure 3.37 Market for lemons with inspection.

pays 15K for the car which is valued at 20K. If player II rejects the deal, player I loses the $200 inspection fee and there is no cost to player II.

If the car is a peach and the seller plays 5i, then her payoff if player II accepts is −5.2K because the seller values the car at 10K but she sells it for only 5K plus the cost of the inspection. The buyer has a payoff of 15K because she values the car at 20K and she has paid 5K.

If the car is a lemon and player I asks 15K without inspection and player II accepts the deal, then player I gains 15K because he values the car at 0 and player II loses 15K because he just bought a car worth nothing. The remaining payoffs are similar.

There are three BEs for this game listed.

1. If at $1:1$ play 15; if at $1:2$ play 15 so offer to sell the car for 15K whether peach or lemon. The Buyer (player II) should play A, i.e., buy the car, giving expected payoff $5\frac{4}{5} + 15\frac{1}{5} = 7$ to the seller, and $5\frac{4}{5} - 15\frac{1}{5} = 1$ to the buyer. For this BE, the consistent beliefs are $\mu_{23H} = \frac{4}{5}, \mu_{23L} = \frac{1}{5}$. Off the equilibrium path the beliefs are indeterminate. This is a pooling BE.

   In words, the seller should offer the car at 15K whether it is High or Low quality. The Buyer should Accept the offer. The option to get it inspected should be declined.

2. If at $1:1$ play 15i with probability $\frac{1}{4}$ and 15 with probability $\frac{3}{4}$; if at $1:2$ play 15. The Buyer (player II) should play A if the offer is 15i. If the offer is 15, the buyer should accept with probability $\frac{24}{25}$.

   For this BE, the beliefs are $\mu_{21H} = 1, \mu_{23H} = \frac{3}{4}$. The expected payoff is $\frac{168}{25}$ to the seller, and 1 to the buyer. This is a separating BE. To see the calculation for the expected payoff, we have for player I, the seller

$$\frac{4}{5}\left(\frac{1}{4}(4.8) + \frac{3}{4}(\frac{24}{25}5 + \frac{1}{25}(0))\right) + \frac{1}{5}\left(1(\frac{24}{25}15 + \frac{1}{25}0)\right) = 6.72.$$

   and for the buyer

$$\frac{4}{5}\left(\frac{1}{4}(5) + \frac{3}{4}(\frac{24}{25}5 + \frac{1}{25}(0))\right) + \frac{1}{5}\left(1(\frac{24}{25}(-15) + \frac{1}{25}0)\right) = 1.$$

   The Seller should always offer the car at 15K, but if it is High quality the seller should also offer the inspection 25% of the time.

3. If at $1:1$ play 15i; if at $1:2$ play 15. The Buyer (player II) should play A if the offer is 15i and R if the offer is 15. The expected payoff to the Seller is $\frac{96}{25}$ and 4 to the Buyer. This is also a separating BE.

   In this BE, the Seller should offer the car at 15K with the inspection option if it is High quality. If it is a Lemon, the Seller should offer the car at 15K without the inspection. The Buyer should Accept the price with the inspection and Reject the deal if the price is 15K and no inspection. In all cases, the Buyer should accept the price with an inspection and reject the car without an inspection. The seller signals it is a lousy car if it is offered without an inspection.

You should work through the calculations to see which if any of these are PBEs. This example shows that there will be a market for peaches if an inspection is offered.

Example 3.28   Three Games. Suppose we are given the following three game matrices. Nature selects the game with the given probability.

Game 1 *Prob $\frac{1}{3}$*				Game 2 *Prob $\frac{1}{3}$*				Game 3 *Prob $\frac{1}{3}$*		
	L	R			L	R			L	R
T	$(0,0)$	$(6,-1)$		T	$(1,3)$	$(0,0)$		T	$(2,-2)$	$(-2,2)$
B	$(-1,6)$	$(4,4)$		B	$(0,0)$	$(3,1)$		B	$(-2,2)$	$(2,-2)$

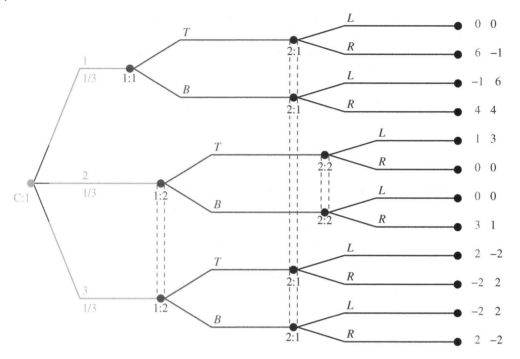

Figure 3.38 3 Games.

The wrinkle in this setup is that player I knows if Nature has selected Game 1 or Game 2 or 3 but not which of Games 2 or 3 if not Game 1. Similarly, player II knows if Nature has selected Game 2 or Game 1 or 3 but not which of the two if not Game 2. The two players make their plays simultaneously. We want to find the NE and verify it is a PBE. First, here is the game tree in Figure 3.38.

To find the PBE, let p =probability player I plays T if at $1:1$, q =probability I plays T if at $1:2$. Let r =probability II plays L if at $2:1$ and s =probability II plays L at $2:2$. We need to find p, q, r, s.

We begin with $2:1$. The beliefs of player II at $2:1$ are

$$\mu_{21}^1 = \frac{p}{p+1-p+q+1-q} = \frac{p}{2}, \quad \mu_{21}^2 = \frac{1-p}{2}, \quad \mu_{21}^3 = \frac{q}{2}, \quad \mu_{21}^4 = \frac{1-q}{2}.$$

and then the expected payoff to II is

$$E_{II}(r, 2:1) = \tfrac{1}{2}r(-p - 8q + 6) + \tfrac{1}{2}(-5p + 4q + 2).$$

This is maximized when $r = \begin{cases} 1, & 6 > p + 8q \\ 0, & 6 < p + 8q, \\ \text{arb.,} & 6 = p + 8q. \end{cases}$

In a similar way we calculate $E_{II}(s, 2:2)$ is maximized when $s = \begin{cases} 1, & q > \frac{1}{4} \\ 0, & q < \frac{1}{4}, \\ \text{arb.,} & q = \frac{1}{4}. \end{cases}$

Next $E_I(q, 1:2)$ is maximized when $q = \begin{cases} 1, & 4s + 8r > 7 \\ 0, & 4s + 84 < 7, \\ \text{arb.,} & 4s + 8r = 7. \end{cases}$

$E_I(p, 1:1)$ is maximized when $p = \begin{cases} 1, & r < 2 \\ 0, & r > 2, \\ \text{arb.}, & r = 2. \end{cases}$

We put these all together. The conditions on r require that $p = 1$ since $0 \le r \le 1$. Replacing p by

1 gives, $r = \begin{cases} 1, & q < \frac{5}{8} \\ 0, & q > \frac{5}{8}, \\ \text{arb.}, & q = \frac{5}{8}. \end{cases}$

If $q > \frac{5}{8} \implies r = 0, s = 1 \implies 4s + 8r = 4 < 7 \implies q = 0$, a contradiction.

If $\frac{1}{4} < q < \frac{5}{8} \implies r = 1, s = 1 \implies 4s + 8r = 12 > 7 \implies q = 1$, a contradiction.

If $q < \frac{1}{4} \implies r = 1, s = 0 \implies 4s + 8r = 8 > 7 \implies q = 1$, a contradiction.

The only possibility left for q is $q = \frac{5}{8} \implies r = \text{arb.}, s = 1 \implies 4s + 8r = 4 + 8r$. If $4 + 8r > 7 \implies r > \frac{3}{8} \implies q = 1$, a contradiction. If $4 + 8r < 7 \implies r < \frac{3}{8} \implies q = 0$, a contradiction. The only possibility is $4 + 8r = 7 \implies r = \frac{3}{8}$. Since $q = \frac{5}{8}$ we also get $s = 1$, and we are done.

The PBE is $p = 1, q = \frac{5}{8}, r = \frac{3}{8}, s = 1$. In words, the PBE says that if Game 1 is played, player I should play T while if Game 2 or 3 is played player 1 should play T with probability $\frac{5}{8}$ and B with probability $\frac{3}{8}$. Player II should play L if Game 2 is played and L with probability $\frac{3}{8}$, R with probability $\frac{5}{8}$ if Game 1 or 3 is played.

Once we have the beliefs it is easy to calculate the expected payoffs. For example, for player I we have

$$(0\frac{3}{8} + 6\frac{5}{8})\frac{1}{2} + ((-1)\frac{3}{8} + 4\frac{5}{8})0 + (2\frac{3}{8} + (-2)\frac{5}{8})\frac{5}{16} + ((-2)\frac{3}{8} + 2\frac{5}{8})\frac{3}{16} + 1\frac{5}{8} = \frac{17}{12}.$$

Problems

3.35 In the Gift Game, Example 3.25 suppose that the probability the sender is a Friend is $\frac{2}{3}$, and suppose

- The sender's payoff is $\begin{cases} 1, & \text{if his gift is accepted,} \\ -1, & \text{if his gift is rejected,} \\ 0, & \text{if he does not give any gift.} \end{cases}$

- If the sender is a friend, then the receiver's payoff is $\begin{cases} 1, & \text{if they accept,} \\ 0, & \text{if they reject.} \end{cases}$

  If the sender is an enemy, then the receiver's payoff is $\begin{cases} 0, & \text{if they accept,} \\ -1, & \text{if they reject.} \end{cases}$

Find all NEs and verify that not all the NEs are PBEs.

3.36 **Signalling Games** Signalling is a two-player game involving the sender and the receiver. In the first stage, Nature chooses the sender's type. In the second stage, the sender knows her type and chooses an action. In the third stage, the receiver observes the sender's action, modifies her beliefs about the sender's type in view of the updated information, and selects an action.

This example is the simplest signalling game giving the main features. The sequence of plays is

- Nature moves first. This determines player I's type is t_1 with probability p and type t_2 with probability $1 - p$.

- Player I makes a move aware of his type, i.e., knowing the choice of Nature. Player I sends a signal by his action.
- Player II observes player I's action, i.e., player II knows which action player I has taken but is unaware of the choice of Nature (and so unaware of the type of player I). The signals player II receives are player I's choice of actions from the info sets.
- Player II makes a move and payoffs determined.

Find the extensive form of this signalling game. Assume Player I has two actions T and B and player II also has two actions L and R. Find all PBEs for this game.

3.37   Beer or Quiche Sequential Equilibrium In the Beer-Quiche game in Figure 3.39

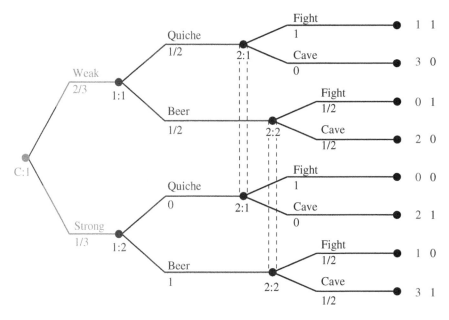

Figure 3.39 Beer or Quiche game.

we have found that the NE is for player I to play Quiche and Beer with probability $\frac{1}{2}$ if player I is weak and to play Beer always if player I is strong. Player II should always Fight if he spots player I with a Quiche, and to Fight or Cave with probability $\frac{1}{2}$ if player I has a Beer. Verify that this is a sequential equilibrium.

3.38   Consider the game with 2 matrices played with equal probability

$\frac{1}{2}$	L	R
T	$(1,1)$	$(0,0)$
B	$(0,0)$	$(0,0)$

$\frac{1}{2}$	L	R
T	$(0,0)$	$(0,0)$
B	$(0,0)$	$(2,2)$

Find the PBE. The game matrix is

	L	R
TT	$\left(\frac{1}{2}, \frac{1}{2}\right)$	$(0, 0)$
TB	$\left(\frac{1}{2}, \frac{1}{2}\right)$	$(1, 1)$
BT	$(0, 0)$	$(0, 0)$
BB	$(0, 0)$	$(1, 1)$

3.39 Find all Bayesian equilibria in the game with two states of nature.

Game 1 Prob $\frac{1}{2}$	L	R
T	$(1, 1)$	$(0, 0)$
B	$(0, 0)$	$(0, 0)$

Game 2 Prob $\frac{1}{2}$	L	R
T	$(0, 0)$	$(0, 0)$
B	$(0, 0)$	$(2, 2)$

Player I knows if Nature has picked Game 1 or game 2, but player II does not. Moves are simultaneous and payoffs are determined by the game drawn by Nature. Produce the game tree.

3.40 Suppose we are given the following three game matrices. Nature selects the game with the given probability.

Game 1 Prob $\frac{9}{13}$	L	R
T	$(2, 2)$	$(0, 0)$
B	$(3, 0)$	$(1, 1)$

Game 2 Prob $\frac{3}{13}$	L	R
T	$(2, 2)$	$(0, 0)$
B	$(0, 0)$	$(1, 1)$

Game 3 Prob $\frac{1}{13}$	L	R
T	$(2, 2)$	$(0, 0)$
B	$(0, 0)$	$(1, 1)$

The players are aware of the following rules.

1. Player I knows if Nature has selected Game 1 or not.
2. Player II knows if Nature has selected Game 3 or not.
3. Players I and II choose simultaneously.

The game tree and matrix form are as follows.

	RR	RL	LR	LL
BB	$(1, 1)$	$\left(\frac{28}{13}, \frac{1}{13}\right)$	$\left(\frac{12}{13}, \frac{12}{13}\right)$	$\left(\frac{27}{13}, 0\right)$
BT	$\left(\frac{4}{13}, \frac{4}{13}\right)$	$\left(\frac{19}{13}, \frac{19}{13}\right)$	$\left(\frac{3}{13}, \frac{3}{13}\right)$	$\left(\frac{18}{13}, \frac{18}{13}\right)$
TB	$\left(\frac{9}{13}, \frac{9}{13}\right)$	$\left(\frac{33}{13}, \frac{6}{13}\right)$	$\left(\frac{11}{13}, \frac{11}{13}\right)$	$\left(\frac{35}{13}, \frac{8}{13}\right)$
TT	$(0, 0)$	$\left(\frac{24}{13}, \frac{24}{13}\right)$	$\left(\frac{2}{13}, \frac{2}{13}\right)$	$(2, 2)$

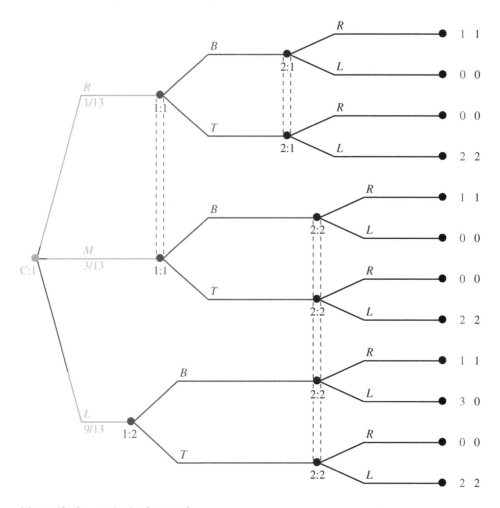

(a) Verify the entries in the matrix.
(b) Find the unique PBE.

3.41 In this version of the end game in poker, there are two players I and II both of whom must ante $1 to play. The deck has 3 cards, (K)ing, (Q)ueen, and (J)ack. A dealer deals out the three cards. PI gets the top card and PII gets the second card. Each player knows their own card but not the other player's card. PI goes first and may (F)old or (B)et an additional $1. If PI folds, PII wins the pot. If PI bets, PII may fold or (C)all the raise. Find the beliefs at each node of each information set and solve this game.

3.42 Consider the following Bayesian game:
1. Nature decides whether the payoffs are as in Matrix I or Matrix II, with equal probabilities.
2. PI is informed of the choice of Nature, PII is not.
3. Choices are made simultaneously.

Solve the game by finding all sequential rational equilibria.

MatrixI	L	R
T	(1,1)	(0,2)
B	(0,2)	(1,1)

MatrixII	L	R
T	(2,2)	(0,1)
B	(4,4)	(2,3)

3.43 **War in WWII.** In 1939 Hitler invaded Czechoslovakia and Britain's prime minister Neville Chamberlain decided to meet with Hitler in Munich in order to discuss Hitler's intentions in the future, in particular Hitler's menacing posture toward Poland. Clearly, Chamberlain had no knowledge of Hitler's payoffs or aims. Hitler can conceal his true intentions to Chamberlain by Lying (and assuring Chamberlain that his territorial ambitions end with Czechoslovakia), or telling the Truth (that his goal is world domination). Chamberlain has the options of believing Hitler and making concessions, or deciding that Hitler is lying and threatening war. Hitler's type is $\{t_1 = L, t_2 = T\}$. Hitler's type is unknown to Chamberlain but he thinks $Prob(Hitler = L) = 0.6$ while $Prob(Hitler = T) = 0.4$. We may model this as an extensive form game in which Nature decides if Hitler is Lying or telling the Truth.

In this example, Chamberlain cannot know if Hitler is lying but Hitler does know if he is lying.

(a) Solve the game in Figure 3.40 and determine what Chamberlain and Hitler should do.

(b) Suppose the probabilities of Hitler lying or telling the truth are reversed. Solve the game. Does anything change?

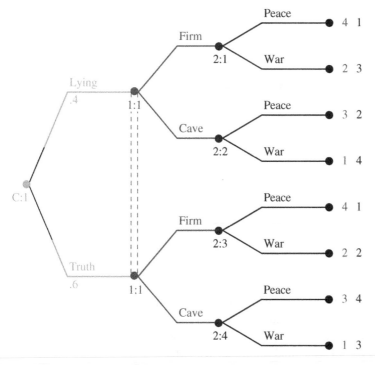

Figure 3.40 Hitler vs Chamberlain.

3.44 The game tree in figure has matrix

	T	B
R	(2, 2)	(2, 2)
M	(1, 1)	(0, 2)
L	(0, 2)	(3, 1)

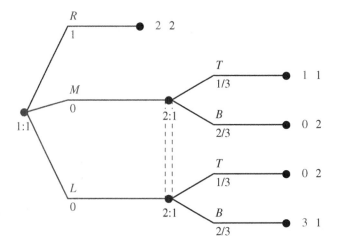

(a) If player I assigns to actions, the probabilities $(\frac{1}{9}, \frac{3}{9}, \frac{5}{9})$ to R, M, L find the belief player II has that player I played M.

(b) What if player I played R with probability 1?

3.45 Consider the following extensive game.

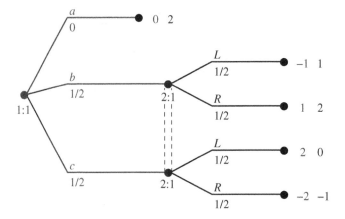

Let μ be the belief of player II at $2:1$ at the top node. Let a be the probability player I plays action 1 and b the probability player I plays action 2 and c the probability she plays action 3, $a + b + c = 1$. Also, let p =probability II plays L. The probabilities below each action represent a particular NE. Find **all** sequential equilibria, i.e., PBEs.

3.46 **Cuban Missile Crisis** In 1962 the USSR was caught placing nuclear missiles in Cuba. JFK had options to use military force or do nothing. It was unknown how the USSR would respond. The matrices are

USSR War	W	D
Invade	$(1, -8)$	$(-10, -4)$
Don't	$(-2, 2)$	$(-2, 2)$

USSR Peace	W	D
Invade	$(1, -4)$	$(-10, -8)$
Don't	$(-2, 2)$	$(-2, 2)$

The probability USSR is of type War is p. Find all PBEs.

3.47 Consider the game in figure:

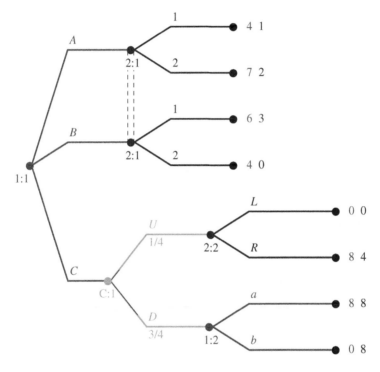

Suppose the beliefs at the information set $2:1$ at the top node is μ and the bottom node is $1 - \mu$. Find all PBEs for this game.

3.48 Suppose a seller wants to sell his used car. The seller knows what is the quality of the car, but the buyer does not. The buyer knows only that the car could be a peach with probability $\frac{2}{3}$ and a lemon with probability $\frac{1}{3}$. If the car is a peach, the buyer's valuation for it is $25,000 and the seller's is $18,000. If it is a lemon, both buyer's and seller's valuations are $0. The seller's asking price is either $10,000 or $18,000. Then, the buyer can accept the offer (buy the car) or reject the offer.
(a) Would there be a separating perfect Bayesian equilibrium in this case?
(b) Find all the pooling perfect Bayesian equilibria if any.

3.49 Consider the Battle of Sexes example.

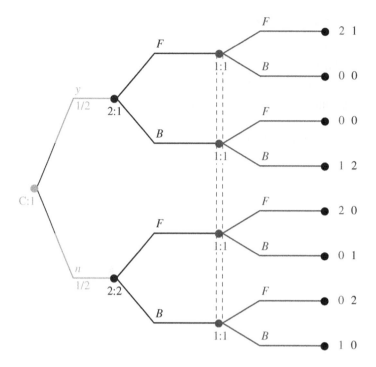

Find all Bayesian Equilibria and the beliefs at each node of information set 1 : 1 for each BE.

3.50 Sometimes when one or both players have more information they may both do worse than lacking the information. Consider the following matrices:

	L	M	R
T	(1, 2)	(1, 0)	(1, 3)
B	(4, 4)	(0, 0)	(0, 5)

	L	M	R
T	(1, 2)	(1, 3)	(1, 0)
B	(4, 4)	(0, 5)	(0, 0)

(a) Suppose that both players only know that each matrix is played with probability $\frac{1}{2}$. Find the Perfect Bayesian Equilibrium and the expected payoffs to each player.

(b) Suppose that player II knows which matrix is played but player I does not. Find the Perfect Bayesian Equilibrium and the expected payoffs and compare to the first part.

3.51 Consider the end games in poker.

(a) **2 Card Poker Endgame** Consider the game with imperfect information where chance deals hands to players 1 and 2. Each player gets two cards, either an A or a K each with probability $\frac{1}{2}$. Each hand therefore has probability $\frac{1}{4}$. Each player knows her own cards but not that of the other player. After each card is dealt, the player can fold or bet. Assume the ante is 4 and the bet to stay in is 6. Also **assume each player always bets if they have an ace.** Let p be the probability for player 1 to bet with a king, and q the probability for player 2 to bet with a king. We will find the PBE (i.e., the sequential equilibrium).

Since a player must bet with an Ace, the only unknown to the opposing player is whether or not the player holds two aces or one ace.

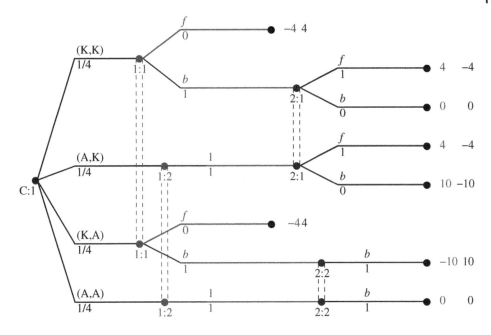

(b) Find the beliefs for player 2 on info set $2:1$ and $2:2$.

(c) Find $E_{II}(q, 2:1), E_{II}(q, 2:2), E_I(p, 1:1)$ and the values of q and p which maximize the expected payoffs.

(d) Find the Perfect Bayesian Equilibrium.

(e) **3 possible cards.** There are two players who each get one card that can be either an ace A, a king K, or a queen Q, so there are 9 equiprobable hands. The ante is 4 and the bet to call is 6 and high card wins the pot (a tie is worth 0). Assume player I bets or folds first. Find the sequential equilibrium.

(f) Now assume that each player will always bet with and ace and always fold with a queen. Find the PBE (sequential equilibrium).

3.52 The market for lemons problem. If the car is a peach, it is worth $3000 to the buyer and $2000 to the seller. If the car is a lemon, it is worth $1000 to the buyer and $0 to the seller. Suppose the probability the car is a peach is $\frac{2}{3}$ and the probability it is a lemon is $\frac{1}{3}$. The payoff matrices for both cases are

Buyer/seller Peach	T	NT
T	$(3000 - p, p - 2000)$	$(0, 2000)$
NT	$(0, 2000)$	$(0, 2000)$

Buyer/seller lemon	T	NT
T	$(1000 - p, p)$	$(0, 0)$
NT	$(0, 0)$	$(0, 0)$

The variable $p \geq 0$ represents the agreed upon price for the car. Find the PBE for a given p and predict the value of p.

3.53 **Two Opponents in Dispute.** The strategies of each player is to Fight or Yield. Player I doesn't know if player II is Strong or Weak and must choose his strategy without that knowledge. Here is the model assuming $Prob(\text{II is Strong}) = p = \frac{1}{3}, Prob(\text{II is Weak}) = 1 - p = \frac{2}{3}$. With $p = \frac{1}{3}$ the Bayesian equilibrium is (F, FY), with expected payoffs $(\frac{1}{3}, \frac{1}{3})$.

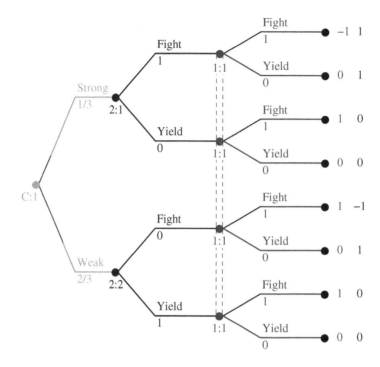

Now take p as a parameter and go through calculating the BE.

3.54 In Example 3.21, we showed in case (1) that there is a unique NE and therefore a PBE which is the same as the NE. Derive the PBE using sequential rationality.

Bibliographic Notes

Extensive games are a major generalization of static games. Without the use of software like Gambit, it is extremely difficult to solve complex games and it is an essential tool in this chapter. The theory of extensive games can get quite complicated and this is where most modern game theory is conducted. Backward induction is also known by the name **dynamic programming**. Bayesian games are also known by the name **stochastic games**. The book by Myerson [29] or [1] is a good reference for further study and of course there are many advanced books on dynamic programming and stochastic games.

Many of the problems of this chapter have appeared at one time or another on exams in game theory given at the London School of Economics. The used car problem is a standard example of how lack of information can impact markets and lead to other mechanisms in economic problems.

4

N-Person Nonzero Sum Games and Games with a Continuum of Strategies

The race is not always to the swift nor the battle to the strong, but that's the way to bet.

–Damon Runyon, More than Somewhat

Infinite striving to be the best is man's duty; it is its own reward. Everything else is in God's hands.

– Mahatma Gandhi

The best weapon against an enemy is another enemy.

– Friedrich Nietzsche

4.1 Motivating Examples

For many problems arising in real-world application, the choices of each player (and there may be more than two) can be any value in an interval, or more generally a continuum set. For example, if firms are competing on the price of a gadget they make, or if two competing firms are deciding where to place their retail outlets, or if countries at war are trying to determine the best time to end the conflict, etc., these are all games with a continuum of strategies. Each player has their own payoff depending on their own and their opponents choice of strategies. Let's start with a specific example.

Example 4.1 The Median Voter Problem. Do politicians pick a position on issues to maximize their votes? Suppose that is exactly what they do (after all, there are plenty of real examples). Suppose that voter preferences on the issue are distributed from $[0, 1]$ according to a continuous probability density function $f(x) > 0$ and $\int_0^1 f(x)\, dx = 1$. The density $f(x)$ approximately represents the percentage of voters who have preference $x \in [0, 1]$ over the issue. The midpoint $x = \frac{1}{2}$ is taken to be middle of the road, while x values in $[0, \frac{1}{2})$ are **leftist or liberal,** and x values in $(\frac{1}{2}, 1]$ are **rightist or conservative.** The question a politician might ask is: "Given f, what position in $[0, 1]$ should I take in order to maximize the votes that I get in an election against my opponent?" The opponent also asks the same question. **We assume that voters will always vote for the candidate nearest to their own positions.**

Let's call the two candidates I and II, and let's take the position of player I to be $q_1 \in [0, 1]$ and for player II, $q_2 \in [0, 1]$. Let V be the random variable that is the position of a randomly chosen voter so that V has continuous density function f.

Game Theory: An Introduction, Third Edition. E. N. Barron.
© 2024 John Wiley & Sons, Inc. Published 2024 by John Wiley & Sons, Inc.

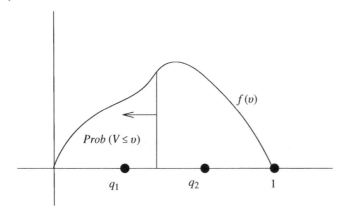

Figure 4.1 Area to the left of V is candidate I's percentage of the vote.

The payoff functions for player I and II is given by

$$
u_1(q_1, q_2) \equiv \begin{cases} Prob(V \le \dfrac{q_1 + q_2}{2}) & \text{if } q_1 < q_2; \\ \dfrac{1}{2} & \text{if } q_1 = q_2; \\ Prob(V > \dfrac{q_1 + q_2}{2}) & \text{if } q_1 > q_2. \end{cases}
$$

$$
u_2(q_1, q_2) \equiv 1 - u_1(q_1, q_2).
$$

Here is the reasoning behind these payoffs. Suppose that $q_1 < q_2$, so that candidate I takes a position to the left of candidate II. The midpoint of

$$
[q_1, q_2] \text{ is } \gamma \equiv \dfrac{q_1 + q_2}{2}.
$$

Now, the voters below q_1 will certainly vote for candidate I because q_2 is farther away. The voters above q_2 will vote for candidate II. The voters with positions in the interval $[q_1, \gamma]$ are closer to q_1 than to q_2, and so they vote for candidate I. Consequently, candidate I receives the total percentage of votes $Prob(V \le \gamma = (q_1 + q_2)/2)$ if $q_1 < q_2$. Similarly, if $q_2 < q_1$, candidate I receives the percentage of votes $Prob(V > \gamma)$. Finally, if $q_1 = q_2$, there is no distinction between the candidates' positions and they evenly split the vote. Recall from elementary probability theory that the probabilities for a given density are given by (see Figure 4.1)

$$
Prob(V \le v) = \int_0^v f(x)\, dx \text{ and } Prob(V > v) = 1 - Prob(V \le v).
$$

The question is what should be the choices of each player in order to maximize their percentage of the vote? In other words, each player wants to choose q_i so as to maximize $u_i(q_1, q_2)$. If we focus on player I, player I chooses q_1 to maximize $u_1(q_1, q_2)$, Clearly, the value of q_1 must depend on q_2. But, player I doesn't know q_2 and player II doesn't want to help maximize u_1 but in fact to minimize it. Here's what we want:

$\quad q_1^*$ should maximize $u_1(q_1, q_2)$,

and

$\quad q_2^*$ should maximize $u_2(q_1, q_2)$.

If we can find such a pair (q_1^*, q_2^*), then

$$u_1(q_1^*, q_2^*) \geq u_1(q_1, q_2^*), \text{ and } u_2(q_1^*, q_2^*) \geq u_2(q_1^*, q_2)$$

for all $q_1 \in [0, 1], q_2 \in [0, 1]$, and it will be called a **Nash Equilibrium** (NE) for the game and it will give us the point from which any deviation by an opposing player will result in a lower voter percentage for that player.

In calculus we know how to find maxima of functions if we can take derivatives of the functions involved. We would like to do that for this problem. This is a problem with a discontinuous payoff pair, and we cannot simply take derivatives and set them to zero to find the equilibrium. But we can take an educated guess as to what the equilibrium should be by the following reasoning. Suppose that I takes a position to the left of II, $q_1 < q_2$. Then she receives $Prob(V \leq \gamma)$ percent of the vote. Because this is the cumulative distribution function of V, we know that it increases as γ (which is the midpoint of q_1 and q_2) increases. Therefore, player I wants γ as large as possible. Once q_1 increases past q_2, then candidate II starts to gain because $u_1(q_1, q_2) = 1 - Prob(V \leq \gamma)$ if $q_1 > q_2$. It seems that player I should not go further to the right than q_2, but should equal q_2. In addition, we should have

$$Prob(V \leq \gamma) = Prob(V \geq \gamma) = 1 - Prob(V \leq \gamma) \implies Prob(V \leq \gamma) = \frac{1}{2}.$$

In probability theory, we know that this defines γ as the **median** of the random variable V. Therefore, the equilibrium γ should be the solution of

$$F_V(\gamma) \equiv Prob(V \leq \gamma) = \int_0^\gamma f(x)\, dx = \frac{1}{2}.$$

Because $F_V'(\gamma) = f(\gamma) > 0$, $F_V(\gamma)$ is strictly increasing, and there can only be one such $\gamma = \gamma^*$ that solves the equation; that is, a random variable has only one median when the density is strictly positive.

On the basis of this reasoning, we now verify that

If γ^* is the median of the voter positions

then

(γ^*, γ^*) is a Nash equilibrium for the candidates.

If this is the case, then $u_i(\gamma^*, \gamma^*) = \frac{1}{2}$ for each candidate and both candidates split the vote.

How do we check this? We need to verify directly that

$$u_1(\gamma^*, \gamma^*) = \frac{1}{2} \geq u_1(q_1, \gamma^*) \text{ and } u_2(\gamma^*, \gamma^*) = \frac{1}{2} \geq u_2(\gamma^*, q_2)$$

for every $q_1, q_2 \in [0, 1]$. If we assume $q_1 > \gamma^*$, then

$$u_1(q_1, \gamma^*) = Prob\left(V \geq \frac{q_1 + \gamma^*}{2}\right) \leq Prob\left(V \geq \frac{\gamma^* + \gamma^*}{2}\right) = Prob(V \geq \gamma^*) = \frac{1}{2}.$$

If, on the other hand, $q_1 < \gamma^*$, then

$$u_1(q_1, \gamma^*) = Prob\left(V \leq \frac{q_1 + \gamma^*}{2}\right) \leq Prob\left(V \geq \frac{\gamma^* + \gamma^*}{2}\right) = Prob(V \geq \gamma^*) = \frac{1}{2}.$$

and we are done.

In the special case, V has a uniform distribution on $[0, 1]$, so that $Prob(V \leq v) = v, 0 \leq v \leq 1$, we have $\gamma^* = \frac{1}{2}$. In that case, each candidate should be in the center. That will also be true if voter positions follow a bell curve (or, more generally, is symmetric). It seems reasonable that in the

United States if we account for all regions of the country, national candidates should be in the center. Naturally, that is where the winner usually is, but not always. For instance, when James Earl Carter was president, the country swung to the right, and so did γ^*, with Ronald Reagan elected as president in 1980. Carter should have moved to the right. Moreover, a model of V involving electoral college votes may lead to a median which is not in the center.

In this chapter we study many examples in many different areas and try to find the Nash equilibrium for the players.

4.2 The Basics

In this section we will precisely define what we mean by a Nash Equilibrium (NE) for a game with N players.

If there are N players in a game, we assume that each player has her own payoff function depending on her choice of strategy and the choices of the other players. Suppose that the strategies must take values in sets $Q_i, i = 1, \ldots, N$ and the payoffs are real-valued functions

$$u_i : Q_1 \times \cdots \times Q_N \to \mathbb{R}, \qquad i = 1, 2, \ldots, N.$$

Here is a formal definition of a pure Nash equilibrium point, keeping in mind that each player wants to maximize their own payoff.

Definition 4.2.1 *A collection of strategies $q^* = (q_1^*, \ldots, q_n^*) \in Q_1 \times \cdots \times Q_N$ is a pure Nash equilibrium for the game with payoff functions $\{u_i(q_1, \ldots, q_n)\}, i = 1, \ldots, N$, if for each player $i = 1, \ldots, N$, we have*

$$u_i(q_1^*, \ldots, q_{i-1}^*, q_i^*, q_{i+1}^*, \ldots, q_N^*)$$
$$\geq u_i(q_1^*, \ldots, q_{i-1}^*, q_i, q_{i+1}^*, \ldots, q_N^*), \text{for all } q_i \in Q_i.$$

A short hand way to write this is $u_i(q_i^*, q_{-i}^*) \geq u_i(q_i, q_{-i}^*)$ for all $q_i \in Q_i$, where q_{-i} refers to all the players except the ith. That is,

$$q_{-i} = (q_1, \ldots, q_{i-1}, q_{i+1}, \ldots, q_n) = \text{all players except } i,$$

$$(q_i, q_{-i}) = (q_1, \ldots, q_{i-1}, q_i, q_{i+1}, \ldots, q_n) = \text{all players including } i.$$

A NE satisfies

$$u_i(q_i^*, q_{-i}^*) = \max_{q_i \in Q_i} u_i(q_i, q_{-i}^*).$$

Best response strategies play a critical role in finding Nash equilibria. The definition for games with a continuum of strategies is given next.

Definition 4.2.2 *Given payoff functions $u_i(q_1, \ldots, q_n)$, a best response of player i, written $q_i = BR_i(q_1, \ldots, q_{i-1}, q_{i+1}, \ldots, q_n)$ is a value so that $u_i(q_i, q_{-i}) = \max_{q_i \in Q_i} u_i(q_i, q_{-i})$. In other words, q_i provides the maximum payoff for player i, given the values of the other player's $q_i's$.*

In general, we say that $q_i = BR_i(q_{-i}) \in \arg\max_{q_i \in Q_i} u_i(q_i, q_{-i})$.[1]

---

1 The arg max of a function is the set of points where the maximum is attained.

A Nash equilibrium consists of strategies that are all best responses to each other. The point is that no player can do better by deviating from a Nash point, assuming that no one else deviates. It doesn't mean that a **group** of players couldn't do better by playing something else.

Remarks

1. Notice that if there are two players and $u_1 = -u_2$, then a point (q_1^*, q_2^*) is a Nash point if

$$u_1(q_1^*, q_2^*) \geq u_1(q_1, q_2^*) \text{ and } u_2(q_1^*, q_2^*) \geq u_2(q_1^*, q_2), \forall (q_1, q_2).$$

But then

$$-u_1(q_1^*, q_2^*) \geq -u_1(q_1^*, q_2) \Longrightarrow u_1(q_1^*, q_2^*) \leq u_1(q_1^*, q_2),$$

and putting these together we see that

$$u_1(q_1, q_2^*) \leq u_1(q_1^*, q_2^*) \leq u_1(q_1^*, q_2), \forall (q_1, q_2).$$

This, of course, says that (q_1^*, q_2^*) is a saddle point of the two-person zero sum game. This would be a pure saddle point if the Q sets were pure strategies and a mixed saddle point if the Q sets were the set of mixed strategies.

2. In many cases, the problem of finding a Nash point can be reduced to a simple calculus problem. To do this we need to have the strategy sets Q_i to be open intervals and the payoff functions to have at least two continuous derivatives because we are going to apply the second derivative test. The steps involved in determining $(q_1^*, q_2^*, \dots, q_n^*)$ as a Nash equilibrium are the following:

(a) Solve

$$\frac{\partial u_i(q_1, \dots, q_n)}{\partial q_i} = 0, \quad i = 1, 2, \dots, n.$$

(b) Verify that q_i^* is the only stationary point of the function

$$q \mapsto u_i(q_1^*, \dots, q_{i-1}^*, q, q_{i+1}^*, \dots, q_n^*) \text{ for } q \in Q_i.$$

(c) Verify

$$\frac{\partial^2 u_i(q_1, \dots, q_n)}{\partial q_i^2} < 0, \quad i = 1, 2, \dots, n,$$

evaluated at $q_1^*, \dots, q_n^*$.

If these three points hold for $(q_1^*, q_2^*, \dots, q_n^*)$, then it must be a Nash equilibrium. The last condition guarantees that the function is concave down in each variable when the other variables are fixed. This guarantees that the critical point (in that variable) is a maximum point. There are many problems where this is all we have to do to find a Nash equilibrium, but remember that this is a sufficient but not necessary set of conditions because many problems have Nash equilibria that do not satisfy any of the three conditions.

3. Carefully read the definition of Nash equilibrium. For the calculus approach, we take the partial of u_i with respect to q_i, **not** the partial of each payoff function with respect to all variables. We are not trying to maximize each payoff function over **all** the variables, but each payoff function to each player as a function only of the variable they control, namely, q_i. That is the difference between a Nash equilibrium and simply the old calculus problem of finding the maximum of a function over a set of variables.

Let's start with a straightforward example.

Example 4.2 We have a two-person game with pure strategy sets $Q_1 = Q_2 = \mathbb{R}$ and payoff functions

$$u_1(q_1, q_2) = -q_1 q_2 - q_1^2 + q_1 + q_2 \text{ and } u_2(q_1, q_2) = -3q_2^2 - 3q_1 + 7q_2.$$

Then

$$\frac{\partial u_1}{\partial q_1} = -q_2 - 2q_1 + 1 \text{ and } \frac{\partial u_2}{\partial q_2} = -6q_2 + 7.$$

There is one and only one solution of these (so only one stationary point), and it is given by $q_1 = -\frac{1}{12}, q_2 = \frac{7}{6}$. Finally, we have

$$\frac{\partial^2 u_1}{\partial q_1^2} = -2 < 0 \text{ and } \frac{\partial^2 u_2}{\partial q_2^2} = -6 < 0,$$

and so $(q_1, q_2) = (-\frac{1}{12}, \frac{7}{6})$ is indeed a Nash equilibrium.

For a three-person example take

$$u_1(q_1, q_2, q_3) = -q_1^2 + 2q_1 q_2 - 3q_2 q_3 - q_1$$
$$u_2(q_1, q_2, q_3) = -q_2^2 - q_1^2 + 4q_1 q_3 - 5q_1 q_2 + 2q_2$$
$$u_3(q_1, q_2, q_3) = -4q_3^2 - (q_1 + q_3)^2 + q_1 q_3 - 3q_2 q_3 + q_1 q_2 q_3.$$

Taking the partials and finding the stationary point gives the unique solution $q_1^* = \frac{1}{7}, q_2^* = \frac{9}{14}$, and $q_3^* = -\frac{97}{490}$. Since $\partial^2 u_1/\partial q_1^2 = -2 < 0$, $\partial^2 u_2/\partial q_2^2 = -2 < 0$, and $\partial^2 u_3/\partial q_3^2 = -10 < 0$, we know that our stationary point is a Nash equilibrium.

This example has found a **pure** Nash equilibrium which exists because the payoff functions are concave in each variable separately. When that is not true we have to deal with mixed strategies which we will discuss below.

Example 4.3 Using Mathematica to Find a Saddle Point. When there are only two players and the game is zero sum we may find a saddle point by finding the critical points and checking the determinant of the Hessian to make sure it is negative. Here's an example. Consider the function graphed in Figure 4.2:

$$u(x, y) = x^3 - 2x + y^2, \quad x \in [-2, 1], y \in [-1, 1].$$

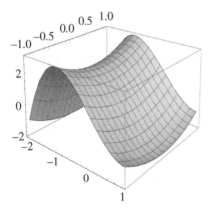

Figure 4.2 Saddle point at $(-1, 0)$.

This function is not concave or convex in x but it is convex in y. A saddle point is not guaranteed by the von Neumann minimax theorem. Nevertheless, we can find a critical point and check the second-order condition using the code

```
f[x_,y_]:=x^3-3x+y^2;
```

```
saddle=NSolve[{D[f[x,y],x]==0,D[f[x,y],y]==0,
D[f[x,y],{x,2}] D[f[x,y],{y,2}]-D[f[x,y],x,y]^2<0},{x,y}];
```

```
Print["The saddle point(s) of f(x,y) is/are ",saddle];
```

```
The saddle point(s) of f(x,y) is/are {{x->-1.,y->0.}}
```

This code finds the critical points which satisfy the condition $f_{xx}f_{yy} - (f_{xy})^2 < 0$ if there are any. It turns out that this function does have a saddle point at $(-1, 0)$ showing that von Neumann's Theorem 1.7.2 provides a sufficient but not necessary condition for a saddle.

Example 4.4 This example will illustrate the use of best response strategies to calculate a Nash equilibrium.

Suppose players I and II have payoff functions

$$u_1(x,y) = x(2 + 2y - x) \qquad u_2(x,y) = y(4 - x - y).$$

The variables must be nonnegative, $x \geq 0, y \geq 0$.

First we find the best response functions for each player, namely, $y(x)$ which satisfies $u_2(x, y(x)) = \max_y u_2(x, y)$, and $x(y)$ which satisfies $u_1(x(y), y) = \max_x u_1(x, y)$.

For our payoffs $\frac{\partial u_1}{\partial x} = 0 \implies x(y) = 1 + y$, and $\frac{\partial u_2}{\partial y} = 0 \implies y(x) = \frac{4-x}{2}$. This requires $4 \geq x \geq 0$. The Nash equilibrium is where the best response curves cross, i.e., where $x(y) = y(x)$. For this example, we have $x^* = 2$ and $y^* = 1$. Taking the second partials show that these are indeed maxima of the respective payoff functions. Furthermore $u_1(2, 1) = 4$ and $u_2(2, 1) = 1$.

Knowing the best response functions allow us to consider what happens if the players choose sequentially instead of simultaneously.

Suppose that player I assumes that II will always use the best response function $y(x) = \frac{4-x}{2}$. Why wouldn't player I then choose x to maximize $u_1(x, y(x)) = x(2 + 2(\frac{4-x}{2}) - x)$? This function has a maximum at $x = \frac{3}{2}$ and $u_1(\frac{3}{2}, y(\frac{3}{2})) = \frac{9}{2} > 4$. Thus player I can do better if she knows that player II will use her best response function. Also, since $y(\frac{3}{2}) = \frac{5}{4}$, $u_2(\frac{3}{2}, \frac{5}{4}) = \frac{25}{16} > 1$, and both players do better. On the other hand, if player I uses $x = \frac{3}{2}$ and player II uses $y = 1$, then $u_1(\frac{3}{2}, 1) = \frac{15}{4} < 4$ and $u_2(\frac{3}{2}, 1) = \frac{3}{2} > 1$.

Similarly, if player II chooses her point y to maximize $u_2(x(y), y)$, she will see that $y = \frac{3}{4}$ and then $x(\frac{3}{4}) = \frac{7}{4}$. Then $u_1(\frac{7}{4}, \frac{3}{4}) = \frac{49}{16} < 4$ and $u_2(\frac{7}{4}, \frac{3}{4}) = \frac{9}{8} > 1$. Only player II does better in this setup.

The conclusion is that both players prefer that player II commits to using her best response and that player I announces she will play $x = \frac{3}{2}$.

Example 4.5 There is another idea for solving nonzero sum games which was introduced by John Roemer[1] in [36] and is called a **Kantian Equilibrium**. The idea is basically that if a player deviates from the Kantian equilibrium, then the other players, instead of not deviating from the

---

1 Elizabeth S. and A. Varick Stout Professor of Political Science and Economics at Yale University.

equilibrium, will in fact deviate from the equilibrium just as the player who first deviated. Here's the definition in the case of two players.

Definition 4.2.3 *Suppose there are two players in a nonzero sum game with payoff functions $u_1(x, y), u_2(x, y)$. Then (x^*, y^*) is a Kantian Equilibrium if it satisfies the condition*

$$u_i(x^*, y^*) \geq u_i(\gamma x^*, \gamma y^*), \text{ for all } \gamma > 0,$$

or equivalently,

$$u_i(x^*, y^*) = \max_{\gamma > 0} u_i(\gamma x^*, \gamma y^*), \quad i = 1, 2,$$

and the maximum must be attained with $\gamma = 1$.

The definition says that if either player decides to deviate from her part of the Kantian equilibrium by some proportion $0 < \gamma \leq 1$, then the other player will also deviate from her part of the equilibrium by the same proportion.

In Example 4.4 , we had the payoff functions

$$u_1(x, y) = x(2 + 2y - x) \qquad u_2(x, y) = y(4 - x - y).$$

To find the Kantian equilibrium, we first find

$$\max_{\gamma > 0} \gamma x(2 + 2\gamma y - \gamma x) \implies \gamma = \frac{1}{x - 2y}$$

and

$$\max_{\gamma > 0} \gamma y(4 - \gamma x - \gamma y) \implies \gamma = \frac{2}{x + y}.$$

In order for (x^*, y^*) to be a Kantian equilibrium, the condition in the definition forces $\gamma = 1$. Therefore, we have

$$\frac{1}{x - 2y} = 1 \text{ and } \frac{2}{x + y} = 1.$$

Solving, we get $x^* = \frac{5}{3}, y^* = \frac{1}{3}$. The Kantian equilibrium is $(\frac{5}{3}, \frac{1}{3})$. In this case the resulting payoffs to each player are $u_1(x^*, y^*) = \frac{5}{3}, u_2(x^*, y^*) = \frac{2}{3}$. Compared to the Nash equilibrium payoffs the Kantian equilibrium provides significantly lower payoffs in this example.

4.2.1   Do We Have Mixed Strategies in Continuous Games

How do we set up the game when we allow mixed strategies and we have a continuum of pure strategies? First note that this involves some concepts on continuous random variables from probability. Here is a sketch of what we have to do. Consider only two players with payoff functions $u_1 : Q_1 \times Q_2 \to \mathbb{R}$ and $u_2 : Q_1 \times Q_2 \to \mathbb{R}$, where $Q_1 \subset \mathbb{R}^n, Q_2 \subset \mathbb{R}^m$ are the pure strategy sets for each player. For simplicity, let's assume the control sets Q_i is some interval like $Q_1 = [a, b]$.

A mixed strategy for player I is now a probability distribution on Q_1, but since Q_1 has a continuum of points, this means that a mixed strategy is a probability density function (pdf) which we label $X(x)$. That is a function $X(x) \geq 0$ and $\int_{Q_1} X(x)\, dx = 1$.[1] Similarly, a mixed strategy for player

---

1 More generally one would have to take a mixed strategy as a cumulative distribution function (cdf) $F_1(x) = Prob(\text{I plays } q \leq x)$ and $F_1'(x) = X(x)$. Not every cdf has a pdf so this is a restrictive definition.

II is a probability density function $Y(y)$ on Q_2 with $Y(y) \geq 0$, $\int_{Q_2} Y(y)\,dy = 1$. Intuitively, $X(x)\,dx$ represents $Prob$(I chooses in $[x, x + dx]$) but more accurately

$$Prob(\text{I plays a number in } [a, b]) = \int_a^b X(x)\,dx.$$

Given choices of mixed strategies X, Y we then calculate the expected payoff for player I as

$$E_{\mathrm{I}}(X, Y) = \int_{Q_1} \int_{Q_2} u_1(x, y) X(x) Y(y)\,dx\,dy$$

and

$$E_{\mathrm{II}}(X, Y) = \int_{Q_1} \int_{Q_2} u_2(x, y) X(x) Y(y)\,dx\,dy$$

for player II.

It may be proved that there is a Nash equilibrium in mixed strategies under mild assumptions on u_1, u_2, Q_1, Q_2. For example, if the payoff functions are continuous and Q_i are convex and compact then a Nash equilibrium is guaranteed to exist. The question is how to find it, because now we are looking for **functions** X, Y, not points or vectors. The only method we present here for finding the Nash equilibrium (or saddle point if $u_2 = -u_1$) is based on the Equality of Payoffs theorem, also known as the Indifference Principle.

Proposition 4.2.4 _Indifference Principle._ _Let_ $X^*(x), Y^*(y)$ _be Nash equilibrium probability density functions. Assume the payoff functions are upper semicontinuous in each variable._[1]
_If_ $x_0 \in Q_1$ _is a point where_ $X^*(x_0) > 0$, _then_

$$E_{\mathrm{I}}(x_0, Y^*) = \int_{Q_2} u_1(x_0, y)\, Y^*(y)\,dy = v_{\mathrm{I}} := E_{\mathrm{I}}(X^*, Y^*).$$

_If_ $y_0 \in Q_2$ _is a point where_ $Y^*(y_0) > 0$, _then_

$$E_{\mathrm{II}}(X^*, y_0) = \int_{Q_1} u_2(x, y_0)\, X^*(x)\,dx = v_{\mathrm{II}} := E_{\mathrm{II}}(X^*, Y^*)$$

In particular, if $X^*(x) > 0$ _for all_ $x \in Q_1$, _then_ $x \mapsto E_{\mathrm{I}}(x, Y^*)$ _is a constant and that constant is_ $E_{\mathrm{I}}(X^*, Y^*)$. _Similarly, if_ $Y^*(y) > 0$ _for all_ $y \in Q_2$, _then_ $y \mapsto E_{\mathrm{II}}(X^*, y)$ _is a constant and that constant is_ $E_{\mathrm{II}}(X^*, Y^*)$.

Proof: We will only show the first statement and we will take for simplicity $Q_1 = [a, b]$. Since X^*, Y^* is a Nash equilibrium, $v_{\mathrm{I}} \geq E_{\mathrm{I}}(x, Y^*)$ for all $x \in Q_1$ and in particular this is true for $x = x_0$. Now suppose that $v_{\mathrm{I}} > E_{\mathrm{I}}(x_0, Y^*)$. Then this must be true on an interval $a < c < x_0 < d < b$. Thus we have,

$$v_{\mathrm{I}}\, length([a, b] \backslash (c, d)) \geq \int_{[a,b] \backslash (c,d)} E_{\mathrm{I}}(x, Y^*)\, X^*(x)\,dx$$

and

$$v_{\mathrm{I}}\,(d - c) > \int_c^d E_{\mathrm{I}}(x, Y^*)\, X^*(x)\,dx.$$

---

1 f is upper semicontinuous if $\{x \mid f(x) < a\}$ is open for each $a \in \mathbb{R}$

The last strict inequality holds because $X^*(x_0) > 0$ and hence $X^*(x) > 0$ on (c, d). Adding the two inequalities, we get

$$v_{\mathrm{I}} = v_{\mathrm{I}} \left((b - d) + (c - a) \right) + v_{\mathrm{I}} (d - c)$$

$$> \int_{[a,b]\backslash(c,d)} E_{\mathrm{I}}(x, Y^*) X^*(x) \, dx + \int_c^d E_{\mathrm{I}}(x, Y^*) X^*(x) \, dx$$

$$= \int_a^b E_{\mathrm{I}}(x, Y^*) X^*(x) \, dx$$

$$= v_{\mathrm{I}}$$

a contradiction. ∎

Now we will consider several examples to see how to compute things.

Example 4.6   War of Attrition. A game called the war of attrition game in which two countries, labeled 1 and 2, compete for a resource which they each value at v. The players incur a cost of c per unit time for continuing the war for the resource but they may choose to quit at any time. For each player, the payoff functions are taken to be for $t_1 \geq 0, t_2 \geq 0$,

$$u_1(t_1, t_2) = \begin{cases} v - c \, t_2, & \text{if } t_1 > t_2; 2 \text{ quits before } 1; \\ -c \, t_1, & \text{if } t_1 \leq t_2; 1 \text{ quits before } 2. \end{cases}$$

and

$$u_2(t_1, t_2) = \begin{cases} v - c \, t_1, & \text{if } t_2 > t_1; 1 \text{ quits before } 2; \\ -c \, t_2, & \text{if } t_2 \leq t_1; 2 \text{ quits before } 1. \end{cases}$$

This means that if, say player 1 quits before player 2, the payoff to 1 is $-ct_1$, whereas if 2 quits before 1, then 1 gets the resource valued at v less the cost of running the war up to the time 2 quits.

One way to find the pure Nash equilibrium, if any, is to find the best response functions (which are sets in general) and see where they intersect.

It is not hard to check that

$$BR_1(t_2) = \begin{cases} 0, & \text{if } t_2 > \frac{v}{c}; \\ (t_2, \infty), & \text{if } t_2 < \frac{v}{c}; \\ \{0\} \cup (t_2, \infty), & \text{if } t_2 = \frac{v}{c}. \end{cases}$$

and

$$BR_2(t_1) = \begin{cases} 0, & \text{if } t_1 > \frac{v}{c}; \\ (t_1, \infty), & \text{if } t_1 < \frac{v}{c}; \\ \{0\} \cup (t_1, \infty), & \text{if } t_1 = \frac{v}{c}. \end{cases}$$

For instance, given a quit time $t_2 > \frac{v}{c}$ for player 2, player 1's best response is to quit immediately, i.e., $t_1 = 0$. If player 2 quits at time $t_2 < \frac{v}{c}$, player 1's best response is to quit at any time after t_2. If player 2 quits at exactly time $t_2 = \frac{v}{c}$, then player 1 should quit either immediately ($t_1 = 0$) or at any time after t_2, in both cases receiving a payoff of 0.

By graphing these set-valued functions on the same set of axes in the t_1-t_2 plane, we obtain

$$BR_1(t_2) \cap BR_2(t_1) = \{(t_1 \geq \frac{v}{c}, t_2 = 0), (t_1 = 0, t_2 \geq \frac{v}{c})\}$$

and these are the pure Nash equilibria. Either player I stops immediately and player II stops at time $t_2^* = \frac{v}{c}$ or vice-versa. The payoffs of the countries are

$$u_1(0, t_2^*) = 0, \quad u_2(0, t_2^*) = v, \quad t_2^* \geq \frac{v}{c}, \quad u_1(t_1^*, 0) = v, \quad u_2(t_1^*, 0) = 0, \quad t_1^* \geq \frac{v}{c}.$$

This means that a pure Nash equilibrium involves one of the two countries stopping immediately. Notice too that if one player stops immediately, any time after $\frac{v}{c}$ is the other player's part of the Nash equilibrium. But this seems a little unrealistic and certainly doesn't seem to be what happens in practice. In Example 4.7 , we will see if there are any mixed Nash equilibria that make more sense.

Example 4.7 Mixed Strategies in the War of Attrition. Consider[1] again the war of attrition game. The pure Nash equilibria indicated that one of the players should quit immediately. If that was realistic there would never be wars.

Next we see if there is a mixed Nash equilibrium. Let $X(t)$ and $Y(t)$ denote density functions for each player assuming that the times of quitting labelled τ_1 and τ_2 are random variables. We use Proposition 4.2.4 and assume that the density function for player I's part of the Nash equilibrium satisfies $X^*(t_1) > 0$. In that case we know that for player II's density function Y^*, we have for all $0 \leq t_1 < \infty$,

$$v_I = \int_0^\infty u_1(t_1, t_2) Y^*(t_2)\, dt_2 = \int_0^{t_1} (-ct_1) Y^*(t_2)\, dt_2 + \int_{t_1}^\infty (v - ct_2) Y^*(t_2)\, dt_2.$$

Since this is true for any t_1, we take the derivative of both sides with respect to t_1 to get using the fundamental theorem of calculus[2]

$$\begin{aligned}
0 &= -ct_1 Y^*(t_1) - c \int_0^{t_1} Y^*(t_2)\, dt_2 - (v - ct_1) Y^*(t_1) \\
&= -v Y^*(t_1) - c \int_0^{t_1} Y^*(t_2)\, dt_2
\end{aligned}$$

Again take a derivative with respect to t_1 to see that

$$\frac{d}{dt_1} Y^*(t_1) + \frac{c}{v} Y^*(t_1) = 0$$

This is a first order ordinary differential equation for Y^* which may be solved by integration to see that

$$Y^*(t_1) = C e^{-\frac{c}{v} t_1}$$

where C is a constant determined by the condition $\int_0^\infty Y^*(y)\, dy = 1$. We determine that $C = \frac{c}{v}$ and hence the density is (replacing the variable name t_1 with player II's time variable t_2)

$$Y^*(t_2) = \frac{c}{v} e^{-\frac{c}{v} t_2}, \quad t_2 \geq 0,$$

which is an exponential density for an exponentially distributed random variable[3] representing the time that player II will quit with mean $\frac{v}{c}$. Since the players in this game are completely symmetric, it is easy to check that

$$X^*(t_1) = \frac{c}{v} e^{-\frac{c}{v} t_1}, \quad t_1 \geq 0.$$

---

1 The solution of this game is due to James Webb [42], a former lecturer at Nottingham Trent University, UK.
2 This is called Leibniz's rule $\frac{d}{dt} \int_{a(t)}^{b(t)} f(x, t)\, dx = f(b(t), t) b'(t) - f(a(t), t) a'(t) + \int_{a(t)}^{b(t)} f_t(x, t)\, dx$.
3 A random variable is exponentially distributed if it has pdf $f(x) = \lambda e^{-\lambda x}, x \geq 0$, where $\lambda > 0$ is a parameter. The mean of such a rv is $\frac{1}{\lambda}$.

Now suppose we want to know the expected length of time the war will last. To answer that question just requires a little probability. First, let τ denote the random variable giving the length of the war and as above let $\tau_i, i = 1, 2$ be the random variable that gives the time each player will quit. The war goes on until time $\tau = \min\{\tau_1, \tau_2\}$. Then from what we just found we know that τ_1 has density $X^*(s) = \frac{c}{v}e^{-\frac{c}{v}s}$ and τ_2 has density $Y^*(s) = \frac{c}{v}e^{-\frac{c}{v}s}$. Therefore,

$$
\begin{aligned}
Prob(\tau \leq t) &= Prob(\min\{\tau_1, \tau_2\} \leq t) \\
&= 1 - Prob(\min\{\tau_1, \tau_2\} > t) \\
&= 1 - Prob(\tau_1 > t \text{ and } \tau_2 > t) \\
&= 1 - Prob(\tau_1 > t)Prob(\tau_2 > t) \quad \text{(by independence)} \\
&= 1 - e^{-\frac{c}{v}t}e^{-\frac{c}{v}t} = 1 - e^{-\frac{2c}{v}t}.
\end{aligned}
$$

This tells us that τ has a density $\frac{2c}{v}e^{-\frac{2c}{v}t}$ and hence is also exponentially distributed with mean $\frac{v}{2c}$. Because each player would quit on average at time $\frac{v}{c}$ you might expect that the war would last the same amount of time, but that is incorrect. On average, the time the **first** player quits is half of the time the war ends.

Next we want to calculate the expected payoffs to each player if they are playing their Nash equilibrium. This is a straightforward calculation

$$
\begin{aligned}
u_1(X^*, Y^*) &= \int_0^\infty \int_0^\infty u_1(t_1, t_2)X^*(t_1)Y^*(t_2)\, dt_1\, dt_2 \\
&= \int_0^\infty \left(\int_0^{t_2} -ct_1 X^*(t_1)\, dt_1 + \int_{t_2}^\infty (v - ct_2)X^*(t_1)\, dt_1 \right) Y^*(t_2)\, dt_2 \\
&= 0.
\end{aligned}
$$

Similarly, the expected payoff to player II is also zero when the optimal strategies are played.

The war of attrition is a particular example of a **game of timing**. In general, we have two functions $\alpha(t), \beta(t)$ and payoff functions depending on the time that each player chooses to quit. In particular, we take the strategy sets to be $[0, 1]$ for both players and

$$
u_1(t_1, t_2) = \begin{cases} \beta(t_2), & \text{if } t_1 > t_2; 2 \text{ quits before } 1; \\ \alpha(t_1), & \text{if } t_1 < t_2; 1 \text{ quits before } 2. \\ \frac{\alpha(t_1) + \beta(t_1)}{2}, & \text{if } t_1 = t_2. \end{cases}
$$

and

$$
u_2(t_1, t_2) = \begin{cases} \beta(t_1), & \text{if } t_2 > t_1; 1 \text{ quits before } 2; \\ \alpha(t_2), & \text{if } t_2 < t_1; 2 \text{ quits before } 1. \\ \frac{\alpha(t_1) + \beta(t_1)}{2}, & \text{if } t_1 = t_2. \end{cases}
$$

If we take α and β to be continuous, decreasing functions with $\alpha(t) < \beta(t)$ and $\alpha(0) > \beta(1)$, then the general war of attrition has two pure Nash equilibria given by $\{(t_1 = \tau^*, t_2 = 0), (t_1 = 0, t_2 = \tau^*)\}$ where $\tau^* > 0$ is the first time where $b(\tau^*) = a(0)$.

Mixed strategies for continuous games can be very complicated. They may not have continuous densities and generally should be viewed as general cumulative distribution functions. This is a topic for a much more advanced course. Now we return to one of our theme games and consider continuous poker.

Example 4.8   A Simplified Model of Poker. Each player antes \$1 to start the game and then is dealt a hand. Model a hand for each player $i =$ I,II as a random number $X_i = x_i \in [0, 1], i = 1, 2$, where X_i has a uniform distribution on $[0, 1]$. Higher numbers are better hands. A player only knows her own hand. Each player can then choose to either **fold** or **bet** c with player I moving first. If I folds, the game is over and player II wins the pot. If I bets the additional c, player II can either fold or call by putting in an additional c. If player II folds, player I wins the pot. If player II **calls**, the player with the highest hand wins the pot, which would be $(c + 2)$, to net $(c + 1)$ to the winner.

Strategies for this game will be functions

$$\alpha_1(x_1) = Prob(\text{I bets if she gets } X_1 = x_1),$$
$$\alpha_2(x_2) = Prob(\text{II bets if she gets } X_2 = x_2).$$

Define the signum function $\mathrm{sgn}(x) = \begin{cases} +1, & x > 0 \\ -1, & x < 0, \\ 0, & x = 0 \end{cases}$. The payoff to player I is then dependent on the two functions α_1, α_2 and is given by

$$P(\alpha_1, \alpha_2) = \begin{cases} -1, & \text{if I folds, with probability } 1 - \alpha_1(x_1), \\ +1, & \text{if I bets and II folds, with probability } \alpha_1(x_1)(1 - \alpha_2(x_2)), \\ (c + 1)\mathrm{sgn}(x_1 - x_2), & \text{if I bets and II calls, with probability } \alpha_1(x_1)\alpha_2(x_2). \end{cases}$$

Since this is a zero sum game, the payoff to player II is $-P(\alpha_1, \alpha_2)$. The expected payoff to player I will be (recall that both random variables are uniform)

$$
\begin{aligned}
E(\alpha_1, \alpha_2) &= \int_0^1 \int_0^1 P(\alpha_1(x_1), \alpha_2(x_2)) \, dx_1 \, dx_2 \\
&= \int_0^1 \int_0^1 (-1)(1 - \alpha_1(x_1)) + (+1)\alpha_1(x_1)(1 - \alpha_2(x_2)) \\
&\qquad\qquad + (c + 1)\mathrm{sgn}(x_1 - x_2)\alpha_1(x_1)\alpha_2(x_2) \, dx_1 \, dx_2 \\
&= \int_0^1 \alpha_1(x_1) \left(1 + \int_0^1 (1 - \alpha_2(x_2)) + (c + 1)\mathrm{sgn}(x_1 - x_2)\alpha_2(x_2) \, dx_2 \right) dx_1 - 1 \\
&= \int_0^1 \alpha_1(x_1)K(x_1) \, dx_1 - 1,
\end{aligned}
$$

where $K(x_1) = 1 + \int_0^1 (1 - \alpha_2(x_2)) + (c + 1)\mathrm{sgn}(x_1 - x_2)\alpha_2(x_2) \, dx_2$.

We factored out α_1 because we want to use it to maximize $E(\alpha_1, \alpha_2)$. For player II we instead write this so that we isolate $\alpha_2(x_2)$ to get another way to express the payoff

$$
\begin{aligned}
E(\alpha_1, \alpha_2) &= \int_0^1 \alpha_2(x_2) \left(\int_0^1 \alpha_1(x_1)((c + 1)\mathrm{sgn}(x_1 - x_2) - 1) \, dx_1 \right) dx_2 \\
&\qquad + \int_0^1 (2\alpha_1(x_1) - 1) \, dx_1 \\
&= \int_0^1 \alpha_2(x_2)M(x_2) \, dx_2 + \int_0^1 (2\alpha_1(x_1) - 1) \, dx_1,
\end{aligned}
$$

where $M(x_2) = \int_0^1 \alpha_1(x_1)((c + 1)\mathrm{sgn}(x_1 - x_2) - 1) \, dx_1$.

Player I wants to choose $\alpha_1(x_1)$ to maximize $E(\alpha_1, \alpha_2)$. Because α_1 is a probability $0 \le \alpha_1 \le 1$ and so

$$\alpha_1(x_1) = \begin{cases} 1, & \text{if } K(x_1) > 0, \\ 0, & \text{if } K(x_1) < 0, \\ \text{undetermined}, & \text{if } K(x_1) = 0. \end{cases}$$

That is, player I will bet if $K(x_1) > 0$ and fold if $K(x_1) < 0$. Since $K(x_1)$ is obviously a non-decreasing function (why?), it is clear that it will not change sign more than once. This means that it is reasonable to assume there is a threshold level for player I, call it $\xi_1 \in [0,1]$ so that when $X_1 > \xi_1$ player I will bet and if $X_1 < \xi_1$ player I should fold. Determining ξ_1 will solve the problem for player I.

Similarly, player II wants to choose $\alpha_2(x_2)$ to minimize $E(\alpha_1, \alpha_2)$. Because α_2 is a probability $0 \le \alpha_2 \le 1$

$$\alpha_2(x_2) = \begin{cases} 0, & \text{if } M(x_2) > 0, \\ 1, & \text{if } M(x_2) < 0, \\ \text{undetermined}, & \text{if } M(x_2) = 0. \end{cases}$$

Player II will fold if $M(x_2) > 0$ and bet if $M(x_2) < 0$. Since $M(x_2)$ is a non-increasing function (why?), it will not change sign more than once. This means that it is reasonable to assume there is also a threshold level for player II, call it $\xi_2 \in [0,1]$ so that when $X_2 > \xi_2$ player II will fold and if $X_2 < \xi_2$ player II should bet. Again, determining ξ_2 will solve the problem for player II. Summarizing we may assume

$$\alpha_1(x_1) = \begin{cases} 1, & x_1 > \xi_1, \\ 0, & x_1 < \xi_1, \end{cases}, \text{ and } \alpha_2(x_2) = \begin{cases} 0, & x_2 > \xi_2, \\ 1, & x_2 < \xi_2, \end{cases}.$$

We have dropped what happens when $x_1 = \xi_1$ or $x_2 = \xi_2$ because the probability this happens is zero. Let's find ξ_1, ξ_2. First we calculate (exercise)

$$M(x_2) = \int_0^1 \alpha_1(x_1)((c+1)\text{sgn}(x_1 - x_2) - 1) \, dx_1$$
$$= \begin{cases} c(1 - \xi_1), & \text{if } x_2 < \xi_1, \\ -2(c+1)x_2 + \xi_1(c+2) + c, & \text{if } x_2 \ge \xi_1. \end{cases}$$

This is a horizontal line when $x_2 < \xi_1$ and it is a line with negative slope when $x_2 > \xi_1$. This line crosses zero when

$$-2(c+1)\xi_2 + \xi_1(c+2) + c = 0 \implies \xi_2 = \frac{\xi_1(c+2) + c}{2(c+1)}.$$

Next we plug this into $E(\alpha_1, \alpha_2)$ and consider it as a function of ξ_1. We get after some algebra (verify this)

$$E(\alpha_1, \alpha_2) = (c+1)\xi_2^2 - \xi_2(\xi_1(c+2) + c) + \xi_1 c$$
$$= \frac{(c+2)^2}{4(c+1)} \left(-\xi_1^2 + (2\xi_1 - 1)\frac{c^2}{(c+2)^2} \right)$$
$$= F(\xi_1).$$

Since player I wants to maximize the expected payoff, we want to choose ξ_1 so as to maximize the function $F(\xi_1)$. Taking the derivative and setting it to zero, we solve to get the optimal thresholds for each player

$$\xi_1^* = \left(\frac{c}{c+2} \right)^2 \text{ and plugging into } \xi_2 \implies \xi_2^* = \frac{c}{c+2}.$$

Once we know the optimal strategies, we may find the value of this game to player I as $E(\alpha_1, \alpha_2) = -\left(\frac{c}{c+2}\right)^2$. Since this is negative, on average player I will lose money using this or any other strategy. In particular, if the bet amount is $c = 5$, player I will have an average loss of $ 0.5102 with threshold levels, $\xi_1 = 0.5102$ and $\xi_2 = 0.714$. Would you rather be player I or player II? Explain why it is reasonable that player I should be at a disadvantage.

Refer to [25] and [20] for much more on continuous models of poker.

It is possible to incorporate bluffing in Example 4.8 by slightly changing the rules so that if player I folds the hands are compared and the higher hand wins the pot. This is instead of player I folding and player II automatically winning the pot. This seemingly small difference will change the strategy of player I so that there are two threshold values, say $0 < a < b < 1$ so that player I will bet if $X_1 < a$ or if $X_1 > b$, i.e., she will bet with high enough and low enough hands. Player I will optimally bluff. We omit the analysis and refer to [25] or [20] for the details.

Example 4.8 exhibits an interesting way to find a Nash equilibrium (actually a saddle point in that case). First we found that the NE should be a threshold-type strategy which reduces the problem from finding functions which maximize to finding a point which maximizes and then we can use calculus. We have already seen it is not true that Nash equilibria can be found only for payoff functions that have derivatives. Here is another example.

Example 4.9 Negotiation with Arbitration. A firm is negotiating with the union for a new contract on wages. As often happens (NBA vs. Player's Union 2011), the negotiations have gone to an arbitrator in an effort to insert a neutral party to decide the issue. The offers are submitted to the arbitrator simultaneously by the firm and the union. Suppose the offers are x by the firm and y by the union and we may assume $y > x$, since if the firm offers at least what the union wants the problem goes away. The arbitrator has in mind a settlement he considers fair, say f and the offer submitted closest to f will be the one the arbitrator selects and will impose this as the settlement. The amount f is known only to the arbitrator but the firm and union guess that f is distributed according to some known cumulative probability distribution function $G(x) = Prob(f \leq x)$.

The parties know that the arbitrator will choose the closest offer to f and so each calculates the expected payoff

$$u(x, y) = xProb(x \text{ is closest}) + yProb(y \text{ is closest})$$

Of course it must be true that $x < f < y$. By drawing a number line it is easy to see that

$$x \text{ is closest to} f \iff f < \frac{x+y}{2}$$

and

$$y \text{ is closest to} f \iff f > \frac{x+y}{2}.$$

Thus

$$Prob(x \text{ is closest}) = Prob\left(f < \frac{x+y}{2}\right) = G\left(\frac{x+y}{2}\right),$$
$$Prob(y \text{ is closest}) = 1 - G\left(\frac{x+y}{2}\right),$$

and

$$u(x, y) = xG\left(\frac{x+y}{2}\right) + y\left(1 - G\left(\frac{x+y}{2}\right)\right).$$

This is a two-person zero sum game. The union is the maximizer and the firm is the minimizer. It is a calculus calculation to check that

$$\det \begin{bmatrix} u_{xx} & u_{xy} \\ u_{xy} & u_{yy} \end{bmatrix} = -\left(G'\left(\frac{x+y}{2}\right)\right)^2 < 0$$

as long as $G'\left(\frac{x+y}{2}\right) \neq 0$. At any such point, if the first derivatives are zero there, the point must be a saddle point.

To find a saddle point, we calculate

$$\frac{\partial u}{\partial x} = \frac{x-y}{2}G'\left(\frac{x+y}{2}\right) + G\left(\frac{x+y}{2}\right) = 0$$

and

$$\frac{\partial u}{\partial y} = \frac{x-y}{2}G'\left(\frac{x+y}{2}\right) + \left(1 - G\left(\frac{x+y}{2}\right)\right) = 0$$

Solving for x^*, y^* we see that $G\left(\frac{x^*+y^*}{2}\right) = \frac{1}{2}$. We interpret this to say that there are many possible offers that each of the two sides can make, but the saddle point must satisfy the requirement that the average of the two offers should be a **median** of the arbitrator's distribution. Of course neither player knows the other's offer but they can calculate the median of G and then propose that offer. The average of the two offers will give the median. Since this is a saddle point, any deviation from that will not be optimal for the deviant player.

Next, since $G(\frac{x+y}{2}) = \frac{1}{2}$ we may substitute this into either $u_x = 0$ or $u_y = 0$ to see that

$$y^* - x^* = \frac{1}{G'\left(\frac{x^*+y^*}{2}\right)}.$$

To illustrate a way to play this, let's take the mediated fair settlement to have a normal distribution with mean μ and variance σ^2. The cumulative distribution function and density are given by

$$G(z) = \frac{1}{\sigma\sqrt{2\pi}}\int_{-\infty}^{z} e^{-\frac{(x-\mu)^2}{2\sigma^2}}\,dx \implies g(z) = G'(z) = \frac{1}{\sigma\sqrt{2\pi}}e^{-\frac{(z-\mu)^2}{2\sigma^2}}.$$

The condition for $u(x,y) = xG\left(\frac{x+y}{2}\right) + y(1 - G\left(\frac{x+y}{2}\right))$ to have a saddle point becomes

$$\det \begin{bmatrix} u_{xx} & u_{xy} \\ u_{xy} & u_{yy} \end{bmatrix} = -\left(G'\left(\frac{x+y}{2}\right)\right)^2 = -\frac{1}{2\pi\sigma^2}e^{-\frac{(x+y-2\mu)^2}{4\sigma^2}} < 0.$$

Any critical point is therefore a saddle point so we may apply the preceding.

The mean μ is the median for this distribution resulting in $\frac{x^*+y^*}{2} = \mu$. The spread between the union offer and the management offer should be

$$y^* - x^* = \frac{1}{g(\mu)} = \sigma\sqrt{2\pi}.$$

Solving the two linear equations $y^* + x^* = 2\mu, y^* - x^* = \sigma\sqrt{2\pi}$ results in the offers by management $x^* = \mu - \sigma\sqrt{\frac{\pi}{2}}$ and the union $y^* = \mu + \sigma\sqrt{\frac{\pi}{2}}$.

4.2.2 Existence of Pure NE

Of course it is not true that every collection of payoff functions will have a pure Nash equilibrium as we have seen, but there is at least one result guaranteeing that one exists. It should remind you of von Neumann's Minimax Theorem, but it is more general than that because it doesn't have to be zero sum. It also generalizes our previous statement of Nash's theorem for bimatrix games.

Theorem 4.2.5 *Let $Q_1 \subset \mathbb{R}^n$ and $Q_2 \subset \mathbb{R}^m$ be compact and convex sets.*
 (a) Suppose that the payoff functions $u_i : Q_1 \times Q_2 \to \mathbb{R}$, $i = 1, 2$, satisfy

1. *u_1 and u_2 are continuous.*
2. *$q_1 \mapsto u_1(q_1, q_2)$ is concave for each fixed q_2.*
3. *$q_2 \mapsto u_2(q_1, q_2)$ is concave for each fixed q_1.*

 Then, there is a Nash equilibrium for (u_1, u_2).
 (b) Let $Q_i \subset \mathbb{R}^{n_i}$ be convex and compact. If we have N payoff functions $u_i : Q_1 \times Q_2 \times \cdots \times Q_N \to \mathbb{R}$, $i = 1, 2, \ldots, N$, which are continuous and $q_i \mapsto u_i(q_i, q_{-i})$ is concave, then there is a Nash equilibrium $(q_1^*, \ldots, q_N^*)$.

We will not prove this theorem which is due to John Nash. It indicates that if we do not have convexity of the strategy sets and concavity of the payoff functions, we may not have an equilibrium. It also says more than that. It gives us a way to set up mixed strategies for games that do not have pure Nash equilibria; namely, we have to convexify the strategy sets, and then make the payoffs concave in each variable. We do that by using **continuous probability distributions** instead of the discrete ones corresponding to mixed strategies for matrix games. We have already seen an example of this in Example 4.8

Getting back to the similarities with von Neumann's minimax Theorem 1.7.2, von Neumann's theorem says roughly that a function $f(x, y)$ that is concave in x and convex in y will have a saddle point. The connection with the Nash theorem is made by noticing that $f(x, y)$ is the payoff for player I and $-f(x, y)$ is the payoff for player II, and so if $y \mapsto f(x, y)$ is convex, then $y \mapsto -f(x, y)$ is concave. So Nash's result is a true generalization of the von Neumann minimax theorem. The Nash theorem is only a sufficient condition, not a necessary condition for a Nash equilibrium. In fact, many of the conditions in the theorem may be weakened, but we will not go into that here.

We conclude this section with several examples for finding pure Nash equilibria in interesting and important N-person games.

Example 4.10   Tragedy of the Commons. There are many players consuming common resources, some of which may be renewable or nonrenewable. Typical examples of common resources include (1) fisheries in international waters, (2) ground water, (3) email, (4) pollution, etc.

Suppose there are N farmers who share grazing land for sheep. Each of the farmers has the option of keeping 1 sheep. The payoff to a farmer for having a sheep is 1, but sheep damage the common grazing land at cost -5 per sheep. Suppose each farmer has the payoff function

$$u_i(x_1, \ldots, x_N) = x_i - 5 \frac{x_1 + \cdots + x_N}{N},$$

$$x_i = \begin{cases} 1, & \text{if } i \text{ has a sheep;} \\ 0, & \text{otherwise.} \end{cases} \quad , \quad i = 1, 2, \ldots, N.$$

The total damage done to the grazing land is -5 times the number of sheep, but the damage is shared by all the farmers.

Now here is an amazing result. If $N \geq 5$ a Nash equilibrium is $(1, 1, \ldots, 1)$, all the farmers should have a sheep and $u_i(1, 1, \ldots, 1) = -4$ for each farmer.

To see why this is true, if i decides to not have a sheep while everybody else sticks with their sheep, we have

$$u_i(0, 1_{-i}) = u_i(1, 1, \ldots, 1, 0, 1, \ldots, 1) = 0 - 5\frac{N-1}{N}$$

$$= -5 + \frac{5}{N} \leq -4 = u_i(1, 1_{-i})$$

if and only if $N \geq 5$.

The result is everybody loses a lot. Is there a way to avoid this outcome? One way is to impose a sheep tax. Let's say the tax is α and this changes the payoffs to

$$u_i(x_1, \ldots, x_N) = x_i - \alpha x_i - 5\frac{x_1 + \cdots + x_N}{N}$$

Then, if everyone has a sheep $u_i(1, \ldots, 1) = -4 - \alpha$. If $\alpha \geq 1$ it is easy to see that $u_i(0, 1_{-i}) > -4 - \alpha$, and so player i is better off getting rid of his sheep. Indeed, the new Nash equilibrium is $(0, 0, \ldots, 0)$, and no farmer has a sheep.

Tragedy of the Commons, a Continuous Version. In this example we again take the common resource to be grazing land for sheep but we assume farmers have a continuous number of sheep. There are two shepherds each of which has $x_i, i = 1, 2$ sheep. Sheep generate wool depending on how much grass they eat. Income for each shepherd is proportional to the wool they can sell so a reasonable payoff function representing the amount of wool generated is

$$u_1(x_1, x_2) = x_1^2 + x_1(100 - 2(x_1 + x_2)),$$
$$u_2(x_1, x_2) = x_2^2 + x_2(100 - 2(x_1 + x_2)).$$

The second term takes into account that the grazing land can accommodate no more than 100 sheep. Taking partials and setting to zero gives

$$2x_1 + 100 - 4x_1 - 2x_2 = 0, \ 2x_2 + 100 - 4x_2 - 2x_1 = 0 \implies x_1 + x_2 = 50$$

and, to be fair, each shepherd should graze 25 sheep, yielding $u_i(25, 25) = 625$.

Suppose that the two shepherds meet in a bar before the grazing begins in order to come to an agreement about how many sheep each shepherd should graze. Naturally, they will assume that $x_1 = x_2$ or they begin with an unfair situation. Consequently, the payoff function of each player becomes

$$u_1(x, x) = x^2 + x(100 - 4x) = -3x^2 + 100x.$$

The maximum occurs at $x = \frac{50}{3} < 25$. Together they should graze a total of about 33 sheep while without an agreement they would graze a total of 50 sheep. The payoff to each shepherd if they follow the agreement will be

$$u_1(\frac{50}{3}, \frac{50}{3}) = \frac{2500}{3}$$

and this is strictly greater than 625, the amount they could get on their own. By cooperating both shepherds get a higher amount of wool. On the other hand, if you have up to 25 sheep to graze, you might think that you could sneak in a few more than $\frac{50}{3}$ to get a few more units of wool. Suppose you assume the other guy, say player 2, will go with the agreed upon $\frac{50}{3}$. What should you do then? Your payoff is

$$u_1(x_1, \frac{50}{3}) = x_1^2 + x_1(100 - 2(x_1 + \frac{50}{3}))$$

which is maximized at $x_1 = \frac{100}{3}$, yielding shepherd 1 a payoff of $\frac{10000}{9} = 1111.11$. Shepherd 1 definitely has an incentive to cheat. So does shepherd 2. Unless there is some penalty for violating the agreement, each player will end up grazing their Nash equilibrium.

If we were to consider a Kantian Equilibrium instead we would look at

$$\max_{\gamma > 0} \gamma^2 x_1^2 + \gamma x_1 (100 - 2\gamma(x_1 + x_2)) \implies \gamma = \frac{100 x_1}{2x_1^2 + 4x_1 x_2}.$$

Since we must have $\gamma = 1$, we get the equation $x_1^2 + 2x_1 x_2 - 50 x_1 = 0$. By symmetry, it seems reasonable that $x_1 = x_2$ and so we get $3x_1^2 - 50x_1 = 0$, which has solution $x_1 = \frac{50}{3}$. This is the same as the NE.

In general, take any game with a profit function of the form

$$u_i(x_1, x_2) = x_i f(x_1 + x_2) - c\, x_i$$

where c is a unit cost and $f(\cdot)$ is a function giving the unit return to grazing. It is assumed that f is a decreasing function of the total sheep grazing to reflect the overgrazing effect. The result will illustrate a tragedy of the commons effect.

Example 4.11 Consumption Over Two Periods. Suppose we have a two-period consumption problem. If we have two players consuming $x_i, i = 1, 2$ units of a resource today, then they have to take into account that they will also consume the resource tomorrow. Suppose the total amount of the resource is $M > 0$. On day one we must have $x_1 + x_2 \leq M$. However, since the players act independently, it is possible they will choose $x_1 + x_2 > M$ in which case we will set $x_1 = x_2 = \frac{M}{2}$.

On day two, the amount of resource available is $M - (x_1 + x_2)$ and we assume that each player will get half of the amount available at that time. Each player only decides how much they consume today but they have to take into account tomorrow as well. Now let's set up each player's payoff function. For player $i = 1, 2$,

$$u_i(x_1, x_2) = \ln x_i + \ln\left(\frac{M - x_1 - x_2}{2}\right)$$

where we measure the utility of consumption using an increasing but concave down function. This is a typical payoff function illustrating diminishing returns – a dollar more when you have a lot of them is not going to bring you a lot of pleasure. Another point to make is that a player chooses to consume x_i in period one, and then equally shares the remaining amount of resource in period 2. Anyone who doesn't like to share has an incentive to consume more in period one.

Let's assume that $0 < x_1 + x_2 < M$. Then

$$\frac{\partial u_i(x_1, x_2)}{\partial x_i} = \frac{1}{x_i} - \frac{1}{M - x_1 - x_2} = 0, \ i = 1, 2.$$

Consider player 1. We solve the condition for x_1 as a function of x_2 to obtain the best response function for player 1 given by

$$x_1^*(x_2) = \frac{M - x_2}{2}.$$

Similarly $x_2^*(x_1) = \frac{M - x_1}{2}$. If we plot these two lines in the $x_1 - x_2$ plane, we see that they cross when $x_1^* = x_2^* = \frac{M}{3}$. This is the Nash equilibrium. They consume together $\frac{2}{3}M$ in the first period and defer one third for the next period. The payoff to each player is then

$$u_i\left(\frac{M}{3}, \frac{M}{3}\right) = \ln\frac{M}{3} + \ln\left(\frac{M}{6}\right).$$

Next, we consider what happens if the players act as one for the entire problem. In other words, they seek to maximize their total payoffs as a function of (x_1, x_2). The problem becomes

$$\text{Maximize } u(x_1, x_2) = u_1(x_1, x_2) + u_2(x_1, x_2), \quad x_1 + x_2 \leq M.$$

Maximizing the sum of the payoffs is called looking for the **social optimum**. Assuming $0 < x_1 + x_2 < M$ we again take derivatives and set to zero to get

$$\frac{1}{x_i} - 2\frac{1}{M - x_1 - x_2} = 0, \quad i = 1, 2$$

which results in

$$x_1 = \frac{M}{4} = x_2$$

as the social optimum amount of consumption. Acting together, they each consume half the resource in period one. Why do they do that?

We can generalize this to $N > 2$ players quite easily.

Suppose that M is the total amount of a common resource available and let $x_i, i = 1, 2, \ldots, N$ denote the amount player i consumes in period one. We must have

$$x_1 + \cdots + x_N \leq M.$$

By convention, if $x_1 + \cdots + x_N > M$, we take $x_i = \frac{M}{N}$. Tomorrow the amount of consumption for each player is set as $\frac{M - (x_1 + \cdots + x_N)}{N}$ which means they get an equal share of what's left over after the first days consumption.

Next, the utility of consumption is given by

$$u_i(x_1, \ldots, x_N) = \ln(x_i) + \ln\left(\frac{M - (x_1 + \cdots + x_N)}{N}\right)$$

Then, player i's goal is to

$$\text{Maximize } u_i(x_1, \ldots, x_N), \quad \text{subject to } 0 \leq x_i \leq M, x_1 + \cdots + x_N \leq M.$$

Assuming we have an interior maximum, we can find the best response function by taking derivatives.

$$\frac{\partial u_i(x_1, \ldots, x_N)}{\partial x_i} = \frac{1}{x_i} - \frac{1}{M - \sum\limits_{k=1}^{N} x_k} = 0$$

which implies $x_i^*(x_1, x_2, \ldots, x_{i-1}, x_{i+1}, \ldots, x_N) = \frac{1}{2}(M - \sum_{k=1, k \neq i}^{N} x_k)$. By symmetry we get

$$x_1^* = x_2^* = \cdots = x_N^* = \frac{M}{N+1}$$

This means that each consumer uses up $\frac{M}{N+1}$ of the resources today, and then $\frac{M}{N(N+1)}$ of the resource tomorrow.

Now we compare what happens if the consumers share the resource agreeing to maximize their benefits as a group instead of independently. In this case, we maximize the social optimum

$$\sum_{i=1}^{N} u_i(x_1, \ldots, x_N) = \sum_{i=1}^{N} \ln(x_i) + N \ln\left(\frac{M - (x_1 + \cdots + x_N)}{N}\right).$$

Solving this system results in the social optimum point $x_1^* = \cdots = x_N^* = \frac{M}{2N}$. Note that as $N \to \infty$ the social optimum amount of consumption in period one approaches zero. If a player gives up a unit

Figure 4.3 Braess paradox.

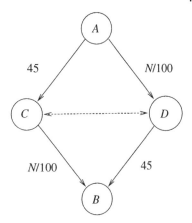

of consumption in period one, then the player has to share that unit of consumption with everyone in period two.

Our next example has had far reaching implications to many areas but especially transportation problems. Did you ever think that adding a new road to an already congested system might makes things worse?

Example 4.12   Braess's Paradox. In this example we will consider an N-person nonzero sum game in which each player wants to minimize a cost. The problem will add a resource which should allow the players to reduce their cost, but in fact makes things worse. This is a classic example called the Braess's paradox[1]. Refer to Figure 4.3 for the network of roads with branches labelled with travel times. The zip road is the road which may be added is the dotted line.

There are N commuters who want to travel from A to B. The travel time on each leg depends on the number of commuters who choose to take that leg. Notice that if a commuter chooses $A \to D \to B$ the initial choice of leg determines automatically the second leg before the zip road is added. The travel times are as follows:

- $A \to D$ and $C \to B$ is $N/100$
- $A \to C$ and $D \to B$ is 45

Each player wants to minimize her own travel time, which is taken as the payoff for each player.

The total travel time for each commuter $A \to D \to B$ is $N/100 + 45$, and the total travel time $A \to C \to B$ is also $N/100 + 45$. If one of the routes took less time, any commuter would switch to the quicker route. Thus the only Nash equilibrium is for exactly $N/2$ players to take each route (assuming N is even, what if N is odd?) For example, if $N = 2000$ then the travel time under equilibrium is 55 and 1000 choose $A \to D \to B$ and 1000 choose $A \to C \to B$. This assumes perfect information for all commuters.

Now if a zip road (the dotted line) from $C \leftrightarrow D$ is built which takes essentially no travel time, then consider what a rational driver would do. We'll take $N = 2000$ for now. Since $N/100 = 20 < 45$ all drivers would choose initially $A \to D$. When the driver gets to D, a commuter would take the zip road to C and then take $C \to B$. This results in a total travel time of $N/100 + N/100 = 40 < 55$, and the zip road saves everyone 15.

But is this always true no matter how many commuters are on the roads? Actually, no. For example, if $N = 4000$ then all commuters would pick $A \to D$ and this would take $N/100 = 40$; then

---

1 Due to Professor Dietrich Braess, Mathematics Department, Ruhr University Bochum, Germany.

they would take the zip road to C and travel from $C \to B$. The total commute time would be 80. If they skip the zip road, their travel time would be $40 + 45 = 85 > 80$ so they will take the zip road. On the other hand, if the zip road did not exist, half the commuters would take $A \to D \to B$ and the other half would take $A \to C \to B$, with a travel time of $45 + 20 = 65 < 80$. Giving the commuters a way to cross over actually makes things worse! Having an option and assuming commuters always do what's best for them, results in longer travel time for everyone.

Braess's paradox has been seen in action around the world and traffic engineers are well aware of the paradox. Moreover, this paradox has applications to much more general networks in computer science and telecommunications. It should be noted that this is not really a paradox since it has an explanation, but it is completely counterintuitive.

Our next two examples show what happens when players think through the logical consequences.

Example 4.13   The Traveler's Paradox.[1] Two airline passengers who have luggage with identical contents are informed by the airline that their luggage has been lost. The airline offers to compensate them if they make a claim in some range acceptable to the airline. Here are the payoff functions for each player

$$u_1(q_1, q_2) = \begin{cases} q_1 & \text{if } q_1 = q_2; \\ q_1 + R & \text{if } q_2 > q_1; \\ q_2 - R & \text{if } q_1 > q_2. \end{cases} \quad \text{and } u_2(q_1, q_2) = \begin{cases} q_2 & \text{if } q_1 = q_2; \\ q_1 - R & \text{if } q_2 > q_1; \\ q_2 + R & \text{if } q_1 > q_2. \end{cases}$$

It is assumed that the acceptable range is $[a, b]$ and $q_i \in [a, b], i = 1, 2$. The idea behind these payoffs is that if the passengers' claims are equal the airline will pay the amount claimed. If passenger I claims less than passenger II, $q_1 < q_2$, then passenger II will be penalized an amount R and passenger I will receive the amount she claimed plus R. Passenger II will receive the lower amount claimed minus R. Similarly, if passenger I claims more than does passenger II, $q_1 > q_2$, then passenger I will receive $q_2 - R$, and passenger II will receive the amount claimed q_2 plus R.

Suppose, to be specific, that $a = 80, b = 200$, so the airline acceptable range is from \$80 to \$200. We take $R = 2$, so the penalty is only 2 dollars for claiming high. Believe it or not, we will show that the Nash equilibrium is $(q_1 = 80, q_2 = 80)$ so both players should claim the low end of the airline range (under the Nash equilibrium concept). To see that, we have to show that

$$u_1(80, 80) = 80 \geq u_1(q_1, 80) \text{ and } u_2(80, 80) = 80 \geq u_2(80, q_2)$$

for all $80 \leq q_1, q_2 \leq 200$. Now

$$u_1(q_1, 80) = \begin{cases} 80 & \text{if } q_1 = 80; \\ 80 - 2 = 78 & \text{if } q_1 > 80. \end{cases}$$

and

$$u_2(80, q_2) = \begin{cases} 80 & \text{if } q_2 = 80; \\ 80 - 2 = 78 & \text{if } q_2 > 80; \end{cases}$$

So indeed $(80, 80)$ is a Nash equilibrium with payoff 80 to each passenger. But clearly, they can do better if they both claim $q_1 = q_2 = \$200$. Why don't they do that? The problem is that there is an incentive to undercut the other traveler. If R is \$2, then, if one of the passengers drops her claim to \$199, this passenger will actually receive \$ 201. This cascades downward, and the undercutting

---

1 Adapted from an example in [10].

disappears only at the lowest range of the acceptable claims. Do you think that the passengers would, in reality, make the lowest claim?

Example 4.14 Guessing Two-thirds of the Average. A common game people play is to choose a number in order to match some objective, usually to guess the number someone is thinking, as close as possible. In a group of people playing this game, the closest number wins. We consider a variation of this game which seems much more difficult for everyone involved.

Suppose we have 3 people who are going to choose an integer from 1 to N. The person who chooses closest to two-thirds of the average of all the numbers chosen wins $1. If two or more people choose the same number closest to two-thirds of the average, they split the $1. We set up the payoff functions as follows.

$$u_i(x_1, x_2, x_3) = \begin{cases} 1 & \text{if } x_1 \neq x_2 \neq x_3, x_i \text{ closest to } \frac{2}{3}\bar{x}, \\ \frac{1}{2} & \text{if } x_i = x_j \text{ for some } j \neq i, \text{ and } x_i \text{ closest to } \frac{2}{3}\bar{x}, \\ \frac{1}{3} & \text{if } x_1 = x_2 = x_3, \\ 0 & \text{otherwise}. \end{cases}$$

where $\bar{x} = \frac{x_1 + x_2 + x_3}{3}$. The possible choices for x_i are $\{1, 2, \ldots, N\}$.

The claim is that $(x_1^*, x_2^*, x_3^*) = (1, 1, 1)$ is a Nash equilibrium in pure strategies. All the players should call 1 and they each receive $\frac{1}{3}$.

To see why this is true, let's consider player 1. The rest would be the same. We must show

$$u_1(1, 1, 1) \geq u_1(x_1, 1, 1), \qquad x_1 \neq 1, x_1 \in \{2, 3, \ldots, N\}.$$

Suppose player 1 chooses $x_1 = k \geq 2$. Then

$$\frac{2}{3}\frac{1 + 1 + k}{3} = \frac{4}{9} + \frac{2}{9}k$$

and the distance between two-thirds of the average and k is

$$|k - \frac{4}{9} - \frac{2}{9}k| = |\frac{7}{9}k - \frac{4}{9}|$$

If player 1 had chosen $x_1 = 1$, the distance from 1 to two-thirds of the average is $|1 - \frac{2}{3}| = \frac{1}{3}$. We ask ourselves if there is a $k > 1$ so that $|\frac{7}{9}k - \frac{4}{9}| < \frac{1}{3}$? The answer to that is no, not for $k > 1$, as you can easily check. In a similar way we show that none of the players do better by switching to another integer and conclude that $(1, 1, 1)$ is a Nash equilibrium.

Example 4.15 This example illustrates a generalization of Nash Equilibrium to games in which the player's choice of strategy may depend on the **other players strategies.** For example, suppose there are two players with payoffs $P_i(x, y), i = 1, 2$ and each player want to maximize their payoff. However, there is also a constraint on the choice of strategies, say of the form $g(x, y) \leq 0$, which may come about because the players share a resource. So the game for each player is

$$\max_{x \mid g(x,y) \leq 0} P_1(x, y) \text{ and } \max_{y \mid g(x,y) \leq 0} P_2(x, y).$$

This is a generalized game and a NE is now called a **Generalized Nash Equilibrium**. Here is a specific example.

There is a triangular plot of land with vertices at $(0, 0), (1, 0)$, and $(1, 1)$. Player I picks a point $0 \leq x \leq 1$ and player II a point in $0 \leq y \leq 1$. However, the point (x, y) must lie in the triangle and so $0 \leq y \leq x \leq 1$. Player I will receive the land in the triangle to the left of x and below y as well as the

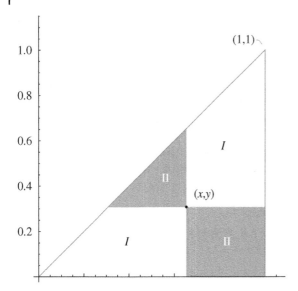

Figure 4.4 Land game.

land to the right of x and above y. Player II receives the remaining land in the triangle. This setup is depicted in Figure 4.4. Find the payoff to each player and a Nash equilibrium.

The area for player I is

$$A_I(x,y) = \frac{1}{2}(2x - y)y + \frac{1}{2}(1 + x - 2y)(1 - x), \quad x \geq y.$$

For player II, $A_{II}(x,y) = \frac{1}{2} - A_I(x,y)$.

This means player I wants to maximize A_I and player II wants to minimize it.

Set $S(y) = \{x \in [0,1] \mid 0 \leq y \leq x\}$. We have

$$\max_{x \in S(y)} A_I(x,y) = f(y) = \begin{cases} y - \frac{y^2}{2}, & \frac{1}{2} \leq y \leq 1 \\ \frac{3y^2}{2} - y + \frac{1}{2}, & 0 \leq y \leq \frac{1}{2} \end{cases}$$

with the maximum achieved at $x^*(y) = \begin{cases} 1, & \frac{1}{2} \leq y \leq 1 \\ 2y, & 0 \leq y \leq \frac{1}{2} \end{cases}$. Then, $\min_{0 \leq y \leq 1} f(y) = \frac{1}{3}$ at $y^* = \frac{1}{3}$. The

max is achieved at $x^*(y^*) = \frac{2}{3}$. The area to each player is $1/3$ to I and $1/6$ to II. We conclude that

$$\min_{0 \leq y \leq 1} \max_{x \in S(y)} A_I(x,y) = \frac{1}{3}.$$

Furthermore, we have $x^*(y^*) = \frac{2}{3}, y^* = \frac{1}{3}$ which, as the reader can check is the **unconstrained** saddle point of $A(x,y)$.

Now let's do it with player I going first. Set $S(x) = \{y \in [0,1] \mid 0 \leq y \leq x\}$. We have

$$\min_{y \in S(x)} A(x,y) = g(x) = \begin{cases} \frac{1}{2}(1 - x^2), & x = 0 \text{ or } \frac{2}{3} \leq x \leq 1 \\ (x-1)x + \frac{1}{2}, & 0 < x < \frac{2}{3} \end{cases}$$

achieved at $y^*(x) = \begin{cases} x, & 0 \leq x < \frac{2}{3} \\ 0, & \frac{2}{3} \leq x \leq 1 \end{cases}$. Next, $\max_{0 \leq x \leq 1} g(x) = \frac{1}{2}$ achieved at $x = 0$. We see that

$$\max_{0 \leq x \leq 1} \min_{y \in S(x)} A_I(x,y) = \frac{1}{2} \implies A_{II}(x^*, y^*) = 0.$$

Consequently, if player I chooses first, there is really nothing player II can do to prevent player I from grabbing all the land. Player I would simply choose $x = 0$ and because $0 \leq y \leq x$ this forces $y = 0$ and player I gets it all. On the other hand, if player II chooses first player II should pick $y = 1/3$ and this prevents player I from playing $x = 0$ because $0 \leq y \leq x$. The constraint makes a huge difference. Observe that this example shows that min max is not always greater than max min when there are constraints between the strategies. Generalized game theory is a very active area of research because of the many practical applications but in many respects it is much more difficult. Think about what it would mean to have a mixed strategy as one example of the difficulties.

Example 4.16   Breaking Up Can Be Hard to Do. Two partners must dissolve their partnership. Partner 1 currently owns share $s \in (0, 1)$ of the partnership, partner 2 owns proportion share $1 - s$. The partners agree to play the following game: partner 1 names a price, p, for the whole partnership, and partner 2 then chooses either to buy 1's share for $p \cdot s$ or to sell her share to 1 for $p(1 - s)$. Partner 1 has the option to buy out partner 2 or sell to partner 2. Suppose it is common knowledge that the partners' valuations for owning the whole partnership are independently and uniformly distributed, but that each partner's valuation is private information. What should each player do?

To answer this question, we see that player 1 has a choice of what price to ask, p. Let V_1, V_2 denote respectively, the valuation that each partner places on the company, with each V_i a random variable which is uniformly distributed on $[0, v_i]$, i.e., $V_i \sim Unif[0, v_i], i = 1, 2$. We assume $p \leq v_1, p \leq v_2$. Player 1 wants to choose p to maximize her profit given by

$$u_1 = \begin{cases} V_1 - (1 - s)p, & \text{if 2 sells to 1} \\ p \cdot s, & \text{if 2 buys out 1.} \end{cases}$$

Player 2's profit is

$$u_2 = \begin{cases} V_2 - p \cdot s, & \text{if 1 sells to 2} \\ (1 - s)p, & \text{if 1 buys out 2.} \end{cases}$$

If $V_2 - p \cdot s \leq p(1 - s)$ this says player 2's valuation less the cost to buy player 1's share is less than the amount player 2 would get from player 1 if 2 sells to 1. This is the condition needed for player 2 to sell. Therefore player 1's expected profit is

$$E(u_1) = E[V_1 - (1 - s)p]Prob(V_2 - p \cdot s \leq p(1 - s)) + p \cdot s \ Prob(V_2 - p \cdot s \geq p(1 - s)).$$

Similarly, player 1 will sell to 2 if and only if $p \cdot s \geq V_1 - (1 - s)p$ and player 2's expected profit is

$$E(u_2) = E[V_2 - p \cdot s]Prob(V_1 - (1 - s)p \leq p \cdot s) + (1 - s) \cdot p \ Prob(V_1 - (1 - s)p \geq p \cdot s).$$

Next, since we are assuming valuations are uniformly distributed,

$$Prob(V_2 - p \cdot s \leq p - p \cdot s) = Prob(V_2 \leq p) = \frac{p}{v_2}, and \qquad Prob(V_1 - (1 - s)p \leq p \cdot s) = \frac{p}{v_1},$$

so that

$$E(u_1)(p) = (\frac{v_1}{2} - (1 - s)p)\frac{p}{v_2} + p \cdot s(1 - \frac{p}{v_2})$$

and

$$E(u_2)(p) = (\frac{v_2}{2} - p \cdot s)\frac{p}{v_1} + (1 - s)p(1 - \frac{p}{v_1}).$$

Now, if player 1 is trying to decide on an offer p, she would want to maximize her expected payoff. Taking a derivative and setting to zero we get the optimal p^* as

$$p^* = \frac{2sv_2 + v_1}{4}.$$

This is dependent on player 1's belief that player 2's valuation is uniformly distributed with value no more than v_2. Player 1 knows his own valuation v_1 exactly. The best response of player 2 is then to sell if $v_2 \leq p^*$ and to buy out player 1 if $v_2 > p^*$.

Example 4.17 Jury Trials. This example which was introduced and analyzed by political scientists David Austen–Smith[1] and Jeffrey Banks[2] [2] in 1996 will illustrate once again the utility of Bayes' rule in game theory and why it is so important.

We have a defendant coming to trial by a jury who must decide to acquit or convict. Before a juror has to decide, the juror gets a signal of guilt, g, or innocence, i, based on the evidence presented during trial. Let G be the event the defendant is actually guilty and $I = G^c$ the event of actual innocence.

The jurors have beliefs about the defendant's guilt before the trial begins and we denote by $\pi = Prob(G)$ this prior belief. Let $p = Prob(g|G)$ and $q = Prob(i|I)$. Then, p is the probability a juror receives evidence of guilt given that the defendant is guilty and q is the probability the juror receives evidence of innocence given that the defendant is innocent. Of course $1 - q = Prob(g|I)$ is the probability of receiving the signal of guilt given that the defendant is innocent. We assume

$$Prob(g|G) > P(g|I), \text{i.e.,} p > 1 - q \text{ and } p > \frac{1}{2}, q > \frac{1}{2}.$$

We also assume that jurors will have signals of innocence or guilt arrived at independently.

Recapping, we have

$$\pi = Prob(G), p = Prob(g|G) = 1 - Prob(i|G), q = Prob(i|I) = 1 - Prob(g|I)$$

Every juror has the same payoff which we take to be as follows. Let $0 < z < 1$.

$$P(z) = \begin{cases} 0, & \text{if a guilty defendant is convicted} \\ & \text{or an innocent defendant is acquitted,} \\ -z, & \text{if an innocent defendant is convicted} \\ -(1 - z), & \text{if a guilty defendant is acquitted.} \end{cases}$$

In other words if the juror reaches the correct decision her payoff is 0. The juror's payoff is $-z < 0$ if an innocent defendant is convicted, and $-(1 - z) < 0$ if a guilty defendant is acquitted.[3]

Each juror wants to maximize their payoff, which is zero, and make a correct decision. If a juror **thinks a defendant is guilty with probability** r, then the expected payoff to the juror by convicting is

$$E(\text{convict}) = r \cdot \underbrace{0}_{\text{correct decision}} + (1 - r) \underbrace{(-z)}_{\text{convict innocent}} = -z(1 - r).$$

The expected payoff to the juror if she acquits is

$$E(\text{acquit}) = r \cdot \underbrace{0}_{\text{correct decision}} + r\underbrace{(-(1 - z))}_{\text{acquit guilty}} = -r(1 - z).$$

The juror, who wants the largest expected payoff, will convict if $-z(1 - r) > -(1 - z)r$, which reduces to $r > z$. Therefore for any **prior belief of guilt** above z a juror will optimally vote to

---

1 Professor of Managerial Economics and Decision Sciences, Northwestern University.
2 1958–2000, was a Professor of Political Science at Cal Tech.
3 In the language of statistics, z is the probability of a type I error, $1 - z$ the cost of a type II error.

convict. This means that z may be interpreted as the **threshold level** above which the juror will believe the defendant is guilty.

Now let's analyze what jurors should do depending on the signals of guilt or innocence.

Case 1: One juror. We know $Prob(i|G) = 1 - p$. By Bayes' rule, the probability the juror thinks the defendant is actually guilty given that she receives an innocent signal is

$$Prob(G|i) = \frac{Prob(i|G)Prob(G)}{Prob(i|G)Prob(G) + Prob(i|I)Prob(I)} = \frac{(1-p)\pi}{(1-p)\pi + q(1-\pi)}.$$

Because of the assumption $p > 1 - q$, it is easy to check that $Prob(G|i) < \pi$. This makes sense because the probability of actual guilt given the juror receives an innocence signal should be less than the juror's prior belief of guilt. Also, her posterior probability the defendant is guilty given a guilty signal is

$$Prob(G|g) = \frac{Prob(g|G)Prob(G)}{Prob(g|G)Prob(G) + Prob(g|I)Prob(I)} = \frac{p\pi}{p\pi + (1-q)(1-\pi)} > \pi.$$

The probability of guilt goes up after a guilty signal and down after an innocence signal. We see that $Prob(G|g) > \pi > Prob(G|i)$, as it should be.

If the juror's conviction threshold z satisfies $z \geq \max\{Prob(G|i), Prob(G|g)\}$, she will acquit no matter what signal she gets because she votes to convict if $r > z$. If $z \leq \min\{Prob(G|i), Prob(G|g)\}$, she will convict no matter what.

Therefore the signals matter to the juror only in the case when $Prob(G|i) < z < Prob(G|g)$, i.e.,

$$\frac{(1-p)\pi}{(1-p)\pi + q(1-\pi)} < z < \frac{p\pi}{p\pi + (1-q)(1-\pi)}. \tag{S}$$

Case 2: Two jurors. Label the 2 jurors j_1, j_2. Suppose j_2 votes to acquit if her signal is i and convict if her signal is g.

If $j_2's$ signal is i (innocent), then $j_1's$ vote is irrelevant because a unanimous verdict is needed for conviction. So j_1 can ignore the possibility $j_2's$ signal may be i and assume it is g, otherwise j_2 has already decided the verdict.

When will j_1 vote to acquit when her signal is i? In the case j_1 is i and j_2 is g, j_1 thinks the probability the defendant is guilty is

$$Prob(G|i, g) = \frac{Prob(i, g|G)Prob(G)}{Prob(i, g|G)Prob(G) + Prob(i, g|I)Prob(I)}$$
$$= \frac{(1-p)p\pi}{(1-p)p\,\pi + q(1-q)(1-\pi)}$$

In Bayes' rule we have used the conditional independence of the jurors. Using the payoffs, j_1 will vote to acquit if $(1 - z)Prob(G|i, g) \leq z(1 - Prob(G|i, g))$ which requires

$$z > Prob(G|i, g) = \frac{(1-p)p\pi}{(1-p)p\,\pi + q(1-q)(1-\pi)}.$$

Similarly, if j_1 receives a signal g, she votes to convict if

$$z < \frac{p^2\pi}{p^2\,\pi + q(1-q)^2(1-\pi)}.$$

As before, we conclude j_1 optimally votes depending on her signal if

$$\frac{(1-p)p\pi}{(1-p)p\,\pi + q(1-q)(1-\pi)} < z < \frac{p^2\pi}{p^2\,\pi + q(1-q)^2(1-\pi)}. \tag{S1}$$

This says that in the case of two jurors, voting according to the signal received is a Nash equilibrium if (S1) holds.

We claim now that a juror is less worried about convicting an innocent defendant when there are two jurors than when there is only one. In fact, since $p > \frac{1}{2} > 1 - q$ by assumption, we compare lower bounds for z,

$$\underbrace{\frac{(1-p)p\pi}{(1-p)p\,\pi + q(1-q)(1-\pi)}}_{\text{2 jurors}} \geq \underbrace{\frac{(1-p)\pi}{(1-p)\pi + q(1-\pi)}}_{\text{one juror}}.$$

For instance, if $p = \frac{3}{4}, q = \frac{7}{8}$, and $\pi = \frac{1}{2}$, we have the left side is 0.34772 and the right side is 0.222. In the critical case, we have $Prob(G|i,g) = 0.34772$ as the juror's posterior probability of guilt.

Case 3: N jurors: Suppose every juror other than j_1 votes according to her signal, i.e., votes to acquit if her signal is i and convict if her signal is g. Assume (S) holds for each juror. We want to see if voting according to signal is a Nash equilibrium.

Due to unanimity, j_1 can ignore the possibility some other juror's signals may be i and just assume they are all g. When will j_1 vote to acquit when her signal is i? When $j_1's$ signal is i and every other juror's signal is g, j_1 thinks the probability the defendant is guilty is

$$Prob(G|i,g,g,\dots,g) = \frac{Prob(i,g,g,\dots,g|G)Prob(G)}{Prob(i,g,g,\dots,g|G)Prob(G) + Prob(i,g,g,\dots,g|I)Prob(I)}$$

$$= \frac{(1-p)p^{N-1}\pi}{(1-p)^{N-1}p\,\pi + q(1-q)^{N-1}(1-\pi)}$$

$$= \frac{1}{1 + \frac{q}{1-p}\left(\frac{1-q}{p}\right)^{N-1}\left(\frac{1-\pi}{\pi}\right)}$$

If the chance of finding a guilty defendant guilty is greater than the chance of not finding an innocent defendant innocent, i.e., $p > 1 - q$ the denominator goes to 1 as $N \to \infty$. So the lower bound on z for which j_1 votes for acquittal when her signal is i goes to 1 as $N \to \infty$. In words, in a large jury, if jurors to any extent do not want to acquit a guilty defendant then a juror who sees evidence of innocence will nevertheless vote for conviction. As a numerical example, if $N = 12, \pi = 0.5, p = q = 0.8$, we already have $Prob(G|i,g,g,\dots,g) = 0.9999$. In a 12 member jury in which guilt or innocence is a toss up at the beginning of the trial, but the probability of guilt goes up to 0.8 if a guilt signal is received and the probability of innocence goes up to 0.8 if an innocence signal is received, the threshold of conviction is almost 100%. Does this make sense? One explanation is that for N large enough people will not vote according to their signals since a juror thinks that their innocent signal cannot be trusted when it is the only one in a large group of jurors voting to convict.

It is possible to show that a juror whose strategy is to vote according to the signal received cannot be a Nash equilibrium as $N \to \infty$. There is always a Nash equilibrium where all jurors vote to acquit. However, there is a mixed Nash equilibrium, which we will not derive, in which jurors vote to convict with probability 1 if they get a guilty signal. If they get an innocence signal, they still vote to convict with some positive probability γ given by

$$\gamma = \frac{p\,B^{\frac{1}{N-1}} - (1-q)}{q - (1-p)B^{\frac{1}{N-1}}} \quad \text{where } B = \frac{(1-z)(1-p)\pi}{(1-\pi)\,z}$$

Notice that $\gamma \to 1$ as $N \to \infty$. Remember that γ is the probability a juror votes to convict even though they get an innocence signal. With $N = 12$ jurors and $\pi = 0.5$ the prior probability of guilt, and $p = Prob(g|G) = q = Prob(I|i) = 0.8$, and $z = 0.7$ as the threshold level of guilt above which a juror votes to convict, we calculate

$$B = \frac{(1-0.7)(1-0.8)0.5}{(1-0.5) \ 0.7} = 0.0857, \text{ and then } \gamma = \frac{0.8(0.0857)^{1/11} - (1 - 0.8)}{0.8 - 0.2(0.0857)^{1/11}} = 0.6873$$

In the mixed Nash equilibrium, a juror in a 12 member jury will optimally vote to convict is almost 70% even though an innocence signal was received.

Problems

4.1 This problem makes clear the connection between a pair of numbers (q_1^*, q_2^*) which maximizes both payoff functions, and a Nash equilibrium. Suppose that we have a two-person game with payoff functions $u_i(q_1, q_2)$, $i = 1, 2$. Suppose there is a pure strategy pair (q_1^*, q_2^*) which maximizes both u_1 and u_2 as functions of the pair (q_1, q_2).
   (a) Verify that (q_1^*, q_2^*) is a Nash equilibrium.
   (b) Construct an example in which a Nash equilibrium (q_1^*, q_2^*) does *not* maximize both u_1 and u_2 as functions of (q_1, q_2).

4.2 Consider the zero sum game with payoff function for player 1 given by $u(x, y) = -2x^2 + y^2 + 3xy - x - 2y, 0 \leq x, y \leq 1$. Show that this function is concave in x and convex in y. Find the saddle point and the value.

4.3 The payoff functions for Curly and Shemp are, respectively,

$$c(x, y) = xy^2 - x^2, \quad s(x, y) = 8y - xy^2,$$

with $x \in [1, 3], y \in [1, 3]$. Curly chooses x and Shemp chooses y.
   (a) Find and graph the rational reaction sets for each player. Recall that the best response for Curly is $BR_c(y) \in \arg\max c(x, y)$ and the rational reaction set for Curly is the graph of $BR_c(y)$. Similarly for Shemp.
   (b) Find a pure Nash equilibrium.

4.4 Consider the game in which two players choose nonnegative integers no greater than 1000. Player 1 must choose an even integer while player 2 must choose an odd integer. When they announce their number, the player who chose the lower number wins the number she announced in dollars. Find the Nash equilibrium.

4.5 There are N players each using r_i units of a resource whose total value is $R = \sum_{i=1}^{N} r_i$. The cost to player i for using r_i units of the resource is $f(r_i) + g(R - r_i)$ where we take $f(x) = 2x^2, g(x) = x^2$. This says that a player's cost is a function of the amount of the total used by that player and a function of the amount used by the other players. The revenue player i receives from using r_i units is $h(r_i)$, where $h(x) = \sqrt{x}$. Assume that the total resources used by all players is not unlimited so that $0 \leq R \leq R_0 < \infty$.
   (a) Find the payoff function for player $i = 1, 2, \ldots, N$.
   (b) Find the Nash equilibrium in general and when $N = 12$.

(c) Now suppose that the total amount of resources is R and each player will use $\frac{R}{N}$ units of the total. Here R is unknown and is chosen so that the *social welfare* is maximized, i.e.,

$$\text{Maximize} \sum_{i=1}^{N} u_i \left(\frac{R}{N}, \ldots, \frac{R}{N} \right),$$

over $R \geq 0$. Practically, this means that the players do not act independently but work together so that the total payoffs to all players is maximized. Find the value of R which provides the maximum, R^s, and find it's value when $N = 12$.

4.6 In the tragedy of the commons Example 4.10, we saw that if $N \geq 5$ everyone owning a sheep is a Nash equilibrium. Analyze the case $N \leq 4$.

4.7 Consider the median voter model Example 4.1. Let X denote the preference of a voter and suppose X has density $f(x) = -1.23x^2 + 2x + 0.41, 0 \leq x \leq 1$. Find the Nash equilibrium position a candidate should take in order to maximize their voting percentage.

4.8 In the arbitration game Example 4.9, we have seen that the payoff function is given by

$$u(x, y) = xG \left(\frac{x+y}{2} \right) + y \left(1 - G \left(\frac{x+y}{2} \right) \right).$$

where G is a cumulative distribution function.

(a) Assume that $G' = g$ is a continuous density function with $g(z) > 0$. Show that any critical point of $u(x, y)$ must be a saddle point of u in the sense that $\det \begin{bmatrix} u_{xx} & u_{xy} \\ u_{xy} & u_{yy} \end{bmatrix} < 0$.

(b) Derive the optimal offers for the exponential distribution with $\lambda > 0$

$$G(z) = 1 - e^{-\lambda z}, z \geq 0 \implies g(z) = \lambda e^{-\lambda z}, z > 0.$$

Assume that the minimum offer must be at least $a > 0$ and find the smallest possible $a > 0$ so that the offers are positive. In other words, the range of offers is $X + a$ and X has an exponential distribution.

4.9 Consider the game in which player I chooses a non-negative number x, and player II chooses a non-negative number y (x and y are not necessarily integers). The payoffs are

$$u_1(x, y) = x(4 + y - x) \text{ for player I}, u_2(x, y) = y(4 + x - y) \text{ for player II}.$$

Determine player I's best response $x(y)$ to a given y, and player II's best response $y(x)$ to a given x. Find a Nash equilibrium and give the payoffs to the two players.

4.10 Two players decide on the amount of effort they each will exert on a project. The effort level of each player is $q_i \geq 0, i = 1, 2$. The payoff to each player is

$$u_1(q_1, q_2) = q_1(c + q_2 - q_1), \quad \text{and} \quad u_2(q_1, q_2) = q_2(c + q_1 - q_2),$$

where $c > 0$ is a constant. This choice of payoff models a synergistic effect between the two players.

(a) Find each player's best response function.

(b) Find the Nash equilibrium.

4.11 Suppose two citizens are considering how much to contribute to a public playground, if anything. Suppose their payoff functions are given by

$$u_i(q_1, q_2) = q_1 + q_2 + w_i - q_i + (w_i - q_i)(q_1 + q_2), i = 1, 2,$$

where w_i =wealth of player i. This payoff represents a benefit from the total provided for the playground $q_1 + q_2$ by both citizens, the amount of wealth left over for private benefit $w_i - q_i$, and an interaction term $(w_i - q_i)(q_1 + q_2)$ representing the benefit of private money with the amount donated. Assume $w_1 = w_2 = w$ and $0 \leq q_i \leq w$. Find a pure Nash equilibrium.

4.12 Consider the following escalating war of attrition. C1 and C2 are at war. On day N of the war, fighting costs each approximately N^2 billion dollars. On day one, both C1 and C2 pay $1B$ each. On day two, each pays $4B$, for a total of $5B$ each after 2 days of fighting. There is a prize worth $500B$ for the winner of the war, and the loser will pay a penalty of an additional $800B$ in reparations to the winner.
  (a) Show that the total cost of fighting for N days is *approximately* $\frac{N^3}{3}$.
      Hint: $\sum_{k=1}^{n} k^2 = \frac{n(n+1)(2n+1)}{6}$.
  (b) What is the payoff to each player if the war ends with C1 giving up on day 5?

4.13 With the same information as in the previous problem, assume C2 is playing a mixed strategy for the time to quit, the random variable T with pdf $p(t)$, and C1 is playing the pure strategy "give up on day 5, if C2 is still fighting." Set up an expression (which will involve some integrals) giving the expected payoff to C1.
  (a) Write down the expression for the payoff if C1 is still playing "give up on day 5, if C2 is still fighting," and C2 uses the mixed strategy $p(t) = 2e^{-2t}$.
  (b) Write down the equation you would use to solve for a totally mixed symmetric Nash $p(t)$ using Proposition 4.2.4.
  (c) Differentiate your equation twice to find a differential equation for $p(\cdot)$.
  (d) Solve the differential equation for $p(t)$ using the condition $\int_0^\infty p(t)\, dt = 1$.

4.14 Consider the war of attrition game in Example 4.6. Verify that

$$BR_1(t_2) = \begin{cases} 0, & \text{if } t_2 > \frac{v}{c}; \\ (t_2, \infty), & \text{if } t_2 < \frac{v}{c}; \\ \{0\} \cup (t_2, \infty), & \text{if } t_2 = \frac{v}{c}. \end{cases}$$

and

$$BR_2(t_1) = \begin{cases} 0, & \text{if } t_1 > \frac{v}{c}; \\ (t_1, \infty), & \text{if } t_1 < \frac{v}{c}; \\ \{0\} \cup (t_1, \infty), & \text{if } t_1 = \frac{v}{c}. \end{cases}$$

  (a) Verify that

$$BR_1(t_2) \cap BR_2(t_1) = \{(t_1 \geq \frac{v}{c}, t_2 = 0), (t_1 = 0, t_2 \geq \frac{v}{c})\}.$$

  (b) Verify that $t_1^* = \frac{v}{c}, t_2^* = 0$ is a pure Nash equilibrium.
  (c) We assumed the cost of continuing the war was $c\, t$. Suppose instead the cost is $c\, t^2$. Find the Nash equilibrium in pure strategies.

(d) Suppose we modify the payoff function to include a cost if country 1 quits after country 2 and vice versa. In addition, we slightly change what happens when they quit at the same time. Take the payoff for country 1 to be

$$u_1(t_1, t_2) = \begin{cases} -c\, t_1, & \text{if } t_1 < t_2; \\ v - c\, t_2 - c\, t_1, & \text{if } t_1 \geq t_2. \end{cases}$$

Find the payoff for country 2 and find the pure Nash equilibrium.

4.15 This problem considers the War of Attrition but with differing costs of continuing the war and differing values placed upon the resource over which the war is fought. Country 1 places value v_1 on the land, and country 2 values it at v_2. The players choose the time at which to concede the land to the other player, but there is a cost for letting time pass. Suppose that the cost to each country is $c_i, i = 1, 2$ per unit of time. The first player to concede yields the land to the other player at that time. If they concede at the same time, each player gets half the land. Determine the payoffs to each player and determine the pure Nash equilibria.

4.16 Consider the general war of attrition as described in this section with given functions α, β. The payoff functions are

$$u_1(t_1, t_2) = \begin{cases} \beta(t_2), & \text{if } t_1 > t_2; 2 \text{ quits before } 1; \\ \alpha(t_1), & \text{if } t_1 < t_2; 1 \text{ quits before } 2. \\ \frac{\alpha(t_1) + \beta(t_1)}{2}, & \text{if } t_1 = t_2. \end{cases}$$

and

$$u_2(t_1, t_2) = \begin{cases} \beta(t_1), & \text{if } t_2 > t_1; 1 \text{ quits before } 2; \\ \alpha(t_2), & \text{if } t_2 < t_1; 2 \text{ quits before } 1. \\ \frac{\alpha(t_1) + \beta(t_1)}{2}, & \text{if } t_1 = t_2. \end{cases}$$

Verify that if we take α and β to be continuous, decreasing functions with $\alpha(t) < \beta(t)$ and $\alpha(0) > \beta(1)$, then the general war of attrition has two pure Nash equilibria given by $\{(t_1 = \tau^*, t_2 = 0), (t_1 = 0, t_2 = \tau^*)\}$ where $\tau^* > 0$ is the first time where $\beta(\tau^*) = \alpha(0)$.

4.17 This is an exercise on Braess's paradox. Consider the traffic system in which travelers want to go from $A \to C$. The travel time for each car depends on the total traffic. In Figure 4.5 the travel time on each leg is given by the formula in which x is the number of cars on the leg.
(a) Suppose the total number of cars is 6 and each car can decide to take $A \to B \to C$ or $A \to D \to C$. Show that 3 cars taking each path is a Nash equilibrium.

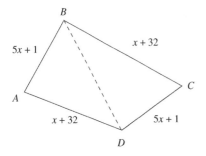

Figure 4.5 Braess' paradox.

(b) Suppose a new road $B \rightarrow D$ is built which has travel time zero no matter how many cars are on it. Cars can now take the paths $A \rightarrow B \rightarrow D \rightarrow C$ and $A \rightarrow D \rightarrow B \rightarrow C$ in addition to the paths from the first part. Find the Nash equilibrium and show that the travel time is now worse for all the players.

4.18 We have a game with $N \geq 2$ players. Each player i must simultaneously decide whether to join the team or not. If player i joins then $x_i = 1$ and otherwise $x_i = 0$. Let $T = \sum_{i=1}^{N} x_i$ denote the size of the team. If player i doesn't join the team, then player i receives a payoff of zero. If player i does join the team then i pays a cost of c. If all N players join the team, so that $T = N$, then each player enjoys a benefit of v. Hence player i's payoff is

$$u_i(x_1, \ldots, x_N) = \begin{cases} v - c, & \text{if } T = \sum_{i=1}^{N} x_i = N; \\ -x_i c, & \text{if } T < N. \end{cases}$$

Suppose that $v > c > 0$.
(a) Find all of the pure-strategy Nash equilibria.
(b) Take $N = 2$. Find a mixed Nash equilibrium.

4.19 Two companies are at war over a market with a total value of $V > 0$. Company 1 allocates an effort $x > 0$, and company 2 allocates an effort $y > 0$ to obtain all or a portion of V. The portion of V won by company 1 if they allocate effort x is $(\frac{x}{x+y})V$ at cost $C_1 x$, where $C_1 > 0$ is a constant. Similarly, the portion of V won by company 2 if they allocate effort y is $(\frac{y}{x+y})V$ at cost $C_2 y$, where $C_2 > 0$ is a constant. The total reward to each company is then

$$u_1(x, y) = V \frac{x}{x + y} - C_1 x \quad \text{and} \quad u_2(x, y) = V \frac{y}{x + y} - C_2 y, \quad x > 0, y > 0.$$

Show that these payoff functions are concave in the variable they control and then find the Nash equilibrium using calculus.

4.20 Suppose that a firm's output depends on the number of laborers they have hired. Let $p =$ number of workers, $w =$ worker compensation (per unit time) and assume

$$u_f(p, w) = f(p) - pw, \quad f(p) = \begin{cases} p(1000 - p), & \text{if } p \leq 500; \\ 25000, & \text{if } p > 500. \end{cases}$$

is the payoff to the firm if they hire p workers and pay them w. Assume that the union payoff is $u_u(p, w) = pw$.
Find the best response for the firm to a wage demand assuming $w_m \leq w \leq W$. Then find w to maximize $u_u(p(w), w)$. This scheme says the firm will hire the number of workers which maximizes its payoff for a given wage. Then the union, knowing that the firm will use its best response, will choose a wage demand that maximizes its payoff assuming the firm uses $p(w)$.

4.21 Suppose that $N > 2$ players choose an integer in $\{1, 2, \ldots, 100\}$. The payoff to each player is 1 if that player has chosen an integer which is closest to $\frac{2}{3}$ of the average of the numbers chosen by all the players. If two or more players choose the closest integer, then they equally split the 1. The payoff of other players is 0. Show that the Nash equilibrium is for each player to choose the number 1.

4.22 Two countries share a long border. The pollution emitted by one country affects the other. If country $i = 1, 2$ pollutes a total of Q_i tons per year, they will emit p_i tons per year into the atmosphere, and clean up $Q_i - p_i$ tons per year (before it is emitted) at cost c dollars per ton. Health costs attributed to the atmospheric pollution are proportional to the square of the total atmospheric pollution in each country. The total atmospheric pollution in country 1 is $p_1 + kp_2$, and in country 2, it is $p_2 + kp_1$. The constant $0 < k < 1$ represents the fraction of the neighboring country's pollution entering the atmosphere. Assume the proportionality constants and cost of cleanup is the same for both countries.

(a) Find a function giving the total cost of polluting for each country as a function of p_1, p_2.

(b) Find the Nash equilibrium level of pollution emitted for each country. Notice here that we want to minimize the payoff functions in this problem, not maximize.

(c) Find the optimal levels of pollution and payoffs if $c = 1000, a = 10, k = \frac{1}{2}$, $Q_1 = Q_2 = 300$.

4.23 Corn is a food product in high demand but also enjoys a government price subsidy. Assume that the demand for corn (in bushels) is given by $D(p) = 150000(15 - p)^+$, where p is the price per bushel. The government program guarantees that $p \geq 2$. Suppose that there are three corn producers who have reaped 1 million bushels each. They each have the choice of how much to send to market and how much to use for feed (at no profit). Find the Nash equilibrium. What happens if one farmer sends the entire crop to market?

4.24 This is known as the **division of land game**. Suppose that there is a parcel of land as in the following Figure 4.6:

Player I chooses a vertical line between $(0, 1)$ on the x axis and player II chooses a horizontal line between $(0, 1)$ on the y axis. Player I gets the land below II's choice and right of I's choice as well as the land above II's choice and left of I's line. Player II gets the rest of the land. Both

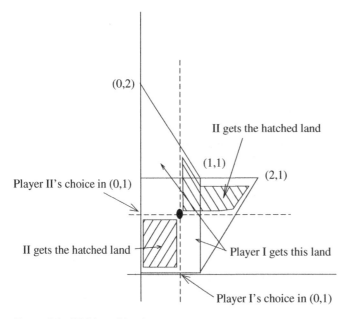

Figure 4.6 Division of land.

players want to choose their line so as to maximize the amount of land they get. Formulate this as a game with continuous strategies and solve it.

4.25 A famous economics problem is called the Samaritan's Dilemma. In one version of this problem, a citizen works in period 1 at a job giving a current annual net income of I. Out of her income she saves x for her retirement and earns $r\%$ interest per period. When she retires, considered to occur at period 2, she will receive an annual amount y from the government. The payoff to the citizen is

$$u_C(x,y) \equiv u_1(I - x) + u_2(x(1 + r) + y)\delta,$$

where u_1 is the first-period utility function and is increasing but concave down, and u_2 is the second-period utility function, that is also increasing and concave down. These utility functions model increasing utility but at a diminishing rate as time progresses. The constant $\delta > 0$, called a discount rate, finds the present value of dollars that are not delivered until the second period. The government has a payoff function

$$u_G(x,y) = u(T - y) + \alpha u_C(x,y) = u(T - y) + \alpha \left(u_1(I - x) + u_2(x(1 + r) + y)\delta \right)$$

where $u(t)$ is the government's utility function for income level T, such as tax receipts, and $\alpha > 0$ represents the factor of benefits received by the government for a happy citizen, called an **altruism factor**.

(a) Assume that the utility functions u_1, u_2, u are all logarithmic functions of their variables. Find a Nash equilibrium.

(b) Given $\alpha = 0.25, I = 22,000, T = 100,000, r = 0.15$ and $\delta = 0.03$, find the optimal amount for the citizen to save and the optimal amount for the government to transfer to the citizen in period 2.

4.26 Suppose we have a zero sum game and mixed densities X_0 for player I and Y_0 for player II.

(a) Suppose there are constants c and b so that $u(X_0, y) = c$ for all pure strategies y for player II and $u(x, Y_0) = b$, for all pure strategies x for player II. Show that (X_0, Y_0) is a saddle point in mixed strategies for the game and we must have $b = c$.

(b) Suppose there is a constant c so that $u(X_0, y) \geq c$, for all pure x, and $u(x, Y_0) \leq c$, for all pure y. Show that (X_0, Y_0) is a saddle point in mixed strategies.

4.27 Two investors choose investment levels from the unit interval. The investor with the highest level of investment wins the game, which has payoff 1 but costs the level of investment. If the same level of investment is chosen, they split the market. Take the investment levels to be $x \in [0, 1]$ for player I and $y \in [0, 1]$ for player II. Investor I's payoff function is

$$u_1(x,y) = \begin{cases} 1 - x, & \text{if } x > y; \\ \frac{1}{2} - x, & \text{if } x = y; \\ -x, & \text{if } x < y. \end{cases}$$

(a) What is player II's payoff function?

(b) Show that there is no pure Nash equilibrium in this game.

(c) Now use mixed strategies. Let $f(x)$ be the density for player I and $g(y)$ the density for player II. The equality of payoffs theorem would say $v_I = E_I(x, Y) = E_I(x', Y)$ for any $x, x' \in [0, 1]$ with $f(x), f(x') > 0$. This says

$$E_I(x, Y) = \int_0^1 u_1(x, y)g(y) \, dy$$

is a constant, v_I, independent of x (assuming $f(x) > 0$). Take the derivative of both sides with respect to x and solve for the density $g(x)$ for player II. Do a similar procedure to find the density for player I. Now find v_I and v_{II}.

4.28 Firm A is taking over another firm B. The true value of B is known to B but unknown to A. Firm A assumes the value of B is $X \sim Unif[0, y]$, where y is the maximum value A will bid for B. If A takes over B, the value of B will become λx where $\lambda > 1$ and x is the value of B before the takeover (remember this is known to B but $X \sim Unif[0, y]$ to A.)

A strategy for A is a bid y. A strategy for B is a yes or no decision, sell or don't sell. So if the type (B, x) accepts, A gets $\lambda X - y$, while (B, x) gets y. If (B, x) rejects, A gets 0 while (B, x) gets x. It follows, therefore, that (B, x) will accept if $x < y$, while she will reject if $x > y$.
(a) Find the expected value of the payoff to A.
(b) When will firm A decide to takeover firm B?

4.29 There are five countries, C_1, C_2, C_3, C_4, C_5 all fishing in the same ocean. They each get a positive payoff $100x_i$ (say in millions of dollars) for the fish they catch (say in millions of tons), and a negative payout of $-T(T - 10)/5$, where T is the total of all the fish caught, due to damaging the future potential of the ocean (you can assume $T \geq 10$, so this really is negative). So, if C_i catches x_i fish, the payoff to, e.g., C_4 is

$$P_4 = 100x_4 - (x_1 + x_2 + x_3 + x_4 + x_5)(x_1 + x_2 + x_3 + x_4 + x_5 - 10).$$

1. Take the partial derivative of P_4 with respect to x_4, and use this to find the possible Nash equilibria. Note: there are a lot! All you can really solve for is the total of the fish caught by all 5 countries.
2. Find the payoff to each country if they play equilibrium strategies where all catch the same amount of fish.
3. Now allow the countries to cooperate and fish the amount that gives the highest total payoff. How much do they catch now? What are the new payoffs to each country (assume they all fish the same amount)?

4.3 Economics Applications of Nash Equilibria

The majority of game theorists are economists (or some combination of economics and mathematics), and in recent years many winners of the Nobel Prize in economics have been game theorists (Aumann[1] Harsanyi[2], Myerson[3], Schelling[4], Selten[5], and, of course, Nash, to mention only a few. Nash, however, would probably consider himself primarily a pure mathematician). Most recently Paul R. Milgrom[6] and Robert B. Wilson[7] have been awarded the prize in 2020 for

---

1 Robert Aumann (born June 8, 1930) won a Nobel Memorial Prize in Economic Sciences, 2005.
2 John Harsanyi (born May 29, 1920 in Budapest, Hungary; died August 9, 2000 in Berkeley, California, United States) won a Nobel Memorial Prize in Economic Sciences, 1994.
3 Roger Myerson (born March 29, 1951) won a Nobel Memorial Prize in Economic Sciences, 2007.
4 Thomas Schelling (born April 14, 1921; died December 13, 2016), won a Nobel Memorial Prize in Economic Sciences, 2005.
5 Reinhard Selten (born October 5, 1930; died August 23, 2016), won the 1994 Nobel Memorial Prize in Economic Sciences (shared with John Harsanyi and John Nash).
6 Born April 20, 1948, Professor at Stanford University.
7 Born May 16, 1937, Professor at Stanford University.

their work in auction theory. In this section we will discuss some of the basic applications of game theory to economics problems.

Cournot Duopoly

Cournot[1] developed one of the earliest economic models of the competition between two firms. Suppose that there are two companies producing the same gadget. Firm $i = 1, 2$ chooses to produce the quantity $q_i \geq 0$, so the total quantity produced by both companies is $q = q_1 + q_2$.

We assume in this simplified model that the price of a gadget is a decreasing function of the total quantity produced by the two firms. Let's take it to be

$$P(q) = (\Gamma - q)^+ = \begin{cases} \Gamma - q & \text{if } 0 \leq q \leq \Gamma; \\ 0 & \text{if } q > \Gamma. \end{cases}$$

Γ represents the price of gadgets beyond which the quantity to produce is essentially zero, and the price a consumer is willing to pay for a gadget if there are no gadgets on the market. Suppose also that to make one gadget costs firm $i = 1, 2$, c_i dollars per unit so the total cost to produce q_i units is $c_i q_i$, $i = 1, 2$.

The total quantity of gadgets produced by the two firms together is $q_1 + q_2$, so that the revenue to firm i for producing q_1 units of the gadget is $q_i P(q_1 + q_2)$. The cost of production to firm i is $c_i q_i$. Each firm wants to maximize its own profit function, which is total revenue minus total costs and is given by

$$u_1(q_1, q_2) = P(q_1 + q_2)q_1 - c_1 q_1 \text{ and } u_2(q_1, q_2) = P(q_1 + q_2)q_2 - c_2 q_2. \tag{4.3.1}$$

Observe that the only interaction between these two profit functions is through the price of a gadget which depends on the total quantity produced.

Let's begin by taking the partials and setting to zero. We assume that the optimal production quantities are in the interval $(0, \Gamma)$:

$$\frac{\partial u_1(q_1, q_2)}{\partial q_1} = 0 \Longrightarrow -2q_1 - q_2 + \Gamma - c_1 = 0,$$

$$\frac{\partial u_2(q_1, q_2)}{\partial q_2} = 0 \Longrightarrow -2q_2 - q_1 + \Gamma - c_2 = 0.$$

Notice that we take the partial of u_i with respect to q_i, not the partial of each payoff function with respect to both variables. We are not trying to maximize each profit function over both variables, but each profit function to each firm as a function only of the variable they control, namely, q_i. That is a Nash equilibrium. Solving each equation gives the best response functions

$$q_1(q_2) = \frac{\Gamma - c_1 - q_2}{2} \text{ and } q_2(q_1) = \frac{\Gamma - c_2 - q_1}{2}$$

Now the intersection of these two best responses gives the optimal production quantities for each firm at

$$q_1^* = \frac{\Gamma + c_2 - 2c_1}{3} \text{ and } q_2^* = \frac{\Gamma + c_1 - 2c_2}{3}.$$

We will have $q_1^* > 0$ and $q_2^* > 0$ if we have $\Gamma > 2c_1, \Gamma > 2c_2$.

---

1 Antoine Augustin Cournot (August 28, 1801–March 31,1877) was a French philosopher and mathematician. One of his students was Auguste Walras, who was the father of Leon Walras. Cournot and Auguste Walras convinced Leon Walras to give economics a try. Walras then came up with his famous equilibrium theory in economics.

Now we have to check the second order conditions to make sure we have a maximum. At these points we have

$$\frac{\partial^2 u_1(q_1^*, q_2^*)}{\partial^2 q_1} = -2 < 0 \text{ and } \frac{\partial^2 u_2(q_1^*, q_2^*)}{\partial q_2^2} = -2 < 0,$$

and so (q_1^*, q_2^*) are values that maximize the profit functions, when the other variable is fixed. The total amount the two firms should produce is

$$q^* = q_1^* + q_2^* = \frac{2\Gamma - c_1 - c_2}{3}$$

and $\Gamma > q^* > 0$ if $\Gamma > 2c_1, \Gamma > 2c_2$. We see that our assumption about where the optimal production quantities would be found was correct if Γ is large enough. If however, say $\Gamma < 2c_1 - c_2$ then the formula for $q_1^* < 0$ and since the payoffs are concave down, the Nash equilibrium point is then $q_1^* = 0, q_2^* = \frac{\Gamma - c_2}{2}$. For the remainder of this section, we assume $\Gamma > 2c_1, \Gamma > 2c_2$.

The price function at the quantity q^* is then

$$P(q_1^* + q_2^*) = \Gamma - q_1^* - q_2^* = \Gamma - \frac{2\Gamma - c_1 - c_2}{3} = \frac{\Gamma + c_1 + c_2}{3}.$$

That is the market price of the gadgets produced by both firms when producing optimally.

Turn it around now and suppose that the price of gadgets is set at

$$P(q_1 + q_2) = p = \frac{\Gamma + c_1 + c_2}{3}.$$

If this is the market price of gadgets how many gadgets should each firm produce? The total quantity that both firms should produce (and will be sold) at this price is $q = P^{-1}(p)$, or

$$q = P^{-1}(p) = \Gamma - p = \frac{2\Gamma - c_1 - c_2}{3}$$
$$= \frac{\Gamma + c_2 - 2c_1}{3} + \frac{\Gamma + c_1 - 2c_2}{3} = q_1^* + q_2^*.$$

We conclude that the quantity of gadgets sold (demanded) will be exactly the total amount that each firm **should** produce at this price. This is called a **market equilibrium** and it turns out to be given by the Nash point equilibrium quantity to produce. In other words, in economics, a market equilibrium exists when the quantity demanded at a price p is $q_1^* + q_2^*$ and the firms will optimally produce the quantities q_1^*, q_2^* at price p. That is exactly what happens when we use a Nash equilibrium to determine q_1^*, q_2^*.

Finally, substituting the Nash equilibrium point into the profit functions gives the equilibrium profits

$$u_1(q_1^*, q_2^*) = \frac{(\Gamma + c_2 - 2c_1)^2}{9} \text{ and } u_2(q_1^*, q_2^*) = \frac{(\Gamma + c_1 - 2c_2)^2}{9}.$$

Notice that the profit of each firm depends on the costs of the other firm. That's a problem because how is a firm supposed to know the costs of a competing firm? The costs can be estimated, but known for sure...? This example is only a first-cut approximation, and we will have a lot more to say about this later.

Example 4.18 We take the inverse demand function to be $p(Q) = 3.94 - Q, Q = q_1 + q_2$ and the profit function for each firm is $u_i(q_1, q_2) = (p(Q) - c_i)q_i, i = 1, 2$ with costs $c_1 = 0.4, c_2 = 1.2$. The Cournot equilibrium is then

$$q_1^* = \frac{\Gamma + c_2 - 2c_1}{3} = 1.45$$
$$q_2^* = \frac{\Gamma + c_1 - 2c_2}{3} = 0.65.$$

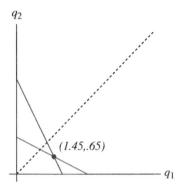

Figure 4.7 Best responses and Cournot equilibrium.

The equilibrium price is 1.85. The Figure 4.7 shows the best responses of each firm and where the lines intersect we have the Cournot equilibrium.

Cournot Model with Uncertain Costs

Now here is a generalization of the Cournot model that is more realistic and also more difficult to solve because it involves a lack of information on the part of at least one player. It is assumed that one firm has no information regarding the other firm's cost function. Here is the model setup.

Assume that both firms produce gadgets at constant unit cost. Both firms know that firm 1's cost is c_1, but firm 1 does not know firm 2's cost of c_2, which is known only to firm 2. Suppose that the cost for firm 2 is considered as a random variable to firm 1, say, C_2. Now firm 1 has reason to believe that

$$Prob(C_2 = c^+) = p \text{ and } Prob(C_2 = c^-) = 1 - p$$

for some $0 < p < 1$ that is known by firm 1.

Again, the payoffs to each firm are its profits. For firm 1 we have the payoff function, assuming that firm 1 makes q_1 gadgets and firm 2 makes q_2 gadgets

$$u_1(q_1, q_2) = q_1[P(q_1 + q_2) - c_1],$$

where $P(q_1 + q_2)$ is the market price for the total production of $q_1 + q_2$ gadgets. Firm 2's payoff function is

$$u_2(q_1, q_2) = q_2[P(q_1 + q_2) - C_2].$$

From firm 1's perspective, this is a random variable because of the unknown cost. The way to find an equilibrium now is the following:

1. Find the optimal production level for firm 2 using the costs c^+ and c^- giving the two numbers q_2^+ and q_2^-.
2. Firm 1 now finds the optimal production levels for the two firm 2 quantities from step 1 using the **expected payoff**

$$E(u_1(q_1, q_2(C_2))) = [u_1(q_1, q_2^-)]Prob(C_2 = c^-) + [u_1(q_1, q_2^+)]Prob(C_2 = c^+)$$
$$= q_1[P(q_1 + q_2^-) - c_1](1 - p) + q_1[P(q_1 + q_2^+) - c_1]p.$$

3. From the previous two steps, you end up with three equations involving q_1, q_2^+, q_2^-. Treat these as three equations in three unknowns and solve.

For example, let's take the price function $P(q) = \Gamma - q, 0 \leq q \leq \Gamma$.

1. Firm 2 has the cost of production $c^+ q_2$ with probability p and the cost of production $c^- q_2$ with probability $1 - p$. Firm 2 will solve the problem for each cost c^+, c^- assuming that q_1 is known:

$$\max_{q_2} q_2(\Gamma - (q_1 + q_2) - c^+) \implies q_2^+ = \frac{1}{2}(\Gamma - q_1 - c^+)$$

$$\max_{q_2} q_2(\Gamma - (q_1 + q_2) - c^-) \implies q_2^- = \frac{1}{2}(\Gamma - q_1 - c^-)$$

2. Next, firm 1 will maximize the expected profit using the two quantities q_2^+, q_2^-. Firm 1 seeks the production quantity q_1, which solves

$$\max_{q_1} q_1[\Gamma - (q_1 + q_2^+) - c_1]p + q_1[\Gamma - (q_1 + q_2^-) - c_1](1 - p).$$

This is maximized at

$$q_1 = \frac{1}{2}[p(\Gamma - q_2^+ - c_1) + (1 - p)(\Gamma - q_2^- - c_1)].$$

3. Summarizing, we now have the following system of equations for the variables q_1, q_2^-, q_2^+:

$$q_2^+ = \frac{1}{2}(\Gamma - q_1 - c^+),$$

$$q_2^- = \frac{1}{2}(\Gamma - q_1 - c^-),$$

$$q_1 = \frac{1}{2}[p(\Gamma - q_2^+ - c_1) + (1 - p)(\Gamma - q_2^- - c_1)].$$

Observe that q_1 is the expected quantity of production assuming that firm 2 produces q_2^+ with probability p and q_2^- with probability $1 - p$.

Solving the equations in (3), we finally arrive at the optimal production levels:

$$q_1^* = \frac{1}{3}[\Gamma - 2c_1 + pc^+ + (1 - p)c^-],$$

$$q_2^{+*} = \frac{1}{3}[\Gamma + c_1] - \frac{1}{6}[(1 - p)c^- + pc^+] - \frac{1}{2}c^+,$$

$$q_2^{-*} = \frac{1}{3}[\Gamma - 2c^- + c_1] + \frac{1}{6}p(c^- - c^+).$$

Notice that if we require that the production levels be nonnegative, we need to put some conditions on the costs and Γ.

What happens if the demands for a gadget are uncertain?

Cournot Duopoly with Uncertain Demands

The inverse demand function is $P(q_1, q_2) = a - q_1 - q_2$, with q_i the production level of firm $i = 1, 2$. Total cost of production is given by $c_i(q_i) = c\, q_i$. The demand, a for widgets is a modified binomial random variable.

$$Prob(a = x) = \begin{cases} \gamma & x = a_H \\ 1 - \gamma, & x = a_L \end{cases}$$

and we assume $a_H > a_L > c$. Let's also assume that firm 1 knows whether demand is high, a_H, or low a_L. Firm 2 does not know demand with certainty but does know the value of $0 < \gamma < 1$. If firm 1 knows that demand is a_H then q_{1H} is the quantity produced, while if demand is a_L, firm 1 produces q_{1L}.

The strategy requirements are $0 \leq q_{1H} \leq a_H, 0 \leq q_{1L} \leq a_L$ for firm 1, and $0 \leq q_2 \leq \gamma a_H + (1 - \gamma)a_L$, for firm 2, where $a_d = \gamma a_H + (1 - \gamma)a_L$ is the expected demand. We will also assume

$$3(a_L - c) > d - c.$$

In the terminology of the definition, $T_1 = \{H, L\}$ is the type space of firm 1 and firm 2 has only one type. For given q_2, firm 1 will maximize

$$\begin{cases} q_{1H}(a_H - c - q_2 - q_{1H}), & \text{if demand is high} \\ q_{1L}(a_L - c - q_2 - q_{1L}), & \text{if demand is low.} \end{cases}$$

The optimal quantity for firm 1 to produce is

$$q_1^* = \begin{cases} q_{1H}^*, & \text{if } a = a_H \\ q_{1L}^*, & \text{if } a = a_L \end{cases} = \begin{cases} \frac{1}{2}(a_H - c - q_2), \\ \frac{1}{2}(a_L - c - q_2). \end{cases}$$

Firm 2 will choose q_2 to maximize

$$\gamma q_2(a_H - c - q_{1H}^* - q_2) + (1 - \gamma)q_2(a_L - c - q_{1L}^* - q_2).$$

The maximum is achieved at

$$q_2^* = \frac{1}{2}(\gamma(a_H - q_{1H}^*) + (1 - \gamma)(a_L - q_{1L}^*) - c).$$

Solving now for $q_{1H}^*, q_{1L}^*, q_2^*$, we get

$$q_{1H}^* = \frac{1}{2}(a_H - c) - \frac{1}{6}(\gamma a_H + (1 - \gamma)a_L - c)$$

$$q_{1L}^* = \frac{1}{2}(a_L - c) - \frac{1}{6}(\gamma a_H + (1 - \gamma)a_L - c)$$

$$q_2^* = \frac{1}{3}(\gamma a_H + (1 - \gamma)a_L - c)$$

The Bertrand Model

Joseph Bertrand [1] didn't like Cournot's model. He thought firms should set prices to accommodate demand not production quantities. Here is the setup he introduced. We again have two companies making identical gadgets. In this model they can set prices, not quantities, and they will only produce the quantity demanded at the given price. So the quantity sold is a function of the price set by each firm, say, $q = \Gamma - p$. This is better referred to as the **demand function** for a given price:

$$D(p) = \Gamma - p, \quad 0 \leq p \leq \Gamma \quad \text{and} \quad D(p) = 0 \quad \text{when} \quad p > \Gamma.$$

In a classic problem the model says that if both firms charge the same price, they will split the market evenly, with each selling exactly half of the total sold. But the company that charges a lower price will capture the entire market. We have to assume that each company has enough capacity to make the entire quantity if it captures the whole market. The cost to make gadgets is still c_i, $i = 1, 2$, dollars per unit gadget. We first assume the firms have differing costs:

$$c_1 \neq c_2 \quad \text{and} \quad \Gamma > \max\{c_1, c_2\}.$$

---

1 Joseph Louis François Bertrand (March 11, 1822–April 5, 1900) was a French mathematician who contributed to number theory, differential geometry, probability theory, thermodynamics, and economics. He reviewed the Cournot competition model, arguing that Cournot had reached a misleading conclusion.

The profit function for firm $i = 1, 2$, assuming that firm 1 sets the price as p_1 and firm 2 sets the price at p_2, is

$$u_1(p_1, p_2) = \begin{cases} p_1(\Gamma - p_1) - c_1(\Gamma - p_1) & \text{if } p_1 < p_2; \\ \dfrac{(p - c_1)(\Gamma - p)}{2} & \text{if } p_1 = p_2 = p \geq c_1; \\ 0, & \text{if } p_1 > p_2. \end{cases}$$

This says that if firm 1 sets the price lower than firm 2, firm 1's profit will be (price – cost) × quantity demanded; if the prices are the same, firm 1's profits will be $\left(\frac{1}{2}\right)$ (price – cost) × quantity demanded; and zero if firm 1's price of a gadget is greater than firm 2's. This assumes that the **lower price captures the entire market**. Similarly, firm 2's profit function is

$$u_2(p_1, p_2) = \begin{cases} p_2(\Gamma - p_2) - c_2(\Gamma - p_2) & \text{if } p_2 < p_1; \\ \dfrac{(p - c_2)(\Gamma - p)}{2} & \text{if } p_1 = p_2 = p \geq c_2; \\ 0 & \text{if } p_2 > p_1. \end{cases}$$

Figure 4.8 is a plot of $u_1(p_1, p_2)$ with $\Gamma = 10, c_1 = 3$.

Now we have to find a Nash equilibrium for these discontinuous (non differentiable) payoff functions. Let's suppose that there is a Nash equilibrium point at (p_1^*, p_2^*). By definition, we have

$$u_1(p_1^*, p_2^*) \geq u_1(p_1, p_2^*) \text{ and } u_2(p_1^*, p_2^*) \geq u_2(p_1^*, p_2), \text{for all } (p_1, p_2).$$

Let's break this down by considering three cases:

Case 1. $p_1^* > p_2^*$. Then it should be true that firm 2, having a lower price, captures the entire market so that for firm 1

$$u_1(p_1^*, p_2^*) = 0 \geq u_1(p_1, p_2^*), \text{ for every } p_1.$$

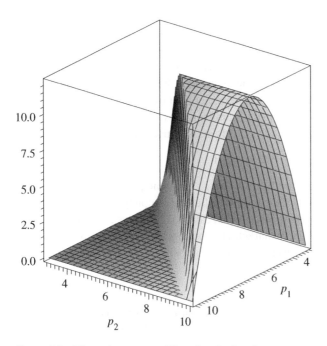

Figure 4.8 Discontinuous payoff function for firm 1.

But if we take any price $c_1 < p_1 < p_2^*$, the right side will be

$$u_1(p_1, p_2^*) = (p_1 - c_1)(\Gamma - p_1) > 0,$$

so $p_1^* > p_2^*$ cannot hold and still have (p_1^*, p_2^*) a Nash equilibrium.

Case 2. $p_1^* < p_2^*$. Then it should be true that firm 1 captures the entire market and so for firm 2

$$u_2(p_1^*, p_2^*) = 0 \geq u_2(p_1^*, p_2), \quad \text{for every } p_2.$$

But if we take any price for firm 2 with $c_2 < p_2 < p_1^*$, the right side will be

$$u_2(p_1^*, p_2) = (p_2 - c_2)(\Gamma - p_2) > 0,$$

that is, strictly positive, and again it cannot be that $p_1^* < p_2^*$ and fulfill the requirements of a Nash equilibrium. So the only case left is the following.

Case 3. $p_1^* = p_2^*$. But then the two firms split the market and we must have for firm 1

$$u_1(p_1^*, p_2^*) = \frac{(p_1^* - c_1)(\Gamma - p_1^*)}{2} \geq u_1(p_1, p_2^*), \quad \text{for all } p_1 \geq c_1.$$

If we take firm 1's price to be $p_1 = p_1^* - \varepsilon < p_2^*$ with really small $\varepsilon > 0$, then firm 1 drops the price ever so slightly below firm 2's price. Under the Bertrand model, firm 1 will capture the entire market at price p_1 so that in this case we have

$$u_1(p_1^*, p_2^*) = \frac{(p_1^* - c_1)(\Gamma - p_1^*)}{2} < u_1(p_1^* - \varepsilon, p_2^*) = (p_1^* - \varepsilon - c_1)(\Gamma - p_1^* + \varepsilon).$$

This inequality won't be true for every ε, but it will be true for small enough $\varepsilon > 0$, (say, $0 < \varepsilon < \frac{p_1^* - c_1}{2}$).

Thus, in all cases, we can find prices so that the condition that (p_1^*, p_2^*) be a Nash point is violated and so there is no Nash equilibrium in pure strategies. But there is one case when there is a Nash equilibrium. In the analysis above, we assumed in several places that prices would have to be above costs. What if we drop that assumption? The first thing that happens is that in case 3 we won't be able to find a positive ε to drop p_1^*.

What's the problem here? It is that neither player has a continuous profit function (as you can see in Figure 4.8). By lowering the price just below the competitor's price, the firm with the lower price can capture the entire market. So the incentive is to continue lowering the price of a gadget. In fact, we are led to believe that maybe $p_1^* = c_1, p_2^* = c_2$ is a Nash equilibrium. Let's check that, and let's just assume that $c_1 < c_2$, because a similar argument would apply if $c_1 > c_2$.

In this case, $u_1(c_1, c_2) = 0$, and if this is a Nash equilibrium, then it must be true that $u_1(c_1, c_2) = 0 \geq u_1(p_1, c_2)$ for all p_1. But if we take any price $c_1 < p_1 < c_2$, then

$$u_1(p_1, c_2) = (p_1 - c_1)(\Gamma - p_1) > 0,$$

and we conclude that (c_1, c_2) also is not a Nash equilibrium. The only way that this could work is if $c_1 = c_2 = c$, so the costs to each firm are the same. In this case, we leave it as an exercise to show that $p_1^* = c, p_2^* = c$ is a Nash equilibrium and optimal profits are zero for each firm. So, what good is this if the firms make no money, and even that is true only when their costs are the same? This leads us to examine assumptions about exactly how profits arise in competing firms. Is it strictly prices and costs, or are there other factors involved?

The Traveler's Paradox Example 4.13 illustrates what goes wrong with the Bertrand model. It also shows that when there is an incentive to drop to the lowest price, it leads to unrealistic expectations. On the other hand, the Bertrand model can be modified to take other factors into account like demand sensitivities to prices or production quantity capacity limits. These modifications are explored in the exercises.

The Stackelberg Model

What happens if two competing firms don't choose the production quantities at the same time, but choose sequentially one after the other? Stackelberg[1] gave an answer to this question. In this model, we will assume that there is a dominant firm, say, firm 1, who will announce its production quantity publicly. Then firm 2 will decide how much to produce. In other words, given that one firm knows the production quantity of the other, determine how much each will or should produce.

Suppose that firm 1 announces that it will produce q_1 gadgets at cost c_1 dollars per unit. It is then up to firm 2 to decide how many gadgets, say, q_2 at cost c_2, it will produce. We again assume that the unit costs are constant. The price per unit will then be considered a function of the total quantity produced so that $p = p(q_1, q_2) = (\Gamma - q_1 - q_2)^+ = \max\{\Gamma - q_1 - q_2, 0\}$. The profit functions will be

$$u_1(q_1, q_2) = (\Gamma - q_1 - q_2)q_1 - c_1 q_1,$$
$$u_2(q_1, q_2) = (\Gamma - q_1 - q_2)q_2 - c_2 q_2.$$

These are the same as in the simplest Cournot model, but now q_1 is fixed as given. It is not variable when firm 1 announces it. So what we are really looking for is the best response of firm 2 to the production announcement q_1 by firm 1. In other words, firm 2 wants to know how to choose $q_2 = q_2(q_1)$ so as to

Maximize over q_2, given q_1, the function $u_2(q_1, q_2(q_1))$.

This is given by calculus as

$$q_2(q_1) = \frac{\Gamma - q_1 - c_2}{2}.$$

This is the amount that firm 2 should produce when firm 1 announces the quantity of production q_1. It is the same best response function for firm 2 as in the Cournot model.

Now, firm 1 has some clever employees who know calculus and game theory and can perform this calculation as well as we can. Firm 1 knows what firm 2's optimal production quantity should be, given its own announcement of q_1. Therefore, firm 1 should choose q_1 to maximize its own profit function knowing that firm 2 will use production quantity $q_2(q_1)$:

$$u_1(q_1, q_2(q_1)) = q_1(\Gamma - q_1 - q_2(q_1)) - c_1 q_1$$
$$= q_1\left(\Gamma - q_1 - \frac{\Gamma - q_1 - c_2}{2}\right) - c_1 q_1$$
$$= q_1\frac{\Gamma - q_1}{2} + q_1\left(\frac{c_2}{2} - c_1\right).$$

Firm 1 wants to choose q_1 to make this as large as possible. By calculus, we find that

$$q_1^* = \frac{\Gamma - 2c_1 + c_2}{2}.$$

The optimal production quantity for firm 2 is

$$q_2^* = q_2(q_1^*) = \frac{\Gamma + 2c_1 - 3c_2}{4}.$$

The equilibrium profit function for firm 2 is then

$$u_2(q_1^*, q_2^*) = \frac{(\Gamma + 2c_1 - 3c_2)^2}{16},$$

---

1 Heinrich Freiherr von Stackelberg (1905–1946) was a German economist who contributed to game theory and oligopoly theory.

and for firm 1, it is

$$u_1(q_1^*, q_2^*) = \frac{\left(\Gamma - 2c_1 + c_2\right)^2}{8}.$$

For comparison, we will set $c_1 = c_2 = c$ and then recall the optimal production quantities for the Cournot model:

$$q_1^c = \frac{\Gamma - 2c_1 + c_2}{3} = \frac{\Gamma - c}{3}, \qquad q_2^c = \frac{\Gamma + c_1 - 2c_2}{3} = \frac{\Gamma - c}{3}.$$

The equilibrium profit functions were

$$u_1(q_1^c, q_2^c) = \frac{(\Gamma + c_2 - 2c_1)^2}{9} = \frac{(\Gamma - c)^2}{9},$$

$$u_2(q_1^c, q_2^c) = \frac{(\Gamma + c_1 - 2c_2)^2}{9} = \frac{(\Gamma - c)^2}{9}.$$

In the Stackelberg model, we have

$$q_1^* = \frac{\Gamma - c}{2} > q_1^c, \qquad q_2^* = \frac{\Gamma - c}{4} < q_2^c.$$

So firm 1 produces more and firm 2 produces less in the Stackelberg model than if firm 2 did not have the information announced by firm 1. For the firm's profits, we have

$$u_1(q_1^c, q_2^c) = \frac{(\Gamma - c)^2}{9} < u_1(q_1^*, q_2^*) = \frac{(\Gamma - c)^2}{8},$$

$$u_2(q_1^c, q_2^c) = \frac{(\Gamma - c)^2}{9} > u_2(q_1^*, q_2^*) = \frac{(\Gamma - c)^2}{16}.$$

Firm 1 makes more money by announcing the production level, and firm 2 makes less with the information.

One last comparison is the total quantity produced

$$q_1^c + q_2^c = \frac{2}{3}(\Gamma - c) < q_1^* + q_2^* = \frac{3\Gamma - 2c_1 - c_2}{4} = \frac{3}{4}(\Gamma - c)$$

and the price at equilibrium (recall that $\Gamma > c$):

$$P(q_1^* + q_2^*) = \frac{\Gamma + 3c}{4} < P(q_1^c + q_2^c) = \frac{\Gamma + 2c}{3}.$$

Entry Deterrence

In this example we ask the following question. If there is currently only one firm producing a gadget, what should be the price of the gadget in order to make it unprofitable for another firm to enter the market and compete with firm 1? This is a famous problem in economics called the **entry deterrence problem**.

Of course, a monopolist may charge any price at all as long as there is a demand for gadgets at that price. But it should also be true that competition should lower prices, implying that the price to prevent entry by a competitor should be lower than what the firm would otherwise set.

We call the existing company firm 1 and the potential challenger firm 2. The demand function is $p = D(q) = (\Gamma - q)^+$.

Now, before the challenger enters the market the profit function to firm 1 is

$$u_1(q_1) = (\Gamma - q_1)q_1 - (aq_1 + b),$$

where we assume that the cost function $C(q) = aq + b$, with $\Gamma > a, b > 0$. This cost function includes a fixed cost of $b > 0$ because even if the firm produces nothing, it still has expenses.

Now firm 1 is acting as a monopolist in our model because it has no competition. So firm 1 wants to maximize profit, which gives a production quantity of

$$q_1^* = \frac{\Gamma - a}{2}$$

and maximum profit for a monopolist of

$$u_1(q_1^*) = \frac{(\Gamma - a)^2}{4} - b.$$

In addition, the price of a gadget at this quantity of production will be

$$p = D(q_1^*) = \frac{\Gamma + a}{2}.$$

Now firm 2 enters the picture and calculates firm 2's profit function knowing that firm 1 will or should produce $q_1^* = (\Gamma - a)/2$ to get firm 2's payoff function

$$u_2(q_2) = \left(\Gamma - \frac{\Gamma - a}{2} - q_2 \right) q_2 - (aq_2 + b).$$

So firm 2 calculates its maximum possible profit and optimal production quantity as

$$u_2(q_2^*) = \frac{(\Gamma - a)^2}{16} - b, \quad q_2^* = \frac{\Gamma - a}{4}.$$

The price of gadgets will now drop (recall $\Gamma > a$) to

$$p = D(q_1^* + q_2^*) = \Gamma - q_1^* - q_2^* = \frac{\Gamma + 3a}{4} < D(q_1^*) = \frac{\Gamma + a}{2}.$$

As long as $u_2(q_2^*) \geq 0$, firm 2 has an incentive to enter the market. If we interpret the constant b as a fixed cost to enter the market, this will require that

$$\frac{(\Gamma - a)^2}{16} > b,$$

or else firm 2 cannot make a profit.

Now here is a more serious analysis because firm 1 is not about to sit by idly and let another firm enter the market. Therefore, firm 1 will now analyze the Cournot model assuming that there is a firm 2 against which firm 1 is competing. Firm 1 looks at the profit function for firm 2:

$$u_2(q_1, q_2) = (\Gamma - q_1 - q_2)q_2 - (aq_2 + b),$$

and maximizes this as a function of q_2 to get

$$q_2^m = \frac{\Gamma - q_1 - a}{2} \quad \text{and} \quad u_2(q_1, q_2^m) = \frac{(\Gamma - q_1 - a)^2}{4} - b$$

as the maximum profit to firm 2 if firm 1 produces q_1 gadgets. Firm 1 reasons that it can set q_1 so that firm 2's profit is zero:

$$u_2(q_1, q_2^m) = \frac{(\Gamma - q_1 - a)^2}{4} - b = 0 \implies q_1^0 = \Gamma - 2\sqrt{b} - a.$$

This gives a zero profit to firm 2. Consequently, if firm 1 decides to produce q_1^0 gadgets, firm 2 has no incentive to enter the market. The price at this quantity will be

$$D(q_1^0) = \Gamma - (\Gamma - 2\sqrt{b} - a) = 2\sqrt{b} + a,$$

and the profit for firm 1 at this level of production will be

$$u_1(q_1^0) = (\Gamma - q_1^0)q_1^0 - (aq_1^0 + b) = 2\sqrt{b}(\Gamma - a) - 5b.$$

This puts a requirement on Γ that $\Gamma > a + \frac{5}{2}\sqrt{b}$, or else firm 1 will also make a zero profit.

Hotelling's Location Model

In this game firms 1 and 2 are located at the end points of the interval $[0, 1]$ with firm 1 at 0 and firm 2 at 1. They set their price p_1, p_2, respectively, for gadgets they produce at the same production cost $c > 0$. Customers for these gadgets are uniformly distributed in the interval $[0, 1]$, and their payoff if located at $0 \le x \le 1$ and shopping at firm $i = 1, 2$, is given by

$$u_i(x) = \begin{cases} v - p_1 - t\,x, & i = 1 \\ v - p_2 - t(1 - x), & i = 2 \end{cases}$$

Here t is the unit cost to travel which will be paid by the customer and v is maximum willingness to pay for a gadget. A customer's utility is the difference between the maximum amount they are willing to pay for a gadget and the costs of obtaining the gadget.

The payoff to each firm is then $P_i(p_1, p_2) = (p_i - c)q_i, i = 1, 2$, where $q_i =$ the quantity of gadgets sold by firm i.

We begin by finding the distance for customers so that they don't care if they buy the gadget from firm 1 or firm 2. That is the position $\hat{x}$ where

$$v - p_1 - t\,\hat{x} = v - p_2 - t(1 - \hat{x}) \implies \hat{x} = \frac{p_2 - p_1 + t}{2t}.$$

It makes sense that if $p_1 = p_2$ then $\hat{x} = \frac{1}{2}$. Economists say the market is covered if all customers will buy a gadget. In order for this to happen in this model, we need

$$u_1(\hat{x}) = v - p_1 - t\left(\frac{p_2 - p_1 + t}{2t}\right) \ge 0 \implies v \ge p_1 + t\left(\frac{p_2 - p_1 + t}{2t}\right).$$

The quantity of gadgets each firm will sell is basically the demand generated by all the customers that will travel to that firm. That means the demand functions for each firm are given by

$$q_1 = D_1(p_1, p_2) = \int_0^{\hat{x}} 1 \; dq = \hat{x} = \frac{p_2 - p_1 + t}{2t} = \frac{p_2 - p_1}{2t} + \frac{1}{2}$$

$$q_2 = D_2(p_1, p_2) = \int_{\hat{x}}^1 1 \; dq = 1 - \hat{x} = 1 - \frac{p_2 - p_1 + t}{2t} = \frac{p_1 - p_2}{2t} + \frac{1}{2}$$

Now we want to maximize the profit for each firm. For firm 1,

$$\max_{p_1} P_1(p_1, p_2) = \max_{p_1} (p_1 - c)D_1(p_1, p_2) = \max_{p_1} (p_1 - c)\frac{p_2 - p_1}{2t} + \frac{1}{2}$$

Taking a partial derivative and setting to zero, we get the critical point

$$\frac{\partial P_1(p_1, p_2)}{\partial p_1} = \frac{p_2 - p_1 + t}{2t} - \frac{1}{2t}(p_1 - c) = 0 \implies p_1^* = \frac{1}{2}(p_2 + t + c).$$

Clearly, because of symmetry $p_1^* = p_2^* = p^*$ and p^* satisfies

$$p^* = \frac{1}{2}(p^* + t + c) \implies p^* = t + c.$$

What if $t = 0$ so there is no cost to travel to the firm? In that case $p^* = c$ and $P_1 = P_2 = 0$, which is the result of the Bertrand pricing model.

Now that we have the equilibrium price for each player, namely $p^* = t + c$, we have $\hat{x} = \frac{1}{2}$, $D_1(p^*, p^*) = \frac{1}{2}$, and $P_1(p^*, p^*) = (p^* - c)D_1(p^*, p^*) = \frac{t}{2}$.

How Should a Seller Set the Buy-it-now Price?

Frequently an item up for sale may solicit offers but the seller also offers the option of setting a price which is at a level in which the seller is willing to take the object off the market and the buyer who is willing to pay the price gets the object. This is a modification of an auction which we consider in more depth later. Now we consider the problem of the seller of the object as to how to set the **buy-it-now** price.

The seller may set a **reserve price** r, which is a nonnegotiable lowest price you must get to consider selling the object. You may also declare a price $p \geq r$, which is your **take-it-or-leave-it price** or **buy-it-now price** and wait for some buyer, who hopefully has a valuation greater than or equal to p to buy the object. The problem is to determine p. Now we get into the analysis.

There are $N > 1$ bidders with valuations of an object for sale $V_i, i = 1, 2, \ldots, N$. The valuations are random variables. The information that we assume is known to the seller is the joint cumulative distribution function

$$F(v_1, v_2, \ldots, v_N) = Prob(V_1 \leq v_1, \ldots, V_n \leq v_N),$$

and each buyer i has knowledge of his or her own distribution function $F_i(v_i) = P(V_i \leq v_i)$.

The solution **involves** calculating the expected payoff from the trade and then maximizing the expected payoff over $p \geq r$. The payoff is the function $U(p)$, which is $p - r$, if there is a buyer with a valuation at least p, and 0 otherwise:

$$U(p) = \begin{cases} p - r. & \text{if } \max\{V_1, \ldots, V_N\} \geq p; \\ 0, & \text{if } \max\{V_1, \ldots, V_N\} < p. \end{cases}$$

This is a random variable because it depends on $V_1, \ldots, V_N$ which are random. The expected payoff is

$$\begin{aligned} u(p) = E[U(p)] &= (p - r) Prob(\max\{V_1, \ldots, V_N\} \leq p) + 0 \cdot Prob(\max\{V_1, \ldots, V_N\} < p) \\ &= (p - r)[1 - Prob(\max\{V_1, \ldots, V_N\} < p)] \\ &= (p - r)[1 - Prob(V_1 < p, \ldots, V_N < p)] \\ &= (p - r)[1 - F(p, p, \ldots p)] \\ &= (p - r)f(p), \end{aligned}$$

where $f(p) = 1 - F(p, p, \ldots, p)$. The seller wants to find p^* such that

$$\max_{p \geq r}(p - r)f(p) = (p^* - r)f(p^*).$$

If there is a maximum $p^* > r$, we could find it by calculus. It is the solution of

$$(p^* - r)f'(p^*) + f(p^*) = 0.$$

This will be a maximum as long as $(p^* - r)f''(p^*) + 2f'(p^*) < 0$. To proceed further, we need to know something about the valuation distribution. Let's take the simplest case that $\{V_i\}$ is a collection of N independent and identically distributed random variables. In this case

$$F(v_1, \ldots, v_N) = G(v_1) \cdots G(v_N), \quad \text{where } G(v) = F_i(v), 1 \leq i \leq N.$$

Then

$$f(p) = 1 - G(p)^N, \quad f'(p) = -NG(p)^{N-1}G'(p), \quad G'(p) = g(p).$$

is the density function of V_i, if it is a continuous random variable. So the condition for a maximum at p^* becomes

$$-(p^* - r)NG(p^*)^{N-1}g(p^*) + 1 - G(p^*)^N = 0.$$

Now we take a particular distribution for the valuations that is still realistic. In the absence of any other information, we might as well assume that the valuations are uniformly distributed over the interval $[r, R]$, remembering that r is the reserve price. For this uniform distribution, we have

$$g(p) = \begin{cases} \frac{1}{R-r}, & r < p < R; \\ 0, & \text{otherwise.} \end{cases} \quad \text{and} \quad G(p) = \begin{cases} 0, & \text{if } p < r; \\ \frac{p-r}{R-r}, & \text{if } r \leq p \leq R; \\ 1, & \text{if } p > R. \end{cases}$$

If we assume that $r < p^* < R$, then we may solve the first-order condition

$$-(p^* - r)N\left(\frac{p^* - r}{R - r}\right)^{N-1}\left(\frac{1}{R - r}\right) + 1 - \left(\frac{p^* - r}{R - r}\right)^N = 0$$

for p^* to get the **take-it-or-leave-it price**

$$p^* = r + (R - r)\left(\frac{1}{N + 1}\right)^{1/N}.$$

For this p^*, we have $f(p^*) = 1 - \left(\frac{p^*-r}{R-r}\right)^N = \frac{N}{N+1}$ and the expected payoff

$$u(p^*) = (p^* - r)f(p^*) = (R - r)N\left(\frac{1}{N + 1}\right)^{\frac{1}{N}+1}.$$

Of particular interest are the cases $N = 1, N = 2$, and $N \to \infty$. We label the take-it-or-leave-it price as $p^* = p^*(N)$. Here are the results:

1. When there is only one potential buyer, the **take-it-or-leave-it price** should be set at

   $$p^*(1) = r + \frac{R - r}{2} = \frac{r + R}{2}$$

   the midpoint of the range $[r, R]$. The expected payoff to the seller is

   $$u(p^*(1)) = \frac{R - r}{4}.$$

2. When there are two potential buyers, the take-it-or-leave-it price should be set at

   $$p^*(2) = r + \frac{R - r}{\sqrt{3}}, \text{ and then } u(p^*(2)) = (R - r)\frac{2\sqrt{3}}{9}.$$

3. As $N \to \infty$, we have the take-it-or-leave-it price should be set at

   $$p^*(\infty) = \lim_{N \to \infty} p^*(N) = \lim_{N \to \infty} r + (R - r)\left(\frac{1}{N + 1}\right)^{1/N} = R,$$

and then the expected payoff is $u(p^*(\infty)) = R - r$. Note that we may calculate $\lim_{N \to \infty}(N + 1)^{1/N} = 1$ using L'Hôpital's rule. We conclude that as the number of potential buyers increases, the price should be set at the upper range of valuations. A totally reasonable result.

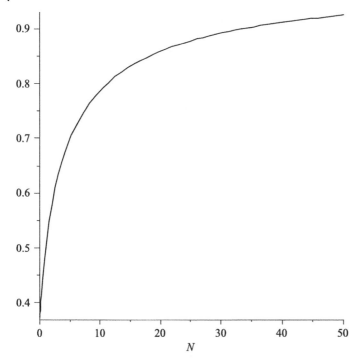

Figure 4.9 Take-it-or-leave-it price versus number of bidders.

You can see from Figure 4.9, which plots $p^*(N)$ as a function of N (with $R = 1, r = 0$), that as the number of buyers increases, there is a rapid increase in the take-it-or-leave-it price before it starts to level off.

Problems

4.30 In the Cournot model we assumed there were two firms in competition. Instead suppose that there is only one firm, a monopolist, producing at quantity q. Formulate a model to determine the optimal production quantity. Compare the optimal production quantity for the monopolist with the quantities derived in the Cournot duopoly assuming $c_1 = c_2 = c$. Does competition produce lower prices?

4.31 The Cournot model assumed the firms acted independently. Suppose instead the two firms collude to set production quantities.
(a) Find the optimal production quantity for each firm assuming the unit costs are the same for both firms, $c_1 = c_2 = c$.
(b) Show that the optimal production quantities in the cartel solution are unstable in the sense that they are not best responses to each other. Assume $c_1 = c_2 = c$.

4.32 Two firms produce identical widgets. The price function is $P(q) = (150 - q)^+$ and the cost to produce q widgets is $C(q) = 120q - \frac{2}{3}q^2$ for each firm.
(a) What are the profit functions for each firm?
(b) Find the Nash equilibrium quantities of production.

(c) Find the price of a widget at the Nash equilibrium level of production as well as the profits for each firm.

(d) If firm 1 assumes firm 2 will use it's best response function, what is the best production level of firm 1 to maximize it's profits assuming firm 1 will then publicly announce the level. In other words, they will play sequentially, not simultaneously.

4.33 Suppose instead of two firms in the Cournot model with payoff functions (4.3.1), that there are N firms. Formulate this model and find the optimal quantities each of the N firms should produce. Instead of a duopoly, this is an oligopoly. What happens when the firms all have the same costs and $N \to \infty$?

4.34 Two firms produce identical products. The market price for total production quantity q is $P(q) = 100 - 2\sqrt{q}$. Firm 1's production cost is $C_1(q_1) = q_1 + 10$, and firm 2's production cost is $C_2(q_2) = 2q_2 + 5$. Find the profit functions and the Nash equilibrium quantities of production and profits.

4.35 Consider the Cournot duopoly model with a somewhat more realistic price function given by

$$P(q) = \begin{cases} \frac{1}{4}q^2 - 5q + 26, & \text{if } 0 \le q \le 10; \\ 1, & \text{if } q > 10. \end{cases}$$

Take the cost of producing a gadget at $c = 1$ for both firms.

(a) We may restrict productions for both firms q_1, q_2 to $[0, 10]$. Why?

(b) Find the optimal production quantity and the profit at that quantity if there is only one firm making the gadget.

(c) Now suppose there are two firms with the same price function. Find the Nash equilibrium quantities of production and the profits for each firm.

4.36 Compare profits for firm 1 in the model with uncertain costs and the standard Cournot model. Assume $\Gamma = 15, c_1 = 4, c^+ = 5, c^- = 1$ and $p = 0.5$.

4.37 Suppose that we consider the Cournot model with uncertain costs but with three possible costs for firm 2, $Prob(C_2 = c^i) = p_i$, $i = 1, 2, 3$, where $p_i \ge 0, p_1 + p_2 + p_3 = 1$.

(a) Solve for the optimal production quantities.

(b) Find the explicit optimal production quantities when $p_1 = \frac{1}{2}, p_2 = \frac{1}{8}, p_3 = \frac{3}{8}, \Gamma = 100$, and $c_1 = 2, c^1 = 1, c^2 = 2, c^3 = 5$.

4.38 Suppose that two firms have constant unit costs $c_1 = 2, c_2 = 1$ and $\Gamma = 19$ in the Stackelberg model.

(a) How much should firm 2 produce as a function of q_1?

(b) How much should firm 1 produce?

(c) How much, then, should firm 2 produce?

4.39 Set up and solve a Stackelberg model given three firms with constant unit costs c_1, c_2, c_3 and firm 1 announcing production quantity q_1.

4.40 Gadget prices range in the interval $(0, 100]$. Assume the demand function is simply $D(p) = 100 - p$. If two firms make the gadget, firm 1's profit function is given by

$$u_1(p_1, p_2) = \begin{cases} p_1(100 - p_1), & p_1 < p_2 \\ \frac{p_1(100 - p_1)}{2}, & p_1 = p_2 \\ 0, & p_1 > p_2 \end{cases}$$

(a) What is firm 2's profit function?
(b) Suppose firm 2 picks a price $p_2 > 50$. What is the best response of firm 1?
(c) Suppose firm 2 picks a price $p_2 \leq 50$. Is there a best response of firm 1?
(d) Consider the game played in rounds. In round one, eliminate all strategies that are *not* best responses for both firms in the two cases considered in earlier parts of this problem. In round two, now find a better response to what's left from round one. If you keep going, what is the conclusion?

4.41 In the Bertrand model of this section shows that if $c_1 = c_2 = c$, then $(p_1^*, p_2^*) = (c, c)$ is a Nash equilibrium but each firm makes zero profit.

4.42 Determine the entry deterrence level of production for the model in this section for firm 1 given $\Gamma = 100, a = 2, b = 10$.
(a) How much profit is lost by setting the price to deter a competitor?

4.43 We could make one more adjustment in the Bertrand model of this section and see what effect it has on the model. What if we put a limit on the total quantity that a firm can produce? This limits the supply and possibly will put a floor on prices. Let $K \geq \frac{\Gamma}{2}$ denote the maximum quantity of gadgets that each firm can produce and recall that $D(p) = \Gamma - p$ is the quantity of gadgets demanded at price p. Find the profit functions for each firm.

4.44 Suppose that the demand functions in the Bertrand model are given by

$$q_1 = D_1(p_1, p_2) = (\Gamma - p_1 + bp_2)^+ \text{ and } q_2 = D_2(p_1, p_2) = (\Gamma - p_2 + bp_1)^+,$$

where $1 \geq b > 0$. This says that the quantity of gadgets sold by a firm will increase if the price set by the opposing firm is too high. Assume that both firms have a cost of production $c \leq \min\{p_1, p_2\}$. Then, since profit is revenue minus costs the profit functions will be given by

$$u_i(p_1, p_2) = D_i(p_1, p_2)(p_i - c), \quad i = 1, 2.$$

(a) Using calculus, show that there is a unique Nash equilibrium at

$$p_1^* = p_2^* = \frac{\Gamma + c}{2 - b}.$$

(b) Find the profit functions at equilibrium.
(c) Suppose the firms have different costs and sensitivities so that

$$q_1 = D_1(p_1, p_2) = (\Gamma - p_1 + b_1p_2)^+ \quad \text{and} \quad q_2 = D_2(p_1, p_2) = (\Gamma - p_2 + b_2p_1)^+,$$

and

$$u_i(p_1, p_2) = D_i(p_1, p_2)(p_i - c_i), \quad i = 1, 2.$$

Find a Nash equilibrium and the profits at equilibrium.
(d) Find the equilibrium prices, production quantities, and profits if $\Gamma = 100, c_1 = 5, c_2 = 1, b_1 = \frac{1}{2}, b_2 = \frac{3}{4}$.

4.45 Suppose that firm 1 announces a price in the Bertrand model with demand functions

$$q_1 = D_1(p_1, p_2) = (\Gamma - p_1 + bp_2)^+ \text{ and } q_2 = D_2(p_1, p_2) = (\Gamma - p_2 + bp_1)^+,$$

where $1 \geq b > 0$. Assume the firms have unit costs of production c_1, c_2.
Construct the Stackelberg model; firm 2 should find the best response $p_2 = p_2(p_1)$ to maximize $u_2(p_1, p_2)$ and then firm 1 should choose p_1 to maximize $u_1(p_1, p_2(p_1))$. Find the equilibrium prices and profits. What is the result when $\Gamma = 100, c_1 = 5, c_2 = 1, b = 1/2$?

4.46 Two companies share the market for some gadget. The market price will be $200 - q_1 - q_2$ where q_1 and q_2 are the two production quantities, in millions of units. The costs for the two companies are as follows:

For company 1, $30 per unit.
For company 2, $40 per unit.

Use the Cournot duopoly model to find the Nash equilibrium production values. Also, find the corresponding profits for the two companies.

4.47 A company has a monopoly on the market for simple wooden chairs. Their annual cost of production is $ 25 thousand plus 10 thousand per thousand chairs produced. The price of chairs is $80 - N$, when N is the number of chairs produced, in thousands. These are simple to make, and any other company could enter the market and have the same production costs.
   (a) Find the optimal number of chairs the company should produce, and their profit, if there is no threat of another company entering the market.
   (b) Find their entry deterrence production and profit. That is, the number of chairs they should produce so that no other company can enter the market and make a profit, and how much profit they are making at that production level (assuming they successfully deter entry, so no other company is making chairs). Do this by setting up an equation in one variable. You can solve by computer if you need to.

4.48 Consider the Hotelling Location problem except that the transportation cost of a customer at position x is given by $t\,x^2$ if the customer buys from firm 1, and $t(1-x)^2$ if they buy from firm 1. Find the Nash equilibrium prices and quantities for each firm.

4.49 Two firms share a resource modeled by the interval $[0, 1]$. A firm can use as much of the resource as desired but there is a penalty for using too much. The firm which uses more than the opponent must pay a fine of $0 \leq b \leq \frac{1}{2}$ to the opponent.
   (a) Find the payoff functions to each firm.
   (b) Suppose player 1 uses the mixed strategy $X(x)$ (i.e., the density function for player 1). Player 2 will use the pure strategy $y \in [0, 1]$. The expected payoff to player 1 should be indifferent to any player 2 strategy, and so independent of y, when player 1 is playing an optimal strategy X. (See Proposition 4.2.4.) Find the expected payoff for player 1 when player 1 uses X and player 2 uses y.
   (c) Using the fact that the expected payoff should be independent of y when X is optimal, set the derivative to zero using Leibniz's rule and solve for X.
   (d) What is the optimal strategy for player 2?

(e) Find the value of the game, i.e., calculate the expected payoff for player 1 if both players are using their optimal mixed strategies.

4.50 Consider the following model of Bertrand duopoly with differentiated products. Demand for firm i is $q_i(p_i, p_j) = a - p_i + b_i p_j$. Costs are zero for both firms. The sensitivity of firm i's demand to firm j's price is either high or low. That is, b_i is either b_H or b_L, where $b_H > b_L > 0$. For each firm, $b_i = b_H$ with probability γ and $b_i = b_L$ with probability $1 - \gamma$. Each firm knows its own b_i but not its competitor's. All of this is common knowledge. Assume that $\gamma b_H + (1 - \gamma)b_L < 2$. Find the pure-strategy Nash equilibria of this game.

4.51 Determine the buy-it-now price when the valuations are independent and follow a triangular distribution with density function

$$
g(p) = \begin{cases} \dfrac{2(p-r)}{(c-r)(R-r)}, & r \le p \le c \\ \dfrac{2(R-p)}{(R-c)(R-r)}, & c < p \le R \end{cases} \quad \text{and} \quad G(p) = \begin{cases} \dfrac{(p-r)^2}{(c-r)(R-r)}, & r \le p \le c \\ 1 - \dfrac{(R-p)^2}{(R-c)(R-r)}, & c < p \le R \\ 1 & p > R \end{cases}
$$

The value of c is the mode of the distribution, i.e., c is the peak value of the density function.

4.4 Duels

Duels are used to model not only the actual dueling situation but also many problems in other fields. For example, a battle between two companies for control of a third company or asset can be regarded as a duel in which the accuracy functions could represent the probability of success. Duels can be used to model competitive auctions between two bidders. So there is ample motivation to study a theory of duels.

We begin with a simple example of a three-person truel with strange rules for a real shooting truel but may be realistic for companies trying to take over other companies.

Example 4.19 A, B, and C each have a gun containing a single bullet. Each person, as long as he is alive, may shoot at any surviving person. A shoots first, then B (if still alive), then C (if still alive).

Denote by p_i the probability that player $i =$ A, B, C, hits her intended target; assume that $0 < p_i < 1$. Assume that each player wishes to maximize her probability of survival. Among outcomes in which her survival probability is the same, she wants to minimize the survival probabilities of others.

Assuming that $p_A, p_B,$ and p_C are all distinct, we will show that

- C is better off if $p_C < p_B$ rather than $p_C > p_B$.
- B is a better target for A than is C when $p_B > p_C$. Always eliminate the more accurate shooter.
- If C is still alive after A and B have taken their shot, C still chooses to take her shot. This entails C's desire to minimize the chances of survival of A and B.

Here's how we solve the problem. We use backward induction and dominance.

1. Start with C, who shoots last. If C chooses to shoot at B, then A will survive (recall each player has one shot).
2. By dominance, B always shoots at C since C poses a threat to B while A has already taken her shot.

3. Suppose C is alive when it is B's turn to shoot, B will shoot at C. Suppose C is dead when it is B's turn to shoot, B will shoot at A.
4. A shoots at B.

Now let's assume that if A and B are both alive when it's C's turn to shoot, then C will shoot at A or B with equal probability 0.5.

Given the above analysis, if A shoots at B, the probability A is killed is

$$Prob(A\text{ is killed}) = \underbrace{p_A p_C}_{\text{A kills B \& C kills A}} + 0.5\, \underbrace{(1 - p_A)}_{\text{A misses B}}\underbrace{(1 - p_B)}_{\text{B misses C}}\underbrace{p_C}_{\text{C kills A}}.$$

The first term corresponds to A kills B and C kills A. The last term corresponds to the scenario where C kills A conditional on both A and B miss the shot. Under this scenario, we assume 50% chance that C chooses to shoot at A.

If A shoots at C the probability A is killed is

$$Prob(A\text{ is l killed}) = \underbrace{p_A p_B}_{\text{B kills C \& C kills A}} + 0.5\, \underbrace{(1 - p_A)}_{\text{A misses C}}\underbrace{(1 - p_B)}_{\text{B misses C}}\underbrace{p_C}_{\text{C kills A}}.$$

If $p_C < p_B$, then the probability A is killed is smaller if A shoots at B than if A shoots at C, so A will shoot at B first. This means that C will only have to deal with the shot coming from B if B survives the shot from A.

Now we consider continuous versions of duels. In Chapter 1 we considered discrete versions of a duel in which the players were allowed to fire only at certain distances . In reality, a player can shoot at any distance (or time) once the duel begins. That was only one of our simplifications. The theory of duels includes multiple bullets, machine gun duels, silent and noisy, and so on.[1]

Here are the precise rules that we use here. There are two participants, I and II, each with a gun, and each has exactly one bullet. They will fire their guns at the opponent at a moment of their own choosing. The players each have functions representing their accuracy or probability of killing the opponent, say, $p_I(x)$ for player I and $p_{II}(y)$ for player II, with $x, y \in [0, 1]$. The choice of strategies is a **time** in [0, 1] at which to shoot. Assume that $p_I(0) = p_{II}(0) = 0$ and $p_I(1) = p_{II}(1) = 1$. So, in the setup here you may assume that they are farthest apart at time 0 or $x = y = 0$ and closest together when $x = y = 1$. It is realistic to assume also that both p_I and p_{II} are continuous, strictly increasing, and have continuous derivatives up to any order needed. Accuracy increases as the distance between the duelists decreases.

Noisy Duels

In a noisy duel each opponent knows when the opponent has taken a shot. If a duelist does not know if the opponent has taken their shot, it is called a silent duel.

Finally, if I hits II, player I receives +1 and player II receives −1, and conversely. If both players miss, the payoff is 0 to both. The payoff functions will be the expected payoff depending on the accuracy functions and the choice of the $x \in [0, 1]$ or $y \in [0, 1]$ at which the player will take the shot. We will consider only the symmetric case with continuous strategies.

We break our problem down into the cases where player I shoots before player II, player II shoots before player I, or they shoot at the same time.

---

1 Refer to Karlin's book [20] for a very nice exposition of the general theory of duels.

If player I shoots before player II, then $x < y$, and

$$u_1(x, y) = (+1)p_I(x) + (-1)(1 - p_I(x))p_{II}(y),$$
$$u_2(x, y) = (-1)p_I(x) + (+1)(1 - p_I(x))p_{II}(y).$$

If player II shoots before player I, then $y < x$ and we have similar expected payoffs:

$$u_1(x, y) = (-1)p_{II}(y) + (+1)(1 - p_{II}(y))p_I(x),$$
$$u_2(x, y) = (+1)p_{II}(y) + (-1)(1 - p_{II}(y))p_I(x).$$

Finally, if they choose to shoot at the same time, then $x = y$ and we have

$$u_1(x, x) = (+1)p_I(x)(1 - p_{II}(x)) + (-1)(1 - p_I(x))p_{II}(x) = -u_2(x, x).$$

In this simplest setup, this is a zero sum game, but, as mentioned earlier, it is easily changed to nonzero sum.

We have set up this duel without consideration yet that the duel is noisy. Each player will hear (or see, or feel) the shot by the other player, so that if a player shoots and misses, the surviving player will know that all she has to do is wait until her accuracy reaches 1. With certainty that occurs at $x = y = 1$, and she will then take her shot. In a **silent** duel the players would not know that a shot was taken (unless they didn't survive). Silent duels are more difficult to analyze, and we will consider a special case later.

Let's simplify the payoffs in the case of a noisy duel. In that case, when a player takes a shot and misses, the other player (if she survives) waits until time 1 to kill the opponent with certainty. So, the payoffs become

$$u_1(x, y) = \begin{cases} (+1)p_I(x) + (-1)(1 - p_I(x)) = 2p_I(x) - 1, & x < y; \\ p_I(x) - p_{II}(x), & x = y; \\ 1 - 2p_{II}(y), & x > y. \end{cases}$$

For player II, $u_2(x, y) = -u_1(x, y)$.

Now, to solve this, we cannot use the procedure outlined using derivatives, because this function has no derivatives exactly at the places where the optimal things happen.

Figure 4.10 below shows a graph of $u_1(x, y)$ in the case when the players have the distinct accuracy functions given by $p_I(x) = x^3$ and $p_{II}(x) = x^2$. Player I's accuracy function increases at a slower rate than that for player II. Nevertheless, we will see that both players will fire at the same time. That conclusion seems reasonable when it is a noisy duel. If one player fires before the opponent, the accuracy suffers, and, if it is a miss, death is certain.

In fact, we will show that there is a unique point $x^* \in [0, 1]$ that is the unique solution of

$$p_I(x^*) + p_{II}(x^*) = 1, \tag{4.4.1}$$

so that

$$u_1(x^*, x^*) \geq u_1(x, x^*) \text{ for all } x \in [0, 1] \tag{4.4.2}$$

and

$$u_2(x^*, x^*) \geq u_2(x^*, y) \text{ for all } y \in [0, 1]. \tag{4.4.3}$$

This says that (x^*, x^*) is a Nash equilibrium for the noisy duel. Of course, since $u_2 = -u_1$, the inequalities reduce to

$$u_1(x, x^*) \leq u_1(x^*, x^*) \leq u_1(x^*, y), \text{ for all } x, y \in [0, 1],$$

so that (x^*, x^*) is a saddle point for u_1.

Figure 4.10 $u_1(x,y)$ with accuracy functions $p_{\mathrm{I}} = x^3, p_{\mathrm{II}} = x^2$.

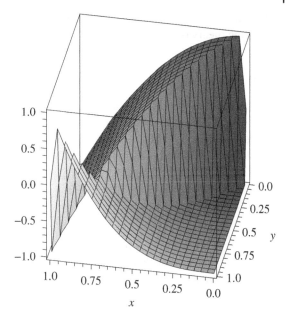

To verify that the inequalities (4.4.2) and (4.4.3) hold for x^* defined in (4.4.1), we have, from the definition of u_1, that

$$u_1(x^*, x^*) = p_{\mathrm{I}}(x^*) - p_{\mathrm{II}}(x^*).$$

Using the fact that both accuracy functions are increasing, we have, by (4.4.1)

$$u_1(x^*, x^*) = p_{\mathrm{I}}(x^*) - p_{\mathrm{II}}(x^*) = 1 - 2p_{\mathrm{II}}(x^*)$$
$$\leq 1 - 2p_{\mathrm{II}}(y) = u_1(x^*, y) \text{ if } x^* > y,$$
$$u_1(x^*, x^*) = p_{\mathrm{I}}(x^*) - p_{\mathrm{II}}(x^*)$$
$$= 2p_{\mathrm{I}}(x^*) - 1 = u_1(x^*, y) \text{ if } x^* < y, \text{ and}$$
$$u_1(x^*, x^*) = p_{\mathrm{I}}(x^*) - p_{\mathrm{II}}(x^*)$$
$$= u_1(x^*, y) \text{ if } x^* = y.$$

So, in all cases $u_1(x^*, x^*) \leq u_1(x^*, y)$ for all $y \in [0, 1]$. We verify (4.4.3) in a similar way and leave that as an exercise. We have shown that (x^*, x^*) is indeed a Nash point.

That x^* exists and is unique is shown by considering the function $f(x) = p_{\mathrm{I}}(x) + p_{\mathrm{II}}(x)$. We have $f(0) = 0, f(1) = 2$, and $f'(x) = p_{\mathrm{I}}'(x) + p_{\mathrm{II}}'(x) > 0$. By the **intermediate value theorem** of calculus, we conclude that there is an x^* satisfying $f(x^*) = 1$. The uniqueness of x^* follows from the fact that p_{I} and p_{II} are strictly increasing.

In the example shown in Figure 4.10 with $p_{\mathrm{I}}(x) = x^3$ and $p_{\mathrm{II}}(x) = x^2$, we have the condition $x^{*3} + x^{*2} = 1$, which has solution at $x^* = 0.754877$. With these accuracy functions, the duelists should wait until less than 25% of the time is left until they **both** fire. The expected payoff to player I is $u_1(x^*, x^*) = -0.1397$, and the expected payoff to player II is $u_2(x^*, x^*) = 0.1397$. It appears that player I is going down. We expect that result in view of the fact that player II has greater accuracy at x^*, namely, $p_{\mathrm{II}}(x^*) = 0.5698 > p_{\mathrm{I}}(x^*) = 0.4301$.

For a more dramatic example, suppose that $p_{\mathrm{I}}(x) = x^3$, $p_{\mathrm{II}}(x) = x$. This says that player I is at a severe disadvantage. His accuracy does not improve until a lot of time has passed (and so the duelists are closer together). In this case $x^* = 0.68232$ and they both fire at time 0.68232 and

$u_1(0.68232, 0.68232) = -0.365$. Player I would be stupid to play this game with real bullets. That is why game theory is so important.

Silent Duel on [0, 1]

In case you are curious as to what happens when we have a silent duel, we will present this example to show that things get considerably more complicated. We take the simplest possible accuracy functions $p_I(x) = p_{II}(x) = x \in [0, 1]$ because this case is already much more difficult than the noisy duel. The payoff of this game to player I is

$$u_1(x, y) = \begin{cases} x - (1 - x)y, & x < y; \\ 0, & x = y; \\ -y + (1 - y)x, & x > y. \end{cases}$$

For player II, since this is zero sum, $u_2(x, y) = -u_1(x, y)$. Now, in the problem with a silent duel, intuitively it seems that there cannot be a pure Nash equilibrium because silence would dictate that an opponent could always take advantage of a pure strategy. But how do we allow mixed strategies in a game with continuous strategies? In a discrete matrix game, a mixed strategy is a probability distribution over the pure strategies. Why not allow the players to choose continuous probability distributions? No reason at all. So we consider the mixed strategy choice for each player

$$X(x) = \int_0^x f(a) \, da,$$

$$Y(y) = \int_0^y g(b) \, db,$$

$$\int_0^1 f(a) \, da = \int_0^1 g(b) \, db = 1.$$

The cumulative distribution function $X(x)$ represents the probability that player I will choose to fire at a point $\leq x$. The expected payoff to player I if he chooses X and his opponent chooses Y is

$$E(u_1(X, Y)) = \int_0^1 \int_0^1 u_1(x, y) f(x) g(y) \, dx \, dy.$$

As in the discrete-game case, we define the value of the game as

$$v \equiv \min_Y \max_X E(u_1(X, Y)) = \max_X \min_Y E(u_1(X, Y)).$$

The equality follows from the existence theorem of a Nash equilibrium (actually a saddle point in this case) because the expected payoff is not only concave–convex, but actually linear in each of the probability distributions X, Y. It is completely analogous to the existence of a mixed strategy saddle point for matrix games. The value of games with a continuum of strategies exists if the players choose from within the class of probability distributions. (Actually, the probability distributions should include the possibility of point masses, but we do not go into this generality here.) A saddle point in mixed strategies has the same definition as before: (X^*, Y^*) is a saddle if

$$E(X, Y^*) \leq E(X^*, Y^*) = v \leq E(X^*, Y), \forall X, Y \text{ probability distributions.}$$

Now, the fact that both players are symmetric and have the same accuracy functions allows us to guess that $v = 0$ for the silent duel. To find the optimal strategies, namely, the density functions

$f(x), g(y)$, we use the necessary condition that if X^*, Y^* are optimal, then using Proposition 4.2.4,

$$E(X, y) = \int_0^1 u_1(x, y)f(x)\, dx = v = 0,$$

$$E(x, Y) = \int_0^1 u_1(x, y)g(y)\, dy = v = 0, \quad \forall x, y \in [0, 1].$$

This is completely analogous to the **equality of payoffs** Theorem 2.2.3 to find mixed strategies in bimatrix games, or to the geometric solution of two person 2×2 games in which the value occurs where the two payoff lines cross. We replace $u_1(x, y)$ to work with the following equation:

$$\int_0^y [x - (1 - x)y]f(x)\, dx + \int_y^1 [-y + (1 - y)x]f(x)\, dx = 0, \quad \forall y \in [0, 1].$$

If we expand this, we get

$$
\begin{aligned}
0 &= \int_0^y [x - (1 - x)y]f(x)\, dx + \int_y^1 [-y + (1 - y)x]f(x)\, dx \\
&= \int_0^y xf(x)\, dx - y \int_0^y (1 - x)f(x)\, dx - y \int_y^1 f(x)dx + (1 - y)\int_y^1 xf(x)\, dx \\
&= \int_0^1 xf(x)\, dx - y \int_0^1 f(x)\, dx + y \int_0^y xf(x)\, dx - y \int_y^1 xf(x)\, dx \\
&= \int_0^1 xf(x)\, dx - y + y \int_0^y xf(x)\, dx - y \int_y^1 xf(x)\, dx.
\end{aligned}
$$

The first term in this last line is actually a constant, and the constant is $E[X] = \int_0^1 xf(x)\, dx$, which is the **mean of the strategy** X.

Now a key observation is that the equation we have should be looked at, not in the unknown function $f(x)$, but in the unknown function $xf(x)$. Let's call it $\varphi(x) \equiv xf(x)$, and we see that

$$E[X] - y + y \int_0^y \varphi(x)\, dx - y \int_y^1 \varphi(x)\, dx = 0.$$

Consider the left side as a function of $y \in [0, 1]$. Call it

$$F(y) \equiv E[X] - y + y \int_0^y \varphi(x)\, dx - y \int_y^1 \varphi(x)\, dx.$$

Then $F(y) = 0, 0 \le y \le 1$. We take a derivative using the fundamental theorem of calculus in an attempt to get rid of the integrals:

$$F'(y) = -1 + \int_0^y \varphi(x)\, dx + y[\varphi(y)] - \int_y^1 \varphi(x)\, dx + y[\varphi(y)] = 0$$

and then another derivative

$$
\begin{aligned}
F''(y) &= \varphi(y) + \varphi(y) + y\varphi'(y) + \varphi(y) + \varphi(y) + y\varphi'(y) \\
&= 4\varphi(y) + 2y\varphi'(y) = 0.
\end{aligned}
$$

So we are led to the differential equation for $\varphi(y)$, which is

$$4\varphi(y) + 2y\varphi''(y) = 0, \quad 0 \le y \le 1.$$

This is a first-order **ordinary differential equation** that will have general solution

$$\varphi(y) = C\frac{1}{y^2},$$

as you can easily check by plugging in. Then $\varphi(y) = yf(y) = \frac{C}{y^2}$ implies that $f(y) = \frac{C}{y^3}$, or $f(x) = \frac{C}{x^3}$ returning to the x variable. We have to determine the constant C.

You might think that the way to find C is to apply the fact that $\int_0^1 f(x) \, dx = 1$. That would normally be correct, but it also points out a problem with our formulation. Look at $\int_0^1 x^{-3} \, dx$. This integral **diverges** (i.e., it is infinite), because x^{-3} is not integrable on $(0, 1)$. This would stop us dead in our tracks because there would be no way to fix that with a constant C unless the constant was zero. That can't be, because then $f = 0$, and it is not a probability density. The way to fix this is to assume that the function $f(x)$ is zero on the starting subinterval $[0, a]$ for some $0 < a < 1$. In other words, we are assuming that the players will not shoot on the interval $[0, a]$ for some unknown time $a > 0$. The lucky thing is that the procedure we used at first, but now repeated with this assumption, is the same and leads to the equation

$$E[X] - y + y \int_a^y \varphi(x) \, dx - y \int_y^1 \varphi(x) \, dx = 0,$$

which is the same as where we were before except that 0 is replaced by a. So, we get the same function $\varphi(y)$ and eventually the same $f(x) = \frac{C}{x^3}$, except we are now on the interval $0 < a \leq x \leq 1$. This idea does not come for free, however, because now we have two constants to determine, C and a. C is easy to find because we must have $\int_a^1 \frac{C}{x^3} \, dx = 1$. This says, $C = \dfrac{2a^2}{(1 - a^2)} > 0$. To find $a > 0$, we substitute $f(x) = \frac{C}{x^3}$ into (recall that $\varphi(x) = xf(x)$)

$$0 = E[X] - y + y \int_a^y \varphi(x) \, dx - y \int_y^1 \varphi(x) \, dx$$

$$= y \left(C + \frac{C}{a} - 1 \right) + C \left(-3 + \frac{1}{a} \right).$$

This must hold for all $a \leq y \leq 1$ which implies that $C + \frac{C}{a} - 1 = 0$. Therefore, $C = \frac{a}{(a+1)}$. But then we must have

$$C = \frac{2a^2}{1 - a^2} = \frac{a}{a + 1} \implies a = \frac{1}{3}, \quad C = \frac{1}{4}.$$

So, we have found $X(x)$. It is the cumulative distribution function of the strategy for player I and has density

$$f(x) = \begin{cases} 0 & \text{if } 0 \leq x < \frac{1}{3}; \\ \dfrac{1}{4x^3} & \text{if } \frac{1}{3} \leq x \leq 1. \end{cases}$$

We know that $\int_a^1 u_1(x, y)f(x) \, dx = 0$ for $y \geq a$, but we have to check that with this $C = \frac{1}{4}$ and $a = \frac{1}{3}$ to make sure that

$$\int_a^1 u_1(x, y)f(x) \, dx > 0 = v, \text{ when } y < a. \tag{4.4.4}$$

That is, we need to check that X played against any pure strategy in $[0, a]$ must give at least the value $v = 0$ if X is optimal. Let's take a derivative of the function $G(y) = \int_a^1 u_1(x, y)f(x) \, dx$, $0 \leq a \leq 1$. We have,

$$G(y) = \int_a^1 u_1(x, y)f(x) \, dx = \int_a^1 (-y + (1 - y)x)f(x) \, dx,$$

which implies that

$$G'(y) = \int_a^1 [-1 - xf(x)] \, dx = -\frac{3}{2} < 0.$$

This means $G(y)$ is decreasing on $[0, a]$. Since $G(a) = \int_a^1 (-a + (1-a)x)f(x)\, dx = 0$, it must be true that $G(y) > 0$ on $[0, a]$, so the condition (4.4.4) checks out. Finally, since this is a symmetric game, it will be true that $Y(y)$ will have the same density as player I.

Problems

4.52 Determine the optimal time to fire for each player in the noisy duel with accuracy functions $p_I(x) = \sin\left(\frac{\pi}{2}x\right)$ and $p_{II}(x) = x^2$, $0 \le x \le 1$.

4.53 **Noisy duel** The difference between a silent and noisy duel is that in a noisy duel each player knows when the opponent has taken the shot. Once again, two gun-fighters are in a duel. Each can choose when to shoot, which is a time in $[0, 1]$. The accuracy functions are:
- $p_I(x) = x^2$.
- $p_{II}(y) = y^3$.

In this version of the noisy duel, assume the players care equally about killing their opponent as surviving, so the payoff to I is

$$\text{Pay}_1 = \text{Probability II gets shot} - \text{Probability I gets shot}.$$

(a) Explain that a pure strategy for I is to choose a time x to shoot, and shoot then unless II has already taken her shot.

(b) Assume the players have chosen strategies x and y. Write down the payoff to I. You will need three cases, depending on whether $x < y, x > y$, or they are equal (in which case both players shoot at the same time).

(c) Give an argument that there can be no pure Nash equilibrium with $x \ne y$.

(d) Show that at a pure equilibrium with $x = y$, the payoff to I assuming x is very slightly less than y should be the same as the payoff with x very slightly greater than y.

(e) Write an equation that lets you solve for the pure equilibrium strategies. **Solve by computer or calculator**.

4.54 **Silent duel** The accuracy functions for both duelists are linear: $p_I(x) = p_{II}(x) = x$. In a silent duel **players cannot tell if their opponent has shot and missed or is still waiting**.
As before, assume the players care equally about killing their opponent as surviving, so the payoff to I is

$$\text{Pay}_1 = \text{Probability II gets shot} - \text{Probability I gets shot}$$

(a) Explain that a pure strategy for I is to choose a time x to shoot, and shoot then unless you are dead.

(b) Assume the players have chosen strategies x and y. Write down the payoff to I. You will still need three cases, depending on whether $x < y, x > y, x = y$.

(c) Give an argument that there can be no pure Nash equilibrium with $x \ne y$.

(d) Give an argument that there can be no pure Nash equilibrium with $x = y$. Hint: if we are shooting at the same time, how would you like to change your strategy?

(e) Assume I is playing a totally mixed equilibrium with a probability density function $f_I(x)$, and I shoots at a random time chosen from that distribution. Write an expression for the payoff if II shoots at (i) $y = 0.3$ and (ii) $y = T$ for a general T.

(f) Explain why, if the players are both playing the same mixed equilibrium $f(x)$, the expression you just found should be constant, so its derivative with respect to T should be 0.

(g) Take the derivative from the previous part, and set it equal to 0. You will need to use the fundamental theorem of calculus. Simplify as much as possible.

(h) There is still an integral, and we only know how to solve differential equations... differentiate again to get rid of it.

(i) Now you should have a differential equation. Solve it! Keep track of the constant of integration C.

(j) You should be able to find C using the fact that $\int_0^1 f(x)\,dx = 1$. Try this, and explain what goes wrong.

(k) Discuss: it is probably a bad idea to shoot very early! Now guess that the strategy is

$$f(x) = \begin{cases} 0 & x < a \\ \text{what you found above} & x \geq a \end{cases}$$

for some a.

(l) Solve for C in terms of a

(m) Show that your answer with $a = \frac{1}{3}$ works. Discuss how you might have figured out $a = \frac{1}{3}$.

4.5 Auctions

There were probably auctions by cavemen for clubs, tools, skins, and so on, but we can be sure (because there is a written record) that there was bidding by Babylonians for both men and women slaves and wives. Auctions today are ubiquitous with many internet auction houses led by eBay, which does more than 6 billion dollars of business a year and earns almost 2 billion dollars as basically an auctioneer. This pales in comparison with United States treasury bills, notes, and bond auctions, which have total dollar values each year in the trillions of dollars. Rights to the airwaves, oil leases, pollution rights, and tobacco, all the way down to auctions for a Mickey Mantle-signed baseball are common occurrences.

There are different types of auctions we study. Their definitions are summarized here.

Definition 4.5.1 *The different types of auctions are:*

- **English Auction.** *Bids are announced publicly, and the bids rise until only one bidder is left. That bidder wins the object at the highest bid.*
- **Sealed Bid, First Price.** *Bids are private and are made simultaneously. The highest sealed bid wins and the winner pays that bid.*
- **Sealed Bid, Second Price.** *Bids are private and made simultaneously. The high bid wins, but the winner pays the second highest bid. This is also called a Vickrey auction after the Nobel Prize winner who studied them.*
- **Dutch Auction.** *The auctioneer (which could be a machine) publicly announces a high bid. Bidders may accept the bid or not. The announced prices are gradually lowered until someone accepts that price. The first bidder who accepts the announced price wins the object and pays that price.*
- **Private Value Auction.** *Bidders are certain of their own valuation of the object up for auction and these valuations (which may be random variables) are independent.*
- **Common Value Auction.** *The object for sale has the same value (that is not known for certain to the bidders) to all the bidders. Each bidder has their own estimate of this value.*

In this section we will present a game theory approach to the theory of auctions and will be more specific about the type of auction as we cover it. Common value auctions require the use of more advanced probability theory and will not be considered further.

Let's start with a simple nonzero sum game that shows why auction firms like eBay even exist (or need to exist).

Example 4.20   In an online auction with no middleman, the seller of the object and the buyer of the object may choose to renege on the deal dishonestly or go through with the deal honestly. How is that carried out? The buyer could choose to wait for the item and then not pay for it. The seller could simply receive payment but not send the item. Here is a possible payoff matrix for the buyer and the seller:

Buyer/Seller	Send	Keep
Pay	$(1,1)$	$(-2,2)$
Don't Pay	$(2,-2)$	$(-1,-1)$

There is only one Nash equilibrium in this problem and it is at (don't pay, keep); neither player should be honest! Amazing, the transaction will never happen and it is all due to either lack of trust on the part of the buyer and seller, or total dishonesty on both their parts. If a buyer can't trust the seller to send the item and the seller can't depend on the buyer to pay for the item, there won't be a deal.

Now let's introduce an auction house that serves two purposes: (1) it guarantees payment to the seller and (2) it guarantees delivery of the item for the buyer. Of course, the auction house will not do that out of kindness but because it is paid by the seller (or the buyer) in the form of a commission. This introduces a third strategy for the buyer and seller to use: Auctioneer. This changes the payoff matrix as follows:

Buyer/Seller	Send	Keep	Auctioneer
Pay	$(1,1)$	$(-2,2)$	$(1, 1-c)$
Don't pay	$(2,-2)$	$(-1,-1)$	$(0,-c)$
Auctioneer	$(1-c,1)$	$(-c,0)$	$(1-c, 1-c)$

The idea is that each player has the **option, but not the obligation,** of using an auctioneer. If somehow they should agree to both be honest, they both get +1. If they both use an auctioneer, the auctioneer will charge a fee of $0 < c < 1$ and the payoff to each player will be $1 - c$.

Observe that $(-1,-1)$ is no longer a pure Nash equilibrium. We use a calculus procedure to find the mixed Nash equilibrium for this symmetric game as a function of c. The result of this calculation is

$$X_c = \left(\frac{1}{2}(1-c), \frac{1}{2}c, \frac{1}{2}\right) = Y_c$$

and (X_c, Y_c) is the unique Nash equilibrium. The expected payoffs to each player are

$$E_I(X_c, Y_c) = 1 - \frac{3}{2}c = E_{II}(X_c, Y_c).$$

As long as $\frac{2}{3} > c > 0$, both players receive a positive expected payoff. Because we want the payoffs to be close to $(1,1)$, which is what they each get if they are both honest and don't use an auctioneer, it will be in the interest of the auctioneer to make $c > 0$ as small as possible because at some point the transaction will not be worth the cost to the buyer or seller.

The Nash equilibrium tells the buyer and seller to use the auctioneer half the time, no matter what value c is. Each player should be dishonest with probability $\frac{c}{2}$, which will increase as c increases.

The players should be honest with probability only $\frac{(1-c)}{2}$. If $c = 1$ they should never play honestly and either play dishonestly or use an auctioneer half the time.

You can see that the existence and only the existence of an auctioneer will permit the transaction to go through. From an economics perspective, this implies that auctioneers **will** come into existence as an economic necessity for online auctions and it can be a very profitable business (which it is). But this is true not only for online auctions. Auction houses for all kinds of specialty and non-specialty items (like horses, tobacco, diamonds, gold, houses, etc.) have been in existence for decades, if not centuries, because they serve the economic function of guaranteeing the transaction.

A common feature of auctions is that the seller of the object may set a price, called the **reserve price,** so that if none of the bids for the object are above the reserve price the seller will not sell the object. One question that arises is whether sellers should use a reserve price. Here is an example to illustrate.

Example 4.21 Assume that the auction has two possible buyers. The seller must have some information, or estimates, about how the buyers value the object. Perhaps these are the seller's own valuations projected onto the buyers. Suppose that the seller feels that each buyer values the object at either $\$s$ (small amount) or $\$L$ (large amount) with probability $\frac{1}{2}$ each. But, assuming bids may go up by a minimum of $1, the winning bids with no reserve price set are $\$s, \$(s + 1), \$(s + 1)$, or $\$L$ each with probability $\frac{1}{4}$. Without a reserve price, the expected payoff to the seller is

$$\frac{s + 2(s + 1) + L}{4} = \frac{3s + 2 + L}{4}.$$

Suppose next that the seller sets a reserve price at the higher valuation $\$ L$, and this is the lowest acceptable price to the seller. Let $B_i, i = 1, 2$ denote the random variable that is the bid for buyer i. Without collusion or passing of information, we may assume that the B_i values are independent. The seller is assuming that the valuations of the bidders are $(s, s), (s, L), (L, s)$ and (L, L), each with probability $\frac{1}{4}$. If the reserve price is set at $\$L$, the sale will not go through 25% of the time, but the expected payoff to the seller will be

$$\left(\frac{3}{4}\right)L + \left(\frac{1}{4}\right)0 = \frac{3L}{4}.$$

The question is whether it can happen that there are valuations s and L, so that

$$\frac{3L}{4} > \frac{3s + 2 + L}{4}.$$

Solving for L, the requirement is that $L > \frac{3s+2}{2}$, and this certainly has solutions. For example, if $L = 100$, any lower valuation $s < 66$ will lead to a higher expected payoff to the seller with a reserve price set at $100. Of course, this result depends on the valuations of the seller.

Example 4.22 If you need more justification for a reserve price, consider that if only one bidder shows up for your auction and you must sell to the high bidder, then you do not make any money at all unless you set a reserve price. Of course, the question is how to set the reserve price. Assume that a buyer has a valuation of your gizmo at $\$ V$, where V is a random variable with cumulative distribution function $F(v)$. Then, if your reserve price is set at $\$p$, the expected payoff (assuming one bidder) will be

$$E[\text{Payoff}] = p\,Prob(V > p) + 0\,Prob(V \le p)$$
$$= p\,Prob(V > p) = p(1 - F(p)).$$

You want to choose p to maximize this function $g(p) \equiv p(1 - F(p))$. Let's find the critical points assuming that $f(p) = F'(p)$ is the probability density function of V:

$$g'(p) = (1 - F(p)) - pf(p) = 0.$$

Assuming that $g''(p) = 2f(p) + pf'(p) \geq 0$, we will have a maximum. Observe that the maximum is not achieved at $p = 0$ because $g'(0) = 1 - F(0) = 1 > 0$. So, we need to find the solution, p^*, of

$$1 - F(p^*) - p^*f(p^*) = 0.$$

For a concrete example, we assume that the random variable V has a normal distribution with mean 0.5 and standard deviation 0.2. The density is

$$f(p) = \frac{2.5}{\sqrt{2\pi}}e^{-12.5(p-0.5)^2}.$$

We need to solve $1 - F(p) - pf(p) = 0$, where $F(p)$ is the cumulative normal distribution with mean 0.5 and standard deviation 0.2. The plot of $b(u) = 1 - F(u) - p\,f(u)$ is shown in the following figure.

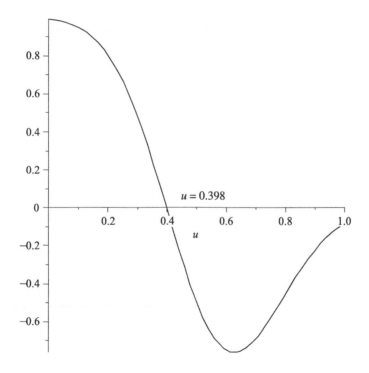

We see that the function $b(u)$ crosses the axis at $u = p^* = 0.398$. The reserve price should be set at 39.8% of the maximum valuation.

Having seen why auction houses exist, let's get into the theory from the bidders' perspective. There are N bidders (=players) in this game. There is one item up for bid and each player **values the object** at $v_1 \geq v_2 \geq \cdots \geq v_N > 0$ dollars. One question we have to deal with is whether the bidders know this ranking of values.

4.5.1 Complete Information

In Section 4.3 we have considered how a seller should set her buy-it-now price for an item she puts up for sale and there are N bidders (but this was an optimization problem and not a game). We assumed their valuations were random variables. Now we consider this problem as a game in which each player wants to maximize their own payoff. In the simplest and almost totally unrealistic model, all the bidders have complete information about the valuations of all the bidders. Now we define the rules of the auctions considered in this section.

Definition 4.5.2   *A first-price, sealed-bid auction is an auction in which each bidder submits a bid $b_i, i = 1, 2, \dots, N$ in a sealed envelope. After all the bids are received the envelopes are opened by the auctioneer and the person with the highest bid wins the object and pays the bid b_i. If there are identical bids, the winner is chosen at random from the identical bidders.*

So what is the payoff to player i if the bids are $b_1, \dots, b_N$? Well, if bidder i doesn't win the object, she pays nothing and gets nothing. That will occur if she is not a high bidder:

$$b_i < \max\{b_1, \dots, b_N\} \equiv M.$$

On the other hand, if she is a high bidder, so that $b_i = M$, then the payoff is the difference between what she bid and what she thinks it's worth (i.e., $v_i - b_i$). If she bids less than her valuation of the object, and wins the object, then she gets a positive payoff, but she gets a negative payoff if she bids more than it's worth to her. To take into account the case when there are k ties in the high bids, she would get the average payoff. Let's use the notation that $\{k\}$ is the set of high bidders. So, in symbols

$$u_i(b_1, \dots, b_N) = \begin{cases} 0 & \text{if } b_i < M, \text{ she is not a high bidder;} \\ v_i - b_i & \text{if } b_i = M, \text{ she is the sole high bidder;} \\ \dfrac{v_i - b_i}{k} & \text{if } i \in \{k\}, \text{she is one of } k \text{ high bidders.} \end{cases}$$

Naturally, bidder i wants to know the amount to bid. That is determined by finding the maximum payoff of player i, assuming that the other players are fixed. We want a Nash equilibrium for this game. This doesn't really seem too complicated. Why not just bid v_i? That would guarantee a payoff of zero to each player, but is that the maximum? Should a player ever bid more than her valuation? These questions are answered in the following rules.

With complete information and a sealed-bid, first-price auction:

1. Each bidder should bid $b_i \le v_i, i = 1, 2, \dots, N$. Never bid more than the valuation. To see this, just consider the following cases. If player i bids $b_i > v_i$ and wins the auction, then $u_i < 0$, even if there are ties. If player i bids $b_i > v_i$ and does not win the auction, then $u_i = 0$. But if player i bids $b_i \le v_i$, in all cases $u_i \ge 0$.

2. In the case when the highest valuation is strictly bigger than the second highest valuation, $v_1 > v_2$, player 1 bids $b_1 \approx v_2, v_1 > b_1 > v_2$; that is, player 1 wins the object with any bid greater than v_2 and so should bid very close to but higher than v_2. Notice that this is an open interval and the maximum is not actually achieved by a bid. If bidding is in whole dollars, then $b_1 = v_2 + 1$ is the optimal bid. There is no Nash equilibrium achieved in the case where the winning bid is in $(v_2, v_1]$ because it is an open interval at v_2.

3. In the case when $v_1 = v_2 = \dots = v_k$, so there are k players with the highest, but equal, value of the object, then player i should bid v_i (i.e, $b_i = v_i, 1 \le i \le N$). So, the bid $(b_1, \dots, b_N) = (v_1, \dots, v_N)$ will be a Nash equilibrium in this case. We leave it as an exercise to verify that.

Problems

4.55 Verify that in the first-price auction $(b_1, \ldots, b_N) = (v_1, \ldots, v_N)$ is a Nash equilibrium assuming $v_1 = v_2$.

4.56 In a second-price sealed-bid auction with complete information the winner is the high bidder but she pays, not the price she bid, but the second highest bid. If there are ties, then the winner is drawn at random from among the high bidders and she pays the highest bid. Formulate the payoff functions and show that the following rules are optimal:
(a) Each player bids $b_i \leq v_i$.
(b) If $v_1 > v_2$, then player 1 wins by bidding any amount $v_2 < b_1 < v_1$.
(c) If $v_1 = v_2 = \cdots v_k$, then $(v_1, v_2, \ldots, v_N)$ is a Nash equilibrium.

4.57 A homeowner is selling her house by auction. Two bidders place the same value on the house at \$100,000, while the next bidder values the house at \$80,000. Should the homeowner use a first-price or second-price auction to sell the house, or does it matter? What if the highest valuation is \$100,000, the next is \$95,000 and the rest are no more than \$90,000?

4.58 Find the optimal reserve price to set in an auction assuming that the density of the value random variable V is $f(p) = 6p(1 - p), 0 \leq p \leq 1$.

4.5.2 Symmetric Independent Private Value Auctions

In this section[1] we take it a step further and model the auction as a game. Again, there is one object up for auction by a seller and the bidders know their own valuations of the object but not the valuations of the other bidders. It is assumed that the unknown valuations $V_1, \ldots, V_N$ are independent and identically distributed continuous random variables. The **symmetric** part of the title of this section comes from assuming that the bidders all have the same valuation distribution.

We will consider two types of auction:

1. English auction, where bids increase until everyone except the highest one or two bidders are gone. In a first-price auction, the high bidder gets the object.
2. Dutch auction, where the auctioneer asks a price and lowers the price continuously until one or more bidders decide to buy the item at the latest announced price. In a tie, the winner is chosen randomly.

Remarks A Dutch auction is equivalent to the first-price sealed-bid auction. Why? A bidder in each type of auction has to decide how much to bid to place in the sealed envelope or when to yell "Buy" in the Dutch auction. That means that the strategies and payoffs are the same for these types of auction. In both cases the bidder must decide the highest price she is willing to pay and submit that bid. In a Dutch auction, the object will be awarded to the highest bidder at a price equal to her bid. That is exactly what happens in a first-price sealed-bid auction. The strategies for making a bid are identical in each of the two seemingly different types of auction.

An English auction can also be shown to be equivalent to a **second-price sealed-bid auction** (as long as we are in the private values set up). Why? As long as bidders do not change their valuations based on the other bidders' bids, which is assumed, then bidders should accept to pay any price up to

---

1 Refer to the article by Milgrom [28] for an excellent essay on auctions from an economic perspective.

their own valuations. A player will continue to bid until the current announced price is greater than how much the bidder is willing to pay. This means that the item will be won by the bidder who has the highest valuation and she will win the object at a price equal to the **second highest valuation**.

The equivalence of the various types of auctions was first observed by William Vickrey.[1] The result of an English auction can be achieved by using a sealed-bid auction in which the item is won by the second-highest submitted bid. A bidder should submit a bid that is equal to her valuation of the object since she is willing to pay an amount less than her valuation but not willing to pay a price greater than her valuation. Assuming that each player does that, the result of the auction is that the object will be won by the bidder with the highest valuation and the price paid will be the second highest valuation.

For simplicity it will be assumed that the reserve price is normalized to $r = 0$. As above, we assume that the joint distribution function of the valuations is given by

$$F_J(v_1, \ldots, v_N) = Prob(V_1 \leq v_1, \ldots, V_N \leq v_N) = F(v_1)F(v_2) \cdots F(v_N),$$

which holds because we assume independence and identical distribution of the bidder's valuations. Suppose that the maximum possible valuation of all the players is a fixed constant $w > 0$. Then, a bidder gets to choose a bidding function $b_i(v)$ that takes points in $[0, w]$ and gives a positive real bid. Because of the symmetry property, it should be the case that all players have the same payoff function and that all optimal strategies for each player are the same for all other players.

Now suppose that we have a player with bidding function $b = b(v)$ for a valuation $v \in [0, w]$. The payoff to the player is given by

$$u(b(v), v) = Prob(b(v) \text{ is high bid})v - E(\text{payment for bid } b(v)).$$

We have to use the expected payment for the bid $b(v)$ because we don't know whether the bid $b(v)$ will be a winning bid. The probability that $b(v)$ is the high bid is

$$Prob(b(v) > \max \{b(V_1), \ldots, b(V_N)\}).$$

We will simplify notation a bit by setting

$$f(b(v)) = Prob(b(v) > \max \{b(V_1), \ldots, b(V_N)\}),$$

and then write the payoff as

$$u(b(v), v) = f(b(v))v - E(b(v)).$$

We want the bidder to maximize this payoff by choosing a bidding strategy which we denote by $\beta(v)$.

One property of $\beta(v)$ should be obvious, namely, as the valuation increases, the bid must increase. The fact that $\beta(v)$ is strictly increasing as a function of v can be proved using some **convex analysis** but we will skip the proof.

Once we know the strict increasing property, the fact that all bidders are essentially the same leads us to the conclusion that the bidder with the highest valuation wins the auction.

Let's take the specific example that bidder's valuations are uniform on $[0, R] = [0, 1]$. Then for each player, we have

$$F(v) = \begin{cases} 0 & \text{if } v < 0; \\ v & \text{if } 0 \leq v \leq 1; \\ 1 & \text{if } v > 1. \end{cases}$$

---

1 William Vickrey (1914–1996) won the 1996 Nobel Prize in Economics primarily for his foundational work on auctions. He earned a BS in mathematics from Yale in 1935 and a PhD in economics from Columbia University.

Remark Whenever we say in the following that we are normalizing to the interval $[0, 1]$, this is not a restriction because we may always transform from an interval $[r, R]$ to $[0, 1]$ and the reverse by the linear transformation $t = (s - r)/(R - r),\ r \le s \le R$, or $s = r + (R - r)t, 0 \le t \le 1$.

We can now establish the following theorem.[1]

Theorem 4.5.3 *Suppose that valuations $V_1, \ldots, V_N$, are uniformly distributed on $[0, 1]$, and the expected payoff function for each bidder is*

$$u(b(v), v) = f(b(v))v - E(b(v)),\ \text{where}$$

$$f(b(v)) = Prob(b(v) > \max\{b(V_1), \ldots, b(V_N)\}). \tag{4.5.1}$$

Then there is a unique Nash equilibrium $(\beta, \ldots, \beta)$ given by

$$\beta(v_i) = v_i,\quad i = 1, 2, \ldots, N$$

in the case when we have an English auction, and

$$\beta(v_i) = \left(1 - \frac{1}{N}\right)v_i,\quad i = 1, 2, \ldots, N,$$

in the case when we have a Dutch auction. In either case the expected payment price for the object is

$$p^* = \frac{N - 1}{N + 1}.$$

Proof: To see why this is true, we start with the Dutch auction result. Since all players are indistinguishable, we might as well say our that guy is player 1. Suppose that player 1 bids $b = \beta(v)$. Then the probability that she wins the object is given by

$$Prob(\beta(\max\{V_2, \ldots, V_N\}) < b).$$

Now here is where we use the fact that β is strictly increasing, because then it has an inverse, $v = \beta^{-1}(b)$, and so we can say

$$f(b) = Prob(\beta(\max\{V_2, \ldots, V_N\}) < b)$$

$$= Prob(\max\{V_2, \ldots, V_N\} < \beta^{-1}(b)) = Prob(V_i < \beta^{-1}(b), i = 2, \ldots, N)$$

$$= F(\beta^{-1}(b))^{N-1} = [\beta^{-1}(b)]^{N-1} = v^{N-1},$$

because all valuations are independent and identically distributed. The next-to-last-equality is because we are assuming a uniform distribution here. The function $f(b)$ is the probability of winning the object with a bid of b. ∎

Now, for the given bid $b = \beta(v)$, player 1 wants to maximize her expected payoff. The expected payoff (4.5.1) becomes

$$u(\beta(v), v) = f(\beta(v))v - E(\beta(v))$$

$$= f(b)v - (b Prob(\text{win}) + 0 \cdot Prob(\text{lose}))$$

$$= f(b)v - bf(b) = (v - b)f(b).$$

Taking a derivative of $u(b, v)$ with respect to b, evaluating at $b = \beta(v)$ and setting to zero, we get the condition

$$f'(\beta)(v - \beta) - f(\beta) = 0. \tag{4.5.2}$$

Since $f(b) = [\beta^{-1}(b)]^{N-1} = v^{N-1}, v = \beta^{-1}(b)$, we have

$$\frac{df(b)}{db} = (N - 1)[\beta^{-1}(b)]^{N-2}\frac{d\beta^{-1}(b)}{db}.$$

---

1 See the article by Wolfstetter [45] for a very nice presentation and many more results on auctions, including common value auctions.

Therefore, after dividing out the term $[\beta^{-1}(b)]^{N-2}$, the condition (4.5.2) becomes

$$(N-1)[\beta^{-1}(b) - b]\frac{d\beta^{-1}(b)}{db} - \beta^{-1}(b) = 0.$$

Let's set $y(b) = \beta^{-1}(b)$ to see that this equation becomes

$$(N-1)[y(b) - b]y'(b) - y(b) = 0. \qquad (4.5.3)$$

This is a **first-order ordinary differential equation** for $y(b)$ with initial condition $y(0) = 0$, that we may solve to get

$$y(b) = \beta^{-1}(b) = v = \frac{N}{N-1}b.$$

The reason $y(0) = 0$ is because when the valuation of the item is zero, the bid must be zero. Solving for b in terms of v we get

$$b = \beta(v) = \left(1 - \frac{1}{N}\right)v,$$

which is the claimed optimal bidding function in a Dutch auction. Note that β is an increasing function of N, with $\beta(v) = \frac{v}{2}$ when $N = 2$, and $\beta = v$ when $N \to \infty$. For a given valuation v known to the bidder, she should choose the bid optimally $\left(1 - \frac{1}{N}\right)v$. She will win the auction if her bid is the highest among all N bidders and she will place the bid when the price is lowered to $(1 - \frac{1}{N})v$. Thus we see that the Dutch auction and the first price sealed-bid auction are strategically equivalent.

Next we calculate the expected revenue received by the seller. In a Dutch auction, we know that the payment will be the highest bid. We know that is going to be the random variable $\beta(\max\{V_1, \dots, V_N\})$, which is the optimal bidding function evaluated at the largest of the random valuations. Then

$$\begin{aligned}
E(\beta(\max\{V_1, \dots, V_N\})) &= E\left[\left(1 - \frac{1}{N}\right)\max\{V_1, \dots, V_N\}\right] \\
&= \left(1 - \frac{1}{N}\right)E[\max\{V_1, \dots, V_N\}] \\
&= \frac{N-1}{N+1}
\end{aligned}$$

because $E[\max\{V_1, \dots, V_N\}] = \frac{N}{(N+1)}$, as we will see next, when the V_i values are uniform on $[0,1]$.

Here is why $E[\max\{V_1, \dots, V_N\}] = N/(N+1)$. The cumulative distribution function of $Y = \max\{V_1, \dots, V_N\}$ is derived as follows. Since the valuations are independent and all have the same distribution,

$$F_Y(x) = Prob(\max\{V_1, \dots, V_N\} \leq x) = P(V_i \leq x)^N = F_V(x)^N.$$

Then the density of Y is

$$f_Y(x) = F_Y'(x) = N(F_V(x))^{N-1}f_V(x).$$

In case V has a uniform distribution on $[0,1]$, $f_V(x) = 1, F_V(x) = x, 0 < x < 1$, and so

$$f_Y(x) = Nx^{N-1}, 0 < x < 1 \implies E[Y] = \int_0^1 xf_Y(x)\,dx = \frac{N}{N+1}.$$

So we have verified everything for the Dutch auction case.

Next we solve the English auction game. We have already discussed informally that in an English auction each bidder should bid his or her true valuation so that $\beta(v) = v$. Here is a formal statement and proof.

Theorem 4.5.4 *In an English auction with valuations $V_1, \ldots, V_N$ and $V_i = v_i$ known to player i, then player i's optimal bid is v_i.*

We will not prove this theorem but give an example to see why it should be true.

Example 4.23 Let's say there are three bidders in an English auction for a painting. The bidders' valuations for the painting are \$100, \$200, and \$300, respectively. The reserve price is \$100.

The Nash equilibrium for this auction is as follows:

Bidder 1 bids \$100. Bidder 2 bids \$200. Bidder 3 bids \$300. This is because each bidder is bidding their true valuation for the painting. If any bidder were to bid less than their true valuation, they would lose the auction and receive no payoff. If any bidder were to bid more than their true valuation, they would lose money.

To see this, let's consider what would happen if bidder 1 bid less than \$100. In this case, bidder 2 would win the auction and bidder 1 would receive no payoff. Similarly, if bidder 2 bid less than \$200, bidder 3 would win the auction and bidder 2 would receive no payoff. Finally, if bidder 3 bid less than \$300, they would lose the auction and receive no payoff.

On the other hand, if any bidder were to bid more than their true valuation, they would lose money. For example, if bidder 1 bid \$150, they would still lose the auction to bidder 2, but they would lose \$50 in the process. Similarly, if bidder 2 bid \$250, they would still lose the auction to bidder 3, but they would lose \$50 in the process. Finally, if bidder 3 bid \$350, they would still win the auction, but they would have to pay \$50 more than they were willing to.

Therefore, the Nash equilibrium for this auction is for each bidder to bid their true valuation.

Given that the optimal bid in an English auction is $b_i = \beta(v_i) = v_i$, we next calculate the expected payment. The winner of the English auction with uniform valuations makes the payment of the second-highest bid, which is given by

$$\text{If } V_j = \max\{V_1, \ldots, V_N\}, \text{ then } E[\max_{i \neq j}\{V_i\}] = \frac{N-1}{N+1}.$$

This follows from knowing the density of the random variable $Y = \max_{i \neq j}\{V_i\}$, the second highest valuation. In the case when V is uniform on $[0, 1]$ the density of Y is

$$f_Y(x) = N(N-1)x(1-x)^{N-1}, 0 < x < 1 \implies E[Y] = \frac{N-1}{N+1}.$$

The derivation of this uses order statistics in elementary probability theory and we omit it. Therefore we have proved all parts of Theorem 4.5.3.

One major difference between English and Dutch auctions is the risk characteristics as measured by the variance of the selling price (see Wolfstetter [45] for the derivation).

1. In an English auction, the selling price random variable is the second highest valuation, that we write as $P_E = \max_2\{V_1, \ldots, V_N\}$. In probability theory this is an **order statistic**, and it is shown that if the valuations are all uniformly distributed on $[0, 1]$, then

$$Var(P_E) = \frac{2(N-1)}{(N+1)^2(N+2)}.$$

2. In a Dutch auction, equivalent to a first-price sealed-bid auction, the selling price is $P_D = \beta(\max\{V_1, \ldots, V_N\})$, and we have seen that with uniform valuations

$$\beta(\max\{V_1, \ldots, V_N\}) = \frac{N-1}{N}\max\{V_1, \ldots, V_N\}.$$

Consequently

$$Var(P_D) = Var(\beta(\max\{V_1, \ldots, V_N\}))$$
$$= \left(\frac{N-1}{N}\right)^2 Var(\max\{V_1, \ldots, V_N\})$$
$$= \frac{(N-1)^2}{N(N+1)^2(N+2)}.$$

We claim that $Var(P_D) < Var(P_E)$. That will be true if

$$\frac{2(N-1)}{(N+1)^2(N+2)} > \frac{(N-1)^2}{N(N+1)^2(N+2)}.$$

After using some algebra, this inequality reduces to the condition $2 > \frac{(N-1)}{N}$, which is absolutely true for any $N \geq 1$. We conclude that Dutch auctions are less risky for the seller than English auctions, as measured by the variance of the payment.

If the valuations are not uniformly distributed, the problem will be much harder to solve explicitly. But there is a general formula for the Nash equilibrium still assuming independence and that each valuation has distribution function $F(v)$. If the distribution is continuous, the Dutch auction will have a unique Nash equilibrium given by

$$\beta(v) = v - \frac{1}{F(v)^{N-1}} \int_r^v F(y)^{N-1} \, dy.$$

The proof of this formula comes basically from having to solve the differential equation that we derived earlier for the Nash equilibrium

$$(N-1)f(y(b))(y(b) - b)y'(b) - F(y(b)) = 0,$$

where we have set $y(b) = \beta^{-1}(b)$, and $F(y)$ is the cumulative distribution function and $f = F'$ is the density function.

The expected payment in a Dutch auction with uniformly distributed valuations was shown to be $E[P_D] = \frac{N-1}{N+1}$. The expected payment in an English auction was also shown to be $E[P_E] = \frac{N-1}{N+1}$. You will see in the problems that the expected payment in an **all pay** auction, in which all bidders will have to pay their bid, is also $\frac{N-1}{N+1}$. What's going on? Is the expected payment for an auction always $\frac{N-1}{N+1}$, at least for valuations that are uniformly distributed? The answer is "Yes," and not just for uniform distributions:

Theorem 4.5.5 *Any symmetric private value auction with identically distributed valuations, satisfying the following conditions, always has the same expected payment to the seller of the object:*

1. *They have the same number of bidders (who are risk-neutral).*
2. *The object at auction always goes to the bidder with the highest valuation.*
3. *The bidder with the lowest valuation has a zero expected payoff.*

This is known as the **revenue equivalence theorem**.

In the remainder of this section, we will verify directly as sort of a review that **linear trading rules** are the way to bid when the valuations are uniformly distributed in the interval $[r, R]$, where r is the reserve price. For simplicity we consider only two bidders who will have payoff functions

$$u_1((b_1, v_1), (b_2, v_2)) = \begin{cases} v_1 - b_1 & \text{if } b_1 > b_2; \\ \dfrac{v_1 - b_1}{2} & \text{if } b_1 = b_2; \\ 0 & \text{if } b_1 < b_2. \end{cases}$$

and

$$u_2((b_1, v_1), (b_2, v_2)) = \begin{cases} v_2 - b_2 & \text{if } b_2 > b_1; \\ \dfrac{v_2 - b_2}{2} & \text{if } b_1 = b_2; \\ 0 & \text{if } b_2 < b_1. \end{cases}$$

We have explicitly indicated that each player has two variables to work with, namely, the bid and the valuation. Of course the bid will depend on the valuation eventually. The independent valuations of each player are random variables V_1, V_2 with identical cumulative distribution function $F_V(v)$. Each bidder knows his or her own valuation but not the opponent's. So the expected payoff to player 1 is

$$U_1(b_1, b_2) \equiv Eu_1(b_1, v_1, b_2(V_2), V_2) = Prob(b_1 > b_2(V_2))(v_1 - b_1)$$

because in all other cases, the expected value is zero. In the case $b_1 = b_2(V_2)$, it is zero since we have continuous random variables. In addition, we write player 1's bid and valuation with lower-case letters because player 1 knows her own bid and valuation with certainty. Similarly

$$U_2(b_1, b_2) \equiv Eu_2(b_1(V_1), V_1, b_2, v_2) = Prob(b_2 > b_1(V_1))(v_2 - b_2).$$

A Nash equilibrium must satisfy

$$U_1(b_1^*, b_2^*) \geq U_1(b_1, b_2^*) \text{ and } U_2(b_1^*, b_2^*) \geq U_1(b_1^*, b_2).$$

In the case that the valuations are uniform on $[r, R]$, we will verify that the bidding rules

$$\beta_1^*(v_1) = \frac{r + v_1}{2} \text{ and } \beta_2^*(v_2) = \frac{r + v_2}{2}$$

constitute a Nash equilibrium. We only need to show it is true for β_1^* because it will be the same procedure for β_2^*.

So, by the assumptions, we have

$$\begin{aligned} Prob(b_1 > \beta_2^*(V_2)) &= Prob\left(b_1 > \frac{r + V_2}{2}\right) \\ &= Prob(2b_1 - r > V_2) \\ &= \frac{(2b_1 - r) - r}{R - r}, \end{aligned}$$

if $\frac{r}{2} < b_1 < \frac{(r+R)}{2}$, and the expected payoff

$$\begin{aligned} U_1(b_1, \beta_2^*(V_2)) &= (v_1 - b_1)Prob(b_1 > \beta_2^*(V_2)) \\ &= (v_1 - b_1)Prob(V_2 < 2b_1 - r) \\ &= \begin{cases} 0 & \text{if } b_1 < \frac{r}{2}; \\ (v_1 - b_1)\dfrac{2b_1 - 2r}{R - r} & \text{if } \frac{r}{2} < b_1 < \frac{(r+R)}{2}; \\ v_1 - b_1 & \text{if } \frac{(r+R)}{2} < b_1. \end{cases} \end{aligned}$$

We want to maximize this as a function of b_1. To do so, let's consider the case $\frac{r}{2} < b_1 < \frac{(r+R)}{2}$ and set

$$g(b_1) = (v_1 - b_1)\frac{2b_1 - 2r}{R - r}.$$

The function g is strictly concave down as a function of b_1 and has a unique maximum at $\beta_1 = \frac{r}{2} + \frac{v_1}{2}$, as the reader can readily verify by calculus. We conclude that $\beta_1^*(v_1) = \frac{(r+v_1)}{2}$ maximizes $U_1(b_1, \beta_2^*)$. This shows that $\beta_1^*(v_1)$ is a best response to $\beta_2^*(v_2)$. We have verified the claim.

Example 4.24 Two players are bidding in a first-price sealed-bid auction for a 1901s United States penny, a very valuable coin for collectors. Each player values it at somewhere between \$750K (where K = 1000) and \$1000K dollars with a uniform distribution (so $r = 750, R = 1000$). In this case, each player should bid $\beta_i(v_i) = \frac{1}{2}(750 + v_i)$. So, if player 1 values the penny at \$800K, she should optimally bid $\beta_1(800) = 775K$. Of course, if bidder 2 has a higher valuation, say, at \$850K, then player 2 will bid $\beta_2(850) = \frac{1}{2}(750 + 850) = 800K$ and win the penny. It will sell for \$800K and player 2 will benefit by $850 - 800 = 50K$.

On the other hand, if this were a second-price sealed-bid auction, equivalent to an English auction, then each bidder would bid their own valuations. In this case $b_1 = \$800K$ and $b_2 = \$850K$. Bidder 2 still gets the penny, but the selling price is the same. On the other hand, if player 1 valued the penny at \$775K, then $b_1 = \$775K$, and that would be the selling price. It would sell for \$25K less than in a first-price auction.

Problems

4.59 In an **all-pay auction** all the bidders must actually pay their bids to the seller, but only the high bidder gets the object up for sale. This type of auction is also called a **charity auction**. By following the procedure for a Dutch auction, show that the equilibrium bidding function for all players is $\beta(v) = \frac{N-1}{N}v^N$, assuming that bidders' valuations are uniformly distributed on the normalized interval $[0, 1]$. Find the expected total amount collected by the seller.

4.6 Stable Matching, Marriage, and Residencies

The stable matching problem is one of the most ubiquitous problems in economics, computer science, and social choice theory. It involves a method to match each member of one group to a member of another group so that they are matched with their most preferred opposing member. Marriages fall into this category but many other matching problems do also. A well-known instance of this problem arises when medical school graduates apply to hospitals for medical residencies. Graduates list their selections of hospitals and rank them. The hospitals interview the prospective residents and also rank them. These ranks are unknown to the opposing entities. The method of matching graduates with hospitals is called the stable matching algorithm.

Here is a precise statement of the problem and definitions of the terms. We have two sets of elements, say M and W referred to simply as **men** and **women**. For simplicity, we assume they have the same number of elements n but this is not necessary. We want each element of M to be paired with an element of W. Each element in each set has a ranked list of members of the other set they would prefer.

A matching must be done in a way such that it is **stable**. A matching is **not stable** if there are elements $A \in M$ and $B \in W$, such that

1. A and B are not currently matched with each other.
2. A prefers B over its current pairing.
3. B also prefers A over its current pairing.

This condition would mean that A and B could run off with each other. This situation is not stable. The stable marriage problem consists of matching each member of M to a member of W so that all matches are stable and each member of M ranks each member of W and vice versa.

Definition 4.6.1 *A **stable matching** of men with women is one in which there does not exist a pair of a woman and a man who prefer each other to the partners with whom they are currently matched.*

It turns out the stable marriage problem has a solution, i.e., everyone has a stable match, and there is an efficient algorithm to find it: the **Gale–Shapley algorithm**[1]. Notice that we did not say there is only one stable match. In fact, the stable marriage problem can have many solutions. However, the Gale–Shapley algorithm will find the matching that is optimal for all elements in the proposer set. That is if the members of M do the proposing for a match with members of W, all proposers will get the best possible partner among all stable matchings. For the members of W, the opposite occurs. That is in the proposed to set W everyone gets the worst possible outcome, not in the sense that they all get their least preferred choice, but among all possible stable matchings, they get the one with their least preferred partner. Therefore, one can consider two versions of the stable marriage problem: the one in which M proposes to W, and the one in which W proposes to M.

Let's start with an example of an unstable matching.

Example 4.25 Suppose there are 5 men labeled $M = \{A, B, C, D, E\}$ and 5 women labeled $W = \{a, b, c, d, e\}$. The preferences for men and women are listed.

Men	Women
$A : cdabe$	$a : ADCEB$
$B : acbde$	$b : ABDCE$
$C : aebdc$	$c : DECAB$
$D : baced$	$d : CBAED$
$E : bcade$	$e : ABDEC$

Consider the matchings $(A, a), (B, b), (C, c), (D, d), (E, e)$. Suppose A has been matched to a woman a and d has been matched to a man D. The pair (A, d) can destabilize the matching if both of them rank each other higher than their current match. In the tables A prefers d to a and d prefers A to D. This means A and d could run off, get matched, and destabilize the matchings.

Consider the pair (A, c). Since c prefers C to A and c is matched to C, even though A may want to match with c, c has no incentive to switch to A. The pair (A, c) cannot destabilize the matchings.

Gale–Shapley Algorithm

The Gale–Shapley algorithm will find a stable marriage. It does matches in stages, or rounds. Here's how it goes.

1. In the first round,
   - Each unmatched man proposes to the woman he prefers most, and then
   - Each woman replies "maybe" to her suitor she most prefers and "no" to all other suitors.
   - She is then provisionally "matched" to the suitor she most prefers so far, and that suitor is likewise provisionally engaged to her.
2. In each subsequent round,
   - Each unmatched man proposes to the most preferred woman to whom he has not yet proposed (regardless of whether the woman is already matched), and then
   - Each woman replies "maybe" if she is currently not matched or if she prefers this man over her current provisional partner (in this case, her current provisional partner becomes unmatched). This allows every already-matched woman to "trade up" (and reject her temporary partner)
3. This process is repeated until everyone is matched.

---

1 **Gale D, Shapley LS.**, *College admissions and the stability of marriage,* Amer Math Monthly 1962; 69(1): 9–15. David Gale (December 13, 1921 – March 7, 2008) was an American mathematician and economist. He was a professor emeritus at the University of California, Berkeley. Lloyd Shapley (June 2, 1923 – March 12, 2016) was an mathematician and economist. With Alvin E. Roth, Shapley won the 2012 Nobel Memorial Prize in Economic Sciences.

This algorithm is guaranteed to produce a stable marriage for all participants in time $O(n^2)$ steps where n is the number of men or women.

Among all possible different stable matchings, it always yields the one that is best for all men among all stable matchings, and worst for all women. One of the exercises illustrates this. No man can get a better matching for himself by misrepresenting his preferences. However, each woman may be able to misrepresent her preferences and get a better match.

Remarks

- As long as the algorithm continues, there is at least one woman without a proposal.
- At least one man is rejected as long as the algorithm continues.
- The algorithm terminates in at most $n(n-1)+1$ rounds. To see that, on the last round, there is at least one woman receiving a proposal. Other than this last woman, the rest of the women will be married in a maximum of $n(n-1)$ days.
- Obviously, instead of starting with men doing the proposing, we could start with women. The algorithm is the same with men and women reversed. Does allowing women to do the proposing first change the matches? The answer is it definitely could.

Example 4.26 The preferences are

Men	Women
$A : badc$	$a : CBDA$
$B : cbad$	$b : DBAC$
$C : bdca$	$c : ACBD$
$D : dbac$	$d : ADCB$

If men do the proposing, the algorithm results in the stable matching

$$(A, a), (B, b), (C, c), (D, d).$$

Here's why. Let's go through the algorithm.

1. $A \to b$
2. $B \to c$
3. $C \to b$ but b is engaged to $A \succ C$ so C moves on to d.
4. $D \to d$ but d is engaged to C but $D \succ C$ so d is engaged to D.
5. C was engaged to d and has to move on to next choice which is c. But c is engaged to B and $C \succ B$ so c is engaged with C and unengaged from B.
6. B was engaged to c and has to move on to next choice which is b. But b was engaged to A and $B \succ A$ so b is engaged to B and is unengaged from A.
7. A was engaged to b and has to move on to next choice which is a. Since a is free, that is a match.

All men and women are now matched according to Gale–Shapley.

If women do the proposing, the algorithm results in the stable matching (verify this!)

$$(A, d), (B, a), (C, c), (D, b).$$

Theorem 4.6.2 *The Gale–Shapley algorithm results in a stable matching.*

Proof: The proof is by contradiction. Assume that the procedure does not result in a stable matching. This implies that there exist two couples (given by our matching procedure), say (A, a) and (B, b), such that A prefers b over a and b prefers A over B. This means that A must have proposed to b before proposing to a but b rejected A because there was somebody b preferred to A in her list at some time. Therefore, it is not possible that later she ends up with B, who she prefers less than A. This is a contradiction and there are no such pairs. ∎

What does game theory have to say about this? Game Theory can be used to analyze if there exists any incentive for a person, or a group of persons to misrepresent their true preferences and end up with a better match. Another way to phrase the question is to ask if the matching given by Gale–Shapley is a Nash equilibrium.

It has been shown that the Gale–Shapley stable matchings do result in a Nash equilibrium and hence there can be no incentive for a person to misrepresent their preferences. But, if people are allowed to be indifferent among various choices of partners the Gale–Shapley algorithm need not lead to a stable matching. Changing $>$ to $\geq$ makes a difference.

Example 4.27   The Stable Roommates Problem. This is a variation on the stable marriage problem which arises in many situations. We are given a set S with an even number of members n. Each member has a preference list of roommates consisting of all the other members of S. We need to match the members of S such that every member has their most preferred roommate. If such a matching exists, we say that it is a stable matching. The stable roommates algorithm was developed by Robert Irving in 1985.

There are significant differences between the stable marriage problem and the stable roommates problem. First, the stable marriage problem matches members of two disjoint sets but in the stable roommates problem, there is only one set and we try to match each member of the set to another member of the same set.

Second, a stable match always exists in the stable marriage problem. That is not true in the stable roommates problem as you will see in the exercises.

Example 4.28   Medical Residency Matching. The algorithm is a slight adjustment of Gale-Shapley because hospitals match with more than one med student. Here is an example how it works. There are 6 students and 3 hospitals but each hospital has 2 positions.

$$A : 123$$
$$B : 213$$
$$C : 321 \qquad 1 : ABEFCD$$
$$D : 231 \qquad 2 : BAFDEC$$
$$E : 132 \qquad 3 : FDCEAB$$
$$F : 123$$

The algorithm has students proposing. So, $A \to 1, B \to 2, C \to 3, D \to 2, E \to 1, F \to 1$. Hospital 1 accepts A and E and rejects F. Hospital 2 accepts B and D. Hospital 3 accepts C. Student F is not matched yet and 3 has only one student. On to the next round.

$F \to 2$ and for 2, $F > D$ so 2 dumps D and accepts F. D is now unmatched. On to round 3.

$D \to 3$. Hospital 3 accepts D because 3 had only one match, namely C. We are done and the stable matching is

$$(1, (A, E)), (2, (B, F)), (3, (C, D)).$$

This algorithm is called the Roth–Peranson algorithm[1] and is actually used by the medical establishment. Some modifications to account for couples can also be implemented. It has also been shown that this algorithm is optimal in that there is essentially no honest strategy which will improve on the stable matching[2].

---

1 Roth AE, Peranson E., *The effects of the change in the NRMP matching algorithm*, JAMA. 1997;278(9):729–732.

2 See Roth AE, Sotomayor MAO. *Two-sided matching: study in game-theoretic modeling and analysis,* Cambridge University Press; 1992.

Stable Marriage and Integer Programming

There is a way to find stable matchings by converting the problem into an integer programming problem. Integer programming is simply a version of linear programming in which the variables must be integers. So let's set it up.

We are given two lists, say men, M, and women, W. Let's take a specific man m and a specific woman w. The set of all women j that m prefers over w is denoted by $j \succ_m w$ where we use the notation $j \succ_m w$ to mean man m prefers woman j over woman w. Similarly, $i \succ_w m$ is the set of all men i that woman w prefers over man m.

Let $x_{m,w} \in \{0,1\}$ for each $m \in M, w \in W$ be a variable so that $x_{m,w} = 1$ if m and w have been matched. If they haven't been paired, then $x_{m,w} = 0$. These variables must satisfy two types of constraints.

- Matching Constraints– Each member of M and W has exactly one match in the opposite set. This means the map must be one-to-one.
- Stability Constraints– If $m \in M$ and $w \in W$ have not been paired, then at least one of them must have been matched with a member which is a preferred match, i.e., a better partner. If m is not matched to w then either m is matched to somebody better than w or w is matched to someone better than m. If this doesn't hold, then it must be that (m, w) blocks the current marriage.

Let's consider the following example.

Example 4.29 Suppose $M = \{A(1), B(2), C(3), D(4)\}$ and $W = \{a(1), b(2), c(3), d(4)\}$. We also have to specify their preference lists:

$$
\begin{array}{ll}
A : cadb & a : DCAB \\
B : acdb & b : CDAB \\
C : dcab & c : CBAD \\
D : cbad & d : DBCA
\end{array}
$$

The first matching constraint is for A. He must be matched with one and only one member of W. This means that we must have

$$x_{11} + x_{12} + x_{13} + x_{14} = 1 \Leftrightarrow \sum_{j \in W} x_{1j} = 1.$$

This is true of all members of M so that

$$\sum_{j \in W} x_{ij} = 1, \text{ for all } i \in M.$$

But it is also true for all members of W so it is also required that

$$\sum_{i \in M} x_{ij} = 1, \text{ for all } j \in W.$$

How do we implement the stability constraint? Let's focus on the pair $(2, 4)$, i.e., (B, d). Stability says that if B and d are not married to each other, then at least one of them will be matched to a better partner. Looking at the preference lists, we have for B, $a > c > d$. For d, we have $D > B$. To be stable either B and d are matched to each other, $x_{24} = 1$, or B is matched with a or c, $x_{21} + x_{23} = 1$, or d is matched to D, $x_{44} = 1$. Since variables are either 0 or 1, we can combine to say

$$\underbrace{x_{24}}_{\text{2 paired with 4}} + \underbrace{x_{21} + x_{23}}_{\text{women preferred to 4}} + \underbrace{x_{44}}_{\text{men preferred to 2}} = 1. \tag{4.6.1}$$

Note that $j \succ_2 4 = \{1, 3\}, i \succ_4 2 = \{4\}$. so we may write this as

$$\sum_{j \succ_2 4} x_{2,j} + \sum_{i \succ_4 2} x_{i,4} + x_{24} = 1,$$

There are 16 such constraints.

Now consider the general formulation integer programming problem:

$$\text{Maximize} \sum_{i,j} x_{ij} \tag{4.6.2}$$

subject to $\tag{4.6.3}$

$$\sum_{j} x_{m,j} = 1, \text{ for all men } m \in M \tag{4.6.4}$$

$$\sum_{i} x_{i,w} = 1, \text{ for all women } w \in W \tag{4.6.5}$$

$$\sum_{j \succ_m w} x_{m,j} + \sum_{i \succ_w m} x_{i,w} + x_{m,w} = 1, \text{ for all pairs } (m, w), \tag{4.6.6}$$

$$x_{m,w} \in \{0, 1\}, \text{ for all pairs } (m, w) \tag{4.6.7}$$

This is a mathematical formulation of the stable matching problem which has the goal of matching men and women in an optimal way. The constraint says that for each pair (m, w), either man m is matched to someone he prefers to woman w, or woman w marries someone she prefers to man m, or they are matched to each other. This constraint is the stability constraint.

4.6.1 Finding a Stable Marriage Using Mathematica

There is a significant connection between the theory of Graphs and the Stable Marriage problem. It would take us too far afield to discuss this connection but simple commands in Mathematica can be used to always find a stable match given preference lists. Here are a few examples.

Example 4.30 The preferences of men and women are

	Men	Women
$m1$:	$w2, w1, w4, w3$	$w1$: $m3, m2, m4, m1$
$m2$:	$w3, w2, w1, w4$	$w2$: $m4, m2, m1, m3$
$m3$:	$w2, w4, w3, w1$	$w3$: $m1, m3, m2, m4$
$m4$:	$w4, w2, w1, w3$	$w4$: $m1, m4, m3, m2$

The first step is to enter the preferences to construct a graph with edges. The edges are specified as follows:

$$\begin{aligned}
\text{edges} = \{ &M1 \to W2, M1 \to W1, M1 \to W4, M1 \to W3, \\
&M2 \to W3, M2 \to W2, M2 \to W1, M2 \to W4, \\
&M3 \to W2, M3 \to W4, M3 \to W3, M3 \to W1, \\
&M4 \to W4, M4 \to W2, M4 \to W1, M4 \to W3, \\
&W1 \to M3, W1 \to M2, W1 \to M4, W1 \to M1, \\
&W2 \to M4, W2 \to M2, W2 \to M1, W2 \to M3, \\
&W3 \to M1, W3 \to M3, W3 \to M2, W3 \to M4, \\
&W4 \to M1, W4 \to M4, W4 \to M3, W4 \to M2\};
\end{aligned}$$

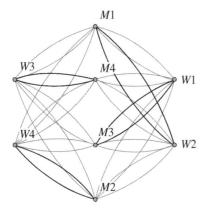

Figure 4.11 Stable match edges in bold.

Now we create a graph with the edges specified in *edges*. Once the graph is created, we can use the Mathematica command to find a stable match. The graph and the stable match are illustrated in Figure 4.11.

$$In[4] := \text{Find Independent Edge Set}[g] \text{ and } Out[4] = \{M1W2, W1M3, W4M2, W3M4\}$$

In general, there may be more than one stable match and they can be found using Mathematica in a tricky way and we will not go into that here. Another feature of Mathematica is that if you want to check if a given match is a stable match you may use the command

$$\text{Independent Edge Set Q}[g, M1 \rightarrow W2, M2 \rightarrow W4, M3 \rightarrow W3, M4 \rightarrow W1]$$

If this is a stable match it will return True, otherwise False.

Problems

4.60 Suppose $M = \{A(1), B(2), C(3), D(4)\}$ and $W = \{a(1), b(2), c(3), d(4)\}$. We also have to specify their preference lists:

$$\begin{array}{ll} A : cadb & a : DCAB \\ B : acdb & b : CDAB \\ C : dcab & c : CBAD \\ D : cbad & d : DBCA \end{array}$$

(a) Find the stable matching if Men propose.
(b) Find the stable matching if Women propose.

4.61 Consider the preference lists for men and women:

$$\begin{array}{ll} \textit{Men} & \textit{Women} \\ m1 : w1, w2, w3 & w1 : m3, m2, m1 \\ m2 : w2, w3, w1 & w2 : m1, m3, m2 \\ m3 : w3, w1, w2 & w3 : m2, m1, m3 \end{array}$$

(a) Find the stable matching if Men propose.
(b) Find the stable matching if Women propose.
(c) Show that $(m1, w1), (m2, w3), (m3, w2)$ is not a stable match.

4.62 Assume there are 5 men labeled $\{1, 2, 3, 4, 5\}$ and 5 women labeled $\{A, B, C, D, E\}$. Their preference lists are

1 : *BAECD*	*A* : 53124
2 : *EACBD*	*B* : 52314
3 : *CBEAD*	*C* : 15234
4 : *CBDAE*	*D* : 53421
5 : *EACBD*	*E* : 13524.

(a) Run the stable marriage algorithm with the men proposing.

(b) Run the stable marriage algorithm with the women proposing. Notice that the answer is different.

(c) Show that $(1, A), (2, C), (3, B), (4, D), (5, E)$ is also stable. Hint: in theory, you need to check 20 possibly new pairs, but many can be eliminated immediately. For instance, no pair $(5, ?)$ is an instability (why?).

Note that it is at least as good for all men as the women-proposing outcome, and at least as good for the women as the men-proposing outcome (why?).

4.63 **Medical residencies** The Gale–Shapely algorithm is used in quite a few real situations. Most famously, it is used to place med school students to hospital medical residencies. From 1952 to 1997, they used a **hospitals proposed** algorithm. Since 1998, they have used an **applicant proposed** algorithm (this change was partly in response to pressure from medical students and perhaps fear of lawsuits). The actual matching is slightly more complex, most importantly because they allow couples to match together, which in theory causes serious problems but in practice has little effect. We will ignore that.

Consider the following preferences, for three hospitals (ABC), which can each accept two residents, and 7 prospective residents (123456):

	1 : *ABC*
	2 : *BCA*
A : 3625741	3 : *BCA*
B : 7415263	4 : *BAC*
C : 7643512	5 : *ACB*
	6 : *BAC*
	7 : *ACB*

(a) Run the Gale–Shapely algorithm with students proposing. Note: each hospital can accept 2 residents, so only rejects if they have 3 or more proposals.

(b) Run the Gale–Shapely algorithm with hospitals proposing.

(c) The applicant proposed algorithm is the best possible for all applicants, but still one applicant fails to match. Explain why the same student will fail to match in all stable matchings. Look at the original lists of hospital preferences. Would you have been able to guess which student was out of luck?

4.64 Four people are trying to be matched as roommates of size 2. This is the stable roommate problem. The 4 people have the following preference lists:

A : *BCD*

B : *CAD*

C : *ABD*

D : *ACB*

There are only 3 possible matchings. Show by directly checking that all three are unstable. That means there is no stable solution!

4.7 Selected Chapter Problems

Problems

4.65 **Trimatrix games** This is an exercise on a three-person game.

(a) Find three pure Nash equilibria for the (non-zero-sum) trimatrix game give by

$$\begin{bmatrix} (0,0,3) & (2,1,1) \\ (7,2,1) & (0,3,1) \end{bmatrix} \quad \begin{bmatrix} (0,1,0) & (6,1,1) \\ (1,2,3) & (1,1,0) \end{bmatrix}$$

(b) Do you think there are any mixed equilibria for this game?

4.66 **Continuous tragedy of the commons.** A group of 3 farmers are sharing grazing grounds. These are large grassy grounds, so they will have a lot of sheep; we can pretend the number of sheep is a continuous variable. If the farmers have x_1, x_2, x_3 sheep respectively, assume the payoff to each farmer is

$$u_i(x_1, x_2, x_3) = \begin{cases} 10x_i, & \text{if } x_1 + x_2 + x_3 < 2100 \\ 10x_i - (x_1 + x_2 + x_3 - 2100)^2/120, & \text{if } x_1 + x_2 + x_3 \geq 2100 \end{cases}$$

(a) Explain why, at a Nash equilibrium, one can assume that $x_1 + x_2 + x_3 \geq 2100$.

(b) Since the Nash equilibrium is likely to occur when $x_1 + x_2 + x_3 \geq 2100$ set up equations to find the maximum of each payoff for each player.

(c) Use equations from part (b) to find some Nash equilibria. Find one that is equitable in the sense that all players get the same payoff.

(d) Now set up and solve an equation to give the best possible strategy if the farmers can agree to cooperate, and trust each other. Hint: this means they are trying to maximize the sum of their payoffs (i.e., the social welfare), not just one payoff. The result should give a maximum social welfare of 22000.

4.67 **Braess' paradox.** Four cities are connected by roads as shown below.

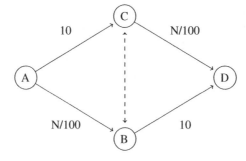

This gives the travel time, where N is the number of people trying to take the road. So each traveler has two choices (go through B or go through C). You can think of this as an N-player game where the pure strategies are the two possible choices.

(a) Assume there are 800 total cars, and pretend the dotted arrow does not exist. Find a Nash equilibrium, and give the expected travel time for each traveler.

(b) Now we add the dotted arrow: this is a high-speed bypass that can take you from B to C or from C to B in only 1 minute, no matter how many people are using it. Find a Nash equilibrium and the travel time(s) for this situation.

(c) Your equilibrium travel time in (b) should be strictly bigger than in (a). Discuss.

4.68 **War of attrition.** Two countries are at war. The war is costing each a billion dollars a day (this is their estimate for the dollar value of the total cost, including lives lost, productivity lost, actual money spent on combat...). The benefit to each side of winning the war is $200 billion, and the cost of losing is $300 billion. At any point, either side can surrender, and the war is over. This can happen any time day or night, so the surrender time is a continuous variable. Assume each side is rich, so the war could potentially go on for a long time. Assume also that they are evenly matched, and that if neither surrenders they will fight until both are destroyed. This is of course a (high stakes) two-player game.

(a) State what the pure strategies are.

(b) Find some Nash equilibria in pure strategies. Hint: they will be very asymmetric.

(c) Discuss what a mixed strategy might be.

(d) Assume PI is playing a totally mixed Nash equilibrium (discuss what that is). Write down an equation giving the expected payoff if PII is playing the strategy (i) quit immediately (ii) quit after 5 days, (iii) quit after 10 days. Hint: two of these will involve integrals.

(e) Show that, if there is a fully mixed Nash equilibrium, then PI's strategy must satisfy:

$$\int_0^T -tp(t)dt - (500 + T)\left(1 - \int_0^T p(t)dt\right) = -500,$$

where this is an equality of functions of the time T that PII chooses to quit. You may end up with a slightly different form of this equation.

(f) Take the derivative of the equation in part (4.68.e) and simplify to get a new equation, which will involve $p(T)$ and an integral.

(g) Take the derivative of the equation in part (4.68.f) and simplify (if needed) to get a new equation. This time there should be no integrals left, but there is a derivative.

(h) Solve the differential equation in (4.68.g) to find the equilibrium strategy...be careful about the constant of integration.

(i) Find the expected payout for the equilibrium strategy, and discuss if you think it is reasonable.

4.69 **Cournot's Duopoly** Boeing and Airbus have close to total control of the market for large commercial aircraft. Lets pretend the planes are identical. It costs Boeing 180M to build a plane, and costs Airbus 150M. Assume the price aircraft will sell for (which depends on the total number produced) is given by $510 - q_1 - q_2$, where q_1 and q_2 are the number of plane produced by Boeing and Airbus respectively. Numbers here are fairly large, so lets assume q_1, q_2 are continuous variables.

(a) Write down an equation given the profit of each manufacturer in terms of q_1, q_2.

(b) At a Nash equilibrium, each firm's profit should be roughly indifferent to making one more or less plane. Write down equations that say this.

(c) Solve your equations to find the equilibrium strategies.

(d) Use this to find the two profits, and decide if this makes sense.

(e) It seems strange that Airbus doesn't just put Boeing out of business. Can you explain in words why Airbus doesn't have an incentive to do this?

(f) Now assume Boeing's cost is $350 and Airbus's is still $150. Solve the problem again. Something different happens. What? What would actually happen?

4.70 Stackelberg model: Sequential choices. Consider the same situation as in Problem 4.69, with costs of 180M and 150M respectively.

(a) Assume that Boeing gets to choose their production level q_1 first, and then Airbus responds by choosing the q_2 that gives them the greatest profit given q_1. At the optimal value, the partial derivative with respect to q_2 should be 0. Use this to solve for q_2 in terms of q_1.

(b) Substitute your answer from (a) into Boeing's payoff function. This now gives their profit as a function of only q_1. Find the optimal value by standard single-variable calculus methods.

(c) Find the profit to Boeing and to Airbus in this setup.

(d) Your answer does not agree with the Nash equilibrium value...and Boeing should not actually be making a (slightly) bigger profit than Airbus! Discuss.

4.71 Entry deterrence This is a classic economics question: A firm (call it firm A) has a monopoly on a market. They produce N gadgets at a total cost of $c = 10N + 400$ million dollars, where N is the number they produce in millions. They can then sell those gadgets for $120 - N$ dollars each. It is well-known how to make these gadgets, and another firm could enter the market at any time and produce them for the same price. The question is, how many gadgets should the firm make in order to maximize their profit without giving another firm an incentive to enter the market?

(a) Calculate the firm's profit as a function of N assuming that no other firm has any interest in entering the market.

(b) How many gadgets should the firm produce if no one else has any interest in entering the market? What will their profit be?

(c) Now, assume another firm (call it firm B) is thinking of entering the market. Firm B has a spy who can tell them how much firm A is planning to produce. Find a function giving firm B's profit as a function of the number of gadgets N_2 they produce, assuming that firm A sticks to its plan and produces the number of gadgets you found in part (b). Maximize this function to determine how firm B should respond.

(d) Now assume firm A also has a spy, and they know in advance that firm B is planning to enter the market. They also know firm B has spies. They decide to increase their production to the point where firm B has no incentive to enter the market. That is, they want to set production so that firm B's maximum profit would be zero. What should their production be? Hint: write down an equation for firm B's maximum profit in terms of firm A's production N.

(e) Find the price of gadgets and firm A's profits if they must worry about entry deterrence. How much lower are they?

(f) Does this always work? That is, can firm A always find a price where they remain profitable but there is no incentive for a firm to enter the market?

Bibliographic Notes

This chapter has many applications of N-person nonzero sum games, especially to economics, duels, auctions, and a tangential connection to the stable matching problem. In addition, there is an application to the theory of juries constructed by political scientists. The juror example in this chapter is representative of the use of game theory in social science. See the excellent book by McCarty and Meirowitz [26] for many applications of game theory to political science.

Example 4.5 introduces another concept of equilibrium distinct from a Nash equilibrium. This idea is due to Roemer [36]

The voter model in Example 4.1 is a standard application of game theory to voting. More than two candidates are allowed for a model to get an $N > 2$ person game. Here we follow the presentation in the reference by Aliprantis and Chakrabarti [1] in the formulation of the payoff functions.

The economics models presented in Section 4.3 have been studied for many years, and these models appear in both the game theory and economic theory literature. The Bertrand model and its ramifications follow Ferguson [6] and Aliprantis and Chakrabarti [1]. The traveler's paradox Example 4.13 (from Reference [10]) is also well known as a critique of both the Nash equilibrium concept and a Bertrand-type model of markets. Entry deterrence is a modification of the same model in Ferguson's notes [6]. The theory of duels as examples of games of timing is due to Karlin [20]. Our derivation of the payoff functions and the solution of the silent continuous duel also follows Karlin [20].

The land division game in Problem 4.24 is a wonderful calculus and game theory problem. For this particular geometry of land, we take the formulation in Jones [16]. Problem 4.19 is formulated as an advertising problem in Ferguson [6]. H. Gintis [9] is an outstanding source of problems in many areas of game theory. Problems 4.25 and 4.23 are modifications of the tobacco farmer problem and Social Security problem in Gintis [9].

The stable matching problem is widely used to match medical graduates to residencies but there are many more applications. The Mathematica commands gives a beautiful way to find any and all matchings.

5

Repeated Games

The same thing happened today that happened yesterday, only to different people
−Walter Winchell

Change begets change as much as repetition reinforces repetition.
−Bill Drayton

The more often a stupidity is repeated, the more it gets the appearance of wisdom.
−Friedrich Nietzsche

Repetition is the mother of learning, but variation is the spice of life.
−George Bernard Shaw

What happens when a game is played more than once? Many real games are played over and over again, such as setting prices or production levels for each of two or more firms. Battles in a war can be considered as two game opponents fighting essentially the same battle again and again. Companies set prices on their products every day to maximize sales. There are many games played over and over again.

Think of a prisoner's dilemma game played more than once. Recall that in the prisoner's dilemma, the course of action predicted by Nash would be to defect, i.e., to rat out the other guy. But both players would do better if they cooperated and clammed up. But what if a prisoner knew that she was going to play again; would it affect the best course of action? Is it possible that knowing a game will be played again, possibly with no end, would lead to cooperation? Would knowledge of the players of how the opponents chose to play make any difference in a prisoner's dilemma game played more than once?

To answer that, let's consider the two-stage extensive form of the prisoner's dilemma given in Figure 5.1.

The information sets are designed so that each player knows her own moves but not that of the opponent. The one-stage game matrix is

$$
\begin{array}{c|cc}
 & C & D \\
\hline
C & (2,2) & (0,3) \\
D & (3,0) & (1,1)
\end{array}
\qquad \text{(PD)}
$$

Game Theory: An Introduction, Third Edition. E. N. Barron.
© 2024 John Wiley & Sons, Inc. Published 2024 by John Wiley & Sons, Inc.

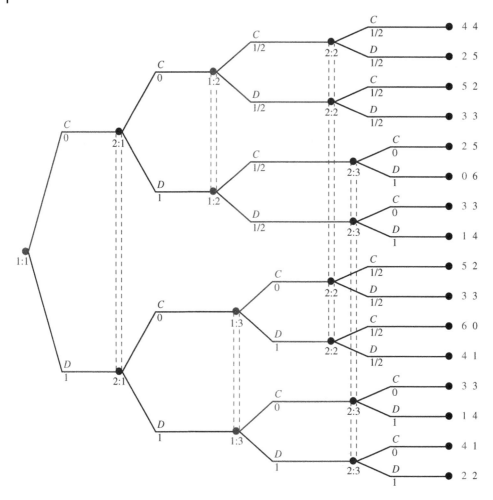

Figure 5.1 2 Stage Prisoner's Dilemma Game

In the one-shot game, the strategies are simply C or D for each player; C means cooperate and D means defect. In the two-stage game, in which the payoff at the end of the second stage is the **sum of the payoffs** at each stage, there are many more strategies. Here is the game matrix for the 2-step Prisoner's dilemma game.

	CC^*	CD^*	D^*C	D^*D
CC^*	$(4,4)$	$(2,5)$	$(2,5)$	$(0,6)$
CD^*	$(5,2)$	$(3,3)$	$(3,3)$	$(1,4)$
D^*C	$(5,2)$	$(3,3)$	$(3,3)$	$(1,4)$
D^*D	$(6,0)$	$(4,1)$	$(4,1)$	$(2,2)$

The Nash equilibrium is at (D^*D, D^*D). It is optimal for each player to play D at each stage. Thus, in the two-round prisoner's dilemma game, the optimal strategies remain to rat out the other guy. You might wonder what happens if we make this a perfect information game instead, i.e., each node is a new information set. This would correspond to each player having knowledge of the other player's choices. The answer is nothing changes as long as the payoffs stay the same as you should verify. Each player will rat out the other player.

What happens if we play this game for more than 2 rounds? Clearly, nothing will change unless new strategies are considered. Let's consider four possible strategies if the game is played for k rounds.

- **cooperate**: play C in every round.
- **defect**: play D in every round.
- **tit-for-tat**: start by cooperating; first play C, then in each round play the last strategy played by the opponent.
- **grim-trigger**: start by cooperating; play C as long as the opponent plays C; if the opponent plays D in a particular round, then play D from this round forward. This is a punishment strategy. Any deviation from cooperation will be punished but if both players use grim-trigger no reason to deviate will ever come up.

We have already seen in Figure 5.1 that playing D in every round is a NE but this is clearly not what the players want. Playing C in every round is obviously not a NE for the two-stage game or for the k-stage game. Will playing the tit-for-tat strategy or the grim-trigger strategy make a difference?
Before we answer that question let's introduce some general terminology.

Terminology. Suppose we denote a game in which the players make simultaneous moves by G. This is called the **stage game** or **one-shot game**. If we play this game multiple times, say T times with strategies consisting of a complete plan of action for the entire sequence of games, this is called a **repeated game**, which we label G^T. If the game is played an undetermined number of times we label the game as G^∞. The payoffs for the games G^T and G^∞ will be defined below.

The question is what kind of behavior can emerge if the game is played a finite or infinite number of times and what happens if we play strategies like tit-for-tat or grim-trigger. We have the following general result for games repeated a finite number of times.

Proposition 5.0.1 *If the one-shot game G has a unique Nash Equilibrium, then the game repeated $T < \infty$ times has a unique subgame perfect Nash equilibrium in which the players play the one-shot Nash equilibrium at each stage.*
If G has a unique Nash equilibrium, then for each Nash equilibrium of the repeated game G^T the corresponding payoff is the same as the payoff of the one-stage Nash equilibrium played T times.

When a game is played repeatedly and has **more than one Nash equilibrium**, it is not true that the only NEs for the repeated game are simply the NE for the one-shot game played in each round. Here is an example where there are two NEs and an NE is found which incorporates plays which are not part of a NE for the one-shot game.

Example 5.1 The one-shot game has matrix

	A	B	C
A	(5, 5)	(0, 0)	(12, 0)
B	(0, 0)	(2, 2)	(0, 0)
C	(0, 12)	(0, 0)	(10, 10)

In the one-shot game, it seems that (A, A) is the best NE. But now, with two NEs what would be an NE when the game is played twice? If the strategy is to play (A, A) in round one and also in round two, the payoff is 10 to each player. If the strategy is to play (B, B) in each round, the payoff is 4 for each player. Can they do better? Clearly, cooperating and playing C in both rounds for each player gets a payoff of 20 but each player can get more by defecting to A as long as the other player sticks with C. It's not clear what to do. Consider the following strategy, call it s, for each player.

- In the first-round play C.
- If the opponent played C in the first round, then play A in the second round. If the opponent did not play C in the first round, play B in the second round.

We claim that this is in fact a NE if the game is played twice. To see this, notice that if they both play s then they play C in the first round and therefore A in the second round resulting in a total of 15 for each player. How can a player, say player I, possibly get more than 15 in the two-stage game? Looking at the numbers in the table, the only way is for player I to play A in the first round (which results in 12 to player I since II will play C). But then player II will play B in the second round and player I will also play B in the second round resulting in a payoff of 2 to each player in that round. The result is the payoff of at most 14 to player I, assuming the strategy s is followed. The same conclusion holds for player II. We conclude that any deviation from s by either player results in a payoff less than 15 to whichever player deviates. That means s is an NE.

It turns out that s is a pretty good strategy and it involves cooperation. Are there other ways this can happen and how do we find good strategies?

5.1 Games Repeated Until ...

Many games continue round after round with no predetermined end, such as price setting, warfare (cold and hot), etc. We may model this situation using the chances the game will be played again after a round ends. The players don't know if the last game they played will in fact be the last, or that the game will be played again.

Let γ = probability of playing another round, $0 \leq \gamma \leq 1$. Then $1 - \gamma$ is the probability the game ends after the current round. Let T be the **random variable representing how many rounds** are played before everything ends. The random variable T could have many distributions but we will choose the most natural one and assume that T is a **Geometric random variable** with

$$Prob(T = k) = \gamma^{k-1}(1 - \gamma), k = 1, 2, \ldots .$$

One of the features of a Geometric random variable is that the probability of playing, say, t future rounds does not depend on how many rounds have been played so far. In other words, T is **memoryless**, which means formally $P(T > t + s | T > t) = P(T > s)$, the probability the game will go on for at least s additional rounds given that it has lasted for at least t rounds, does not depend on the fact it has lasted for at least t rounds. It's as if the game has just started anew.

The **expected number of rounds** in an indefinitely repeated game is[1]

$$E[T] = \sum_{k=1}^{\infty} k Prob(T = k) = \sum_{k=1}^{\infty} k\gamma^{k-1}(1 - \gamma) = \frac{1}{1 - \gamma}.$$

Next suppose that the payoff to a player at round k is given by a_k. Then the expected total payoff to the player is given by

$$\boxed{\sum_{k=1}^{\infty} a_k Prob(T = k) = \sum_{k=1}^{\infty} (1 - \gamma)\gamma^{k-1} a_k.}$$

---

1 Throughout this chapter we use the formulas from calculus

$$\sum_{k=1}^{r} a^k = \frac{a(1 - a^r)}{1 - a}, \ \sum_{k=1}^{\infty} a^k = \frac{a}{1 - a}, \ \sum_{k=1}^{\infty} k\gamma^{k-1} = \frac{1}{(1 - \gamma)^2} \text{ and } \sum_{k=r}^{\infty} a^k = \frac{a^r}{1 - a}.$$

For the infinite series we need $|a| < 1$ for convergence.

There is another interpretation of the parameter γ. It is based on the time value of money in that a dollar today is worth more than a dollar tomorrow. We then consider γ **as the discount factor**. The total payoff to a player in the game G^T is then

$$
\begin{cases}
\dfrac{\gamma(1-\gamma^t)}{1-\gamma}\displaystyle\sum_{k=1}^{t}\gamma^{k-1}a_k, & \text{if } T = t < \infty, \\[2ex]
(1-\gamma)\displaystyle\sum_{k=1}^{\infty}\gamma^{k-1}a_k, & \text{if } T = t = \infty.
\end{cases}
$$

In the $T = \infty$ case, because $0 < \gamma < 1$, it is always the case that the infinite series converges. The factor $1 - \gamma$ in the payoff says that if the players use the same strategies in every round so that $a_k = a$ for all $k = 1, 2, \ldots$, then the payoff in the repeated game is the same as the payoff in the one-shot game. It is a normalizing factor which does not affect game theory considerations.

Example 5.2 This example illustrates γ as an interest rate and strategies for setting prices. Suppose there are two corporations, PI and PII, producing widgets. If they collude (i.e., cooperate) on price, they can each earn C per quarter. Under this cooperation, if the discount rate (i.e., interest rate) is γ, then the income stream to, say PI is

$$
P1 = C + \gamma C + \gamma^2 C + \cdots = C\sum_{k=0}^{\infty}\gamma^k = C\frac{1}{1-\gamma}.
$$

If PII also sticks to the collusion, PII will also receive this same amount, $P2 = C\frac{1}{1-\gamma}$. On the other hand, if PII deviates from this strategy and instead decides to raise the price in quarter $k_1 - 1$, then in that period PII will earn $D > C$. Then PI cuts the price in quarter $k_1 + 1$ and earns $C_1 < C < D$ from that quarter on. In quarter $k_1 + 1$ PII also cuts the price and earns $D_1 < C_1$ from that quarter on. The payoff to PI under this scenario will be

$$
P1_1 = \sum_{k=0}^{k_1}C\gamma^k + C_1\sum_{k=k_1+1}^{\infty}\gamma^k = C\frac{1-\gamma^{k_1+1}}{1-\gamma} + C_1\frac{\gamma^{k_1+1}}{1-\gamma}.
$$

The payoff to PII will be

$$
P2_1 = \sum_{k=0}^{k_1-1}C\gamma^k + D\gamma^{k_1} + D_1\sum_{k=k_1+1}^{\infty}\gamma^k = C\frac{1-\gamma^{k_1}}{1-\gamma} + D\gamma^{k_1} + D_1\frac{\gamma^{k_1+1}}{1-\gamma}.
$$

It will not be profitable for PII to deviate from the cooperation strategy if and only if

$$
C\frac{1}{1-\gamma} > C\frac{1-\gamma^{k_1}}{1-\gamma} + D\gamma^{k_1} + D_1\frac{\gamma^{k_1+1}}{1-\gamma}
$$

or,

$$
C > C(1 - \gamma^{k_1}) + D\gamma^{k_1}(1-\gamma) + D_1(\gamma^{k_1+1}).
$$

Simplifying, we see that deviating from cooperating is not profitable if and only if

$$
\gamma > \frac{D-C}{D-D_1} > 0,
$$

and $\gamma < 1$, since $D_1 < C_1 < C < D$. This makes sense since it says that if the interest rate is large enough the future loss of profit from raising the price in period k_1 is enough to dissuade such action.

This example provides some insight into why unilaterally raising prices by a company generally doesn't work, especially if doing so starts a price war. But it may work if interest rates are high enough.

Remark The payoff in G^∞ is not the only possible payoff we could use. One other possible payoff is the long-run average: $\lim_{n\to\infty} \frac{1}{n} \sum_{k=1}^{n} a_k$. This has the disadvantage that the limit may not actually exist. In this chapter we will only consider the discounted payoff.

For repeated games one way to verify a strategy is a subgame perfect NE is to check what happens if a player deviates in one stage. Suppose s is a strategy for players in a repeated game, so s specifies what move each player makes at each stage of the game, depending on the prior history of moves in the game.

Definition 5.1.1 *We say the strategy s satisfies the **One-Stage Deviation Principle** if no player can gain by deviating from s in a single round and sticking to s otherwise.*

The next theorem which we will not prove but should be almost obvious simplifies verifying a subgame perfect NE.

Theorem 5.1.2 *Suppose we have the one-shot game G with bounded payoffs to each player. A strategy s for a repeated game with $0 < \gamma \le 1$ is a subgame perfect Nash equilibrium if and only if it satisfies the one-stage deviation principle.*

We will use this theorem in what follows to check that a strategy is a subgame perfect NE.

We have seen that in the Prisoner's Dilemma game (PD), if we play the game a finite number of times $G^T, T < \infty$, in each round each player will play D. There is no way finite repetition of the game without some sort of major modification in the strategies will lead to the players choosing C. That doesn't mean there aren't NEs for a finitely repeated game which involves cooperation, as we have already seen.

In an indefinitely repeated game cooperation can be enforced, for instance by using a grim-trigger strategy which penalizes an opponent who does not cooperate. This is only true if the chances the game continues are large enough (or if the discount rate is large enough). That's what the next proposition considers.

Proposition 5.1.3 *Suppose the one-shot game is G which is the Prisoner's Dilemma (PD). For any (discount factor) $\frac{1}{2} < \gamma \le 1$, if each player in the game G^∞ plays a grim-trigger strategy, it will be a subgame perfect NE. That is, if the PD game has a greater than 50% chance of being played again, then grim trigger is a subgame perfect NE.*

Once again, recall that the **grim-trigger strategy is to play C as long as the opponent plays C; if the opponent plays D in a particular round, then play D from this round forward.** Therefore, if they both play the grim-trigger strategy they choose (C, C) in each round and forever, but if a player deviates in a round, they both pay the price. To show that grim-trigger is a subgame perfect NE we will use the one-stage deviation principle.

To verify the proposition, we need to show that if a player deviates from the grim-trigger, then they do worse. Suppose player I deviates at some round and player II uses the grim-trigger. Because I deviates, there must be a first round, say k_1, where instead of playing C, player I chooses D.

1. Play C in rounds $1, 2, \ldots, k_1 - 1$, payoff to I and II is 2.

2. Play D in round k_1, payoff to I is 3, to II is 0.
3. Rounds $k_1 + 1, k_1 + 2, \ldots$, payoff to I and II is 1 since II will then play D over C.

If we calculate the largest possible payoff to I, it is then[1]

$$P_I = (1 - \gamma)\left(\sum_{k=1}^{k_1} 2\gamma^{k-1} + 3\gamma^{k_1} + \sum_{k=k_1+2}^{\infty} 1\gamma^{k-1}\right)$$

$$= 2 + (1 - 2\gamma)\gamma^{k_1}.$$

If both players used their grim-trigger strategies we know that the payoff to player I and II is exactly 2 because they would both play C forever. Are there values of $0 < \gamma \le 1$ so that $P_I < 2$? In order for that to happen we need

$$2 + (1 - 2\gamma)\gamma^{k_1} < 2 \Leftrightarrow \frac{1}{2} < \gamma.$$

Consequently, deviation from the grim-trigger strategy will result in a payoff to player I of less than she could get by sticking to it, as long as $\frac{1}{2} < \gamma \le 1$. By the one-stage deviation principle, this shows that the grim-trigger strategies form a subgame perfect NE.

We may interpret this result by saying that if the discount factor is large enough and a dollar today is worth much more than a dollar tomorrow, the players should cooperate. An alternative interpretation is that if the probability the game will continue for another round is more than 50% at each round, then the players should cooperate. We see that the grim-trigger strategy is indeed a NE and it does lead to cooperation by the players. However, in case you are overly optimistic, it is also true that each player choosing D at every stage is also a NE and it is not clear that without some coordination the grim-trigger strategy would be played.

Example 5.3 Consider the game,

	C	D	P
C	$(-1, -1)$	$(-10, 0)$	$(-15, -10)$
D	$(0, -10)$	$\boxed{(-9, -9)}$	$(-15, -10)$
P	$(-10, -15)$	$(-10, -15)$	$\boxed{(-12, -12)}$

This game is a version of Prisoner's Dilemma with punishment. It has three Nash equilibria, namely $(D, D), (P, P)$, and the mixed Nash $((\frac{3}{4}D, \frac{1}{4}P), (\frac{3}{4}D, \frac{1}{4}P))$.

Clearly in the one stage game (C, C) is not a Nash equilibrium even though both players would like to receive those payoffs. However, we will show that if the discount factor $\gamma \ge \frac{1}{12}$, it will be the case that (C, C) will be part of a subgame perfect equilibrium in the indefinitely repeated game. Basically, they start out cooperating and if either player drops out and plays D in an effort to do better, then the opponent switches to P which will punish both players. This only works if the chances the game does not end at any round are sufficiently high. Here are the details.

Consider the modified grim-trigger strategy for each player.

- In the first-round play C.
- If D is ever played by the opponent, then switch to P and play P forever.

---

1 The general formula which may be helpful is

$$(1 - \gamma)\left(\sum_{k=1}^{k_1} a\gamma^{k-1} + b\gamma^{k_1} + \sum_{k=k_1+2}^{\infty} c\gamma^{k-1}\right) = a + \gamma^{k_1}(-a + b - b\gamma + c\gamma).$$

The claim is that this is a subgame perfect NE for each player giving payoff -1 to each player and C is played by each player forever.

If player I deviates from this trigger strategy and plays D in some round k_1, while player II sticks to the strategy, then the payoff to player I is

$$P_I = (1 - \gamma) \left(\sum_{k=1}^{k_1} \underbrace{(-1)}_{\text{I \& II play C}} \gamma^{k-1} + \underbrace{(0)}_{\text{I plays D, II plays C}} \gamma^{k_1} + \sum_{k=k_1+2}^{\infty} \underbrace{(-12)}_{\text{I \& II play P}} \gamma^{k-1} \right)$$

$$= -1 + (1 - 12\gamma)\gamma^{k_1}.$$

When is this lower than simply playing C at every round?

$$-1 > -1 + (1 - 12\gamma)\gamma^{k_1} \Leftrightarrow \gamma > \frac{1}{12}.$$

You can check the same conclusion results if player II deviates from the modified grim-trigger as long as $\gamma > \frac{1}{12}$. Therefore, no player has an incentive to switch to D if $\gamma > \frac{1}{12}$ and (C, C) should be played forever. By the one-stage deviation principle, the modified grim-trigger is a subgame perfect NE.

The grim-trigger strategy is aptly labeled grim because everyone loses unless they cooperate. Is there a somewhat milder way to ensure cooperation? Let's see what happens with the Tit-for-Tat strategy.

Example 5.4 *Tit-for-Tat Strategy*

The donation game has two players. Each player may choose to make a donation of c or nothing. If a player makes the donation and the opposing player chooses not to make a donation, i.e., chooses to pass, the passing player will receive $b > c$. If both players choose to donate, they each receive $b - c$. The game matrix is therefore,

	Donate	Pass
Donate	$(b - c, b - c)$	$(-c, b)$
Pass	$(b, -c)$	$(0, 0)$

The NE strategy for each player is to Pass. Even if the game is played multiple times, the best strategy for each player is to always Pass. Is this realistic? What is the tit-for-tat strategy for this and is it a NE if the game is played repeatedly?

The tit-for-tat strategy is to first play Donate, then in each round, play the last strategy played by the opponent. First, if they play Donate at each stage, the payoff will be $b - c$ to each player. If player I deviates from playing Donate in some round k_1 then player II will play Pass in round $k_1 + 1$, player I will Donate in round $k_1 + 1$ and they will alternate from round to round. It looks like this

Player	$\cdots$	$k_1 - 1$	k_1	$k_1 + 1$	$k_1 + 2$	$k_1 + 3$	$k_1 + 4$	$k_1 + 5 \cdots$
I	$\cdots$	Donate	Pass	Donate	Pass	Donate	Pass	Donate
II	$\cdots$	Donate	Donate	Pass	Donate	Pass	Donate	Pass

Remember that the tit-for-tat strategy has each player doing the same as the opponent in the **last round**. To make things a little easier, let's take $k_1 = 1$ so player I chooses to Pass in the first round, player II will Donate, and then they alternate in every round. If the players always play Donate, their payoff is $b - c$. When can they do better by playing tit-for-tat? To do so the payoff from tit-for-tat should be larger than $b - c$:

$$b - c < (1 - \gamma)\left(b + (-c)\gamma + b\gamma^2 + (-c)\gamma^3 + b\gamma^4 + (-c)\gamma^5 + \cdots\right)$$

$$= (1 - \gamma)\left(b\sum_{k=1}^{\infty}(\gamma^2)^{k-1} + (-c)\sum_{k=1}^{\infty}\gamma^{2k-1}\right)$$

$$= (1 - \gamma)\left(b\frac{1}{1 - \gamma^2} + (-c)\frac{\gamma}{1 - \gamma^2}\right)$$

$$= \frac{b}{1 + \gamma} - \frac{c\gamma}{1 + \gamma} = \frac{b - c\gamma}{1 + \gamma}.$$

This inequality will be true for all $0 \leq \gamma < \frac{c}{b}$. This means that for any $\gamma < \frac{c}{b}$, the tit-for-tat strategy will be better than always playing Donate. Another way to say this is that for $\gamma > \frac{c}{b}$ the tit-for-tat strategy will not be better than always playing Donate. If the chance the game is repeated for another round is large enough, player I is better off always playing Donate, whereas if the probability the current round is the last round is smaller than $\frac{c}{b}$ the player could do better with tit-for-tat. Notice that because of symmetry the same result is true for player II.

Next we want to compare the use of the grim-trigger strategy for this game. Each player will play Donate until someone deviates to Pass and then they will play Pass forever. To not deviate we need

$$b - c > (1 - \gamma)\left(\sum_{k=1}^{k_1}\gamma^{k-1}(b - c) + \gamma^{k_1}(b) + \sum_{i=k_1+2}^{\infty}\gamma^{k-1}(0)\right) \Leftrightarrow$$

$$b - c > (b - c) + \gamma^{k_1}(c - b\gamma) \Leftrightarrow \gamma > \frac{c}{b}.$$

If $1 \geq \gamma > \frac{c}{b}$, deviation from grim-trigger will not be profitable and the strategy to always Donate will ensue. By the one-round deviation principle, both players using grim-trigger is a subgame perfect NE.

Example 5.5   *Tit-for-Tat Is a NE but Not Always Subgame Perfect*
Once again we consider a prisoner's dilemma game with matrix

	C	D
C	(5, 5)	(0, 6)
D	(6, 0)	(1, 1)

The tit-for-tat strategy says that each player will begin by cooperating, i.e., play C, but then play the same strategy as the opponent in the last round if the opponent deviates in any round. We will first show that if both players use tit-for-tat, it is a NE.

If at any round, say k_1 a player, say player II deviates to D then player I will play C in round k_1 and D in round $k_1 + 1$ and then both players will play D forever. This is the same as grim-trigger. The payoff to player II is

$$(1 - \gamma)\left(5\sum_{k=1}^{k_1}\gamma^{k-1} + 1\gamma^{k_1} + 6\sum_{k=k_1+2}^{\infty}\gamma^{k-1}\right) = 5 + (1 - 5\gamma)\gamma^{k_1}.$$

The payoff to player I, who is not the player who deviates from C but is following the rules of tit-for-tat, is

$$(1 - \gamma)\left(5\sum_{k=1}^{k_1}\gamma^{k-1} + 0\gamma^{k_1} + \sum_{k=k_1+2}^{\infty}\gamma^{k-1}\right) = 5 + (-5 + \gamma)\gamma^{k_1} \leq 5.$$

Clearly, if $\gamma > \frac{1}{5}$ deviating from playing C by player II in any round yields a lower payoff than not deviating, i.e., $5 > 5 + (1 - 5\gamma)\gamma^{k_1}$ if $\gamma > \frac{1}{5}$, and hence, (tit-for-tat, tit-for-tat) is indeed a Nash equilibrium.

If at the outset both players have played (C, C), the payoff to each player is 5 and this is their payoff if they never deviate. So suppose player II sticks with tit-for-tat and let's consider what player I can do if there is a round, say $k_1 - 1$ ending in each of one of the following outcomes.

1. (C, C): If the outcome is (C, C) up to round $k_1 - 1$ and player I sticks to tit-for-tat, the outcome is (C, C) in every round. The payoff in the subgame is

$$(1 - \gamma)(5\gamma^{k_1} + 5\gamma^{k_1+1} + \cdots) = (1 - \gamma)5\gamma^{k_1}\frac{1}{1 - \gamma} = 5\gamma^{k_1}.$$

   If she chooses D in round k_1 of the subgame and then sticks to tit-for-tat, the outcome alternates between (D, C) and (C, D). Player I's payoff is then

$$(1 - \gamma)(6\gamma^{k_1} + 0\gamma^{k_1+1} + 6\gamma^{k_1+2} + 0\gamma^{k_1+3} + \cdots) = (1 - \gamma)6\gamma^{k_1}\sum_{k=0}^{\infty}(\gamma^2)^k$$

$$= (1 - \gamma)\frac{6\gamma^{k_1}}{1 - \gamma^2} = \frac{6\gamma^{k_1}}{1 + \gamma}.$$

   Therefore, to be profitable to do this, I needs $\frac{6\gamma^{k_1}}{1+\gamma} > 5\gamma^{k_1} \implies \frac{1}{5} > \gamma$. Alternatively, to not be profitable for player I we need $\gamma \geq \frac{1}{5}$.

2. (C, D): If I sticks to tit-for-tat the outcomes alternates between (D, C) and (C, D). In this case, the payoff to I will be $\frac{6\gamma^{k_1}}{1+\gamma}$. If player I deviates to C in round k_1 of subgame and then sticks to tit-for-tat, the outcome is (C, C) in every round and payoff is $5\gamma^{k_1}$. So we need $\frac{6\gamma^{k_1}}{1+\gamma} \geq 5\gamma^{k_1} \implies \gamma \leq \frac{1}{5}$, in order for a one round deviation from tit-for-tat to not be profitable for I is.

3. (D, C): If I sticks to tit-for-tat the outcomes alternates between (C, D) and (D, C) so payoff to I

$$(1 - \gamma)(0\gamma^{k_1} + 6\gamma^{k_1+1} + 0\gamma^{k_1+2} + 6\gamma^{k_1+3} + \cdots) = (1 - \gamma)6\gamma^{k_1+1}\sum_{k=0}^{\infty}(\gamma^2)^k$$

$$= (1 - \gamma)\frac{6\gamma^{k_1+1}}{1 - \gamma^2} = \frac{6\gamma^{k_1+1}}{1 + \gamma}.$$

   If player I deviates to D in round k_1 and then sticks to tit-for-tat the outcome is (D, D) in every round and the payoff in the subgame is $1\gamma^{k_1}$. Therefore, for a one-period deviation to not be profitable to player I we need $\frac{6\gamma^{k_1+1}}{1+\gamma} \geq 1\gamma^{k_1} \implies \gamma \geq \frac{1}{5}$.

4. (D, D): If I sticks to tit-for-tat, the outcome is (D, D) in every round so the payoff is 1. If she deviates to C in round k_1 of the subgame and then sticks to tit-for-tat, the outcomes alternate from (C, D) and (D, C). The payoff to player I is then

$$(1 - \gamma)(0\gamma^{k_1} + 6\gamma^{k_1+1} + 0\gamma^{k_1+2} + 6\gamma^{k_1+3} + \cdots) = (1 - \gamma)6\gamma^{k_1+1}\sum_{k=0}^{\infty}(\gamma^2)^k$$

$$= (1 - \gamma)\frac{6\gamma^{k_1+1}}{1 - \gamma^2} = \frac{6\gamma^{k_1+1}}{1 + \gamma}.$$

Thus we need $\frac{6\gamma^{k_1+1}}{1+\gamma} \leq 1\gamma^{k_1} \implies \gamma \leq \frac{1}{5}$ for the deviation to not be profitable for player I.

Similarly, for player II the same conclusions hold. Therefore (tit-for-tat, tit-for-tat) is subgame perfect if and only if $\gamma = \frac{1}{5}$.

5.2 Grim-Trigger in General

We let (s_1, s_2) be a NE of the one-shot game with associated payoffs (P_1, P_2). Suppose that the choice of strategies (s_1^*, s_2^*) would produce better payoffs (P_1^*, P_2^*) where $P_i^* > P_i$ for each player $i = 1, 2$. The players would definitely prefer to play (P_1^*, P_2^*) but we assume it is **not a NE**. When an opposing player, say player II chooses s_2^*, we assume the **maximal payoff** that player I can achieve by changing her strategy away from s_1^* is $d_1 > P_1^* > P_1$.

If we consider (s_1^*, s_2^*) as a cooperative strategy we will determine if grim-trigger can be used to enforce it, for some range of γ's. Here is grim-trigger specifically.

- play $s_i^*, i = 1, 2$ to start the game and play it as long as both players play (s_1^*, s_2^*). Begin by cooperating and cooperate as long as the opponent cooperates.
- if any player ever deviates from the pair (s_1^*, s_2^*) then switch to the NE strategy s_i for every round thereafter. Switch to the NE if cooperation fails.

To summarize the notation,

- P_i is the payoff to player $i = 1, 2$ of the one-shot game if the NE is played.
- P_i^* is the payoff to player $i = 1, 2$ if some other strategy is played, and we are assuming $P_i^* > P_i$.
- d_i is the **largest** payoff player $i = 1, 2$ can achieve by deviating from s_i^* if the opponent still uses s_{-i}^*, and we assume $d_i > P_i^* > P_i$.

We will see that the inequality necessary and sufficient for player $i = 1, 2$ to prefer to not deviate from the trigger strategy is:

$$\boxed{\gamma \geq \frac{d_i - P_i^*}{d_i - P_i}.}$$

The derivation is similar to the previous examples and goes like this .

$$P_i^* \geq (1 - \gamma) \left(\underbrace{\sum_{k=1}^{k_1} \gamma^{k-1} P_i^*}_{\text{cooperative}} + \underbrace{\gamma^{k_1} d_i}_{\text{max payoff}} + \underbrace{\sum_{k=k_1+2}^{\infty} \gamma^{k-1} P_i}_{\text{NE payoff}} \right) \Leftrightarrow$$

$$P_i^* \geq P_i^* + \gamma^{k_1}(-P_i^* + d_i(1 - \gamma) + P_i \gamma) \Leftrightarrow$$

$$\gamma \geq \frac{d_i - P_i^*}{d_i - P_i} = \gamma_i^*.$$

We have a lower bound on γ which ensures that player i will not deviate from his grim-trigger strategy given that the other player uses his grim-trigger strategy.

Remarks

- Everything works in a similar way if there are $N > 2$ players.
- Since each player does their own calculation of γ we actually have a threshold γ_i^* for each player i, $\gamma = \gamma_i \geq \gamma_i^*$. The threshold $\gamma^* = \max\{\gamma_1^*, \ldots, \gamma_N^*\}$ works for all the players.
- Any player who deviates from the better strategies (s_1^*, s_2^*) triggers the switch by both players to the Nash equilibrium strategies (s_1, s_2). If any player deviates, all players suffer the consequences.
- If $d_i \leq P_i^*$, then player i has no incentive to deviate from s_i^*.

Example 5.6 The game which illustrates the calculations is

	L	C	R
T	$(1,-1)$	$(2,1)$	$(1,0)$
M	$(3,4)$	$(0,1)$	$(-3,2)$
B	$(4,-5)$	$(-1,3)$	$(1,1)$

The unique pure NE is (T,C) so player I gets 2 and player II gets 1. Clearly, (M,L) is better for both players but this is not a NE. Can the players get to (M,L) in an infinitely repeated game for some discount factor (i.e., probability that the game continues for another round)? The grim-trigger strategy is the following for each player:

- Player I will play M in the first round and continue to do so as long as (M,L) is played. If in some round (M,L) is not played then I will switch to T, her part of the NE, for all future rounds.
- Player II will play L in the first round and continue to do so as long as (M,L) is played. If in some round (M,L) is not played then II will switch to C, her part of the NE, for all future rounds.

In this example, we have $(P_1, P_2) = (2,1)$, $(P_1^* = 3, P_2^* = 4)$. The maximum possible payoff for player I is $d_1 = 4$ and for player II is $d_2 = 4$.

If the players use their trigger strategies, the payoff to each player will be greater than the payoff from using the NE if the discount factor satisfies for each player,

$$\gamma_1 \geq \frac{d_1 - P_1^*}{d_1 - P_1} = \frac{4-3}{4-2} = \frac{1}{2} \text{ and } \gamma_2 \geq \frac{4-4}{4-1} = 0.$$

If $\gamma_1 \geq \frac{1}{2}$, then player I's trigger strategy is a best response to player II's trigger strategy. Given that player I plays M, player II's best response is L since that is the best choice. Player II's trigger strategy is thus a best response to player I's trigger strategy for all values of γ_1 and γ_2 plays no role. In sum, the trigger strategies support cooperation of both players if the probability the game continues for another round at the end of any round is at least 50%.

The next example shows that these ideas work for continuous games as well.

Example 5.7 *Repeated Cournot Duopoly*

Refer back to the Cournot Duopoly game Section 4.3 in which two firms are setting their production quantities q_i to maximize their profit. The price of a gadget if the production level is q is $P(q)$ and the cost to produce q gadgets is $c_i q$. Let

$$q = q_1 + q_2, \quad p(q) = \Gamma - q, \quad C_i(q_i) = c_i q_i, \quad P_i(q_1, q_2) = q_i p(q) - c_i q_i.$$

The NE quantities to produce for each firm is

$$q_1^* = \frac{1}{3}(\Gamma + c_2 - 2c_1), \quad q_2^* = \frac{1}{3}(\Gamma + c_1 - 2c_2).$$

For simplicity, let $c_1 = c_2 = c$. Then, the NE and the profits to each firm is

$$q_1^* = q_2^* = \frac{1}{3}(\Gamma - c), \quad P_1(q_1^*, q_2^*) = P_2(q_1^*, q_2^*) = \frac{1}{9}(\Gamma - c)^2.$$

Now suppose this game is played over many rounds, indefinite in number, and each firm needs to set a production quantity for each round. We would like cooperation between the firms so we look at how to set the q_i's if there was cooperation.

Let's use the values of $q = q_1 + q_2$ which maximizes the total profit of both firms (this is called the social welfare profit),

$$P(q) = P_1(q_1, q_2) + P_2(q_1, q_2) = q(\Gamma - q) - c q.$$

The maximum is achieved at $\hat{q} = \frac{1}{2}(\Gamma - c)$ and we assume each firm produces half the optimal widgets $\hat{q}_i = \frac{1}{4}(\Gamma - c), i = 1, 2$. The profit to each firm for these quantities is

$$\hat{P}_i = \frac{1}{4}(\Gamma - c)(\Gamma - \frac{1}{2}(\Gamma - c)) - c\frac{1}{4}(\Gamma - c) = \frac{1}{8}(\Gamma - c)^2.$$

Notice that $P_i(q_1^*, q_2^*) = \frac{1}{9}(\Gamma - c)^2 < \hat{P}_i = P_i(\hat{q}_1, \hat{q}_2)$.

So, the grim-trigger strategies each player will use is to produce the cooperation levels $\hat{q}_i$ in each period and if any player deviates from this in any period then the other firm will switch to playing the Nash equilibrium quantity q_i^*. We will show that if the discount factor is large enough, the grim-trigger strategy is a subgame perfect NE and cooperation will ensue.

Suppose player II at some stage switches to $\hat{q}_2 = \frac{1}{4}(\Gamma - c)$. What will player I do? Player I could switch to the value which maximizes $P_1(q_1, \frac{1}{4}(\Gamma - c))$, and that value is given by $q_1^\% = \frac{3}{8}(\Gamma - c)$. The resulting price of a widget is then

$$P = \Gamma - \frac{1}{4}(\Gamma - c) - \frac{3}{8}(\Gamma - c) = \frac{3\Gamma + 5c}{8}.$$

How much profit does firm make by doing this? Here it is:

$$P_1\left(\frac{3}{8}(\Gamma - c), \frac{1}{4}(\Gamma - c)\right) = \frac{3\Gamma + 5c}{8}\frac{3}{8}(\Gamma - c) - c\frac{3}{8}(\Gamma - c) = \frac{9}{64}(\Gamma - c)^2.$$

It's important to notice that $\frac{9}{64}(\Gamma - c)^2 > \frac{1}{8}(\Gamma - c)^2$ so the profit to firm I from deviating is greater than the profit from colluding with player II. Since there is an incentive to deviate from the collusive quantities there is no way that $(\hat{q}_1, \hat{q}_2)$ can be a NE.

The strategy to be used by each player is now clear. Each player will produce the quantity $\hat{q}_i = \frac{1}{4}(\Gamma - c)$ as long as they both play it. But, if a player deviates (and she would deviate to $\frac{3}{8}(\Gamma - c)$), then each player should switch to the NE quantities $q_i^* = \frac{1}{3}(\Gamma - c)$.

Start with the discounted profits if a player uses the proposed strategy and a player deviates from the collusive amounts. Remember that deviation would be to $q_1^\%$ or $q_2^\%$. Suppose deviation occurs in round k. The profit to the deviating player is then

$$Q_D = (1 - \gamma)\left(\underbrace{\sum_{i=1}^{k}\gamma^{i-1}\frac{1}{8}(\Gamma - c)^2}_{\text{coop quantity}} + \underbrace{\gamma^k\frac{9}{64}(\Gamma - c)^2}_{\text{deviate}} + \underbrace{\sum_{i=k+2}^{\infty}\gamma^{i-1}\frac{1}{9}(\Gamma - c)^2}_{\text{penalty}}\right).$$

If they do not deviate from the collusive quantities, the profit to each player is

$$Q_{ND} = (1 - \gamma)\sum_{i=1}^{\infty}\gamma^{i-1}\frac{1}{8}(\Gamma - c)^2 = \frac{1}{8}(\Gamma - c)^2.$$

What prevents deviating? Obviously, if the profit from not deviating is at least as large as that from deviating. This gives us the requirement for not deviating $Q_{ND} \geq Q_D$ and that will give us a condition on γ. Here's how it goes.[1]

$$\frac{1}{8}(\Gamma - c)^2 \geq \sum_{i=1}^{k}\gamma^{i-1}\frac{1}{8}(\Gamma - c)^2 + \gamma^k\frac{9}{64}(\Gamma - c)^2 + \sum_{i=k+2}^{\infty}\gamma^{i-1}\frac{1}{9}(\Gamma - c)^2$$

$$= \frac{1}{8}(\Gamma - c)^2 + \gamma^k(\Gamma - c)^2(\frac{1}{64} - \frac{17}{576}\gamma).$$

---

1 Use the formula $\sum_{i=0}^{n}r^i = \frac{1-r^{n+1}}{1-r}$.

Simplifying this we get that this inequality is true if and only if $\gamma \geq \frac{9}{17} = \gamma^*$. This says that if the probability the game will continue is at least $\frac{9}{17}$, deviating will be less profitable than not deviating from the trigger strategy. Thus, by the one-stage deviation principle, the grim-trigger strategy is a subgame perfect NE.

It may have occurred to you to ask the question about what happens if we consider the strategies to always do the same thing, tit-for-tat, and grim-trigger, as strategies in a game played against each other. We will illustrate what happens in a general prisoner's dilemma game.

Example 5.8   Consider the Prisoner's Dilemma game G with two players

	C	D
C	(a, a)	(s, t)
D	(t, s)	(d, d)

To make it a model for prisoner's dilemma we assume that $t > a > d > s$, so that (D, D) is the NE in the one-shot game. Consider three strategies played in the repeated game G^∞.

(C) Always play C in every round.
(D) Always play D in every round.
(TFT) Play tit-for-tat strategy: each player starts with C and then plays in the next round the same choice as the opponent.

We consider the game in which each player's pure strategies are (C), (D), and (TFT). We need to calculate the payoffs of each strategy played against the opponent who is also playing C, D, or TFT. Recall that the payoff in a game G^∞ with payoff a_k at the kth round is $(1 - \gamma) \sum_{k=1}^{\infty} a_k \gamma^{k-1}$.

	(C)	(D)	(TFT)
(C)	(a, a)	(s, t)	(a, a)
(D)	(t, s)	(d, d)	$(t + (d - t)\gamma, s + (d - s)\gamma)$
(TFT)	(a, a)	$(s + (d - s)\gamma, t + (d - t)\gamma)$	(a, a)

For example, if a_{ij} is the i, j entry of the matrix for player I,

$$a_{23} = (1 - \gamma)(P(D, C) + \sum_{k=2}^{\infty} P(D, D)\gamma^{k-1}) = (1 - \gamma)\left(t + d\sum_{k=2}^{\infty} \gamma^{k-1}\right) = t + \gamma(d - t),$$

where $P(i, j)$ is the payoff to player I in the one-shot game G.

Clearly, player I would never play strategy (C) because (TFT) is always better. Similarly, player II would never play (C) for a similar reason. So now we are left with the reduced 2×2 game

	(D)	(TFT)
(D)	(d, d)	$(t + (d - t)\gamma, s + (d - s)\gamma)$
(TFT)	$(s + (d - s)\gamma, t + (d - t)\gamma)$	(a, a)

Recall that $t > a > d > s$. Since $d > s$ we have $d \geq s + (d - s)\gamma$ so $((D), (D))$ is a pure NE. If $\gamma \leq \frac{t-a}{t-d}$ then (TFT, TFT) is also a NE, as you can verify, which leads to a cooperative outcome (a, a).

5.2.1 A Better Estimate for the Discount Factor

In the previous section we have shown that the grim-trigger strategy consists of the players begin by cooperating but, if a player deviates from cooperation, then both players switch to their NE. Playing the grim-trigger strategy is a subgame perfect NE as long as the discount factor is sufficiently large, i.e., the probability the game will be played again is sufficiently large. Now we will modify the grim-trigger so that if a player unilaterally deviates from cooperation the players will switch to their minimax strategies instead of their Nash equilibrium strategies. We'll see that this results in a better lower bound on the discount rate and the trigger strategy will still be a NE for the infinitely repeated game, but it may not be subgame perfect.

Suppose we have a bimatrix game (A, B). Set

$$m_1 = \min_Y \max_X X\, A\, Y^T = value(A), \text{ and } m_2 = \min_X \max_Y X\, B\, Y^T.$$

Notice that m_2 is not $value(B)$ although it looks like it is. In fact,

$$m_2 = \min_X \max_Y X\, B\, Y^T = \min_X \max_Y Y\, B^T\, X^T = \min_Y \max_X X\, B^T\, Y^T = value(B^T),$$

where we simply relabel the strategies X, Y in the last equality to conform with previous notation. We see that m_1, m_2 are the safety values of the game. Set $(\hat{X}, \hat{Y})$ strategies so that

$$m_1 = \max_X X\, A\, \hat{Y}^T \text{ and } m_2 = \max_Y \hat{X} B Y^T$$

These are the strategies at which the minimums are attained in m_1, m_2. Choosing $\hat{X}$ is the worst that player I can do to player II and choosing $\hat{Y}$ is the worst player II can do to player I.

In the G^∞ game we want to show that the grim-trigger strategy in which the players begin by cooperating but then switch to $(\hat{X}, \hat{Y})$ if a player deviates is a NE. The difference here in what we did earlier is that the players switched to a NE of the one-shot game G instead of the minimax strategies, which may not be a NE.

We denote the players by 1 and 2 and their minimax strategies by $\hat{X}_1 = \hat{X}, \hat{X}_2 = \hat{Y}$. Let's label the cooperating strategies for each player by X_1^* and X_2^* for players $i = 1, 2$. If player i deviates from the cooperation strategy (X_1^*, X_2^*) with payoffs (v_1, v_2) then player i will play the punishment strategy $\hat{X}_i$, the minimax strategy. We call this **grim-trigger with punishment**. Player $i = 1, 2$ compares her payoff v_i if she follows this trigger strategy to her best possible payoff $d_i > v_i$.

Suppose player i deviates in round k_1. The payoff to player i is

- v_i, the best payoff achievable by both players cooperating in rounds $1, 2, \ldots, k_1 - 1$.
- d_i, the best payoff to player i if she deviates in round k_1.
- m_i, the minimax payoff for player i in rounds $k_1 + 1, k_1 + 2, \ldots,$

The payoff is compared to v_i to make sure deviation is not profitable.

$$v_i \geq (1 - \gamma)\left(v_i \sum_{j=1}^{k_1} \gamma^{j-1} + d_i \gamma^{k_1} + m_i \sum_{j=k_1+2}^{\infty} \gamma^{j-1} \right)$$

$$= v_i + \gamma^{k_1}(-v_1 + (1 - \gamma)d_i + \gamma m_i).$$

This holds if

$$\gamma > \frac{d_i - v_i}{d_i - m_i} = \gamma_i^*.$$

Since $d_i > v_i > m_i$, $0 < \gamma_i^* < 1$. This means that for $\gamma > \gamma_i^*$ and i deviates from the strategy, then player i's payoff will be smaller than v_i. When she sticks to the strategy, her payoff is v_i. We conclude that for large enough γ, grim-trigger with punishment is a NE.

One thing to notice is that when the players switch to a NE strategy as they did before instead of their minimax strategy we compare lower bounds on γ:

$$\frac{d_i - v_i}{d_i - m_i} \leq \frac{d_i - v_i}{d_i - P_i}$$

where P_i is the payoff if the NE is played in the one-shot game. This is due to $P_i \geq m_i$. Consequently, the bound for the discount rate will be lower using m_i.

Why isn't the grim-trigger strategy with punishment subgame perfect? The answer is because the pair of minimax strategies played by each player to get the payoffs m_i may not be a NE. There is no reason to expect that the strategy for player 1 is a best response to that for player 2 and vice versa. Therefore, in the subgame in which the minimax strategies are played, there is no reason to believe that $(\hat{X}_1, \hat{X}_2)$ is a NE.

Example 5.9 Here's an example to show that the bound on the discount rate improves using grim-trigger with punishment.

	L	C	R
T	$(1, -1)$	$\boxed{(2, 1)}$	$(1, 0)$
M	$(3, 4)$	$(0, 1)$	$(-3, 2)$
B	$(4, -5)$	$(-1, 3)$	$(1, 1)$

This game has one pure NE at (T, C). There are two other mixed NEs, $X_1 = (\frac{3}{5}, \frac{2}{5}, 0), Y_1 = (\frac{1}{2}, \frac{1}{2}, 0)$ with payoffs $(\frac{3}{2}, 1)$ and $X_2 = \frac{2}{7}, \frac{4}{7}, \frac{1}{7}), Y_2 = (\frac{1}{2}, \frac{1}{2}, 0)$ with payoffs $(\frac{3}{2}, \frac{9}{7})$.

The matrices involved are

$$A = \begin{bmatrix} 1 & 2 & 1 \\ 3 & 0 & -3 \\ 4 & -1 & 1 \end{bmatrix} \text{ and } B = \begin{bmatrix} -1 & 1 & 0 \\ 4 & 1 & 2 \\ -5 & 3 & 1 \end{bmatrix}, B^T = \begin{bmatrix} -1 & 4 & -5 \\ 1 & 1 & 3 \\ 0 & 2 & 1 \end{bmatrix}.$$

We have $m_1 = value(A) = 1$ and $m_2 = value(B^T) = 1$. Suppose we want to implement the cooperative payoff $(v_1, v_2) = (3, 4)$.

There is a pure NE at (T, C) with payoffs $(P_1, P_2) = (2, 1)$. The maximum possible payoffs for each player are $d_1 = d_2 = 4$. If the players implement the trigger strategy with punishment, the discount factor must satisfy

$$\gamma \geq \max\left\{\frac{d_1 - v_1}{d_1 - m_1}, \frac{d_2 - v_2}{d_2 - m_2}\right\} = \frac{4 - 3}{4 - 1} = \frac{1}{3}.$$

If, on the other hand, the players revert to their NE strategies the discount factor must satisfy

$$\gamma \geq \max\left\{\frac{d_1 - v_1}{d_1 - P_1}, \frac{d_2 - v_2}{d_2 - P_2}\right\} = \frac{4 - 3}{4 - 2} = \frac{1}{2}.$$

The trigger minimax strategy player 2 should use if she wants to punish player I is $\hat{Y} = (0, 0, 1)$, i.e., play R. The minimax strategy player 1 should use to punish player 2 is $\hat{X} = (0, 1, 0)$, i.e., play M. Notice that $(\hat{X}, \hat{Y})$ is not a NE so reverting to this pair of strategies shows that the grim-trigger with punishment may not be subgame perfect.

5.2.2 Folk Theorems

There are a bunch of theorems referred to as Folk Theorems because apparently they were known to be true before any particular person published them. These folk theorems say that as long as there is a larger cooperative payoff than the NE of the one-shot game, and the cooperative payoff is a (convex) combination of all the reasonable payoffs possible, then there is a NE of the infinitely repeated game G^∞ which achieves that payoff as long as the discount factor is large enough. Here is a precise statement which, in fact, says there is a subgame perfect NE achieving the cooperative payoff.

Theorem 5.2.1 [1] *Let the one-shot game G have a pure NE labeled s^* with payoff vector P^*. Let P be a payoff vector of some strategy vector s of G such that $P_i \geq P_i^*$, for all players i. Then there is a $\gamma_0 > 0$ such that for $1 \geq \gamma > \gamma_0$ there exists a subgame perfect equilibrium of G^∞ with payoff vector P.*

Once again, the idea is all players start off by cooperating by playing s but if anyone deviates, they all switch to playing s^*. Since s^* is a NE there is no further incentive to switch.

More can be said if the game is two-player. No matter what payoff in a two-player game one-shot game G is achievable, there is a subgame perfect NE of G^∞ which gives a payoff as close as we want to the given payoff as long as the discount factor is large enough[2].

Problems

5.1 Consider the PD game

	C	D
C	(2, 2)	(−3, 3)
D	(3, −3)	(−2, −2)

There is a unique subgame perfect NE in the finite game G^T, $T < \infty$, and it is (D, D) at every stage. Even in the infinitely repeated game, it is still a subgame perfect NE. Now we ask the question if it is possible that (C, C) played at every stage of the game G^∞ is a NE? Write down the grim-trigger strategy involving cooperation and find the range of γ's so that it is a subgame perfect NE.

5.2 Consider the following game:

	K	M
K	(2, 2)	(6, 0)
M	(0, 6)	(4, 4)

This game has a unique NE at (K, K) and it is strictly dominant. Consider now the game G^∞. Find and show that the grim-trigger strategy is a subgame perfect NE for some range of the discount factor γ.

5.3 Consider the game

	C	D
C	(5, 5)	(−3, 8)
D	(8, −3)	(0, 0)

1 J.W. Friedman, *A non-cooperative equilibrium for supergames*, The Review of Economic Studies, 38(1), pp. 1–12, 1971.
2 For further information and more theorems refer to Fudenberg and Maskin, *The folk theorem in repeated games with discounting or with incomplete information*, Econometrica, 54(3), pp. 533–554, 1986.

a) Suppose both players use their trigger strategies so that they play C unless a player defects to D and then plays D forever. Find γ^* so that if $1 > \gamma \geq \gamma^*$ the pair of trigger strategies is a subgame perfect NE.

b) Consider the following strategy for player I. Play $C, D, C, \ldots$ as long as player II plays $D, C, D, \ldots$. If player II deviates from this, then player I will play D forever. The strategy for player II is to play $D, C, D, \ldots$, as long as player I plays $C, D, C, \ldots$, and if player I deviates from this, then player II will play D forever. Find γ^* so that this pair of strategies is a subgame perfect NE.

5.4 Consider the game

	L	R
U	$(5, 1)$	$(1, 2)$
D	$(4, 2)$	$(2, 4)$

and the pair of strategies: player I plays D and then D as long as (D, L) is played on even-numbered rounds and (D, R) is played on odd-numbered rounds. For any deviation, play U forever. Player II plays L on even rounds and R at odd rounds as long as (D, L) was played on even rounds and (D, R) at odd rounds. For any deviation play R forever.

a) Show that if these strategies are played the long-term average payoffs to each player is 3.

b) Find the smallest γ^* so that $1 > \gamma \geq \gamma^*$ implies these strategies are a NE.

5.5 The game

	L	R
U	$(4, 4)$	$(0, 2)$
D	$(3, 6)$	$(1, 8)$

has pure NEs at (U, L) and (D, R) and a mixed NE $X^* = (\frac{1}{2}, \frac{1}{2}) = Y^*$.

a) Find the minimax values v_1, v_2.

b) Consider the strategies

- player I: Start with D and play it as long as (D, L) has been played so far. Upon a deviation, play $X = (\frac{1}{2}, \frac{1}{2})$ forever.
- player II: Start with L and play it as long as (D, L) has been played so far. Upon a deviation, play $Y = (\frac{1}{2}, \frac{1}{2})$ forever.

Find the payoff to each player if these strategies are played and show that the pair of strategies is subgame perfect NE.

5.6 Find conditions on γ so that Example 5.8 has a mixed NE.

5.7 Suppose the inverse demand curve in the Cournot Model is $q = \Gamma - p$ and the payoff for firm i in the one-shot game is $q_i(p_i, p_{-i}) = \begin{cases} \Gamma - p_i, & p_i < p_j \\ \frac{1}{2}(\Gamma - p_i), & p_i = p_j \\ 0, & p_i > p_j. \end{cases}$ In the one-shot game the NE is $p_1^* = p_2^* = c$ where c is the common marginal cost of the firms.

The profit-maximizing price is $p^* = \frac{\Gamma + c}{2}$. The optimal collusion price is $p_1 = p_2 = p^*$. and the quantity of gadgets they make is $q_1 = q_2 = \frac{\Gamma - c}{4}$. This gives a profit for each firm of $\pi^* = \frac{(\Gamma - c)^2}{8}$. Consider now the infinitely repeated game and the strategies for each firm: Set the price $p^* = \frac{\Gamma + c}{2}$ as long as the opponent also uses this price. If a firm deviates from producing the collusive output for the price p^*, then set the price $p_1 = p_2 = c$. Show that if $\gamma \geq \frac{1}{2}$ then the pair of strategies is subgame perfect NE.

5.8 **Repeated Cournot Duopoly.** Consider a Cournot Duopoly game. As before q_i is the output of firm i and total output of the two firms is $q = q_1 + q_2$. The price of a widget for total output q is $p(q) = 14 - q$ and the cost function for each firm is given by $C(q_i) = \frac{q_i^2}{4}, i = 1, 2$. With these price and cost functions, the payoffs to each firm are given by

$$P_i(q_1, q_2) = q_i p(q) - C(q_i) = q_i(14 - (q_1 + q_2)) - \frac{q_i^2}{4}, \quad i = 1, 2.$$

As usual, we find the NE production quantities by taking derivatives and setting to zero.
a) What are the NE production quantities?
b) Instead of using the NE quantities, suppose the firms cooperated. Calculate the quantities and resulting payoffs that maximize the social welfare

$$P_1(q_1, q_2) + P_2(q_1, q_2) = (q_1 + q_2)(14 - (q_1 + q_2)) - \frac{1}{4}q_1^2 - \frac{1}{4}q_2^2.$$

c) What are the trigger strategies each firm should use?
d) Find the lower bound on the discount rate for each player so that the trigger strategy is a subgame perfect NE.

5.9 Consider the following indefinitely repeated Prisoner's Dilemma game

	C	D
C	(x, x)	$(0, y)$
D	$(y, 0)$	$(1, 1)$

Show that (tit-for-tat,tit-for-tat) is a subgame perfect NE for this game with discount factor γ if and only if $y - x = 1$ and $\gamma = 1/x$.

5.10 Consider the one-shot games:
a)

	C	D
C	$(3, 3)$	$(0, 5)$
D	$(5, 0)$	$(1, 1)$

Which of the payoffs (i) $(3, 3)$; (ii) $(5/4, 5/4)$; (iii) $(1, 1)$; (iv) $(5, 0)$, are not both feasible and enforceable? Feasible means there are strategies which achieve the payoff and enforceable means that any deviation from the strategies by a player results in a lower payoff to the deviating player.

b)

	C	D
C	$(4, 4)$	$(0, 5)$
D	$(5, 0)$	$(3, 3)$

Which of the payoffs (i) $(6, 6)$; (ii) $(5, 2)$; (iii) $(5/4, 5/4)$; $(4, 4)$, are both feasible and enforceable?

5.11 Consider a model on workings of OPEC. Countries aim to collude on production, drive up prices and profits and return to equilibrium if someone deviates. Let $P = 300 - 5Q$ world demand for oil (where $Q = \sum_i q_i$ is total production and q_i be the production of country i). Let marginal cost for production be c for all countries.

    a) What is a payoff function that gives the profit to each country?

    b) Using the payoff function for country i find a static NE (where each country tries to maximize its profit). Assume there are 4 countries and $c = 20$.

5.12 Now suppose countries try to enforce a Grim-Trigger strategy by keeping $q_i = 7$ each, unless someone deviates. If someone deviates, production is increased to $q_i = 11.2$ each forever.

    a) Is it enforceable and feasible?

    b) Suppose a country deviates as follows: $q_i = 28 - 21/2 = 17.5$, thus earning a profit of 1531.25. Calculate the threshold γ.

5.13 Consider the following game involving N countries. The per period payoff for country i is

$$(P - c)q_i = (300 - 5(q_{-i} + q_i) - 20)q_i$$

and γ is the discount factor. Consider a grim-trigger threat as part of a strategy: if there is a deviation from the prescribed production, go to producing $q = 11.2$ forever after. If each produces $q_i = 10$, which of the following sentences is wrong:

1. the resulting price is $300 - 5(40) = 100$;
2. The profit for each country is $(100 - 20)10 = 800$ M\$/day;
3. The Nash Equilibrium profit is 627.2 M\$/day.
4. Producing $q = 10$ is not sustainable with the grim-trigger threat described above for any γ.

5.14 Consider the matrix for the one-shot game G

	M	F
M	$(4,4)$	$(-1,5)$
F	$(5,-1)$	$(1,1)$

    a) Consider the payoffs of $(2, 2)$. Show that for a discount factor γ that is close enough to 1, we can get payoffs (v_1, v_2) that are arbitrarily close to $(2, 2)$. Start with the equations

$$a_1(4, 4) + a_2(-1, 5) + a_3(5, -1) + a_4(1, 1) = (2, 2),$$
$$a_1 + a_2 + a_3 + a_4 = 1$$

    Choose the solution with $a_1 = a_4 = 0, a_2 = a_3 = \frac{1}{2}$. Half of the time we want to play $X = (1, 0), Y = (0, 1)$ and half the time we want $X = (0, 1), Y = (1, 0)$.

    b) Take the following strategies:

    Player 1: in period 1 play F; in every even period play M if the pattern of play was $(F, M), (M, F), (F, M)$ ...; otherwise play F. In every odd period after period 1 play F.

    Player 2: in period 1 play M; in every odd period play F; in every odd period after period 1 play M if the pattern was $(F, M), (M, F), (F, M)$ ...; otherwise play F.

    These strategies will cause the payoff realizations to alternate between $(5, -1)$ and $(-1, 5)$. Find the resulting payoffs for the two players.

    c) Show that $\lim_{\gamma \to 1} v_1 = 2 = \lim_{\gamma \to 1} v_2$ so that payoffs arbitrarily close to $(2, 2)$ can be obtained for γ close enough to 1.

    d) Are these strategies subgame perfect?

5.15 J. Harrington in [12] and [13] analyzed the problem of price fixing among the auction houses Christies and Sothebys. It is assumed that they charge a percentage of the sold auction price of an item at the auction. They compete for items put up for auction by the percentage they charge. Suppose they charge 2, 4, 6, or 8 percent. The game matrix is

C/S	2	4	6	8
2	$(-20, -20)$	$(60, 0)$	$(140, -60)$	$(220, -200)$
4	$(0, 60)$	$(100, 100)$	$(220, 60)$	$(140, -60)$
6	$(-60, 140)$	$(60, 220)$	$(180, 180)$	$(320, 80)$
8	$(-200, 220)$	$(-60, 140)$	$(80, 320)$	$(230, 230)$

Each pair in the matrix represents the net profit to each auction house if they charge the percentage in the row and column heading. The pure NE is boxed and says they should each charge 4 percent. But clearly, they can do better if they both charge 6 or even 8 percent. This is a prisoner's dilemma game. As a one-shot game, i.e., if they compete only once or a finite number of times it is clear they should play the NE. There is no direct communication between the houses and collusion is illegal. They are going to compete forever on many items coming up for auction. Assuming they play indefinitely what happens if Sothebys decides to unilaterally raise the price to 8%? Christie's best response is to charge 6%, to which Sothebys will respond with 4% to which Christies will respond with 4% and we are at the NE. How can they raise rates optimally and permanently? Consider the grim-trigger strategy and the grim-trigger with punishment strategy and find the discount factor which works in both cases.

Bibliographic Notes

Repeated games are known as either **finite horizon** or **infinite horizon**, meaning either finitely repeated or of indefinite repetition. One of the problems with infinite horizon games is that the extensive form basically cannot be established. Gambit is of no help with these games and even finitely repeated games are difficult to analyze with Gambit. The textbooks in the references have much more on repeated games and especially the text [1]. See also the books [7] and [3].

6

Cooperative Games

Government and cooperation are in all things the laws of life; anarchy and competition the laws of death.

–John Ruskin, Unto this Last

We must all hang together, or assuredly we will all hang separately.
–Benjamin Franklin, at the signing of the Declaration of Independence

Talent wins games, but teamwork and intelligence wins championships.

–Michael Jordan

Teamwork means never having to take all the blame yourself.

–Stephen Hawking

There are two kinds of people: Those who say to God "Thy will be done," and those to whom God says, "All right, then, have it your way."

–C.S. Lewis

6.1   What Is a Cooperative Game?

Cooperative game theory is a branch of game theory that studies situations in which players can make binding agreements about how to cooperate. In contrast, non-cooperative game theory studies situations in which players cannot make binding agreements and must rely on self-interest to motivate their behavior.

Cooperative game theory has a wide range of applications, including economics, business, and political science. It can be used to analyze situations such as labor negotiations, corporate mergers, and international treaties.

One of the main goals of cooperative game theory is to find fair ways to allocate the payoffs from a game among the players. There are many different fairness criteria that can be used, such as the Shapley value, the nucleolus, and the core.

Here are some of the key concepts in cooperative game theory:

- Coalition: A group of players who have agreed to cooperate.
- Characteristic function: A function that assigns a payoff to each coalition.

Game Theory: An Introduction, Third Edition. E. N. Barron.
© 2024 John Wiley & Sons, Inc. Published 2024 by John Wiley & Sons, Inc.

- Payoff vector: A vector of payoffs that is assigned to the players.
- Fairness criterion: A rule for allocating the payoffs from a game among the players in a fair way.
- The nucleolus: The nucleolus is a payoff vector that is Pareto efficient and individually rational. In other words, no coalition can improve its payoff without making at least one player worse off, and no player can improve its payoff without making at least one coalition worse off.
- The Shapley value: The Shapley value is a payoff vector that is calculated using the concept of marginal contribution. The Shapley value assigns to each player a payoff that is equal to the average of the marginal contributions that the player makes to all coalitions that the player can join.

One example of a payoff for a coalition of players is a measure of power in an organization, such as a parliament, congress, or board of directors of a company. How coalitions can form in order to maximize the power of the coalition is an example to be studied in cooperative game theory.

6.2 Coalitions and Characteristic Functions

There are $n > 1$ players numbered $1, 2, \ldots, n$. **In this chapter we use the letter n to denote the number of players and N to denote the set of all the players $N = \{1, 2, \ldots, n\}$.** We consider a game in which the players may choose to cooperate by forming coalitions. A **coalition** is any subset $S \subset N$, or numbered collection of the players. Since there are 2^n possible subsets of N, there are 2^n possible coalitions. Coalitions form in order to benefit every member of the coalition so that all members might receive more than they could individually on their own. In this section we try to determine a **fair allocation** of the benefits of cooperation among the players to each member of a coalition. A major problem in cooperative game theory is to precisely define what **fair** means. The definition of **fair** of course determines how the allocations to members of a coalition are made.

First, we need to quantify the benefits of a coalition through the use of a real-valued function, called the **characteristic function**. The characteristic function of a coalition $S \subset N$ is the **largest** guaranteed payoff to the coalition.

Definition 6.2.1 *Let 2^N denote the set of all possible coalitions for the players N. If $S = \{i\}$ is a coalition containing the single member i, we simply denote S by i.*

Any function $v : 2^N \to \mathbb{R}$ satisfying

$$v(\emptyset) = 0 \quad and \quad v(N) \geq \sum_{i=1}^{n} v(i)$$

is a characteristic function (of an n-person cooperative game).

In other words, the only condition placed on a characteristic function is that the benefit of the empty coalition be zero and the benefit of the **grand coalition** N, consisting of all the players, be at least the sum of the benefits of the individual players if no coalitions form. This means that everyone pulling together should do better than each player on his or her own. With that much flexibility, games may have more than one characteristic function. Let's start with some simple examples.

Example 6.1

1. Suppose that there is a factory with n workers each doing the same task. If each worker earns the same amount b dollars, then we can take the characteristic function to be $v(S) = b|S|$, where $|S|$ is the number of workers in S. Clearly, $v(\emptyset) = b|\emptyset| = 0$, and $v(N) = b|N| = bn = b\sum_{i=1}^{n} v(i)$.

2. Suppose that the owner of a car, labeled player 1, offers it for sale for $M. There are two customers interested in the car. Customer C, labeled player 2, values the car at c and customer D, labeled player 3, values it at d. Assume that the price is nonnegotiable. This means that if $M > c$ and $M > d$, then no deal will be made. We will assume then that $M < \min\{c, d\}$, and, for definiteness we may assume $M < c \le d$. The set of possible coalitions are $2^N \equiv \{123, 12, 13, 23, 1, 2, 3, \emptyset\}$. For simplicity we are dropping the braces in the notation for any individual coalition.

   It requires a seller and a buyer to reach a deal. Therefore, we may define the characteristic function as follows:

$$v(123) = d, \quad v(1) = M, \quad v(\emptyset) = 0$$
$$v(13) = d, \quad v(12) = c, \quad v(23) = 0,$$
$$v(2) = v(3) = 0.$$

   Why? Well, $v(123) = d$ because the car will be sold for d, $v(1) = M$ because the car is worth M to player 1, $v(13) = d$ because player 1 will sell the car to player 3 for $d > M$, $v(12) = c$ because the car will be sold to player 2 for $c > M$, and so on. The reader can easily check that v is a characteristic function.

3. A small airport has a single runway and all of the different planes have to use it. Assume there are three different types of planes that use the runway. The largest plane needs 1000 feet, the medium plane needs 750 feet, and the small planes need 500 feet. The runway must be built to serve the biggest plane. The runway's cost is directly proportional to length. If there are 3 users of the runway, one of each type with 1=small, 2=medium, and 3=large, then a possible characteristic function is

$$v(1) = 500, \quad v(2) = 750, \quad v(3) = 1000,$$
$$v(12) = 750, \quad v(13) = 1000, \quad v(23) = 1000$$
$$v(123) = 1000, \quad v(\emptyset) = 0.$$

4. Suppose a family-owned corporation has 5 shareholders, Family member $i = 1, 2, 3, 4, 5$ holds $11, 10, 10, 20, 30$ shares, respectively. In order for a resolution to pass, a simple majority of shares must be in favor of the resolution. However, player 5 as the patriarch of the family has veto power. A possible characteristic function is $v(S) = 1$ for any coalition consisting of 51 or more shares and with player 5 as a member of S, and $v(S) = 0$, otherwise.

5. A **simple game** is one in which $v(S) = 1$ or $v(S) = 0$ for all coalitions S. A coalition with $v(S) = 1$ is called a **winning coalition** and one with $v(S) = 0$ is a **losing coalition**. For example, if we take $v(S) = 1$ if $|S| > \frac{n}{2}$ and $v(S) = 0$ otherwise, we have a simple game that is a model of majority voting. If a coalition contains more than half of the players, it has the majority of votes and is a winning coalition.

6. In any bimatrix (A, B) nonzero sum game we may obtain a characteristic function by taking $v(1) = value(A)$, $v(2) = value(B^T)$, and $v(12) =$ sum of largest payoff pair in (A, B). Checking that this is a characteristic function is skipped. The next example works one out.

Example 6.2 In this example we will construct a characteristic function for a version of the prisoner's dilemma game in which we assumed that there was no cooperation. Now we will assume that the players may cooperate and negotiate. One form of the prisoner's dilemma is with the bimatrix

$$\begin{bmatrix} (8, 8) & (0, 10) \\ (10, 0) & (2, 2) \end{bmatrix}.$$

Here $N = \{1, 2\}$ and the possible coalitions are $2^N = \{\emptyset, 1, 2, 12\}$. If the players do not form a coalition, they are playing the nonzero sum noncooperative game. Each player can guarantee only that they receive their **safety level**. For player I that is the value of the zero sum game with matrix $A = \begin{bmatrix} 8 & 0 \\ 10 & 2 \end{bmatrix}$, which is $value(A) = 2$. For player II the safety level is the value of the game with matrix $B^T = \begin{bmatrix} 8 & 0 \\ 10 & 2 \end{bmatrix}$. Again $value(B^T) = 2$.

Thus we could define $v(1) = v(2) = 2$ as the characteristic function for single-member coalitions. Now, if the players cooperate and form the coalition $S = \{12\}$, can they do better? Figure 6.1 shows what is going on. The parallelogram is the boundary of the set of all possible payoffs to the two players when they use all possible mixed strategies.

The vertices are the pure payoff pairs in the bimatrix. You can see that without cooperation they are each at the lower left vertex point $(2, 2)$. Any point in the parallelogram is attainable with some suitable selection of mixed strategies if the players cooperate. Consequently, the maximum benefit to cooperation for both players results in the payoff pair at vertex point $(8, 8)$, and so we set $v(12) = 16$ as the **maximum sum of the benefits** awarded to each player. With $v(\emptyset) = 0$, the specification is complete.

As an aside, notice that $(8, 8)$ is Pareto-optimal as is any point on the two lines connecting $(0, 10)$ and $(8, 8)$ and $(10, 0)$ with $(8, 8)$. This is the Pareto-optimal boundary of the payoff set. This is clear from Figure 6.1 because if you take any point on the lines, you cannot simultaneously move up and right and remain in the set.

Example 6.3 Here is a much more complicated but systematic way to create a characteristic function given any n-person, noncooperative, nonzero sum game. The idea is to create a two-person zero sum game in which any given coalition is played against a pure opposing coalition consisting of everybody else. The two players are the coalition S versus all the other players, which is also a coalition $N - S$. The characteristic function will be the value of the game associated with each coalition S. The way to set this up will become clear if we go through an example.

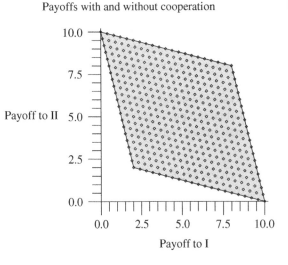

Payoffs with and without cooperation

Figure 6.1 Payoff to player I versus payoff to player II.

Let's work out a specific example using a three-player game. Suppose that we have a three-player nonzero sum game with the following matrices:

3 plays A		player 2	
		A	*B*
player 1	*A*	$(1, 1, 0)$	$(4, -2, 2)$
	B	$(1, 2, -1)$	$(3, 1, -1)$

3 plays B		player 2	
		A	*B*
player 1	*A*	$(-3, 1, 2)$	$(0, 1, 1)$
	B	$(2, 0, -1)$	$(2, 1, -1)$

Each player has the two pure strategies *A* and *B*. Because there are three players, in matrix form this could be represented in three dimensions. That is a little hard to write down, so instead we have broken this into two 2×2 matrices. Each matrix assumes that player 3 plays one of the two strategies that is fixed. Now we want to find the characteristic function of this game.

We need to consider all of the zero sum games which would consist of the two-player coalitions versus each player, and the converse, which will switch the roles from maximizer to minimizer and vice versa. The possible two-player coalitions are $\{12\}, \{13\}, \{23\}$ versus the single-player coalitions, $\{1, 2, 3\}$, and conversely. For example, one such possible game is $S = \{12\}$ versus $N - S = 3$, in which player $S = \{12\}$ is the row player and player 3 is the column player. We also have to consider the game 3 versus $\{12\}$, in which player 3 is the row player and coalition $\{12\}$ is the column player. So now we go through the construction.

1. **Play $S = \{12\}$ versus $\{3\}$.** Players 1 and 2 team up against player 3. We first write down the associated matrix game.

12 versus 3		player 3	
		A	*B*
player 12	*AA*	2	-2
	AB	2	1
	BA	3	2
	BB	4	3

For example, if 1 plays A, 2 plays A, and 3 plays B, the payoffs in the nonzero sum game are $(-3, 1, 2)$ and so the payoff to player 12 is $-3 + 1 = -2$, the **sum of the payoff to player 1 and player 2**, which is our coalition. Now we calculate the value of the zero sum two-person game with this matrix to get the *value*(12 versus 3) = 3 and we write $v(12) = 3$. This is the maximum possible guaranteed benefit to coalition $\{12\}$ because it even assumes that player 3 is actively working against the coalition.

In the game $\{3\}$ versus $\{12\}$, we have $\{3\}$ as the row player and players $\{12\}$ as the column player. We now want to know the maximum possible payoff to player 3 assuming that the coalition $\{12\}$ is actively working against player 3. The matrix is

3 versus 12		12			
		AA	*AB*	*BA*	*BB*
3	*A*	0	2	-1	-1
	B	2	1	-1	-1

The value of this game is −1. Consequently, in the game {3} versus {12} we would get $v(3) = -1$. Observe that the game matrix for 3 versus 12 is **not** the transpose of the game matrix for 12 versus 3.

2. **Play** $S = \{13\}$ **versus** $\{2\}$. The game matrix is

13 versus 2		player 2	
		A	*B*
player 13	*AA*	1	6
	AB	−1	1
	BA	0	2
	BB	1	1

We see that the value of this game is 1 so that $v(13) = 1$. In the game {2} versus {13}, we have {2} as the row player and the matrix

2 versus 13		13			
		AA	*AB*	*BA*	*BB*
2	*A*	1	1	2	0
	B	−2	1	1	1

The value of this game is $\frac{1}{4}$, and so $v(2) = \frac{1}{4}$.

Continuing in this way, we summarize that the characteristic function for this three-person game is

$$v(1) = 1, \ v(2) = \frac{1}{4}, \ v(3) = -1,$$

$$v(12) = 3, \ v(13) = 1, \ v(23) = 1,$$

$$v(123) = 4, \ v(\emptyset) = 0.$$

The value $v(123) = 4$ is obtained by taking the largest sum of the payoffs that they would achieve if they all cooperated. This number is obtained from the pure strategies: 3 plays A, 1 plays A, and 2 plays B with payoffs $(4, -2, 2)$. Summing these payoffs for all the players gives $v(123) = 4$. This is the most the players can get if they form a grand coalition, and they can get this only if all the players cooperate. The central question in cooperative game theory is how to allocate the reward of 4 to the three players. In this example, player 2 contributes a payoff of −2 to the grand coalition, so should player 2 get an equal share of the 4? On the other hand, the 4 can only be obtained if player 2 agrees to play strategy B, so player 2 does have to be induced to do this. What would be a **fair allocation**?

One more observation is that player 3 seems to be in a bad position. On her own she can be guaranteed to get only $v(3) = -1$, but with the assistance of player 1 $v(13) = 1$. Since $v(1) = 1$, this gain for $S = 13$ is solely attributed to 1. Note however that $v(2) = \frac{1}{4}$, while $v(23) = 1$. Player 3 can definitely add value by joining with player 2.

Remark There is a general formula for the characteristic function obtained by converting an n-person nonzero sum game to a cooperative game. Given any coalition $S \subset N$, the characteristic function is

$$v(S) = \max_{X \in X_S} \min_{Y \in Y_{N-S}} \sum_{i \in S} E_i(X, Y) = \min_{Y \in Y_{N-S}} \max_{X \in X_S} \sum_{i \in S} E_i(X, Y),$$

where X_S is the set of mixed strategies for the coalition S, Y_{N-S} is the set of mixed strategies for the coalition $N - S$, $E_i(X, Y)$ is the expected payoff to player $i \in S$, and $\sum_{i \in S} E_i(X, Y)$ is the total payoff for each player in $i \in S$ and represents the payoff to the coalition S. The set of pure strategies for coalition S is the set of all combinations of pure strategies for the members of S. This definition of characteristic function satisfies the requirements to be a characteristic function and the property of superadditivity discussed below.

It is important to not be confused about the definition of characteristic function. A characteristic function is **any** function that satisfies $v(\emptyset) = 0, v(N) \geq \sum v(i)$. It does not have to be defined as we did in the examples with the matrices but that is a convenient way of obtaining one that will work.

Here are some additional observations and definitions.

Remarks on Characteristic Functions

1. A very desirable property of a characteristic function is that it satisfy

$$v(S \cup T) \geq v(S) + v(T) \quad \text{for all } S, T \subset N, S \cap T = \emptyset.$$

   This is called **superadditivity**. It says that the benefits of the larger consolidated coalition $S \cup T$ of the two separate coalitions S, T must be at least the total benefits of the individual coalitions S and T. Many results on cooperative games do not need superadditivity, but we will take it as an **axiom that our characteristic functions in all that follows must be superadditive**. With the assumption of superadditivity, the players have the incentive to form and join the grand coalition N.

2. A game is **inessential** if and only if $v(N) = \sum_{i=1}^{n} v(i)$. An **essential** game therefore is one with $v(N) > \sum_{i=1}^{n} v(i)$. The word **inessential** implies that these games are not important. That turns out to be true. They turn out to be easy to analyze, as we will see.

3. Any game with $v(S \cup T) = v(S) + v(T)$, for all $S, T \subset N, S \cap T = \emptyset$, is called an **additive** game. A game is inessential if and only if it is additive.

To see why a characteristic function for an inessential game must be additive, we simply write down the definitions. In fact, let $S, T \subset N, S \cap T = \emptyset$. Then

$$v(N) = \sum_{i=1}^{n} v(i) \qquad \text{(inessential game)}$$

$$= \sum_{i \in S} v(i) + \sum_{i \in T} v(i) + \sum_{i \in N - (S \cup T)} v(i)$$

$$\leq v(S) + v(T) + v(N - (S \cup T)) \qquad \text{(superadditivity)}$$

$$\leq v(S \cup T) + v(N - (S \cup T)) \qquad \text{(superadditivity)}$$

$$\leq v(N) \qquad \text{(superadditivity again)}.$$

Since we now have equality throughout

$$v(S) + v(T) + v(N - (S \cup T)) = v(S \cup T) + v(N - (S \cup T)),$$

and so $v(S) + v(T) = v(S \cup T)$.

We need a basic definition regarding the allocation of rewards to each player. Recall that $v(N)$ represents the reward available if all players cooperate.

Definition 6.2.2 *Let x_i be a real number for each $i = 1, 2, \ldots, n$, with $\sum_{i=1}^{n} x_i \leq v(N)$. A vector $\vec{x} = (x_1, \ldots, x_n)$ is an* **imputation** *if*

- $x_i \geq v(i)$ **(individual rationality)**
- $\sum_{i=1}^{n} x_i = v(N)$ **(group rationality)**

Each x_i represents the share of the value of $v(N)$ received by player i. The imputation $\vec{x}$ is also called a **payoff** **vector** *or an* **allocation**, *and we will use these words interchangeably.*

Remarks

1. It is possible for x_i to be a negative number! That allows us to model coalition members that do not benefit and may be a detriment to a coalition.
2. Individual rationality means that the share received by player i should be at least what he could get on his own. Each player must be individually rational, or else why join the grand coalition?
3. Group rationality means any increase of reward to a player must be matched by a decrease in reward for one or more other players. Why is group rationality reasonable? Well, we know that $v(N) \geq \sum_i x_i \geq \sum_i v(i)$, just by definition. If in fact $\sum_i x_i < v(N)$, then each player could actually receive a bigger share than simply x_i; in fact, one possibility is an additional amount $(v(N) - \sum_i x_i)/n$. This says that the allocation x_i would be rejected by each player, so it must be true that $\sum_i x_i = v(N)$ for any reasonable allocation. Nothing should be left over.
4. Any inessential game, i.e., $v(N) = \sum_{i=1}^{n} v(i)$, has one and only one imputation that satisfies all of the requirements and it is $\vec{x} = (v(1), \dots, v(n))$. The verification is a simple exercise (see the problems). These games are uninteresting because there is no incentive for any of the players to form any sort of coalition and there is no wiggle room in finding a better allocation.

The main objective in cooperative game theory is to determine the imputation that results in a **fair** allocation of the total rewards. Of course, this will depend on the definition of **fair**, as we mentioned earlier. That word is not at all precise. If you change the meaning of **fair** you will change the imputation.

We begin by presenting a way to transform a given characteristic function for a cooperative game to one which is frequently easier to work with. It is called the **(0,1) normalization of the original game**. This is not strictly necessary, but it does simplify the computations in many problems. The normalized game will result in a characteristic function with $v(i) = 0, i = 1, 2 \dots, n$ and $v(N) = 1$. In addition, any two n-person cooperative games may be compared by comparing their normalized characteristic functions. If they are the same, the two games are said to be **strategically equivalent**.

The lemma will show how to make the conversion to $(0, 1)$ normalized.

Lemma 6.2.3 *Any essential game with characteristic function v has a $(0, 1)$ normalization with characteristic function v'. That is, given the characteristic function $v(\cdot)$ set*

$$v'(S) = \frac{v(S) - \sum_{i \in S} v(i)}{v(N) - \sum_{i=1}^{n} v(i)}, \quad S \subseteq N.$$

Then $v'(\cdot)$ is a characteristic function that satisfies $v'(N) = 1, v'(i) = 0, i = 1, 2, \dots, n$.

It should be obvious from the way v' is defined that v' satisfies the conclusions and so is a normalized characteristic function.

Remark: How Does Normalizing Affect Imputations? If we have an imputation for an unnormalized game, what does it become for the normalized game? Conversely, if we have an imputation for the normalized game, how do we get the imputation for the original game?

The set of imputations for the original game is

$$X = \left\{ \vec{x} = (x_1, \ldots, x_n) \mid x_i \geq v(i), \sum_{i=1}^{n} x_i = v(N) \right\}.$$

For the normalized game, indeed, for any game with $v(i) = 0, v(N) = 1$, the set of all possible imputations is given by

$$X' = \left\{ \vec{x}' = (x_1', \ldots, x_n') \mid x_i' \geq 0, \sum_{i=1}^{n} x_i' = 1 \right\}.$$

That is because any imputation $\vec{x}' = (x_1', \ldots, x_n')$ must satisfy $x_i' \geq v'(i) = 0$ and $\sum_i x_i' = v'(N) = 1$.

If $\vec{x}' = (x_1', \ldots, x_n') \in X'$ is an imputation for v' then the imputation for v becomes for $\vec{x} = (x_1, \ldots, x_n) \in X$,

$$\boxed{x_i = x_i' \left(v(N) - \sum_{i=1}^{n} v(i) \right) + v(i), i = 1, 2, \ldots, n.} \tag{6.2.1}$$

This shows that x_i is the amount he could get on his own, $v(i)$, plus the fraction of the left over rewards he is entitled to by his worth to coalitions.

Conversely, if $\vec{x} = (x_1, \ldots, x_n) \in X$ is an imputation for the original game, then $\vec{x}' = (x_1', \ldots, x_n')$ is the imputation for the normalized game, where

$$\boxed{x_i' = \frac{x_i - v(i)}{v(N) - \sum_{i=1}^{n} v(i)}, i = 1, 2, \ldots, n.}$$

We can make the computations simpler by setting

$$c = \frac{1}{v(N) - \sum_{i=1}^{n} v(i)} \quad \text{and} \quad a_i = -c\, v(i), \quad i = 1, 2, \ldots, n,$$

and then

$$x_i = \frac{(x_i' - a_i)}{c} \quad \text{and} \quad x_i' = c\, x_i + a_i, \quad i = 1, 2, \ldots, n.$$

Example 6.4 In the three-person nonzero sum game considered above we found the (unnormalized) characteristic function to be

$$v(1) = 1, \quad v(2) = \frac{1}{4}, \quad v(3) = -1$$

$$v(12) = 3, \quad v(13) = 1, \quad v(23) = 1$$

$$v(123) = 4.$$

To normalize this game we compute the normalized characteristic function by v' as

$$v'(i) = 0$$

$$v'(12) = \frac{v(12) - \sum_{i \in 12} v(i)}{v(123) - \sum_{i=1}^{3} v(i)} = \frac{3 - 1 - 1/4}{4 - 1 - 1/4 + 1} = \frac{7}{15}$$

$$v'(13) = \frac{v(13) - \sum_{i \in 13} v(i)}{v(123) - \sum_{i=1}^{3} v(i)} = \frac{4}{15}$$

$$v'(23) = \frac{v(23) - \sum_{i \in 23} v(i)}{v(123) - \sum_{i=1}^{3} v(i)} = \frac{7}{15}$$

$$v'(123) = 1$$

In the rest of this section we let X denote the set of imputations $\vec{x}$. We look for an allocation $\vec{x} \in X$ as a **solution** to the game. The problem is the definition of the word **solution**. It is as vague as the word **fair**. We are seeking an imputation that, in some sense, is fair and allocates a fair share of the payoff to each player. To get a handle on the idea of **fair**, we introduce the following subset of X.

Definition 6.2.4 *The **reasonable allocation set** of a cooperative game is a set of imputations $R \subset X$ given by*

$$R \equiv \{\vec{x} \in X \mid x_i \le \max_{T \in \Pi^i} (v(T) - v(T - i)), i = 1, 2, \dots, n\},$$

where Π^i is the set of all coalitions for which player i is a member. If $T \in \Pi^i$, then $i \in T \subset N$, and $T - i$ denotes the coalition T without the player i.

In other words, the reasonable set is the set of imputations so that the amount allocated to each player is no greater than the maximum benefit that the player brings to any coalition of which the player is a member. The difference $v(T) - v(T - i)$ is the measure of the rewards for coalition T due to player i. The reasonable set gives us a first way to reduce the size of X and try to focus in on a solution.

If the reasonable set has only one element, which is extremely unlikely for most games, then that is our solution. If there are many elements in R, we need to cut it down further. In fact, we need to cut it down to the **core** imputations, or even further. Here is the definition.

Definition 6.2.5 *Let $S \subset N$ be a coalition and let $\vec{x} \in X$. The **excess** of coalition $S \subset N$ for imputation $\vec{x} \in X$ is defined by*

$$e(S, x) = v(S) - \sum_{i \in S} x_i.$$

It is the amount by which the rewards allocated to the coalition S differs from the benefits associated with S.

*The **core of the game** is*

$$C(0) = \{\vec{x} \in X \mid e(S, \vec{x}) \le 0, \ \forall S \subset N\} = \left\{ \vec{x} \in X \mid v(S) \le \sum_{i \in S} x_i, \ \forall S \subset N \right\}.$$

The core is the set of allocations so that each coalition receives at least the rewards associated with that coalition. The core may be empty.

*The ε-**core**, for $-\infty < \varepsilon < +\infty$, is*

$$C(\varepsilon) = \{\vec{x} \in X \mid e(S, \vec{x}) \le \varepsilon, \; \forall S \subset N, S \neq N, S \neq \emptyset\}.$$

*Let $\varepsilon^1 \in (-\infty, \infty)$ be the first ε for which $C(\varepsilon) \neq \emptyset$. The **least core**, labeled X^1, is $C(\varepsilon^1)$. It is possible for ε^1 to be positive, negative, or zero.*

Since any allocation must satisfy $v(N) = \sum_{i=1}^{n} x_i$, the excess function with the grand coalition $e(N, \vec{x}) = v(N) - \sum_{i=1}^{n} x_i = 0$. The grand coalition is excluded in the requirements for $C(\varepsilon)$ because if N were an eligible coalition, then $e(N, \vec{x}) = 0 \le \varepsilon$, and it would force ε to be nonnegative. That would put too strict a requirement on ε in order for $C(\varepsilon)$ to be nonempty. That is why N is excluded. In fact, we could exclude it in the definition of $C(0)$ because any imputation must satisfy $e(N, \vec{x}) = 0$, and it automatically satisfies the requirement $e(N, \vec{x}) \le 0$. Similarly, $e(\emptyset, \vec{x}) = 0$ and we exclude $S = \emptyset$.

We will use the notation that for a given imputation $\vec{x} = (x_1, \ldots, x_n)$ and a given coalition $S \subset N$

$$\vec{x}(S) = \sum_{i \in S} x_i,$$

the total amount allocated to coalition S. The excess function then becomes $e(S, \vec{x}) = v(S) - \vec{x}(S)$.

Remarks

1. The idea behind the definition of the core is that an imputation $\vec{x}$ is a member of the core if no matter which coalition S is formed, the total payoff given to the members of S, namely, $\vec{x}(S) = \sum_{i \in S} x_i$, must be at least as large as $v(S)$, the maximum possible benefit of forming the coalition. If $e(S, \vec{x}) > 0$, this would say that the maximum possible benefits of joining the coalition S are greater than the total allocation to the members of S using the imputation $\vec{x}$. But then the members of S would not be very happy with $\vec{x}$ since the amount available is not actually allocated. In that sense, if $\vec{x} \in C(0)$, then $e(S, \vec{x}) \le 0$ for every coalition S, and there would be no *a priori* incentive for any coalition to try to use a different imputation. An imputation is in the core of a game if it is acceptable to all coalitions. The excess function $e(S, \vec{x})$ is a **measure of dissatisfaction** of a particular coalition S with the allocation $\vec{x}$. Consequently, $\vec{x}$ is in the core if all coalitions are satisfied with $\vec{x}$. If the core has only one allocation, that is our solution.

2. Likewise, if $\vec{x} \in C(\varepsilon)$, then the measure of dissatisfaction of a coalition with $\vec{x}$ is limited to ε. The size of ε determines the measure of dissatisfaction because $e(S, \vec{x}) \le \varepsilon$.

3. It should be clear, since $C(\varepsilon)$ is just a set of linear inequalities, as ε increases, $C(\varepsilon)$ gets bigger, and as ε decreases, $C(\varepsilon)$ gets smaller. In other words, $\varepsilon < \varepsilon' \implies C(\varepsilon) \subset C(\varepsilon')$. The idea is that we should shrink (or expand if necessary) $C(\varepsilon)$ by adjusting ε until we get one and only one imputation in it, if possible.

4. If $\varepsilon > 0$ we have $C(0) \subset C(\varepsilon)$. If $\varepsilon < 0, C(\varepsilon) \subset C(0)$. If the core of the game $C(0)$ is empty, then there will be a positive $\varepsilon > 0$ such that $C(0) \subset C(\varepsilon) \neq \emptyset$ and the first such $\varepsilon = \varepsilon^1$ gives us the least core $C(\varepsilon^1)$. If $C(0)$ is not empty but contains more than one allocation, then the first $\varepsilon = \varepsilon^1$ which will make $C(\varepsilon^1)$ nonempty will be negative, $\varepsilon^1 < 0$. There will always be some $\varepsilon \in (-\infty, \infty)$ so that $C(\varepsilon) \neq \emptyset$. The least core uses the first such ε. Make a note that $\varepsilon^1 > 0$ means that $C(0) = \emptyset$.

5. We will see shortly that $C(0) \subset R$, every allocation in the core is always in the reasonable set.

6. The definition of **solution** for a cooperative game we are going to use in this section is that an imputation should be a fair allocation if it is the allocation which minimizes the maximum dissatisfaction for all coalitions. This is von Neumann and Morgenstern's idea.

Example 6.5 Let's give an example of a calculation of $C(0)$. Take the three-person game $N = \{1, 2, 3\}$, with characteristic function

$$v(1) = 1, \ v(2) = 2, \ v(3) = 3,$$

$$v(23) = 6, v(13) = 5, v(12) = 4, v(\emptyset) = 0, v(N) = 8.$$

The excess functions for a given imputation $\vec{x} = (x_1, x_2, x_3) \in C(0)$ must satisfy

$$e(1, \vec{x}) = 1 - x_1 \le 0, \ e(2, \vec{x}) = 2 - x_2 \le 0, \ e(3, \vec{x}) = 3 - x_3 \le 0$$

$$e(12, \vec{x}) = 4 - x_1 - x_2 \le 0, \ e(13, \vec{x}) = 5 - x_1 - x_3 \le 0,$$

$$e(23, \vec{x}) = 6 - x_2 - x_3 \le 0,$$

and we must have $x_1 + x_2 + x_3 = 8$. These inequalities imply that $x_1 \ge 1, x_2 \ge 2, x_3 \ge 3$, and

$$x_1 + x_2 \ge 4, x_1 + x_3 \ge 5, x_2 + x_3 \ge 6.$$

If we use some algebra and the substitution $x_3 = 8 - x_1 - x_2$ to solve these inequalities, we see that

$$C(0) = \{(x_1, x_2, 8 - x_1 - x_2) \mid 1 \le x_1 \le 2, 2 \le x_2 \le 3, 4 \le x_1 + x_2 \le 5\}.$$

If we plot this region in the (x_1, x_2) plane, we get the following diagram

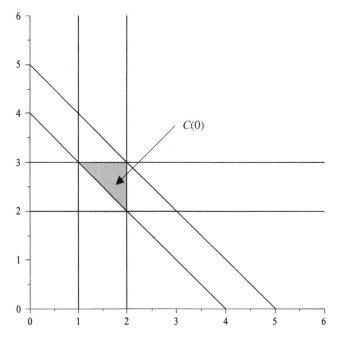

The problem with the core is that it contains more than one allocation. The problem of allocating the rewards $v(N)$ has many reasonable allocations but not a single well-defined allocation. For this we next need to calculate the least core.

We will show that the first ε for which $C(\varepsilon) \ne \emptyset$ is $\varepsilon = \varepsilon^1 = -\frac{1}{3}$. In fact, the least core is the single imputation $C\left(-\frac{1}{3}\right) = \left\{ \left(\frac{5}{3}, \frac{8}{3}, \frac{11}{3} \right) \right\}$. Indeed, the imputations in $C(\varepsilon)$ must satisfy $e(S, \vec{x}) \le \varepsilon$ for all nonempty coalitions $S \subsetneq N$. Written out, these inequalities become

$$1 - \varepsilon \leq x_1 \leq 2 + \varepsilon, \quad 2 - \varepsilon \leq x_2 \leq 3 + \varepsilon,$$
$$4 - \varepsilon \leq x_1 + x_2 \leq 5 + \varepsilon,$$

where we have eliminated $x_3 = 8 - x_1 - x_2$. Adding the inequalities involving only x_1, x_2 we see that $4 - \varepsilon \leq x_1 + x_2 \leq 5 + 2\varepsilon$, which implies that $\varepsilon \geq -\frac{1}{3}$. You can check that this is the first ε for which $C(\varepsilon) \neq \emptyset$. With $\varepsilon = -\frac{1}{3}$, it follows that $x_1 + x_2 = \frac{13}{3}$ and $x_1 \leq \frac{5}{3}, x_2 \leq \frac{8}{3}$. Then,

$$\left(\frac{5}{3} - x_1 \right) + \left(\frac{8}{3} - x_2 \right) = 0,$$

which implies that $x_1 = \frac{5}{3}, x_2 = \frac{8}{3}$ because two nonnegative terms adding to zero must each be zero. This is one technique for finding ε^1 and $C(\varepsilon^1)$.

We now have determined the single allocation $\left\{ \vec{x} = \left(\frac{5}{3}, \frac{8}{3}, \frac{11}{3} \right) \right\} = C \left(-\frac{1}{3} \right) \subset C(0)$ and this is our solution to the cooperative game. Of the $8 = v(N)$ units available for distribution to the members, player 1 should get $\frac{5}{3}$, player 2 should get $\frac{8}{3}$, and player 3 should get $\frac{11}{3}$.

Now we will formalize some properties of the core. We begin by showing that the core must be a subset of the reasonable set.

Lemma 6.2.6 $C(0) \subset R$.

Proof: We may assume the game is in normalized form because we can always transform it to one that is and then work with that one. So $v(N) = 1, v(i) = 0, i = 1, \ldots, n$. Let $\vec{x} \in C(0)$. If $\vec{x} \notin R$ there is some player j such that

$$x_j > \max_{T \in \mathbb{P}^j} v(T) - v(T - j).$$

This means that for every $T \subset N$ with $j \in T, x_j > v(T) - v(T - j)$, and so the amount allocated to player j is larger than the amount of her benefit to any coalition of which she is a member. Take $T = N$ since she certainly is a member of N. Then

$$x_j > v(N) - v(N - j) = 1 - v(N - j).$$

But then, $v(N - j) > 1 - x_j = \sum_{i \neq j} x_i = \vec{x}(N - j)$, and so $e(N - j, \vec{x}) = v(N - j) - \vec{x}(N - j) > 0$, which means $\vec{x} \notin C(0)$. ∎

Example 6.6 In this example we will normalize the given characteristic function, find the reasonable set, and find the core of the game. Finally, we will find the least core, show that it has only one allocation, and then convert back to the unnormalized imputation.

We have the characteristic function in the three-player game from Example 6.3:

$$\bar{v}(1) = 1, \bar{v}(2) = \frac{1}{4}, \bar{v}(3) = -1, \bar{v}(12) = 3, \bar{v}(13) = 1, \bar{v}(23) = 1, \bar{v}(123) = 4.$$

This is an essential game that we normalized in Example 6.4 to obtain the characteristic function that we will use:

$$v(i) = 0, \quad v(123) = 1, \quad v(12) = \frac{7}{15}, \quad v(13) = \frac{4}{15}, \quad v(23) = \frac{7}{15}.$$

The normalization constants are $c = \frac{4}{15}$, and $a_1 = -\frac{4}{15}, a_2 = -\frac{1}{15}$, and $a_3 = \frac{4}{15}$. Since this is now in normalized form, the set of imputations is

$$X = \left\{ \vec{x} = (x_1, x_2, x_3) \mid x_i \geq 0, \sum_{i=1}^{3} x_i = 1 \right\}.$$

The reasonable set is easy to find and here is the result:

$$R = \{ \vec{x} = (x_1, x_2, x_3) \in X \mid x_i \leq \max_{T \in \Pi^i} \{ v(T) - v(T - i) \}, i = 1, 2, 3 \}$$

$$= \left\{ (x_1, x_2, 1 - x_1 - x_2) \mid x_1 \leq \frac{8}{15}, x_2 \leq \frac{11}{15}, \frac{7}{15} \leq x_1 + x_2 \leq 1 \right\}.$$

To see how to get this let's consider

$$x_1 \leq \max_{T \in \Pi^1} v(T) - v(T - 1).$$

The coalitions containing player 1 are $\{1, 12, 13, 123\}$, so we are calculating the maximum of

$$v(1) - v(\emptyset) = 0, \quad v(12) - v(2) = \frac{7}{15}, \quad v(13) - v(3) = \frac{4}{15},$$

$$v(123) - v(23) = 1 - \frac{7}{15} = \frac{8}{15}.$$

Hence $0 \leq x_1 \leq \frac{8}{15}$. Similarly, $0 \leq x_2 \leq \frac{11}{15}$. We could also show $0 \leq x_3 \leq \frac{8}{15}$, but this isn't good enough because we can't ignore $x_1 + x_2 + x_3 = 1$. That is where we use

$$0 \leq 1 - x_1 - x_2 = x_3 \leq \frac{8}{15} \implies \frac{7}{15} \leq x_1 + x_2 \leq 1.$$

Another benefit of replacing x_3 is that now we can draw the reasonable set in (x_1, x_2) space. Figure 6.2 below is a plot of R.

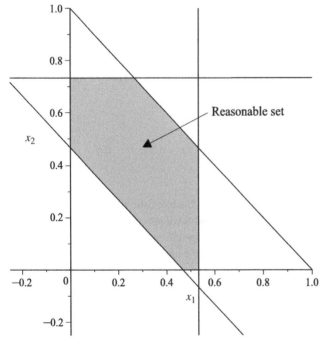

Figure 6.2 The set of reasonable imputations.

You can see from Figure 6.2 that there are lots of reasonable imputations. This is a starting point. We would like to find next the point (or points) in the reasonable set which is acceptable to all coalitions. That is the core of the game:

$$C(0) = \{\vec{x} = (x_1, x_2, 1 - x_1 - x_2) \in X \mid e(S, x) \le 0, \ \forall S \subsetneq N, S \ne \emptyset\}$$
$$= \left\{ x_1 \ge 0, x_2 \ge 0, \frac{7}{15} - x_1 - x_2 \le 0, -\frac{11}{15} + x_2 \le 0, -\frac{8}{15} + x_1 \le 0, \right.$$
$$\left. x_1 + x_2 \le 1 \right\}.$$

Unfortunately, this gives us exactly the same set as the reasonable set, $C(0) = R$ in this example, and that is too big a set.

Now let's calculate the ε-core for any $\varepsilon \in (-\infty, \infty)$. The ε-core is, by definition

$$C(\varepsilon) = \{\vec{x} \in X \mid e(S, \vec{x}) \le \varepsilon, \ \forall S \subset N, S \ne N, S \ne \emptyset\}$$
$$= \left\{ x_1 \ge 0, x_2 \ge 0, \frac{7}{15} - x_1 - x_2 \le \varepsilon, -\frac{11}{15} + x_2 \le \varepsilon, -\frac{8}{15} + x_1 \le \varepsilon, \right.$$
$$\left. -x_1 \le \varepsilon, -x_2 \le \varepsilon, -1 + x_1 + x_2 \le \varepsilon \right\}.$$

We have used the fact that $x_1 + x_2 + x_3 = 1$ to substitute $x_3 = 1 - x_1 - x_2$.

By working with the inequalities in $C(\varepsilon)$, we can find the least core X^1. We verify that the first ε so that $C(\varepsilon) \ne \emptyset$ is $\varepsilon^1 = -\frac{4}{15}$. The procedure is to add the inequality for x_2 with the one for x_1 and then use the first inequality:

$$-\frac{11}{15} + x_2 - \frac{8}{15} + x_1 = -\frac{19}{15} + x_1 + x_2 \le 2\varepsilon,$$

but $x_1 + x_2 \ge \frac{7}{15} - \varepsilon$, so that

$$-\frac{12}{15} - \varepsilon = -\frac{19}{15} + \frac{7}{15} - \varepsilon \le -\frac{19}{15} + x_1 + x_2 \le 2\varepsilon$$

which can be satisfied if and only if $\varepsilon \ge -\frac{4}{15} = \varepsilon^1$.

If we replace ε by $\varepsilon^1 = -\frac{4}{15}$, the **least core** is the set

$$C(\varepsilon^1) = X^1 = \left\{ (x_1, x_2, 1 - x_1 - x_2) \mid \right.$$
$$\left. \frac{4}{15} \ge x_1 \ge 0, \frac{7}{15} \ge x_2 \ge 0, x_1 + x_2 = \frac{11}{15} \right\}$$
$$= \left\{ \left(\frac{4}{15}, \frac{7}{15}, \frac{4}{15} \right) \right\}.$$

The least core is exactly the single point $(x_1 = \frac{4}{15}, x_2 = \frac{7}{15}, x_3 = \frac{4}{15})$ because if $x_1 + x_2 = \frac{11}{15}$, then $\left(\frac{4}{15} - x_1 \right) + \left(\frac{7}{15} - x_2 \right) = 0$ and each of the two terms is nonnegative, and therefore both zero or they couldn't add to zero. This says that there is one and only one allocation that gives all coalitions the smallest dissatisfaction.

If we want the imputation of the original unnormalized game, we use $\bar{x}_i = (x_i - a_i)/c$ and obtain

$$\bar{x}_1 = \frac{\frac{4}{15} + \frac{4}{15}}{\frac{4}{15}} = 2, \quad \bar{x}_2 = \frac{\frac{7}{15} + \frac{1}{15}}{\frac{4}{15}} = 2, \quad \bar{x}_3 = \frac{\frac{4}{15} - \frac{4}{15}}{\frac{4}{15}} = 0.$$

It is an exercise to directly get this imputation for the original game without going through the normalization and solving the inequalities from scratch.

Now here is a surprising (?) conclusion. Remember from Example 6.3 that the payoff of 4 for the grand coalition $\{123\}$ is obtained only by cooperation of the three players in the game in which the payoffs to each player was $(4, -2, 2)$. But remember also that player 3 was in a very weak position according to the characteristic function we derived. The conclusion of the imputation we derived is that players 1 and 2 will split the 4 units available to the grand coalition and player 3 gets nothing. That is still better than what she could get on her own because $\bar{v}(3) = -1$. So that is our fair allocation.

We have already argued that the core $C(0)$ should consist of the good imputations and so would be considered the solution of our game. If in fact $C(0)$ contained exactly one point, then that would be true. Unfortunately, the core may contain many points, as in Example 6.6, or may even be empty. Here is an example of a game with an empty core.

Example 6.7   Suppose that the characteristic function of a three-player game is given by

$$v(123) = 1 = v(12) = v(13) = v(23) \quad \text{and} \quad v(1) = v(2) = v(3) = 0.$$

Since this is already in normalized form, the set of imputations is

$$X = \left\{ \vec{x} = (x_1, x_2, x_3) \mid x_i \geq 0, \sum_{i=1}^{3} x_i = 1 \right\}.$$

To calculate the reasonable set R, we need to find

$$x_i \leq \max_{T \in \Pi^i} \{v(T) - v(T - i)\}, i = 1, 2, 3.$$

Starting with $\Pi^1 = \{1, 12, 13, 123\}$, we calculate

$$v(1) - v(\emptyset) = 0, \ v(12) - v(2) = 1, v(13) - v(3) = 1, v(123) - v(23) = 0,$$

so $x_1 \leq \max \{0, 1, 1, 0\} = 1$. This is true for x_2 as well as x_3. So all we get from this is $R = X$, all the imputations are reasonable.

Next we have

$$C(0) = \left\{ \vec{x} \in X \mid v(S) \leq \sum_{i \in S} x_i, \ \forall S \subsetneq N \right\}.$$

If $\vec{x} \in C(0)$, we calculate

$$e(i, \vec{x}) = v(i) - x_i = -x_i \leq 0, \ e(12, \vec{x}) = 1 - (x_1 + x_2) \leq 0$$

and, in likewise fashion

$$e(13, \vec{x}) = 1 - (x_1 + x_3) \leq 0, \ e(23, \vec{x}) = 1 - (x_2 + x_3) \leq 0.$$

The set of inequalities we have to solve are

$$x_1 + x_2 \geq 1, \ x_1 + x_3 \geq 1, \ x_2 + x_3 \geq 1, \ x_1 + x_2 + x_3 = 1, \ x_i \geq 0.$$

But clearly there is no $\vec{x} \in X$ that can satisfy these inequalities, because it is impossible to have three positive numbers, any two of which have sum at least 1, which can add up to 1. In fact, adding the three inequalities gives us $2 = 2x_1 + 2x_2 + 2x_3 \geq 3$. Therefore $C(0) = \emptyset$.

As mentioned earlier, when $C(0) = \emptyset$, it will be true that the least core $C(\varepsilon)$ must have $\varepsilon > 0$.

In the next proposition we will determine a necessary and sufficient condition for any cooperative game with three players to have a nonempty core.

We take $N = \{1, 2, 3\}$ and a characteristic function in normalized form

$$v(i) = v(\emptyset) = 0, \quad i = 1, 2, 3, \quad v(123) = 1,$$
$$v(12) = a_{12}, \quad v(13) = a_{13}, \quad v(23) = a_{23}.$$

Of course, we have $0 \le a_{ij} \le 1$. We can state the proposition.

Proposition 6.2.7 *For the three-person cooperative game with normalized characteristic function v we have $C(0) \ne \emptyset$ if and only if*

$$a_{12} + a_{13} + a_{23} \le 2.$$

Proof: We have

$$C(0) = \{(x_1, x_2, 1 - x_1 - x_2) | x_i \ge 0, a_{12} \le x_1 + x_2,$$
$$a_{13} \le x_1 + (1 - x_1 - x_2) = 1 - x_2, \quad \text{and} \quad a_{23} \le 1 - x_1\}.$$

So, $x_1 + x_2 \ge a_{12}, x_2 \le 1 - a_{13}$, and $x_1 \le 1 - a_{23}$. Adding the last two inequalities says $x_1 + x_2 \le 2 - a_{23} - a_{13}$ so that with the first inequality $a_{12} \le 2 - a_{13} - a_{23}$. Consequently, if $C(0) \ne \emptyset$, it must be true that $a_{12} + a_{13} + a_{23} \le 2$.

For the other side, if $a_{12} + a_{13} + a_{23} \le 2$, we define the imputation

$$\vec{x} = (x_1, x_2, x_3)$$
$$= \left(\frac{1 - 2a_{23} + a_{13} + a_{12}}{3}, \frac{1 + a_{23} - 2a_{13} + a_{12}}{3}, \frac{1 + a_{23} + a_{13} - 2a_{12}}{3} \right).$$

Then $x_1 + x_2 + x_3 = 1 = v(123)$. Furthermore,

$$v(23) - x_2 - x_3 = a_{23} - x_2 - x_3 = a_{23} - x_2 - (1 - x_1 - x_2)$$
$$= a_{23} - 1 + x_1$$
$$= a_{23} - 1 + \frac{1 - 2a_{23} + a_{13} + a_{12}}{3}$$
$$= \frac{a_{23} + a_{13} + a_{12} - 2}{3} \le 0.$$

Similarly, $v(12) - x_1 - x_2 \le 0$ and $v(13) - x_1 - x_3 \le 0$. Hence $\vec{x} \in C(0)$ and so $C(0) \ne \emptyset$. ∎

Remark: An Automated Way to Determine Whether $C(0) = \emptyset$

With the use of a computer algebra system, there is a simple way of determining whether the core is empty. Consider the linear program:

$$\text{Minimize } z = x_1 + \cdots + x_n$$
$$\text{subject to } v(S) \le \sum_{i \in S} x_i \text{ for every } S \subsetneq N.$$

It is not hard to check that $C(0)$ is not empty if and only if the linear program has a minimum, say, z^*, and $z^* \le v(N)$. If the game is normalized, then we need $z^* \le 1$. When this condition is not satisfied, $C(0) = \emptyset$. For instance, in Example 6.7 the Mathematica commands are

$$obj = x + y + z, \quad cnsts = \{1 - x - z \le 0, 1 - y - z \le 0, 1 - x - y \le 0\}$$
$$\text{Minimize}[\{obj, cnsts\}, \{x, y, z\}].$$

The result is $\left\{ x = \frac{1}{2}, y = \frac{1}{2}, z = \frac{1}{2} \right\}$ as the allocation and $obj = \frac{3}{2}$ as the sum of the allocation components. Since this is a game in which the allocation components must sum to 1, because $v(N) = 1$, we see that the core must be empty.

Remark: Cost Allocation Games

In many games we are interested in allocating costs rather than rewards to members of a coalition. In this case the appropriate characteristic function is $c : 2^N \to \mathbb{R}$ which is subadditive

$$c(S \cup T) \le c(S) + c(T), \quad \forall \, S, T \subset N, S \cap T = \emptyset.$$

This says that it costs less for two coalitions to act together than to operate independently.

Next, a cost allocation is a vector $\vec{x} = (x_1, \dots, x_n)$ which satisfies

- $x_i \ge 0, i = 1, 2, \dots, n, \sum_{i=1}^{n} x_i = c(N)$.
- $x_i \le c(i), \ i = 1, 2, \dots, n$ (**individual rationality**).
- $\sum_{i \in S} x_i \le c(S)$ (**collective rationality**)

We may either work with $c(S)$ directly, or define the ordinary characteristic function

$$v(S) = \sum_{i \in S} c(i) - c(S),$$

which represents the savings that the coalition S can achieve. If we obtain a solution of the game with v, say $\vec{y} = (y_1, \dots, y_n)$, then the cost allocation game has solution $\vec{x} = (x_1, \dots, x_n), x_i c(i) = y_i$, $i = 1, 2, \dots, n$. The basic difference is that each player wants to minimize its cost allocation but maximize its savings allocation.

Problems

6.1 A Stag Hunt game has characteristic function $v(S) = \alpha |S|, S \subset N, v(N) = 1$, where $\frac{1}{n} > \alpha > 0$.
(a) Find the normalized Stag Hunt characteristic function.
(b) Find $C(0)$ using the normalization.

6.2 A customer wants to buy a bolt and a nut for the bolt. There are three players but player 1 owns the bolt and players 2 and 3 each own a nut. A bolt together with a nut is worth 5 but is worthless otherwise. Also, a nut without a bolt is worthless. Define a characteristic function for this game and verify that it is superadditive.

6.3 A river has n pollution producing factories dumping water into the river. Assume that the factory does not have to pay for the water it uses but it may need to expend money to clean the water before it is suitable for use. Assume the cost of a factory to clean polluted water before it can be used is proportional to the number of polluting factories. Let $c =$cost per factory. Assume also that a factory may choose to clean the water it dumps into the river at a cost of b per factory. We assume the inequalities $0 < c < b < nc$.
If a coalition S forms, all of its members could agree to pollute with a payoff of $-|S|(nc)$. The other possibility is all of its members could agree to clean the water and the factories not in the coalition pollute the river, which results in a total payoff to coalition S of

$-|S|(n - |S|)c - |S|b$. Hence, the characteristic function is

$$v(S) = \begin{cases} \max\{-|S|(nc), -|S|(n-|S|)c - |S|b\}, & \text{if } S \subset N; \\ \max\{-n^2c, -nb\}, & \text{if } S = N. \end{cases}$$

Show that $\vec{x} = (-b, -b, \dots, -b) \in C(0)$, which means $C(0) \neq \emptyset$. The allocation in which every factory cleans the water before it is dumped in the river is in the core.

6.4 A small drug research company, labeled 1, has developed a drug. It does not have the resources to get FDA (Food and Drug Administration) approval or to market the drug, so it considers selling the rights to the drug to a big drug company. Drug companies 2 and 3 are interested in buying the rights but only if both companies are involved in order to spread the risks. Suppose that the drug research company wants $1 billion, but will take $100 million if only one of the two big drug companies are involved. The profit to a participating drug company 2 or 3 is $5 billion, which they split. Here is a possible characteristic function with units in billions:

$$v(1) = v(2) = v(3) = 0, v(12) = 0.1, v(13) = 0.1, v(23) = 0, v(123) = 5,$$

because any coalition which doesn't include player 1 will be worth nothing.

(a) Find the normalized characteristic function and find the core using the normalized characteristic function.

(b) The least core using normalized allocation is $C(-\frac{1}{3}) = \{(\frac{1}{3}, \frac{1}{3}, \frac{1}{3})\}$. Find the least core in unnormalized allocations.

6.5 Look back at Example 6.5. Find the normalized characteristic function and the normalized element in the least core.

6.6 Consider the bimatrix game with

$$A = \begin{bmatrix} 4 & 1 \\ 0 & 2 \end{bmatrix} \quad \text{and} \quad B = \begin{bmatrix} 2 & 1 \\ 0 & 4 \end{bmatrix}.$$

(a) Find the characteristic function of this game.

(b) Find the core of the game $C(0)$.

(c) Find the least core.

6.7 Given the characteristic function $v(i) = 0, i = 1, 2, 3, 4$, and

$$v(12) = 4, v(13) = 4, v(14) = 3, v(23) = 6, v(24) = 2, v(34) = 2$$

$$v(123) = 10, v(124) = 7, v(134) = 7, v(234) = 8, v(1234) = 13,$$

find the normalized characteristic function. Given the fair allocation

$$\vec{x} = \left(\frac{1}{4}, \frac{33}{104}, \frac{33}{104}, \frac{3}{26}\right)$$

for the normalized game find the unnormalized allocation.

6.8 Odd Man out is a three player coin toss game in which each player chooses H or T. If all 3 make the same choice, the house pays each player 1; otherwise the odd man out pays the other players 1. Consider this as a 3 player nonzero sum game.

(a) Find the characteristic function for the game.

(b) Find the core.

6.9 Find the characteristic function for the following three–player game. Each player has two strategies, A, B. If player 1 plays A the matrix is

$$\begin{bmatrix} (1,2,1) & (3,0,1) \\ (-1,6,-3) & (3,2,1) \end{bmatrix},$$

while if player 1 plays B the matrix is

$$\begin{bmatrix} (-1,2,4) & (1,0,3) \\ (7,5,4) & (3,2,1) \end{bmatrix}.$$

In each matrix, player 2 is the row player and player 3 is the column player. Next find the normalized characteristic function.

6.10 Derive the least core for the game with

$$v(123) = 1 = v(12) = v(13) = v(23) \quad \text{and} \quad v(1) = v(2) = v(3) = 0.$$

6.11 Given the characteristic function

$$v(1) = 1, v(2) = \frac{1}{4}, v(3) = -1, v(12) = 3, v(13) = -1, v(23) = 1, v(123) = 4,$$

find the least core without normalizing.

6.12 Larger amounts of money invested in things like CDs (certificates of deposit) get a better rate of return. Suppose the rates of return depend on the amount invested as follows:

Invested amount	Rate of return
0–1,000,000	4%
1,000,000–3,000,000	5%
> 3,000,000	5.5%

Three companies are going to form an investment partnership to pool their money. Suppose Company 1 will invest $1,800,000, Company 2, $900,000, and Company 3, $300,000. How should the net amount earned on the total investment be split among the three companies? Define an appropriate characteristic function. Find the core.

6.13 A **constant sum game** is one in which $v(S) + v(N - S) = v(N)$ for all coalitions $S \subset N$. Show that any *essential* constant sum game must have empty core $C(0) = \emptyset$.

6.14 In this problem you will see why inessential games are of no interest. Show that an **inessential** game has one and only one imputation and is given by

$$\vec{x} = (x_1, \ldots, x_n) = (v(1), v(2), \ldots, v(n));$$

that is, each player is allocated exactly the benefit of the one-player coalition.

6.15 A player i is a **dummy** if $v(S) = v(S \cup i)$, for every $S \subset N$ with $i \notin S$. It looks like a dummy contributes nothing. Show that if i is a dummy and $v(i) = 0$, then for any $\vec{x} \in C(0)$, it must be true that $x_i = 0$.

6.16 Show that a vector $\vec{x} = (x_1, x_2, \ldots, x_n)$ is an imputation if and only if there are nonnegative constants $a_i \geq 0, i = 1, 2, \ldots, n$, such that $\sum_{i=1}^{n} a_i = v(N) - \sum_{i=1}^{n} v(i)$, and $x_i = v(i) + a_i$ for each $i = 1, 2, \ldots, n$.

6.17 Let $\delta_i = v(N) - v(N - i)$. Show that $C(0) = \emptyset$ if $\sum_{i=1}^{n} \delta_i < v(N)$.

6.18 Verify the statement: $C(0) \neq \emptyset$ if and only if the linear program

$$\text{Minimize } z = x_1 + \cdots + x_n$$
$$\text{subject to } v(S) \leq \sum_{i \in S} x_i \text{ for every } S \subsetneq N$$

has a finite minimum, say z^*, and $z^* \leq v(N)$.

6.19 Show that in any $n \geq 2$ player cooperative game, each player belongs to exactly 2^{n-1} coalitions.

6.2.1 More on the Core and Least Core

Theorem 6.2.9 formalizes the idea above that when $e(S, \vec{x}) \leq 0$ for all coalitions, then the player should be happy with the imputation $\vec{x}$ and would not want to switch to another one.

One way to describe the fact that one imputation is better than another is the concept of domination.

Definition 6.2.8 *If we have two imputations $\vec{x} \in X, \vec{y} \in X$, and a nonempty coalition $S \subset N$, then $\vec{x}$ **dominates** $\vec{y}$ (for the coalition S) if $x_i > y_i$ for all members $i \in S$, and $\vec{x}(S) = \sum_{i \in S} x_i \leq v(S)$. If $\vec{x}$ dominates $\vec{y}$ for the coalition S, we write $\vec{x} >_S \vec{y}$.*

If $\vec{x}$ dominates $\vec{y}$ for the coalition S, then members of S prefer the allocation $\vec{x}$ to the allocation $\vec{y}$, because they get more $x_i > y_i$, for each $i \in S$, and the coalition S can actually achieve the allocation because $v(S) \geq \sum_{i \in S} x_i$. The next result is another characterization of the core of the game.

Theorem 6.2.9 *The core of a game is the set of all undominated imputations for the game; that is,*

$$C(0) = \{\vec{x} \in X \mid \text{there is no } \vec{z} \in X \text{ and } S \subset N \text{ such that } \vec{z} >_S \vec{x}\}.$$
$$= \left\{ \vec{x} \in X \mid \text{there is no } \vec{z} \in X \text{ and } S \subset N \text{ such that } z_i > x_i, \forall \, i \in S, \text{ and } \sum_{i \in S} z_i \leq v(S) \right\}.$$

Proof: Call the right-hand side the set B. We have to show $C(0) \subset B$ and $B \subset C(0)$.

We may assume that the game is in $(0, 1)$ normalized form.

Let $\vec{x} \in C(0)$ and suppose $\vec{x} \notin B$. Since $\vec{x} \notin B$ that means that $\vec{x}$ must be dominated by another imputation for at least one nonempty coalition $S \subset N$; that is, there is $\vec{y} \in X$ and $S \subset N$ such that $y_i > x_i$ for all $i \in S$ and $v(S) \geq \sum_{i \in S} y_i$. Summing on $i \in S$, this shows

$$v(S) \geq \sum_{i \in S} y_i > \sum_{i \in S} x_i \Longrightarrow e(S, \vec{x}) > 0,$$

contradicting the fact that $\vec{x} \in C(0)$. Therefore $C(0) \subset B$.

Now let $\vec{x} \in B$. If $\vec{x} \notin C(0)$, there is a nonempty coalition $S \subset N$ so that $e(S, \vec{x}) = v(S) - \sum_{i \in S} x_i > 0$. Let

$$\varepsilon = v(S) - \sum_{i \in S} x_i > 0 \quad \text{and} \quad \alpha = 1 - v(S) \geq 0.$$

Let $s = |S|$, the number of players in S, and

$$z_i = \begin{cases} x_i + \dfrac{\varepsilon}{s} & \text{if } i \in S; \\ \dfrac{\alpha}{n-s} & \text{if } i \notin S. \end{cases}$$

If $i \in S$ z_i is an allocation which is x_i plus the (positive) excess equally divided among members of S. If $i \notin S$, z_i allocates the total rewards minus the rewards for S equally among the members not in S.

We will show that $\vec{z} = (z_1, \dots, z_n)$ is an imputation and $\vec{z}$ dominates $\vec{x}$ for the coalition S; that is, that $\vec{z}$ is a better allocation for the members of S than is $\vec{x}$.

First $z_i \geq 0$ and

$$\sum_{i=1}^{n} z_i = \sum_{i \in S} x_i + \sum_{i \in S} \frac{\varepsilon}{s} + \sum_{i \notin N - S} \frac{\alpha}{n-s} = \sum_{i \in S} x_i + \varepsilon + \alpha = v(S) + 1 - v(S) = 1.$$

Therefore $\vec{z}$ is an imputation.

Next, we show $\vec{z}$ is a better imputation than is $\vec{x}$ for the coalition S. If $i \in S$ $z_i = x_i + \varepsilon/s > x_i$ and $\sum_{i \in S} z_i = \sum_{i \in S} x_i + \varepsilon = v(S)$. Therefore $\vec{z}$ dominates $\vec{x}$. But this says $\vec{x} \notin B$ and that is a contradiction. Hence $B \subset C(0)$. ∎

Example 6.8 This example [1] will present a game with an empty core. We will see again that when we calculate the least core $X^1 = C(\varepsilon^1)$, where ε^1 is the first value for which $C(\varepsilon^1) \neq \emptyset$, we will obtain a reasonable fair allocation (and hopefully only one). Recall the fact that $C(0) = \emptyset$ means that when we calculate ε^1 it must be the case that $\varepsilon^1 > 0$ because if $\varepsilon^1 < 0$, by the definition of ε^1 as the first ε making $C(\varepsilon) \neq \emptyset$, we know immediately that $C(0) \neq \emptyset$ because $C(\varepsilon)$ increases as ε gets bigger.

Suppose that Curly has 150 sinks to give away to whomever shows up to take them away. Aggie(1), Maggie(2), and Baggie(3) simultaneously show up with their trucks to take as many of the sinks as their trucks can haul. Aggie can haul 45, Maggie 60, and Baggie 75, for a total of 180, 30 more than the maximum of 150. The wrinkle in this problem is that the sinks are too heavy for any one person to load onto the trucks so they must cooperate in loading the sinks. The question is: How many sinks should be allocated to each person?

Define the characteristic function $v(S)$ as the number of sinks the coalition $S \subset N = \{1, 2, 3\}$ can load. We have $v(i) = 0, i = 1, 2, 3$, since they must cooperate to receive any sinks at all, and

$$v(12) = 105, \quad v(13) = 120, \quad v(23) = 135, \quad v(123) = 150.$$

The set of imputations will be $X = \{(x_1, x_2, x_3) \mid x_i \geq 0, \sum x_i = 150\}$. To see if the core is nonempty, note that the inequalities

$$x_1 + x_2 \geq 105, x_1 + x_3 \geq 120, \text{ and } x_2 + x_3 \geq 135$$

imply that $2(x_1 + x_2 + x_3) = 2(150) = 300 \geq 360$, which is impossible, so $C(0) = \emptyset$.

---

1 Due to Mesterton-Gibbons [27].

A second way to see $C(0) = \emptyset$ is to use Proposition 6.2.7. If we do that, we normalize v to get $v'(12) = \frac{105}{150}, v'(13) = \frac{120}{150}, v'(23) = \frac{135}{150}$. Then, since $v'(12) + v'(13) + v'(23) = \frac{360}{150} > 2$, the proposition tells us that $C(0) = \emptyset$.

Finally, a third way to see $C(0) = \emptyset$ is to use Problem (6.17). We calculate

$$\delta_i = v(N) - v(N - i) \implies$$
$$\delta_1 = 150 - v(23) = 15,$$
$$\delta_2 = 150 - 120 = 30,$$
$$\delta_3 = 150 - 105 = 45.$$

Then $\delta_1 + \delta_2 + \delta_3 = 90 < 150$, and the core is empty.

Since $C(0) = \emptyset$ we know that no matter what allocation we use, there will be some coalition S and some allocation $\vec{z}$ so that $\vec{z} >_S \vec{x}$. For example, if $\vec{x} = (45, 50, 55)$ we could take $S = 23$ and $\vec{z} = (43, 51, 56)$. Since $z_2 > x_2, z_3 > x_3$ and $z_2 + z_3 \leq v(23)$, clearly $\vec{z} >_S \vec{x}$. No matter what allocation you want, it would be dominated by some other allocation for some coalition.

Now that we know the core is empty, the next step is to calculate the least core. We also know that the first ε for which $C(\varepsilon) \neq \emptyset$ must be positive (since otherwise $C(\varepsilon) \subset C(0) = \emptyset$). Begin with the definition:

$$C(\varepsilon) = \{\vec{x} \in X \mid e(S, \vec{x}) \leq \varepsilon, \forall \emptyset \neq S \subsetneq N\}$$
$$= \left\{ \vec{x} \in X \mid v(S) - \sum_{i \in S} x_i \leq \varepsilon \right\}$$
$$= \{\vec{x} \in X \mid 105 \leq x_1 + x_2 + \varepsilon, 120 \leq x_1 + x_3 + \varepsilon,$$
$$135 \leq x_2 + x_3 + \varepsilon, -x_i \leq \varepsilon, x_i \geq 0\}.$$

We know that $x_1 + x_2 + x_3 = 150$, so by replacing $x_3 = 150 - x_1 - x_2$, we obtain as conditions on ε that

$$120 \leq 150 - x_2 + \varepsilon, \ 135 \leq 150 - x_1 + \varepsilon, \ 105 \leq x_1 + x_2 + \varepsilon.$$

We see that $45 \geq x_1 + x_2 - 2\varepsilon \geq 105 - 3\varepsilon$, implying that $\varepsilon \geq 20$. This is in fact the first $\varepsilon^1 = 20$, making $C(\varepsilon) \neq \emptyset$. Using $\varepsilon^1 = 20$, we calculate

$$C(20) = \{(x_1 = 35, x_2 = 50, x_3 = 65)\}.$$

Hence the fair allocation is to let Aggie have 35 sinks, Maggie 50, and Baggie 65 sinks, and they all cooperate.

We conclude that our fair allocation of sinks is as follows:

Player	Truck capacity	Allocation
Aggie	45	35
Maggie	60	50
Baggie	75	65
Total	180	150

Observe that each player in the fair allocation gets 10 less than the capacity of her truck. It seems that this is certainly a reasonably fair way to allocate the sinks; that is, there is an undersupply of 30 sinks so each player will receive $\frac{30}{3} = 10$ less than her truck can haul. You might think of other ways in which you would allocate the sinks (e.g., along the percentage of truck capacity to the total), but the solution here minimizes the maximum dissatisfaction over any other allocation for all coalitions.

We have already noted that $C(0) = \emptyset$. Therefore, no matter what allocation you might use, there will always be a level of dissatisfaction with that allocation for at least one coalition. The least core allocation provides the lowest level of dissatisfaction for all coalitions.

Naturally, a different characteristic function changes the solution as you will see in the exercises.

The least core plays a critical role in solving the problem when $C(0) = \emptyset$ or there is more than one allocation in $C(0)$. Here is the precise description of the first ε making the set $C(\varepsilon)$ nonempty.

Lemma 6.2.10 *Let*

$$\varepsilon^1 = \min_{\vec{x} \in X} \max_{S \subsetneq N} e(S, \vec{x}).$$

Then the least core $X^1 = C(\varepsilon^1) \neq \emptyset$. *If* $\varepsilon < \varepsilon^1$ *then* $C(\varepsilon) = \emptyset$. *If* $\varepsilon > \varepsilon^1$, *then* $C(\varepsilon^1) \subsetneq C(\varepsilon)$.

Proof: Since the set of imputations is compact (=closed and bounded) and $\vec{x} \mapsto \max_S e(S, \vec{x})$ is at least lower semicontinuous, there is an allocation $\vec{x}_0$ so that the minimum in the definition of ε^1 is achieved, namely, $\varepsilon^1 = \max_S e(S, \vec{x}_0) \geq e(S, \vec{x}_0), \forall S \subsetneq N$. This is the very definition of $\vec{x}_0 \in C(\varepsilon^1)$ and so $C(\varepsilon^1) \neq \emptyset$.

On the other hand, if we have a smaller $\varepsilon < \varepsilon^1 = \min_{\vec{x}} \max_{S \subsetneq N}$, then for every allocation $\vec{x} \in X$, we have $\varepsilon < \max_S e(S, \vec{x})$. So, for any allocation, there is at least one coalition $S \subsetneq N$ for which $\varepsilon < e(S, \vec{x})$. This means that for this ε, no matter which allocation is given, $\vec{x} \notin C(\varepsilon)$. Thus, $C(\varepsilon) = \emptyset$. As a result, ε^1 is the first ε so that $C(\varepsilon) \neq \emptyset$. ∎

Remarks

These remarks summarize the ideas behind the use of the least core.

1. For a given grand allocation $\vec{x}$, the coalition S_0 that most objects to $\vec{x}$ is the coalition giving the largest excess and so satisfies

$$e(S_0, \vec{x}) = \max_{S \subsetneq N} e(S, \vec{x}).$$

For each fixed coalition S, the allocation giving the minimum dissatisfaction for that coalition is

$$e(S, \vec{x}_0) = \min_{\vec{x} \in X} e(S, \vec{x}).$$

The value of ε giving the least ε-core is

$$\varepsilon^1 \equiv \min_{\vec{x} \in X} \max_{S \subsetneq N} e(S, \vec{x}).$$

The procedure is to find the coalition giving the largest excess for a given allocation and then find the allocation which minimizes the maximum excess.

2. If $\varepsilon^1 = \min_{\vec{x}} \max_{S \subsetneq N} e(S, \vec{x}) < 0$, then there is at least one allocation $\vec{x}^*$ that satisfies $\max_S e(S, \vec{x}^*) < 0$. That means that $e(S, \vec{x}^*) < 0$ for every coalition $S \subsetneq N$. Every coalition is satisfied with $\vec{x}^*$ because $v(S) < \vec{x}^*(S)$, so that every coalition is allocated at least its maximum rewards. Obviously, this says $\vec{x}^* \in C(0)$ and is undominated.

If $\varepsilon^1 = \min_{\vec{x}} \max_{S \subsetneq N} e(S, \vec{x}) > 0$, then for every allocation $\vec{x} \in X$, $\max_S e(S, \vec{x}) > 0$. Consequently, there is at least one coalition S so that $e(S, \vec{x}) = v(S) - \vec{x}(S) > 0$. For any allocation, there is at least one coalition that will not be happy with it.

3. The excess function $e(S, \vec{x})$ is a measure of dissatisfaction of S with the imputation $\vec{x}$. It makes sense that the best imputation would minimize the largest dissatisfaction over all the coalitions. This leads us to one possible definition of a solution for the n-person cooperative game. An allocation $\vec{x}^* \in X$ is a solution to the cooperative game if

$$\varepsilon^1 = \min_{\vec{x} \in X} \max_S e(S, \vec{x}) = \max_S e(S, \vec{x}^*),$$

so that $\vec{x}^*$ minimizes the maximum excess for any coalition S. When there is only one such allocation $\vec{x}^*$, it is the fair allocation. The problem is that there may be more than one $\vec{x}^*$ providing the minimum, then we still have a problem as to how to choose among them.

Remark: Calculation of the Least Core The point of calculating the ε-core is that the core is not a sufficient set to ultimately solve the problem in the case when the core $C(0)$ is (1) empty or (2) consists of more than one point. In case (2) the issue, of course, is which point should be chosen as the fair allocation. The ε-core seeks to address this issue by shrinking the core at the same rate from each side of the boundary until we reach a single point. We can use Mathematica to do this.

The calculation of the least core is equivalent to the linear programming problem

Minimize z

subject to

$$v(S) - \vec{x}(S) = v(S) - \sum_{i \in S} x_i \leq z, \text{ for all } S \subsetneq N.$$

The characteristic function need not be normalized. So all we really need to do is to formulate the game using characteristic functions, write down the constraints, and plug them into Mathematica. The result will be the first $z = \varepsilon^1$ that makes $C(\varepsilon^1) \neq \emptyset$, as well as an imputation which provides the minimum.

For example, let's suppose we start with the characteristic function

$$v(i) = 0, \ i = 1, 2, 3, \ v(12) = 2, v(23) = 1, v(13) = 0, \ v(123) = \frac{5}{2}.$$

The constraint set is the ε-core

$$C(\varepsilon) = \{\vec{x} = (x_1, x_2, x_3) \mid v(S) - x(S) \leq \varepsilon, \emptyset \neq S \subsetneq N\}$$
$$= \Big\{ -x_i \leq \varepsilon, i = 1, 2, 3, 2 - x_1 - x_2 \leq \varepsilon, 1 - x_2 - x_3 \leq \varepsilon,$$
$$0 - x_1 - x_3 \leq \varepsilon, x_1 + x_2 + x_3 = \frac{5}{2}, x_i \geq 0 \Big\}.$$

We get

$$x_1 = \frac{5}{4}, \ x_2 = 1, \ x_3 = \frac{1}{4}, \ z = -\frac{1}{4}.$$

Hence the first $\varepsilon^1 = z$ for which the ε-core is nonempty is $\varepsilon^1 = -\frac{1}{4}$. Now, Mathematica also gives us the allocation $\vec{x} = \left(\frac{5}{4}, 1, \frac{1}{4} \right)$ which will be in $C\left(-\frac{1}{4} \right)$, but we don't know if that is the **only point** in $C\left(-\frac{1}{4} \right)$.

Figure 6.3 shows the core $C(0)$.

You can even see how the core shrinks to the ε-core using an animation:

Figure 6.4 results from the animation at $z = -0.18367$ with the dark region constituting the core $C(-0.18367)$. You will see that at $z = -\frac{1}{4}$ the dark region becomes the line segment. Hence $C\left(-\frac{1}{4} \right)$ is certainly not empty, but it is also not just one point.

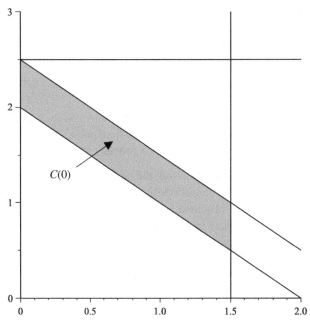

Figure 6.3 Graph of $C(0)$.

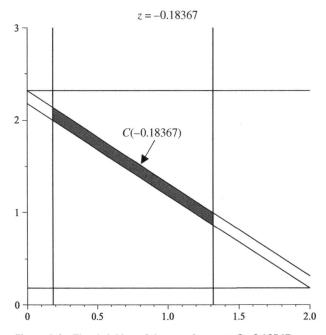

Figure 6.4 The shrinking of the core frozen at $C(-0.18367)$.

In summary, we have solved any cooperative game if the least core contains exactly one point. But when $C(\varepsilon^1) = X^1$ has more than one point, we still have a problem, and that leads us in Section 6.3 to the **nucleolus**.

Problems

6.20 In a three-player game, each player has two strategies A, B. If player 1 plays A, the matrix is

$$\begin{bmatrix} (1, 2, 1) & (3, 0, 1) \\ (-1, 6, -3) & (3, 2, 1) \end{bmatrix},$$

while if player 1 plays B, the matrix is

$$\begin{bmatrix} (-1, 2, 4) & (1, 0, 3) \\ (7, 5, 4) & (3, 2, 1) \end{bmatrix}.$$

In each matrix, player 2 is the row player and player 3 is the column player. The characteristic function is $v(\emptyset) = 0, v(1) = \frac{3}{5}, v(2) = 2, v(3) = 1, v(12) = 5, v(13) = 4, v(23) = 3$, and $v(123) = 16$. Verify that and then find the core and the least core.

6.21 In the sink problem, we took the characteristic function $v(i) = 0, v(12) = 105, v(13) = 120, v(23) = 135$, and $v(123) = 150$, which models the fact that the players get the number of sinks each coalition can load and anyone not in the coalition would get nothing. The truck capacity and least core allocation in this case was

Player	Truck Capacity	$C(20)$ Allocation
Aggie	45	35
Maggie	60	50
Baggie	75	65
Total	180	150

(a) Suppose that each coalition now gets the sinks left over after anyone not in the coalition gets their full truck capacity met first. Curly will help load any one player coalition. What is the characteristic function?

(b) Show that the core of this game is not empty and find the least core.

6.22 A **weighted majority game** has a characteristic function of the form

$$v(S) = \begin{cases} 1, & \text{if } \sum_{i \in S} w_i > q, \\ 0, & \text{otherwise}, \end{cases}$$

where $w_i \geq 0$ are called weights and $q > 0$ is called a quota. Take $q = \frac{1}{2} \sum_{i \in N} w_i$. Suppose that there is one large group with two-fifths of the votes and two equal-sized groups with three-tenths of the vote each.

(a) Find the characteristic function.

(b) Find the core and the least core.

6.23 A classic game is the **garbage game**. Suppose that there are n property owners, each with one bag of garbage that needs to be dumped on somebody's property (one of the n). If n bags of garbage are dumped on a coalition S of property owners, the coalition receives a reward of $-n$. The characteristic function is taken to be the best that the members of a coalition S can do, which is to dump all their garbage on the property of the owners not in S.

(a) Explain why the characteristic function should be $v(S) = -(n - |S|)$, $v(N) = -n$, $v(\emptyset) = 0$, where $|S|$ is the number of members in S.

(b) Show that the core of the game is empty if $n > 2$.

(c) Recall that an imputation $\vec{y}$ dominates an imputation $\vec{x}$ through the coalition S if $e(S, \vec{y}) \geq 0$ and $y_i > x_i$ for each member $i \in S$. Take $n = 4$ in the garbage game. Find a coalition S so that $\vec{y} = (-1.5, -0.5, -1, -1)$ dominates $\vec{x} = (-2, -1, -1, 0)$.

6.24 In the pollution game we have the characteristic function

$$v(S) = \begin{cases} \max\{-|S|(nc), -|S|(n - |S|)c - |S|b\}, & \text{if } S \subset N; \\ \max\{-n^2c, -nb\}, & \text{if } S = N. \end{cases}$$

where $0 < c < b < nc$, and b is the cost of cleaning water before dumping while c is the cost of cleaning the water after dumping. We have seen that $C(0) \neq \emptyset$ since $\vec{x} = (-b, -b, \dots, -b) \in C(0)$. Find the least core if $c = 1, b = 2, n = 3$.

6.3 The Nucleolus

The core $C(0)$ might be empty, but we can find an ε so that $C(\varepsilon)$ is not empty. We can fix the **empty** problem. Even if $C(0)$ is not empty, it may contain more than one point and again we can use $C(\varepsilon)$ to **maybe** shrink the core down to one point or, if $C(0) = \emptyset$, to expand the core until we get it nonempty. The problem is what happens when the least core $C(\varepsilon)$ itself has too many points.

In this section we will address the issue of what to do when the least core $C(\varepsilon)$ contains more than one point. Remember that $e(S, \vec{x}) = v(S) - \sum_{i \in S} x_i = v(S) - \vec{x}(S)$, and the larger the excess, the more unhappy the coalition S is with the allocation $\vec{x}$. So, no matter what, we want the excess to be as small as possible for all coalitions and we want the imputation which achieves that.

In Section 6.2 we saw that we should shrink $C(0)$ to $C(\varepsilon^1)$, so if $C(\varepsilon^1)$ has more than one allocation, why not shrink that also? No reason at all.

Let's begin by working through an example to see how to shrink the ε^1-core.

Example 6.9 Let us take the normalized characteristic function for the three-player game

$$v(12) = \frac{4}{5}, v(13) = \frac{2}{5}, v(23) = \frac{1}{5}, \quad \text{and} \quad v(123) = 1, v(i) = 0, \quad i = 1, 2, 3.$$

Step 1: Calculate the least core. We have the ε-core

$$C(\varepsilon) = \{(x_1, x_2, x_3) \in X \mid e(S, x) \leq \varepsilon, \forall \emptyset \neq S \subsetneq N\}$$

$$= \left\{ (x_1, x_2, 1 - x_1 - x_2) \mid -\varepsilon \leq x_1 \leq \frac{4}{5} + \varepsilon, \right.$$

$$\left. -\varepsilon \leq x_2 \leq \frac{3}{5} + \varepsilon, \quad \frac{4}{5} - \varepsilon \leq x_1 + x_2 \leq 1 + \varepsilon \right\}.$$

We then calculate that the first ε for which $C(\varepsilon) \neq \emptyset$ is $\varepsilon^1 = -\frac{1}{10}$, and then

$$C\left(\varepsilon^1 = -\frac{1}{10}\right) = \left\{ (x_1, x_2, 1 - x_1 - x_2) \mid x_1 \in \left[\frac{2}{5}, \frac{7}{10}\right], \right.$$

$$\left. x_2 \in \left[\frac{1}{5}, \frac{1}{2}\right], \quad x_1 + x_2 = \frac{9}{10} \right\}.$$

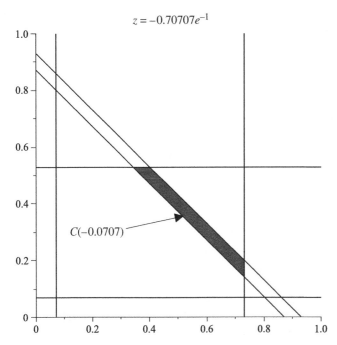

$z = -0.70707e^{-1}$

$C(-0.0707)$

Figure 6.5 $C(-0.07)$: shrinking down to a line segment.

This is a line segment in the (x_1, x_2) plane as we see in Figure 6.5, which is obtained from the Mathematica animation shrinking the core down to the line frozen at $z = -0.07$.

So we have the problem that the least core does not have only one imputation that we would be able to call our solution. What is the fair allocation now? We must shrink the line down somehow. Here is what to do.

Step 2: Calculate the next least core. The idea is, **restricted to the allocations in the first least core**, minimize the maximum excesses over all the allocations in the least core. We must take allocations with $\vec{x} = (x_1, x_2, x_3)$, with $\frac{2}{5} \le x_1 \le \frac{7}{10}$, $\frac{1}{5} \le x_2 \le \frac{1}{2}$, and $x_1 + x_2 = \frac{9}{10}$. This last equality then requires that $x_3 = 1 - x_1 - x_2 = \frac{1}{10}$.

If we take any allocation $\vec{x} \in C(\varepsilon^1)$, we want to calculate the excesses for each coalition:

$$e(1, \vec{x}) = -x_1 \qquad\qquad e(2, \vec{x}) = -x_2$$
$$e(13, \vec{x}) = x_2 - \frac{3}{5} \qquad\qquad e(23, \vec{x}) = x_1 - \frac{4}{5}$$
$$e(12, \vec{x}) = \frac{4}{5} - x_1 - x_2 = -\frac{1}{10} \quad e(3, \vec{x}) = -x_3 = -\frac{1}{10}$$

Since $\vec{x} \in C(\varepsilon^1)$, we know that these are all $\le -\frac{1}{10}$. Observe that the excesses $e(12, \vec{x}) = e(3, \vec{x}) = -\frac{1}{10}$ do not depend on the allocation $\vec{x}$ as long as it is in $C(\varepsilon^1)$. But then, there is nothing we can do about those coalitions by changing the allocation. Those coalitions will always have an excess of $-\frac{1}{10}$ as long as the imputations are in $C(\varepsilon^1)$, and they cannot be reduced. Therefore, we may eliminate those coalitions from further consideration.

Now we set

$$\Sigma^1 \equiv \{S \subsetneq N \mid e(S, \vec{x}) < \varepsilon^1, \text{ for some } \vec{x} \in C(\varepsilon^1)\}.$$

This is a set of coalitions with excesses for some imputation smaller than ε^1. These coalitions can use some imputation that gives a **better** allocation for them, as long as the allocations used are also in $C\left(-\frac{1}{10}\right)$. For our example, we get

$$\Sigma^1 = \{1, 2, 13, 23\}.$$

The coalitions $\{12\}$ and $\{3\}$ are out because the excesses of those coalitions cannot be dropped below $-\frac{1}{10}$ no matter what allocation we use in $C(-\frac{1}{10})$. Their level of dissatisfaction cannot be dropped any further.

Now pick any allocation in $C\left(-\frac{1}{10}\right)$ and calculate the smallest level of dissatisfaction for the coalitions in Σ^1:

$$\varepsilon^2 \equiv \min_{\vec{x} \in X^1} \max_{S \in \Sigma^1} e(S, \vec{x}).$$

The number ε^2 is then the first maximum excess over all allocations in $C\left(-\frac{1}{10}\right)$. It is defined just as is ε^1 except we restrict to the coalitions that can have their dissatisfaction reduced. Finally, set

$$X^2 \equiv \{\vec{x} \in X^1 = C(\varepsilon^1) \mid e(S, \vec{x}) \leq \varepsilon^2, \ \forall S \in \Sigma^1\}.$$

The set X^2 is the subset of allocations from X^1 that are preferred by the coalitions in Σ^1. It plays exactly the same role as the least core $C(\varepsilon^1)$, but now we can only use the coalitions in Σ^1. If X^2 contains exactly one imputation, then that point is our fair allocation, namely, the solution to our problem.

In our example, we now use allocations $\vec{x} \in X^1$ so that $x_1 + x_2 = \frac{9}{10}$, $\frac{2}{5} \leq x_1 \leq \frac{7}{10}$, and $\frac{1}{5} \leq x_2 \leq \frac{1}{2}$. The next least core is

$$C(\varepsilon^2) \equiv X^2 = \{\vec{x} \in X^1 \mid e(1, \vec{x}) \leq \varepsilon^2, e(2, \vec{x}) \leq \varepsilon^2, e(13, \vec{x}) \leq \varepsilon^2, e(23, \vec{x}) \leq \varepsilon^2\}$$

$$= \left\{\vec{x} \in X^1 \mid -x_1 \leq \varepsilon^2, -x_2 \leq \varepsilon^2, x_2 - \frac{3}{5} \leq \varepsilon^2, x_1 - \frac{4}{5} \leq \varepsilon^2\right\}.$$

We need to find the first ε^2 for which X^2 is nonempty. We do this by hand as follows. Since $x_1 + x_2 = \frac{9}{10}$, we get rid of $x_2 = \frac{9}{10} - x_1$. Then

$$-x_1 \leq \varepsilon^2 - x_2 = x_1 - \frac{9}{10} \leq \varepsilon^2 \implies -\varepsilon^2 - \frac{9}{10} \leq \varepsilon^2 \implies \varepsilon^2 \geq -\frac{9}{20}$$

$$\frac{9}{10} - x_1 - \frac{3}{5} \leq \varepsilon^2, \ x_1 - \frac{4}{5} \leq \varepsilon^2 \implies -\frac{5}{10} - \varepsilon^2 \leq \varepsilon^2 \implies \varepsilon^2 \geq -\frac{5}{20},$$

and so on.

The first ε^2 satisfying all the requirements is then $\varepsilon^2 = -\frac{5}{20} = -\frac{1}{4}$. Next, we replace ε^2 by $-\frac{1}{4}$ in the definition of X^2 to get

$$C\left(-\frac{1}{4}\right) \equiv X^2 = \left\{\vec{x} \in X^1 \mid -x_1 \leq -\frac{1}{4}, -x_2 \leq -\frac{1}{4}, x_2 - \frac{3}{5} \leq -\frac{1}{4}, x_1 - \frac{4}{5} \leq -\frac{1}{4}\right\}$$

$$= \left\{\vec{x} \in X^1 \mid \frac{1}{4} \leq x_1 \leq \frac{11}{20}, \frac{1}{4} \leq x_2 \leq \frac{7}{20}, x_1 + x_2 = 18/20\right\}.$$

The last equality gives us

$$0 = \left(\frac{11}{20} - x_1\right) + \left(\frac{7}{20} - x_2\right) \implies x_1 = \frac{11}{20}, \ x_2 = \frac{7}{20},$$

since both terms are nonnegative and cannot add up to zero unless they are each zero. We have found $x_1 = \frac{11}{20}, x_2 = \frac{7}{20}$, and, finally, $x_3 = \frac{2}{20}$. We have our second least core

$$X^2 = \left\{ \left(\frac{11}{20}, \frac{7}{20}, \frac{2}{20} \right) \right\},$$

and X^2 consists of exactly one point. That is our solution to the problem. Notice that for this allocation

$$e(13, \vec{x}) = x_2 - \frac{3}{5} = \frac{7}{20} - 12/20 = -\frac{1}{4}$$

$$e(23, \vec{x}) = x_1 - \frac{4}{5} = 11/20 - \frac{4}{5} = -\frac{1}{4}$$

$$e(1, \vec{x}) = -11/20, \quad \text{and} \quad e(2, x) = -\frac{7}{20},$$

and each of these is a constant smaller than $-\frac{1}{10}$. Because they are all independent of any specific allocation, we know that they cannot be reduced any further by adjusting the imputation. Since X^2 contains only one allocation, no further adjustments are possible in any case. This is the allocation that minimizes the maximum dissatisfaction of all coalitions. In this example X^2 is the **nucleolus** for our problem.

The most difficult part of this procedure is finding ε^1, ε^2, and so on. For instance, we can find $\varepsilon^2 = -\frac{1}{4}$ very easily if we use the Mathematica commands

$$cnsts = \left\{ -x_1 \le z, -x_2 \le z, x_2 - \frac{3}{5} \le z, x_1 - \frac{4}{5} \le z, x_1 + x_2 = \frac{9}{10} \right\}$$

Minimize[{z, cnsts}, {x, y}]

We find that $z = -1/4, x_1 = 11/20, x_2 = 7/20$. You have to be careful of the x_1 and x_2 because these are points providing the minimum but you don't know whether they are the only such points. That is what you must verify.

In general, we would need to continue this procedure if X^2 also contained more than one point. Here are the sequence of steps to take in general until we get down to one point:

1. **Step 0: Initialize.** We start with the set of all possible imputations X and the coalitions excluding N and $\emptyset$:

   $$X^0 \equiv X, \quad \Sigma^0 \equiv \{ S \subsetneq N, S \neq \emptyset \}.$$

2. **Step $k \ge 1$: Successively calculate**
   (a) The minimum of the maximum dissatisfaction

   $$\varepsilon^k \equiv \min_{\vec{x} \in X^{k-1}} \max_{S \in \Sigma^{k-1}} e(S, \vec{x}).$$

   (b) The set of allocations achieving the minimax dissatisfaction

   $$X^k \equiv \{ \vec{x} \in X^{k-1} \mid \varepsilon^k = \min_{\vec{x} \in X^{k-1}} \max_{S \in \Sigma^{k-1}} e(S, \vec{x}) = \max_{S \in \Sigma^{k-1}} e(S, \vec{x}) \}$$

   $$= \{ \vec{x} \in X^{k-1} \mid e(S, \vec{x}) \le \varepsilon^k, \ \forall S \subsetneq \Sigma^{k-1} \}.$$

Alternatively, calculate the first ε^k so that

$$C(\varepsilon) = \{\vec{x} \in X^{k-1} \mid e(S, \vec{x}) \leq \varepsilon, \, \forall S \subsetneq \Sigma^{k-1}\} \neq \emptyset$$

and then $X^k = C(\varepsilon^k)$.

(c) The set of coalitions achieving the minimax dissatisfaction is

$$\Sigma_k = \{S \in \Sigma^{k-1} \mid e(S, \vec{x}) = \varepsilon^k, \, \forall \, \vec{x} \in X^k\}.$$

(2) Delete these coalitions from the previous set

$$\Sigma^k \equiv \Sigma^{k-1} - \Sigma_k.$$

3. **Step: Test if Done.** If $\Sigma^k = \emptyset$ we are done; otherwise set $k = k + 1$ and go to Step (2) with the new k.

When this algorithm stops at, say, $k = m$, then X^m is the **nucleolus** of the core and will satisfy the relationships

$$X^m \subset X^{m-1} \subset \cdots \subset X^1 = C(\varepsilon^1) \subset X^0 = X.$$

Also, $\Sigma^0 \supset \Sigma^1 \supset \Sigma^2 \cdots \Sigma^{m-1} \supset \Sigma^m = \emptyset$. The allocation sets decrease down to a single point, **the nucleolus,** and the unhappiest coalitions decrease down to the empty set. The nucleolus is guaranteed to contain only one allocation $\vec{x}$, and this is the solution of the game. In fact, the following theorem can be proved.[1]

Theorem 6.3.1 *The nucleolus algorithm stops in a finite number of steps $m < \infty$ and for each $k = 1, 2, \ldots, m$, we have*

1. $-\infty < \varepsilon_k < \infty$.
2. $X^k \neq \emptyset$ *are convex, closed, and bounded.*
3. $\Sigma_k \neq \emptyset$ *for $k = 1, 2, \ldots, m - 1$.*
4. $\varepsilon_{k+1} < \varepsilon_k$.

In addition, X^m is a single point, called the **nucleolus of the game**:

$$Nucleolus = X^m = \bigcap_{k=1}^{m} X^k.$$

The nucleolus algorithm stops when all coalitions have been eliminated, but when working this out by hand you don't have to go that far. When you see that X^k is a single point you may stop.

The procedure to find the nucleolus can be formulated as a sequence of linear programs that can be solved using Mathematica.

To begin, set $k = 1$ and calculate the constraint set

$$X^1 = \{\vec{x} \in X \mid e(S, \vec{x}) \leq \varepsilon, \, \forall S \subsetneq N\}.$$

The first ε that makes this nonempty is ε^1, given by

$$\varepsilon^1 = \min_{\vec{x} \in X} \max_{S \in \Sigma^0} e(S, \vec{x}), \quad \Sigma^0 = \{S \mid S \subsetneq N, \emptyset\}.$$

---

1 See, for example, the book by Wang [41].

The first linear programming problem that will yield $\varepsilon^1, X^1, \Sigma^1$ is

> Minimize ε
> subject to $v(S) - \vec{x}(S) \le \varepsilon$, $\vec{x} \in X^0 = X$.

The set of $\vec{x}$ values that provide the minimum in this problem is labeled X^1 (this is the least core). Now we take

$$\Sigma_1 = \{S \in \Sigma^0 \mid e(S, \vec{x}) = \varepsilon^1, \forall \vec{x} \in X^1\},$$

which is the set of coalitions that give excess ε^1 for any allocation in X^1. Getting rid of those gives us the next set of coalitions that we have to deal with, $\Sigma^1 = \Sigma^0 - \Sigma_1$.

The next linear programming problem can now be formulated:

> Minimize ε
> subject to $v(S) - \vec{x}(S) \le \varepsilon$, $\vec{x} \in X^1, S \in \Sigma^1 = \Sigma^0 - \Sigma_1$.

The minimum such ε is ε^2, and we set X^2 to be the set of allocations in X^1 at which $\varepsilon^2 = \max_{S \in \Sigma^1} e(S, \vec{x})$. Then

$$\Sigma_2 = \{S \in \Sigma^1 \mid e(S, \vec{x}) = \varepsilon^2, \forall \vec{x} \in X^2\}.$$

Set $\Sigma^2 = \Sigma^1 - \Sigma_2$ and see if this is empty. If so, we are done; if not, we continue until we get our solution.

Basically, the algorithm says: (1) Calculate the least core; (2) eliminate coalitions whose excesses cannot be reduced any further; (3) calculate the next least core using the remaining coalitions, and so on. When you get down to the last least core, you have the nucleolus.

Example 6.10 Three hospitals, A, B, C, want to have a proton therapy accelerator (PTA) to provide precise radiological cancer therapy. These are very expensive devices because they are subatomic particle accelerators. The hospitals can choose to build their own or build one, centrally located, PTA to which they may refer their patients. The costs for building their own PTA are estimated at $50, 30, 50$, for A, B, C, respectively, in millions of dollars. If A and B cooperate to build a PTA, the total cost will be 60 because of land costs for the location, coordination, and so on. If B and C cooperate, the cost will be 70; if A and C cooperate, the cost will be 110. Because the cost for cooperation between A and C is greater than what it would cost if they built their own, they would decide to build their own, so the cost is still 100 for AC cooperation. Finally, the cost to build one PTA for all three hospitals A, B, C is 105.

In this setup the players want to minimize their costs, but since our previous theory is based on maximizing benefits instead of minimizing, we reformulate the problem by looking at the **amount saved** by each player and for each coalition. The characteristic function is then

> $v(S)$ = total cost if each $i \in S$ builds its own − cost if members in S cooperate.

With A = player 1, B = player 2, C = player 3, we get

$$v(1) = v(2) = v(3) = v(13) = 0, v(12) = 20, v(23) = 10, v(123) = 25.$$

For instance, $v(123) = 50 + 30 + 50 - 105 = 25$. We are looking for the fair allocation of the savings to each hospital that we can then translate back to costs. This game is not in normalized form and need not be.

Once we find the fair allocation of the cost savings game, say $\vec{x} = (x_1, x_2, \ldots, x_n)$, we allocate the costs to each player with the cost allocation vector

$$\vec{y} = (y_1, y_2, \ldots, y_n), \quad y_i = x_i - c(i).$$

The first linear program finds the least core:

Minimize ε

subject to

$$\{-x_i \le \varepsilon, i = 1, 2, 3, -(x_1 + x_3) \le \varepsilon, \ 20 - (x_1 + x_2) \le \varepsilon,$$
$$10 - (x_2 + x_3) \le \varepsilon, \ x_1 + x_2 + x_3 = 25, x_i \ge 0\}.$$

This gives us $\varepsilon^1 = -\frac{5}{2}$. Replacing ε by $\varepsilon^1 = -\frac{5}{2}$ and simplifying, we see that the least core will be the set

$$X^1 = \left\{ (x_1, x_2, 25 - x_1 - x_2) \mid \frac{5}{2} \le x_1 \le \frac{25}{2}, x_1 + x_2 = \frac{45}{2} \right\}.$$

Notice that $x_3 = 25 - \frac{45}{2} = \frac{5}{2}$. Next, we have to calculate

$$\Sigma_1 = \{ S \in \Sigma^0 \mid e(S, \vec{x}) = \varepsilon^1, \forall \vec{x} \in X^1 \}.$$

Calculate the excesses for all the coalitions except $N, \emptyset$, assuming that the allocations are in X^1:

$$e(1, \vec{x}) = v(1) - x_1 = -x_1, \ \ e(2, \vec{x}) = -x_2, \ \ e(3, \vec{x}) = -x_3 = -\frac{5}{2},$$

$$e(12, \vec{x}) = v(12) - x_1 - x_2 = 20 - \frac{45}{2} = -\frac{5}{2},$$

$$e(13, \vec{x}) = v(13) - x_1 - x_3 = 0 - x_1 - \frac{5}{2} = -x_1 - \frac{5}{2},$$

$$e(23, \vec{x}) = v(23) - x_2 - x_3 = 10 - x_2 - \frac{5}{2} = \frac{15}{2} - x_2.$$

Thus the coalitions that give $\varepsilon^1 = -\frac{5}{2}$ (and have no further dependence on any allocation) are $\Sigma_1 = \{12, 3\}$, and so these two coalitions are dropped from consideration in the next step. The remaining coalitions are

$$\Sigma^1 = \Sigma^0 - \{12, 3\} = \{1, 2, 13, 23\}.$$

This is going to lead to the constraint set for the next linear program:

Minimize ε,

subject to $\vec{x} \in X^2$,

$$X^2 = \{ \vec{x} \in X^1 \mid v(S) - \vec{x}(S) \le \varepsilon, \ \forall S \in \Sigma^1 \}$$

$$= \left\{ \left(x_1, x_2, \frac{5}{2} \right) \mid x_1 + x_2 = \frac{45}{2}, -x_1 \le \varepsilon, -x_2 \le \varepsilon - \left(x_1 + \frac{5}{2} \right) \le \varepsilon, \ 10 - \left(x_2 + \frac{5}{2} \right) \le \varepsilon \right\}.$$

We get the solution of this linear program as

$$\varepsilon^2 = -\frac{15}{2}, \ x_1 = \frac{15}{2}, x_2 = 15, \text{ and } x_3 = \frac{5}{2}.$$

Furthermore, it is the one and only solution, so we should be done, but we will continue with the algorithm until we end up with an empty set of coalitions.

Calculate the excesses for all the coalitions excluding $N, \emptyset$, assuming that the allocations are in X^2:

$$e(1, \vec{x}) = -x_1 = -\frac{15}{2}, \ \ e(2, \vec{x}) = -x_2 = -15, \ \ e(3, \vec{x}) = -x_3 = -\frac{5}{2},$$

$$e(12, \vec{x}) = v(12) - x_1 - x_2 = 20 - \frac{45}{2} = -\frac{5}{2},$$

$$e(13, \vec{x}) = v(13) - x_1 - x_3 = 0 - x_1 - \frac{5}{2} = -\frac{15}{2} - \frac{5}{2} = -\frac{20}{2},$$

$$e(23, \vec{x}) = v(23) - x_2 - x_3 = 10 - x_2 - \frac{5}{2} = \frac{15}{2} - x_2 = \frac{15}{2} - 15 = -\frac{15}{2},$$

so that $e(23, \vec{x}) = e(1, \vec{x}) = -\frac{15}{2}$. Now we can get rid of coalitions $\Sigma_2 = \{1, 3, 12, 13, 23\}$ because none of the excesses for those coalitions can be further reduced by changing the allocations. Then

$$\Sigma^2 = \Sigma^1 - \Sigma_2 = \emptyset.$$

We are now done, having followed the algorithm all the way through. We conclude that

$$\text{Nucleolus} = \left\{ \left(\frac{15}{2}, 15, \frac{5}{2} \right) \right\}.$$

Therefore, the fair allocation of savings to hospital A is $\frac{15}{2}$, the savings to B is 15, and the savings to C is $\frac{5}{2}$. Consequently, the costs allocated to each player if all the hospitals cooperate are as follows: A pays $50 - 7.5 = 42.5$, B pays $30 - 15 = 15$, and C pays $50 - 2.5 = 47.5$. Hospital B saves the most and pays the least.

6.3.1 An Exact Nucleolus for Three Player Games

Finding the nucleolus for a game can be a formidable task. It turns out that there is a **formula** derived by Leng and Parlar [22] giving the nucleolus for any 3 person cooperative game of the type we have been considering. Let's get to the results. They are split into two cases: the case when $C(0) = \emptyset$ and the case when $C(0) \neq \emptyset$.

The simplest result is the case when the core of the game is empty. We have seen that an empty core would imply that the least core $X^1 = C(\varepsilon^1)$ must have $\varepsilon^1 > 0$ and X^1 is nonempty. Of course, it might then have more than one point and might have to be cut down even further as we did earlier. The end result of all that work is the following general formula when $C(0) = \emptyset$.

For the characteristic function $v(\cdot)$, the theorems **assume that v is superadditive and one-player coalitions have $v(i) = 0, i = 1, 2, \dots, n$. Further, it is assumed that $v(S) \geq 0$ for all two-player coalitions.** If these assumptions are not satisfied, then we work with the $0 - 1$ normalized game instead.

Theorem 6.3.2 [22] *Suppose we have the three-player cooperative game with players $N = \{1, 2, 3\}$ and characteristic function $v(S), S \subset N$ with $v(i) = 0, i = 1, 2, 3, v(S) \geq 0$. If $C(0) = \emptyset$, the nucleolus allocation is given by*

$$x_i = \frac{v(N) + v(ij) + v(ik) - 2v(jk)}{3}, \quad i, j, k = 1, 2, 3 \text{ and } i \neq j \neq k. \tag{6.3.1}$$

Proof: The core is assumed empty, i.e., the set of individually rational and group rational allocations

$$C(0) = \{\vec{x} = (x_1, x_2, x_3) \mid e(S, \vec{x}) = v(S) - \vec{x}(S) \leq 0, \forall \, S \subset N\} = \emptyset.$$

Claim: At least one of $e(12, \vec{x}), e(13, \vec{x}), e(23, \vec{x})$ must be > 0 for every $\vec{x}$.

Suppose that this is not true. Then for some allocation $\vec{x}_0$ these excesses are ≤ 0. Since $e(i, \vec{x}) = v(i) - x_i = 0 - x_i \leq 0, i = 1, 2, 3$, for any allocation, it must be the case that $\vec{x}_0 \in C(0)$, contradicting the fact that $C(0)$ is assumed empty.

From the claim we may conclude that for every allocation $\vec{x}$

$$\max_{S \subset N} e(S, \vec{x}) = \max \{e(12, \vec{x}), e(13, \vec{x}), e(23, \vec{x})\} = f(\vec{x}) > 0. \tag{6.3.2}$$

We want to choose $\vec{x}$ in order to minimize $f(\vec{x})$.

Claim: The minimum of $f(\vec{x})$ must occur at an allocation $\vec{x}_0$ where $e(12, \vec{x}_0) = e(13, \vec{x}_0) = e(23, \vec{x}_0)$. Suppose the coalition $S = \{12\}$ provides the maximum in (6.3.2), $f(\vec{x}) = e(12, \vec{x})$. We seek to minimize

$$f(\vec{x}) = \max \{v(12) - x_1 - x_2, v(13) - v(N) + x_2, v(23) - v(N) + x_1\} \tag{6.3.3}$$

$$= v(12) - x_1 - x_2 > 0.$$

Now in order to decrease $f(\vec{x})$, we need to increase x_1 or x_2. But if we do that the other two terms in (6.3.3) increase. The most we can increase either x_1 or x_2 is until have at least two equal terms in (6.3.3) and then

$$v(12) - x_1 - x_2 = \max \{v(13) - v(N) + x_2, v(23) - v(N) + x_1\} > 0.$$

Say the larger of the two is $v(13) - v(N) + x_2$. But then we can decrease x_2 and increase x_1 to keep the left side constant, and this can be done until

$$v(13) - v(N) + x_2 = v(23) - v(N) + x_1 = v(12) - x_1 - x_2,$$

but that is exactly what the claim says.

Using the claim we obtain the two equations for two unknowns

$$x_1 + 2x_2 = v(12) - v(13) + v(N), \quad \text{and} \quad 2x_1 + x_2 = v(12) - v(23) + v(N),$$

which are solved to obtain the formulas in the statement of the theorem. ∎

Next, we write down the result in case we begin with a nonempty core. The proof is skipped.[1] The inequalities of the theorem are not easy to check but the implementation is easy using Mathematica.

Theorem 6.3.3 [22] *Let* $\vec{x}^* = (x_1, x_2, x_3)$ *denote the imputation in the nucleolus of a three player cooperative game with nonempty core. Assume that* $v(i) = 0, i = 1, 2, 3, v(S) \geq 0, C(0) \neq \emptyset$. *Then*

1. *If* $i, j = 1, 2, 3$ *are such that* $v(N) \geq 3v(ij), i \neq j \implies x_1 = x_2 = x_3 = \dfrac{v(N)}{3}$.

2. *If* $i, j, k = 1, 2, 3, i \neq j \neq k$ *are such that*

$$3v(ij) \geq v(N) \begin{cases} \geq v(ij) + 2v(ik), & \text{if } i, j, k = 1, 2, 3, i \neq j \neq k; \\ \geq v(ij) + 2v(jk), & \text{if } i, j, k = 1, 2, 3, i \neq j \neq k; \end{cases}$$

Then $x_i = x_j = \dfrac{v(N) + v(ij)}{4}, x_k = \dfrac{v(N) - v(ij)}{2}$.

3. *If* $i, j, k = 1, 2, 3, i \neq j \neq k$ *are such that*

$$v(ij) \geq v(ik) \text{ and } v(N) \begin{cases} \leq v(ij) + 2v(ik), & \text{if } i, j, k = 1, 2, 3, i \neq j \neq k; \\ \geq v(ij) + 2v(jk), & \text{if } i, j, k = 1, 2, 3, i \neq j \neq k; \end{cases}$$

Then $x_i = \dfrac{v(ij) + v(ik)}{2}, x_j = \dfrac{v(N) - v(ik)}{2}, x_k = \dfrac{v(N) - v(ij)}{2}$.

---

1 Refer to the paper by Leng and Parlar [22] for the proof.

4. *If for* $i, j, k = 1, 2, 3, i \neq j \neq k$

$$v(N) + v(ij) \geq 2(v(ik) + v(jk)) \text{ and } v(N) \begin{cases} \leq v(ij) + 2v(ik), \\ \leq v(ij) + 2v(jk), \end{cases}$$

Then

$$x_i = \frac{v(N) + v(ij) + 2(v(ik) - v(jk))}{4},$$

$$x_j = \frac{v(N) + v(ij) + 2(v(jk) - v(ik))}{4},$$

$$x_k = \frac{v(N) - v(ij)}{2}.$$

5. *If for* $i, j, k = 1, 2, 3, i \neq j \neq k$

$$v(N) + v(ij) \leq 2(v(ik) + v(jk))$$

Then

$$x_i = \frac{v(N) + v(ij) + v(ik) - 2v(jk)}{3},$$

$$x_j = \frac{v(N) + v(ij) + v(jk) - 2v(ik)}{3},$$

$$x_k = \frac{v(N) + v(ik) + v(jk) - 2v(ij)}{3}.$$

Of course remembering these formulas is more difficult than deriving the nucleolus from scratch, but the formulas are very helpful in writing computer code to find the nucleolus (see the end of this chapter).

Example 6.11 Here's an application of the formulas.

1. Empty Core. Let $v(12) = 5, v(13) = 6, v(23) = 8, v(123) = 9$. All the others are zero. First, the core is

$$C(0) = \{(x_1, x_2, 9 - x_1 - x_2) \mid x_i \geq 0, 5 \leq x_1 + x_2, x_2 \leq 3, x_1 \leq 1\}$$

which is clearly empty since $5 \leq x_1 + x_2 \leq 4$. So we can use the empty core formulas to get

$$x_1 = \frac{v(123) + v(12) + v(13) - 2v(23)}{3} = \frac{4}{3}$$

Similarly, $x_2 = \frac{10}{3}, x_3 = 9 - x_1 - x_2 = \frac{13}{3}$.

2. Nonempty Core. Take $v(i) = 0, i = 1, 2, 3$, and $v(12) = 1, v(13) = 4, v(23) = 3, v(123) = 6$. Using Proposition 6.2.7 we have for the normalized characteristic function $v'(S) = \frac{v(S)}{6}$,

$$v'(12) + v'(13) + v'(23) = (1 + 4 + 3)/6 \leq 2 \implies C(0) \neq \emptyset.$$

Next, we have to verify one of the sets of conditions in the theorem. Since

$$v(123) = 6 \leq v(13) + 2v(23) = 10,$$

$$v(123) = 6 \geq v(13) + 2v(12) = 6,$$

$$v(13) = 4 \geq v(23) = 3,$$

the third case of the theorem (with $i = 3, j = 1, k = 2$) is satisfied. Using the formulas there, we have

$$x_1 = \frac{[v(123) - v(23)]}{2} = 1.5,$$

$$x_2 = \frac{[v(123) - v(13)]}{2} = 1,$$

$$x_3 = \frac{[v(13) + v(23)]}{2} = 3.5.$$

The next example illustrates the application of cost allocation in a municipal water problem.

Example 6.12 Three cities are to be connected to a water tower at a central location. Label the three cities $1, 2, 3$ and the water tower as 0. The cost to lay pipe connecting location i with location j is denoted as $c_{ij}, i \neq j$. Figure 6.6 contains the data for our problem.

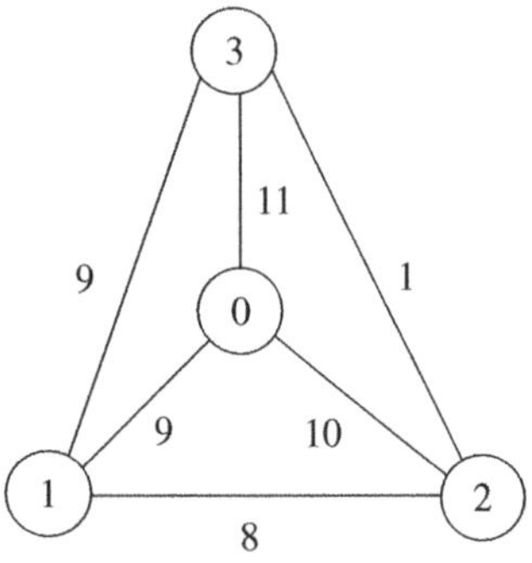

Figure 6.6 Three cities and a water tower.

Coalitions among cities can form for pipe to be laid to the water tower. For example, it is possible for city 1 and city 3 to join up so that the cost to the coalition $\{13\}$ would be the sum of the cost of going from 1 to 3 and then 3 to 0. It may be possible to connect from 1 to 3 to 0 but not from 3 to 1 to 0 depending on land conditions. We have the following costs in which we do not treat the water tower as a player:

$$c(1) = 9, c(2) = 10, c(3) = 11, c(123) = 18, c(12) = 17, c(13) = 18, c(23) = 11.$$

The single-player coalitions correspond to hooking up that city directly to location 0. Converting this to a savings game, we let $c(S)$ be the total cost for coalition S and

$$v(S) = \sum_{i \in S} c(i) - c(S) = \text{amount saved by coalition } S.$$

This is a three-player game and we calculate the characteristic function as

$$v(i) = 0, \ i = 1, 2, 3, \ \ v(123) = 30 - 18 = 12,$$

$$v(12) = 19 - 17 = 2, \ v(13) = 2, \ v(23) = 10.$$

We find the nucleolus of this game using the theorem. First, let's check to see if the core is empty. Using Proposition 6.2.7 we see that for the normalized characteristic function $v'(S) = \frac{v(S)}{12}, v'(12) + v'(13) + v'(23) = (2 + 2 + 10)/12 \leq 2$ and so we have a nonempty core.

Since we have a nonempty core, we must check the cases in the theorem. We check

$$v(123) = 12 \leq v(23) + 2\, v(13) = 10 + 2 \cdot 2 = 14$$

$$v(123) = 12 \leq v(23) + 2\, v(12) = 10 + 2 \cdot 2 = 14$$

$$v(123) + v(23) = 12 + 10 \geq 2(v(13) + v(12)) = 2(2 + 2) = 8$$

This implies

$$x_1 = (v(123) - v(23))/2 = 1,$$

$$x_2 = (v(123) + v(23) - 2(v(13) - v(12)))/4 = \frac{11}{2},$$
$$x_3 = (v(123) + v(23) + 2(v(13) - v(12)))/4 = \frac{11}{2}.$$

The conclusion is that the nucleolus is $\vec{x} = \{(1, \frac{11}{2}, \frac{11}{2})\}$. Our conclusion is that of the 12 units of savings possible for all the cities to cooperate, city 1 only gets 1 unit, while cities 2 and 3 get $\frac{11}{2}$ units each. It makes sense that city 1 would get the least because city 1 brings the least benefit to any coalition compared with 2 and 3.

Example 6.13 shows how cooperative game theory can be applied to important transport problems with applications in many areas including computer science.

Example 6.13 Shortest Path Routing. Let $c(S)$ denote the total mileage driven in order to inspect the machinery of towns in S situated as in Figure 6.7.

Figure 6.7 Shortest path routing

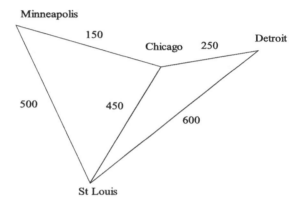

The home headquarters is Chicago. We have the choice of going to each city in a round trip or combining cities. We have (with Minneapolis=1, St Louis=2, Detroit=3)

$$c(1) = 300, c(2) = 900, c(3) = 500, c(12) = 1100, c(13) = 800, c(23) = 1300$$

and $c(123) = 1500$. We are ignoring things like hotels and other expenses but these too can be taken into account. The characteristic function for a cost savings game is $v(S) = \sum_{i \in S} c(i) - c(S)$ and we get

$$v(i) = 0, i = 1, 2, 3, v(12) = 100, v(13) = 0, v(23) = 100, v(123) = 200.$$

Using Proposition 6.2.7 we have that $v'(12) + v'(13) + v'(23) = 1 \leq 2$, where $v'(S) = \frac{v(S)}{200}$ is the normalized characteristic function, and hence we have a nonempty core. Since

$$v(123) = 200 \leq v(12) + 2v(23) = 300,$$
$$v(123) = 200 \geq v(12) + 2v(13) = 100,$$
$$v(23) = 100 \leq v(12) = 100,$$

We use the formulas from the theorem and get the allocation of cost savings

$$x_1 = (v(123) - v(23))/2 = 50,$$
$$x_2 = (v(12) + v(23))/2 = 100,$$
$$x_3 = (v(123) - v(12))/2 = 50.$$

Remark: The examples make it look like finding which set of conditions in the theorem is easy. It is easy but extremely time consuming if you do it by hand. Leng and Parlar [22] wrote a Maple procedure to do the calculations. A Mathematica translation is provided at the end of this chapter.

Problems

6.25 There are 3 types of planes (1, 2, and 3) which use an airport runway. Plane 1 needs a 100-yard runway, 2 needs a 150-yard runway, and 3 needs a 400-yard runway. The cost of maintaining a runway is equal to its length. Suppose this airport has one 400-yard runway used by all 3 types of planes and assume also that only one plane of each type will land at the airport on a given day. We want to know how much of the $400 cost should be allocated to each plane.
   (a) Find the characteristic function.
   (b) Find the least core and show it has only one allocation.

6.26 In a glove game with 3 players, player 1 can supply one left glove and players 2 and 3 can supply one right glove each. The value of a coalition is the number of paired gloves in the coalition.
   (a) Find the characteristic function.
   (b) Find $C(0)$.

6.27 Consider the normalized characteristic function for a three-person game

$$v(12) = \frac{4}{5}, v(13) = \frac{2}{5}, v(23) = \frac{1}{5}.$$

Find the core, the least core X^1, and the next least core X^2. X^2 will be the nucleolus.

6.28 Find the fair allocation in the nucleolus for the three-person characteristic function game with

$$v(i) = 0, i = 1, 2, 3,$$
$$v(12) = v(13) = 2, v(23) = 10,$$
$$v(123) = 12.$$

6.29 In Problem 6.12 we considered the problem in which companies can often get a better cash return if they invest larger amounts. There are 3 companies who may cooperate to invest money in a venture that pays a rate of return as follows:

Invested amount	Rate of return
0–1,000,000	4%
1,000,000–3,000,000	5%
> 3,000,000	5.5%

Suppose Company 1 will invest $1,800,000, Company 2 will invest $900,000, and Company 3 will invest $400,000. This problem was considered earlier but with a different amount of cash for player 3. How should the net amount earned on the total investment be split among the three companies? Define an appropriate characteristic function. Find the nucleolus.

6.30 There are three ambitious computer science students, named 1, 2, and 3, with a lucrative idea and they wish to start a business. None of the students can start a business on their own but any two of them can.
- If 1 and 2 form a business their total salary will be 120,000.
- If 1 and 3 form a business their total salary will be 100,000.
- If 2 and 3 form a business their total salary will be 80,000.
- If all three go into business together they will pay themselves a total salary of 150,000.
  (a) Find an appropriate characteristic function and find the core.
  (b) Suppose the 3 together will pay themselves 140,000. Find the least core.
  (c) Find the nucleolus if they will pay themselves a total of 300,000.

6.31 Four doctors, Moe, Larry, Curly, and Shemp,[1] are in partnership and cover hours for each other. At most one doctor needs to be in the office to see patients at any one time. They advertise their office hours as follows:

(1)	Shemp	12:00–5:00
(2)	Curly	9:00–4:00
(3)	Larry	10:00–4:00
(4)	Moe	2:00–5:00

A coalition is an agreement by one or more doctors as to the times they will really be in the office to see everybody's patients. The characteristic function should be the amount of time saved by a given coalition. Note that $v(i) = 0$, and $v(1234) = 13$ hours. This problem is an example of the use of cooperative game theory in a scheduling problem and is an important application in many disciplines.
(a) Find the characteristic function for all coalitions.
(b) Find X^1, X^2. When you get to X^2, you should have the fair allocation in terms of hours saved.
(c) Find the exact times of the day that each doctor will be in the office according to the allocation you found in X^3.

6.32 Use software to solve the four-person game with unnormalized characteristic function

$$v(i) = 0, i = 1, 2, 3, 4,$$
$$v(12) = v(34) = v(14) = v(23) = 1,$$
$$v(13) = \frac{3}{4}, v(24) = 0,$$
$$v(123) = v(124) = v(134) = v(234) = 1,$$
$$v(1234) = 3.$$

6.33 We have four players involved in a game to minimize their costs. We have transformed such games to savings games by defining $v(S) = \sum_{i \in S} c(i) - c(S)$, the total cost if each player is

---

1 This scheduling problem is adapted from an example in Mesterton-Gibbons [27].

in a one-player coalition, minus the cost involved if players form the coalition S. Find the nucleolus allocation of costs for the four-player game with costs

$$c(1) = 7, \ c(2) = 6, \ c(3) = 4, \ c(4) = 5,$$

$$c(12) = 7.5, \ c(13) = 7, \ c(14) = 7.5, \ c(23) = 6.5, \ c(24) = 6.5, \ c(34) = 5.5,$$

$$c(123) = 7.5, \ c(124) = 8, \ c(134) = 7.5, \ c(234) = 7,$$

$$c(1234) = 8.5, \ c(\emptyset) = 0.$$

6.34 Show that in a cost-saving game if we define the cost-saving characteristic function $v(S) = \sum_{i \in S} c(i) - c(S)$, if $v(S)$ is superadditive, then $c(S)$ is subadditive, and conversely.

6.35 Consider the three-player cooperative game with characteristic function

$$v(1) = 4, v(2) = 1, v(3) = 6, v(12) = 7, v(13) = 13, v(23) = 10, v(123) = 17$$

1. Graph the zero core.
2. Graph the -1 core.
3. Find the least core by solving a system of three equations and three unknowns.

6.36 Consider the three-player cooperative game with characteristic function

$$v(1) = 1, v(2) = 2, v(3) = 2, v(12) = 8, v(13) = 5, v(23) = 6, v(123) = 12$$

1. Graph the zero core.
2. Graph the -1 core.
3. What is the least core in this case? This shouldn't involve doing any math, but it is confusing.
4. What imputation (i.e. solution) would you use?

6.4 The Shapley Value

In an entirely different approach to deciding a fair allocation in a cooperative game, we change the definition of **fair** from minimizing the maximum dissatisfaction to **allocating an amount proportional to the benefit each coalition derives from having a specific player as a member**. The fair allocation would be the one recognizing the amount that each member adds to a coalition. Players who add nothing should receive nothing, and players who are indispensable should be allocated a lot. The question is how do we figure out how much benefit each player adds to a coalition. Lloyd Shapley[1] came up with a way.

Definition 6.4.1 *An allocation $\vec{x} = (x_1, \ldots, x_n)$ is called the* **Shapley value** *if*

$$x_i = \sum_{\{S \in \Pi^i\}} [v(S) - v(S - i)] \frac{(|S| - 1)!(|N| - |S|)!}{|N|!}, \quad i = 1, 2, \ldots, n,$$

---

[1] Lloyd Shapley (1926–2016) was Emeritus Professor of Mathematics at UCLA and a member of the greatest generation. He and John Nash had the same PhD advisor, A. W. Tucker, at Princeton. He has received many honors, not the least of which is a Bronze Star from the US Army for his service in World War II in breaking a Soviet code.

where Π^i is the set of all coalitions $S \subset N$ containing i as a member (i.e., $i \in S$), $|S| =$ number of members in S, and $|N| = n$.

If players join the grand coalition one after the other, we know that there are $n!$ possible sequences. Assume the probability of any particular sequence forming is then $\frac{1}{n!}$. When player i shows up and finds $S - i$ players already there, then $i's$ contribution is to the coalition S is $v(S) - v(S - i)$. The Shapley allocation is each player's expected contribution to any possible sequencing of players joining the grand coalition.

More precisely, fix a player, say, i, and consider the random variable Z_i, which takes its values in the set of all possible coalitions 2^N. Z_i is the coalition S in which i is the last player to join S and $n - |S|$ players join the grand coalition after player i. Diagrammatically, if i joins the coalition S on the way to the formation of the grand coalition, we have

$$\underbrace{(1)(2)\cdots(|S| - 2)(|S| - 1)}_{|S| - 1 \text{ arrive}} \quad \underbrace{(i)}_{i \text{ arrives}} \quad \underbrace{(n - |S|)(n - |S| - 1)\cdots(2)(1)}_{n - |S| \text{ remaining arrive}}$$

Remember that because a characteristic function is superadditive, the players have the incentive to form the grand coalition.

For a given coalition S, by elementary probability, there are $(|S| - 1)!(n - |S|)!$ ways i can join the grand coalition N, joining S first. With this reasoning, we assume that Z_i has the probability distribution

$$Prob(Z_i = S) = \frac{(|S| - 1)!(n - |S|)!}{n!}.$$

We choose this distribution because $|S| - 1$ players have joined before player i, and this can happen in $(|S| - 1)!$ ways; and $n - |S|$ players join after player i, and this can happen in $(n - |S|)!$ ways. The denominator is the total number of ways that the grand coalition can form among n players. Any of the $n!$ permutations has probability $\frac{1}{n!}$ of actually being the way the players join. This distribution assumes that they are **all equally likely**. One could debate this choice of distribution, but this one certainly seems reasonable. Also, see Example 6.4.17 below for a direct example of the calculation of the arrival of a player to a coalition and the consequent benefits.

Therefore, for the fixed player i, the benefit player i brings to the coalition Z_i is $v(Z_i) - v(Z_i - i)$. It seems reasonable that the amount of the total grand coalition benefits that should be allocated to player i should be the expected value of $v(Z_i) - v(Z_i - i)$. This gives,

$$x_i \equiv E[v(Z_i) - v(Z_i - i)] = \sum_{\{S \in \Pi_i\}} [v(S) - v(S - i)]Prob(Z_i = S)$$
$$= \sum_{\{S \in \Pi^i\}} [v(S) - v(S - i)]\frac{(|S| - 1)!(n - |S|)!}{n!}.$$

The **Shapley value (or vector)** is then the allocation $\vec{x} = (x_1, \dots, x_n)$. At the end of this chapter you can find the Mathematica code to find the Shapley value.

Example 6.14 Two players have to divide \$M, but they each get zero if they can't reach an agreement as to how to dived it. What is the fair division? Obviously, without regard to the benefit derived from the money the allocation should be $M/2$ to each player. Let's see if Shapley gives that.

Define $v(1) = v(2) = 0, v(12) = M$. Then

$$x_1 = [v(1) - v(\emptyset)]\frac{0!1!}{1!}2! + [v(12) - v(2)]\frac{1!0!}{2!} = \frac{M}{2}.$$

Note that if we solve this problem using the least core approach, we get

$$C(\varepsilon) = \{(x_1, x_2) \mid e(S, x) \le \varepsilon, \forall S \subsetneq N\}$$
$$= \{(x_1, x_2) \mid -x_1 \le \varepsilon, -x_2 \le \varepsilon, \ x_1 + x_2 = M\}$$
$$= \left\{(x_1, x_2) \mid x_1 = x_2 = \frac{M}{2}\right\}.$$

The reason for this is that if $x_1 \ge -\varepsilon$, $x_2 \ge -\varepsilon$, then adding, we have $-2\varepsilon \le x_1 + x_2 = M$. This implies that $\varepsilon \ge -M/2$ is the restriction on ε. The first ε that makes $C(\varepsilon) \ne \emptyset$ is then $\varepsilon^1 = -M/2$. Then $x_1 \ge M/2$, $x_2 \ge M/2$, and, since they add to M, it must be that $x_1 = x_2 = M/2$. So the least core allocation and the Shapley value coincide in the problem.

Before we continue, let's verify that in fact the Shapley allocation actually satisfies the required conditions.

First, we show that $x_i \ge v(i), i = 1, 2, \ldots, n$, i.e., individual rationality is satisfied. To see this, since $v(S) \ge v(S - i) + v(i)$ by superadditivity, we have

$$x_i = \sum_{\{S \in \Pi^i\}} [v(S) - v(S - i)] \frac{(|S| - 1)!(n - |S|)!}{n!}$$
$$\ge \sum_{\{S \in \Pi^i\}} v(i) \frac{(|S| - 1)!(n - |S|)!}{n!}$$
$$= v(i) \sum_{\{S \in \Pi^i\}} \frac{(|S| - 1)!(n - |S|)!}{n!}$$
$$= v(i).$$

The last equality follows from the fact that

$$\sum_{\{S \in \Pi^i\}} P(Z_i = S) = \sum_{\{S \in \Pi^i\}} \frac{(|S| - 1)!(n - |S|)!}{n!} = 1.$$

Second, we have to show the Shapley allocation satisfies group rationality, i.e., $\sum_{i=1}^{n} x_i = v(N)$. Let's add up the $x_i's$,

$$\sum_{i=1}^{n} x_i = \sum_{i=1}^{n} \sum_{\{S \in \Pi^i\}} [v(S) - v(S - i)] \frac{(|S| - 1)!(n - |S|)!}{n!}.$$

By carefully analyzing the coefficients in the double sum, we can show that group rationality holds. Instead of going through that let's just look at the case $n = 2$. We have

$$\sum_{\{S \in \Pi^1\}} [v(S) - v(S - 1)] \frac{(|S| - 1)!(2 - |S|)!}{2!}$$
$$+ \sum_{\{S \in \Pi^2\}} [v(S) - v(S - 2)] \frac{(|S| - 1)!(2 - |S|)!}{2!}$$
$$= \frac{1}{2}[v(1) - v(\emptyset)] + \frac{1}{2}[v(12) - v(2)] + \frac{1}{2}[v(2) - v(\emptyset)] + \frac{1}{2}[v(12) - v(1)]$$
$$= v(12).$$

Example 6.15 Let's go back to the sink allocation (Example 6.4.8) with Amy, Agnes, and Agatha. Using the core concept, we obtained

Player	Truck Capacity	Allocation
Amy	45	35
Agnes	60	50
Agatha	75	65
Total	180	150

Let's see what we get for the Shapley value. Recall that the characteristic function was $v(i) = 0, v(13) = 120, v(12) = 105, v(23) = 135, v(123) = 150$. In this case, $n = 3, n! = 6$, and for player $i = 1, \Pi^1 = \{1, 12, 13, 123\}$, so

$$
\begin{aligned}
x_1 &= \sum_{\{S \in \Pi^1\}} [v(S) - v(S-1)] \frac{(|S|-1)!(3-|S|)!}{3!} \\
&= [v(1) - v(\emptyset)] \frac{2!0!}{3!} + [v(12) - v(2)] \frac{1!1!}{3!} \\
&\quad + [v(13) - v(3)] \frac{1!1!}{3!} + [v(123) - v(23)] \frac{2!0!}{3!} \\
&= 0 + 105\frac{1}{6} + 120\frac{1}{6} + [150 - 135]\frac{2}{6} \\
&= 42.5.
\end{aligned}
$$

Similarly, with $\Pi^2 = \{2, 12, 23, 123\}$,

$$
\begin{aligned}
x_2 &= \sum_{\{S \in \Pi^2\}} [v(S) - v(S-2)] \frac{(|S|-1)!(3-|S|)!}{6} \\
&= [v(2) - v(\emptyset)]Prob(Z_2 = 2) + [v(12) - v(1)]Prob(Z_2 = 12) \\
&\quad + [v(23) - v(3)]Prob(Z_2 = 23) + [v(123) - v(13)]Prob(Z_2 = 123) \\
&= 0 + 105\frac{1}{6} + 135\frac{1}{6} + [150 - 135]\frac{2}{6} \\
&= 50,
\end{aligned}
$$

and with $\Pi^3 = \{3, 13, 23, 123\}$

$$
\begin{aligned}
x_3 &= \sum_{\{S \in \Pi^3\}} [v(S) - v(S-3)] \frac{(|S|-1)!(3-|S|)!}{6} \\
&= [v(3) - v(\emptyset)]Prob(Z_3 = 3) + [v(13) - v(1)]Prob(Z_3 = 13) \\
&\quad + [v(23) - v(2)]Prob(Z_3 = 23) + [v(123) - v(12)]Prob(Z_3 = 123) \\
&= 0 + 120\frac{1}{6} + 135\frac{1}{6} + [150 - 105]\frac{2}{6} \\
&= 57.5.
\end{aligned}
$$

Consequently, the Shapley vector is $\vec{x} = (42.5, 50, 57.5)$, or, since we can't split sinks $\vec{x} = (43, 50, 57)$, an allocation quite different from the nucleolus solution of $(35, 50, 65)$.

Example 6.16 A typical and interesting fair allocation problem involves a debtor who owes money to more than one creditor. The problem is that the debtor does not have enough money to pay off the entire amount owed to all the creditors. Consequently, the debtor must negotiate with the creditors to reach an agreement about what portion of the assets of the debtor will be paid to each creditor. Usually, but not always, these agreements are imposed by a bankruptcy court.

Let's take a specific problem. Suppose that debtor D has exactly \$100,000 to pay off three creditors A, B, C. Debtor D owes A \$50,000; D owes B \$65,000, and D owes C \$10,000.

Now it is possible for D to split up the $100K (K=1000) on the basis of percentages; that is, the total owed is $125,000 and the amount owed to A is 40 of that, to B is 52%, and to C about 8%, so A would get $40K, B would get $52K and C would get $8K. What if the players could form coalitions to try to get more?

Let's take the characteristic function as follows. The three players are A, B, C and (with amounts in thousands of dollars)

$$v(A) = 25, \ v(B) = 40, \ v(C) = 0,$$
$$v(AB) = 90, \ v(AC) = 35, \ v(BC) = 50, \ v(ABC) = 100.$$

To explain this choice of characteristic function, consider the coalition consisting of just A If we look at the worst that could happen to A, it would be that B and C get paid off completely and A gets what's left, if anything. If B gets $65K and C gets $10K, then $25K is left for A, and so we take $v(A) = 25$. Similarly, if A and B get the entire $100K, then C gets $0. If we consider the coalition AC, they look at the fact that in the worst case B gets paid $65K and they have $35K left as the value of their coalition. This characteristic function is a little pessimistic since it is also possible to consider that AC would be paid $75K and then $v(AC) = 75$. So other characteristic functions are certainly possible. On the other hand, if two creditors can form a coalition to freeze out the third creditor not in the coalition, then the characteristic function we use here is exactly the result.

Now we compute the Shapley values. For player A, we have

$$x_A = [v(A) - v(\emptyset)]\frac{1}{3} + [v(AB) - v(B)]\frac{1}{6}$$
$$+ [v(AC) - v(C)]\frac{1}{6} + [v(ABC) - v(BC)]\frac{1}{3}$$
$$= \frac{25}{3} + \frac{50}{6} + \frac{35}{6} + \frac{50}{3} = \frac{235}{6} = 39.17K.$$

Similarly, for players B and C

$$x_B = [v(B) - v(\emptyset)]\frac{1}{3} + [v(AB) - v(A)]\frac{1}{6}$$
$$+ [v(BC) - v(C)]\frac{1}{6} + [v(ABC) - v(AC)]\frac{1}{3}$$
$$= 40\frac{1}{3} + 65\frac{1}{6} + 50\frac{1}{6} + 65\frac{1}{3} = \frac{325}{6} = 54.17K$$

$$x_C = [v(C) - v(\emptyset)]\frac{1}{3} + [v(BC) - v(B)]\frac{1}{6}$$
$$+ [v(AC) - v(A)]\frac{1}{6} + [v(ABC) - v(AB)]\frac{1}{3}$$
$$= 0\frac{1}{3} + 10\frac{1}{6} + 10\frac{1}{6} + 10\frac{1}{3} = \frac{40}{6} = 6.67K,$$

where again K=1000. The Shapley allocation is $\vec{x} = (39.17, 54.17, 6.67)$ compared to the allocation by percentages of $(40, 52, 8)$. Player B will receive more under the Shapley allocation at the expense of players A and C, who are owed the least.

A reasonable question to ask is why is the Shapley allocation any better than the percentage allocation? After all, the percentage allocation gives a perfectly reasonable answer–or does it? Actually, if players can combine to freeze out another player (especially in a court of law with lawyers involved), or if somehow coalitions among the creditors can form, then the percentage allocation ignores the power of the coalition. The players do not all have the same negotiating power. The Shapley allocation takes that into account through the characteristic function, while the percentage allocation does not.

Example 6.17 A typical problem facing small biotech companies is that they can discover a new drug but they don't have the resources to manufacture and market it. Typically the solution is to team up with a large partner. Let's say that A is the biotech firm and B and C are the candidate big pharmaceutical companies. If B or C teams up with A, the big firm will split $1 billion with A. Here is a possible characteristic function:

$$v(A) = v(B) = v(C) = v(BC) = 0, \quad v(AB) = v(AC) = v(ABC) = 1.$$

We will indicate a quicker way to calculate the Shapley allocation when there are a small number of players. We make a table indicating the value brought to a coalition by each player on the way to formation of the grand coalition:

Order of arrival	Player A	Player B	Player C
ABC	0	1	0
ACB	0	0	1
BAC	1	0	0
BCA	1	0	0
CAB	1	0	0
CBA	1	0	0
Total	4	1	1

The numbers in the table are the amount of value added to a coalition when that player arrives. For example, if A arrives first, no benefit is added; then, if B arrives and joins A, player B has added 1 to the coalition AB; finally, when C arrives (so we have the coalition ABC), C adds no additional value. Since it is assumed in the derivation of the Shapley value that each arrival sequence is **equally likely** we calculate the average benefit brought by each player as the total benefit brought by each player (the sum of each column), divided by the total number of possible orders of arrival. We get

$$x_A = \frac{4}{6}, \ x_B = \frac{1}{6}, \ \text{and} \ x_C = \frac{1}{6}.$$

So company A, the discoverer of the drug should be allocated two-thirds of the $1 billion and the big companies split the remaining third.

It is interesting to compare this with the nucleolus. The core, which will be the nucleolus for this example, is

$$C(0) = \{\vec{x} = (x_A, x_B, 1 - x_A - x_B) \mid -x_A \le 0, -x_B \le 0, -(x_B + 1 - x_A - x_B) \le 0, 1 - x_A - x_B \le 0,$$
$$1 - x_A - (1 - x_A - x_B) \le 0, x_A + x_B \le 1\}$$
$$= \{(x_A, x_B, x_C) = (1, 0, 0)\}.$$

This says that A gets the entire $1 billion and the other companies get nothing. The Shapley value is definitely more realistic.

Shapley vectors can also quickly analyze the winning coalitions in games where winning or losing is all we care about: who do we team up with to win. Here are the definitions.

Definition 6.4.2 *Suppose that we are given a characteristic function v(S) that satisfies that for every S ⊂ N, either v(S) = 0 or v(S) = 1. This is called a* **simple game**. *If v(S) = 1, the coalition S is said to be a* **winning coalition**. *If v(S) = 0, the coalition S is said to be a* **losing coalition**. *Let*

$$W^i = \{S \in \Pi^i \mid v(S) = 1, v(S - i) = 0\},$$

denote the set of coalitions who win with player i and lose without player i. These are the winning coalitions for which player i is **critical**.

Simple games are very important in voting systems. For example, a game in which the coalition with a majority of members wins has $v(S) = 1$, if $|S| > n/2$, as the winning coalitions. Losing coalitions have $|S| \leq n/2$ and $v(S) = 0$. If only unanimous votes win, then $v(N) = 1$ is the only winning coalition. Finally, if there is a certain player who has dictatorial power, say, player 1, then $v(S) = 1$ if $1 \in S$ and $v(S) = 0$ if $1 \notin S$.

In the case of a simple game for player i, we need only consider coalitions $S \in \Pi^i$ for which S is a winning coalition, but $S - i$, that is, S without i, is a losing coalition. We have denoted that set by W^i. We need only consider $S \in W^i$ because $v(S) - v(S - i) = 1$ only when $v(S) = 1$, and $v(S - i) = 0$. In all other cases $v(S) - v(S - i) = 0$. In particular, it is an exercise to show that $v(S) = 0$ implies that $v(S - i) = 0$ and $v(S - i) = 1$ implies that $v(S) = 1$. Hence, the Shapley value for a simple game is

$$x_i = \sum_{\{S \in \Pi^i\}} [v(S) - v(S - i)] \frac{(|S| - 1)!(n - |S|)!}{n!}$$

$$= \sum_{\{S \in W^i\}} \frac{(|S| - 1)!(n - |S|)!}{n!}.$$

The Shapley allocation for player i represents the power that player i holds in a game. You can think of it as the probability that a player can tip the balance. It is also called the **Shapley–Shubik index**.

Example 6.18 A corporation has four stockholders (with 100 total shares) who all vote their own individual shares on any major decision. The majority of shares voted decides an issue. A majority consists of more than 50 shares. Suppose that the holdings of each stockholder are as follows:

Player	1	2	3	4
Shares	10	20	30	40

The winning coalitions, that is, with $v(S) = 1$ are

$$W = \{24, 34, 123, 124, 234, 1234\}.$$

We find the Shapley allocation. For x_1, it follows that $W^1 = \{123\}$ because $S = \{123\}$ is winning but $S - 1 = \{23\}$ is losing. Hence

$$x_1 = \frac{(4 - 3)!(3 - 1)!}{4!} = \frac{1}{12}.$$

Similarly, $W^2 = \{24, 123, 124\}$, and so

$$x_2 = \frac{1}{12} + \frac{1}{12} + \frac{1}{12} = \frac{1}{4}.$$

Also, $x_3 = \frac{1}{4}$ and $x_4 = \frac{5}{12}$. We conclude that the Shapley allocation for this game is $\vec{x} = \left(\frac{1}{12}, \frac{3}{12}, \frac{3}{12}, \frac{5}{12}\right)$. Notice that player 1 has the least power, but players 2 and 3 have the same power even though player 3 controls 10 more shares than does player 2. Player 4 has the most power, but a coalition is still necessary to constitute a majority.

Continue this example, but change the shares as follows

Player	1	2	3	4
Shares	10	30	30	40

Computing the Shapley value as $x_1 = 0, x_2 = x_3 = x_4 = \frac{1}{3}$, we see that player 1 is completely marginalized as she contributes nothing to any coalition. She has no power. In addition, player 4's additional shares over players 2 and 3 provide no advantage over those players since a coalition is essential to carry a majority in any case.

Example 6.19 The United Nations Security Council has 15 members, five of whom are permanent (Russia, Great Britain, France, China, and the United States). These five players have veto power over any resolution. To pass a resolution requires all five permanent member's votes and four of the remaining 10 nonpermanent member's votes. This is a game with fifteen players, and we want to determine the Shapley–Shubik index of their power. We label players 1,2,3,4,5 as the permanent members.

Instead of the natural definition of a winning coalition as one that can pass a resolution, it is easier to use the definition that a winning coalition is one that can **defeat** a resolution. So, for player 1 the winning coalitions are those for which $S \in \Pi^1$, and $v(S) = 1$, $v(S-1) = 0$; that is, player 1, or player 1 and any number up to six nonpermanent members can defeat a resolution, so that the winning coalitions for player 1 is the set

$$W^1 = \{1, 1a, 1ab, 1abc, 1abcd, 1abcde, 1abcdef\},$$

where the letters denote distinct nonpermanent members. The number of distinct two-player winning coalitions which have player 1 as a member is $10 = \binom{10}{1}$,[1] three-player coalitions is $\binom{10}{2}$, four-player coalitions is $\binom{10}{3}$, and so on, and each of these coalitions will have the same coefficients in the Shapley value. So we get

$$x_1 = \frac{0!14!}{15!} + \binom{10}{1}\frac{1!13!}{15!} + \binom{10}{2}\frac{2!12!}{15!} + \cdots + \binom{10}{6}\frac{6!8!}{15!}.$$

We can use Mathematica to give us the result with this command:

```
tot:=0; For[ k=0,k<=6,k++,tot=tot+Binomial[10,k] k! (14-k)!/15!];
Print[tot]
```

We get $x_1 = \frac{421}{2145} = 0.1963$. Obviously, it must also be true that $x_2 = x_3 = x_4 = x_5 = 0.19623$. The five permanent members have a total power of $5 \times 0.19623 = 0.9812$ or 98.12% of the power, while the nonpermanent members have $x_6 = \cdots = x_{15} = 0.0019$ or 0.19% each, or a total power of 1.88%.

Example 6.20 In this example[2] we show how cooperative game theory can determine a fair allocation of taxes to a community. For simplicity, assume that there are only four households and that the community requires expenditures of $100,000. The question is how to allocate the cost of the $100,000 among the four households.

---

1 Recall that the binomial coefficient is $\binom{n}{k} = \frac{n!}{k!(n-k)!}$.
2 Adapted from Aliprantis and Chakrabarti [[1], p. 232].

As in most communities, we consider the wealth of the households as represented by the value of their property. Suppose the wealth of household i is w_i. Our four households have specific wealth values

$$w_1 = 75, \ w_2 = 175, \ w_3 = 200, \ w_4 = 300,$$

again with units in thousands of dollars. In addition, suppose that there is a cap on the amount that each household will have to pay (on the basis of age, income, or some other factors) that is independent of the value of their property value. In our case, we take the maximum amount each of the four households will be required to pay as

$$u_1 = 25, \ u_2 = 30, \ u_3 = 20, \ u_4 = 80.$$

What is the fair allocation of expenses to each household?

Let's consider the general problem first. Define the variables

T	Total costs of community
u_i	Maximum amount i will have to pay
w_i	Net worth of player i
z_i	Amount player i will have to pay
$u_i - z_i$	Surplus of the cap over the assessment

The quantity $u_i - z_i$ is the difference between the maximum amount that household i would ever have to pay and the amount household i actually pays. It represents the amount household i does not have to pay.

We will assume that the total wealth of all the players is greater than T, and that the total amount that the players are willing (or are required) to pay is greater than T, but the total actual amount that the players will have to pay is exactly T:

$$\sum_{i=1}^{n} w_i > T, \ \sum_{i=1}^{n} u_i > T, \ \text{and} \ \sum_{i=1}^{n} z_i = T. \tag{6.4.1}$$

This makes sense because "you can't squeeze blood out of a turnip." Here is the characteristic function we will use:

$$v(S) = \begin{cases} \max \left(\sum_{i \in S} u_i - T, 0 \right) & \text{if } \sum_{i \in S} w_i \geq T; \\ 0 & \text{if } \sum_{i \in S} w_i < T. \end{cases}$$

In other words, $v(S) = 0$ in two cases: (1) if the total wealth of the members of coalition S is less than the total cost, $\sum_{i \in S} w_i < T$, or (2) if the total maximum amount coalition S is required to pay is less than T, $\sum_{i \in S} u_i < T$. If a coalition S cannot afford the expenditure T, then the characteristic function of that coalition is zero.

The Shapley value involves the expression $v(S) - v(S - j)$ in each term. Only the terms with $v(S) - v(S - j) > 0$ need to be considered.

Suppose first that the coalition S and player $j \in S$ **satisfies** $v(S) > 0$ and $v(S - j) > 0$. That means the coalition S and the coalition S without player j can finance the community. We compute

$$v(S) - v(S - j) = \sum_{i \in S} u_i - T - \left(\sum_{i \in S, i \neq j} u_i - T \right) = u_j.$$

Next, suppose that the coalition S can finance the community, but not without j: $v(S) > 0$, $v(S - j) = 0$. Then

$$v(S) - v(S - j) = \sum_{i \in S} u_i - T.$$

Summarizing the cases, we have

$$v(S) - v(S - j) = \begin{cases} u_j & \text{if } v(S) > 0, v(S - j) > 0; \\ \sum_{i \in S} u_i - T & \text{if } v(S) > 0, v(S - j) = 0; \\ 0 & \text{if } v(S) = v(S - j) = 0. \end{cases}$$

Notice that if $j \in S$ and $v(S - j) > 0$, then automatically $v(S) > 0$. We are ready to compute the Shapley allocation. For player $j = 1, \ldots, n$, we have,

$$x_j = \sum_{\{S \in \Pi^j\}} [v(S) - v(S - j)] \frac{(|S| - 1)!(n - |S|)!}{n!}$$

$$= \sum_{\{S | j \in S, v(S-j) > 0\}} u_j \frac{(|S| - 1)!(n - |S|)!}{n!} + \sum_{\{S | j \in S, v(S) > 0, v(S-j) = 0\}} \left(\sum_{i \in S} u_i - T \right) \frac{(|S| - 1)!(n - |S|)!}{n!}$$

By our definition of the characteristic function for this problem, the allocation x_j is the portion of the surplus $\sum_i u_i - T > 0$ that will be assessed to household j. Consequently, the amount player j will be billed is actually $z_j = u_j - x_j$.

For the four-person problem data above, we have $T = 100$, $\sum w_i = 750 > 100$, $\sum_i u_i = 155 > 100$, so all our assumptions in (6.4.1) are verified. Remember that the units are in thousands of dollars. Then we have

$$v(i) = 0, \ v(12) = v(13) = v(23) = 0, \ v(14) = 5, v(24) = 10, v(34) = 0,$$

$$v(123) = 0, v(134) = 25, v(234) = 30, v(124) = 35, v(1234) = 55.$$

For example, $v(134) = \max(u_1 + u_3 + u_4 - 100, 0) = 125 - 100 = 25$. We compute

$$x_1 = \sum_{\{S \ | \ 1 \in S, v(S-1) > 0\}} u_1 \frac{(|S| - 1)!(4 - |S|)!}{4!}$$

$$+ \sum_{\{S \ | \ 1 \in S, v(S) > 0, v(S-1) = 0\}} \left(\sum_{i \in S} u_i - T \right) \frac{(|S| - 1)!(4 - |S|)!}{4!}$$

$$= \frac{2!1!}{4!} \cdot u_1 + \frac{3!0!}{4!} u_1 + \frac{1!2!}{4!} \left([u_1 + u_4 - 100] \right) + \frac{2!1!}{4!} [u_1 + u_3 + u_4 - 100]$$

$$= \frac{65}{6}.$$

The first term comes from coalition $S = 124$; the second term, from coalition $S = 1234$; the third term comes from coalition $S = 14$; and the last term from coalition $S = 134$.

As a result, the amount player 1 will be billed will be $z_1 = u_1 - x_1 = 25 - \frac{65}{6} = \frac{85}{6}$ thousand dollars. In a similar way, we calculate

$$x_2 = \frac{40}{3}, \ x_3 = \frac{25}{3}, \ \text{and } x_4 = \frac{45}{2},$$

so that the actual bill to each player will be

$$z_1 = 25 - \tfrac{65}{6} = 14.167,$$
$$z_2 = 30 - \tfrac{40}{3} = 16.667,$$
$$z_3 = 20 - \tfrac{25}{3} = 11.667,$$
$$z_4 = 80 - \tfrac{45}{2} = 57.5.$$

For comparison purposes it is not too difficult to calculate the nucleolus for this game to be $\left(\tfrac{25}{2}, 15, 10, \tfrac{35}{2}\right)$, so that the payments using the nucleolus will be

$$z_1 = 25 - \tfrac{25}{2} = \tfrac{25}{2} = 12.5,$$
$$z_2 = 30 - 15 = 15,$$
$$z_3 = 20 - 10 = 10,$$
$$z_4 = 80 - \tfrac{35}{2} = \tfrac{125}{2} = 62.5.$$

There is yet a third solution, the straightforward solution that assesses the amount to each player in proportion to each household's maximum payment to the total assessment. For example, $u_1/(\sum_i u_i) = 25/155 = 0.1613$ and so player 1 could be assessed the amount $0.1613 \times 100 = 16.13$.

We end this section by explaining how Shapley actually came up with his fair allocation, because it is very interesting in its own right.

First, we separate players who don't really matter. A player i is a **dummy** if for any coalition S in which $i \notin S$, we have

$$v(S \cup i) = v(S).$$

So dummy player i contributes nothing to any coalition. The players who are not dummies are called the **carriers** of the game. Let's define C = set of carriers.

Shapley now looked at things this way. Given a characteristic function v, we should get an allocation as a function of v, $\varphi(v) = (\varphi_1(v), \ldots, \varphi_n(v))$, where $\varphi_i(v)$ will be the allocation or worth or value of player i in the game, and this function φ should satisfy the following properties:

1. $v(N) = \sum_{i=1}^n \varphi_i(v)$. (Group rationality).
2. If players i and j satisfy $v(S \cup i) = v(S \cup j)$ for any coalition with $i \notin S, j \notin S$, then $\varphi_i(v) = \varphi_j(v)$. If i and j provide the same marginal contribution to any coalition, they should have the same worth.
3. If i is a dummy player, $\varphi_i(v) = 0$. Dummies should be allocated nothing.
4. If v_1 and v_2 are two characteristic functions, then $\varphi(v_1 + v_2) = \varphi(v_1) + \varphi(v_2)$.

The last property is the strongest and most controversial. It essentially says that the allocation to a player using the sum of characteristic functions should be the sum of the allocations corresponding to each characteristic function.

Now these properties, if you agree that they are reasonable, leads to a surprising conclusion. There is one and only one function φ that satisfies them! It is given by $\varphi(v) = (\varphi_1(v), \ldots, \varphi_n(v))$, where

$$\varphi_i(v) = \sum_{\{S \in \Pi^i\}} [v(S) - v(S - i)] \frac{(|S| - 1)!(|N| - |S|)!}{|N|!}, i = 1, 2, \ldots, n.$$

This is the **only** function satisfying the properties, and, sure enough, it is the Shapley value.

Remark: This is the Mathematica code to calculate the Shapley Value given $S = 1, 2, \ldots n$ and $v[A]$, for $A \subset S$. For an example v we take $S = \{1, 2, 3\}$ and

```
v[{1}]=25;
v[{2}]=40;
v[{3}]=0;
v[{1,2}]=90;
v[{1,3}]=35;
v[{2,3}]=50;
v[{1,2,3}]=100;
v[{}]=0;
numPlayers=Length[S];
L=Subsets[S];
K=Length[L];
```

The code is then

```
For[i=1,i<=numPlayers,i++,
{x[i]=0;
For[k=1,k<=K,k++,
If[MemberQ[L[[k]],i] \&\& Length[L[[k]]]>=1,
x[i]=x[i]+(v[L[[k]]]-v[DeleteCases[L[[k]],i]])*
 (Factorial[Length[L[[k]]]-1]*Factorial[numPlayers-Length[L[[k]]]])/
 Factorial[numPlayers]]],Print[x[i]]}]
```

```
The output is
235/6
325/6
20/3
```

Problems

6.37 The formula for the Shapley value is

$$x_i = \sum_{S \in \Pi^i} [v(S) - v(S - i)] \frac{(|S| - 1)!(n - |S|)!}{n!}, \quad i = 1, 2, \ldots, n,$$

where Π^i is the set of all coalitions $S \subset N$ containing i as a member (i.e., $i \in S$). Show that an equivalent formula is

$$x_i = \sum_{T \in 2^N \cdot \Pi^i} [v(T \cup i) - v(T)] \frac{|T|!(n - |T| - 1)!}{n!}, \quad i = 1, 2, \ldots, n.$$

6.38 Let v be a superadditive characteristic function for a simple game. Show that if $v(S) = 0$ and $A \subset S$, then $v(A) = 0$, and if $v(S) = 1$ and $S \subset A$, then $v(A) = 1$.

6.39 Moe, Larry, and Curly have banded together to form a leg, back, and lip waxing business, LBLWax, Inc. The overhead to the business is 40K per year. Each stooge brings in annual business and incurs annual costs as follows: Moe-155K revenue, 40K costs; Larry-160K revenue, 35K costs; Curly-140K revenue, 38K costs. Costs include, wax, flame throwers, antibiotics, etc. Overhead includes rent, secretaries, insurance, etc. At the end of each year they take out all the profit and allocate it to the partners.

(a) Find a characteristic function which can be used to determine how much each waxer should be paid.

(b) Find the nucleolus.

(c) Find the Shapley allocation.

6.40 In Problems 6.12 and 6.29, we considered the problem in which 3 investors can earn a greater rate of return if they invest together and pool their money. Find the Shapley values in both cases discussed in those exercises.

6.41 Three chiropractors, Moe, Larry, and Curly are in partnership and cover hours for each other. At most one chiropractor needs to be in the office to see patients at any one time. They advertise their office hours as follows:

(1)	Moe	$2:00 - 5:00$
(2)	Larry	$11:00 - 4:00$
(3)	Curly	$9:00 - 1:00$

A coalition is an agreement by one or more doctors as to the times they will really be in the office to see everyone's patients. The characteristic function should be the amount of time **saved** by a given coalition.

Find the Shapley value using the table of order of arrival.

6.42 In the figure, three locations are connected as in the network. The numbers on the branches represent the cost to move one unit along that branch. The goal is to connect each location to the main trunk and to each other in the most economical way possible. The benefit of being connected to the main trunk is shown next to each location. Branches can be traversed in both directions. Take the minimal cost path to the main trunk in all cases.

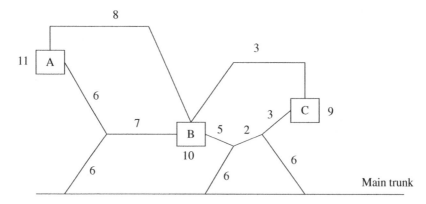

(a) Find an appropriate characteristic function and be sure it is superadditive.

(b) Find the nucleolus.

(c) Find the Shapley value.

6.43 Suppose that the seller of an object (which is worthless to the seller) has two potential buyers who are willing to pay $100 or $130, respectively.
(a) Find the characteristic function and then the core and the Shapley value. Show that the Shapley value is **not** in the core.
(b) Show that the Shapley value is individually and group rational.

6.44 Consider the characteristic function in Example 6.4.16 for the creditor–debtor problem. By writing out the table of the order of arrival of each player versus the benefit the player brings to a coalition when the player arrives as in Example 6.4.17, calculate the Shapley value.

6.45 Find the Shapley allocation for the three-person characteristic function game with

$$v(i) = 0, i = 1, 2, 3,$$
$$v(12) = v(13) = 2, v(23) = 10,$$
$$v(123) = 12.$$

6.46 Once again, we consider the four doctors, Moe(4), Larry(3), Curly(2), and Shemp(1), and their problem to minimize the amount of hours they work as in Problem (6.31). The characteristic function is the number of hours saved by a coalition. We have $v(i) = 0$ and

$$v(12) = 4, v(13) = 4, v(14) = 3, v(23) = 6, v(24) = 2, v(34) = 2,$$
$$v(123) = 10, v(124) = 7, v(134) = 7, v(234) = 8, v(1234) = 13.$$

Find the Shapley allocation.

6.47 **Garbage Game.** Suppose that there are four property owners each with one bag of garbage that needs to be dumped on somebody's property (one of the four). Find the Shapley value for the garbage game with $v(N) = -4$ and $v(S) = -(4 - |S|)$.

6.48 A farmer (player 1) owns some land which he values at $100K. A speculator (player 2) feels that if she buys the land, she can subdivide it into lots and sell the lots for a total of $150K. A home developer (player 3) thinks that he can develop the land and build homes that he can sell. So the land to the developer is worth $160K.
(a) Find the characteristic function and the Shapley allocation.
(b) Compare the Shapley allocation with the nucleolus allocation.

6.49 Find the Shapley allocation for the cost game in Problem 6.33.

6.50 Consider the five player game with $v(S) = 1$ if $1 \in S$ and $|S| \geq 2, v(S) = 1$ if $|S| \geq 4$ and $v(S) = 0$ otherwise. Player 1 is called Mr BIG, and the others are called Peons. Find the Shapley value of this game. (Hint: Use symmetry to simplify.)

6.51 Consider the glove game in which there are 4 players; player 1 can supply one left glove and players 2, 3, and 4 can supply one right glove each.
(a) Find $v(S)$ for this game if the value of a coalition is the number of paired gloves in the coalition.
(b) Find $C(0)$.

(c) What is the Shapley allocation?

(d) Now we change the problem a bit. Suppose there are two players, each with 3 gloves. Player 1 has 2 left hand gloves and 1 right hand glove; player 2 has 2 right hand gloves and 1 left hand glove. They can sell a pair of gloves for 10 dollars. How should they split the proceeds. Determine the nucleolus and the Shapley allocation.

6.52 In Problem 6.25, we considered that there are 3 types of planes (1, 2, and 3) which use an airport. Plane 1 needs a 100-yard runway, 2 needs a 150-yard runway, and 3 needs a 400-yard runway. The cost of maintaining a runway is equal to its length. Suppose this airport has one 400-yard runway used by all 3 types of planes and assume that for the day under study, only one plane of each type lands at the airport. We want to know how much of the $400 cost should be allocated to each plane. We showed that the nucleolus is $C(\varepsilon = -50) = \{x_1 = x_2 = -50, x_3 = -300\}$.

(a) Find the Shapley cost allocation

(b) Calculate the excesses for both the nucleolus and Shapley allocations.

6.53 A river has n pollution producing factories dumping water into the river. Assume that the factory does not have to pay for the water it uses but it may need to expend money to clean the water before it can use it. Assume the cost of a factory to clean polluted water before it can be used is proportional to the number of polluting factories. Let $c =$cost per factory. Assume also that a factory may choose to clean the water it dumps into the river at a cost of b per factory. We take the inequalities $0 < c < b < nc$. The characteristic function is

$$v(S) = \begin{cases} \max\{|S|(-nc), |S|(-(n - |S|)c) - |S|b\}, & \text{if } S \subset N; \\ \max\{-n^2c, -nb\}, & \text{if } S = N. \end{cases}$$

Take $n = 5, b = 3, c = 2$. Find the Shapley allocation.

6.54 Suppose we have a game with characteristic function v which satisfies the property $v(S) + v(N - S) = v(N)$ for all coalitions $S \subset N$. These are called constant sum games.

(a) Show that for a two-person constant sum game the nucleolus and the Shapley value are the same.

(b) Show that the nucleolus and the Shapley value are the same for a three person constant sum game.

(c) Check the result of the preceding part if $v(i) = i, i = 1, 2, 3, v(12) = 5, v(13) = 6, v(23) = 7$, and $v(123) = 8$.

6.55 Three plumbers, Moe Howard(1), Larry Fine(2), and Curly Howard(3), work at the same company and at the same times. Their houses are located as in Figure 6.8 and they would like to carpool to save money.[1] Once they reach the expressway the distance to the company is 12 miles. They drive identical Chevy Corvettes so each has the identical cost of driving to work of 1 dollar per mile. Assume that for any coalition, the route taken is always the shortest distance to pick up the passengers (doubling back may be necessary to pick someone up). It doesn't matter whose car is used. Only the direction from home to work is to be considered. Shemp is not to be considered in this part of the problem.

(a) By considering $v(S) = \sum_{i \in S} c(i) - c(S)$, where $c(S)$ is the cost of driving to work if S forms a carpool, what is the characteristic function of this game.

---

1 Based on a similar problem in Mesterton-Gibbons [27].

Figure 6.8 Moe, Larry, and Curly Carpool to Work

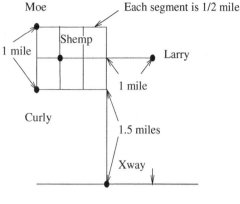

3 and 4 player Car Pool

(b) Find the least core.

(c) Assuming that Moe is the driver, how much should Larry and Curly pay Moe if they all carpool?

(d) Find the Shapley allocation of savings for each player.

(d) The problem is the same as before but now Shemp Howard(4) wants to join the plumbers Howard, Fine, and Howard at the company. Answer all of the questions posed above.

6.56 An alternative to the Shapley–Shubik index is called the **Banzhaf–Coleman index**, which is the imputation $\vec{b} = (b_1, b_2, \dots, b_n)$ defined by

$$b_i = \frac{|W^i|}{|W^1| + |W^2| + \cdots + |W^n|}, i = 1, 2, \dots, n,$$

where W^i is the set of coalitions which win with player i and lose without player i. We also refer to W^i as the coalitions for which player i is **critical** for passing a resolution. It is primarily used in **weighted voting systems** in which player i has w_i votes, and a resolution passes if $\sum_j w_j \geq q$, where q is known as the **quota**.

(a) Why is this a reasonable definition as an index of power?

(b) Consider the 4-player weighted voting system in which a resolution passes if it receives at least 10 votes. Player 1 has 6 votes, player 2 has 5 votes, player 3 has 4 votes, and player 1 has 2 votes.

(c) Find the Shapley–Shubik index.

(d) Find the Banzhaf–Coleman index.

6.57 Suppose a game has 4 players with votes $4, 2, 1, 1$, respectively, for each player $i = 1, 2, 3, 4$. The quota is $q = 5$. Show that the Banzhaf–Coleman index for player 1 is more than twice the index for player 2 even though player 2 has exactly half the votes player 1 does.

6.58 The Senate of the 112th Congress has 100 members of whom 53 are Democrats and 47 are Republicans. Assume that there are 3 types of Democrats and 3 types of Republicans–Liberals, Moderates, Conservatives. Assume that these types vote as a block. For the Democrats, Liberals(1) have 20 votes, Moderates(2) have 25 votes, Conservatives(3) have 8 votes. Also,for Republicans, Liberals(4) have 2 votes, Moderates(5) have 15 votes, and Conservatives(6) have 30 votes. A resolution requires 60 votes to pass.

(a) Find the Shapley–Shubik index and the total power of the Republicans and Democrats.
(b) Find the Banzhaf–Coleman index.
(c) What happens if the Republican Moderate votes becomes 1, while the Republican Conservative votes becomes 44.

6.59 Consider the cooperative game with characteristic function

$$v(1) = 0, v(2) = 3, v(3) = 2, v(12) = 8, v(13) = 7, v(23) = 6, v(123) = 11$$

(a) Graph $C(0)$.
(b) Find the least core, and plot it on your graph from (a)
(c) Find the Shapely value, and plot it on your graph from (a). Comment on why this is interesting.

Bibliographic Notes

The pioneers of the theory of cooperative games include L. Shapley, W. F. Lucas, M. Shubik, and many others, but may go back to Francis Edgeworth in the 1880s. The theory received a huge boost in the publication in 1944 of the seminal work by von Neumann and Morgenstern [31] and then again in a 1953 paper by L. Shapley in which he introduced the Shapley value of a cooperative game.

There are many very good discussions on cooperative game theory, and they are listed in the references. The conversion of any N-person nonzero sum game to characteristic form is due to von Neumann and Morgenstern, which we follow, as presented in references by Wang [41] and Jones [16]. Example 6.8 (due to Mesterton-Gibbons) is called the "log hauling problem" by Mesterton-Gibbons [27] as a realistic example of a game with empty core. It is a good candidate to illustrate how the least core with a positive ε^1 results in a fair allocation in which all the players are dissatisfied with the allocation. The use of Mathematica to plot and animate $C(\varepsilon)$ as ε varies is a great way to show what is happening with the level of dissatisfaction and the resulting allocations. For the concept of the nucleolus, we follow the sequence in Wang's book [41], but this is fairly standard. The allocation of costs and savings games can be found in the early collection of survey papers in reference [23]. Problem 6.31 is a modification of a scheduling problem known as the "antique dealer's problem" in Mesterton-Gibbon's fine book [27], in which we may consider saving games in **time** units rather than monetary units.

The explicit solution of the nucleolus for three-player games was only recently carried out by Leng and Parlar [22] on which the discussion in Section 6.3.1 is based. The Maple software for carrying out the nonempty core calculation can be obtained from Prof. Leng's website. Here we translate that into a Mathematica program.

The Shapley value is popular because it is relatively easy to compute but also because, for the most part, it is based on a commonly accepted set of economic principles. The United Nations Security Council example (Example 6.19) has been widely used as an illustration of quantifying the power of members of a group. The solution given here follows the computation by Jones [16]. Example 6.20

is adapted from an example due to Aliprantis and Chakrabarti [1] and gives an efficient way to compute the Shapley allocation of expenses to multiple users of a resource and taking into account the ability to pay and requirement to meet the expenditures. Similarly, Problem 6.55 is a cost-saving game which arises in everyday life due to Mesterton-Gibbons [27].

We have only scratched the surface of the theory of cooperative games. Refer to the previously mentioned references and the books by Gintis [9], Rasmussen [35], and especially the book by Osborne [32], for many more examples and further study of cooperative games.

7

Bargaining

A good rule to remember for life is that when it comes to plastic surgery and sushi, never be attracted by a bargain.

– Graham Norton

While money doesn't buy love, it puts you in a great bargaining position.

– Christopher Marlowe

Man is an animal that makes bargains: no other animal does this–no dog exchanges bones with another.

–Adam Smith

Bargaining is the third stage of the grief process.

–David Levithan

7.1   Introduction

Bargaining theory is a branch of game theory that studies how two or more parties can reach an agreement on how to divide a limited resource. The theory assumes that the parties are rational and self-interested and that they want to maximize their own gain from the bargaining process.

There are many different bargaining models, but they all share some common features. First, each party has a **threat point**, which is the outcome they would receive if the bargaining process failed and they had to go their separate ways. Second, each party has a **reserve value**, which is the minimum outcome they are willing to accept in order to reach an agreement.

The bargaining process begins with each party making an offer. The parties then continue to make offers and counteroffers until they reach an agreement or one of the parties decides to walk away. The outcome of the bargaining process will depend on the parties' threat points, reserve values, and the bargaining skills of the negotiators.

Bargaining theory has been used to study a wide range of phenomena, including wage negotiations, labor disputes, and international relations. The theory can be used to predict the outcome of bargaining processes and to develop strategies for negotiators.

Game Theory: An Introduction, Third Edition. E. N. Barron.
© 2024 John Wiley & Sons, Inc. Published 2024 by John Wiley & Sons, Inc.

Here are some of the key concepts in bargaining theory:

- **Threat point**: The outcome a party would receive if the bargaining process failed and they had to go their separate ways.
- **Reserve value**: The minimum outcome a party is willing to accept in order to reach an agreement.
- **Asymmetric bargaining**: A bargaining situation in which the parties have different threat points or reservation values.
- **Bargaining power**: The ability of a party to influence the outcome of a bargaining process.
- **Bargaining tactics**: The strategies that negotiators use to reach an agreement.

Let us start with a simple example to illustrate the benefits of bargaining and cooperation. Consider the symmetric two-player nonzero-sum game with bimatrix

	II_1	II_2
I_1	$(2,1)$	$(-1,-1)$
I_2	$(-1,-1)$	$(1,2)$

You can easily check that there are three Nash equilibria given by $X_1 = (0,1) = Y_1, X_2 = (1,0) = Y_2$, and $X_3 = (\frac{3}{5}, \frac{2}{5}), Y_3 = (\frac{2}{5}, \frac{3}{5})$. Now consider Figure 7.1.

The points represent the possible pairs of payoffs to each player $(E_1(x,y), E_2(x,y))$ given by

$$E_1(x,y) = (x, 1-x)A \begin{bmatrix} y \\ 1-y \end{bmatrix}, \qquad E_2(x,y) = (x, 1-x)B \begin{bmatrix} y \\ 1-y \end{bmatrix}.$$

The horizontal axis (abscissa) is the payoff to player I, and the vertical axis (ordinate) is the payoff to player II. Any point in the parabolic region is achievable for some $0 \leq x \leq 1, 0 \leq y \leq 1$.

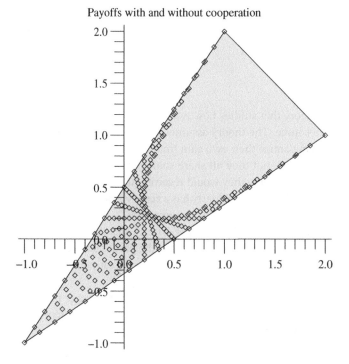

Figure 7.1 Payoff I versus payoff II.

The parabola is given by the implicit equation $5(E_1 - E_2)^2 - 2(E_1 + E_2) + 1 = 0$. If the players play pure strategies, the payoff to each player will be at one of the vertices. The pure Nash equilibria yield the payoff pairs $(E_1 = 1, E_2 = 2)$ and $(E_1 = 2, E_2 = 1)$. The mixed Nash point gives the payoff pair $(E_1 = \frac{1}{5}, E_2 = \frac{1}{5})$, which is strictly inside the region of points, called the **noncooperative payoff set**.

Now, if the players do not cooperate, they will achieve one of two possibilities: (1) The vertices of the figure if they play pure strategies, or (2) any point in the region of points bounded by the two lines and the parabola, if they play mixed strategies. The portion of the triangle outside the parabolic region is **not** achievable simply by the players using mixed strategies. However, if the players agree to cooperate, then any point on the boundary of the triangle, the entire shaded region,[1] including the boundary of the region, are achievable payoffs, which we will see shortly. Cooperation here means an agreement as to which combination of strategies each player will use and the proportion of time that the strategies will be used.

Player I wants a payoff as large as possible and thus as far to the right on the triangle as possible. Player II wants to go as high on the triangle as possible. So player I wants to get the payoff at $(2, 1)$, and player II wants the payoff at $(1, 2)$, but this is possible if and only if the opposing player agrees to play the correct strategy. In addition, it seems that nobody wants to play the mixed Nash equilibrium because they can both do better, but they have to cooperate to achieve a higher payoff.

Here is another example illustrating the achievable payoffs.

Example 7.1

	II$_1$	II$_2$	II$_3$
I$_1$	$(1, 4)$	$(-2, 1)$	$(1, 2)$
I$_2$	$(0, -2)$	$(3, 1)$	$(\frac{1}{2}, \frac{1}{2})$

We will draw the pure payoff points of the game as the vertices of the graph and connect the pure payoffs with straight lines, as in Figure 7.2.

The vertices of the polygon are the payoffs from the matrix. The solid lines connect the pure payoffs. The dotted lines extend the region of payoffs to those payoffs that **could** be achieved if both players cooperate. For example, suppose that player I always chooses row 2, I$_2$, and player II plays the mixed strategy $Y = (y_1, y_2, y_3)$, where $y_i \geq 0, y_1 + y_2 + y_3 = 1$. The expected payoff to I is then

$$E_1(2, Y) = 0 \, y_1 + 3y_2 + \frac{1}{2} \, y_3,$$

and the expected payoff to II is

$$E_2(2, Y) = -2y_1 + 1y_2 + \frac{1}{2} \, y_3.$$

Hence

$$(E_1, E_2) = y_1(0, -2) + y_2(3, 1) + y_3 \left(\frac{1}{2}, \frac{1}{2} \right),$$

which, as a linear combination of the three points $(0, -2)$, $(3, 1)$, and $(\frac{1}{2}, \frac{1}{2})$, is in the convex hull of these three points. This means that if players I and II can agree that player I will always play row 2, then player II can choose a (y_1, y_2, y_3) so that the payoff pair to each player will be in the triangle bounded by the lower dotted line in Figure 7.2 and the lines connecting $(0, -2)$ with $(\frac{1}{2}, \frac{1}{2})$

---

1 This region is called the **convex hull** of the pure payoff pairs. The convex hull of a set of points is the smallest convex set containing all the points.

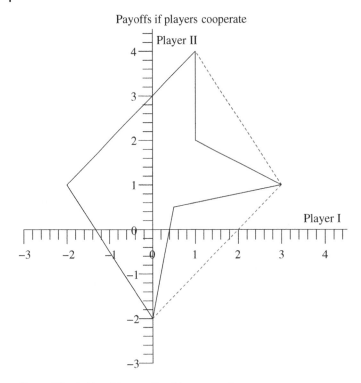

Payoffs if players cooperate

Figure 7.2 Achievable payoffs with cooperation.

with (3, 1). The conclusion is that any point in the convex hull of all the payoff points is achievable if the players agree to cooperate.

One thing to be mindful of is that Figure 7.2 does not show the actual payoff pairs that are achievable in the noncooperative game as we did for the 2×2 symmetric game (Figure 7.1) because it is too involved. The boundaries of that region may not be straight lines or parabolas.

The entire four-sided region in Figure 7.2 is called the **feasible set** for the problem. The precise definition, in general, is as follows.

Definition 7.1.1 *The* **feasible set** *is the convex hull of all the payoff points corresponding to pure strategies of the players.*

The objective of player I in Example 7.1 is to obtain a payoff as far to the right as possible in Figure 7.2, and the objective of player II is to obtain a payoff as far up as possible in Figure 7.2. Player I's ideal payoff is at the point (3, 1), but that is attainable only if II agrees to play II$_2$. Why would he do that? Similarly, II would do best at (1, 4), which will happen only if I plays I$_1$, and why would she do that? There is an incentive for the players to reach a compromise agreement in which they would agree to play in such a way so as to obtain a payoff along the line connecting (1, 4) and (3, 1). That portion of the boundary is known as the **Pareto-optimal boundary** because it is the edge of the set and has the property that if either player tries to do better (say, player I tries to move further right), then the other player will do worse (player II must move down to remain feasible). That is the definition. We have already defined what it means to be Pareto-optimal, but it is repeated here for convenience.

Definition 7.1.2 *The Pareto-optimal boundary of the feasible set is the set of payoff points in which no player can improve his payoff without at least one other player decreasing her payoff.*

The point of this discussion is that there is an incentive for the players to cooperate and try to reach an agreement that will benefit both players. The result will always be a payoff pair occurring on the Pareto-optimal boundary of the feasible set.

In any bargaining problem there is always the possibility that negotiations will fail. Hence, each player must know what the payoff would be if there were no bargaining. This leads us to the next definition.

Definition 7.1.3 *The* **status quo payoff point**, *or* **safety point**, *or* **security point** *in a two-person game is the pair of payoffs* (u^*, v^*) *that each player can achieve if there is no cooperation between the players.*

The safety point usually is, but does not have to be, the same as the safety levels defined earlier. Recall that the safety levels we used in previous sections were defined by the pair $(value(A), value(B^T))$. In the context of bargaining it is simply a noncooperative payoff to each player if no cooperation takes place. For most problems considered in this section, the status quo point **will** be taken to be the values of the zero-sum games associated with each player, because those values can be guaranteed to be achievable, no matter what the other player does.

Example 7.2 We will determine the security point for each player in Example 7.1. In this example we take the security point to be the values that each player can guarantee receiving no matter what. This means that we take it to be the value of the zero-sum games for each player.

Consider the payoff matrix for player I:

$$A = \begin{bmatrix} 1 & -2 & 1 \\ 0 & 3 & \frac{1}{2} \end{bmatrix}$$

We want the value of the game with matrix A. By the methods of Chapter 1 we find that $v(A) = \frac{1}{2}$ and the optimal strategies are $Y = \left(\frac{5}{6}, \frac{1}{6}, 0 \right)$ for player II and $X = \left(\frac{1}{2}, \frac{1}{2} \right)$ for player I.

Next, we consider the payoff matrix for player II. We call this matrix B but since we want to find the value of the game from player II's perspective, we actually need to work with B^T since it is always the row player who is the maximizer (and II is trying to maximize his payoff). Now

$$B^T = \begin{bmatrix} 4 & -2 \\ 1 & 1 \\ 2 & \frac{1}{2} \end{bmatrix}.$$

For this matrix $v(B^T) = 1$, and we have a saddle point at row 2 column 2.

We conclude that the status quo point for this game is $(E_1, E_2) = \left(\frac{1}{2}, 1 \right)$ since that is the guaranteed payoff to each player without cooperation or negotiation. This means that any bargaining must begin with the guaranteed payoff pair $\left(\frac{1}{2}, 1 \right)$. This cuts off the feasible set as in Figure 7.3.

The new feasible set consists of the points in Figure 7.3 above and to the right of the lines emanating from the security point $\left(\frac{1}{2}, 1 \right)$. It is like moving the origin to new point $\left(\frac{1}{2}, 1 \right)$.

Notice that in this problem, the Pareto-optimal boundary is the line connecting $(1, 4)$ and $(3, 1)$ because no player can get a bigger payoff on this line without forcing the other player to get a smaller payoff. A point in the set can't go to the right and stay in the set without also going down; a point in the set can't go up and stay in the set without also going to the left.

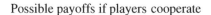

Possible payoffs if players cooperate

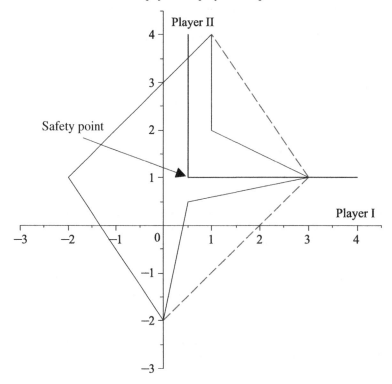

Figure 7.3 The reduced feasible set; safety at $\left(\frac{1}{2}, 1\right)$.

The question now is to find the cooperative, negotiated best payoff for each player. How does cooperation help? Well, suppose, for example, that the players agree to play as follows: I will play row I_1 half the time and row I_2 half the time as long as II plays column II_1 half the time and column II_2 half the time. This is not optimal for player II in terms of his safety level. But, if they agree to play this way, they will get $\frac{1}{2}(1, 4) + \frac{1}{2}(3, 1) = \left(2, \frac{5}{2}\right)$. So player I gets $2 > \frac{1}{2}$ and player II gets $\frac{5}{2} > 1$, a big improvement for each player over his or her own individual safety level. So, they definitely have an incentive to cooperate.

Example 7.3 Here is another example. The bimatrix is

	II_1	II_2
I_1	$(2, 17)$	$(-10, -22)$
I_2	$(-19, -7)$	$(17, 2)$

The reader can calculate that the safety level is given by the point

$$(value(A), value(B^T)) = \left(-\frac{13}{4}, -\frac{5}{2}\right),$$

and the optimal strategies that will give these values are $X_A = \left(\frac{3}{4}, \frac{1}{4}\right)$, $Y_A = \left(\frac{9}{16}, \frac{7}{16}\right)$, and $X_B = \left(\frac{1}{2}, \frac{1}{2}\right)$, $Y_B = \left(\frac{3}{16}, \frac{13}{16}\right)$. Negotiations start from the safety point. Figure 7.4 shows the safety point and the associated feasible payoff pairs above and to the right of the dark lines.

The shaded region in Figure 7.4 is the convex hull of the pure payoffs, namely, the feasible set, and is the set of all possible negotiated payoffs. The region of dot points is the set of noncooperative

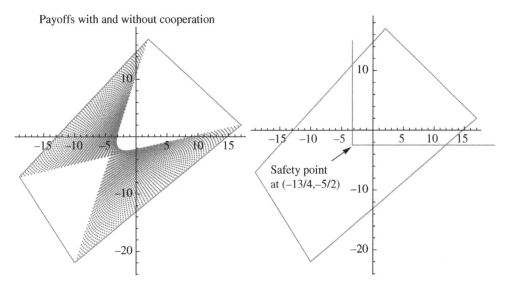

Payoffs with and without cooperation

Figure 7.4 Achievable payoff pairs with cooperation; safety point $= \left(-\frac{13}{4}, -\frac{5}{2} \right)$.

payoff pairs if we consider the use of all possible mixed strategies. The set we consider is the shaded region above and to the right of the safety point. It appears that a negotiated set of payoffs will benefit both players and will be on the line farthest to the right, which is the Pareto-optimal boundary. Player I would love to get $(17, 2)$, while player II would love to get $(2, 17)$. That probably won't occur, but they could negotiate a point along the line connecting these two points and compromise on obtaining, say, the midpoint

$$\frac{1}{2}(2, 17) + \frac{1}{2}(17, 2) = (9.5, 9.5).$$

So they could negotiate to get 9.5 each if they agree that each player would use the pure strategies $X = (1, 0) = Y$ half the time and play pure strategies $X = (0, 1) = Y$ exactly half the time. They have an incentive to cooperate.

Now, suppose that player II threatens player I by saying that she will always play strategy II_1 unless I cooperates. Player II's goal is to get the 17 if and when I plays I_1, so I would receive 2. Of course, I does not have to play I_1, but if he doesn't, then I will get -19, and II will get -7. So if I does not cooperate and II carries out her threat, they will both lose, but I will lose much more than II. Therefore, II is in a much stronger position than I in this game and can essentially **force** I to cooperate. This implies that the safety level of $(-\frac{13}{4}, -\frac{5}{2})$ loses its effect here because II has a credible threat that she can use to force player I to cooperate. This also seems to imply that maybe player II should expect to get more than 9.5 to reflect her stronger bargaining position from the start.

The preceding example indicates that there may be a more realistic choice for a safety level than the values of the associated games, taking into account various threat possibilities. We will see later that this is indeed the case.

7.2 The Nash Model with Security Point

We start with any old security status quo point (u^*, v^*) for a two-player cooperative game with matrices A, B. This leads to a feasible set of possible negotiated outcomes depending on the point we start from (u^*, v^*). This may be the safety point $u^* = value(A), v^* = value(B^T)$, or not. For any given such point and feasible set S, we are looking for a negotiated outcome, call it $(\overline{u}, \overline{v})$. This point will depend on (u^*, v^*) and the set S, so we may write

$$(\overline{u}, \overline{v}) = f(S, u^*, v^*).$$

The question is how to determine the point $(\overline{u}, \overline{v})$? John Nash proposed the following requirements for the point to be a negotiated solution:

- **Axiom 1**: We must have $\overline{u} \geq u^*$ and $\overline{v} \geq v^*$. Each player must get at least the status quo point.
- **Axiom 2**: The point $(\overline{u}, \overline{v}) \in S$, that is, it must be a feasible point.
- **Axiom 3**: If (u, v) is any point in S, so that $u \geq \overline{u}$ and $v \geq \overline{v}$, then it must be the case that $u = \overline{u}, v = \overline{v}$. In other words, there is no other point in S, where **both** players receive more. This is **Pareto-optimality**.
- **Axiom 4**: If $(\overline{u}, \overline{v}) \in T \subset S$ and $(\overline{u}, \overline{v}) = f(T, u^*, v^*)$ is the solution to the bargaining problem with feasible set T, then for the larger feasible set S, either $(\overline{u}, \overline{v}) = f(S, u^*, v^*)$ is the bargaining solution for S, or the actual bargaining solution for S is in $S - T$. We are assuming that the security point is the same for T and S. So, if we have more alternatives, the new negotiated position can't be one of the old possibilities.
- **Axiom 5**: If T is an affine transformation of S, $T = aS + b = \varphi(S)$ and $(\overline{u}, \overline{v}) = f(S, u^*, v^*)$ is the bargaining solution of S with security point (u^*, v^*), then $(a\overline{u} + b, a\overline{v} + b) = f(T, au^* + b, av^* + b)$ is the bargaining solution associated with T and security point $(au^* + b, av^* + b)$. This says that the solution will not depend on the scale or units used in measuring payoffs.
- **Axiom 6**: If the game is symmetric with respect to the players, then so is the bargaining solution. In other words, if $(\overline{u}, \overline{v}) = f(S, u^*, v^*)$ and (i) $u^* = v^*$, and (ii) $(u, v) \in S \Rightarrow (v, u) \in S$, then $\overline{u} = \overline{v}$. So, if the players are essentially interchangeable they should get the same negotiated payoff.

The amazing thing that Nash proved is that if we assume these axioms, there is one and only one solution to the bargaining problem. In addition, the theorem gives a constructive way of finding the bargaining solution.

Theorem 7.2.1 *Let the set of feasible points for a bargaining game be nonempty and convex, and let $(u^*, v^*) \in S$ be the security point. Consider the nonlinear programming problem*

$$\text{Maximize } g(u, v) := (u - u^*)(v - v^*)$$

$$\text{subject to } (u, v) \in S, u \geq u^*, v \geq v^*.$$

Assume that there is at least one point $(u, v) \in S$ with $u > u^*, v > v^*$. Then there exists one and only one point $(\overline{u}, \overline{v}) \in S$ that solves this problem, and this point is the unique solution of the bargaining problem $(\overline{u}, \overline{v}) = f(S, u^*, v^*)$ that satisfies the axioms $1 - 6$. If, in addition, the game satisfies the symmetry assumption, then the conclusion of axiom 6 tells us that $\overline{u} = \overline{v}$

Example 7.4 In an earlier example we considered the game with bimatrix

	II$_1$	II$_2$
I$_1$	(2, 17)	(−10, −22)
I$_2$	(−19, −7)	(17, 2)

The safety levels are $u^* = value(A) = -\frac{13}{4}, v^* = value(B^T) = -\frac{5}{2}$, Figure 7.5 for this problem shows the safety point and the associated feasible payoff pairs above and to the right. Now, we need to describe the Pareto-optimal boundary of the feasible set. We need the equation of the lines

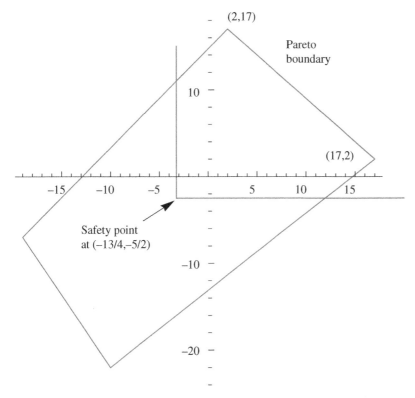

Figure 7.5 Pareto-optimal boundary is line connecting $(2, 17)$ and $(17, 2)$.

forming the Pareto-optimal boundary. In this example it is simply $v = -u + 19$, which is the line with negative slope to the right of the safety point. It is the only place where both players cannot simultaneously improve their payoffs. (If player I moves right, to stay in the feasible set, player II must go down.)

To find the bargaining solution for this problem, we have to solve the nonlinear programming problem

$$\text{Maximize } \left(u + \frac{13}{4}\right)\left(v + \frac{5}{2}\right)$$

$$\text{subject to } u \geq -\frac{13}{4}, \quad v \geq -\frac{5}{2}, \quad v \leq -u + 19.$$

The Mathematica commands used to solve this are

> FindMaximum[(u + 13/4)*(v + 5/2), u >= −13/4, v >= −5/2, u + v <= 19, u, v]

This gives the optimal bargained payoff pair ($\bar{u} = \frac{73}{8} = 9.125, \bar{v} = \frac{79}{8} = 9.875$). The maximum of g is $g(\bar{u}, \bar{v}) = 153.14$, which we do not really use or need.

The bargained payoff to player I is $\bar{u} = 9.125$, and the bargained payoff to player II is $\bar{v} = 9.875$. We do **not** get the point we expected, namely, $(9.5, 9.5)$; that is due to the fact that the security point is not symmetric. Player II has a small advantage.

You can see in Figure 7.6 that the solution of the problem occurs just where the level curves, or contours of g are tangent to the boundary of the feasible set. Since the function g has concave up contours and the feasible set is convex, this must occur at exactly one point.

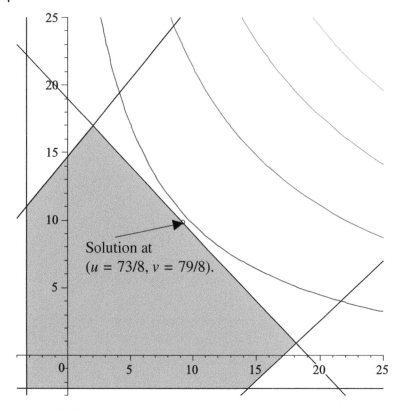

Figure 7.6 Bargaining solution where curves just touch Pareto boundary at $(9.125, 9.875)$.

Finally, knowing that the optimal point must occur on the Pareto-optimal boundary means we could solve the nonlinear programming problem by calculus. We want to maximize

$$f(u) = g(u, -u + 19) = \left(u + \frac{13}{4}\right)\left(-u + 19 + \frac{5}{2}\right), \text{ on the interval } 2 \le u \le 17.$$

This is an elementary calculus maximization problem.

Example 7.5 We will work through another example from scratch. We start with the following bimatrix:

	II_1	II_2
I_1	$(1, 3)$	$(-4, -2)$
I_2	$(-1, -3)$	$(2, 1)$

1. **Find the security point:** To begin, we find the values of the associated matrices

$$A = \begin{bmatrix} 1 & -4 \\ -1 & 2 \end{bmatrix}, \quad B^T = \begin{bmatrix} 3 & -3 \\ -2 & 1 \end{bmatrix}.$$

Then, $value(A) = -\frac{1}{4}$ and $value(B^T) = -\frac{1}{3}$. Hence the security point is $(u^*, v^*) = \left(-\frac{1}{4}, -\frac{1}{3}\right)$.

2. **Find the feasible set:** The feasible set, taking into account the security point, is

$$S^* = \{(u, v) \mid u \ge -\frac{1}{4}, v \ge -\frac{1}{3}, 0 \le 10 + 5u - 5v, 0 \le 10 + u + 3v,$$
$$0 \le 5 - 4u + 3v, 0 \le 5 - 2u - v\}.$$

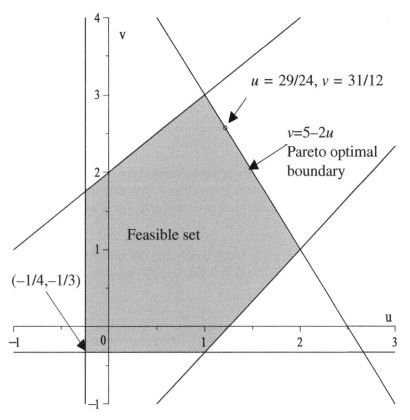

Figure 7.7 Security point $\left(-\frac{1}{4}, -\frac{1}{3}\right)$, Pareto boundary $v = -2u + 5$, solution $(1.208, 2.583)$.

3. **Set up and solve the nonlinear programming problem:** The nonlinear programming problem is then

$$\text{Maximize } g(u, v) \equiv \left(u + \frac{1}{4}\right)\left(v + \frac{1}{3}\right)$$

$$\text{subject to } (u, v) \in S^*.$$

Mathematica gives us the solution $\overline{u} = \frac{29}{24} = 1.208, \overline{v} = \frac{31}{12} = 2.583$. If we look at Figure 7.7 for S^*, we see that the Pareto-optimal boundary is the line $v = -2u + 5, 1 \leq u \leq 2$.

The solution with the safety point given by the values of the zero-sum games is at point $(\overline{u}, \overline{v}) = (1.208, 2.583)$. The conclusion is that with this security point, player I receives the negotiated solution $\overline{u} = 1.208$ and player II the amount $\overline{v} = 2.583$. Again, we do not need Mathematica to solve this problem if we know the line where the maximum occurs, which here is $v = -2u + 5$, because then we may substitute into g and use calculus:

$$f(u) = g(u, -2u + 5) = \left(u + \frac{1}{4}\right)\left(-2u + \frac{16}{3}\right)$$

$$\implies f'(u) = -4u + \frac{29}{6} = 0$$

$$\implies u = \frac{29}{24}.$$

So this gives us the solution as well.

4. **Find the strategies giving the negotiated solution:** How should the players cooperate in order to achieve the bargained solutions we just obtained? To find out, the only points in the bimatrix that are of interest are the endpoints of the Pareto-optimal boundary, namely, $(1, 3)$ and $(2, 1)$. he cooperation must be a linear combination of the strategies yielding these payoffs. Solve

$$\left(\frac{29}{24}, \frac{31}{12}\right) = \lambda(1, 3) + (1 - \lambda)(2, 1),$$

to get $\lambda = \frac{19}{24}$. This says that (I, II) must agree to play (row 1, col 1) with probability $\frac{19}{24}$ and (row 2, col 2) with probability $\frac{5}{24}$.

The Nash bargaining theorem also applies to games in which the players have payoff functions $u_1(x, y), u_2(x, y)$, where x, y are in some interval and the players have a continuum of strategies. As long as the feasible set contains some security point u_1^*, u_2^*, we may apply Nash's theorem. Here is an example.

Example 7.6 Suppose that two persons are given \$1000, which they can split if they can agree on how to split it. If they cannot agree, they each get nothing. One player is rich, so her payoff function is

$$u_1(x, y) = \frac{x}{2}, \quad 0 \le x + y \le 1000$$

because the receipt of more money will not mean that much. The other player is poor, so his payoff function is

$$u_2(x, y) = \ln(y + 1), \quad 0 \le x + y \le 1000,$$

because small amounts of money mean a lot, but the money has less and less impact as he gets more but no more than \$1000. We want to find the bargained solution. The safety points are taken as $(0, 0)$ because that is what they get if they can't agree on a split. The feasible set is $S = \{(x, y) \mid 0 \le x, y \le 1000, x + y \le 1000\}$.

Figure 7.8 illustrates the feasible set and the contours of the objective function.

The Nash bargaining solution is given by solving the problem

$$\text{Maximize} \, (u_1 - 0)(u_2 - 0) = \left(\frac{x}{2} - 0\right)(\ln(y + 1) - 0)$$

subject to

$$0 \le x \le 1000, 0 \le y \le 1000, x + y \le 1000.$$

Since the solution will occur where $x + y = 1000$, substitute $x = 1000 - y$. If we take a derivative of $f(y) = \frac{1}{2}(1000 - y)\ln(y + 1)$ and set it to zero, we obtain that we have to solve the equation $\frac{1000 - y}{y + 1} = \ln(y + 1)$, which with the aid of a calculator is $y = 163.09$.

The solution is obtained using Mathematica as follows.

Find Maximum[f[x, y], x >= 0, y >= 0, x <= 1000, y <= 1000, x + y <= 1000, x, y]

Mathematica tells us that the maximum is achieved at $x = 836.91, y = 163.09$, so the poor man gets \$163, while the rich woman gets \$837. The utility (or value of this money) to each player is $u_1 = 418.5$ to the rich guy and $u_2 = 5.10$ to the poor guy. Figure 7.8 shows the feasible set as well as several level curves of $f(x, y) = k$. The optimal solution is obtained by increasing k until the curve is tangent to the Pareto-optimal boundary. That occurs here at the point $(836.91, 163.09)$. The actual value of the maximum is of no interest to us.

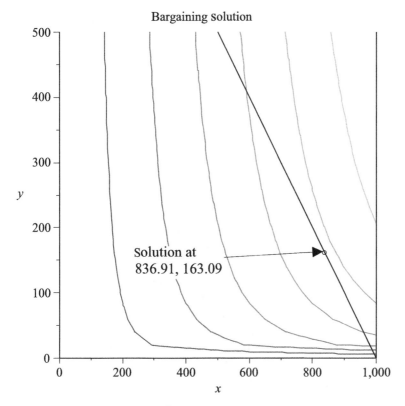

Figure 7.8   Rich and poor split $1000: solution at $(836.91, 163.09)$.

7.3   Threats

Negotiations of the type considered in the previous section do not take into account the relative strength of the positions of the players in the negotiations. As mentioned earlier, a player may be able to force the opposing player to play a certain strategy by threatening to use a strategy that will be very detrimental to the opponent. These types of threats will change the bargaining solution. Let's start with an example.

Example 7.7   We will consider the two-person game with bimatrix

	II_1	II_2
I_1	$(2, 4)$	$(-3, -10)$
I_2	$(-8, -2)$	$(10, 1)$

Player I's payoff matrix is

$$A = \begin{bmatrix} 2 & -3 \\ -8 & 10 \end{bmatrix}$$

and for II

$$B = \begin{bmatrix} 4 & -10 \\ -2 & 1 \end{bmatrix}, \text{ so we look at } B^T = \begin{bmatrix} 4 & -2 \\ -10 & 1 \end{bmatrix}.$$

It is left to the reader to verify that $value(A) = -\frac{4}{23}$, $value(B^T) = -\frac{16}{17}$ so the security point is $(u^*, v^*) = (-\frac{4}{23}, -\frac{16}{17})$.

With this security point, we solve the problem

$$\text{Maximize } g(u, v) = \left(u + \frac{4}{23} \right) \left(v + \frac{16}{17} \right)$$

$$\text{subject to } u \ge -\frac{4}{23}, v \ge -\frac{16}{17}, v \ge \frac{11}{13}u - \frac{97}{13},$$

$$v \le -\frac{3}{8}u + \frac{38}{8}, v \le \frac{6}{10}u + \frac{28}{10}.$$

In the usual way, we get the solution $\bar{u} = 7.501$, $\bar{v} = 1.937$. This is achieved by players I and II agreeing to play the pure strategies (I_1, II_1) 31.2% of the time and pure strategies (I_2, II_2) 68.8% of the time. So with the safety levels as the value of the games, we get the bargained payoffs to each player as 7.501 to player I and 1.937 to player II. Figure 7.9 is a three-dimensional diagram of the contours of $g(u, v)$ over the shaded feasible set. The dot shown on the Pareto boundary is the solution to our problem.

The problem is that this solution is not realistic for this game. Why? The answer is that player II is actually in a much stronger position than player I. In fact, player II can threaten player I with always playing II_1. If player II does that and player I plays I_1, then I gets $2 < 7.501$, but II gets $4 > 1.937$. So, why would player I do that? Well, if player I instead plays I_2, in order to avoid getting less, then player I actually gets -8, while player II gets -2. So both will lose, but I loses much more. Player II's threat to always play II_1, if player I doesn't cooperate on II's terms, is a credible and serious threat that player I cannot ignore. Next, we consider how to deal with this problem.

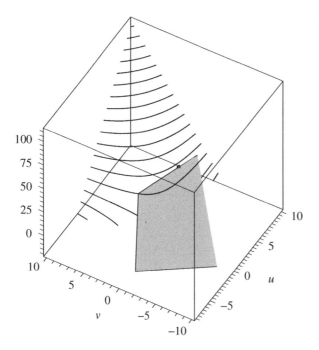

Figure 7.9 The feasible set and level curves in three dimensions. Solution is at $(7.5, 1.93)$ for security point $\left(-\frac{4}{23}, -\frac{16}{17} \right)$.

7.3.1 Finding the Threat Strategies

In a threat game we replace the security levels (u^*, v^*), which we have so far taken to be the value of the associated games $u^* = value(A), v^* = value(B^T)$, with the expected payoffs to each player if threat strategies are used. Why do we focus on the security levels? The idea is that the players will threaten to use their threat strategies, say (X_t, Y_t), and if they do not come to an agreement as to how to play the game, then they will use the threat strategies and obtain payoffs $E_I(X_t, Y_t)$ and $E_{II}(X_t, Y_t)$, respectively. This becomes their security point if they don't reach an agreement.

In Example 7.7, player II seemed to have a pure threat strategy. In general, both players will have a mixed-threat strategy, and we have to find a way to determine them. For now, suppose that in the bimatrix game player I has a threat strategy X_t and player II has a threat strategy Y_t. The new status quo or security point will be the expected payoffs to the players if they both use their threat strategies:

$$u^* = E_A(X_t, Y_t) = X_t A Y_t^T \text{ and } v^* = E_B(X_t, Y_t) = X_t B Y_t^T.$$

Then we return to the cooperative bargaining game and apply the same procedure as before but with the new threat security point; that is, we seek to

$$\text{Maximize } g(u, v) := (u - X_t A Y_t^T)(v - X_t B Y_t^T)$$

$$\text{subject to } (u, v) \in S, u \geq X_t A Y_t^T, v \geq X_t B Y_t^T.$$

Notice that we are not using B^T for player II's security point but the matrix B because we only need to use B^T when we consider player II as the row maximizer and player I as the column minimizer.

In Example 7.7, let's suppose that the threat strategies are $X_t = (0, 1)$ and $Y_t = (1, 0)$. Then the expected payoffs give us the safety point $u^* = X_t A Y_t^T = -8$ and $v^* = X_t B Y_t^T = -2$ (see Figure 7.10). Changing to this security point increases the size of the feasible set and changes the objective function to $g(u, v) = (u + 8)(v + 2)$.

When we solved this example with the security point $(-\frac{4}{23}, -\frac{16}{17})$ we obtained the payoffs 7.501 for player I and 1.937 for player II. The solution of the threat problem is $\overline{u} = 5 < 7.501, \overline{v} = 2.875 > 1.937$. This reflects the fact that player II has a credible threat and, therefore, should get more than if we ignore the threat.

The question now is how to pick the threat strategies. How do we know in the previous example that the threat strategies we chose were the best ones? We continue our example to see how to solve this problem. This method follows the procedure as presented by A. J. Jones [16].

We look for a different security point associated with threats that we call the **optimal threat security point**. Let u^t denote the payoff to player I if both players use their optimal threat strategies. Similarly, v^t is the payoff to player II if they both play their optimal threat strategies. At this point, we don't know what these payoffs are, and we don't know what the optimal threat strategies are, but (u^t, v^t) will be what they each get if their threats are carried out. This should be our threat security point.

The Pareto-optimal boundary for our problem is the line segment $v = -\frac{3}{8}u + \frac{38}{8}, 2 \leq u \leq 10$. This line has slope $m_p = -\frac{3}{8}$. Consider now a line with slope $-m_p = \frac{3}{8}$ through **any possible threat security point in the feasible set** (u^t, v^t). Referring to Figure 7.11, the line will intersect the Pareto-optimal boundary line segment at some possible negotiated solution $(\overline{u}, \overline{v})$. The line with slope $-m_p$ through (u^t, v^t), whatever the point is, has the equation

$$v - v^t = -m_p(u - u^t).$$

It is a fact (see Lemma 7.3.1) that **this line *must* pass through the optimal threat security point.**

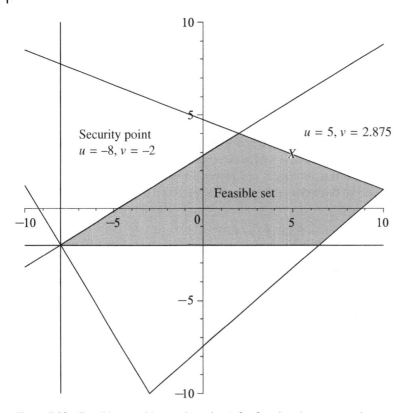

Figure 7.10 Feasible set with security point (−8, −2) using threat strategies.

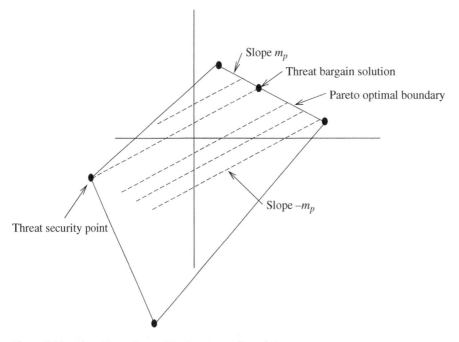

Figure 7.11 Lines through possible threat security points.

The equation of the Pareto-optimal boundary line is

$$v = m_p u + b = -\frac{3}{8}u + \frac{38}{8},$$

so the intersection point of the two lines will be at the coordinates

$$\bar{u} = \frac{m_p\, u^t + v^t - b}{2m_p} = \frac{3u^t - 8v^t + 38}{6},$$

$$\bar{v} = \frac{1}{2}(m_p u^t + v^t + b) = \frac{-(3u^t - 8v^t) + 38}{16}.$$

Now, remember that we are trying to find the best threat strategies to use, but the primary objective of the players is to maximize their payoffs $\bar{u}, \bar{v}$. This tells us exactly what to do to find the optimal threat security point.

- Player I will **maximize** $\bar{u}$ if she chooses threat strategies to maximize the quantity $-m_p\, u^t - v^t = \frac{3}{8}u^t - v^t$.
- Player II will maximize $\bar{v}$ if she chooses threat strategies to **minimize** the same quantity $-m_p\, u^t - v^t$ because the Pareto-optimal boundary will have $m_p < 0$, so the sign of the term multiplying u^t will be opposite in $\bar{u}$ and $\bar{v}$.

Putting these two goals together, it seems that we need to solve a game with some matrix. The rules following will show exactly what we need to do.

7.3.1.1 Summary Approach for Bargaining with Threat Strategies

Here is the general procedure for finding u^t, v^t and the optimal threat strategies as well as the solution to the bargaining game:

1. Identify the Pareto-optimal boundary of the feasible payoff set and find the slope of that line, call it m_p. This slope should be < 0. The equation of the Pareto-optimal boundary is $v = m_p u + b$, so b is the v-intercept.

2. Construct the new matrix for a zero-sum game

$$-m_p u^t - v^t = -m_p(X_t A Y_t^T) - X_t B Y_t^T = X_t(-m_p A - B)Y_t^T$$

with matrix $-m_p A - B$.

3. Find the optimal strategies X_t, Y_t for that game and compute $u^t = X_t A Y_t^T$ and $v^t = X_t B Y_t^T$. This (u^t, v^t) is the threat security point to be used to solve the bargaining problem.

4. Once we know (u^t, v^t), we may use the following formulas for $(\bar{u}, \bar{v})$:

$$\boxed{\bar{u} = \frac{m_p\, u^t + v^t - b}{2m_p}, \quad \bar{v} = \frac{1}{2}(m_p u^t + v^t + b).} \tag{7.3.1}$$

Alternatively, we may apply the nonlinear programming method with security point (u^t, v^t) to find $(\bar{u}, \bar{v})$.

Example 7.7 *continued.*

Carrying out these steps for our example, $m_p = -\frac{3}{8}, b = \frac{38}{8}$, we find

$$\frac{3}{8}A - B = \begin{bmatrix} -\frac{26}{8} & \frac{71}{8} \\ -1 & \frac{22}{8} \end{bmatrix}$$

We find $value(\frac{3}{8}A - B) = -1$ and, because there is a saddle point at the second row and first column, optimal threat strategies $X_t = (0, 1)$, $Y_t = (1, 0)$. Then $u^t = X_t A Y_t^T = -8$, and $v^t = X_t B Y_t^T = -2$. Once we know that, we can use the formulas above to get

$$\bar{u} = \frac{-\frac{3}{8}(-8) + (-2) - \frac{38}{8}}{2(-\frac{3}{8})} = 5,$$

$$\bar{v} = \frac{1}{2}\left(-\frac{3}{8}(-8) + (-2) + \frac{38}{8}\right) = 2.875.$$

This matches with our previous solution in which we simply took the threat security point to be $(-8, -2)$. Now we see that $(-8, -2)$ is indeed the **optimal threat** security point.

7.3.1.2 Another Way to Derive the Threat Strategies Procedure

The preceding derivation leads to concise formulas for the bargained solution using optimal threat strategies. To further explain how this solution comes from the Nash bargaining theorem, we apply it directly and justify the fact that the Pareto line, with slope m_p, and the line through $(\bar{u}, \bar{v})$ and (u^t, v^t), will have slope $-m_p$.

Assume the Pareto optimal set is the straight-line segment

$$P = \{(u, v) \mid \alpha \leq u \leq \beta, v = m_p u + b\}$$

for some constants $\alpha, \beta, m_p < 0, b$. The negotiated payoffs should end up on P. Let X_t, Y_t denote any (to be determined) threat strategies.

Let's start by finding the optimal threat for player I. Nash proves that player I will get an eventual payoff u that maximizes the function

$$f(u) = (u - X_t A Y_t^T)(m_p u + b - X_t B Y_t^T)$$

where we have used the fact that $v = m_p u + b$. This function has a maximum in $[\alpha, \beta]$. Let's assume it is an interior maximum, so we can use calculus to find the critical point.

$$f'(u) = 2m_p u + X_t(-m_p A - B)Y_t^T + b = 0$$

which, solving for u, gives us

$$\bar{u} = \frac{X_t(-m_p A - B)Y_t^T + b}{-2m_p}$$

Since $f''(u) = 2m_p < 0$, we know $\bar{u}$ does provide a maximum of f. Since $\bar{v} = m_p \bar{u} + b$, we also get

$$\bar{v} = \frac{1}{2}(b - X_t(-m_p A - B)Y_t^T). \tag{7.3.2}$$

Now, we analyze the goals for each player. Player I wants to make $\bar{u}$, the payoff that player I should receive, as large as possible. This says, player I should choose X_t to maximize the term $X_t(-m_p A - B)Y_t^T$ against all possible threats Y_t for player II.

From II's point of view, player II wants $\bar{v}$ also large as possible. Looking at the formula for $\bar{v}$ in (7.3.2), we see that player II, who controls the choice of Y_t, will choose the threat strategy Y_t so as to maximize $-X_t(-m_p A - B)Y_t^T$, which is equivalent to *minimize* $X_t(-m_p A - B)Y_t^T$ against any threat X_t for player I.

Once again we arrive at the problem that we must solve the zero-sum matrix game with matrix $-m_p A - B$. Von Neumann's minimax theorem guarantees there is a value and saddle point for this game, denoted as (X_t, Y_t), and they are the optimal threat strategies for each player in the original nonzero-sum game.

Let $value(-m_pA - B) = X_t(-m_pA - B)Y_t^T$ denote the value of the game with matrix $-m_pA - B$. The payoffs for the bargained solution will then be given by

$$\overline{u} = \frac{1}{-2m_p}(b + value(-m_pA - B)), \quad \overline{v} = \frac{1}{2}(b - value(-m_pA - B))$$

Assuming that $\alpha < \overline{u} < \beta$, we have completely solved the problem assuming we have an interior maximum $\alpha < \overline{u} < \beta$.

The next lemma proves that if the Pareto optimal boundary has slope m_p, then the slope of the line through $(u^t = X_tAY_t^T, v^t = X_tBY_t^T)$ and the bargaining solution $(\overline{u}, \overline{v})$ must have slope $-m_p$. This lemma is essential in seeing what happens if the Pareto optimal boundary is not just a line segment but consists of several line segments.

Lemma 7.3.1 *If $(\overline{u}, \overline{v})$ is the solution of the Nash bargaining problem with any security point (u_0, v_0) and the Pareto optimal boundary through $(\overline{u}, \overline{v})$ is a straight line with slope m_p and $(\overline{u}, \overline{v})$ is not at an endpoint, then*

$$\frac{\overline{v} - v_0}{\overline{u} - u_0} = -m_p.$$

That is, the slope of the line through (u_0, v_0) and $(\overline{u}, \overline{v})$ must be the negative of the slope of the Pareto optimal boundary at the point $(\overline{u}, \overline{v})$.

Proof: We have already seen that if $\overline{u}$ is interior to (α, β) then $(\overline{u}, \overline{v})$ by Nash's theorem, maximizes $f(u) = (u - u_0)(m_pu + b - v_0)$. Taking the derivative and setting it to zero, we see that

$$b - v_0 + m_pu^t + m_p(u^t - u_0) = 0 \implies u^t = \frac{b - v_0 - m_pu_0}{-2m_p}, v^t = \frac{b + v_0 + m_pu_0}{-2m_p}$$

For arbitrary security point (u_0, v_0) the maximizing point is given by

$$\overline{u} = \frac{-m_pu_0 + b - v_0}{-2m_p}, \quad \overline{v} = m_p\overline{u} + b = \frac{b + m_pu_0 + v_0}{2}$$

Now, we calculate the slope of the line through (u_0, v_0) and $(\overline{u}, \overline{v})$.

$$\frac{\overline{v} - v_0}{\overline{u} - u_0} = \frac{\frac{b + m_pu_0 + v_0}{2} - v_0}{\frac{-m_pu_0 + b - v_0}{-2m_p} - u_0}$$

$$= -m_p\frac{b + m_pu_0 - v_0}{m_pu_0 + b - v_0}$$

$$= -m_p. \qquad \blacksquare$$

The preceding discussion gives us a general procedure to solve for the threat strategies. Notice, however, that several things can make this procedure more complicated. First, the determination of the Pareto-optimal boundary of S is of critical importance. In Figure 7.10 it consisted of only one line segment, but in practice there may be many such line segments and we have to work separately with each segment. That is because we need the slopes of the segments. This means that the threat strategies and the threat point u^t, v^t could change from segment to segment. An example below will illustrate this.

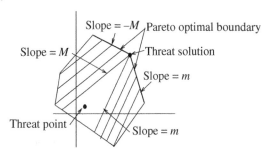

Slope = −M — Pareto optimal boundary

Slope = M

Threat solution

Slope = m

Threat point

Slope = m

Figure 7.12 Bargaining solution for threats when threat point is in the cone is the vertex.

Threat solution for vertex

Another problem is that the Pareto-optimal boundary could be a point of intersection of two segments, so there is no slope for the point. Then, what do we do? The answer is that when we calculate the threat point (u^t, v^t) for each of the two line segments that intersect at a vertex, if this threat point is in the cone emanating from this vertex with the slopes shown in Figure 7.12, then the threat solution of our problem is in fact at the vertex.

Example 7.8 Consider the cooperative game with bimatrix

	II$_1$	II$_2$
I$_1$	(−1, −1)	(1, 1)
I$_2$	(2, −2)	(−2, 2)

The individual matrices are

$$A = \begin{bmatrix} -1 & 1 \\ 2 & -2 \end{bmatrix}, \qquad B = \begin{bmatrix} -1 & 1 \\ -2 & 2 \end{bmatrix}.$$

It is easy to calculate that $value(A) = 0, value(B^T) = 1$ and so the status quo security point for this game is at $(u^*, v^*) = (0, 1)$. The problem we then need to solve is

Maximize $u(v − 1)$,

subject to $(u, v) \in S^*$,

where

$$S^* = \{(u, v) \mid v \le \left(-\frac{1}{3}\right) u + \frac{4}{3}, v \le -3u + 4, u \ge 0, v \ge 1\}.$$

The solution of this problem is at the unique point $(\bar{u}, \bar{v}) = (\frac{1}{2}, \frac{7}{6})$, which you can see in Figure 7.13. The solution of the problem is given by the Mathematica commands:

```
FindMaximum[{u(v-1), u >= 0, v >= 1, -3 u - v <= 4, u + 3 v <= 4,
              3 u + v <= 4, -u - 3 v <= 4}, {u, v}]
```

We get from these commands that $z = 0.083, u = \bar{u} = 0.5, v = \bar{v} = 1.167$. As mentioned earlier, you may also get this by hand using calculus. You need to find the maximum of $g(u, v) = u(v − 1)$ subject to $u \ge 0, v \ge 1$, and $v = -\frac{u}{3} + \frac{4}{3}$. So, $f(u) = g(u, v) = u(-\frac{u}{3} + \frac{4}{3})$ is the function to maximize. Since $f'(u) = -\frac{2u}{3} + \frac{1}{3} = 0$ at $u = \frac{1}{2} \ge 0$, we have that $\bar{u} = \frac{1}{2} > 0, \bar{v} = \frac{7}{6} > 1$ as our interior feasible maximum.

Next, to find the threat strategies, we note that we have two possibilities because we have two line segments in Figure 7.13 as the Pareto-optimal boundary. We have to consider both $m_p = -\frac{1}{3}, b = \frac{4}{3}$ and $m_p = -3, b = 4$.

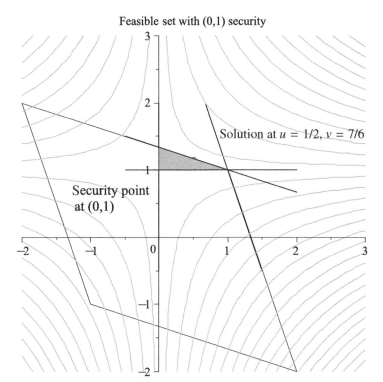

Figure 7.13 Security point $(0, 1)$; cooperative solution $(\overline{u} = \frac{1}{2}, \overline{v} = \frac{7}{6})$.

Let's use $m_p = -3, b = 4$. We look for the value of the game with matrix $3A - B$:

$$3A - B = \begin{bmatrix} -2 & 2 \\ 8 & -8 \end{bmatrix}$$

Then $value(3A - B) = 0$, and the optimal threat strategies are $X_t = (\frac{1}{2}, \frac{1}{2}) = Y_t$. Then the security threat points are

$$u^t = X_t A Y_t^{~T} = 0 \text{ and } v^t = X_t B Y_t^{~T} = 0.$$

This means that each player threatens to use (X_t, Y_t) and receive 0 rather than cooperate and receive more.

Now the maximization problem becomes

> Maximize uv,
>
> subject to $(u, v) \in S^t$,

where

$$S^t = \{(u, v) \mid v \le \left(-\frac{1}{3}\right) u + \frac{4}{3}, v \le -3u + 4, u \ge 0, v \ge 0\}.$$

The solution of this problem is at the unique point $(\overline{u}, \overline{v}) = (1, 1)$. You can see in Figure 7.14 how the level curves have bent over to touch at the vertex.

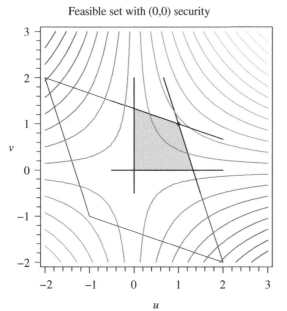

Feasible set with (0,0) security

Figure 7.14 Security point $(0, 0)$; cooperative solution $(\overline{u} = 1, \overline{v} = 1)$.

What would have happened if we used the slope of the other line of the Pareto-optimal boundary? Let's look at $m_p = -\frac{1}{3}, b = \frac{4}{3}$. The matrix is

$$\frac{1}{3} A - B = \begin{bmatrix} \frac{2}{3} & -\frac{2}{3} \\ \frac{8}{3} & -\frac{8}{3} \end{bmatrix}$$

Then $value(\frac{1}{3} A - B) = -\frac{2}{3}$, and the optimal threat strategies are $X_t = (1, 0)$, $Y_t = (0, 1)$. The security threat points are

$$u^t = X_t A Y_t^T = 1 \text{ and } v^t = X_t B Y_t^T = 1.$$

This point is exactly the vertex of the feasible set.

Now the maximization problem becomes

> Maximize $(u - 1)(v - 1)$,
>
> subject to $(u, v) \in S^t$,

where

$$S^t = \{(u, v) | v \le \left(-\frac{1}{3}\right) u + \frac{4}{3}, \ v \le -3u + 4, \ u \ge 1, \ v \ge 1\}.$$

But this set has exactly one point (as you should verify), and it is $(1, 1)$, so we immediately get the solution $(\overline{u} = 1, \overline{v} = 1)$, the same as what we got earlier.

What happens if we try to use the formulas (7.3.1) for the threat problem? This question arises now because the contours of g are hitting the feasible set right at the point of intersection of two lines. The two lines have the equations

$$v = -3u + 4 \text{ and } v = -\frac{1}{3} u + \frac{4}{3}.$$

So, do we use $m_p = -3, b = 4$, or $m_p = -\frac{1}{3}, b = \frac{4}{3}$? Let's calculate for both. For $m_p = -3, b = 4, u^t = v^t = 0$, we have

$$\bar{u} = \frac{m_p u^t + v^t - b}{2m_p} = \frac{-3(0) + (0) - 4}{2(-3)} = \frac{2}{3},$$

$$\bar{v} = \frac{1}{2}(m_p u^t + v^t + b) = \frac{1}{2}(-3(0) + (0) + 4) = 2.$$

The point $(\frac{2}{3}, 2)$ is not in S^t because $(-\frac{1}{3})(\frac{2}{3}) + \frac{4}{3} = \frac{10}{9} < 2$. So we no longer consider this point. However, because the point $(u^t, v^t) = (0, 0)$ is inside the cone region formed by the lines through $(1, 1)$ with slopes $\frac{1}{3}$ and 3, we know that the threat solution should be $(1, 1)$.

For $m_p = -\frac{1}{3}, b = \frac{4}{3}, u^t = v^t = 1$,

$$\bar{u} = \frac{m_p u^t + v^t - b}{2m_p} = \frac{-\frac{1}{3}(1) + (1) - \frac{4}{3}}{2(-\frac{1}{3})} = 1,$$

$$\bar{v} = \frac{1}{2}(m_p u^t + v^t + b) = \frac{1}{2}\left(-\frac{1}{3}(1) + (1) + \frac{4}{3}\right) = 1.$$

This gives $(\bar{u} = 1, \bar{v} = 1)$, which is the correct solution.

7.4 The Kalai–Smorodinsky Bargaining Solution

Nash's bargaining solution does not always give realistic solutions because the axioms on which it is based may not be satisfied. The next example illustrates what can go wrong.

Example 7.9 At the risk of undermining your confidence, this example will show that the Nash bargaining solution can be totally unrealistic and in an important problem. Suppose that there is a person, Moe, who owes money to two creditors, Larry and Curly. He owes more than he can pay. Let's say that he can pay at most \$100, but he owes a total of \$ 150 > \$100 dollars, \$90 to Curly and \$60 to Larry. The question is how to divide the \$100 among the two creditors. We set this up as a bargaining game and use Nash's method to solve it.

First, the feasible set is

$$S = \{(u, v) \mid u \le 60, v \le 90, u + v \le 100\},$$

where u is the amount Larry gets, and v is the amount Curly will get.

The objective function we want to maximize at first is $g(u, v) = uv$ because if Larry and Curly can't agree on the split, then we assume that they each get nothing (because they have to sue and pay lawyers, etc.).

For the solution, we want to maximize $g(u, v)$ subject to $(u, v) \in S, u \ge 0, v \ge 0$. It is straightforward to show that the maximum occurs at $\bar{u} = \bar{v} = 50$, as shown in Figure 7.15.

In fact, if we take any safety point of the form $u^* = a = v^*$, we would get the exact same solution. This says that even though Moe owes Curly \$90 and Larry \$60, they both get the same amount as a settlement. That doesn't seem reasonable, and I'm sure Curly would be very upset.

Now let's modify the safety point to $u^* = -60$ and $v^* = -90$, which is still feasible and reflects the fact that the players actually lose the amount owed in the worst case, that is, when they are left holding the bag. This case is illustrated in Figure 7.16.

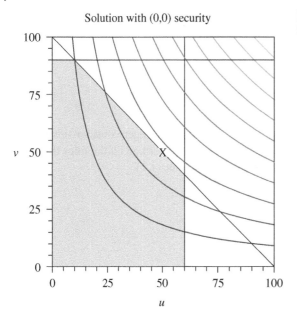

Solution with (0,0) security

Figure 7.15 Moe pays both Curly and Larry $50 each.

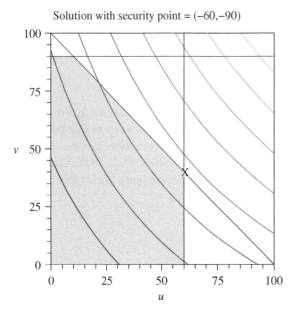

Solution with security point = (−60,−90)

Figure 7.16 Moe pays Larry $60 and pays Curly $40.

The solution is now obtained from maximizing $g(u, v) = (u + 60)(v + 90)$ subject to $(u, v) \in S, u \geq -60, v \geq -90$, and results in $\bar{u} = 60$ and $\bar{v} = 40$. This is ridiculous because it says that Larry should be paid off in full while Curly, who is owed more, gets less than half of what he is owed.

Of course, it is also possible to set this game up as a cooperative game and find the allocation. In this case we take the characteristic function to be $v(L) = 60, v(C) = 90, v(LC) = 100$. Note that Moe is not a player in this game–he gives up the $100 no matter what. In this case, the least core is

$$C(\varepsilon) = \{60 - x_L \leq \varepsilon, 90 - x_C \leq \varepsilon, x_L + x_C = 100\}.$$

The smallest ε so that $C(\varepsilon) \neq \emptyset$ is $\varepsilon^1 = 25$, and then we get $C(25) = \{(35, 65)\}$. The nucleolus solution is that Larry gets \$35, and Curly gets \$65. It is also easy to calculate that this is the same as the Shapley value.

The problem with the Nash bargaining solution occurs whenever the players are not even close to being symmetric, as in this example. In this case, there is an idea for solution due to Kalai and Smorodinsky [18] that gives an alternate approach to solving the bargaining problem.

We will give a brief description of Kalai and Smorodinsky's idea.[1] The KS solution is based on the idea that each player should get an amount proportional to the player's contribution. How to carry that out is given in the next steps.

1. Given the feasible set S, find

$$a = \max_{(u,v) \in S} u \text{ and } b = \max_{(u,v) \in S} v.$$

Essentially, a is the maximum possible payoff for player I and b is the maximum possible payoff for player II.

2. Given any status quo security point (u^*, v^*), consider the line that has equation

$$v - v^* = k(u - u^*), \quad k = \frac{b - v^*}{a - u^*}. \tag{7.4.1}$$

This line passes through (u^*, v^*) and (a, b), and is called the **KS line**, after Kalai and Smorodinsky.

3. The KS solution to the bargaining problem is the highest exit point on line KS from the feasible set. It roughly allocates to each player an amount proportional to the ratio of their maximum possible payoffs.

For the Moe–Larry–Curly problem, with $u^* = -60, v^* = -90$, we calculate

$$a = 60, b = 90 \implies k = \frac{180}{120} = \frac{3}{2}, \text{ and } v + 90 = \frac{3}{2}(u + 60).$$

This KS line will exit the feasible set where it crosses the line $u + v = 100$. This occurs at $\bar{u} = 40$, and so $\bar{v} = 60$. Consequently, now Larry gets \$40 and Curly gets \$60. This is much more reasonable because almost everyone would agree that each creditor should be paid the same percentage of the amount Moe has as his percentage of the total owed. In this case $90/(90 + 60) = \frac{3}{5}$, so that Curly should get three-fifths of the \$100 or \$60. That is the KS solution as well.

The KS solution gives another alternative for solution of the bargaining problem. However, it should be noted that the KS solution does not satisfy all the axioms for a desirable solution. In particular it does not satisfy an economic axiom called the **independence of irrelevant alternatives axiom**.

7.5 Sequential Bargaining

In most bargaining situations there are offers and counteroffers. Generally this can go several rounds until an agreement is reached or negotiations break down. In this section we will present a typical example to see how to analyze these problems.

It will be helpful to think about the problem in which there is a buyer and a seller. Player 1 is the buyer and player 2 is the seller. Suppose the item up for sale is a house.

---

1 Refer to the discussion in Aliprantis and Chakrabarti [1] for more details and further results and examples.

Now suppose that the seller has a bottom line absolutely lowest price below which he will not sell the house. Let's call that the **reserve price**. The seller offers the house for sale at a price called the **ask price**. The difference between the two prices is called the **spread=asked-reserve**. The spread is the negotiating range. Let x denote the fraction of the spread going to the buyer and $1 - x$ the fraction of the spread going to the seller. If we determine $x, 0 \leq x \leq 1$, the buyer and seller will have settled on a price and the transaction will take place at price reserve$+(1 - x)$spread.

In a one shot bargaining problem the buyer and seller are aware of the reserve price and the ask price and the offer from the buyer is a take-it-or-leave-it deal. Let's solve this problem for the one shot bargain.

We take the payoff for each player as

$$u_1(x, 1 - x) = x, \qquad u_2(x, 1 - x) = 1 - x$$

which is the fraction of the spread going to each player. Each player wants to make his fraction as large as possible. Let's take any safety point (d_1, d_2) which determines the worth to each player if no deal is made.

The Nash bargaining problem then consists of maximizing

$$g(x, 1 - x) = (x - d_1)(1 - x - d_2) \text{ over the set } 0 \leq x \leq 1.$$

The solution is

$$x^* = \frac{1 + d_1 - d_2}{2}$$

and this works as long as $x^* > d_1$ which means we must have $d_1 + d_2 \leq 1$. Note that if $d_1 = d_2$, the optimal split is $\frac{1}{2}$, so the transaction takes place at the midpoint of the spread.

To contrast this with the KS solution, we first calculate

$$s_1 = \max_{0 \leq x \leq 1} u_1(x, 1 - x) = 1, \text{ and } s_2 = \max_{0 \leq x \leq 1} u_2(x, 1 - x) = 1.$$

and the KS line with security point $(d_1, d_2), 0 \leq d_i < 1$ is $y - 1 = \frac{1-d_2}{1-d_1}(x - 1)$. This line intersects the boundary of the feasible set $x + y = 1$ when

$$x^{KS} = \frac{1 - d_2}{2 - d_1 - d_2}.$$

When $d_1 = d_2 = 0$, the split is at the midpoint of the spread.

Now consider what happens if sequential bargaining is possible. Player 1 begins by offering the split at $1 - x$. Player 2 then can choose to either accept the split or reject it. Then player 1 can accept or reject that. This can go on presumably forever. What prevents it from endlessly going through offer-counter offer, is the time value of money. Let $r > 0$ and $\delta = \frac{1}{1+r}$. Then r represents the interest rate on money and δ is the discount rate. What that means is that \$1 today is worth \$1 $(1 + r)$ in the next period, conversely, \$1 in the next period is worth \$1 δ today. We assume that each player has their own discount rate δ_1, δ_2 if we think of the discount rate as representing a discount one will offer today to make the deal now.

The payoff functions for each player are now given by

$$u_1(x_n, 1 - x_n) = \delta_1^{n-1} x_n, \text{ and } u_2(x_n, 1 - x_n) = \delta_2^{n-1}(1 - x_n), \ n = 1, 2.$$

Let's consider the two stage bargaining problem. Consider the tree in Figure 7.17.

The game begins with an offer x_1 from player 1. Player 2 may accept or reject the offer. If it is accepted, the game is over and player 1 gets x_1, player 2 gets $1 - x_1$. If the first offer by player 1 is rejected, player 2 makes a counter offer $y_2 \in [0, 1]$. If player 1 rejects the counter offer the game

Figure 7.17 Two stage bargaining.

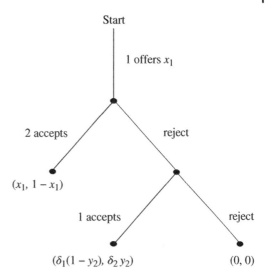

is over, no deal is reached, and the payoffs are 0 to each player. If player 1 accepts the offer, the game ends and the payoff to player 1 is $\delta_1(1 - y_2)$ while the payoff to player 2 is $\delta_2 y_2$.

This is an extensive game with perfect information and can be solved by looking for a subgame perfect equilibrium via backward induction. Starting at the end of the second period, player 1 makes the final decision. If $\delta_1(1 - y_2) > 0$, i.e., if $y_2 \neq 1$, then player 1 receives the larger payoff $\delta_1(1 - y_2)$. Naturally, player 2 will choose y_2 extremely close to 1 since 2 knows that *any* $y_2 < 1$ will give 1 a positive payoff. Thus at the beginning of the last stage where player 2 offers $y_2 \approx 1$, player 2 gets $\approx \delta_2$ and player 1 gets ≈ 0 (but positive).

Player 2, at the start of the first stage, will compare $1 - x_1$ with $\delta_2 y_2$. If $1 - x_1 > \delta_2 y_2 \approx \delta_2$, player 2 will accept player 1's first offer. Player 1's offer needs to satisfy $x_1 < 1 - \delta_2$, but as large as possible. That means player 1 will play $x_1 = 1 - \delta_2$, and that should be the offer 1 makes at the start of the game.

So here is the result giving the subgame perfect equilibrium for each player.

- Player 1 begins the game by offering $x_1 = 1 - \delta_2$. Then, if player 2 accepts the offer, the payoff to 1 is $1 - \delta_2$ and the payoff to 2 is δ_2. If player 2 rejects and counter offers $y_2 \approx 1, y_2 < 1$, then player 1 will accept the offer, receiving $\delta_1(1 - y_2) > 0$. If player 2 counters with $y_2 = 1$, then player 1 is indifferent between acceptance or rejection (and so will reject).

Now let's analyze the bargaining game with three rounds in Figure 7.18. To simplify the details we will assume that both players have the same discount factor $\delta = \delta_1 = \delta_2$.

As before, player 1 begins by offering to take the fraction x_1 of the spread, and 2 is left with $1 - x_1$ if 2 accepts. If 2 rejects 1's offer, then two counters with an offer of $y_2 \in [0, 1]$, and $1 - y_2$ for player 1. Finally, player 1 may either accept player 2's offer and the game ends, or reject the offer and counter with an offer of x_3 for player 1, and $1 - x_3$ for player 2. Taking into account the discount factor δ, the payoffs are as follows:

1. If player 2 accepts the first offer x_1, player 1 gets $u_1(x_1, 1 - x_1) = x_1$, and player 2 gets $u_2(x_1, 1 - x_1) = 1 - x_1$, otherwise ...
2. If player 1 accepts player 2's offer of y_2, then player 1 gets $u_1(1 - y_2, y_2) = \delta(1 - y_2)$ and player 2 gets $u_2(1 - y_2, y_2) = \delta y_2$, otherwise ...

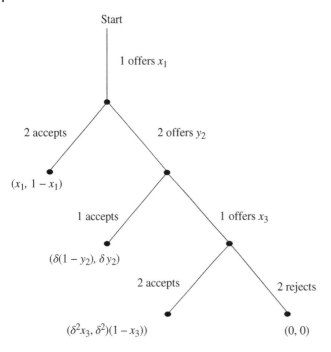

Figure 7.18 Three stage bargaining.

3. If player 2 accepts player 1's offer of x_3, then player 1 gets $u_1(x_3, 1 - x_3) = \delta^2 x_3$ and player 2 gets $u_2(x_3, 1 - x_3) = \delta^2(1 - x_3)$, otherwise player 1 gets $u_1(x_3, 1 - x_3) = 0$ and player 2 gets $u_2(x_3, 1 - x_3) = 0$.

Now let's calculate the subgame perfect equilibrium using backward induction.

1. In the final stage where player 2 decides, if $1 > x_3 \approx 1$, then player 2 will accept 1's offer of x_3 since otherwise player 2 gets 0.
2. Player 1 will accept player 2's offer of y_2 if $\delta(1 - y_2) > \delta^2 x_3^* \approx \delta^2$ (since $x_3^* \approx 1$ otherwise). That is, if $1 - y_2 > \delta$, or $1 - \delta > y_2$, then 1 will accept. Thus player 2 should offer the largest possibility, or $y_2^* = 1 - \delta$ in the second stage.
3. In the first stage, player 2 will accept player 1's offer of x_1 if $1 - x_1 > 1 - y_2^* = \delta$, i.e., $1 - \delta > x_1$. Consequently, player 1 should offer $x_1^* = 1 - \delta$.

What happens if this game can continue possibly forever? Well, we can figure that out by what we have already calculated. First we need to make a change in the third stage game if the final offer is rejected. **Instead of** $(0, 0)$ **going to each player, let** s **denote the total spread available and assume that when player 2 rejects player 1's counter at stage 3, the payoff to each player is** $(\delta^2 s, \delta^2(1 - s))$. We go through the analysis again.

Player 2 in stage 3 will accept the counter of player 1, x_3, if $\delta^2(1 - x_3) \geq \delta^2(1 - s)$, which is true if $x_3 \leq s$. Since player 1 gets $\delta^2 x_3$ if the offer is accepted, player 1 makes the offer x_3 as large as possible and so $x_3^* = s$.

Working back to the second stage, player 2's offer of y_2 will be accepted if $\delta(1 - y_2) \geq \delta^2 s$, i.e., if $y_2 \leq 1 - \delta s$. Since player 2 receives δy_2, player 2 wants y_2 to be as large as possible and hence offers $y_2 = 1 - \delta s$ in the second stage.

In the first stage, player 2 will accept player 1's offer of x_1 if $1 - x_1 > \delta y_2 = \delta(1 - \delta s)$. Simplifying, this requires $x_1 < 1 - \delta(1 - \delta s)$, or, making x_1 as large as possible, player 1 should offer $x_1 = 1 - \delta + \delta^2 s$ in the first stage.

We may summarize the subgame perfect equilibrium by writing this down in reverse order.

Player 1	**Player 2**
1. Offer player 2 $x_1 = 1 - \delta + \delta^2 s$.	1. Accept if $x_1 = 1 - \delta + \delta^2 s$.
2. If player 2 offers $y_2 = 1 - \delta s$, accept.	2. Offer $y_2 = 1 - \delta s$.
3. Offer $x_3 = s$.	3. Accept if $x_3 = s$.

What should s be? If the bargaining goes on for third stages and ends, s is 0 if no agreement can be achieved. In theory, the bargaining can go back and forth for quite a long time. The way to choose s is to observe that when we are in a third stage game, at the third stage we are back to the original conditions at the start of the game for both players except for the fact that time has elapsed. In other words, player 1 will now begin bargaining just as she did in the beginning of the game. That implies that for the original third stage game, s should be chosen so that the offer at the first stage x_1 is the same as what she should offer at stage 3. This results in

$$1 - \delta + \delta^2 s = s \implies s = \frac{1}{1 + \delta}.$$

Remark It may make more sense in many sequential bargaining problems to view the discount factor not as the time value of money, but as the probability the bargaining problem will end with a rejection of the latest offer. This would make δ very subjective, which is often the case with bargaining. In particular, when you are negotiating to buy a car, or house, you have to assess the chances that an offer will be rejected, and with no possibility of revisiting the offer. The next example illustrates this.

Example 7.10 Suppose you want to sell your car. You know it is worth at least $2000 and won't take a penny under that. On the other hand, you really want to unload the car so you advertise it for sale at $2800. The market for your used car is not very good but eventually a buyer shows up. The buyer looks like will give an offer but may not negotiate if the offer is turned down. The buyer thinks the same of you but he knows the car is worth at least $2000. Let's take $\delta = 0.5$ to account for the uncertainty in the continuation of bargaining. In this case, assuming indefinite stages of bargaining, the buyer should offer

$$x^* = 2000 + \frac{1}{1 + \delta} 800 = 2000 + \frac{2}{3} 800 = 2533.33$$

and the seller should accept this offer.

The next proposition tells us what to do if there is no limit to the number of periods bargaining can go on. We'll allow separate discount factors for each player.

Proposition 7.5.1 *There is a unique subgame perfect equilibrium, called V^*, in the perpetual sequential bargaining game described as follows: We assume that player 1 is the player making the first offer.*

Whenever player 1 makes an offer, she suggests a split $(x^*, 1 - x^*)$ with $x^* = \frac{1 - \delta_2}{1 - \delta_1 \delta_2}$. Player 2 accepts any offer giving her at least $1 - x^*$.

Whenever player 2 makes an offer, she suggests a split $(y^*, 1 - y^*)$ with $y^* = \frac{\delta_1(1 - \delta_2)}{1 - \delta_1 \delta_2} = \delta_1 x^*$. Player 1 accepts any offer giving her at least y^*.

The bargaining ends immediately with a split $(x^*, 1 - x^*)$.

Remarks

- Player 1's payoff is $v_1(\delta_1, \delta_2) = \frac{1-\delta_2}{1-\delta_1\delta_2}$ is an increasing function of δ_1 and a decreasing function of δ_2, (take the derivative in δ_2). This means that waiting to come to a deal is better for player 1.
- There is a first-mover advantage for the first offer. If the discount factors are the same, the split is $(\frac{1}{1+\delta}, \frac{\delta}{1+\delta})$ which is better for player 1. But, if $\delta \to 1$ the split goes to $(\frac{1}{2}, \frac{1}{2})$ and the advantage disappears.
- Optimally, there is no second round. Player 2 accepts player 1's first offer.

Example 7.11 Suppose a buyer and a seller are 10K apart in a transaction. Suppose the buyer is player 1 and $\delta_1 = 0.5$. Player 2 is the seller and $\delta_2 = 0.8$. Then buyer 1 should retain

$$x = \frac{1-\delta_2}{1-\delta_1\delta_2} = \frac{0.2}{0.6} = \frac{1}{3}$$

of the 10K, or \$3333.33 and offer the seller \$6666.66, which the seller should accept.

On the other hand, if $\delta_1 = \delta_2 = 0.8$, then the buyer should retain

$$x = \frac{1}{1+\delta} = \frac{1}{18} = 0.556$$

of the 10K or \$5560 and the buyer should get \$4440. You can see that if the buyer and seller believe that the bargaining will continue indefinitely with probability 0.8, then the buyer does better than if he believes there is only a 50% chance the bargaining will continue.

Problems

7.1 A British game show involves a final prize. The two contestants may each choose either to Split the prize, or Claim the prize. If they Split the prize, they each get $\frac{1}{2}$; if they each Claim the prize, they each get 0. If one player Splits, and the other player Claims, the player who Claims the prize gets 1 and the other player gets 0. The game matrix is

$$\begin{bmatrix} (\frac{1}{2},\frac{1}{2}) & (0,1) \\ (1,0) & (0,0). \end{bmatrix}.$$

a) Find the Nash bargaining solution without threats.
b) Apparently, each player has a credible threat to always Claim the prize. Find the optimal threat strategies and the Nash solution as well as the combination of pure strategies which the players should agree to in the threat game.
c) If the game will only be played one time and one player announces that he will definitely Claim the prize and then split the winnings after the show is over, what must the other player do?

7.2 Find the solution to the Nash bargaining problem for the game

$$\begin{bmatrix} (1,4) & (-1,-4) \\ (-4,-1) & (4,1) \end{bmatrix}.$$

7.3 Find the Nash bargaining solution, the threat solution, and the KS solution to the battle of the sexes game with matrix

$$\begin{bmatrix} (4,2) & (2,-1) \\ (-1,2) & (2,4) \end{bmatrix}.$$

Compare the solutions with the solution obtained using the characteristic function approach.

7.4 Find the Nash bargaining solution and the threat solution to the game with bimatrix

$$\begin{bmatrix} (-2,5) & (-7,3) & (3,4) \\ (4,-3) & (6,1) & (-6,-6) \end{bmatrix}.$$

Find the KS line and solution.

7.5 Find the Nash bargaining solution and the threat solution to the game with bimatrix

$$\begin{bmatrix} (-3,-1) & (0,5) & (1,\frac{19}{4}) \\ (2,\frac{7}{2}) & (\frac{5}{2},\frac{3}{2}) & (-1,-3) \end{bmatrix}.$$

Find the KS line and solution.

7.6 Consider the sequential bargaining problem. Suppose that each player has their own discount factor $\delta_1, \delta_2, 0 < \delta_i < 1$. Find the subgame perfect equilibrium for each player assuming the bargaining has 3 stages and ends, as well as assuming the stages could continue forever.

7.7 You want to buy a seller's condo. You look up what the seller paid for the condo two years ago and find that she paid \$305,000. You figure that she will absolutely not sell below this price. The seller has listed the condo at \$350,000. Assume that the sequential bargaining problem will go at most three rounds before everyone is fed up with the process and the deal falls apart. Take the discount factors for each player to be $\delta = 0.99$. What should the offers be at each stage? What if the process could go on indefinitely?

7.8 The Nash solution also applies to payoff functions with a continuum of strategies. For example, suppose that two investors are bargaining over a piece of real estate and they have payoff functions $u(x,y) = x + y$, while $v(x,y) = x + \sqrt{y}$, with $x, y \geq 0$, and $x + y \leq 1$. Both investors want to maximize their own payoffs. The bargaining solution with safety point $u^* = 0, v^* = 0$ (because both players get zero if negotiations break down) is given by the solution of the problem

$$\text{Maximize} (u(x,y) - u^*)(v(x,y) - v^*) = (x+y)(x + \sqrt{y})$$

$$\text{subject to} x, y \geq 0, x + y \leq 1.$$

Solve this problem to find the Nash bargaining solution.

7.9 A classic bargaining problem involves a union and management in contract negotiations. If management hires $w \geq 0$ workers, the company produces $f(w)$ revenue units, where f is a continuous, increasing function. The maximum number of workers who are represented by the union is W. A person who is not employed by the company gets a payoff $p_0 \geq 0$, which is either unemployment benefits or the pay at another job. In negotiations with the union, the firm agrees to the pay level p and to employ $0 \leq w \leq W$ workers. We may consider the payoff functions as

$$u(p,w) = f(w) - pw \quad \text{to the company}$$

and

$$v(p, w) = pw + (W - w)p_0 \quad \text{to the union.}$$

Assume the safety security point is $u^* = 0$ for the company and $v^* = Wp_0$ for the union.

a) What is the nonlinear program to find the Nash bargaining solution?

b) Assuming an interior solution (which means you can find the solution by taking derivatives), show that the solution (p^*, w^*) of the Nash bargaining solution satisfies

$$f'(w^*) = p_0 \text{ and } p^* = \frac{w^* p_0 + f(w^*)}{2w^*}.$$

c) Find the Nash bargaining solution for $f(w) = \ln(w + a) + b, a > 0, \frac{1}{a} > p_0, b > -\ln a.$

7.10 Consider the two player matrix game with bi-matrix

$$\begin{bmatrix} (-5, 0) & (1, 3) \\ (4, 1) & (-8, -1) \end{bmatrix}$$

1. Graph the feasible set.
2. Indicate the Pareto-optimal boundary on your graph (it should be a single line segment).
3. Find the Nash bargaining solution using the unique mixed Nash-equilibrium of the game as the status-quo point. Plot this on your feasible set graph.
4. Solve a zero-sum game to find optimal threat strategies and the corresponding status-quo point.
5. Compare the threat payouts to the Nash equilibrium payouts.
6. Graphically find the Nash Bargaining solution using the threat status-quo point from part (4) (that is plot it on the graph but don't solve the algebra). Which player's payoff has increased?
7. Find the Nash bargaining solution with each of the following security points.

   1. s(−4, 1)   4. (−2, −2)
   2. (−2, 2)   5. (2, −3)
   3. (−3, −1)   6. (4, −4)

7.11 Consider the feasible set and security point shown below:

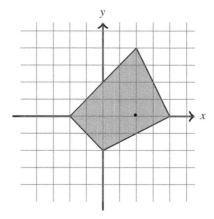

a) Indicate the Pareto-optimal boundary.

b) Explain that the Nash bargaining solution is the solution to

$$\text{Maximize } \{(x - 2)y \mid 2x + y = 8, x \geq 2, y \geq 0\}.$$

c) Remember Lagrange multipliers! If the solution to the above is not at one of the ends of the Pareto-optimal boundary, it should be a solution to

$$\nabla((x - 2)y)) = \lambda \nabla(2x + y),$$

where $\nabla = (\frac{\partial}{\partial x}, \frac{\partial}{\partial y})$ is the vector derivative.

d) Solve!

e) Connect the solution to the security point. The line should have slope 2. First make sure it does, then think about why this makes sense.

f) Using the observation in the previous part, find the Nash bargaining solution if the security point is instead
   1. The origin.
   2. $(-2, 0)$.

7.12 Consider the negotiation game related to the bimatrix game $A = \begin{bmatrix} (7,1) & (-6,-1) \\ (1,4) & (3,2) \end{bmatrix}$.

a) Sketch the feasible set and Pareto-optimal boundary.

b) Find the Nash equilibrium.

c) Using the Nash equilibrium as the security point, what is the Nash bargaining solution?

d) Assume the security point was instead $(1, 1)$. Find the Nash bargaining solution now. Hint: by analogy to the last question, you should draw a line from $(1, 1)$ with slope $\frac{1}{2}$.

e) Explain why the zero-sum game with matrix $M = \begin{bmatrix} 7 & -6 \\ 1 & 3 \end{bmatrix} - 2 \begin{bmatrix} 1 & -1 \\ 4 & 2 \end{bmatrix}$ is relevant.

f) Solve the game from the previous part. Find a new security point by letting players play the resulting (mixed) strategies.

g) Find the new bargaining solution.

7.13 Analyse the Nash bargaining solution with/without threats for the game $\begin{bmatrix} (0,8) & (5,4) \\ (6,0) & (-10,-8) \end{bmatrix}$.
What is different?

Bibliographic Notes

The theory of bargaining presented in this Chapter 7 has two primary developers: Nash and Shapley. Our presentation for finding the optimal threat strategies follows that in Jones' book [16]. The alternative method of bargaining using the KS solution is from Aliprantis and Chakrabarti [1], where more examples and much more discussion can be found. Our union versus management problem (Problem 7.9) is a modification of an example due to Aliprantis and Chakrabarti [1]. The sequential bargaining discussion also follows [1] where much more can be found.

8

Evolutionary Stable Strategies and Population Games

I was a young man with uninformed ideas. I threw out queries, suggestions, wondering all the time over everything; and to my astonishment the ideas took like wildfire. People made a religion of them.

–Charles Darwin

All the ills from which America suffers can be traced to the teaching of evolution.
–William Jennings Bryan

If automobiles had followed the same development cycle as the computer, a Rolls-Royce would today cost $100, get a million miles per gallon, and explode once a year, killing everyone inside.

–Robert Cringely

Every great cause begins as a movement, becomes a business, and eventually degenerates into a racket.

–Eric Hoffer

8.1 Evolution

A major application and extension of game theory to evolutionary theory was initiated by Maynard-Smith and Price.[1] They had the idea that if you looked at interactions among players as a game, then better strategies would eventually evolve and dominate among the players. They introduced the concept of an **evolutionary stable strategy (ESS)** as a good strategy that would not be invaded by any mutants so that bad mutations would not overtake the population. These concepts naturally apply to biology but can be used to explain and predict many phenomena in economics, finance, and other areas in social and political arenas. These applications make finding an ESS an important way to distinguish among Nash equilibria. In this chapter we will present a brief introduction to this important concept.

---

1 John Maynard-Smith, F. R. S. (January 6, 1920–April 19, 2004) was a British biologist and geneticist. George R. Price (1922–January 6, 1975) was an American population geneticist. He and Maynard-Smith introduced the concept of the evolutionary stable strategy (ESS).

Game Theory: An Introduction, Third Edition. E. N. Barron.
© 2024 John Wiley & Sons, Inc. Published 2024 by John Wiley & Sons, Inc.

Consider a population with many members. Whenever two players encounter each other, they play a symmetric bimatrix game with matrices (A, B). Symmetry in this setup means that $B = A^T$.

Let's work with an example.

I/II	Little	Big
Little	$(5, 5)$	$(1, 8)$
Big	$(8, 1)$	$(3, 3)$

The idea is that two organisms of different sizes are competing for resources. If a Little guy meets a Big guy, the Little guy is sure to eke out a barely livable subsistence, but a Big guy does very well. Two Big guys who meet will barely get enough to satisfy them, while two little guys will thrive with the resources available. Our problem is to determine if being little will prevail over the generations or will Big guys eventually take over, or is it the other way around, or can they live side by side.

Begin by observing that only $(3, 3)$ is a Nash equilibrium. Now let's assume that a small percentage p of the population is Big, while a large majority $1 - p$ fraction is Little. Let's consider the pure strategy corresponding to player I always plays Little. When two organisms meet there is probability p a Little guy meets a Big guy, and probability $1 - p$ a Little guy meets another Little guy. This means the expected payoff to a Little guy is

$$FL(p) = 5(1 - p) + 1p = 5 - 4p.$$

On the other hand, if player I is a Big guy, he meets another Big with probability p and a Little with probability $1 - p$. The expected payoff for a Big guy is

$$FB(p) = 8(1 - p) + 3p = 8 - 5p.$$

Would player I prefer to be Big or Little? Which option gives player I a higher expected payoff, i.e., when is $FB(p) > FL(p)$? Obviously, if $5 - 4p < 8 - 5p$, then all this requires is $p < 3$, which is always true. What we can conclude is that playing **Little should not be evolutionary stable** because a Big guy does better.

Is playing Big evolutionary stable? Let's apply the same analysis. Now we assume that a small percentage of the population p is Little while a large percentage $1 - p$ is Big. Player I will play Big and receive expected payoff

$$FB(p) = 8p + 3(1 - p) = 5p + 3.$$

If player I plays Little the expected payoff is

$$FL(p) = 5p + (1 - p) = 4p + 1.$$

Now ask when $FB(p) = 5p + 3 > FL(p) = 4p + 1$, and you see that it would be true if $p > -2$, and that is always true. This implies that Big is indeed evolutionary stable.

As a consequence, if we have a population of Little guys and a small percentage of Big guys show up, either by mutation or invasion from somewhere else, the Big guys will have a small chance of running into another Big but will do extremely well against a Little. Bigs will thrive and cannot be overcome by the Littles. By the same token, introducing a small percentage of Littles into a population of Bigs will not materially affect the Bigs. In evolutionary terms, the Littles will not be able to invade the Bigs, but Bigs can invade Littles and so Big is evolutionary stable.

Now we know how to define what it means to have evolutionary, stable, pure strategies.

Definition 8.1.1 *Given the symmetric game (A, A^T), pure strategy i^* is evolutionary stable (ESS) if there is a $0 < p^* < 1$ so that*

$$(1 - p)a_{i^*i^*} + p\,a_{i^*j} > (1 - p)a_{ji^*} + p\,a_{jj}, \text{ for any } j \neq i^* \text{ and all } 0 < p < p^* \tag{8.1.1}$$

This does not have to be true for all p but for all small enough p. If we think of p as the probability a player does not use i^* (but uses some other strategy j) and $1 - p$ as the probability a player does use i^* then the expected payoff to a player using i^* is

$$\text{Expected payoff using } i^* = (1 - p)a_{i^*i^*} + p\,a_{i^*j}, \quad j \neq i.$$

The expected payoff to a player using j against a player using i^* is

$$\text{Expected payoff using } j = (1 - p)a_{ji^*} + p\,a_{jj}, \quad j \neq i.$$

The definition of ESS says that the expected payoff using i^* is bigger than the expected payoff for a player using any other j, for all percentages of the players deviating from i^* small enough.

Now here is an equivalent and easier to check condition. It has the advantage of not involving p. Furthermore it will show why the fact that $(3, 3)$ is a strict Nash equilibrium is not coincidental for an ESS.

Proposition 8.1.2   *A pure strategy i^* is an ESS if and only if*

1. *(i^*, i^*) is a pure Nash equilibrium, AND,*
2. *$a_{jj} < a_{i^*j}$ for every $j \neq i^*$ which is a best response to i^*.*

Proof: Suppose that i^* is an ESS. In this case (i^*, i^*) must be a Nash equilibrium. In fact if it wasn't, then there is another row j so that $a_{ji^*} > a_{i^*i^*}$. By the definition,

$$(1 - p)(a_{i^*i^*} - a_{ji^*}) > p(a_{jj} - a_{i^*j}), \quad \text{for all } 0 < p < p^* \text{ for some } p^*.$$

Sending $p \to 0$ in this inequality we see that the left side is strictly negative, while the right converges to zero, a contradiction. Hence (i^*, i^*) must be a Nash.

Next, if (i^*, i^*) is a strict Nash, then it must be an ESS. To see why, since i^* is strict, we must have $a_{i^*i^*} > a_{ji^*}$ for any j. But then the condition (8.1.1) is true with $p = 0$ and, by continuity, will also be true for all small $p > 0$.

The final thing we have to answer is the case when (i^*, i^*) is not a strict Nash. When that happens, it means that there is a $j \neq i^*$ which is a best response to i^* so that $a_{i^*i^*} = a_{ji^*}$. But then (8.1.1) is the condition $a_{i^*j} > a_{jj}$. That is

$$(1 - p)a_{i^*i^*} + p\,a_{i^*j} > (1 - p)a_{j,i^*} + p\,a_{jj} \Leftrightarrow a_{i^*j} > a_{jj}$$

for every j which is a best response to i^*. ∎

Example 8.1

1. Consider the symmetric game with

$$\begin{bmatrix} (2,2) & (2,1) \\ (1,2) & (1,1) \end{bmatrix}, \quad A = \begin{bmatrix} 2 & 2 \\ 1 & 1 \end{bmatrix}.$$

   Then $(i^*, i^*) = (1, 1)$ is a pure Nash equilibrium and (1) of the proposition is satisfied. The best response to $i^* = 1$ is only $j = 1$ which means that condition (2) is satisfied vacuously. We conclude that $(1, 1)$ is an ESS. That is, the payoffs $(2, 2)$ as a strict NE, is the ESS.
2. Consider the symmetric game with

$$\begin{bmatrix} (2,2) & (1,1) \\ (1,1) & (2,2) \end{bmatrix}.$$

There are two pure strict Nash equilibria and they are both ESS's. Again condition (2) is satisfied vacuously. (Later we will see the third Nash equilibrium is mixed and it is not an ESS.)

3. Consider the symmetric game with

$$\begin{bmatrix} (1,1) & (2,2) \\ (2,2) & (1,1) \end{bmatrix}.$$

This game also has two pure Nash equilibria neither of which is an ESS since a nonsymmetric NE cannot be an ESS. One of the Nash equilibria is at row 2, column 1, $a_{21} = 2$. If we use $i^* = 2$, row 2 column 2 is NOT a Nash equilibrium. Condition (1) is not satisfied. The pure Nash equilibrium must be symmetric. (We will see later that the mixed Nash equilibrium is symmetric and it is an ESS.)

Now, we consider the case when we have mixed ESSs. We'll start from the beginning.

To make things simple, we assume for now that A is a 2×2 matrix. Then, for a strategy $X = (x, 1-x)$ and $Y = (y, 1-y)$, we will work with the expected payoffs

$$E_{\mathrm{I}}(X, Y) = XAY^T = u(x, y) \quad \text{and} \quad E_{\mathrm{II}}(X, Y) = XA^TY^T = v(x, y).$$

Because of symmetry, we have $u(x, y) = v(y, x)$, so we really only need to talk about $u(x, y)$ and focus on the payoff to player I.

Suppose that there is a strategy $X^* = (x^*, 1 - x^*)$ that is used by most members of the population and s fixed strategies $X_1, \ldots, X_s$, which will be used by **deviants**, a small proportion of the population. Again, we will refer only to the first component of each strategy, $x_1, x_2, \ldots, x_s$.

Suppose that we define a random variable Z, which will represent the strategy played by player I's next opponent. The discrete random variable Z takes on the possible values $x^*, x_1, x_2, \ldots, x_s$ depending on whether the next opponent uses the **usual** strategy, x^*, or a deviant (or mutant) strategy, $x_1, \ldots, x_s$. We assume that the distribution of Z is given by

$$Prob(Z = x^*) = 1 - p, \; Prob(Z = x_j) = p_j, \; j = 1, 2, \ldots, s, \; \sum_{j=1}^{s} p_j = p.$$

It is assumed that X^* will be used by most opponents, so we will take $0 < p \approx 0$, to be a small, but positive number. If player I uses X^* and player II uses Z, the expected payoff, also called the **fitness of the strategy** X^*, is given by

$$F(x^*) \equiv E(u(x^*, Z)) = u(x^*, x^*)(1 - p) + \sum_{j=1}^{s} u(x^*, x_j)p_j. \tag{8.1.2}$$

Similarly, the expected payoff to player I if she uses one of the deviant strategies $x_k, \; k = 1, 2, \ldots, s$, is

$$F(x_k) = E(u(x_k, Z)) = u(x_k, x^*)(1 - p) + \sum_{j=1}^{s} u(x_k, x_j)p_j. \tag{8.1.3}$$

Subtracting (8.1.3) from (8.1.2), we get

$$F(x^*) - F(x_k) = (u(x^*, x^*) - u(x_k, x^*))(1 - p) + \sum_{j=1}^{s} [u(x^*, x_j) - u(x_k, x_j)]p_j. \tag{8.1.4}$$

We want to know what would guarantee $F(x^*) > F(x_k)$ so that x^* is a strictly better, or more **fit**, strategy against any deviant strategy. If

$$u(x^*, x^*) > u(x_k, x^*), k = 1, 2, \ldots, s, \text{ and } p \text{ is a **small enough** positive number,}$$

then, from (8.1.4) we will have $F(x^*) > F(x_k)$, and so no player can do better by using any deviant strategy. This defines x^* as an **uninvadable,** or an **evolutionary stable strategy.** In other words, the strategy X^* resists being overtaken by one of the mutant strategies $X_1, \ldots, X_s$, in case X^* gives a higher payoff when it is used by both players than when any one player decides to use a mutant strategy, and the proportion of deviants in the population p is small enough. We need to include the requirement that p is small enough because $p = \sum_j p_j$ and we need

$$u(x^*, x^*) - u(x_k, x^*) + \frac{1}{1-p} \sum_{j=1}^{s} [u(x^*, x_j) - u(x_k, x_j)] p_j > 0.$$

We can arrange this to be true if p is small but not for all $0 < p < 1$.

Now, the other possibility in (8.1.4) is that the first term could be zero for even one x_k. In that case, $u(x^*, x^*) - u(x_k, x^*) = 0$, and the first term in (8.1.4) drops out for that k. In order for $F(x^*) > F(x_k)$ we would now need $u(x^*, x_j) > u(x_k, x_j)$ for all $j = 1, 2, \ldots, s$. This has to hold for any x_k such that $u(x^*, x^*) = u(x_k, x^*)$. In other words, if there is even one deviant strategy that is as good as x^* when played against x^*, then, in order for x^* to result in a higher average fitness, we must have a bigger payoff when x^* is played against x_j than any payoff with x_k played against x_j. Thus, x^* played against a deviant strategy must be better than x_k against any other deviant strategy x_j.

We summarize these conditions as a definition.

Definition 8.1.3 *A strategy X^* is an ESS against (deviant strategy) strategies $X_1, \ldots, X_s$ if either of (1) or (2) holds:*

(1) $u(x^*, x^*) \geq u(x_k, x^*)$, *for each* $k = 1, 2, \ldots, s$,

(2) *for any* x_k *such that* $u(x^*, x^*) = u(x_k, x^*)$,

we must have $u(x^*, x_j) > u(x_k, x_j)$, *for all* $j = 1, 2, \ldots, s$.

*In the case when there is only one deviant strategy, $s = 1$, and we label $X_1 = X = (x, 1 - x), 0 < x < 1$. That means that every player in the population must use X^* or X with X as the deviant strategy. The proportion of the population using X^* is $1 - p$. The proportion of the population that uses the deviant strategy is p. In this case, the definition reduces to: X^* **is an ESS if and only if either (1) or (2) holds:***

(1) $u(x^*, x^*) > u(x, x^*)$, $\quad \forall \, 0 \leq x \leq 1, x \neq x^*$

(2) $u(x^*, x^*) = u(x, x^*) \implies u(x^*, x) > u(x, x)$, $\quad \forall \, x \neq x^*$.

In the rest of this chapter, we consider only the case $s = 1$.

Remark This definition says that either X^* played against X^* is strictly better than any X played against X^*, or if X against X^* is as good as X^* against X^*, then X^* against any other X must be strictly better than X played against X.

Notice that if X^* and X are any mixed strategies and $0 < p < 1$, then $(1 - p)X + pX^*$ is also a mixed strategy and can be used in an encounter between two players. Here, then, is another definition of an ESS that we will show shortly is equivalent to the first definition.

Definition 8.1.4 *A strategy $X^* = (x^*, 1 - x^*)$ is an evolutionary stable strategy if for every strategy $X = (x, 1 - x)$, with $x \neq x^*$, there is some $p_x \in (0, 1)$, which depends on the particular choice x, such that*

$$u(x^*, px + (1 - p)x^*) > u(x, px + (1 - p)x^*), \quad \textit{for all } 0 < p < p_x. \tag{8.1.5}$$

Remark This definition says that X^* should be a good strategy if and only if this strategy played against the mixed strategy $Y_p \equiv pX + (1-p)X^*$ is better than any deviant strategy X played against $Y_p = pX + (1-p)X^*$, given that the probability p, that a member of the population will use a deviant strategy is sufficiently small.

The left side of (8.1.5) is

$$u(x^*, px + (1-p)x^*) = (x^*, 1-x^*)A \begin{bmatrix} px + (1-p)x^* \\ p(1-x) + (1-p)(1-x^*) \end{bmatrix}$$

$$= X^*A[pX + (1-p)X^*]^T$$

$$= pX^*AX^T + (1-p)X^*AX^{*T}$$

$$= pu(x^*, x) + (1-p)u(x^*, x^*).$$

The right side of (8.1.5) is

$$u(x, px + (1-p)x^*) = (x, 1-x)A \begin{bmatrix} px + (1-p)x^* \\ p(1-x) + (1-p)(1-x^*) \end{bmatrix}$$

$$= XA[pX + (1-p)X^*]^T$$

$$= pXAX^T + (1-p)XAX*^T$$

$$= pu(x, x) + (1-p)u(x, x^*).$$

Putting them together yields that X^* is an ESS according to this definition if and only if

$$pu(x^*, x) + (1-p)u(x^*, x^*) > pu(x, x) + (1-p)u(x, x^*) \quad \text{for } 0 < p < p_x.$$

But this condition is equivalent to

$$p[u(x^*, x) - u(x, x)] + (1-p)[u(x^*, x^*) - u(x, x^*)] > 0 \quad \text{for } 0 < p < p_x. \tag{8.1.6}$$

Now, we can show the two definitions of ESS are equivalent.

Proposition 8.1.5 X^* *is an ESS according to Definition 8.1.3 if and only if X^* is an ESS according to Definition 8.1.4.*

Proof: Suppose X^* satisfies Definition 8.1.3. We will see that inequality (8.1.6) will be true. Now either

$$u(x^*, x^*) > u(x, x^*), \text{ for all } x \neq x^*, \text{ or} \tag{8.1.7}$$

$$u(x^*, x^*) = u(x, x^*) \implies u(x^*, x) > u(x, x) \quad \text{for all } x \neq x^*. \tag{8.1.8}$$

If we suppose that $u(x^*, x^*) > u(x, x^*)$, for all $x \neq x^*$, then for each x there is a small $\gamma_x > 0$ so that

$$u(x^*, x^*) - u(x, x^*) > \gamma_x > 0.$$

For a fixed $x \neq x^*$, since $\gamma_x > 0$, we can find a small enough $p_x > 0$ so that for $0 < p < p_x$ we have

$$u(x^*, x^*) - u(x, x^*) > \frac{1}{1-p}[\gamma_x - p(u(x^*, x) - u(x, x))] > 0.$$

This says that for all $0 < p < p_x$, we have

$$p(u(x^*, x) - u(x, x)) + (1-p)(u(x^*, x^*) - u(x, x^*)) > \gamma_x > 0,$$

which means that (8.1.6) is true. A similar argument shows that if we assume $u(x^*, x^*) = u(x, x^*) \implies u(x^*, x) > u(x, x)$ for all $x \neq x^*$, then (8.1.6) holds. So, if X^* is an ESS in the sense of Definition 8.1.3, then it is an ESS in the sense of Definition 8.1.4.

Conversely, if X^* is an ESS in the sense of Definition 8.1.4, then for each x there is a $p_x > 0$ so that

$$p(u(x^*, x) - u(x, x)) + (1 - p)(u(x^*, x^*) - u(x, x^*)) > 0, \quad \forall \, 0 < p < p_x.$$

If $u(x^*, x^*) - u(x, x^*) = 0$, then since $p > 0$, it must be the case that $u(x^*, x) - u(x, x) > 0$. In case $u(x^*, x^*) - u(x', x^*) \neq 0$ for some $x' \neq x^*$, sending $p \to 0$ in

$$p(u(x^*, x') - u(x', x')) + (1 - p)(u(x^*, x^*) - u(x', x^*)) > 0, \quad \forall \, 0 < p < p_{x'},$$

we conclude that $u(x^*, x^*) - u(x', x^*) > 0$, and this is true for every $x' \neq x^*$. But that says that X^* is an ESS in the sense of Definition 8.1.3. ∎

8.1.1 Properties of an ESS

1. **If X^* is an ESS, then (X^*, X^*) is a Nash equilibrium.** Why? Because if it isn't a Nash equilibrium, then there is a player who can find a strategy $Y = (y, 1 - y)$ such that $u(y, x^*) > u(x^*, x^*)$. Then, for all small enough $p = p_y$, we have

$$p(u(x^*, y) - u(y, y)) + (1 - p)(u(x^*, x^*) - u(y, x^*)) < 0, \quad \forall \, 0 < p < p_y,$$

This is a contradiction of the Definition 8.1.4 of ESS. One consequence of this is that **only the symmetric Nash equilibria** of a game are candidates for ESSs.

2. **If (X^*, X^*) is a strict Nash equilibrium, then X^* is an ESS.** Why? Because if (X^*, X^*) is a strict Nash equilibrium, then $u(x^*, x^*) > u(y, x^*)$ for any $y \neq x^*$. But then, for every small enough p, we would obtain

$$pu(x^*, y) + (1 - p)u(x^*, x^*) > pu(y, y) + (1 - p)u(y, x^*), \text{ for } 0 < p < p_y,$$

for some $p_y > 0$ and all $0 < p < p_y$. But this defines X^* as an ESS according to Definition 8.1.4.

3. **A symmetric Nash equilibrium X^* is an ESS for a symmetric game if and only if $u(x^*, y) > u(y, y)$ for every strategy $y \neq x^*$ that is the best response strategy** to x^*. Recall that $Y = (y, 1 - y)$ is a best response to $X^* = (x^*, 1 - x^*)$ if

$$v(x^*, y) = X^* A^T Y^T = Y A X^{*T} = u(y, x^*) = \max_{W \in S_2} u(w, x^*) = \max_{W \in S_2} W A X^{*T}$$

because of symmetry. In short, $u(y, x^*) = \max_{0 \leq w \leq 1} u(w, x^*)$.

4. **A symmetric 2×2 game with $A = (a_{ij})$ and $a_{11} \neq a_{21}, a_{12} \neq a_{22}$, must have an ESS.**

Let's verify property 4.

Case 1: Suppose $a_{11} > a_{21}$. This implies row 1, column 1 is a strict Nash equilibrium, so immediately, it is an ESS. We reach the same conclusion by the same argument if $a_{22} > a_{12}$.

Case 2: Suppose $a_{11} < a_{21}$ and $a_{22} < a_{12}$. In this case, there is no pure Nash but there is a symmetric mixed Nash given by

$$X^* = (x^*, 1 - x^*), \quad x^* = \frac{a_{12} - a_{22}}{a_{12} - a_{11} + a_{21} - a_{22}} = \frac{a_{12} - a_{22}}{2a_{12} - a_{11} - a_{22}}.$$

Under this case $0 < x^* < 1$. Now, direct calculation shows that

$$u(x^*, x^*) = \frac{a_{11}a_{22} - a_{12}^2}{a_{11} + a_{22} - 2a_{12}} = u(y, x^*), \quad \forall \, Y = (y, 1 - y) \neq X^*.$$

Also,

$$u(y, y) = y^2(a_{11} + a_{22} - 2a_{12}) + 2y(a_{12} - a_{22}) + a_{22}.$$

Using algebra we get

$$u(x^*, y) - u(y, y) = -\frac{(y(a_{11} + a_{22} - 2a_{12}) + a_{12} - a_{22})^2}{a_{11} + a_{22} - 2a_{12}} > 0,$$

since the denominator is negative. Thus X^* satisfies the conditions to be an ESS.

We now present a series of examples illustrating the concepts and calculations.

Example 8.2   In a simplified model of the evolution of currency, suppose that members of a population have currency in either euros or dollars. When they want to trade for some goods, the transaction must take place in the same currency. Here is a possible matrix representation

I/II	Euros	Dollars
Euros	$(1, 1)$	$(0, 0)$
Dollars	$(0, 0)$	$(1, 1)$

Naturally, there are three symmetric Nash equilibria $X_1 = (1, 0) = Y_1, X_2 = (0, 1) = Y_2$ and one mixed Nash equilibrium at $X_3 = (\frac{1}{2}, \frac{1}{2}) = Y_3$. These correspond to everyone using euros, everyone using dollars, or half the population using euros and the other half dollars. We want to know which, if any, of these are ESSs. We have for X_1:

$$u(1, 1) = 1 \quad \text{and} \quad u(x, 1) = x, \text{ so that } u(1, 1) > u(x, 1), \ x \neq 1.$$

This says X_1 is ESS.

Next, for X_2 we have

$$u(0, 0) = 1 \quad \text{and} \quad u(x, 0) = x, \text{ so that } u(0, 0) > u(x, 0), \ x \neq 1.$$

Again, X_2 is an ESS. Note that both X_1 and X_2 are strict Nash equilibria and so ESSs according to Properties (8.1.1), (2).

Finally for X_3, we have

$$u\left(\frac{1}{2}, \frac{1}{2}\right) = \frac{1}{2} \quad \text{and} \quad u\left(x, \frac{1}{2}\right) = \frac{1}{2},$$

so $u(x^*, x^*) = u(\frac{1}{2}, \frac{1}{2}) = \frac{1}{2} = u(x, x^*) = u(x, \frac{1}{2})$, for all $x \neq x^* = \frac{1}{2}$. We now have to check the second possibility in Definition 8.1.3:

$$u(x^*, x^*) = u(x, x^*) = \frac{1}{2} \implies$$
$$u(x^*, x) = \frac{1}{2} > u(x, x) = x^2 + (1 - x)^2 \quad \text{for all } x \neq \frac{1}{2}.$$

But that is false because there are plenty of x values for which this inequality does not hold. We can take, for instance $x = \frac{1}{3}$, to get $\frac{1}{9} + \frac{4}{9} = \frac{5}{9} > \frac{1}{2}$. Consequently, the mixed Nash equilibrium X_3 in which a player uses dollars and euros with equal likelihood is not an ESS. The conclusion is that eventually, the entire population will evolve to either all euros or all dollars, and the other currency will disappear. This seems to predict the future eventual use of one currency worldwide, just as the euro has become the currency for all of Europe.

Example 8.3 Consider the symmetric game with

$$\begin{bmatrix} (2,2) & (1,1) \\ (1,1) & (2,2) \end{bmatrix}.$$

Earlier we saw that there are two pure Nash equilibria and they are both ESS's. There is also a mixed ESS given by $X^* = (\frac{1}{2}, \frac{1}{2})$ but it is not an ESS. In fact, since $u(\frac{1}{2}, y) = \frac{3}{2}$, no matter which $Y = (y, 1-y)$ is used, any mixed strategy is a best response to X^*. If X^* is an ESS it must be true that $u(\frac{1}{2}, y) > u(y, y)$ for all $0 < y < 1, y \neq \frac{1}{2}$. However,

$$u\left(\frac{1}{2}, y\right) = \frac{3}{2} < u(y, y) = 2\left(y - \frac{1}{2}\right)^2 + \frac{3}{2}$$

and X^* is not an ESS.

Now consider the symmetric game with

$$\begin{bmatrix} (1,1) & (2,2) \\ (2,2) & (1,1) \end{bmatrix}.$$

We have seen that this game has two pure Nash equilibria, neither of which is an ESS. There is a mixed symmetric Nash equilibrium given by $X^* = (\frac{1}{2}, \frac{1}{2})$, and it is an ESS. In fact, we will show that

$$u\left(\frac{1}{2}, px + (1-p)\frac{1}{2}\right) > u(x, px + (1-p)\frac{1}{2}), \quad \forall x \neq \frac{1}{2}$$

for small enough $0 < p < 1$. By calculation

$$u\left(\frac{1}{2}, px + (1-p)\frac{1}{2}\right) = \frac{3}{2}, \quad u\left(x, px + (1-p)\frac{1}{2}\right) = \frac{3}{2} - \frac{p}{2}(1-2x)^2,$$

and indeed for any $x \neq \frac{1}{2}$ we have $\frac{3}{2} > \frac{3}{2} - \frac{p}{2}(1-2x)^2$ for any $0 < p < 1$. Hence by (8.1.6) X^* is an ESS.

Example 8.4 Hawk–Dove Game. Each player can choose to act like a hawk or act like a dove. They can either fight or yield when they meet over some roadkill. The payoff matrix is

H/D	Fight	Yield
Fight	$\left(\dfrac{v-c}{2}, \dfrac{v-c}{2}\right)$	$(v, 0)$
Yield	$(0, v)$	$\left(\dfrac{v}{2}, \dfrac{v}{2}\right)$

, so $A = B^T = \begin{bmatrix} \frac{(v-c)}{2} & v \\ 0 & \frac{v}{2} \end{bmatrix}$

The reward for winning a fight is $v > 0$, and the cost of losing a fight is $c > 0$. Each player has an equal chance of winning a fight. The payoff to each player if they both fight is thus $v\frac{1}{2} + (-c)\frac{1}{2}$. If hawk fights, and dove yields, hawk gets v while dove gets 0. If they both yield they both receive $v/2$. This is a symmetric two-person game.

Consider the following cases:

Case 1: $v > c$. When the reward for winning a fight is greater than the cost of losing a fight, there is a unique symmetric Nash equilibrium at (fight,fight). In addition, it is strict, because $\frac{(v-c)}{2} > 0$, and so it is the one and only ESS. Fighting is evolutionary stable, and you will end up with a population of fighters.

Case 2: $v = c$. When the cost of losing a fight is the same as the reward of winning a fight, there are two nonstrict nonsymmetric Nash equilibria at (yield,fight) and (fight,yield). But there is only one symmetric, nonstrict Nash equilibrium at (fight,fight). Since u(fight, yield) $= v >$ u(yield, yield) $= v/2$, the Nash equilibrium (fight,fight) satisfies the conditions in the definition of an ESS.

Case 3: $c > v$. Under the assumption $c > v$, we have two pure nonsymmetric Nash equilibria at $X_1 = (0, 1), Y_1 = (1, 0)$ with payoff $u(0, 1) = v/2$ and $X_2 = (1, 0), Y_2 = (0, 1)$, with payoff $u(1, 0) =$ $(v - c)/2$. It is easy to calculate that $X_3 = \left(\frac{v}{c}, 1 - \frac{v}{c}\right) = Y_3$ is the symmetric mixed strategy Nash equilibrium. This is a symmetric game, and so we want to know which, if any, of the three Nash points are evolutionary stable in the sense of Definition 8.1.3. But we can immediately eliminate the nonsymmetric equilibria.

Before we consider the general case let's take for the time being specifically $v = 4, c = 6$. Then $\frac{v}{c} = \frac{2}{3}, \frac{(v-c)}{2} = -1$, so

$$u(1, 1) = -1, \quad u(0, 0) = 2, \quad \text{and} \quad u\left(\frac{2}{3}, \frac{2}{3}\right) = \frac{2}{3}.$$

Let's consider the mixed strategy $X_3 = Y_3 = \left(\frac{2}{3}, \frac{1}{3}\right)$. We have

$$u\left(\frac{2}{3}, \frac{2}{3}\right) = \frac{2}{3} = u\left(x, \frac{2}{3}\right),$$
$$u\left(\frac{2}{3}, x\right) = -4x + \frac{10}{3} \quad \text{and} \quad u(x, x) = -3x^2 + 2.$$

Since $u\left(\frac{2}{3}, \frac{2}{3}\right) = u\left(x, \frac{2}{3}\right)$ we need to show that the second case in Definition 8.1.3 holds, namely, $u\left(\frac{2}{3}, x\right) = -4x + \frac{10}{3} > -3x^2 + 2 = u(x, x)$ for all $0 \le x \le 1, x \ne \frac{2}{3}$. By algebra, this is the same as showing that $\left(x - \frac{2}{3}\right)^2 > 0$, for $x \ne \frac{2}{3}$, which is obvious. We conclude that $X_3 = Y_3 = \left(\frac{2}{3}, \frac{1}{3}\right)$ is evolutionary stable.

For comparison purposes, let's try to show that $X_1 = (0, 1)$ is not an ESS directly from the definition. Of course, we already know it isn't because (X_1, X_1) is not a Nash equilibrium. So if we try the definition for $X_1 = (0, 1)$, we would need to have that

$$u(0, 0) = 2 \ge u(x, 0) = 2x + 2,$$

which is clearly false for any $0 < x \le 1$. So X_1 is not evolutionary stable.

From a biological perspective, this says that only the mixed strategy Nash equilibrium is evolutionary stable, and so always fighting, or always yielding is not evolutionary stable. Hawks and doves should fight two-thirds of the time when they meet.

In the general case with $v < c$ we have the mixed symmetric Nash equilibrium $X^* = \left(\frac{v}{c}, 1 - \frac{v}{c}\right) =$ Y^*, and since

$$u\left(\frac{v}{c}, \frac{v}{c}\right) = u\left(x, \frac{v}{c}\right), \quad x \ne \frac{v}{c},$$

we need to check whether $u(x^*, x) > u(x, x)$. An algebra calculation shows that

$$u\left(\frac{v}{c}, x\right) - u(x, x) = \frac{c}{2}\left(\frac{v}{c} - x\right)\left(\frac{v}{c} - x\right) > 0, \quad \text{if } x \ne \frac{v}{c}.$$

Consequently, X^* is an ESS.

It is not true that every symmetric game will have an ESS. The game in the next example will illustrate that.

Example 8.5 We return to consideration of the rock–paper–scissors game with a variation on what happens when there is a tie. Consider the matrix

I/II	Rock	Paper	Scissors
Rock	$\left(\frac{1}{2}, \frac{1}{2}\right)$	$(-1, 1)$	$(1, -1)$
Paper	$(1, -1)$	$\left(\frac{1}{2}, \frac{1}{2}\right)$	$(-1, 1)$
Scissors	$(-1, 1)$	$(1, -1)$	$\left(\frac{1}{2}, \frac{1}{2}\right)$

A tie in this game gives each player a payoff of $\frac{1}{2}$, so this is not a zero sum game, but it is symmetric. You can easily verify that there is one Nash equilibrium mixed strategy, and it is $X^* = (\frac{1}{3}, \frac{1}{3}, \frac{1}{3}) = Y^*$, with expected payoff $\frac{1}{6}$ to each player. (If the diagonal terms gave a payoff to each player ≥ 1, there would be many more Nash equilibria, both mixed and pure.) We claim that (X^*, X^*) is not an ESS.

If X^* is an ESS, it would have to be true that either

$$u(X^*, X^*) > u(X, X^*) \quad \text{for all } X \neq X^* \tag{8.1.9}$$

or

$$u(X^*, X^*) = u(X, X^*) \implies u(X^*, X) > u(X, X) \quad \text{for all } X \neq X^*. \tag{8.1.10}$$

Notice that this is a 3×3 game, and so we have to use strategies $X = (x_1, x_2, x_3) \in S_3$. Now $u(X^*, X^*) = X^* A X^{*T} = \frac{1}{6}$, where

$$A = \begin{bmatrix} \frac{1}{2} & -1 & 1 \\ 1 & \frac{1}{2} & -1 \\ -1 & 1 & \frac{1}{2} \end{bmatrix}$$

and

$$u(X, X^*) = \frac{x_1}{6} + \frac{x_2}{6} + \frac{x_3}{6} = \frac{1}{6}, \quad u(X, X) = \frac{(x_1^2 + x_2^2 + x_3^2)}{2}.$$

The first possibility (8.1.9) does not hold.

For the second possibility condition (8.1.10), we have

$$u(X^*, X^*) = \frac{1}{6} = u(X, X^*),$$

but

$$u(X^*, X) = \frac{x_1 + x_2 + x_3}{6} = \frac{1}{6}$$

is **not** greater than $u(X, X) = (x_1^2 + x_2^2 + x_3^2)/2$ for all mixed $X \neq X^*$. For example, take the pure strategy $X = (1, 0, 0)$ to see that it fails. Therefore, neither possibility holds, and X^* is not an ESS.

Our conclusion can be phrased in this way. Suppose that one player decides to use $X^* = \left(\frac{1}{3}, \frac{1}{3}, \frac{1}{3}\right)$. The best response to X^* is the strategy $\overline{X}$ so that

$$u(\overline{X}, X^*) = \max_{Y \in S_3} u(Y, X^*) = \max_{Y \in S_3} \frac{y_1 + y_2 + y_3}{6} = \frac{1}{6},$$

and any pure strategy, for instance $\overline{X} = (1, 0, 0)$, will give that. But then $u(\overline{X}, \overline{X}) = \frac{1}{2} > \frac{1}{6}$, so that any deviant using any of the pure strategies can do better and will eventually invade the population. There is no **uninvadable** strategy.

Remark There is much more to the theory of evolutionary stability and many extensions of the theory, as you will find in the references. In the next section, we will present one of these extensions because it shows how the stability theory of ordinary differential equations enters game theory.

Problems

8.1 In the currency game (Example 8.2), derive the same result we obtained but using the equivalent definition of ESS: X^* is an evolutionary stable strategy if for every strategy $X = (x, 1 - x)$, with $x \neq x^*$, there is some $p_x \in (0, 1)$, which depends on the particular choice x, such that

$$u(x^*, px + (1 - p)x^*) > u(x, px + (1 - p)x^*) \quad \text{for all } 0 < p < p_x.$$

Find the value of p_x in each case an ESS exists.

8.2 It is possible that there is an economy that uses a dominant currency in the sense that the matrix becomes

I/II	Euros	Dollars
Euros	$(1, 1)$	$(0, 0)$
Dollars	$(0, 0)$	$(2, 2)$

Find all Nash equilibria and determine which are ESSs.

8.3 The decision of whether or not each of two admirals should attack an island was studied in Problem (3.22). The analysis resulted in the following matrix.

Fr/Brit	11	12
11	$\left(-\frac{5}{2}, -\frac{5}{2}\right)$	$\left(\frac{7}{4}, \frac{1}{2}\right)$
12	$\left(\frac{1}{2}, \frac{7}{4}\right)$	$\left(\frac{5}{4}, \frac{7}{4}\right)$

Notice that this is a symmetric game. There are 3 Nash equilibria
1. $X_1 = Y_1 = (\frac{1}{7}, \frac{6}{7})$, payoff French $= \frac{8}{7}$, British $= \frac{8}{7}$.
2. $X_2 = (1, 0), Y_2 = (0, 1)$, payoff French $= \frac{1}{2}$, British $= \frac{7}{4}$.
3. $X_3 = (0, 1), Y_3 = (1, 0)$, payoff French $= \frac{7}{4}$, British $= \frac{1}{2}$.
Determine which, if any, of these are ESSs.

8.4 Analyze the Nash equilibria for a version of the prisoner's dilemma game:

I/II	Confess	Deny
Confess	$(4, 4)$	$(1, 6)$
Deny	$(6, 1)$	$(1, 1)$

8.5 Determine the Nash equilibria for rock–paper–scissors with matrix

I/II	Rock	Paper	Scissors
Rock	$(2, 2)$	$(-1, 1)$	$(1, -1)$
Paper	$(1, -1)$	$(2, 2)$	$(-1, 1)$
Scissors	$(-1, 1)$	$(1, -1)$	$(2, 2)$

There are three pure Nash equilibria and four mixed equilibria (all symmetric). Determine which are evolutionary stable strategies, if any, and if an equilibrium is not an ESS, show how the requirements fail.

8.6 In Problem (2.18), we considered the game of the format to be used for radio stations WSUP and WHAP. The game matrix is

WSUP/WHAP	RB	EM	AT
RB	$(25, 25)$	$(50, 30)$	$(50, 20)$
EM	$(30, 50)$	$(15, 15)$	$(30, 20)$
AT	$(20, 50)$	$(20, 30)$	$(10, 10)$

Determine which, if any of the three Nash equilibria are ESSs.

8.7 Consider a game with matrix $A = \begin{bmatrix} a & 0 \\ 0 & b \end{bmatrix}$. Suppose that $ab \neq 0$.

(a) Show that if $ab < 0$, then there is exactly one ESS. Find it.

(b) Suppose that $a > 0, b > 0$. Then there are three symmetric Nash equilibria. Show that the Nash equilibria which are evolutionary stable are the pure ones and that the mixed Nash is not an ESS.

(c) Suppose that $a < 0, b < 0$. Show that this game has two pure nonsymmetric Nash equilibria and one symmetric mixed Nash equilibrium. Show that the mixed Nash is an ESS.

8.8 Verify that $X^* = (x^*, 1 - x^*)$ is an ESS if and only if (X^*, X^*) is a Nash equilibrium and $u(x^*, x) > u(x, x)$ for every $X = (x, 1 - x) \neq X^*$ that is a best response to X^*.

8.2 Population Games

One important idea introduced by considering evolutionary stable strategies is the idea that eventually, players will choose strategies that produce a better-than-average payoff. The clues for this section are the words **eventually** and **better-than-average**. The word **eventually** implies a time dependence and a limit as time passes, so it is natural to try to model what is happening by introducing a time-dependent equation and let $t \to \infty$. That is exactly what we study in this section.

Here is the setup. There are N members of a population. At random two members of the population are chosen to play a certain game against each other. It is assumed that N is a really big number, so that the probability of a faceoff between two of the same members is virtually zero.

We assume that the game they play is a symmetric two-person bimatrix game. **Symmetry** here again means that $B = A^T$. Practically, it means that the players can switch roles without change. In a symmetric game it doesn't matter who is player I and who is player II.

Assume that there are many players in the population. Any particular player is chosen from the population, chooses a mixed strategy, and plays in the bimatrix game with matrix A against any other player chosen from the population. This is called a **random contest**. Notice that all players will be using the same payoff matrix.

The players in the population may use pure strategies $1, 2, \ldots, n$. Suppose that the **percentage of players in the population using strategy j** is

$$P(\text{player uses } j) = p_j, \quad p_j \geq 0, \quad \sum_{j=1}^{n} p_j = 1.$$

Set $\pi = (p_1, \ldots, p_n)$. These p_i components of π are called the **frequencies** and represent the probability a randomly chosen individual in the population will use the strategy i. Denote by

$$\Pi = \left\{ \pi = (p_1, p_2, \ldots, p_n) \mid p_j \geq 0, j = 1, 2, \ldots, n, \sum_{j=1}^{n} p_j = 1 \right\}$$

the **set of all possible frequencies**.

If two players are chosen from the population and player I chooses strategy i and player II chooses strategy j, we calculate the payoffs from the matrix A.

We define the **fitness** of a player playing strategy $i = 1, 2, \ldots, n$ as

$$E(i, \pi) = \sum_{k=1}^{n} a_{i,k} p_k = {}_i A \pi.$$

This is the expected payoff of a random contest to player I who uses strategy i against the other possible strategies $1, 2, \ldots n$, played with probabilities $p_1, \ldots, p_n$. It measures the worth of strategy i in the population. You can see that the π looks just like a mixed strategy and we may identify $E(i, \pi)$ as the expected payoff to player I if player I uses the pure strategy i and the opponent (player II) uses the mixed strategy π.

Next, we calculate the expected fitness of the entire population as

$$E(\pi, \pi) := \sum_{i=1}^{n} p_i \left[\sum_{k=1}^{n} a_{i,k} p_k \right] = \pi A \pi^T.$$

Now suppose that the frequencies $\pi = (p_1, \ldots, p_n) = \pi(t) \in \Pi$ can change with time. This is where the evolutionary characteristics are introduced. We need a model describing how the frequencies can change in time and we use the **frequency dynamics** as the following system of differential equations:

$$\frac{dp_i(t)}{dt} = p_i(t) \left[E(i, \pi(t)) - E(\pi(t), \pi(t)) \right] \tag{8.2.1}$$

$$= p_i(t) \left[\sum_{k=1}^{n} a_{i,k} p_k(t) - \pi(t) A \pi(t)^T \right], \quad i = 1, 2, \ldots, n,$$

or, equivalently,

$$\frac{dp_i(t)}{p_i(t)} = \left[\sum_{k=1}^{n} a_{i,k} p_k(t) - \pi(t) A \pi(t)^T \right] dt. \tag{8.2.2}$$

This is also called the **replicator dynamics**. The idea is that the growth rate at which the population percentage using strategy i changes is measured by how much greater (or less) the expected payoff (or fitness) using i is compared with the expected fitness using all strategies in the population. Better strategies should be used with increasing frequency and worse strategies with decreasing frequency. In the one-dimensional case the right side will be positive if the fitness using pure strategy i is better than average and negative if worse than average. That makes the derivative $dp_i(t)/dt$ positive if better than average, which causes $p_i(t)$ to increase as time progresses, and strategy i will be used with increasing frequency.

We are not specifying at this point the initial conditions $\pi(0) = (p_1(0), \ldots, p_n(0))$, but we know that $\sum_i p_i(0) = 1$. We also note that any realistic solution of Eq. (8.2.1) must have $0 \leq p_i(t) \leq 1$ as well as $\sum_i p_i(t) = 1$ for all $t > 0$. In other words, we must have $\pi(t) \in \Pi$ for all $t \geq 0$. Here is one way to check that: add up the equations in (8.2.1) to get

$$\sum_{i=1}^{n} \frac{dp_i(t)}{dt} = \sum_{i=1}^{n}\sum_{k=1}^{n} p_i(t)a_{i,k}p_k(t) - \sum_{i=1}^{n} p_i(t)\pi(t)A\pi(t)^T$$

$$= \pi(t)A\pi(t)^T \left[1 - \sum_{i=1}^{n} p_i(t)\right]$$

or, setting $\gamma(t) = \sum_{i=1}^{n} p_i(t)$,

$$\frac{d\gamma}{dt} = \pi(t)A\pi(t)^T[1 - \gamma(t)].$$

If $\gamma(0) = 1$, the **unique solution** of this equation is $\gamma(t) \equiv 1$, as you can see by plugging in $\gamma(t) = 1$. Consequently, $\sum_{i=1}^{n} p_i(t) = 1$ for all $t \geq 0$. By uniqueness it is the one and only solution. In addition, assuming that $p_i(0) > 0$, if it ever happens that $p_i(t_0) = 0$ for some $t_0 > 0$, then, by considering the trajectory with that new initial condition, we see that $p_i(t) \equiv 0$ for all $t \geq t_0$. This conclusion also follows from the fact that (8.2.1) has one and only one solution through any initial point, and zero will be a solution if the initial condition is zero. Similar reasoning, which we skip, shows that for each $i, 0 \leq p_i(t) \leq 1$. Therefore $\pi(t) \in \Pi$, for all $t \geq 0$, if $\pi(0) \in \Pi$.

Notice that when the right side of (8.2.1)

$$p_i(t) \left[\sum_{k=1}^{n} a_{i,k}p_k(t) - \pi(t)A\pi(t)^T\right] = 0,$$

then $dp_i(t)/dt = 0$ and $p_i(t)$ is not changing in time. In differential equations a solution that doesn't change in time will be a **steady-state**, **equilibrium**, or **stationary** solution. So, if there is a constant solution $\pi^* = (p_1^*, \ldots, p_n^*)$ of

$$p_i^* \left[\sum_{k=1}^{n} a_{i,k}p_k^* - \pi^*A\pi^{*T}\right] = 0,$$

then if we start at $\pi(0) = \pi^*$, we will stay there for all time and $\lim_{t\to\infty}\pi(t) = \pi^*$ is a steady state solution of (8.2.1).

Remark If the symmetric game has a completely mixed Nash equilibrium $X^* = (x_1^*, \ldots, x_n^*)$, $x_i^* > 0$, then $\pi^* = X^*$ is a stationary solution of (8.2.1). The reason is provided by the equality of payoffs Theorem 2.2.3, which guarantees that $E(i, X^*) = E(j, X^*) = E(X^*, X^*)$ for any pure strategies i, j played with positive probability. So, if the strategy is completely mixed, then $x_i^* > 0$ and the payoff using row i must be the average payoff to the player. But, then if $E(i, X^*) = E(i, \pi^*) = E(\pi^*, \pi^*) = E(X^*, X^*)$, we have the right side of (8.2.1) is zero and then that $\pi^* = X^*$ is a stationary solution.

Example 8.6 Consider the symmetric game with

$$A = \begin{bmatrix} 1 & 3 \\ 2 & 0 \end{bmatrix}, \quad B = A^T.$$

Suppose that the players in the population may use the pure strategies $k \in \{1, 2\}$. The frequency of using k is $p_k(t)$ at time $t \geq 0$, and it is the case that $\pi(t) = (p_1(t), p_2(t)) \in \Pi$. Then, the fitness of a player using $k = 1, 2$ becomes

$$E(1, \pi) = {}_1A \cdot \pi = \sum_{j=1}^{2} a_{1,j}p_j = p_1 + 3p_2,$$

$$E(2, \pi) = {}_2A \cdot \pi = \sum_{j=1}^{2} a_{2j} p_j = 2p_1,$$

and the average fitness in the population is

$$E(\pi, \pi) = \pi A \pi^T = (p_1, p_2) A \begin{bmatrix} p_1 \\ p_2 \end{bmatrix} = p_1^2 + 5p_1 p_2.$$

We end up with the following system of equations:

$$\frac{dp_1(t)}{dt} = p_1(t) \left[p_1 + 3p_2 - (p_1^2 + 5p_1 p_2) \right],$$

$$\frac{dp_2(t)}{dt} = p_2(t) \left[2p_1 - (p_1^2 + 5p_1 p_2) \right].$$

We are interested in the long-term behavior $\lim_{t \to \infty} p_i(t)$ of these equations because the limit is the eventual evolution of the strategies. The steady-state solution of these equations occurs when $dp_i(t)/dt = 0$, which, in this case implies that

$$p_1 \left[p_1 + 3p_2 - (p_1^2 + 5p_1 p_2) \right] = 0 \quad \text{and} \quad p_2 \left[2p_1 - (p_1^2 + 5p_1 p_2) \right] = 0.$$

Since both p_1 and p_2 cannot be zero (because $p_1 + p_2 = 1$), you can check that we have steady-state solutions

$$(p_1 = 0, p_2 = 1), \quad (p_1 = 1, p_2 = 0), \quad \text{and} \quad \left(p_1 = \frac{3}{4}, p_2 = \frac{1}{4} \right).$$

The arrows in Figure 8.1 show the direction a trajectory (=solution of the frequency dynamics) will take as time progresses depending on the starting condition. You can see in the figure that no matter where the initial condition is **inside** the square $(p_1, p_2) \in (0, 1) \times (0, 1)$, (since $p_1 + p_2 = 1$ we are really only on the line from $(0, 1)$ to $(1, 0)$) the trajectories will be sucked into the solution at the point $(p_1 = \frac{3}{4}, p_2 = \frac{1}{4})$ as the equilibrium solution.

Remark Whenever there are only two pure strategies used in the population, we can simplify down to one equation for $p(t)$ using the substitutions $p_1(t) = p(t), p_2(t) = 1 - p(t)$. Since

$$p_1(E(1, \pi) - E(\pi, \pi)) = p(1 - p)(E(1, \pi) - E(2, \pi)),$$

Eq. (8.2.1) then becomes

$$\frac{dp(t)}{dt} = p(t)(1 - p(t))(E(1, \pi) - E(2, \pi)), \quad \pi = (p, 1 - p), \tag{8.2.3}$$

and we must have $0 \leq p(t) \leq 1$. The equation for p_2 in (8.2.1) is redundant.

For Example 8.6 the equation reduces to

$$\frac{dp(t)}{dt} = p(t)(1 - p(t))(-4p(t) + 3),$$

which is no easier to solve exactly. This equation can be solved implicitly using integration by parts to give the implicitly defined solution

$$\ln \left[p^{1/3} |p - 1| (4p - 3)^{-4/3} \right] = t + C.$$

This is valid only away from the stationary points $p = 0, 1, \frac{3}{4}$, because, as you can see, the logarithm is a problem at those points. Figure 8.2 shows the direction field of $p(t)$ versus time on the left and the direction field superimposed with the graphs of four trajectories starting from four different

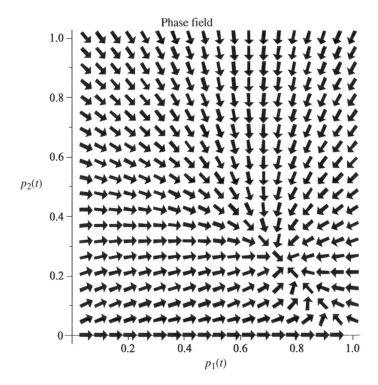

Figure 8.1 Stationary solution at $\left(\frac{3}{4}, \frac{1}{4}\right)$.

initial conditions, two on the edges, and two in the interior of $(0, 1)$. The interior trajectories get pulled quickly as $t \to \infty$ to the steady-state solution $p(t) = \frac{3}{4}$, while the trajectories that start on the edges stay there forever.

In the long run, as long as the population is not using strategies on the edges (i.e., pure strategies), the population will eventually use the pure strategy $k = 1$ exactly 75% of the time and strategy $k = 2$ exactly 25% of the time.

Now, the idea is that the limit behavior of $\pi(t)$ will result in conclusions about how the population will evolve regarding the use of strategies. But that is what we studied in the previous section on evolutionary stable strategies. There must be a connection.

Example 8.7 In Problem (8.7) we looked at the game with matrix $A = \begin{bmatrix} a & 0 \\ 0 & b \end{bmatrix}$, and assumed $ab \neq 0$. You showed that

1. If $ab < 0$, then there is exactly one ESS. It is $X^* = (0, 1) = Y^*$ if $a < 0, b > 0$, and $X^* = (1, 0) = Y^*$ if $b < 0, a > 0$.
2. If $a > 0, b > 0$, then there are three symmetric Nash equilibria. The Nash equilibria that are evolutionary stable are $X_1^* = (1, 0), X_2^* = (0, 1)$, and the mixed Nash $X_3^* = \left(\frac{b}{(a+b)}, \frac{a}{(a+b)}\right)$ is not an ESS.
3. If $a < 0, b < 0$, this game has two pure nonsymmetric Nash equilibria and one symmetric mixed Nash equilibrium. The mixed Nash $X^* = \left(\frac{b}{(a+b)}, \frac{a}{(a+b)}\right)$ is an ESS.

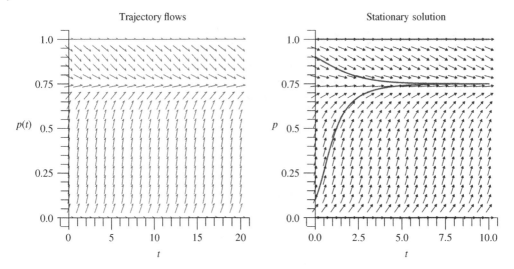

Figure 8.2 Direction field and trajectories for $dp(t)/dt = p(t)(1 - p(t))(-4p(t) + 3)$, $p(t)$ versus time with four initial conditions.

The system of equations for this population game becomes

$$\frac{dp_1(t)}{dt} = p_1(t)\left[\sum_{k=1}^{2} a_{1,k}p_k(t) - (p_1(t), p_2(t))A\begin{bmatrix} p_1(t) \\ p_2(t) \end{bmatrix}\right]$$

$$= p_1(t)[ap_1(t) - ap_1^2(t) - bp_2^2(t)],$$

$$\frac{dp_2(t)}{dt} = p_2(t)\left[\sum_{k=1}^{2} a_{2,k}p_k(t) - (p_1(t), p_2(t))A\begin{bmatrix} p_1(t) \\ p_2(t) \end{bmatrix}\right]$$

$$= p_2(t)[bp_2(t) - ap_1^2(t) - bp_2^2(t)].$$

We can simplify using the fact that $p_2 = 1 - p_1$ to get

$$\frac{dp_1(t)}{dt} = p_1(t)[ap_1(t)(1 - p_1(t)) - bp_2^2(t)]. \tag{8.2.4}$$

Now, we can see that if $a > 0$ and $b < 0$, the right side is always > 0, and so $p_1(t)$ increases while $p_2(t)$ decreases. Similarly, if $a < 0, b > 0$, the right side of (8.2.4) is always < 0 and so $p_1(t)$ decreases while $p_2(t)$ increases. In either case, as $t \to \infty$ we converge to either $(p_1, p_2) = (1, 0)$ or $(0, 1)$, which is at the unique ESS for the game. When $p_1(t)$ must always increase, for example, but cannot get above 1, we know it must converge to 1.

In the case when $a > 0, b > 0$ there is only one mixed Nash (which is not an ESS) occurring with the mixture $X^* = (\frac{b}{(a+b)}, \frac{a}{(a+b)})$. It is not hard to see that in this case, the trajectory $(p_1(t), p_2(t))$ will converge to one of the two pure Nash equilibria that are the ESSs of this game. In fact, if we integrate the differential equation, which is (8.2.4), replacing $p_2 = 1 - p_1$ and $p = p_1$, we have using integration by partial fractions

$$\frac{dp}{p(1 - p)[ap - b(1 - p)]} = dt \implies$$

$$\ln\left[|1 - p|^{-1/a}p^{-1/b}(ap - b(1 - p))^{1/a+1/b}\right] = t + C$$

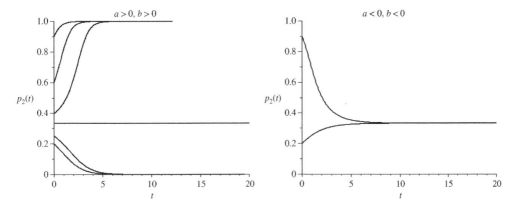

Figure 8.3 Left: $a > 0, b > 0$; Right: $a < 0, b < 0$.

or

$$\frac{(ap - b(1-p))^{1/a+1/b}}{|1 - p|^{1/a} p^{1/b}} = Ce^t.$$

As $t \to \infty$, assuming $C > 0$, the right side goes to ∞. There is no way that could happen on the left side unless $p \to 0$ or $p \to 1$. It would be impossible for the left side to become infinite if $\lim_{t\to\infty} p(t)$ is strictly between 0 and 1.

We illustrate the case $a > 0, b > 0$ in Figure 8.3 on the left for choices $a = 1, b = 2$ and several distinct initial conditions. You can see that $\lim_{t\to\infty} p_2(t) = 0$ or $= 1$, so it is converging to a pure ESS as long as we do not start at $p_1(0) = \frac{2}{3}, p_2(0) = \frac{1}{3}$.

In the case $a < 0, b < 0$ the trajectories converges to the mixed Nash equilibrium, which is the unique ESS. This is illustrated in Figure 8.3 on the right with $a = -1, b = -2$, and you can see that $\lim_{t\to\infty} p_2(t) = \frac{1}{3}$ and so $\lim_{t\to\infty} p_1(t) = \frac{2}{3}$. The phase portraits in Figure 8.4 show clearly what is happening if you follow the arrows.

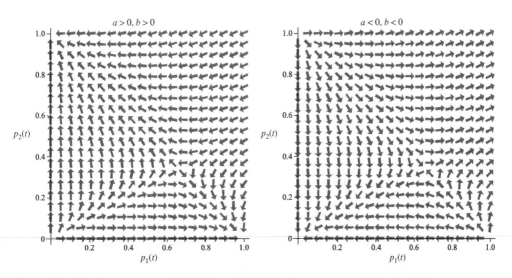

Figure 8.4 Left: Convergence to ESS $(1, 0)$ or $(0, 1)$; Right: Convergence to the mixed ESS.

Remark   Example 8.7 is not as special as it looks because if we are given any two-player symmetric game with matrix

$$A = \begin{bmatrix} a_{11} & a_{12} \\ a_{21} & a_{22} \end{bmatrix}, \quad B = A^T,$$

then the game is equivalent to the symmetric game with matrix

$$A = \begin{bmatrix} a_{11} - a & a_{12} - b \\ a_{21} - a & a_{22} - b \end{bmatrix}, \quad B = A^T$$

for any a, b, in the sense that they have the same set of Nash equilibria. This was Problem (2.28). In particular, if we take $a = a_{21}$ and $b = a_{12}$, we have the equivalent matrix game

$$\bar{A} = \begin{bmatrix} a_{11} - a_{21} & a_{12} - a_{12} \\ a_{21} - a_{21} & a_{22} - a_{21} \end{bmatrix} = \begin{bmatrix} a_{11} - a_{21} & 0 \\ 0 & a_{22} - a_{21} \end{bmatrix}, \quad \bar{B} = \bar{A}^T$$

and so we are in the case discussed in the example.

Let's step aside and write down some results we need from ordinary differential equations. Here is a theorem guaranteeing existence and uniqueness of solutions of differential equations.

Theorem 8.2.1   *Suppose that you have a system of differential equations*

$$\frac{d\pi}{dt} = f(\pi(t)), \quad \pi = (p_1, \ldots, p_n). \tag{8.2.5}$$

Assume that $f : \mathbb{R}^n \to \mathbb{R}^n$ and $\partial f / \partial p_i$ are continuous. Then for any initial condition $\pi(0) = \pi_0$, there is a unique solution up to some time $T > 0$.

Definition 8.2.2   *A steady-state (or stationary, or equilibrium, or fixed-point) solution of the system of ordinary differential equations (8.2.5) is a constant vector π^* that satisfies $f(\pi^*) = 0$. It is (locally)* **stable** *if for any $\varepsilon > 0$ there is a $\delta > 0$ so that every solution of the system with initial condition π_0 satisfies*

$$|\pi_0 - \pi^*| < \delta \implies |\pi(t) - \pi^*| < \varepsilon, \quad \forall t > 0.$$

A stationary solution of (8.2.5) is (locally) **asymptotically stable** *if it is locally stable and if there is $\rho > 0$ so that*

$$|\pi_0 - \pi^*| < \rho \implies \lim_{t \to \infty} |\pi(t) - \pi^*| = 0.$$

The set

$$B_{\pi^*} = \{\pi_0 \mid \lim_{t \to \infty} \pi(t) = \pi^*\}$$

is called the **basin of attraction** *of the steady state π^*. Here $\pi(t)$ is a trajectory through the initial point $\pi(0) = \pi_0$. If every initial point that is possible is in the basin of attraction of π^*, we say that the point π^* is* **globally asymptotically stable***.*

Stability means that if you have a solution of Eq. (8.2.5) that starts **near** the stationary solution, then it will stay near the stationary solution as time progresses. Asymptotic stability means that if a trajectory starts near enough to π^*, then it must eventually converge to π^*. For a system of two equations, such as those that arise with 2×2 games, or to which a 3×3 game may be reduced, we have the criterion given in the following theorem.

Theorem 8.2.3 *A steady-state solution (p_1^*, p_2^*) of the system*

$$\left.\begin{array}{l} \dfrac{dp_1(t)}{dt} = f(p_1(t), p_2(t)) \\[2mm] \dfrac{dp_2(t)}{dt} = g(p_1(t), p_2(t)) \end{array}\right\} \qquad\qquad (8.2.6)$$

is asymptotically stable if

$$f_{p_1}(p_1^*, p_2^*) + g_{p_2}(p_1^*, p_2^*) < 0$$

and

$$\det J(p_1^*, p_2^*) = \det \begin{bmatrix} f_{p_1}(p_1^*, p_2^*) & f_{p_2}(p_1^*, p_2^*) \\ g_{p_1}(p_1^*, p_2^*) & g_{p_2}(p_1^*, p_2^*) \end{bmatrix} > 0.$$

If either $f_{p_1} + g_{p_2} > 0$ or $\det J(p_1^*, p_2^*) < 0$, the steady-state solution $\pi^* = (p_1^*, p_2^*)$ is unstable (i.e., not stable).

The J matrix in the proposition is known as the **Jacobian matrix** of the system. To go a little further into this we need the definition

Definition 8.2.4 *A point (p_1^*, p_2^*) is called an equilibrium point or critical point of (8.2.6) if $f(p_1^*, p_2^*) = g(p_1^*, p_2^*) = 0$. An equilibrium point (p_1^*, p_2^*) is called a **hyperbolic equilibrium point** of (8.2.6) if none of the eigenvalues of the matrix $J(p_1^*, p_2^*)$ have real part which is zero.*

Recall that the eigenvalues of a matrix A are the solutions of $det(A - \lambda I) = 0$. If A is 2×2 there are always two eigenvalues but they may be complex numbers, so a hyperbolic equilibrium point requires the real part of these eigenvalues to be nonzero.

Definition 8.2.5 *An equilibrium point (p_1^*, p_2^*) is called a **sink** if all of the eigenvalues of the matrix $J(p_1^*, p_2^*)$ have negative real part; it is called a **source** if all of the eigenvalues of $J(p_1^*, p_2^*)$ have positive real part; and it is called a **saddle** if it is a hyperbolic equilibrium point and $J(p_1^*, p_2^*)$ has at least one eigenvalue with a positive real part and at least one with a negative real part. It is called a **center** if the pair of eigenvalues have zero real part.*

Remark The stability of any hyperbolic equilibrium point (p_1^*, p_2^*) of (8.2.6) depends on the signs of the real parts of the eigenvalues $J(p_1^*, p_2^*)$. A hyperbolic equilibrium point (p_1^*, p_2^*) is asymptotically stable if (p_1^*, p_2^*) is a sink. A hyperbolic equilibrium point (p_1^*, p_2^*) is unstable if it is either a source or a saddle. Therefore, instead of using Theorem 8.2.3 we can check the eigenvalues of $J(p_1^*, p_2^*)$.

Example 8.8 If there is only one equation $p' = f(p)$, (where $' = d/dt$) then a steady state p^* is asymptotically stable if $f'(p^*) < 0$ and unstable if $f'(p^*) > 0$. It is easy to see why that is true in the one-dimensional case. Set $x(t) = p(t) - p^*$, so now we are looking at $\lim_{t\to\infty} x(t)$ and testing whether that limit is zero. By Taylor's theorem

$$\frac{dx}{dt} = f(x + p^*) = f(p^*) + f'(p^*)x(t) = f'(p^*)x(t)$$

should be true up to first-order terms. The solution of this linear equation is $x(t) = C \exp[f'(p^*)t]$. If $f'(p^*) < 0$, we see that $\lim_{t\to\infty} x(t) = 0$, and if $f'(p^*) > 0$, then $\lim_{t\to\infty} x(t)$ does not exist. So that is why stability requires $f'(p^*) < 0$.

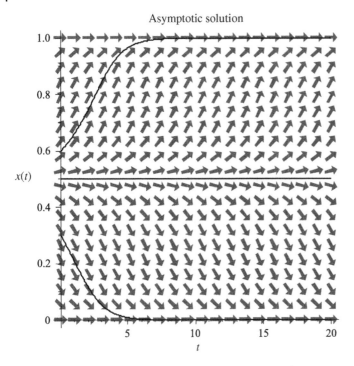

Figure 8.5 Trajectories of $dx/dt = -x(1-x)(1-2x)$; $x = \frac{1}{2}$ is unstable.

Consider the differential equation

$$\frac{dx}{dt} = -x(1-x)(1-2x) = f(x).$$

The stationary states are $x = 0, 1, \frac{1}{2}$, but only two of these are asymptotically stable. These are the pure states $x = 0, 1$, while the state $x = \frac{1}{2}$ is unstable because no matter how small you perturb the initial condition, the solution will eventually be drawn to one of the other asymptotically stable solutions. As $t \to \infty$, the trajectory moves away from $\frac{1}{2}$ unless you start at exactly that point. Figure 8.5 shows this and shows how the arrows lead away from $\frac{1}{2}$.

Checking the stability condition for $f'(0)$, $f'(1)$, $f'(\frac{1}{2})$ we have $f'(x) = -1 + 6x - 6x^2$, and $f'(0) = f'(1) = -1 < 0$, while $f'\left(\frac{1}{2}\right) = \frac{1}{2} > 0$, and so $x = 0, 1$ are asymptotically stable, while $x = \frac{1}{2}$ is not.

Now, here is the connection for evolutionary game theory.

Theorem 8.2.6 *In any 2×2 game, a strategy $X^* = (x_1^*, x_2^*)$ is an ESS if and only if the system (8.2.1) has $p_1^* = x_1^*, p_2^* = x_2^*$ as an asymptotically stable steady state.*

Before we indicate the proof of this, let's recall the definition of what it means to be an ESS. X^* is an ESS if and only if either $E(X^*, X^*) > E(Y, X^*)$, for all strategies $Y \neq X^*$ or $E(X^*, X^*) = E(Y, X^*) \implies E(X^*, Y) > E(Y, Y), \forall Y \neq X^*$.

Proof: Now, we will show that any ESS of a 2×2 game must be asymptotically stable. We will use the stability criterion in Theorem 8.2.3 to do that. Because there are only two pure strategies, we have the equation for $\pi = (p, 1 - p)$:

$$\frac{dp(t)}{dt} = p(t)(1 - p(t))(E(1, \pi) - E(2, \pi)) \equiv f(p(t)).$$

Because of an earlier Problem 8.7 we will consider only the case where

$$A = \begin{bmatrix} a & 0 \\ 0 & b \end{bmatrix}, \quad E(1, \pi) = ap, \quad E(2, \pi) = b(1 - p),$$

and, in general, $a = a_{11} - a_{21}, b = a_{22} - a_{12}$. Then

$$f(p) = p(1 - p)(E(1, \pi) - E(2, \pi)) = p(1 - p)[ap - b(1 - p)]$$

and

$$f'(p) = p(2a + 4b) + p^2(-3a - 3b) - b. \qquad \blacksquare$$

The three steady-state solutions where $f(p) = 0$ are $p^* = 0, 1, \frac{b}{(a+b)}$. Now consider the following cases:

Case 1: $ab < 0$. In this case there is a unique strict symmetric Nash equilibrium and so the ESSs are either $X = (1, 0)$ (when $a > 0, b < 0$), or $X = (0, 1)$ (when $a < 0, b > 0$). We look at $a > 0, b < 0$ and the steady-state solution $p^* = 1$. Then, $f'(1) = 2a + 4b - 3a - 3b - b = -a < 0$, so that $p^* = 1$ is asymptotically stable. Similarly, if $a < 0, b > 0$, the ESS is $X = (0, 1)$ and $p^* = 0$ is asymptotically stable. For an example, if we take $a = 1, b = -2$, the following Figure 8.6 shows convergence to $p^* = 1$ for trajectories from fours different initial conditions:

Case 2: $a > 0, b > 0$. In this case there are three symmetric Nash equilibria: $X_1 = (1, 0), X_2 = (0, 1)$, and the mixed Nash $X_3 = (\gamma, 1 - \gamma), \gamma = b/(a + b)$. The two pure Nash X_1, X_2 are strict and thus are ESSs by the Properties 8.1.9. The mixed Nash is **not** an ESS because $E(X_3, X_3) = a\gamma$ is not larger than $E(Y, X_3) = a\gamma, \forall Y \neq X_3$ (they are equal), and taking $Y = (1, 0)$, $E(Y, Y) = E(1, 1) = a > a\gamma = E(X_3, 1)$. Consequently, X_3 does not satisfy the criteria to be an

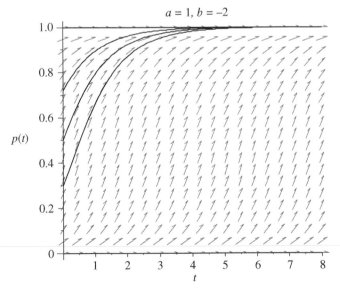

Figure 8.6 Convergence to $p^* = 1$.

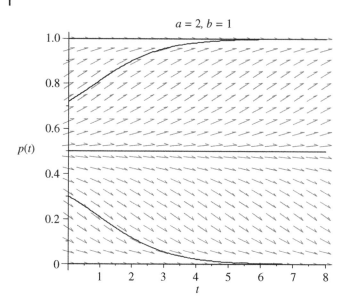

Figure 8.7 Mixed Nash not stable.

ESS. Hence we only need consider the stationary solutions $p^* = 0, 1$. But then, in the case $a > 0, b > 0$, we have $f'(0) = -b < 0$ and $f'(1) = -a < 0$, so they are both asymptotically stable. This is illustrated in the Figure 8.7 with $a = 2, b = 1$.

You can see that from any initial condition $x_0 < \frac{1}{2}$, the trajectory will converge to the stationary solution $p^* = 0$. For any initial condition $x_0 > \frac{1}{2}$, the trajectory will converge to the stationary solution $p^* = 1$, and only for $x_0 = \frac{1}{2}$, will the trajectory stay at $\frac{1}{2}$; $p^* = \frac{1}{2}$ is unstable.

Case 3: $a < 0, b < 0$. In this case, there are two strict but asymmetric Nash equilibria and one symmetric Nash equilibrium $X = (\gamma, 1 - \gamma)$. This symmetric X is an ESS because

$$E(Y, X) = ay_1\gamma + by_2(1 - \gamma) = \frac{ab}{a + b} = E(X, X),$$

and for every $Y \neq X$

$$E(Y, Y) = ay_1^2 + by_2^2 < \frac{ab}{a + b} = E(X, X),$$

since $a < 0, b < 0$. Consequently the only ESS is $X = (\gamma, 1 - \gamma)$, and so we consider the steady state $p^* = \gamma$. Then

$$f'(\gamma) = \frac{ab}{a + b} < 0,$$

and so $p^* = \gamma$ is asymptotically stable. For an example, if we take $a = -2, b = -1$, we get Figure 8.8, in which all interior initial conditions lead to convergence to $\frac{1}{3}$.

It is important to remember that if the trajectory starts exactly at one of the stationary solutions, then it stays there forever. It is starting nearby and seeing where it goes that determines stability.

We have proved that any ESS of a 2×2 game must be asymptotically stable. We skip the opposite direction of the proof and end this section by reviewing the remaining connections between Nash equilibria, ESSs, and stability. The first is a summary of the connections with Nash equilibria.

1. If X^* is a Nash equilibrium for a symmetric game with matrix A, then it is a stationary solution of (8.2.1).

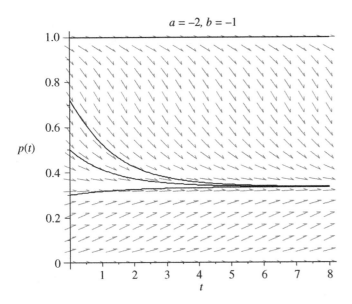

Figure 8.8 Convergence to the mixed ESS $X^* = (b/(a + b), a/(a + b))$, in case $a < 0, b < 0$.

2. If X^* is a strict Nash equilibrium, then it is locally asymptotically stable.
3. If X^* is a stationary solution and if $\lim_{t \to \infty} p(t) = X^*$, where $p(t)$ is a solution of (8.2.1) such that each component $p_i(t)$ of $p(t) = (p_1(t), \dots, p_n(t))$ satisfies $0 < p_i(t) < 1$, then X^* is a Nash equilibrium.
4. If X^* is a locally asymptotically stable stationary solution of (8.2.1), then it is a Nash equilibrium.

The converse statements do not necessarily hold. The verification of all these statements, and much more, can be found in the book by Weibull [43]. Now here is the main result for ESSs and stability.

Theorem 8.2.7 *If X^* is an ESS, then X^* is an asymptotically stable stationary solution of (8.2.1). In addition, if X^* is completely mixed, then it is globally asymptotically stable.*

Again, the converse does not necessarily hold.

Example 8.9 In this example we consider a 3×3 symmetric game so that the frequency dynamics is a system in the three variables (p_1, p_2, p_3). Let's look at the game with matrix

$$A = \begin{bmatrix} 0 & -2 & 1 \\ 1 & 0 & 1 \\ 1 & 3 & 0 \end{bmatrix}.$$

This game has Nash equilibria

X	Y
$(1, 0, 0)$	$(0, 0, 1)$
$(0, 1, 0)$	$(0, 0, 1)$
$(0, 0, 1)$	$(1, 0, 0)$
$(0, 0, 1)$	$(0, 1, 0)$
$(0, \frac{1}{4}, \frac{3}{4})$	$(0, \frac{1}{4}, \frac{3}{4})$

There is only one symmetric Nash equilibrium $X^* = (0, \frac{1}{4}, \frac{3}{4}) = Y^*$. It will be an ESS because we will show that $u(Y, Y) < u(X^*, Y)$ for every best response strategy $Y \neq X^*$ to the strategy X^*. By Properties 8.1.1(3), we then know that X^* is an ESS.

First, we find the set of best response strategies Y. To do that, calculate

$$u(Y, X^*) = y_1 \left(\frac{1}{4} \right) + (y_2 + y_3)\frac{3}{4}.$$

It is clear that u is maximized for $y_1 = 0, y_2 + y_3 = 1$. This means that any best response strategy must be of the form $Y = (0, y, 1 - y)$, where $0 \leq y \leq 1$.

Next, taking any best response strategy, we have

$$u(Y, Y) = 4y - 4y^2 \quad \text{and} \quad u(X^*, Y) = X^* A Y^T = 2y + \frac{1}{4}.$$

We have to determine whether $4y - 4y^2 < 2y + \frac{1}{4}$ for all $0 \leq y \leq 1, y \neq \frac{1}{4}$. This is easy to do by calculus since $f(y) = 4y^2 - (2y + \frac{1}{4})$ has a minimum at $y = \frac{1}{4}$ and $f(\frac{1}{4}) = 0$.

The last plot, exhibited in Figure 8.9, shows that $u(Y, Y) < u(X^*, Y)$ for best response strategies, Y. The replicator dynamics (8.2.1) for this game are

$$\frac{dp_1(t)}{dt} = p_1 \left[-2p_2 + p_3 - (-p_1 p_2 + 2p_1 p_3 + 4p_2 p_3) \right],$$

$$\frac{dp_2(t)}{dt} = p_2 \left[p_1 + p_3 - (-p_1 p_2 + 2p_1 p_3 + 4p_2 p_3) \right],$$

$$\frac{dp_3(t)}{dt} = p_3 \left[p_1 + 3p_2 - (-p_1 p_2 + 2p_1 p_3 + 4p_2 p_3) \right].$$

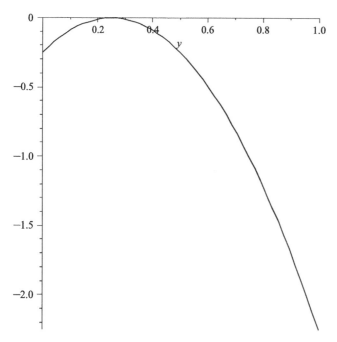

Figure 8.9 Plot of $f(y) = u(Y, Y) - u(X^*, Y) = 4y^2 - 2y - \frac{1}{4}$.

Since $p_1 + p_2 + p_3 = 1$, they can be reduced to two equations to which we may apply the stability theorem. The equations become

$$\frac{dp_1(t)}{dt} = f(p_1, p_2) \equiv p_1 \left[-7p_2 + 1 - 3p_1 + 7p_1p_2 + 2p_1^2 + 4p_2^2 \right],$$

$$\frac{dp_2(t)}{dt} = g(p_1, p_2) \equiv p_2 \left[1 - 5p_2 + 7p_1p_2 - 2p_1 + 2p_1^2 + 4p_2^2 \right].$$

The steady-state solutions are given by the solutions of the pair of equations $f(p_1, p_2) = 0$, $g(p_1, p_2) = 0$, which make the derivatives zero and are given by

$$a = [p_1 = 0, p_2 = 0], \quad b = \left[p_1 = 0, p_2 = \frac{1}{4} \right],$$

$$c = [p_1 = 0, p_2 = 1], \quad d = \left[p_1 = \frac{1}{2}, p_2 = 0 \right], \quad e = [p_1 = 1, p_2 = 0],$$

and we need to analyze each of these. We start with the condition from Theorem 8.2.3 that $f_{p_1} + g_{p_2} < 0$. Calculation yields,

$$k(p_1, p_2) = f_{p_1}(p_1, p_2) + g_{p_2}(p_1, p_2)$$
$$= -17p_2 + 2 - 8p_1 + 28p_1p_2 + 8p_1^2 + 16p_2^2.$$

Directly plugging our points into $k(p_1, p_2)$, we see that

$$k(0,0) = 2 > 0, \quad k\left(0, \frac{1}{4}\right) = -\frac{5}{4} < 0, \quad k(0,1) = 1 > 0,$$

$$k\left(\frac{1}{2}, 0\right) = 0, \quad k(1,0) = 2 > 0,$$

and since we only need to consider the negative and zero terms as possibly stable, we are left with the possibly asymptotically stable value $b = (0, \frac{1}{4})$ and $d = (\frac{1}{2}, 0)$.

Next, we check the Jacobian at that point $b = (0, \frac{1}{4})$ and get

$$\det \begin{bmatrix} -\frac{1}{2} & 0 \\ -\frac{1}{16} & -\frac{3}{4} \end{bmatrix} = \frac{3}{8} > 0.$$

By the stability Theorem 8.2.3 $p_1^* = 0, p_2^* = \frac{1}{4}, p_3^* = \frac{3}{4}$, is indeed an asymptotically stable solution, and hence a Nash equilibrium and an ESS. Figure 8.10 shows three trajectories $(p_1(t), p_2(t))$ starting at time $t = 0$ from the initial conditions $(p_1(0), p_2(0)) = (0.1, 0.2), (0.6, 0.2), (0.33, 0.33)$.

We see the asymptotic convergence of $(p_1(t), p_2(t))$ to $(0, \frac{1}{4})$. Since $p_1 + p_2 + p_3 = 1$, this means $p_3(t) \to \frac{3}{4}$. It also shows a trajectory starting from $p_1 = 0, p_2 = 0$ and shows that the trajectory stays there for all time.

Finally, let's talk about the case $k(\frac{1}{2}, 0) = 0$. Let's check the Jacobian at this point. We have

$$\det \begin{bmatrix} -\frac{1}{2} & -\frac{7}{4} \\ 0 & \frac{1}{2} \end{bmatrix} = -\frac{1}{4} < 0.$$

By the stability theorem, $p_1^* = \frac{1}{2}, p_2^* = 0, p_3^* = \frac{1}{2}$ is not stable.

Example 8.10 This example also illustrates instability with cycling around the Nash mixed strategy as shown in Figure 8.11. The game matrix is

$$A = \begin{bmatrix} 0 & 1 & -1 \\ -1 & 0 & 1 \\ 1 & -1 & 0 \end{bmatrix}.$$

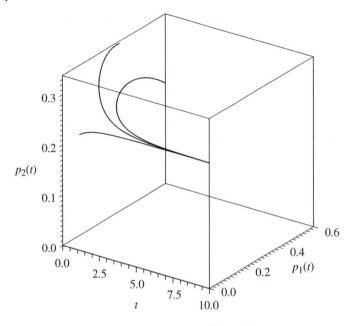

Figure 8.10 Convergence to $p_1^* = 0, p_2^* = \frac{1}{4}, p_3^* = \frac{3}{4}$ from three different initial conditions.

Cycle around the steady state

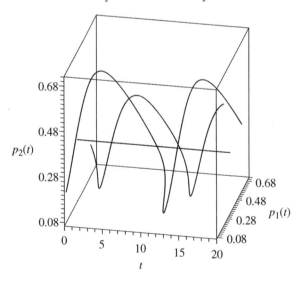

Figure 8.11 $\left(\frac{1}{3}, \frac{1}{3}, \frac{1}{3}\right)$ is a center.

The system of frequency equations, after making the substitution $p_3 = 1 - p_1 - p_2$, becomes

$$\frac{dp_1(t)}{dt} = f(p_1, p_2) \equiv p_1 \left[2p_2 - 1 + p_1 \right],$$

$$\frac{dp_2(t)}{dt} = g(p_1, p_2) \equiv -p_2 \left[2p_1 - 1 + p_2 \right].$$

The steady-state solutions are $(0, 0), (0, 1), (1, 0)$, and $\left(\frac{1}{3}, \frac{1}{3} \right)$. Then

$$k(p_1, p_2) \equiv f_{p_1}(p_1, p_2) + g_{p_2}(p_1, p_2) = 0.$$

Since $\det J(0, 1) = \det J(1, 0) = \det J(0, 0) = -1$, these points are all unstable equilibria. For the critical point $(\frac{1}{3}, \frac{1}{3})$ we have $k(\frac{1}{3}, \frac{1}{3}) = 0$ and $\det J(\frac{1}{3}, \frac{1}{3}) = \frac{1}{3} > 0$. This means we have to calculate the eigenvalues of $J(\frac{1}{3}, \frac{1}{3})$. There are two eigenvalues given by

$$\lambda_1 = \frac{i}{\sqrt{3}} \implies Re(\lambda_1) = 0, \qquad \lambda_2 = -\frac{i}{\sqrt{3}} \implies Re(\lambda_1) = 0.$$

In this case, the system (8.2.6) is said to have a **center** at $(\frac{1}{3}, \frac{1}{3})$. What this means is that the trajectories basically rotate around the point forever. This is illustrated in Figure 8.11.

Three trajectories are shown starting from the points $(0.1, 0.2), (0.6, 0.2)$, and $(0.33, 0.33)$. You can see that unless the starting position is exactly at $(0.33, 0.33)$, the trajectories will cycle around and not converge to the mixed strategy.

8.3 The Von Neumann Minimax Theorem from Replicator Dynamics

In this last section we will present a new proof of the Von Neumann minimax theorem for matrix games using replicator dynamics. This proof is due to J. Hofbauer who constructed this argument in 2018 [14]. The only requirement to follow this is a little advanced calculus.

We will show that for a zero-sum game with matrix $A_{n \times m}$ we have

$$\min_{Y \in S_m} \max_{X \in S_n} X \, A \, Y^T = \max_{X \in S_n} \min_{Y \in S_m} X \, A \, Y^T. \tag{8.3.1}$$

For any bimatrix game (A, B) the replicator equations are

$$\frac{dp_i}{dt} = p_i(t)[E_I[i, q(t)] - E_I[p(t), q(t)]], \qquad i = 1, 2, \ldots, n, \tag{8.3.2}$$

$$\frac{dq_j}{dt} = q_j(t)[E_{II}[p(t), j] - E_{II}[p(t), q(t)]], \qquad j = 1, 2, \ldots, m, \tag{8.3.3}$$

where $p(t) \in S_n$ is a strategy for player I for each $t \in [0, T]$ and $q(t) \in S_m$ is a strategy for player II for each $t \in [0, T]$. Recall that $E_I[p, q] = p A q^T$ and $E_{II}[p, q] = p B q^T$. In particular, for a zero sum game $B = -A$ and the equations become

$$\frac{dp_i}{dt} = p_i(t)[_i A \, q(t) - p(t) \cdot A \, q^T(t)], \qquad i = 1, 2, \ldots, n,$$

$$\frac{dq_j}{dt} = q_j(t)[-p(t)A_j + p(t) \cdot A \, q^T(t)], \qquad j = 1, 2, \ldots, m.$$

Rewrite as

$$\frac{dp_i}{p_i} = [{}_iA\,q(t) - p(t) \cdot A\,q^T(t)]\,dt, \qquad i = 1, 2, \ldots, n,$$

$$\frac{dq_j}{q_j} = [-p(t)A_j + p(t) \cdot A\,q^T(t)]\,dt, \qquad j = 1, 2, \ldots, m.$$

We may eliminate $p(t)\,A\,q^T(t)$ by adding the equations to get the single equation

$$\frac{dp_i}{p_i} + \frac{dq_j}{q_j} = [{}_iAq^T(t) - p(t)A_j]\,dt, \qquad i = 1, 2, \ldots, n, \quad j = 1, 2, \ldots, m.$$

Recall that if there is a pure saddle point of the game, then (8.3.1) will hold. Consequently, we may assume without loss of generality that the strategies must be mixed, i.e., that $0 < p_i(t) < 1$ for all $t > 0$. Next, we integrate this equation on $[0, T]$ and rearrange slightly to get

$$\frac{1}{T}[\ln p_i(T) - \ln p_i(0)] + \frac{1}{T}[\ln q_j(T) - \ln q_j(0)] = {}_iA\overline{q}(T)^T - \overline{p}(T)A_j, \tag{8.3.4}$$

where we set as the average over the time interval $[0, T]$ of the frequencies for each player,

$$\overline{p}(T) = \frac{1}{T}\int_0^T p(t)\,dt \qquad \overline{q}(T) = \frac{1}{T}\int_0^T q(t)\,dt.$$

Because $\overline{p}(T) \in S_n, \overline{q}(T) \in S_m$, which are convex, closed and bounded sets, for any sequence $T_k \to \infty$, there will be a subsequence, still denoted $\overline{p}(T_k)$ and $\overline{q}(T_k)$ and points in $\overline{p} \in S_n$ and $\overline{q} \in S_m$ respectively, such that $\overline{p}(T_k) \to \overline{p}$ and $\overline{q}(T_k) \to \overline{q}$ as $k \to \infty$. Since all frequencies satisfy $0 \leq p_i \leq 1, 0 \leq q_j \leq 1$, we have (remember $0 < p_i(t) < 1$ and $0 < q_j(t) < 1$ for all $t \in [0, 1]$)

$$\lim_{k \to \infty} \frac{1}{T_k}[\ln p_i(T_k) + \ln q_j(T_k)] - \frac{1}{T_k}[\ln p_i(0) + \ln q_j(0)] \leq 0.$$

It follows that, writing $E(X, Y) = X\,A\,Y^T$, from (8.3.4),

$$\lim_{k \to \infty}({}_iA\overline{q}(T_k)^T - \overline{p}(T_k)\,A_j) = {}_iA\overline{q}^T - \overline{p}A_j = E[i, \overline{q}] - E[\overline{p}, j] \leq 0.$$

Let $X = (x_1, \ldots, x_n) \in S_n, Y = (y_1, \ldots, y_m) \in S_m$ be arbitrary. Multiply by x_i and y_j and sum, we have

$$\sum_{j=1}^m \sum_{i=1}^n x_i E[i, \overline{q}] - E[\overline{p}, j]y_j = E[X, \overline{q}] - E[\overline{p}, Y] \leq 0.$$

We have achieved the inequality that for any $X \in S_n$ and $Y \in S_m$, $E[X, \overline{q}] \leq E[\overline{p}, Y]$. Consequently,

$$\min_{Y \in S_m} \max_{X \in S_n} E[X, Y] \leq \max_{X \in S_n} E[X, \overline{q}] \leq \min_{Y \in S_m} E[\overline{p}, Y] \leq \max_{X \in S_n} \min_{Y \in S_m} E[X, Y].$$

Since the reverse inequality is always true, we have proved (8.3.1). With two more lines we can also show that $(\overline{p}, \overline{q})$ is actually a saddle point of the game:

$$\overline{p}A\overline{q}^T \leq \max_{X \in S_n} X\,A\overline{q}^T \leq \min_{Y \in S_m} \overline{p}A\,Y^T \leq \overline{p}\,A\,\overline{q}^T.$$

It must be that the inequalities become equalities. Then for any $X \in S_n, Y \in S_m$,

$$X A \, \bar{q}^T \le \max_{X \in S_n} X A \, \bar{q}^T = \bar{p} A \, \bar{q}^T = \min_{Y \in S_m} \bar{p} A \, Y^T \le \bar{p} A \, Y^T.$$

Which means $(\bar{p}, \bar{q})$ is a saddle point. ∎

That's a very clever argument!

Problems

8.9 Consider a game in which a seller can be either honest or dishonest and a buyer can either inspect or trust (the seller). One game model of this is the matrix $A = \begin{bmatrix} 3 & 2 \\ 4 & 1 \end{bmatrix}$, where the rows are inspect and trust, and the columns correspond to dishonest and honest.
  (a) Find the replicator dynamics for this game.
  (b) Find the Nash equilibria and determine which, if any, are ESSs.
  (c) Analyze the stationary solutions for stability.

8.10 Analyze all stationary solutions for the game with matrix $A = \begin{bmatrix} 1 & 1 \\ 1 & 3 \end{bmatrix}$.

8.11 Consider the symmetric game with matrix

$$A = \begin{bmatrix} 2 & 1 & 5 \\ 5 & 1 & 0 \\ 1 & 4 & 0 \end{bmatrix}.$$

  (a) Find the one and only Nash equilibrium.
  (b) Determine whether the Nash equilibrium is an ESS.
  (c) Reduce the replicator equations to two equations and find the stationary solutions. Check for stability using the stability Theorem 8.2.3.

8.12 Consider the symmetric game with matrix

$$A = \begin{bmatrix} 1 & 1 & 0 \\ 1 & 0 & 1 \\ 0 & 0 & 0 \end{bmatrix}.$$

Show that $X = (1, 0, 0)$ is an ESS that is asymptotically stable for (8.2.1).

8.13 The simplest version of the rock–paper–scissors game has matrix

$$A = \begin{bmatrix} 0 & -1 & 1 \\ 1 & 0 & -1 \\ -1 & 1 & 0 \end{bmatrix}.$$

  (a) Show that there is one and only one completely mixed Nash equilibrium but it is not an ESS.

(b) Show that the statement in the first part is still true if you replace 0 in each row of the matrix by $a > 0$.

(c) Analyze the stability of the stationary points for the replicator dynamics for $a > 0$ and (8.2.1) by reducing to two equations and using the stability Theorem 8.2.3.

8.14 Find the frequency dynamics (8.2.1) for the game

$$A = \begin{bmatrix} 0 & 3 & 1 \\ 3 & 0 & 1 \\ 1 & 1 & 1 \end{bmatrix}.$$

Find the steady-state solutions and investigate their stability, then reach a conclusion about the Nash equilibria and the ESSs.

8.15 Consider the symmetric game with matrix $A = \begin{bmatrix} -a & 1 & -1 \\ -1 & -a & 1 \\ 1 & -1 & -a \end{bmatrix}$, where $a > 0$. Show that $X^* = \left(\frac{1}{3}, \frac{1}{3}, \frac{1}{3}\right)$ is an ESS for any $a > 0$. Compare with Problem (8.13b).

8.16 Consider the symmetric game with matrix $A = \begin{bmatrix} 2 & 1 & 5 \\ 5 & \alpha & 0 \\ 1 & 4 & 3 \end{bmatrix}$ where α is a real number. This problem shows that a stable equilibrium of the replicator dynamics need not be an ESS.

(a) First consider the case $-8 < \alpha < 8.5$. Find the symmetric Nash equilibrium and check if it is an ESS.

(b) Take $\alpha = 2$. Analyze the mixed Nash equilibrium in this case and determine if it is asymptotically stable.

8.17 Find all ESSs, in pure or mixed strategies, in the following game:

I/II	a	b	c
a	$(1,1)$	$(1,1)$	$(0,0)$
b	$(1,1)$	$(0,0)$	$(1,0)$
c	$(0,0)$	$(0,1)$	$(0,0)$

Is there a strategy that cannot be invaded by any combination of mutant strategies?

8.18 Our first Chicken game, with bimatrix $\begin{bmatrix} (0,0) & (-5,5) \\ (5,-5) & (-20,-20) \end{bmatrix}$, is symmetric. Hence it can be turned into an evolutionary game.

(a) Write down a differential equation in one variable that describes the related evolutionary system.

(b) We know that $\left(\frac{3}{4}, \frac{1}{4}\right)$ is the unique symmetric Nash equilibrium, so it should be a fixed point. Verify this by substituting $p = \frac{3}{4}$ into your differential equation. You should find $p' = 0$.

(c) Is this solution an ESS? Hint: you might want to sketch the slope field.

8.19 Consider the symmetric matrix game where $A = B^T = \begin{bmatrix} 10 & 2 & 3 \\ 4 & 5 & 6 \\ 9 & 6 & 3 \end{bmatrix}.$

Gambit gives the following Nash equilibria

	X*			Y*			(E.I,E.II)
a	1	0	0	1	0	0	(10,10)
b	4/5	1/5	0	1/3	0	2/3	(16/3,42/5)
c	1/3	0	2/3	4/5	1/5	0	(42/5,16/3)
d	1/3	1/12	7/12	1/3	1/12	7/12	(21/4,21/4)
e	0	1	0	0	0	1	(6,6)
f	0	0	1	0	1	0	(6,6)
g	0	3/4	1/4	0	3/4	1/4	(21/4,21/4)

Mathematica gives the following stream-plot for the related system of differential equations:

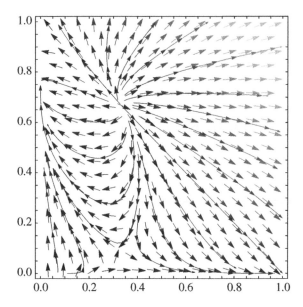

(a) Indicate where each symmetric Nash equilibrium lies in the stream-plot.
(b) Which Nash equilibria are evolutionary stable?
(c) What will happen after a long period of time if the initial populations are in proportions
   1. $(0.6, 0.2, 0.2)$
   2. $(0.3, 0.4, 0.3)$
   3. $(0.41, 0.46, 0.13)$

8.20 Consider the symmetric matrix game defined by $A = \begin{bmatrix} 3 & 1 \\ 2 & 5 \end{bmatrix}$. Think of this as a population game.
   (a) Write down the differential equation describing how the proportion of the population of type 1 changes over time.
   (b) Check that the constant functions $p = 0$ and $p = 1$ are both solutions to the differential equation. Thinking about biology, explain why this should always be the case.

(c) The vector field for this differential equation is shown below. Calculating dp/dt for $p = 0.1$, and make sure this agrees with the plot.

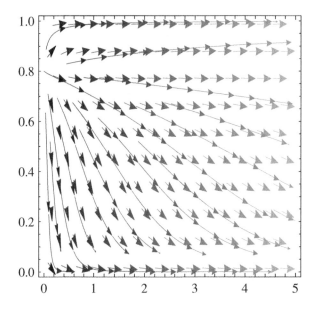

(d) You should be able to immediately see that this game has a unique mixed Nash equilibrium. Can you see what it is from the plot? Is it an ESS?

8.21 The symmetric matrix game $\begin{bmatrix} 1 & 4 \\ 7 & 2 \end{bmatrix}$ corresponds to the differential equation

$$\dot{p} = 8p^3 - 10p^2 + 2p$$

and the slope field

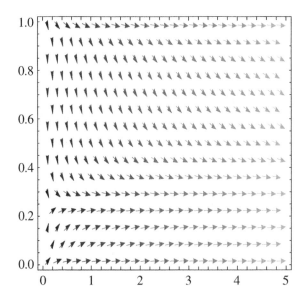

(a) The unique symmetric Nash equilibrium is $X^* = (\frac{1}{4}, \frac{3}{4})$. Is this evolutionary stable?

(b) $p = 1$ is a solution to the differential equation, but $(1, 0)$ is not a symmetric Nash equilibrium. What's going on?

8.22 Consider the symmetric matrix game with matrix $A = \begin{bmatrix} 4 & -1 & -2 \\ -3 & 1 & 4 \\ 3 & 2 & 3 \end{bmatrix}$.

The Nash equilibria are:

- $X = (0, \frac{1}{2}, \frac{1}{2}), Y = (0, \frac{1}{2}, \frac{1}{2})$ with payoffs $P_1 : \frac{5}{2}, P_2 : \frac{5}{2}$.
- $X = (0, 1, 0), Y = (0, 0, 1)$ with payoffs $P_1 : 4, P_2 : 2$.
- $X = (\frac{5}{6}, 0, \frac{1}{6}), Y = (\frac{5}{6}, 0, \frac{1}{6})$ with payoffs $P_1 : 3, P_2 : 3$.
- $X = (1, 0, 0), Y = (1, 0, 0)$ with payoffs $P_1 : 4, P_2 : 4$.
- $X = (0, 0, 1), Y = (0, 1, 0)$ with payoffs $P_1 : 2, P_2 : 4$.

The corresponding differential equations lead to the stream-plot

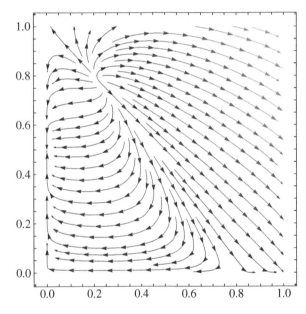

1. Does the Nash $X = (0, 1, 0), Y = (0, 0, 1)$ mean anything for the evolutionary game?
2. Indicate all the symmetric Nash equilibria on the streamplot. Which are ESSs?
3. Looking at the vector plot, there is a fixed-point of the system of differential equations at approximately $(0.2, 0.8)$. However, there is no Nash equilibrium near $(0.2, 0.8, 0)$. How is this possible?
4. If you start with a random population, and let the population game run, what is likely to happen?

8.23 The symmetric matrix game with matrix $A = \begin{bmatrix} 1 & 2 & -2 \\ -2 & 1 & 2 \\ 2 & -2 & 1 \end{bmatrix}$ gives the stream plot

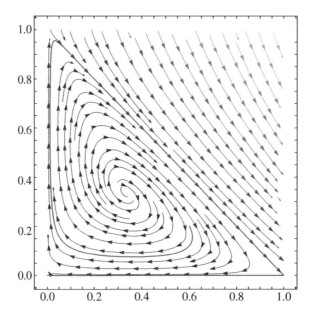

The heavy curve on the plot is for the initial condition $p_1(0) = 0.6, p_2(0) = 0.2$. Are there any ESSs? Discuss. If this was a real biological system, what do you think would happen?

Remark Do populations always choose a better strategy in the long run? There is a fictional story that implies that may not always be the case. The story starts with a measurement. In the United States the distance between railroad rails is 4 feet 8.5 inches. Isn't that strange? Why isn't it 5 feet, or 4 feet? The distance between rails determines how rail cars are built, so wouldn't it be easier to build a car with exact measurements, that determine all the parts in the drive train?

Why is that the measurement chosen?
Well, that's the way they built them in Great Britain, and it was immigrants from Britain who built the US railroads.
Why did the English choose that measurement for the railroads in England?
Because, before there were railroads there were tramways, and that's the measurement they chose for the tramways. (A tramway is a light-rail system for passenger trams.)
Why?
Because they used the same measurements as the people who built wagons for the wheel spacing.
And why did the wagon builders use that measurement?
Because the spacing of the wheel ruts in the old roads had that spacing, and if they used another spacing, the wagon wheels would break apart.
How did the road ruts get that spacing?
The Romans' chariots for the legions made the ruts, and since Rome conquered most of the known world, the ruts ended up being the same almost everywhere they traveled because they were made by the same version of General Motors then, known as Imperial Chariots, Inc.
And why did they choose the spacing that started this story?
Well, that is exactly the width the chariots need to allow two horse's rear ends to fit. And that is how the world's most advanced transportation system is based on 4 feet 8.5 inches.

Bibliographic Notes

Evolutionary games can be approached with two distinct goals in mind. The first is the matter of determining a way to choose a **correct** Nash equilibrium in games with multiple equilibria. The second is to model biological processes in order to determine the eventual evolution of a population. This was the approach of Maynard-Smith and Price (see Vincent and Brown's book [39] for references and biological applications) and has led to a significant impact on biology, both experimental and theoretical. Our motivation for the definition of evolutionary stable strategy is the very nice derivation in the treatise by Mesterton-Gibbons [27]. The equivalent definitions, some of which are easier to apply, are standard and appear in the literature (for example, in references [39] and [43]), as are many more results. The hawk–dove game is a classic example appearing in all books dealing with this topic. The rock–paper–scissors game is an illustrative example of a wide class of games (see the book by Weibull [43] for a lengthy discussion of rock–paper–scissors).

Population games are a natural extension of the idea of evolutionary stability by allowing the strategies to change with time and letting time become infinite. The stability theory of ordinary differential equations can now be brought to bear on the problem (see the book by Scheinermann [38] for the basic theory of stability). The first derivation of the replicator equations seems to have been due to Taylor and Jonker in the article in Reference [17]. Hofbauer and Sigmund, who have made important contributions to evolutionary game theory, have published recently an advanced survey of evolutionary dynamics in Reference [15]. Refer to the book by Gintis [9] for an exercise-based approach to population games and extensions.

While there are many clever proofs of the von Neumann minimax theorem, some of which use very deep mathematical theorems, the recent proof by Hofbauer [15] stands out. It is accessible to any undergraduate student who is familiar with elementary differential equations. It was too nice not to include in a game theory book as a beautiful application of population games.

Problems 8.18–8.23 are exercises adapted from Prof. Peter Tingley assigned in his game theory class.

Appendix A

The Essentials of Matrix Analysis

A matrix is a rectangular collection of numbers. If there are n rows and m columns, we write the matrix as $A_{n \times m}$, and the numbers of the matrix are a_{ij}, where i gives the row number and j gives the column number. These are also called the **dimensions** of the matrix. We compactly write $A = (a_{ij})_{i=1, j=1}^{i=n, j=m}$. In rectangular form

$$A = \begin{bmatrix} a_{11} & a_{12} & \cdots & a_{1m} \\ a_{21} & a_{22} & \cdots & a_{2m} \\ \vdots & \vdots & \cdots & \vdots \\ a_{n1} & a_{n2} & \cdots & a_{nm} \end{bmatrix}.$$

If $n = m$, we say that the matrix is square. The square matrix in which there are all 1s along the diagonal and 0s everywhere else is called the **identity** matrix:

$$I_n := \begin{bmatrix} 1 & 0 & \cdots & 0 \\ 0 & 1 & \cdots & 0 \\ \vdots & \vdots & \cdots & \vdots \\ 0 & 0 & \cdots & 1 \end{bmatrix}.$$

Here are some facts about algebra with matrices:

1. We add two matrices that have the same dimensions, $A + B$, by adding the respective components $A + B = (a_{ij} + b_{ij})$, or

$$A + B = \begin{bmatrix} a_{11} + b_{11} & a_{12} + b_{12} & \cdots & a_{1m} + b_{1m} \\ a_{21} + b_{21} & a_{22} + b_{22} & \cdots & a_{2m} + b_{2m} \\ \vdots & \vdots & \cdots & \vdots \\ a_{n1} + b_{n1} & a_{n2} + b_{n2} & \cdots & a_{nm} + b_{nm} \end{bmatrix}.$$

2. A matrix may be multiplied by a scalar c by multiplying every element of A by c; that is, $cA = (ca_{ij})$ or

$$cA = \begin{bmatrix} ca_{11} & ca_{12} & \cdots & ca_{1m} \\ ca_{21} & ca_{22} & \cdots & ca_{2m} \\ \vdots & \vdots & \cdots & \vdots \\ ca_{n1} & ca_{n2} & \cdots & ca_{nm} \end{bmatrix}.$$

3. We may multiply two matrices $A_{n \times m}$ and $B_{m \times k}$ only if the number of columns of A is exactly the same as the number of rows of B. You have to be careful because not only is $A \cdot B \neq B \cdot A$;

Game Theory: An Introduction, Third Edition. E. N. Barron.
© 2024 John Wiley & Sons, Inc. Published 2024 by John Wiley & Sons, Inc.

in general, it is not even defined if the rows and columns don't match up. So, if $A = A_{n \times m}$ and $B = B_{m \times k}$, then $C = A \cdot B$ is defined and $C = C_{n \times k}$, and is given by

$$
A \cdot B =
\begin{bmatrix}
a_{11} & a_{12} & \cdots & a_{1m} \\
a_{21} & a_{22} & \cdots & a_{2m} \\
\vdots & \vdots & \cdots & \vdots \\
a_{n1} & a_{n2} & \cdots & a_{nm}
\end{bmatrix}
\begin{bmatrix}
b_{11} & b_{12} & \cdots & b_{1k} \\
b_{21} & b_{22} & \cdots & b_{2k} \\
\vdots & \vdots & \cdots & \vdots \\
b_{m1} & b_{m2} & \cdots & b_{mk}
\end{bmatrix}.
$$

We multiply each row $i = 1, 2, \ldots, n$, of A by each column $j = 1, 2, \ldots, k$, of B in this way

$$
_iA \cdot B_j =
\begin{bmatrix}
a_{i1} & a_{i2} & \cdots & a_{im}
\end{bmatrix}
\cdot
\begin{bmatrix}
b_{1j} \\
b_{2j} \\
\vdots \\
b_{mj}
\end{bmatrix}
= a_{i1}b_{1j} + a_{i2}b_{2j} + \cdots + a_{im}b_{mj} = c_{ij}.
$$

This gives the (i, j)th element of the matrix C. The matrix $C_{n \times k} = (c_{ij})$ has elements written compactly as

$$
c_{ij} = \sum_{r=1}^{m} a_{ir}b_{rj}, \quad i = 1, 2, \ldots, n, \quad j = 1, 2, \ldots, k.
$$

4. As special cases of multiplication that we use throughout this book

$$
X_{1 \times n} \cdot A_{n \times m} =
\begin{bmatrix}
x_1 & x_2 & \cdots & x_n
\end{bmatrix}
\begin{bmatrix}
a_{11} & a_{12} & \cdots & a_{1m} \\
a_{21} & a_{22} & \cdots & a_{2m} \\
\vdots & \vdots & \cdots & \vdots \\
a_{n1} & a_{n2} & \cdots & a_{nm}
\end{bmatrix}
$$

$$
= \left[\sum_{i=1}^{n} x_i a_{i1} \quad \sum_{i=1}^{n} x_i a_{i2} \quad \cdots \quad \sum_{i=1}^{n} x_i a_{im} \right].
$$

Each element of the result is $E(X, j), \quad j = 1, 2, \ldots, m$.

5. If we have any matrix $A_{n \times m}$, the **transpose** of A is written as A^T and is the $m \times n$ matrix, which is A with the rows and columns switched:

$$
A^T =
\begin{bmatrix}
a_{11} & a_{21} & \cdots & a_{n1} \\
a_{12} & a_{22} & \cdots & a_{n2} \\
\vdots & \vdots & \cdots & \vdots \\
a_{1n} & a_{2n} & \cdots & a_{nm}
\end{bmatrix}.
$$

If $Y_{1 \times m}$ is a row matrix, then Y^T is an $m \times 1$ column matrix, so we may multiply $A_{n \times m}$ by Y^T on the right to get

$$
A_{n \times m} Y^T_{m \times 1} =
\begin{bmatrix}
a_{11} & a_{12} & \cdots & a_{1m} \\
a_{21} & a_{22} & \cdots & a_{2m} \\
\vdots & \vdots & \cdots & \vdots \\
a_{n1} & a_{n2} & \cdots & a_{nm}
\end{bmatrix}
\begin{bmatrix}
y_1 \\
y_2 \\
\vdots \\
y_m
\end{bmatrix}
=
\begin{bmatrix}
\sum_{j=1}^{m} a_{1j}y_j \\
\sum_{j=1}^{m} a_{2j}y_j \\
\vdots \\
\sum_{j=1}^{m} a_{nj}y_j
\end{bmatrix}.
$$

Each element of the result is $E(i, Y), \quad i = 1, 2, \ldots, n$.

6. A square matrix $A_{n \times n}$ has an **inverse** A^{-1} if there is a matrix $B_{n \times n}$ that satisfies $A \cdot B = B \cdot A = I$, and then B is written as A^{-1}. Finding the inverse is computationally tough, but luckily you can determine whether there is an inverse by finding the determinant of A. The linear algebra theorem says that A^{-1} exists if and only if $\det(A) \neq 0$. The determinant of a 2×2 matrix is easy to calculate by hand:

$$\det(A_{2 \times 2}) = \begin{vmatrix} a_{11} & a_{12} \\ a_{21} & a_{22} \end{vmatrix} = a_{11}a_{22} - a_{12}a_{21}$$

One way to calculate the determinant of a larger matrix is **expansion** by minors which we illustrate for a 3×3 matrix:

$$\det(A_{3 \times 3}) = \begin{vmatrix} a_{11} & a_{12} & a_{13} \\ a_{21} & a_{22} & a_{23} \\ a_{31} & a_{32} & a_{33} \end{vmatrix}$$

$$= a_{11} \begin{vmatrix} a_{22} & a_{23} \\ a_{32} & a_{33} \end{vmatrix} - a_{12} \begin{vmatrix} a_{21} & a_{23} \\ a_{31} & a_{33} \end{vmatrix} + a_{13} \begin{vmatrix} a_{21} & a_{22} \\ a_{31} & a_{32} \end{vmatrix}.$$

This reduces the calculation of the determinant of a 3×3 matrix to the calculation of the determinants of three 2×2 matrices, which are called the **minors** of A. They are obtained by crossing out the row and column of the element in the first row (other rows may also be used). The determinant of the minor is multiplied by the element and the sign $+$ or $-$ alternates starting with a $+$ for the first element. Here is the determinant for a 4×4 reduced to four 3×3 determinants:

$$\det(A_{4 \times 4}) = \begin{vmatrix} a_{11} & a_{12} & a_{13} & a_{14} \\ a_{21} & a_{22} & a_{23} & a_{24} \\ a_{31} & a_{32} & a_{33} & a_{34} \\ a_{41} & a_{42} & a_{43} & a_{44} \end{vmatrix}$$

$$= a_{11} \begin{vmatrix} a_{22} & a_{23} & a_{24} \\ a_{32} & a_{33} & a_{34} \\ a_{42} & a_{43} & a_{44} \end{vmatrix} - a_{12} \begin{vmatrix} a_{21} & a_{23} & a_{24} \\ a_{31} & a_{33} & a_{34} \\ a_{41} & a_{43} & a_{44} \end{vmatrix}$$

$$+ a_{13} \begin{vmatrix} a_{21} & a_{22} & a_{24} \\ a_{31} & a_{32} & a_{34} \\ a_{41} & a_{42} & a_{44} \end{vmatrix} - a_{14} \begin{vmatrix} a_{21} & a_{22} & a_{23} \\ a_{31} & a_{32} & a_{33} \\ a_{41} & a_{42} & a_{43} \end{vmatrix}.$$

7. A system of linear equations for the unknowns $\vec{y} = (y_1, \ldots, y_m)$ may be written in matrix form as $A_{n \times m} \vec{y} = \vec{b}$, where $\vec{b} = (b_1, b_2, \ldots, b_n)$. This is called an **inhomogeneous system** if $\vec{b} \neq \vec{0}$ and a **homogeneous** system if $\vec{b} = \vec{0}$. If A is a square matrix and is invertible, then $\vec{y} = A^{-1}\vec{b}$ is the one and only solution. In particular, if $\vec{b} = \vec{0}$, then $\vec{y} = \vec{0}$ is the only solution.

8. A matrix $A_{n \times m}$ has associated with it a number called the **rank** of A, which is the largest square submatrix of A that has a nonzero determinant. So, if A is a 4×4 invertible matrix, then $rank(A) = 4$. Another way to calculate the rank of A is to **row**-reduce the matrix to row-reduced echelon form. The number of nonzero rows is $rank(A)$.

9. The rank of a matrix is intimately connected to the solution of equations $A_{n \times m} y_{m \times 1} = b_{n \times 1}$. This system will have a unique solution if and only if $rank(A) = m$. If $rank(A) < m$, then $A y_{m \times 1} = b_{n \times 1}$ has an infinite number of solutions. For the homogeneous system $A y_{m \times 1} = 0_{n \times 1}$, this has a unique solution, namely, $y_{m \times 1} = 0_{m \times 1}$, if and only if $rank(A) = m$. In any other case $A y_{m \times 1} = 0_{n \times 1}$ has an infinite number of solutions.

Appendix B

The Essentials of Probability

In this section we give a brief review of the basic probability concepts and definitions used or alluded to in this book. For further information, there are many excellent books on probability (e.g., see the book by Ross [37]).

The space of all possible outcomes of an experiment is labeled Ω. Events are subsets of the space of all possible outcomes, $A \subset \Omega$. Given two events A, B

1. The event $A \cup B$ is the event **either A occurs or B occurs, or they both occur.**
2. The event $A \cap B$ is the event **both A and B occur.**
3. The event $\emptyset$ is called the **impossible event**, while the event Ω itself is called the **sure event**.
4. The event $A^c = \Omega \setminus A$ is called the **complement of** A, so either A occurs or, if A does not occur, then A^c occurs.

The probability of an event A is written as $P(A)$ or $Prob(A)$. It must be true that for any event A, $0 \le P(A) \le 1$. In addition,

1. $P(\emptyset) = 0, P(\Omega) = 1$.
2. $P(A^c) = 1 - P(A)$.
3. $P(A \cup B) = P(A) + P(B) - P(A \cap B)$.
4. $P(A \cap B) = P(A)P(B)$ if A and B are **independent**.

Given two events A, B with $P(B) > 0$, the **conditional probability of A given the event B has occurred**, is defined by

$$P(A|B) = \frac{P(A \cap B)}{P(B)}.$$

Two events are therefore independent if and only if $P(A|B) = P(A)$ so that the knowledge that B has occurred does not affect the probability of A. The very important **laws of total probability** are frequently used:

1. $P(A) = P(A \cap B) + P(A \cap B^c)$
2. $P(A) = P(A|B)P(B) + P(A|B^c)P(B^c)$

These give us a way to calculate the probability of A by breaking down cases. Using the formulas, if we want to calculate $P(A|B)$ but know $P(B|A)$ and $P(B|A^c)$, we can use **Bayes' rule**:

$$P(A|B) = \frac{P(A \cap B)}{P(B)} = \frac{P(B|A)P(A)}{P(B|A)P(A) + P(B|A^c)P(A^c)}.$$

A random variable is a function $X : \Omega \to \mathbb{R}$, that is, a real-valued function of outcomes of an experiment. As such, this random variable takes on its values by chance. We consider events of the

Game Theory: An Introduction, Third Edition. E. N. Barron.
© 2024 John Wiley & Sons, Inc. Published 2024 by John Wiley & Sons, Inc.

form $\{\omega \in \Omega : X(\omega) \leq x\}$ for values of $x \in \mathbb{R}$. For simplicity, this event is written as $\{X \leq x\}$ and the function

$$F_X(x) = P(X \leq x), \quad x \in \mathbb{R}$$

is called the **cumulative distribution function** of X. We have

1. $P(X > a) = 1 - F_X(a)$.
2. $P(a < X \leq b) = F_X(b) - F_X(a)$.

A random variable X is said to be **discrete** if there is a finite or countable set of numbers $x_1, x_2, \ldots$, so that X takes on only these values with $P(X = x_i) = p_i$, where $0 \leq p_i \leq 1$, and $\sum_i p_i = 1$. The cumulative distribution function is a step function with a jump of size p_i at each x_i, $F_X(x) = \sum_{x_i \leq x} p_i$.

A random variable is said to be **continuous** if $P(X = x) = 0$ for every $x \in \mathbb{R}$, and the cumulative distribution function $F_X(x) = P(X \leq x)$ is a continuous function. The probability that a continuous random variable is any particular value is always zero. A probability density function for X is a function

$$f_X(x) \geq 0 \text{ and } \int_{-\infty}^{+\infty} f_X(x)\, dx = 1.$$

In addition, we have

$$F_X(x) = \int_{-\infty}^{x} f_X(y)\, dy \text{ and } \quad f_X(x) = \frac{d}{dx} F_X(x).$$

The cumulative distribution up to x is the area under the density from $-\infty$ to x. With an abuse of notation, we often see $P(X = x) = f_X(x)$, which is clearly nonsense, but it gets the idea across that the density at x is roughly the probability that $X = x$.

Two random variables X and Y are **independent** if $P(X \leq x \text{ and } Y \leq y) = P(X \leq x)P(X \leq y)$ for all $x, y \in \mathbb{R}$. If the densities exist, this is equivalent to $f(x, y) = f_X(x)f_Y(y)$, where $f(x, y)$ is the joint density of (X, Y).

The **mean** or **expected value** of a random variable X is

$$E[X] = \sum_i x_i P(X = x_i) \text{ if } X \text{ is discrete}$$

and

$$E[X] = \int_{-\infty}^{+\infty} x f_X(x)\, dx \text{ if } X \text{ is continuous.}$$

In general, a much more useful measure of X is the **median** of X, which is any number satisfying

$$P(X \geq m) = P(X \leq m) = \frac{1}{2}.$$

Half the area under the density is to the left of m and half is to the right.

The mean of a function of X, say $g(X)$, is given by

$$E[g(X)] = \sum_i g(x_i)P(X = x_i) \text{ if } X \text{ is discrete,}$$

and

$$E[g(X)] = \int_{-\infty}^{+\infty} g(x)f_X(x)\, dx \text{ if } X \text{ is continuous.}$$

With the special case $g(x) = x^2$, we may get the **variance of X** defined by

$$Var(X) = E[X^2] - (E[X])^2 = E[X - E[X]]^2.$$

This gives a measure of the spread of the values of X around the mean defined by the **standard deviation of** X

$$\sigma(X) = \sqrt{Var(X)}.$$

We end this appendix with a list of properties of the main discrete and continuous random variables.

Discrete Random Variables

1. **Bernoulli**: Consider an experiment in which the outcome is either success, with probability $p > 0$ or failure, with probability $1 - p$. The random variable

$$X = \begin{cases} 1 & \text{if success;} \\ 0 & \text{if failure.} \end{cases}$$

is Bernoulli with parameter p. Then $E[X] = p, Var(X) = p(1 - p)$.
2. Suppose that we perform an independent sequence of Bernoulli trials, each of which has probability p of success. Let X be a count of the number of successes in n trials. X is said to be a **binomial random variable** with distribution

$$P(X = k) = \binom{n}{k} p^k(1 - p)^{n-k}, \ k = 0,1, 2, \ldots, n, \ \binom{n}{k} = \frac{n!}{k!(n - k)!}.$$

Then $E[X] = np, Var(X) = np(1 - p)$.
3. In a sequence of independent Bernoulli trials the number of trials until the first success is called a **geometric random variable**. If X is geometric, it has distribution

$$P(X = k) = (1 - p)^{k-1}p, \ \ k = 1,2, \ldots$$

with mean $E[X] = \frac{1}{p}$, and variance $Var(X) = (1 - p)/p^2$.
4. A random variable has a **Poisson distribution with parameter** λ if X takes on the values $k = 0,1, 2, \ldots$, with probability

$$P(X = k) = e^{-\lambda}\frac{\lambda^k}{k!}, \ \ k = 0,1, 2 \ldots.$$

It has $E[X] = \lambda$, and $Var(X) = \lambda$. It arises in many situations and is a limit of binomial random variables. In other words, if we take a large number n of Bernoulli trials with p as the probability of success on any trial, then as $n \to \infty$ and $p \to 0$ but np remaining the constant λ, the total number of successes will follow a Poisson distribution with parameter λ.

Continuous Distributions

1. X is **uniformly distributed on the interval** $[a, b]$ if it has the density

$$f_X(x) = \begin{cases} \dfrac{1}{b - a} & \text{if } a < x < b; \\ 0 & \text{otherwise.} \end{cases}$$

It is a model of picking a number at random from the interval $[a, b]$ in which every number has equal likelihood. The cdf is $F_X(x) = (x - a)/(b - a), a \le x \le b$. The mean is $E[X] = (a + b)/2$, the midpoint, and variance is $Var(X) = (b - a)^2/12$.
2. X has a **normal distribution** with mean μ and standard deviation σ if it has the density

$$f_X(x) = \frac{1}{\sigma\sqrt{2\pi}} \exp\left(-\frac{(x - \mu)^2}{2\sigma^2}\right), \ -\infty < x < \infty,$$

Then the mean is $E[X] = \mu$, and the variance is $Var(X) = \sigma^2$. The graph of f_X is the classic bell-shaped curve centered at μ. The central limit theorem makes this the most important distribution because it roughly says that sums of independent random variables normalized by $\sigma \sqrt{n}$, converge to a normal distribution, no matter what the distribution of the random variables in the sum. More precisely

$$\lim_{n \to \infty} P\left(\frac{\sum_{i=1}^{n} X_i - n\mu}{\sigma \sqrt{n}} \leq x \right) = \frac{1}{\sqrt{2\pi}} \int_{-\infty}^{x} e^{-y^2/2} \, dy.$$

Here $E[X_i] = \mu, Var(X_i) = \sigma^2$ are the mean and variance of the arbitrary members of the sum.
3. The random variable $X \geq 0$ is said to have an **exponential distribution** if

$$P(X \leq x) = F_X(x) = 1 - e^{-\lambda x}, x \geq 0.$$

The density is $f_X(x) = F'_X(x) = \lambda e^{-\lambda x}, x > 0$. Then $E[X] = 1/\lambda$ and $Var(X) = 1/\lambda^2$.

Appendix C

The Mathematica Commands

In this appendix, we will summarize many of the Mathematica constructs and commands used in the book.

C.1 The Upper and Lower Values of a Game

We begin by showing how to use Mathematica to calculate the lower value

$$v^- = \max_{1 \le i \le n} \min_{1 \le j \le m} a_{i,j}$$

and the upper value

$$v^+ = \min_{1 \le j \le m} \max_{1 \le i \le n} a_{i,j}.$$

of the matrix A.

Enter the matrix

```
A = {{1,4,7},{-1,3,5},{2,-6,1.4}}

rows = Dimensions[A][[1]]
      cols = Dimensions[A][[2]]
        a = Table[Min[A[[i]]], {i, rows}]
          b = Max[a[[]]]
            Print["The lower value of the game is= ", b]
            c = Table[Max[A[[All, j]]], {j, cols}]
              d = Min[c[[]]]
              Print["The upper value is=", d]
```

These commands will give the upper and lower values of A. Observe that we do not need to load any packages and we do not need to end a statement with a semicolon. On the other hand, to execute a statement we need to push Shift-Enter at the same time.

Game Theory: An Introduction, Third Edition. E. N. Barron.
© 2024 John Wiley & Sons, Inc. Published 2024 by John Wiley & Sons, Inc.

C.2 The Value of an Invertible Matrix Game with Mixed Strategies

The value and optimal strategies of a game with an invertible matrix (or one which can be made invertible by adding a constant) are calculated in the following. The formulas we use are

$$v(A) = \frac{1}{J_n A^{-1} J_n^T},$$

$$X = v(A)(J_n A^{-1}), \text{ and } Y = v(A)(A^{-1} J_n^T).$$

If these are legitimate strategies, then the whole procedure works; that is, the procedure is actually calculating the saddle point and the value:

```
A = {{4, 2, -1}, {-4, 1, 4}, {0, -1, 5}}
        {{4, 2, -1}, {-4, 1, 4}, {0, -1, 5}}
Det[A]
      72
A1 = A + 3
        {{7, 5, 2}, {-1, 4, 7}, {3, 2, 8}}
B = Inverse[A1]
        {{2/27, -4/27,1/9}, {29/243, 50/243, -17/81},
        {-14/243,1/243,11/81}}
J = {1, 1, 1}
        {1, 1, 1}
v = 1/J.B.J
      81/19
X = v*(J.B)
        {11/19, 5/19, 3/19}
Y = v*(B.J)
        {3/19, 28/57,20/57}
```

At the end of each statement, having pushed Shift+Enter, we are showing the Mathematica result. Remember that you have to subtract the constant you added to the matrix at the beginning in order to get the value for the original matrix. So the actual value of the game is $v(A) = \frac{81}{19} - 3 = \frac{24}{19}$.

C.3 Solving Matrix Games

In this subsection we will present the use of Mathematica to solve a matrix game using the methods presented in the text.

Here is an example using the Maximize and Minimize functions:

```
(*Define the matrix A*)A={{4,0,2,1},{0,4,1,2},{1,-1,3,0},{1,1,0,3},
    {-2,-2,2,2}};

(*Define the variable X*)
n=Length[A];
m=Length[A[[1]]];
X=Array[x,n];
J=Array[1&,n];
K=Array[1&,m];
```

```
Y=Array[y,m];
Z=Append[X,v];
W=Append[Y,v];

(*Define the objective function*)
objectiveFunction=v;

(*Define the constraints for PI and PII*)
constraintsX={X.A>=v, X>=0,X.J==1} ;
constraintsY={A.Y<=v,Y>=0,Y.K==1};

(*Solve*)
Maximize[{objectiveFunction,constraintsX},Z]
Out= {14/9,{x[1]->4/9,x[2]->4/9,x[3]->0,x[4]->0,x[5]->1/9,v->14/9}}
Minimize[{objectiveFunction,constraintsY},W]
Out= {14/9,{y[1]->1/30,y[2]->7/90,y[3]->8/15,y[4]->16/45,v->14/9}}
```

C.4   Interior Nash Points

The system of equations for an interior mixed Nash equilibrium is given by

$$\sum_{j=1}^{m} y_j[a_{kj} - a_{nj}] = 0, \qquad k = 1,2,\dots,n-1$$

$$\sum_{i=1}^{n} x_i[b_{is} - b_{im}] = 0, \qquad s = 1,2,\dots,m-1.$$

$$x_n = 1 - \sum_{i=1}^{n-1} x_i \quad \text{and} \quad y_m = 1 - \sum_{j=1}^{m-1} y_j.$$

The game has matrices (A, B) and the equations, if they have a solution that are strategies, will yield $X^* = (x_1,\dots,x_n)$ and $Y^* = (y_1,\dots,y_m)$.

The following example shows how to set up and solve using Mathematica:

```
A = {{-2, 5, 1}, {-3, 2, 3}, {2, 1, 3}}
B = {{-4, -2, 4}, {-3, 1, 4}, {3, 1, -1}}
rowdim = Dimensions[A][[1]]
coldim = Dimensions[B][[2]]
Y = Array[y, coldim]
X = Array[x, rowdim]
EQ1 = Table[Sum[y[j](A[[k, j]]-A[[rowdim, j]]),
                {j, coldim}], {k, rowdim-1}]
EQ2 = Table[Sum[x[i](B[[i, s]]-B[[i, coldim]]),
                {i, rowdim}], {s, coldim-1}]
Solve[{EQ1[[1]] == 0, EQ1[[2]] == 0, y[1] + y[2] + y[3] == 1},
                {y[1],y[2],y[3]}]
Solve[{EQ2[[1]] == 0, EQ2[[2]] == 0, x[1] + x[2] + x[3] == 1},
                {x[1],x[2],x[3]}]
```

Mathematica gives the output

$$y[1] = \frac{1}{14}, y[2] = \frac{5}{14}, y[3] = \frac{8}{14}, \text{ and } x[1] = \frac{1}{14}, x[2] = \frac{2}{7}, x[3] = \frac{9}{14}.$$

C.5 Lemke–Howson Algorithm for Nash Equilibrium

A Nash equilibrium of a bimatrix game with matrices (A, B) is found by solving the nonlinear programming problem.

$$\text{Maximize} f(X, Y, p, q) = X\,AY^T + X\,BY^T - p - q$$

subject to

$$AY^T \leq p, XB \leq q, \sum_i x_i = 1, \sum_j y_j = 1, X \geq 0, Y \geq 0.$$

Here is the use of Mathematica to solve for a particular example:

```
A = {{-1, 0, 0}, {2, 1, 0}, {0, 1, 1}}
B = {{1, 2, 2}, {1, -1, 0}, {0, 1, 2}}
f[X_] = X.B
g[Y_] = A.Y
FindMaximum[{f[{x, y, z}].{a, b, c} + {x, y, z}.g[{a, b, c}] - p - q,
  f[{x, y, z}][[1]] - q <= 0,
  f[{x, y, z}][[2]] - q <= 0, f[{x, y, z}][[3]] - q <= 0,
  g[{a, b, c}][[1]] - p <= 0, g[{a, b, c}][[2]] - p <= 0,
  g[{a, b, c}][[3]] - p <= 0, x + y + z == 1, x >= 0, y >= 0, z >= 0,
  a >= 0, b >= 0, c >= 0, a + b + c == 1}, {x, y, z, a, b, c, p, q}]
```

The command `FindMaximum` will seek a maximum. This command produces the Mathematica output (in which we round very small numbers to 0)

$$0.0, \{a = 0.333, b = 0.0, c = 0.666\},$$

$$p = 0.666, q = 0.666,$$

$$\{x = 0.0, y = 0.666, z = 0.333\}.$$

The first number verifies that the maximum of f is indeed zero. The optimal strategies for each player are $Y^* = (\frac{1}{3}, 0, \frac{2}{3})$, $X^* = (0, \frac{2}{3}, \frac{1}{3})$. The expected payoff to player I is $p = \frac{2}{3}$, and the expected payoff to player II is $q = \frac{2}{3}$. To adjust the starting positions of the variables to find other NEs, adjust the FindMaximum command to

$$\text{FindMaximum}[..., \{\{x, x_0\}, \{y, y_0\}.z, a, b, c, p, q\}]$$

Here the starting search positions of x is x_0 and y is y_0. You may adjust the positions of any of the variables.

C.6 Is the Core Empty?

A simple check to determine if the core is empty uses a simple linear program. The core is

$$C(0) = \{\vec{x} = (x_1, \ldots, x_n) \mid \sum_{i=1}^{n} x_i = 1, x_i \geq 0, v(S) - \sum_{i \in S} x_i \leq 0, \forall S \subsetneq N\}.$$

Convert to a linear program

$$\text{Minimize} \, x_1 + \cdots + x_n$$

$$\text{subject to} \, v(S) - \sum_{i \in S} x_i \leq 0, \, \forall S \subsetneq N.$$

If the solution of this program produces a minimum of $x_1 + \cdots + x_n > v(N)$, we know that the core is empty; otherwise it is not. For example, if $N = \{1,2,3\}$, and

$$v(i) = 0, v(12) = 105, v(13) = 120, v(23) = 135, \text{and} \, v(123) = 150,$$

we may use the Mathematica command

```
Minimize[{x1+x2+x3,x1+x2>=105,x1+x3>=120,
          x2+x3>=135,x1>=0,x2>=2},{x1,x2,x3}].
```

Mathematica tells us that the minimum of $x_1 + x_2 + x_3$ is $180 > v(123) = 150$. So $C(0) = \emptyset$. The other way to see this is with the Mathematica command

```
LinearProgramming[{1, 1, 1}, {{1, 1, 0}, {1, 0, 1},
                    {0, 1, 1}}, {105, 120,135}]
```

which gives the output $\{x_1 = 45, x_2 = 60, x_3 = 75\}$ and minimum 180. The vector $c = (1,1,1)$ is the coefficient vector of the objective $c \cdot (x_1, x_2, x_3)$; the matrix

$$m = \begin{bmatrix} 1 & 1 & 0 \\ 1 & 0 & 1 \\ 0 & 1 & 1 \end{bmatrix}$$

is the matrix of constraints of the form $\geq b$; and $b = (105, 120, 135)$ is the right-hand side of the constraints.

C.7 Find and Plot the Least Core

The ε-core is

$$C(\varepsilon) = \{\vec{x} = (x_1, \ldots, x_n) \mid v(S) - \sum_{i \in S} x_i \leq \varepsilon, \, \forall \, S \subsetneq N\}.$$

You need to find the smallest $\varepsilon^1 \in \mathbb{R}$ for which $C(\varepsilon^1) \neq \emptyset$.

For example, we use the characteristic function

```
v1 = 0; v2 = 0; v3 = 0;
v12 = 2; v13 = 0; v23 = 1; v123 = 5/2;
```

and then the Mathematica command

```
Minimize[{z, v1 - x1 <= z, v2 - x2 <= z, v3 - x3 <= z,
          v12 - x1 - x2 <= z, v13 - x1 - x3 <= z, v23 - x2 - x3 <= z,
          x1 + x2 + x3 == v123}, {z, x1, x2, x3}]
```

This gives the output

$$z = -\frac{1}{4}, \, x_1 = \frac{5}{4}, \, x_2 = 1, x_3 = \frac{1}{4}.$$

The important quantity is $z = \varepsilon^1 = -\frac{1}{4}$ because we do not know yet whether $C(\varepsilon^1)$ contains only one point. At this stage, we know that $C(-\frac{1}{4}) \neq \emptyset$, and if we take any $\varepsilon < -\frac{1}{4}$, then $C(\varepsilon) = \emptyset$.

Here is how you find the least core and get a plot of $C(0)$ in Mathematica:

```
Core[z] := {v1 - x1 <= z, v2 - x2 <= z, v3 - x3 <= z,
            v12 - x1 - x2 <= z, v13 - x1 - x3 <= z,
            v23 - x2 - x3 <= z, x1 + x2 + x3 == v123}
```

```
A[z] := {z, Core[z]}  Minimize[A[z], {z, x1, x2, x3}]
```

```
Output:-1/4, {z -> -1/4, x1 -> 5/4, x2 -> 1, x3 -> 1/4}
```

```
S = Simplify[Core[z] /. {z -> 0, x3 -> v123 - x1 - x2}]
```

Substitute z=0, x3=v123-x1-x2 in C[z]

```
Reduce[S, {x1, x2}]
```

Output: 0 <= x1 <= 3\/2 && 2 - x1 <= x2 <= 1/2 (5 - 2 x1)
Solve inequalities in C[0] to see if one point.

```
RegionPlot[S, {x1, -2, 2}, {x2, -2, 2}]
```

Plot the core C[0] (see below)

```
F = Simplify[Core[z] /. {z -> -1/4, x3 -> v123 - x1 - x2}]
```

Substitute z=-1/4 for least core.

```
Reduce[F, {x1, x2}]
```

Output: 1/4 <= x1 <= 5/4 && x2 == 1/4 (9 - 4 x1)
Solve inequalities in C[-1/4] to see if one point.

```
Plot[(9 - 4 x1)/4, {x1, 1/4, 5/4}]
```

C[-1/4] is a line segment.

Here is $C(0)$ for the example, generated with Mathematica:

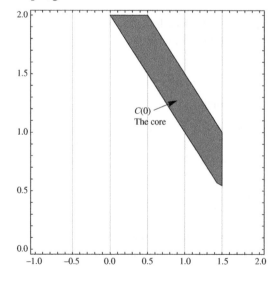

$C(0)$
The core

C.8   Nucleolus Procedure and Shapley Value

This section will give the Mathematica procedure to find the nucleolus. This is presented as a three-player example. The Mathematica commands are interspersed with output and is self-documenting.

```
Clear the variables:
Clear[x1,x2,x3,v1,v2,v3,v12,v13,v23,v123,z,w,S,S2,A,Core]
Define the characteristic function:
v1=0;v2=0;v3=0;v12=1/3;v13=1/6;v23=5/6;v123=1;

Core[z]:={v1-x1<=z,v2-x2<=z,v3-x3<=z,v12-x1-x2<=z,
          v13-x1-x3<=z,v23-x2-x3<=z, x1+x2+x3==v123}

A[z]:={z,Core[z]}

Minimize[A[z],{z,x1,x2,x3}]

Output:-1/12, {z=-1/12, x1=1/12, x2=1/3, x3=7/12}

S=Simplify[Core[z]/.{z->-1/12,x3->v123-x1-x2}]

Reduce[S,{x1,x2}]

Output:
x1=1/12 && 1/3<= x2 <= 3/4

Assign the known variables.

x1=1/12;z=-1/12;x3=v123-x1-x2

Now check the excesses to see which coalitions can be dropped:

v1-x1=-1/2
v2-x2=-x2
v3-x3=-11/12+x2
v12-x1-x2=1/4-x2
v13-x1-x3=-5/6+x2
v23-x2-x3=-1/12

We may drop coalitions {1} and {23} and then recalculate

B[w]={v2-x2<=w,v3-x3<=w,v12-x1-x2<=w,
                v13-x1-x3<=w,x1+x2+x3==v123}

Find the smallest w which makes B[w] nonempty
Minimize[{w,B[w]},{x2,w}]

Output:-7/24, {x2 = 13/24, w = -7/24}
```

```
Check to see if we are done:
S2=Simplify[B[w]/.w->-7/24]

Reduce[S2,{x2}]

Output:x2= 13/24
```

This is the end because we now know x1=1/12,x2=13/24,x3=9/24.

The final result gives the nucleolus as $(x_1 = \frac{1}{12}, x_2 = \frac{13}{24}, x_3 = \frac{9}{24})$.

Shapley Value Procedure:
Here is an example.

```
S = { 1, 2, 3}
numPlayers = Length[S]
L = Subsets[S]
K = Length[L]
v[{1}] = 25;
v[{2}] = 40;
v[{3}] = 0;
v[{1, 2}] = 90;
v[{1, 3}] = 35;
v[{2, 3}] = 50;
v[{1, 2, 3}] = 100;
v[{}] = 0;
For[i = 1, i <= numPlayers, i++,
 { x[i] = 0;
   For[k = 1, k <= K, k++,
    If[MemberQ[L[[k]], i] && Length[L[[k]]] >= 1,
     x[i] =
      x[i] + (v[L[[k]]] -
          v[DeleteCases[L[[k]], i]])*(Factorial[Length[L[[k]]] - 1]
           Factorial[numPlayers - Length[L[[k]]]])/
         Factorial[numPlayers]], Print[x[i]]}]
This gives the output x[1]=235/6, x[2]=325/6,x[3]=20/3.
```

C.9 Mathematica Code for Three-Person Nucleolus

First, we need to check if the core is empty to see if we can use the formulas. The first snippet actually finds the least $\varepsilon = \varepsilon^1$ for the least core.

```
 Clear["Global'*"]
 (* This program implements the theorem with explicit formulas for \
3-person games*)
 (* If the returned value of z is <0, the core is nonempty; if z>0 \
then the core is empty*)
 IsCoreEmpty[v1_, v2_, v3_, v12_, v13_, v23_, v123_] :=
 Module[{x1, x2, x3}, obj = z;
```

```
cnsts = {v1 - x1 <= z, v2 - x2 <= z, v3 - x3 <= z,
v12 -x1-x2<=z,v13 -x1-x3<=z,v23 -x2-x3<=z,
x1 + x2 + x3 == v123} ; Minimize[{obj, cnsts}, {x1, x2, x3} ]]
```
For example, suppose $v(i) = 0$, $v(12) = 100$, $v(13) = 0$, $v(23) = 100$, $v(123) = 200$. The command is:
```
IsCoreEmpty[0, 0, 0, 100, 0, 100, 200]
(* This produces z=-50 so C(0) not empty *)
```
If we take $v(i) = 0$, $v(12) = v(13) = v(23) = v(123) = 1$, the command is:
```
IsCoreEmpty[0, 0, 0, 1, 1, 1, 1]
(* This produces z=1/3 so C(0)= empty *)
```
Assuming the condition $C(0) \neq \emptyset$ the following code (due to Leng and Parlar [22]) will quickly check the conditions of the theorem, print which case holds, and compute the nucleolus.
```
CheckCore[v12_, v13_, v23_, v123_] :=
Module[{y1, y2, y3},
Which [v123 >= 3*v12 && v123 >= 3*v13 && v123 >= 3*v23,
{y1 = v123/3, y2 = y1, y3 = y1,
Print["Case 1"]},
v123 >= v12 + 2*v23 && v123 >= v12 + 2*v13 && v123<= 3* v12,
{y1 = (v123 + v12)/4,
y2 = y1,
y3 = (v123 - v12)/2,
Print["Case 2"]},
v123 >= v12 + 2*v23 && v123 <= v12 + 2*v13 && v12 >= v13,
{y1 = (v12 + v13)/2,
y2 = (v123 - v13)/2,
y3 = (v123 - v12)/2,
Print["Case 3"]},
v123 <= v12 + 2*v23 && v123 >= v12 + 2*v13 && v23 <= v12,
{y1 = (v123 - v23)/2,
y2 = (v12 + v23)/2,
y3 = (v123 - v12)/2,
Print["Case 4"]},
v123 <= v12 + 2*v23 && v123 <= v12 + 2*v13 && v123 + v12 >= 2*(v13
+ v23),
{y1 = (v123 + v12 - 2*(v23 - v13))/4,
y2 = (v123 + v12 + 2*(v23 - v13))/4,
y3 = (v123 - v12)/2,
Print["Case 5"]},
v123 >= v13 + 2*v23 && v123 >= v13 + 2*v12 &&v123 <= 3*v13,
{y1 = (v123 + v13)/4,
y2 = (v123 - v13)/2,
y3 = y1,
Print["Case 6"]},
v123 >= v13 + 2*v23 && v123 <= v13 + 2*v12 && v13 >= v12,
{y1 = (v12 + v13)/2,
y2 = (v123 - v13)/2,
y3 = (v123 - v12)/2,
Print["Case 7"]},
```

```
v123 <= v13 + 2*v23 && v123 >= v13 + 2*v12 && v13 >= v23,
{y1 = (v123 - v23)/2,
y2 = (v123 - v13)/2,
y3 = (v13 + v23)/2,
Print["Case 8"]},
v123 <= v13 + 2*v23 && v123 <= v13 + 2*v12 &&
v123 + v13 >= 2*(v12 + v23),
{y1 = (v123 + v13 - 2*(v23 - v12))/4,
y2 = (v123 - v13)/2,
y3 = (v123 + v13 + 2*(v23 - v12))/4,
Print["Case 9"]},
v123 >= v23 + 2*v13 && v123 >= v23 + 2*v12 && v123 <=3*v23,
{y1 = (v123 - v23)/2,
y2 = (v123 + v23)/4,
y3 = y2,
Print["Case 10"]},
v123 >= v23 + 2*v13 && v123 <= v23 + 2*v12 && v23 >= v12,
{y1 = (v123 - v23)/2,
y2 = (v12 + v23)/2,
y3 = (v123 - v12)/2,
Print["Case 11"]},
v123 <= v23 + 2*v13 && v123 >= v23 + 2*v12 && v23 >= v13,
{y1 = (v123 - v23)/2,
y2 = (v123 - v13)/2,
y3 = (v13 + v23)/2,
Print["Case 12"]},
v123 <= v23 + 2*v13 && v123 <= v23 + 2*v12 &&
v123 + v23 >= 2*(v13 + v12),
{y1 = (v123 - v23)/2,
y2 = (v123 + v23 - 2*(v13 - v12))/4,
y3 = (v123 + v23 + 2*(v13 - v12))/4,
Print["Case 13"]},
v123 + v12 <= 2*(v13 + v23) && v123 + v13 <= 2*(v12+ v23) &&
v123 + v23 <= 2*(v13 + v12),
{y1 = (v123 + v12 + v13 - 2*v23)/3,
y2 = (v123 + v12 + v23 - 2*v13)/3,
y3 = (v123 + v13 + v23 - 2*v12)/3,
Print["Case 14"]} ]
]
```

For example, the command `CheckCore[100,0,100,200]` produces the output `Case 4, x_1=50,x_2=100,x_3=50`.

C.10   Plotting the Payoff Pairs

Given a bimatrix game with matrices (A, B) that are each 2×2 matrices, the expected payoff to player I is

$$E_1(x, y) = (x, 1 - x)A \begin{bmatrix} y \\ 1 - y \end{bmatrix},$$

and for player II

$$E_{II}(x, y) = (x, 1 - x)B \begin{bmatrix} y \\ 1 - y \end{bmatrix}.$$

We want to get an idea of the shape of the set of payoff pairs $(E_I(x, y), E_{II}(x, y))$. In Mathematica we can do that with the following commands:

```
A = {{2, -1}, {-1, 1}}
B = {{1, -1}, {-1, 2}}
f[x_, y_] = {x, 1 - x}.A.{y, 1 - y}
g[x_, y_] = {x, 1 - x}.B.{y, 1 - y}

h[x_, y_] = {f[x, y], g[x, y]}

values = Table[Table[h[x, y], {x, 0, 1, .025}], {y, 0, 1, 0.025}];

s = Flatten[values, 1];
```

```
ListPlot[s]
```

Observe that a function is defined in Mathematica as, for example,

$$f[x\_,y\_] = \{x, 1 - x\}.A.\{y, 1 - y\},$$

which gives $E_I(x, y)$. Mathematica uses the symbol x_ to indicate that x is a variable. The result of these commands is the following graph:

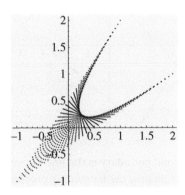

C.11 Bargaining Solutions

We will use Mathematica to solve the following bargaining problem with bimatrix:

$$(A, B) = \begin{bmatrix} (1, 4) & (-\frac{4}{3}, -4) \\ (-3, -1) & (4, 1) \end{bmatrix}.$$

First, we use the safety point $u^* = value(A)$, $v^* = value(B^T)$. Here are the commands to find (u^*, v^*):

```
The matrix is:
A = {{1,-4/3},{-3,4}}
```

```
player I's problem is:
```

```
Maximize[{v, {x, y}.A[[All, 1]] >= v, {x, y}.A[[All, 2]] >= v,
    x + y == 1, x >= 0, y >= 0}, {x, y, v}]
```

player II's problem is:
```
Minimize[{v, A[[1]].{x, y} <= v, A[[2]].{x, y} <= v,
    x + y == 1, x >= 0, y >= 0}, {x, y, v}]
```

This finds $u^* = value(A) = 0$, and, even though we don't need it, $X^* = (\frac{3}{4}, \frac{1}{4}), Y^* = (\frac{4}{7}, \frac{3}{7})$. For v^* we use

```
B = {{4, -4}, {-1, 1}}
BT = Transpose[B]
Minimize[{v, BT[[1]].{x, y} <= v, BT[[2]].{x, y} <= v,
    x + y == 1, x >= 0, y >= 0}, {x, y, v}]
```

which tells us that $v^* = value(B^T) = 0$ and $Y^* = (\frac{1}{5}, \frac{4}{5})$.

So our safety point is $(u^*, v^*) = (0,0)$. A plot of the feasible set is found from

```
ListPlot[{{1, 4}, {-3, -1}, {-4/3, -4}, {4, 1}, {1, 4}},
                    PlotJoined -> True]
```

which produces

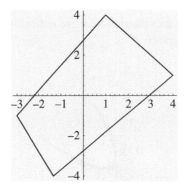

You can see that the Pareto-optimal boundary is the line through (4,1) and (1,4) which has the equation given by $v - 1 = -(u - 4)$. So now the bargaining problem with safety point (0,0) is

$$\text{Maximize } g(u, v) = (u - u^*)(v - v^*) = uv,$$

subject to the constraints

$$(u, v) \in S = \{v \le -u + 5, v \le \frac{5}{4}u - 4, v \ge -\frac{9}{5}u - \frac{32}{5}, v \ge \frac{15}{16}u - \frac{11}{16}\}.$$

The Mathematica command is

```
Maximize[{u v, v <= -u + 5, v <= 5/4 u + 11/4, v >= -9/5 u - 32/5,
    v >= 15/16 u - 11/16, u >= 0, v >= 0}, {u, v}]
```

and gives the output $\bar{u} = \frac{5}{2}, \bar{v} = \frac{5}{2}$ and the maximum is $g = \frac{25}{4}$. This is achieved if the two players agree to play and receive (1,4) and (4,1) exactly half the time. Then they each get $\frac{5}{2}$.

Next, we find the optimal threat strategies.

First, we have seen that the Pareto-optimal boundary is the line $v = -u + 5$, which has slope $m_p = -1$. So we look at the game with matrix $-m_pA - B$, or

$$A - B = \begin{bmatrix} -3 & \frac{8}{3} \\ -2 & 3 \end{bmatrix}$$

It is the optimal strategy that we need for this game. However, in this case it is easy to see that there is a pure saddle point $X^t = (0,1), Y^t = (1,0)$. Consequently, the threat safety point is

$$u^t = X_t A Y_t^{\ T} = -3, v^t = X_t B Y_t^{\ T} = -1.$$

The Mathematica command to solve the bargaining problem with this safety point is

```
Maximize[{ (u + 3) (v + 1), v <= -u + 5, v <= 5/4 u + 11/4,
            v >= -9/5 u - 32/5,
            v >= 15/16 u - 11/16, u >= -3, v >= -1}, {u, v}],
```

which gives the output $\bar{u} = \frac{3}{2}, \bar{v} = \frac{7}{2}$ and maximum $g = \frac{81}{4}$. In this solution player I gets less and player II gets more to reflect the fact that player II has a credible and powerful threat. So the payoffs are obtained by each agreeing to receive (1,4) exactly five-sixths of the time and payoff (4,1) exactly one-sixth of the time (player I throw player II a bone once every 6 plays).

C.12 Mathematica for Replicator Dynamics

Naturally, the solution of the replicator dynamics equations (8.2.1) depends on the matrices involved so we will only give some sample Mathematica commands to produce graphs of the vector fields and trajectories.

For example:

```
a = 2
b = 1
Here is the system we consider:
e1 = p1'[t] == p1[t] (a p1[t] - a p1[t]^2 - b p2[t]^2)
e2 = p2'[t] == p2[t] (b p2[t] - a p1[t]^2 - b p2[t]^2)
This command gives a plot of the vector field in the figure below:
field = VectorPlot[{x (a x - a x^2 - b y^2),
        y (b y - a x^2 - b y^2)}, {x, 0, 1}, {y, 0, 1}];
```

```
This command solves the system for the initial point p1[0],p2[0]:
nsoln = NDSolve[{e1, e2, p1[0] == .3,
                 p2[0] == .7}, {p1, p2}, {t, 0, 20}];
```

```
This plots the trajectory (p1[t],p2[t]) with p1 versus p2:
trajectory =
    ParametricPlot[{p1[t], p2[t]} /. nsoln[[1]], {t, 0, 20},
            PlotRange -> All,
            PlotStyle -> {Hue[1]}, Frame -> True];
```

```
This command plots the trajectory superimposed on the vector field:
```

```
Show[trajectory, field];
```

If you want a graph of the individual functions use:

```
trajectory1 = ParametricPlot[{t, p1[t]} /. nsoln[[1]],
                    {t, 0, 20}, PlotRange -> All,
                    PlotStyle -> {Hue[1]}, Frame -> True];

trajectory2 = ParametricPlot[{t, p2[t]} /. nsoln[[1]],
                    {t, 0, 20}, PlotRange -> All,
                    PlotStyle -> {Hue[1]}, Frame -> True];
Show[trajectory1, trajectory2];
```

This last command produces the following plot of both functions on one graph:

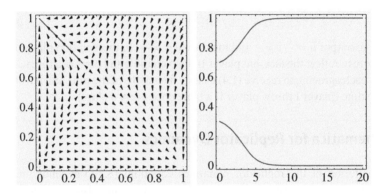

Appendix D

Biographies

D.1 John Von Neumann

John von Neumann was born to an aristocratic family on December 28, 1903, in Budapest, Austria-Hungary and died on February 8, 1957, in Washington, DC, of brain cancer. Von Neumann earned two doctorates at the same time, a PhD in mathematics from the University of Budapest, and a PhD in chemistry from the University of Zurich. He joined the faculty of the prestigious University of Berlin in 1927. Von Neumann made fundamental contributions to quantum physics, functional analysis, economics, computer science, numerical analysis, and many other fields. There was hardly any subject that came to his attention that he didn't revolutionize. Some areas, especially in mathematics, he invented. At age 19, he applied abstract operator theory, much of which he developed, to the brand new field of quantum mechanics. He was an integral member of the Manhattan Project and made fundamental contributions to the development of the hydrogen bomb and the development of the Mutual Assured Destruction policy of the United States. In 1932, he was appointed one of the original and youngest permanent members of the Institute for Advanced Study at Princeton New Jersey (along with A. Einstein) and helped to make it the most prestigious research institute in the world.

The von Neumann minimax theorem was proved in 1928 and was a major milestone in the theory of games. Von Neumann continued to think about games and wrote the classic *Theory of Games and Economic Behavior* [31] (written with economist Oskar Morgenstern[1]) in 1944. It was a breakthrough in the development of economics using mathematics and in the mathematical theory of games. His contributions to pure mathematics fill volumes. The cleverness and ingenuity of his arguments amaze mathematicians to this day.

Von Neumann was one of the most creative mathematicians of the twentieth century. In a century in which there were many geniuses and many breakthroughs, von Neumann contributed more than his fair share. He ranks among the greatest mathematicians of all time for his depth, breadth, and scope of contributions. In addition, and perhaps more importantly, von Neumann was famous for the parties he hosted throughout his lifetime and in the many places he lived. He was an aristocratic *bon vivant* who managed several careers even among the political sharks of the cold-war era without amassing enemies. He was well-liked by all of his colleagues and lived a contributory life.

---

1 Born on January 24, 1902, in Germany and died on July 26, 1977, in Princeton, NJ. Morgenstern's mother was the daughter of the German emperor Frederick III. He was a professor at the University of Vienna when he came to the United States on a Rockefeller Foundation fellowship. In 1938, while in the United States, he was dismissed from his post in Vienna by the Nazis and became a professor of economics at Princeton University where he remained until his death.

If you want to read a very nice biography of John von Neumann, read MacRae's excellent book [24].

D.2   John Forbes Nash

John Forbes Nash Jr. was an American mathematician who made fundamental contributions to game theory, real algebraic geometry, differential geometry, and partial differential equations. He was awarded the 1994 Nobel Memorial Prize in Economics (formally the 1994 Bank of Sweden Prize in Economic Sciences), along with John Harsanyi and Reinhard Selten, for their work on the theory of Nash equilibrium. In 2015, he and Louis Nirenberg were awarded the Abel Prize for their contributions to the field of partial differential equations.

Nash was born in Bluefield, West Virginia, in 1928. He attended Carnegie Institute of Technology (now Carnegie Mellon University) and Princeton University, where he received his doctorate in mathematics in 1950. After a brief stint at the Massachusetts Institute of Technology, he joined the faculty of Princeton in 1951.

As a graduate student at Princeton University in the 1950s, Nash developed the theory of non-cooperative equilibrium which became known as Nash equilibrium. His PhD thesis introduced and developed this theory. Nash's work has had a major impact on economics, political science, military science, artificial intelligence, and many other fields. His work on differential geometry and partial differential equations was ground breaking in mathematics.

In the late 1950s, Nash was diagnosed with paranoid schizophrenia. He spent several years in and out of psychiatric hospitals. However, he eventually recovered and returned to his work in mathematics. In the 1990s, he was the subject of a best-selling biography by Sylvia Nasar, which was later adapted into an Academy Award-winning film.

John Nash died on May 23, 2015, at the age of 86. He was killed in a car crash along with his wife Alicia in New Jersey after returning from Oslo Norway. He had returned from Oslo after attending the ceremony in which he was awarded the Abel Prize.

For further information about Nash, read the book by Nasar [30] and see the movie.

Selected Problem Solutions

Solutions for Chapter 1

1.1 Think of this as a 100×100 game matrix and we are looking for the upper and lower values except that we are really doing it for the transpose of the matrix.

If we take the maximum in each row and then Javier is the minimum maximum, Javier is v^+. If we take the minimum in each column, and Raoul is the maximum of those, then Raoul is the maximum minimum, or v^-. Thus Javier is richer.

Another way to think of this is that the poorest rich guy is wealthier than the richest poor guy. Common sense.

1.3.a The game tree is

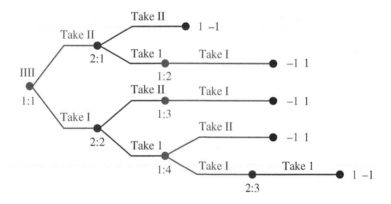

1.3.b Using the notation from the figure we may list the strategies for player I
  1. Go to 2:1; if at 1:2 take 1. [Same as Take 2. (there are no more choices for I after that)]
  2. Go to 2:2; if at 1:3 take 1; if at 1:4 take 2. [Same as Take 1, then if 2 are left, take 1.]
  3. Go to 2:2; if at 1:3 take 1; if at 1:4 take 1. [Same as Take 1, then if 2 are left, take 2.]
  For player II the strategies are
  1. If at 2:1 take 2; if at 2:2 take 2. [Same as If there are 3 left, take 2.]
  2. If at 2;1 take 2; if at 2:2 take 1.
  3. If at 2:1 take 1; if at 2:2 take 2.
  4. If at 2:1 take 1; if at 2:2 take 1; if at 2:3 take 1.

Game Theory: An Introduction, Third Edition. E. N. Barron.
© 2024 John Wiley & Sons, Inc. Published 2024 by John Wiley & Sons, Inc.

The game matrix is

I/II	1	2	3	4
1	1	1	−1	−1
2	−1	−1	−1	−1
3	−1	1	−1	1

1.3.c Since $v^+ = -1, v^- = -1$, this game has a value of -1. Player II can always win by playing as follows: If player I takes 2 pennies, then II should take 1. If player I takes 1 penny, then II should take 2 pennies. No matter what player I does, player II wins.

1.5.a The game matrix is

I/II	1	2	3	4	5
1	0	2	−1	−1	−1
2	−2	0	2	−1	−1
3	1	−2	0	2	−1
4	1	1	−2	0	2
5	1	1	1	−2	0

1.5.b $v^+ = 1$, $v^- = -1$, no pure saddle point.

1.7 Consider first the game with A. To calculate v^- we take the minimum of x and 0, written $\min\{x, 0\}$, and the minimum of 1, 2 which is 1. Then

$$v^- = \max\{\min\{x, 0\}, 1\} = 1$$

since $\min\{x, 0\} \leq 0$ no matter what x is. Similarly,

$$v^+ = \min\{1, \max\{x, 2\}\} = 1 \text{ since } \max\{x, 2\} \geq 2.$$

Thus $v^- = v^+ = 1$ and there is a pure saddle at row 2, column 1.

For matrix B, we have $v^- = \max\{1, \min\{x, 0\}\} = 1, v^+ = \min\{\max\{2, x\}, 1\} = 1$, and there is a pure saddle at row 1, column 2, no matter what x is.

1.9 We have

$$v^+(A) = \min_{1 \leq j \leq n} \max_{1 \leq i \leq n} (i - j) = \min_{1 \leq j \leq n} (n - j) = n - n = 0$$

and

$$v^-(A) = \max_{1 \leq i \leq n} \min_{1 \leq j \leq n} (i - j) = \max_{1 \leq i \leq n} (i - n) = n - n = 0.$$

Thus, $v = 0$ with a pure saddle point at (n, n).

1.11 $v^- = 0.28$ and $v^+ = 0.30$. So baseball does **not** have a saddle point in pure strategies. That shouldn't be a surprise because if there were such a saddle, baseball would be a very dull game, which nonfans say is true anyway.

1.13 Denote the game matrix as

$$A = \begin{bmatrix} a_{11} & a_{12} & a_{13} \\ a_{21} & a_{22} & a_{23} \end{bmatrix}$$

Without loss of generality we may as well assume that the saddle is at row 1 column 1, $v^- = v^+ = a_{11}$. Since $(1, 1)$ is a saddle, we have

$$a_{i1} \leq a_{11} \leq a_{1j}, i = 2, j = 2, 3.$$

In particular $a_{21} \leq a_{11}$. If it is also true that $a_{22} \leq a_{12}$ and $a_{23} \leq a_{13}$ then row 1 dominates row 2 and we are done. Thus we need only suppose that $a_{22} > a_{12}$. Then, from the saddle point inequalities,

$$a_{22} > a_{12} \geq a_{11} \geq a_{21}.$$

But then $a_{12} \geq a_{11}$ and $a_{22} > a_{21}$ says that column 2 is dominated by column 1.

Now consider the 3×3 matrix

$$A = \begin{bmatrix} 4 & -2 & 0 \\ 3 & 1 & 1 \\ 0 & 2 & \frac{1}{2} \end{bmatrix}$$

Then $v^- = v^+ = 1$ and there is a saddle at row 2, column 3, but no row or column dominates another.

1.19.a The matrix is

You/Stranger	H	T
H	3	-2
T	-2	1

The saddle point is $X^* = (\frac{3}{8}, \frac{5}{8}) = Y^*$. This is obtained from solving $3x - 2(1-x) = -2x + (1-x)$, and $3y - 2(1-y) = -2y + (1-y)$. The value of this game is $v = -\frac{1}{8}$, so this is definitely a game you should not play.

1.19.b If the stranger plays the strategy $\tilde{Y} = (\frac{1}{3}, \frac{2}{3})$, then your best response strategy is $\tilde{X} = (0, 1)$ which means you will call Tails all the time. The reason is because

$$E(X, \tilde{Y}) = (x, 1-x)A\,\tilde{Y}^T = -\frac{1}{3}x$$

which is maximized at $x = 0$. This results in an expected payoff to you of zero. $\tilde{Y}$ is not a winning strategy for the stranger.

1.21.a We can find the first derivatives and set to zero:

$$\frac{\partial f}{\partial x} = C - 2x - y = 0 \implies x^*(y) = \frac{C-y}{2},$$

and

$$\frac{\partial g}{\partial y} = D - x - 2y = 0 \implies y^*(x) = \frac{D-x}{2}.$$

These are the best responses since the second partials $f_{xx} = g_{yy} = -2 < 0$.

1.21.b Best responses are $x = \frac{C-y}{2}, y = \frac{D-x}{2}$, which can be solved to give $x^* = \frac{(2C-D)}{3}, y^* = \frac{(2D-C)}{3}$. Next,

$$f(x^*, y^*) = \frac{(D-2C)^2}{9}, \quad \text{and} \quad f(x, y^*) = \frac{(4C-2D-3x)x}{3}.$$

The maximum of $f(x, y^*)$ is $\frac{(D-2C)^2}{9}$, achieved at $x = \frac{2C-D}{3}$, which means it is true that $f(x^*, y^*) \geq f(x, y^*)$ for all x.

1.23 If we use the 2×2 formulas, we get $\det(A) = -80$, $A^* = \begin{bmatrix} 0 & -8 \\ -10 & 1 \end{bmatrix}$, so

$$X^* = \frac{(1\ 1)A^*}{(1\ 1)A^*\begin{bmatrix}1\\1\end{bmatrix}} = (\frac{10}{17}, \frac{7}{17}), Y^{*T} = \frac{A^*\begin{bmatrix}1\\1\end{bmatrix}}{(1\ 1)A^*\begin{bmatrix}1\\1\end{bmatrix}} = (\frac{8}{17}, \frac{9}{17})$$

$$value(A) = \frac{\det(A)}{(1\ 1)A^*\begin{bmatrix}1\\1\end{bmatrix}} = \frac{80}{17}$$

If we use the graphical method, the lines cross where $x + 10(1-x) = 8x \implies x = \frac{10}{17}$. The rest also checks.

1.25 The new matrix is $A = \begin{bmatrix} 1 & 6 \\ 10 & 0 \end{bmatrix}$ and the optimal strategies using the formulas become

$$X^* = (\frac{2}{3}, \frac{1}{3}), Y^* = (\frac{2}{5}, \frac{3}{5}), v(A) = 4.$$

The defense will guard against the pass **more**.

1.27 (a) $X^* = (\frac{15}{22}, \frac{7}{22}), Y^* = (\frac{9}{22}, \frac{13}{22}), v = -\frac{3}{22}.$

(b) $X^* = (\frac{23}{114}, \frac{91}{114}), Y^* = (\frac{31}{38}, \frac{7}{38}), v = \frac{941}{38}.$

(c) $X^* = (\frac{101}{137}, \frac{36}{137}), Y^* = (\frac{31}{137}, \frac{106}{137}), v = -\frac{568}{137}.$

1.29 Consider cases. If $a_{12} > a_{22}$, then because $a_{11} + a_{22} = a_{12} + a_{21}$, it must be that $a_{11} > a_{21}$ since $a_{12} - a_{22} = a_{11} - a_{21} > 0$ so there is a saddle point in the first row. The rest of the cases are similar.

1.31 If we calculate $f(k) = \frac{k-p}{G_k}$ we get $f(1) = 4.5, f(2) = 5.90, f(3) = 5.94$ and $f(4) = 2.46$. This gives us $k^* = 3$. Then we calculate $x_i = \frac{1}{a_i G_3}$ and get

$$X^* = (\frac{14}{53}, \frac{18}{53}, \frac{21}{53}, 0) = (.264, .339, .3962, 0),$$

and $v(A) = \frac{3-p}{G_3} = 5.943$. Finally, to calculate Y^* we have

$$y_j = \begin{cases} \frac{1}{p}\left(1 - \frac{k^*-p}{a_j G_{k^*}}\right), & \text{if } j = 1, 2, \dots, k^*; \\ 0, & \text{otherwise.} \end{cases}$$

which results in $Y^* = (.6792, .30188, .01886, 0)$. Thus player I will attack target 3 about 40% of the time, while player II will defend target 3 with probability only about 2%.

1.33 (a) The matrix has a saddle at row 1, column 2; formulas do not apply.

(b) The inverse of A is

$$A^{-1} = \begin{bmatrix} \frac{8}{35} & -\frac{1}{5} & \frac{1}{35} \\ \frac{1}{70} & \frac{3}{10} & -\frac{13}{70} \\ -\frac{1}{10} & -\frac{1}{10} & \frac{3}{10} \end{bmatrix}$$

The formulas then give $X^* = (\frac{1}{2}, 0, \frac{1}{2}), Y^* = (\frac{1}{5}, \frac{9}{20}, \frac{7}{20}), v = \frac{7}{2}.$ There is another optimal $Y^* = (\frac{1}{2}, 0, \frac{1}{2})$ which we can find by noting that the second column is dominated by a convex combination of the first and third columns (with $\lambda = \frac{2}{3}$). This optimal strategy for player II is not obtained using the formulas. It is in fact optimal since

$$A. \begin{bmatrix} \frac{1}{2} \\ 0 \\ \frac{1}{2} \end{bmatrix} = \begin{bmatrix} \frac{7}{2} \\ \frac{7}{2} \\ \frac{7}{2} \end{bmatrix} \leq \begin{bmatrix} \frac{7}{2} \\ \frac{7}{2} \\ \frac{7}{2} \end{bmatrix}$$

Since $E(2, Y) = 2 < v(A)$, we know the second component of an optimal strategy for player I must have $x_2 = 0$.

(c) Since

$$A^{-1} = \begin{bmatrix} \frac{8}{35} & -\frac{1}{5} & \frac{1}{35} \\ \frac{1}{70} & \frac{3}{10} & -\frac{13}{70} \\ -\frac{1}{10} & -\frac{1}{10} & \frac{3}{10} \end{bmatrix}$$

the formulas give the optimal strategies

$$X^* = \left(\frac{11}{19}, \frac{5}{19}, \frac{3}{19}\right), Y^* = \left(\frac{3}{19}, \frac{28}{57}, \frac{20}{57}\right), v = \frac{24}{19}.$$

(**d**) The matrix has the last column dominated by column 1. Then player I has row 2 dominated by column 1. The resulting matrix is $\begin{bmatrix} -4 & 2 \\ 0 & -1 \end{bmatrix}$. This game can be solved with 2×2 formulas or graphically. The solution for the original game is then

$$X^* = (\tfrac{1}{7}, 0, \tfrac{6}{7}), \ Y^* = (\tfrac{3}{7}, \tfrac{4}{7}, 0), v(A) = -\tfrac{4}{7}.$$

1.35 Use the definition of value and strategy to see that

$$v(A + b) = \min_Y \ \max_X X (A + b) Y^{*T}$$

$$= \min_Y \ \max_X \left(X A Y^{*T} + \sum_{i=1}^{n} \sum_{j=1}^{m} b x_i y_j \right)$$

$$= \min_Y \ \max_X \left(X A Y^{*T} + b \sum_{i=1}^{n} x_i \sum_{j=1}^{m} y_j \right)$$

$$= \min_Y \ \max_X \left(X A Y^{*T} \right) + b$$

$$= v(A) + b.$$

We also see that $X (A + b) Y^T = X A Y^T + b$

If (X^*, Y^*) is a saddle for A, then

$$X (A + b) Y^{*T} = X A Y^{*T} + b \le X^* A Y^{*T} + b = X^* (A + b) Y^{*T}$$

and

$$X^* (A + b) Y^T = X^* A Y^T + b \ge X^* A Y^{*T} + b = X^* (A + b) Y^{*T}$$

for any X, Y. Together this says that X^*, Y^* is also a saddle for $A + b$. The converse is similar.

1.37 By the invertible matrix theorem we have

$$X^* = \frac{J_5 A^{-1}}{J_5 A^{-1} J_5^T} = \left(\frac{1}{5}, \frac{1}{5}, \frac{1}{5}, \frac{1}{5}, \frac{1}{5}\right).$$

Using the formulas we also have $v(A) = 13$ and $Y^* = X^*$. This leads us to suspect that each row and column, in the general case, is played with probability $\frac{1}{n}$.

Now, let S denote the common sum of the rows and columns in an $n \times n$ magic square. For any optimal strategy X for player I and Y for player II, we must have

$$E(i, Y) \le v(A) \le E(X, j), i, j = 1, 2, \ldots, n.$$

Since $E(X, j) = \sum_{i=1}^{n} a_{ij} x_i \ge v$, adding for $j = 1, 2, \ldots, n$, we get

$$\sum_{j=1}^{n} \sum_{i=1}^{n} a_{ij} x_i = \sum_{i=1}^{n} \sum_{j=1}^{n} a_{ij} x_i = S \ge n v(A).$$

Similarly, since $E(i, Y) = \sum_{j=1}^{n} a_{ij} y_j \le v$, adding for $i = 1, 2, \ldots, n$, we get

$$\sum_{i=1}^{n} \sum_{j=1}^{n} a_{ij} y_j = \sum_{j=1}^{n} \sum_{i=1}^{n} a_{ij} y_j = S \le n v(A).$$

Putting them together, we have $v(A) = \frac{S}{n}$.

Finally, we need to verify that $X^* = Y^* = (\frac{1}{n}, \dots, \frac{1}{n})$ is optimal. That follows immediately from the fact that

$$E(X^*, j) = \sum_{i=1}^{n} a_{ij} \frac{1}{n} = \frac{S}{n} = v(A),$$

and similarly $E(i, Y^*) = v(A), i = 1, 2, \dots, n$. We conclude using Theorem 1.4.6.

1.39 Since $\det A = \frac{n!}{(n+1)!} = \frac{1}{n+1} > 0$ the matrix has an inverse. The inverse matrix is

$$B_{n \times n} = A^{-1} = (b_{ij}), \text{ where } b_{ij} = \begin{cases} \frac{i+1}{i}, & i = j, i = 1, 2, \dots, n \\ 0, & i \neq j. \end{cases}$$

Then, setting $q = \sum_{i=1}^{n} \frac{i+1}{i}$ we have

$$v(A) = \frac{1}{q} \quad \text{and} \quad X^* = Y^* = \frac{1}{q} (2, \frac{3}{2}, \frac{4}{3}, \dots, \frac{n+1}{n}).$$

If $n = 5$ we have $q = \frac{437}{60}$ so

$$v(A) = \frac{60}{437}, \quad X^* = Y^* = \frac{60}{437} (2, \frac{3}{2}, \frac{4}{3}, \frac{5}{4}, \frac{6}{5}) = (0.27, 0.21, 0.18, 0.17, 0.16).$$

Notice that you search with decreasing probabilities as the box number increases.

1.41.a The matrix has an inverse if $\det(A) \neq 0$. Since $\det(A) = a_{11}a_{22} \dots a_{nn}$, no diagonal entries may be zero.

1.41.b The inverse matrix is

$$A^{-1} = \begin{bmatrix} 1 & \frac{3}{4} & \frac{1}{4} & -\frac{1}{200} \\ 0 & \frac{1}{4} & -\frac{1}{8} & \frac{7}{400} \\ 0 & 0 & \frac{1}{8} & -\frac{3}{400} \\ 0 & 0 & 0 & \frac{1}{50} \end{bmatrix}.$$

Then, we have

$$v(A) = \frac{40}{91}, \quad X^* = vJ_5 A^{-1} = \frac{1}{91}(40, 40, 10, 1), \quad Y^* = vA^{-1}J_5 = \frac{1}{910}(798, 57, 47, 8),$$

1.43 The game matrix with player I as the row player is

I/II	Right	Left	Center
Right	−0.8	0.6	0.6
Left	0.6	−0.2	0.6
Center	0.6	0.6	−0.4

For example, if player I hits the ball Left, and player II anticipates Left, the expected payoff to player I is $0.6(-1) + 0.4(+1) = -0.2$. If player I hits Right, and II anticipates Right, the payoff is $0.9(-1) + 0.1(+1) = -0.8$. If player I hits Center and II anticipates Center, the expected payoff to I is $0.7(-1) + 0.3(+1) = -0.4$. Finally, if I hits the ball and II anticipates incorrectly, the payoff to I is $0.2(-1) + 0.8(+1) = 0.6$.

The matrix has an inverse given by

$$A^{-1} = \frac{1}{109} \begin{bmatrix} -35 & 75 & 60 \\ 75 & -5 & 105 \\ 60 & 105 & -25 \end{bmatrix}.$$

By the invertible formulas we have $v(A) = 0.2626$, and optimal strategies $X^* = (0.24, 0.42, 0.34) = Y^*$. Player I should aim for the center 42% of the time and II should anticipate a ball to the Center 42% of the time (not 100% of the time).

1.45 The duelists shoot at 10, 6 or 2 paces with accuracies $0.2, 0.4, 1.0$, respectively. The accuracies are the same for both players. The game matrix becomes

B/H	10	6	2
10	0	−0.6	−0.6
6	0.6	0	−0.2
2	0.6	0.2	0

For example, if Burr decides to fire at 10 paces, and Hamilton fires at 10 also, then the expected payoff to Burr is zero since the duelists have the same accuracies and payoffs (it's a draw if they fire at the same time). If Burr decides to fire at 10 paces and Hamilton decides to not fire, then the outcome depends only on whether or not Burr kills Hamilton at 10. If Burr misses, he dies. The expected payoff to Burr is

$$Prob(\text{B kills H at 10})(+1) + Prob(\text{B misses})(−1)$$

$$= 0.2 − 0.8 = −0.6.$$

Similarly, Burr fires at 2 paces and Hamilton fired at 6 paces, then Hamilton's survival depends on whether or not he missed at 6. The expected payoff to Burr is

$$Prob(\text{H kills B at 6})(−1) + Prob(\text{H misses B at 6})(+1)$$

$$= −0.4 + 0.6 = 0.2.$$

Calculating the lower value we see that $v^- = 0$ and the upper value is $v^+ = 0$, both of which are achieved at row 3, column 3. We have a pure saddle point and both players should wait until 2 paces to shoot. Makes perfect sense that a duelist would not risk missing.

1.47 The matrix B is given by

$$B = \begin{bmatrix} 0 & 0 & 5 & 2 & 6 & -1 \\ 0 & 0 & 1 & \frac{7}{2} & 2 & -1 \\ -5 & -1 & 0 & 0 & 0 & 1 \\ -2 & -\frac{7}{2} & 0 & 0 & 0 & 1 \\ -6 & -2 & 0 & 0 & 0 & 1 \\ 1 & 1 & -1 & -1 & -1 & 0 \end{bmatrix}.$$

Now we use the method of symmetry to solve the game with matrix B. Obviously, $v(B) = 0$, and if we let $P = (p_1, \ldots, p_6)$ denote the optimal strategy for player I, we have

$$PA \geq (0, 0, 0, 0, 0, 0) \implies$$

$$-5p_3 - 2p_4 - 6p_5 + p_6 \geq 0$$

$$-p_3 - \frac{7}{2}p_4 - 2p_5 + p_6 \geq 0$$

$$5p_1 + p_2 - p_6 \geq 0$$

$$2p_1 + \frac{7}{2}p_2 - p_6 \geq 0$$

$$6p_1 + 2p_2 - p_6 \geq 0$$

$$-p_1 - p_2 + p_3 + p_4 + p_5 \geq 0$$

$$p_1 + p_2 + p_3 + p_4 + p_5 + p_6 = 1.$$

This system of inequalities has one and only one solution given by

$$P = \left(\frac{5}{53}, \frac{6}{53}, \frac{3}{53}, \frac{8}{53}, 0, \frac{31}{53}\right)$$

As far as the solution of the game with matrix B we have $v(B) = 0$, and P is the optimal strategy for both players I and II.

Now to find the solution of the original game, we have (now with a new meaning for the symbols p_i) we have $n = 2, m = 3$ and

$$p_1 = \frac{5}{53}, p_2 = \frac{6}{53}, q_1 = \frac{3}{53}, q_2 = \frac{8}{53}, q_3 = 0, \gamma = \frac{31}{53}.$$

Then $b = p_1 + p_2 = \frac{11}{53}, b = q_1 + q_2 + q_3 = \frac{3+8}{53} = \frac{11}{53}$. Now define the strategy for the original game with A,

$$X^* = (x_1, x_2), \; x_1 = \frac{p_1}{b} = \frac{5}{11}, \; x_2 = \frac{p_2}{b} = \frac{6}{11}$$

and

$$Y^* = (y_1, y_2, y_3), \; y_j = \frac{q_j}{b} \implies y_1 = \frac{3}{11}, y_2 = \frac{8}{11}, y_3 = 0.$$

Finally, $v(A) = \dfrac{\gamma}{b} = \dfrac{\frac{31}{53}}{\frac{11}{53}} = \dfrac{31}{11}$.

The problem is making the statement that this is indeed the solution of our original game. We verify that by calculating using the original matrix A,

$$E(i, Y^*) = \frac{31}{11} = v(A), \; i = 1, 2 \quad \text{and} \quad E(X^*, j) = \frac{31}{11}, j = 1, 2, \; E(X^*, 3) = \frac{42}{11} \geq v(A)$$

By Theorem 1.4.6 we conclude that X^*, Y^* is optimal and $v(A) = \frac{31}{11}$.

1.49 For (a), the lines for player I cross where $x - (1 - x) = 2(1 - x)$, which gives $x^* = \frac{3}{4}$. For player II, the lines cross where $y = -y + 2(1 - y)$ which gives $y^* = \frac{1}{2}$. Therefore the solution of the game in mixed strategies is

$$X^* = (\frac{3}{4}, \frac{1}{4}), \; Y^* = (\frac{1}{2}, \frac{1}{2}), \; value(A) = \frac{1}{2}.$$

For part (b), the matrix has a saddle point at row 2, column 1, so the optimal strategies won't be a mixed strategies. If we didn't spot the pure saddle point and applied the graphical method anyway, we would get for player II the two lines cross where $3y + 1 - y = 5y + 7(1 - y)$, which gives $y^* = \frac{3}{2} > 1$. The second line lies above the first line for the range $0 \leq y \leq 1$.

1.51 Let Curly be the row player and the thief be the column player. The matrix is

I/II	Home	Office
Home	−1	1
Office	1	0.7

The payoff of 0.7 to Curly if he puts the gold bar at the office and the thief hits the office is obtained by calculating $(+1) \times 0.85 + (-1) \times 0.15 = 0.7$. The two lines for the expected payoffs of player I cross where $E((x, 1 - x), 1) = E((x, 1 - x), 2)$, which become $-x + (1 - x) = x + 0.7(1 - x)$. The solution is $x^* = \frac{3}{23}$.

Similarly for player II the lines cross where $E(1, (y, 1 - y)) = E(2, (y, 1 - y))$, or $-y + (1 - y) = y + 0.7(1 - y)$.

The mixed strategy solution is $X^* = Y^* = (\frac{3}{23}, \frac{20}{23}), v = \frac{17}{23}$.

1.53 Column 2 may be eliminated by dominance: Any $\frac{9}{13} \leq \lambda \leq \frac{3}{4}$ will make

$$13\lambda + 8(1 - \lambda) \leq 29$$

$$18\lambda + 31(1 - \lambda) \leq 22$$

$$23\lambda + 19(1 - \lambda) \leq 22.$$

Once column 2 is gone, row 1 may be dropped. Then we apply the graphical method to get $X^* = (0, \frac{4}{17}, \frac{13}{17})$ and $Y^* = (\frac{12}{17}, 0, \frac{5}{17})$. The value of the game is $v = \frac{37}{17}$.

1.55 The game matrix is

player I/player II	1	2	3	4	5	6
1	1	1	−1	1	1	−1
2	−1	1	−1	−1	1	−1
3	−1	−1	−1	1	1	1

Column 3 immediately dominates every other column. Then it doesn't matter what row player I chooses because the payoff is always −1.

1.57 **(a)** $X^* = (\frac{15}{22}, \frac{7}{22})$; **(b)** $Y^* = (\frac{7}{9}, \frac{2}{9})$; **(c)** $Y^* = (\frac{6}{10}, \frac{4}{10})$.

To see where these come from since they are all 2×2 games without pure saddle points, simply find where the two payoff lines cross for each player.

For (a) we must solve $4x - 9(1 - x) = -3x + 6(1 - x) \implies x = \frac{15}{22}$, and $4y - 3(1 - y) = -9y + 6(1 - y)$ which gives $y = \frac{9}{22}$. Then plugging in to either payoff line we get $v = -\frac{3}{22}$. The other parts are similar.

1.59 Any $\frac{3}{8} \leq \lambda \leq \frac{7}{16}$ will work for a convex combination of columns 2 and 1. To see why

$$0 \cdot \lambda + 8 \cdot (1 - \lambda) \leq 5 \implies \frac{3}{8} \leq \lambda$$

$$8 \cdot \lambda + 4 \cdot (1 - \lambda) \leq 6 \implies \lambda \leq \frac{1}{2}$$

$$12 \cdot \lambda + (-4) \cdot (1 - \lambda) \leq 3 \implies \lambda \leq \frac{7}{16}.$$

The reduced matrix is $\begin{bmatrix} 0 & 8 \\ 8 & 4 \\ 12 & -4 \end{bmatrix}$. The graph for this matrix for player II is

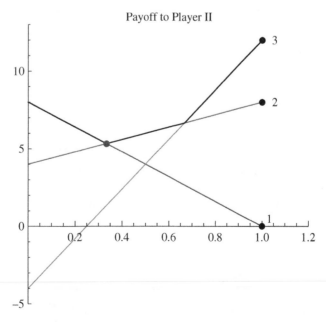

Payoff to Player II

The solution of the original game is therefore $X^* = (\frac{1}{3}, \frac{2}{3}, 0)$, $Y^* = (\frac{1}{3}, \frac{2}{3}, 0)$, and $v(A) = \frac{16}{3}$.

1.61 Column 2 is dominated by column 3; then row 3 is dominated by row 1. The reduced matrix is $\begin{bmatrix} a_4 & a_3 \\ a_1 & a_5 \end{bmatrix}$, which does not have a pure saddle. The resulting solution is obtained where the payoff lines cross; for example $a_4 x + a_1(1-x) = a - 3x + a_5(1-x) \implies x^* = \frac{a_5 - a_1}{a_4 + a_5 - a_1 - a_3}$. We have,

$$v(A) = \frac{a_4 a_5 - a_1 a_3}{g}, X^* = (\frac{a_5 - a_1}{g}, \frac{a_4 - a_3}{g}, 0), Y^* = (\frac{a_5 - a_3}{g}, 0, \frac{a_4 - a_1}{g}),$$

where $g = a_4 + a_5 - a_1 - a_3$.

1.63.a The given strategies are not optimal because $\max_i E(i, Y) = \frac{31}{9}$ and $\min_j E(X, j) = -\frac{42}{9}$. Another way to see it, is to note that since rows 1 and 3 are used with positive probability, Theorem 1.4.6 tells us that if X^* is optimal we must have $E(1, Y^*) = E(3, Y^*) = v$, which you can check easily is not true. Similarly, since columns 2 and 3 are used with positive probability, it must be true that $E(X^*, 2) = E(X^*, 3)$ for X^* to be optimal. But that fails also. Neither X^* nor Y^* are optimal.

1.63.b The optimal Y^* is $Y^* = (\frac{52}{99}, \frac{8}{33}, 0, \frac{23}{99})$. This is obtained from solving the equations

$$E(1, Y^*) = -2y_1 + 3y_2 + 5y_3 - 2y_4 = -\frac{26}{33}$$

$$E(2, Y^*) = 3y_1 - 4y_2 + y_3 - 6y_4 = -\frac{26}{33}$$

$$E(4, Y^*) = -y_1 - 3y_2 + 2y_3 + 2y_4 = -\frac{26}{33}$$

$$y_1 + y_2 + y_3 + y_4 = 1$$

We use the fact that $E(i, Y^*) = v$ if $x_i^* > 0$.

1.65.a $X^* = (\frac{8}{11}, \frac{3}{11}), Y^* = (\frac{6}{11}, \frac{5}{11}), v(A) = \frac{48}{11}$.

1.65.b Solving for the offense, we see that $3x + 12(1-x) = 6x$, which gives $x^* = \frac{4}{5}$, so the optimal strategy is $X^* = (\frac{4}{5}, \frac{1}{5})$. The value of the game is $v(A) = \frac{24}{5}$ and the optimal strategy for the defense is $Y^* = (\frac{2}{5}, \frac{3}{5})$. If the offense gets a better quarterback, the team should Run more!

1.67.a The game matrix is

$$A = \begin{bmatrix} a_1 & 0 & 0 & \cdots \\ 0 & a_2 & 0 & \cdots \\ 0 & 0 & a_3 & \cdots \\ \vdots & \vdots & \vdots & \vdots \end{bmatrix}.$$

Then since there is a 0 in every row and all the $a_k's > 0$, the maximum minimum must be $v^- = 0$. Since there is an $a_k > 0$ in every column the maximum in every column is a_k. The minimum of those is a_1 and so $v^+ = a_1$.

1.67.b We use $E(i, Y^*) \le v \le E(X^*, j), \forall i, j = 1, 2, \ldots$. We get for $X^* = (x_1, x_2, \ldots), E(X^*, j) = a_j x_j \ge v \implies x_j \ge \frac{v}{a_j}$. Similarly, for $Y^* = (y_1, y_2, \ldots)$, implies $y_j \le \frac{v}{a_j}$. Adding these inequalities results in

$$1 = \sum_i x_i \ge v \sum_i \frac{1}{a_i}, \text{ and } 1 = \sum_j y_j \le v \sum_j \frac{1}{a_j}$$

and we conclude that $v = \frac{1}{\sum_i \frac{1}{a_i}}$. We needed the facts that $\sum_i \frac{1}{a_i} < \infty$, and the fact $\sum_i \frac{1}{a_i} \ne 0$, since $a_k > 0$ for all $k = 1, 2, \ldots$.

Next $x_i \geq \frac{v}{a_i} > 0$, and

$$1 = \sum_{i=1}^{\infty} x_i \geq v \sum_{i=1}^{\infty} \frac{1}{a_i} = 1 \implies \sum_{i=1}^{\infty} [x_i - \frac{v}{a_i}] = 0$$

which means it must be true since each term is nonnegative $x_i = \frac{v}{a_i}, i = 1, 2, \dots$. Similarly, $X^* = Y^*$, and the components of both optimal strategies are $\frac{v}{a_i}, i = 1, 2, \dots$.

1.67.c Just as in the second part, we have for any integer $n > 1$, $a_j x_j \geq v$ implies

$$1 \geq \sum_{j=1}^{n} x_j \geq v \sum_{j=1}^{n} \frac{1}{a_j}.$$

Then

$$\frac{1}{\sum_{j=1}^{n} \frac{1}{a_j}} \geq v \geq 0.$$

Sending $n \to \infty$ on the left side and using the fact $\sum_{j=1}^{\infty} \frac{1}{a_j} = \infty$, we get $v = 0$. Notice that we know ahead of time that $v \geq 0$ since

$$v \geq v^- = \max_{X \in S_\infty} \min_{j=1,2,\dots} E(X, j) = \max_{X \in S_\infty} \min_{j=1,2,\dots} x_j a_j \geq 0.$$

Let $X = (x_1, x_2, \dots)$ be any mixed strategy for player I. Then, it is always true that $E(X, j) = x_j a_j \geq v = 0$ for any column j. By Theorem 1.4.6 this says that X is optimal for player I. On the other hand, if Y^* is optimal for player II, then $E(i, Y^*) = a_i y_i \leq v = 0$, $i = 1, 2, \dots$. Since $a_i > 0$, this implies $y_i = 0$ for every $i = 1, 2, \dots$. But then $Y = (0, 0, \dots)$ is not a strategy and we conclude player II does not have an optimal strategy. Since the space of strategies in an infinite sequence space is not closed and bounded, we are not guaranteed that an optimal mixed strategy exists by the minimax theorem.

1.69 Let $\max_i b_i = b_k$. Then $\sum_i x_i b_i - b_k = \sum_i x_i (b_i - b_k) = z$ since $\sum_i x_i = 1$. Now $b_i \leq b_k$ for each i, so $z \leq 0$. Its maximum value is achieved by taking $x_k = 1$ and $x_i = 0, i \neq k$. Hence $\max_X \sum_i x_i b_i - b_k = 0$, which says $\max_X \sum_i x_i b_i = b_k = \max_i b_i$.

1.71 By definition of saddle

$$E(X^0, Y^*) \leq E(X^*, Y^*) \leq E(X^*, Y^0)$$

and

$$E(X^*, Y^0) \leq E(X^0, Y^0) \leq E(X^0, Y^*).$$

Now put them together to get

$$E(X^*, Y^0) \leq E(X^0, Y^0) \leq E(X^0, Y^*) \leq E(X^*, Y^*) \leq E(X^*, Y^0)$$

and so all of them are equal. This implies, for example that (X^*, Y^0) is also a saddle point since

$$E(X, Y^0) \leq E(X^0, Y^0) = E(X^*, Y^0) = E(X^*, Y^*) \leq E(X^*, Y), \quad \forall X, Y.$$

It is similar to see that (X^*, Y_β) and (X_λ, Y^*) are also saddle points.

Let (X, Y) be arbitrary strategies. Then using the bilinearity of $E(X, Y)$ and what we just showed,

$$E(X_\lambda, Y_\beta) = \lambda E(X^*, Y_\beta) + (1 - \lambda)E(X^0, Y_\beta)$$
$$\leq \lambda E(X^*, Y) + (1 - \lambda)E(X^0, Y) = E(X_\lambda, Y), \quad \forall\, Y \in S_m$$

and

$$E(X_\lambda, Y_\beta) = \beta E(X_\lambda, Y^*) + (1 - \beta)E(X_\lambda, Y^0)$$
$$\geq \beta E(X, Y^*) + (1 - \beta)E(X, Y^0) = E(X, Y_\beta), \quad \forall\, X \in S_n$$

1.73 We have

$$v^+ = \min_{-1 \leq y \leq 1}\; \max_{-1 \leq x \leq 1} (x^2 + y^2) = \min_{-1 \leq y \leq 1} (1 + y^2) = 1,$$

and

$$v^- = \max_{-1 \leq x \leq 1}\; \min_{-1 \leq y \leq 1} (x^2 + y^2) = \max_{-1 \leq x \leq 1} (x^2 + 0) = 1.$$

1.75 $v^+ = 1, v^- = 0$. Here's why. For $v^- = \max_x \min_y (x - y)^2$, y can be chosen to be $y = x$ to get a minimum of zero. For $v^+ = \min_y \max_x (x - y)^2$, x wants to be as far away from y as possible. So, if $y < 0$, then $x = 1$, and if $y > 0$, then $x = -1$, so

$$\max_{-1 \leq x \leq 1} (x - y)^2 = \begin{cases} (1 + y)^2 & \text{if } y > 0; \\ (1 - y)^2 & \text{if } y \leq 0. \end{cases}$$

The minimum of this over $y \in [-1, 1]$ is 1, so $v^+ = 1$. You can see this with the Maple commands

```
> f:=y->piecewise(y<0,(1-y)^2,y>=0,(1+y)^2);
> plot(f(y),y=-1..1,view=[-1..1,0..3]);
```

Observe that the function $f(x, y)$ is not concave–convex.

1.77.a The matrix is

Bomber/Sub	12	23	36	69	98	87	74	41	25	65	85	45
1	1	0	0	0	0	0	0	1	0	0	0	0
2	1	1	0	0	0	0	0	0	1	0	0	0
3	0	1	1	0	0	0	0	0	0	0	0	0
4	0	0	0	0	0	0	1	1	0	0	0	1
5	0	0	0	0	0	0	0	0	1	1	1	1
6	0	0	1	1	0	0	0	0	0	1	0	0
7	0	0	0	0	0	1	1	0	0	0	0	0
8	0	0	0	0	1	1	0	0	0	0	1	0
9	0	0	0	1	1	0	0	0	0	0	0	0

We use Linear Programming to solve this game with Mathematica. The linear programming problems are

Player I	Player II
Max v	**Min v**

$x_1 + x_2 \geq v$ $y_1 + y_8 \leq v$

$x_2 + x_3 \geq v$ $y_1 + y_2 + y_9 \leq v$

$x_3 + x_6 \geq v$ $y_2 + y_3 \leq v$

$x_6 + x_9 \geq v$ $y_7 + y_8 + y_{12} \leq v$

$x_8 + x_9 \geq v$ $y_9 + y_{10} + y_{11} + y_{12} \leq v$

$x_7 + x_8 \geq v$ $y_3 + y_4 + y_{10} \leq v$

$x_4 + x_7 \geq v$ $y_6 + y_7 \leq v$

$x_1 + x_4 \geq v$ $y_5 + y_6 + y_{11} \leq v$

$x_2 + x_5 \geq v$ $y_4 + y_5 \leq v$

$x_5 + x_6 \geq v$ $y_j \geq 0$

$x_5 + x_8 \geq v$ $\sum y_j = 1$

$x_4 + x_5 \geq v$

$x_i \geq 0$

$\sum x_i = 1$

This problem has solution $v(A) = \frac{1}{4}$ with optimal strategies

$$X^* = (0, \frac{1}{4}, 0, \frac{1}{4}, 0, \frac{1}{4}, 0, \frac{1}{4}, 0),$$

and

$$Y^* = (0, \frac{1}{4}, 0, \frac{1}{4}, 0, \frac{1}{4}, 0, 0, 0, 0, 0, \frac{1}{4}).$$

1.77.b The sub's strategies can be reduce to (S1) hide in a pair of squares which include the center square (5); (S2) hide in a pair which does not include the center square (5). The bomber's strategies can be reduced to (B1) fire at a corner, (B2) fire at a square in the middle of each side, (B3) fire at the center square. The payoff matrix is then reduced to

Bomber/Sub	(S1)	(S2)
(B1)	0	$\frac{1}{4}$
(B2)	$\frac{1}{4}$	$\frac{1}{4}$
(B3)	1	0

For example if we play *B*1 against *S*2 this means that the bomber will fire at 1 of 4 corners while the sub will hide along the edges. because of the configuration of the grid, that will result in a one in four chance of hitting the sub. That's where the $\frac{1}{4}$ comes from.

This reduced game has a pure saddle point at (*B*2, *S*2). Then $v(A) = \frac{1}{4}$; the bomber fires at one of the four middle side squares with equal probability; the sub hides in one of the eight locations which does not include the center square with equal probability. This results in the same strategies without using symmetry but in compact form.

1.79 The game matrix is

$$\begin{bmatrix} 0.0 & -0.12 & -0.28 & -0.44 & -0.6 \\ 0.12 & 0.0 & 0.04 & -0.08 & -0.2 \\ 0.28 & -0.04 & 0.0 & 0.28 & 0.2 \\ 0.44 & 0.08 & -0.28 & 0.0 & 0.6 \\ 0.6 & 0.2 & -0.2 & -0.6 & 0 \end{bmatrix}$$

The game is symmetric and has solution by linear programming given by

$$X^* = (0, \frac{5}{11}, \frac{5}{11}, 0, \frac{1}{11}) = Y^*, \quad v = 0.$$

1.81 The pure strategies are labeled plane(P), highway(H), roads(R), for each player. The drug runner chooses one of those to try to get to New York, and the cops choose one of those to patrol. The game matrix in which the drug runner is the row player, becomes

$$A = \begin{bmatrix} -18 & 150 & 150 \\ 100 & 24 & 100 \\ 80 & 80 & 35 \end{bmatrix}.$$

For example, if drug runner plays H and cops patrol H, the drug runner's expected payoff is $(-90)(0.4) + (100)(0.6) = 24$. The saddle point is

$$X^* = (0.144, 0.3183, 0.5376) \quad \text{and} \quad Y^* = (0.4628, 0.3651, 0.1721).$$

The drug runners should use the back roads more than half the time, but the cops should patrol the back roads only about 17% of the time.

1.83.a According to the description of the problem, the matrix is

$$A = \begin{bmatrix} 1 & 0 & 0 & 0 & \cdots & 0 \\ b & 1 & 0 & 0 & \cdots & 0 \\ b^2 & b & 1 & 0 & \cdots & 0 \\ \vdots & \vdots & \vdots & \vdots & \vdots & \vdots \\ b^{n-1} & b^{n-2} & b^{n-3} & b^{n-4} & \cdots & 1 \end{bmatrix}$$

1.83.b Assuming that each row is played with positive probability we may set up the system of equations $XA = vJ_n$ to get

$$\begin{aligned} x_1 + bx_2 + \cdots + b^{n-1}x_n &= v \\ x_2 + bx_3 + \cdots + b^{n-2}x_n &= v \\ x_3 + \cdots + b^{n-3}x_n &= v \\ \vdots \quad &\vdots \\ x_{n-1} + bx_n &= v \\ x_n &= v \end{aligned}$$

Using reverse substitution we get

$$\begin{aligned} x_n &= v \\ x_{n-1} + bv &= v & \implies x_{n-1} &= v(1-b) \\ x_{n-2} + bx_{n-1} + b^2x_n &= v & \implies x_{n-2} &= v(1-b) \\ \vdots \\ x_1 + bx_2 + \cdots b^{n-1}x_n &= v & \implies x_1 &= v(1-b) \end{aligned}$$

We have $X = v(1-b, 1-b, \ldots, 1-b, 1)$. To find v, all the components must add to 1 and

$$\sum_{i=1}^{n} x_i = (n-1)v(1-b) + v = 1 \implies v = \frac{1}{1 + (n-1)(1-b)}.$$

The computation for the optimal Y is similar. We get

$$Y = v(1, 1-b, 1-b, \ldots, 1-b), \quad v = \frac{1}{1 + (n-1)(1-b)}.$$

1.85 To find the game matrix with player I as the row player, if I locates the store in town 1, and II locates the store in town 4, for example, then the payoff to I, in terms of market share, is

$$(0.9)(15) + (0.9)(30) + (0.4)(20) + (0.9)(35) = 52\%.$$

Similarly, if store I is in town 1 and store II is in town 2, then the expected payoff to I is

$$(0.9)(15) + (0.1)(30) + (0.4)(20) + (0.4)(35) = 38.5\%.$$

The game matrix becomes

$$A = \begin{bmatrix} 65 & 38.5 & 41.5 & 52 \\ 78 & 65 & 56.5 & 62 \\ 78 & 58.5 & 65 & 62 \\ 63 & 48.5 & 51.5 & 65 \end{bmatrix}$$

The saddle point is $X^* = (0, 0.43, 0.57, 0)$ and $Y^* = (0, 0.57, 0.43, 0)$.

1.87 The saddle point is

$$X^* = \left(\frac{23}{226}, \frac{165}{452}, \frac{73}{452}, \frac{42}{113} \right),$$
$$= (0.102, 0.365, 0.162, 0.372)$$
$$Y^* = \left(\frac{169}{452}, \frac{55}{226}, 0, \frac{61}{452}, \frac{28}{113} \right)$$
$$= (0.374, 0.243, 0.0, 0.135, 0.245).$$

The value is $v = -\frac{431}{226} = -1.907$. According to this, team P should never play a short pass but team B should defend against SP about 16% of the time. Also, on average, P will lose about 2 yards per play. They should pack it in.

1.89.a If we solve this by linear programming, we have the problems

Player I	**Player II**
Max v	**Min v**
$x_1 + 2x_2 + 3x_3 + 4x_4 + 5x_5 \geq v$	$y_1 + 2y_2 + 3y_3 + 4y_4 + 5y_5 \leq v$
$2x_1 + 4x_2 + 5x_3 + x_4 + 3x_5 \geq v$	$2y_1 + 4y_2 + y_3 + 5y_4 + 3y_5 \leq v$
$3x_1 + x_2 + 4x_3 + 5x_4 + 2x_5 \geq v$	$3y_1 + 5y_2 + 4y_3 + 2y_4 + y_5 \leq v$
$4x_1 + 5x_2 + 2x_3 + 3x_4 + x_5 \geq v$	$4y_1 + y_2 + 5y_3 + 3y_4 + 2y_5 \leq v$
$5x_1 + 3x_2 + x_3 + 2x_4 + 4x_5 \geq v$	$5y_1 + 3y_2 + 2y_3 + y_4 + 4y_5 \leq v$
$x_i \geq 0$	$y_j \geq 0$
$\sum x_i = 1$	$\sum y_j = 1$

If we add the inequalities for player I, we see that

$$15(x_1 + x_2 + x_3 + x_4 + x_5) \geq 5v \implies 15 \geq 5v \implies 3 \geq v.$$

Now add the inequalities for player II and get $v \geq 3$, and we conclude that $v(A) = 3$.

We could use software to solve the programs, but instead we will make the judicious guess that $X^* = Y^* = (\frac{1}{5}, \frac{1}{5}, \frac{1}{5}, \frac{1}{5}, \frac{1}{5})$, since all rows and columns contain the same numbers in the same proportions. Now we only need to verify.

First, a direct computation shows $E(X^*, Y^*) = 3$.

Second, $E(i, Y^*) = E(X^*, j) = 3$ for all rows i and columns j. By the conditions to be a saddle point, this is enough to verify that (X^*, Y^*) is a saddle point.

1.83.b Based on the previous part, it is a reasonable guess that $X^* = (\frac{1}{n}, \dots, \frac{1}{n}) = Y^*$ is a saddle point. If that is the case, then

$$v = E(X^*, Y^*) = X^* A Y^{*T} = X^* \begin{bmatrix} \frac{n(n+1)}{2} \frac{1}{n} \\ \vdots \\ \frac{n(n+1)}{2} \frac{1}{n} \end{bmatrix} = \frac{1}{n} \cdot n \cdot \frac{n(n+1)}{2} \frac{1}{n} = \frac{(n+1)}{2}$$

Now we may verify all of this by checking $E(i, Y^*) = E(X^*, j) = \frac{(n+1)}{2} = v$.

1.91 Shemp has the two strategies to (1) tell the truth or (2) call H no matter what. Curly also has two strategies: (1) believe a call of tails or heads, or (2) believe a call of tails and challenge a call of heads. The game matrix to Shemp as the row player is

Shemp/Curly	(1)	(2)
(1)	0	$\frac{1}{2}$
(2)	1	0

For example, if Shemp tells the truth and Curly believes tails but challenges heads, the expected payoff to Shemp is

$$(-1)\frac{1}{2} + (2)\frac{1}{2} = \frac{1}{2}.$$

Similarly, if Shemp always calls heads and Curly believes the call no matter what, Shemp's expected payoff is

$$(-1)\frac{1}{2} + (+1)\frac{1}{2} = 0.$$

The remaining entries are similar.

Since $1 - x = \frac{1}{2}x \implies x = \frac{2}{3}$, and $\frac{1}{2}(1 - y) = y \implies y = \frac{1}{3}$, Shemp should call the actual toss two-thirds of the time and lie one-third of the time. Curly should believe the call one-third of the time and believe tails called but challenge heads called the rest of the time.

1.93.a The game matrix is

L/R	1	2
1	1	-1
2	-2	2

Remember that the numbers represent the payoffs (=winnings) for Left and do not include the money that Left puts into the pot.

Since $x - 2(1 - x) = -x + 2(1 - x) \implies x = \frac{2}{3}$, and $y - (1 - y) = -2 + 2(1 - y) \implies y = \frac{1}{2}$, Left should bet \$1 with probability 2/3 and Right should bet \$1 with probability 1/2. We have $v = 0, X^* = (\frac{2}{3}, \frac{1}{3}), Y^* = (\frac{1}{2}, \frac{1}{2})$.

1.93.b In the first case the game matrix is

L/R	1	2	3	4	5	6
1	1	-1	3	-1	5	-1
2	-2	2	-2	4	-2	6
3	1	-3	3	-3	5	-3
4	-4	2	-4	4	-4	6
5	1	-5	3	-5	5	-5
6	-6	2	-6	4	-6	6

Start with eliminating dominated rows and columns. For example, row 1 dominates row 3 and 5; column 2 dominates columns 4 and 6; then row 2 dominates rows 4 and 6, and column 2 dominates columns 3 and 5. The matrix reduces to the 2×2 case $A = \begin{bmatrix} 1 & -1 \\ -2 & 2 \end{bmatrix}$ which has the same solution as the first part. Left should bet 1 two-thirds of the time; Right should bet 1 half the time, and it is a fair game.

For the second case the matrix is

L/R	2	4	6	8	9	13
2	2	4	6	8	−2	−2
4	2	4	6	8	−4	−4
6	2	4	6	8	−6	−6
8	2	4	6	8	−8	−8
9	−9	−9	−9	−9	9	13
13	−13	−13	−13	−13	9	13

By dominance this reduces to the game

L/R	2	9
2	2	−2
9	−9	9

This game has solution $v = 0, X^* = (\frac{9}{11}, \frac{2}{11}), Y^* = (\frac{7}{22}, \frac{15}{22})$. Left should bet 2 with probability $\frac{9}{11}$ and bet 9 with probability $\frac{2}{11}$. Right should bet the same amounts but with probabilities $\frac{7}{22}$ and $\frac{15}{22}$, respectively.

1.93.c Left and Right each have 4 strategies. The game matrix is

L/R	2	5	16	17
1	−3	6	17	−18
2	4	−7	−18	19
31	33	−36	47	−48
32	−34	37	−48	49

This game is completely mixed and the matrix has an inverse so we could use the invertible matrix game formulas, or we could use the linear programming method. We get the solution

$$v(A) = -\frac{1}{152}, \quad X^* = (50, 48, 181, 153)\frac{1}{456}, \quad Y^* = (10, 10, 67, 65)\frac{1}{152}.$$

In decimal terms,

$$v(A) = -.0066, \quad X^* = (.109, .105, .397, .336), \quad Y^* = (.066, .066, .441, .428)$$

The game is not quite fair to Left, and both players prefer to play the bigger numbers.

1.95.a Your strategies are $H^U H^D, H^U C^D, C^U H^D, C^U C^D$, where H = honest, C = cheat and the superscript refers to the company. The IRS has two strategies: investigate Uno, investigate Due. Here's the matrix:

IRS/You	$H^U H^D$	$H^U C^D$	$C^U H^D$	$C^U C^D$
U	4	3	$4 + 3p$	$3 + 3p$
D	4	$4 + p$	1	$1 + p$

The IRS is the row player because the IRS wants to maximize the tax receipts, while you want to minimize them. For example, if the IRS investigates Uno and you play $H^U H^D$, you simply pay your total bill of 4 million. If, however, you play $C^U H^D$, then you are caught and you have to pay the 4 million you owe as well as a penalty of $3p$, a percentage of the 3 million owed on Uno. The rest of the entries are similar.

1.95.b If $p = \frac{1}{2}$ the matrix is

IRS/You	$H^U H^D$	$H^U C^D$	$C^U H^D$	$C^U C^D$
U	4	3	$\frac{11}{2}$	$\frac{8}{2}$
D	4	$\frac{11}{2}$	1	$\frac{3}{2}$

The solution of this game is $v(A) = \frac{7}{2}, X^* = (\frac{2}{3}, \frac{1}{3})$ and $Y^* = (0, \frac{2}{3}, 0, \frac{1}{3})$. The IRS should audit Uno $\frac{2}{3}$ of the time (because it is worth more to the IRS) and you should play $H^U C^D$ $\frac{2}{3}$ of the time and cheat on both of them $\frac{1}{3}$ of the time.

1.95.c We want $H^U H^D$ to be an optimal pure strategy and we want to know what p should be for that to occur. If it is an optimal strategy, then it must be the case that $v(A) = 4$.

Consider the graphical method for solving the game. The first step is to graph all the lines

$$z = E(X, 1) = 4x + 4(1 - x)$$
$$z = E(X, 2) = 3x + (4 + p)(1 - x)$$
$$z = E(X, 3) = (4 + 3p)x + (1 - x)$$
$$z = E(X, 4) = (3 + 3p)x + (1 + p)(1 - x).$$

Since we want the value of the game to be 4, if all the columns are used with positive probability in an optimal strategy for You, then in all 4 equations $z = 4$. Solve for x in terms of p to get the 4 possible solutions

$$0 \leq x_1 \leq 1, \quad x_2 = \frac{p}{1 + p}, \quad x_3 = \frac{1}{1 + p}, \quad x_4 = \frac{1}{2}\frac{3 - p}{1 + p}$$

The first x_1 is any number in $[0, 1]$. Clearly $x_2, x_3 \in [0, 1]$ for any $0 \leq p \leq 1$. But $x_4 \in [0, 1]$ requires that $3 - p \leq 2 + 2p \implies p \geq 1$.

If the value of the game must be 4, and the optimal strategy for You must be $Y^* = (1, 0, 0, 0)$, then the two lines involved in finding Y^* must include the line corresponding to column 1. This will be true if $p \geq 1$, but is not true for $p < 1$. We conclude that complete honesty will be optimal if the fine is a minimum of 100% of the tax owed.

The figure shows what happens when $p = 1.3$. In this case $X^* = (0.565, 0.435)$, $Y^* = (1, 0, 0, 0), v = 4$.

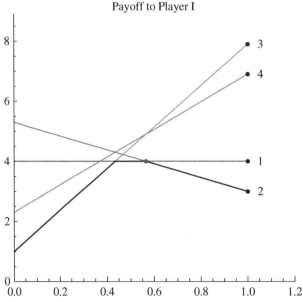

Payoff to Player I

1.97 In order for (X^*, Y^*) to be a saddle point and v to be the value, it is necessary and sufficient that $E(X^*, j) \geq v$, for all $j = 1, 2, \ldots, m$ and $E(i, Y^*) \leq v$ for all $i = 1, 2, \ldots, n$.

1.99 $E(k, Y^*) = v(A)$. If $x_k = 0$, $E(k, Y^*) \leq v(A)$.

1.105 True.

1.107 $x_i = 0$.

1.109 True.

1.111 False. Only if you know $x_1 = 0$ for *every* optimal strategy for player I.

1.113 $X^* = (0, 1, 0)$, $Y^* = (0, 0, 1, 0)$.

1.115 False. There's more to it than that.

1.117 True! This is a point of logic. Since any skew symmetric game must have $v(A) = 0$, the sentence has a false premise. A false premise always implies anything, even something that is false on it's own (since $-Y^*$ can't even be a strategy).

Solutions for Chapter 2

2.1 Use the definitions. For example, if X^*, Y^* is a Nash equilibrium, then $X^* A Y^{*T} \geq XAY^{*T}$, $\forall X$, and $X^*(-A)Y^{*T} \geq X^*(-A)Y^T$, $\forall Y$. Then

$$X^* A Y^T \geq X^* A Y^{*T} \geq XAY^{*T}, \quad \forall X \in S_n, Y \in S_m.$$

2.3.a The two pure Nash equilibria are at (*Turn, Straight*) and (*Straight, Turn*).

2.3.b To verify that $X^* = (\frac{3}{52}, \frac{49}{52})$, $Y^* = (\frac{3}{52}, \frac{49}{52})$ is a Nash equilibrium we need to show that

$$E_I(X^*, Y^*) \geq E_I(X, Y^*), \quad \text{and} \quad E_{II}(X^*, Y^*) \geq E_{II}(X^*, Y),$$

for all strategies X, Y. Calculating the payoffs we get

$$E_I(X, Y^*) = (x, 1 - x) \begin{bmatrix} 19 & -42 \\ 68 & -45 \end{bmatrix} \begin{bmatrix} 3 \\ 49 \end{bmatrix} \frac{1}{52} = -\frac{2001}{52}$$

for any $X = (x, 1 - x), 0 \leq x \leq 1$. Similarly,

$$E_{II}(X^*, Y) = \frac{1}{52}(3, 49) \begin{bmatrix} 19 & 68 \\ -42 & -45 \end{bmatrix} \begin{bmatrix} y \\ 1 - y \end{bmatrix} = -\frac{2001}{52}.$$

Since $E_I(X^*, Y^*) = E_{II}(X^*, Y^*) = -\frac{2001}{52}$, we have shown (X^*, Y^*) is a Nash equilibrium.

2.3.c For player I, $A = \begin{bmatrix} 19 & -42 \\ 68 & -45 \end{bmatrix}$. Then $v(A) = -42$ and that is the safety level for I. The maxmin strategy for player I is $X = (1, 0)$. For player II, $B^T = \begin{bmatrix} 19 & -42 \\ 68 & -45 \end{bmatrix}$ so the safety level for II is also $v(B^T) = -42$. The maxmin strategy for player II is $X^{B^T} = (1, 0)$.

2.5.a Weak dominance by row 1 and then strict dominance by column 2 gives the Nash equilibrium at $(5, 1)$.

2.5.b Player II drops the 3rd column by strict dominance. This gives the matrix

$$\begin{bmatrix} (10, 0) & (5, 1) \\ (10, 1) & (5, 0) \end{bmatrix}$$

Then we have two pure Nash equilibria at $(10, 1)$ and $(5, 1)$.

2.7 There is one pure Nash at Low,Low. Both airlines should set their fares low.

2.9.a Graph $x = BR_1(y)$ and $y = BR_2(x)$ on the same set of axes. Where the graphs intersect is the Nash equilibrium (they could intersect at more than one point). The unique Nash equilibrium is $X^* = (\frac{1}{2}, \frac{1}{2})$, $Y^* = (\frac{1}{4}, \frac{3}{4})$.

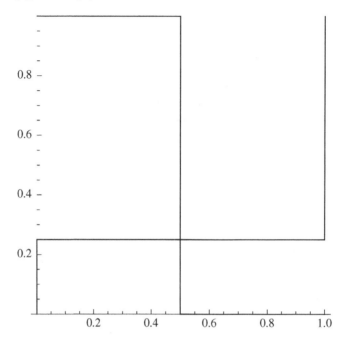

2.9.b Here is a plot of the four lines bounding the best response sets.

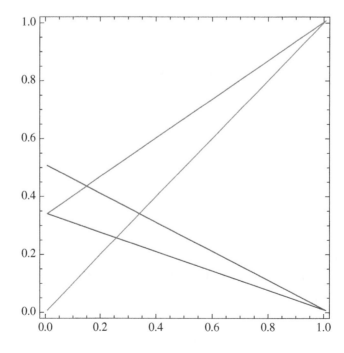

The intersection of the two sets constitutes all the Nash equilibria. Mathematica characterizes the intersection as

$$\left(y = \frac{1}{4}, x = \frac{1}{4}\right) \bigcup \left(\frac{1}{4} < y \le \frac{1}{3}, 1 - 3y \le x \le y\right)$$
$$\bigcup \left(\frac{1}{3} < y < \frac{3}{7}, \frac{(-1 + 3y)}{2} \le x \le 1 - 2y\right)$$
$$\bigcup \left(y = \frac{3}{7}, x = \frac{1}{7}\right)$$

You can see from the figure that there are lots of Nash equilibria. For example $x^* = \frac{1}{4}, y^* = \frac{1}{4}$, is one.

2.11 We calculate the best response functions

$$x = BR_1(y) = \arg\max_{0 \le x \le 1} (x, 1 - x) \begin{bmatrix} 2 & 3 \\ 1 & 4 \end{bmatrix} \begin{bmatrix} y \\ 1 - y \end{bmatrix} = \begin{cases} 0 & 0 \le y < \frac{1}{2} \\ [0,1] & y = \frac{1}{2} \\ 1 & \frac{1}{2} < y \le 1 \end{cases}$$

and

$$y = BR_2(x) = \arg\max_{0 \le y \le 1} (x, 1 - x) \begin{bmatrix} 2 & 1 \\ 3 & 4 \end{bmatrix} \begin{bmatrix} y \\ 1 - y \end{bmatrix} = \begin{cases} 0 & 0 \le x < \frac{1}{2} \\ [0,1] & x = \frac{1}{2} \\ 1 & \frac{1}{2} < x \le 1 \end{cases}$$

Graphing these two functions shows they intersect at $(0,0), (\frac{1}{2}, \frac{1}{2}), (1,1)$.

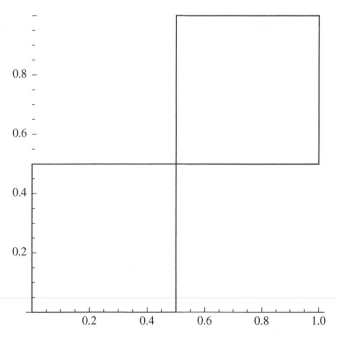

The Nash equilibria are $X_1 = (\frac{1}{2}, \frac{1}{2}), Y_1 = (\frac{1}{2}, \frac{1}{2}), X_2 = (0, 1) = Y_2, X_3 = (1, 0) = Y_3$.

2.13 To find the best response functions calculate

$$x = BR_1(y) = \arg\max_{0 \le x \le 1}(x, 1-x)\begin{bmatrix} -10 & 2 \\ 1 & -1 \end{bmatrix}\begin{bmatrix} y \\ 1-y \end{bmatrix} = \begin{cases} 0 & \frac{3}{14} < y \le 1 \\ [0,1] & y = \frac{3}{14} \\ 1 & 0 \le y < \frac{3}{14} \end{cases}$$

and

$$y = BR_2(x) = \arg\max_{0 \le y \le 1}(x, 1-x)\begin{bmatrix} -5 & -2 \\ -1 & 1 \end{bmatrix}\begin{bmatrix} y \\ 1-y \end{bmatrix} = \begin{cases} 0 & 0 \le x < \frac{2}{9} \\ [0,1] & x = \frac{2}{9} \\ 1 & 0\frac{2}{9} < x \le 1. \end{cases}$$

Here's the figure:

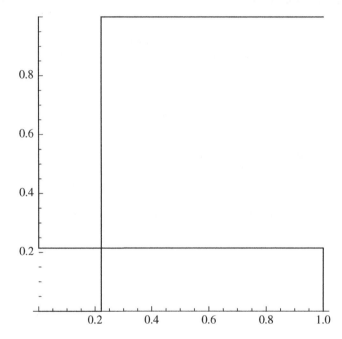

The only Nash is $X = (\frac{2}{9}, \frac{7}{9})$, $Y = (\frac{3}{14}, \frac{11}{14})$, with payoffs $-\frac{4}{7}, \frac{1}{3}$. The rational reaction sets intersect at only this Nash.

2.15.a Let $X^* = (x, 1-x)$. By Equality of Payoffs, assuming both columns are played with positive probability, we get the equations

$$E_{II}(X^*, 1) = E_{II}(X^*, 2) \implies ax + c(1-x) = bx + d(1-x) \implies x = \frac{d-c}{a-b+d-c}.$$

Then $X^* = (\frac{d-c}{a-b+d-c}, \frac{a-b}{a-b+d-c})$. In order for this to work X^* must be a legitimate strategy with positive components.

2.15.b Let $Y^* = (y, 1-y)$. Assuming both rows are played with positive probability,

$$E_I(1, Y^*) = E_I(2, Y^*) \implies Ay + B(1-y) = Cy + D(1-y) \implies y = \frac{D-B}{A-C+D-B}.$$

This requires that $A - C + D - B \ne 0$ and $D \ne B$. Then $Y^* = (\frac{D-B}{A-C+D-B}, \frac{A-C}{A-C+D_B})$ and the components must be positive.

2.17 We need to check that $E_I(X^*, Y^*) \geq E_I(i, Y^*), i = 1, 2, 3$, and $E_{II}(X^*, Y^*) \geq E_{II}(X^*, j), j = 1, 2, 3$, Calculating, we get

$$E_I(X^*, Y^*) = X^* A Y^{*T} = \frac{5}{3}, \quad E_I(i, Y^*) = (\frac{5}{3}(i = 1), \frac{4}{3}(i = 2), \frac{5}{3}(i = 3)).$$

Each of these numbers is $\leq \frac{5}{3}$. Next,

$$E_{II}(X^*, Y^*) = X^* B Y^{*T} = \frac{5}{2}, \quad E_{II}(X^*, j) = (\frac{5}{2}(j = 1), \frac{5}{2}(j = 2), \frac{5}{2}(j = 3)).$$

Each of these are actually equal to $\frac{5}{2}$. We conclude that X^*, Y^* is indeed a Nash equilibrium.

2.19.a The pure strategies for each son is the share of the estate to claim. The matrix is

I/II	0.25	0.5	0.75
0.25	$(0.25, 0.25)$	$(0.25, 0.50)$	$\boxed{(0.25, 0.75)}$
0.5	$(0.50, 0.25)$	$\boxed{(0.50, 0.50)}$	$(0, 0)$
0.75	$\boxed{(0.75, 0.25)}$	$(0, 0)$	$(0, 0)$

There are 3 pure Nash equilibria: (1) row 1, column 3, (2) row 2, column 2, and (3) row 3, column 1.

2.19.b The table contains the pure and mixed Nash points as well as the payoffs.

X	Y	E_I	E_{II}
$(\frac{1}{3}, \frac{1}{6}, \frac{1}{2})$	$(\frac{1}{3}, \frac{1}{6}, \frac{1}{2})$	$\frac{1}{4}$	$\frac{1}{4}$
$(\frac{2}{3}, \frac{1}{3}, 0)$	$(0, \frac{1}{2}, \frac{1}{2})$	$\frac{1}{4}$	$\frac{1}{2}$
$(\frac{1}{3}, 0, \frac{2}{3})$	$(\frac{1}{3}, 0, \frac{2}{3})$	$\frac{1}{4}$	$\frac{1}{4}$
$(0, \frac{1}{2}, \frac{1}{2})$	$(\frac{2}{3}, \frac{1}{3}, 0)$	$\frac{1}{2}$	$\frac{1}{4}$
$(1, 0, 0)$	$(0, 0, 1)$	$\frac{1}{4}$	$\frac{3}{4}$
$(0, 1, 0)$	$(0, 1, 0)$	$\frac{1}{2}$	$\frac{1}{2}$
$(0, 0, 1)$	$(1, 0, 0)$	$\frac{3}{4}$	$\frac{1}{4}$

To use the equality of payoffs theorem to find these you must consider all possibilities in which at least two rows (or columns) are played with positive probability. For example, suppose that columns 2 and 3 are used by player II with positive probability. Equality of payoffs then tells us

$$E_{II}(X^*, 2) = E_{II}(X^*, 3) \implies 0.5x_1 + 0.5x_2 = 0.75x_1 \implies x_2 = 0.5x_1.$$

Hence $X = (x_1, 0.5x_1, 1 - 1.5x_1)$ is a solution as long as $0 < x_1 < \frac{2}{3}$.

If $x_1 = 0$, we get $X = (0, 0, 1)$, which is part of a pure Nash but is not obtained from equality of payoffs. If $x_1 = \frac{2}{3}$ then $X = (\frac{2}{3}, \frac{1}{3}, 0)$. By equality of payoffs we then obtain

$$E_I(1, Y^*) = E_I(2, Y^*) \implies 0.25y_1 + 0.25y_2 + 0.25y_3 = 0.5y_1 + 0.5y_2 \implies y_1 + y_2 = 0.5.$$

Then, $Y = (y_1, \frac{1}{2} - y_1, \frac{1}{2})$ is the solution if $0 \leq y_1 < \frac{1}{2}$. Observe that $y_1 = \frac{1}{2}$ is not consistent with our original assumption that $y_2 > 0, y_3 > 0$, but $y_1 = 0$ is. In that case $Y = (0, \frac{1}{2}, \frac{1}{2})$ and we have the Nash equilibrium $X = (\frac{2}{3}, \frac{1}{3}, 0), Y = (0, \frac{1}{2}, \frac{1}{2})$. Any of the remaining Nash equilibria in the tables are obtained similarly.

If we assume $Y = (y_1, y_2, y_3)$ is part of a mixed Nash with $y_j > 0, j = 1, 2, 3$, then equality of payoffs gives the equations

$$0.25x_1 + 0.25x_2 + 0.25x_3 = v_I$$
$$0.5x_1 + 0.5x_2 = v_I$$
$$0.75x_1 = v_I$$
$$x_1 + x_2 + x_3 = 1$$

The first and fourth equation give $v_I = 0.25$ Then the third equation gives $x_1 = \frac{1}{3}$, then $x_2 = \frac{1}{6}$, and finally $x_3 = \frac{1}{2}$. Since $X = (\frac{1}{3}, \frac{1}{6}, \frac{1}{2})$ we then use equality of payoffs to find Y. The equations are

$$0.25y_1 + 0.25y_2 + 0.25y_3 = v_{II}$$

$$0.5y_1 + 0.5y_2 = v_{II}$$

$$0.75y_1 = v_{II}$$

$$y_1 + y_2 + y_3 = 1$$

This has the same solution as before $Y = X, v_I = v_{II} = 0.25$, and this is consistent with our original assumption that all components of Y are positive.

2.21 We find the best response functions. For the government,

$$x = BR_2(y) = \arg \max_{0 \le x \le 1} (x, 1-x) \begin{bmatrix} 3 & -1 \\ -1 & 0 \end{bmatrix} \begin{bmatrix} y \\ 1-y \end{bmatrix} = \begin{cases} 0, & \text{if } 0 \le y < \frac{1}{5}; \\ [0,1], & \text{if } y = \frac{1}{5}; \\ 1, & \text{if } \frac{1}{5} < y \le 1. \end{cases}$$

Similarly,

$$y = BR_1(x) = \arg \max_{0 \le y \le 1} (x, 1-x) \begin{bmatrix} 2 & 3 \\ 1 & 0 \end{bmatrix} \begin{bmatrix} y \\ 1-y \end{bmatrix} = \begin{cases} 0, & \text{if } 0 \le x < \frac{1}{2}; \\ [0,1], & \text{if } x = \frac{1}{2}; \\ 0, & \text{if } \frac{1}{2} < x \le 1. \end{cases}$$

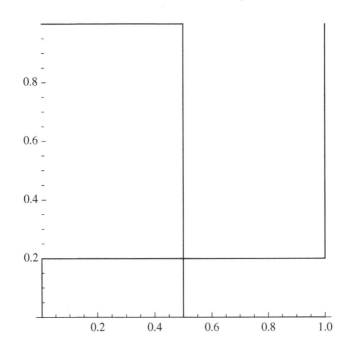

The Nash equilibrium is $X = (\frac{1}{2}, \frac{1}{2})$, $Y = (\frac{1}{5}, \frac{4}{5})$ with payoffs $E_I = -\frac{1}{5}$, $E_{II} = \frac{3}{2}$. The government should aid half the paupers, but 80% of paupers should be bums (or the government predicts that 80% of welfare recipients will not look for work). The rational reaction sets intersect only at the mixed Nash point.

2.23 The system is $0 = 4y_1 - y_2$, $y_1 + y_2 = 1$ which gives $y_1 = \frac{1}{5}$, $y_2 = \frac{4}{5}$.

2.25.a Assuming the formulas actually give strategies we will show that the strategies are a Nash equilibrium. We have

$$X^* A Y^{*T} = \frac{J_n B^{-1}}{J_n B^{-1} J_n^T} A \frac{A^{-1} J_n^T}{J_n A^{-1} J_n^T} = \frac{J_n B^{-1} J_n^T}{J_n B^{-1} J_n^T} \frac{1}{J_n A^{-1} J_n^T} = \frac{1}{J_n A^{-1} J_n^T}.$$

and for any strategy $X \in S_n$,

$$X A Y^{*T} = XA \frac{A^{-1} J_n^T}{J_n A^{-1} J_n^T} = \frac{X J_n^T}{J_n A^{-1} J_n^T} = \frac{1}{J_n A^{-1} J_n^T} = v_I.$$

Similarly,

$$X^* B Y^{*T} = \frac{J_n B^{-1}}{J_n B^{-1} J_n^T} B \frac{A^{-1} J_n^T}{J_n A^{-1} J_n^T} = \frac{1}{J_n B^{-1} J_n^T} \frac{J_n A^{-1} J_n^T}{J_n A^{-1} J_n^T} = \frac{1}{J_n B^{-1} J_n^T}.$$

and for any $Y \in S_n$,

$$X^* B Y^T = \frac{J_n B^{-1}}{J_n B^{-1} J_n^T} B Y^T = \frac{J_n Y^T}{J_n B^{-1} J_n^T} = \frac{1}{J_n B^{-1} J_n^T} = v_{II}.$$

2.25.b We have

$$A^{-1} = \begin{bmatrix} -\frac{1}{16} & \frac{27}{688} & \frac{29}{344} \\ \frac{1}{8} & -\frac{35}{344} & \frac{7}{172} \\ 0 & \frac{5}{43} & -\frac{2}{43} \end{bmatrix}, \quad B^{-1} = \frac{1}{604} \begin{bmatrix} 67 & -3 & -22 \\ -64 & 84 & 12 \\ 43 & -47 & 58 \end{bmatrix},$$

$$X^* = \frac{J_n B^{-1}}{J_n B^{-1} J_n^T} = \left(\frac{23}{64}, \frac{17}{64}, \frac{3}{8} \right)$$

and

$$Y^{*T} = \frac{A^{-1} J_n^T}{J_n A^{-1} J_n^T} = \left(\frac{21}{67}, \frac{22}{67}, \frac{24}{67} \right).$$

These are both legitimate strategies. Finally, $v_I = \frac{344}{67}$, $v_{II} = \frac{151}{32}$.

2.25.c Set $A^* = \begin{bmatrix} a_{22} & -a_{12} \\ -a_{21} & a_{11} \end{bmatrix}$ and $B^* = \begin{bmatrix} b_{22} & -b_{12} \\ -b_{21} & b_{11} \end{bmatrix}$. Then, from the formulas derived in Problem (2.15) we have, assuming both rows and columns are played with positive probability,

$$X^* = \left(\frac{b_{22} - b_{21}}{b_{11} - b_{12} + b_{22} - b_{21}}, \frac{b_{11} - b_{12}}{b_{11} - b_{12} + b_{22} - b_{21}} \right) = \frac{J_n B^*}{J_n B^* J_n^T},$$

and

$$Y^{*T} = \left(\frac{a_{22} - a_{12}}{a_{11} - a_{21} + a_{22} - a_{12}}, \frac{a_{11} - a_{21}}{a_{11} - a_{21} + a_{22} - a_{12}} \right) = \frac{A^* J_n^T}{J_n A^* J_n^T}.$$

(X^*, Y^*) is a Nash equilibrium if they are strategies, and then

$$v_I = \frac{\det(A)}{J_n A^* J_n^T} = E_I(X^*, Y^*), \quad v_{II} = \frac{\det(B)}{J_n B^* J_n^T} = E_{II}(X^*, Y^*).$$

Of course, if A or B does not have an inverse, then $\det(A) = 0$ or $\det(B) = 0$.

For the game in (1), $A^* = \begin{bmatrix} 4 & 1 \\ 2 & 1 \end{bmatrix}$ and $B^* = \begin{bmatrix} 0 & 2 \\ 1 & 3 \end{bmatrix}$. We get

$$X^* = \frac{J_n B^*}{J_n B^* J_n^T} = (\frac{1}{6}, \frac{5}{6}), \quad Y^{*T} = \frac{A^* J_n^T}{J_n A^* J_n^T} = (\frac{5}{8}, \frac{3}{8}).$$

and

$$v_I = \frac{\det(A)}{J_n A^* J_n^T} = \frac{1}{4}, \quad v_{II} = \frac{\det(B)}{J_n B^* J_n^T} = -\frac{1}{3}.$$

There are also two pure Nash equilibria $X_1^* = (0,1), Y_1^* = (0,1)$ and $X_2^* = (1,0), Y_2^* = (1,0)$.

For the game in (2), $A^* = \begin{bmatrix} 4 & 2 \\ 2 & 1 \end{bmatrix}$ and $B^* = \begin{bmatrix} 0 & 2 \\ 1 & 3 \end{bmatrix}$. We get

$$X^* = \frac{J_n B^*}{J_n B^* J_n^T} = (\frac{1}{6}, \frac{5}{6}), \quad Y^{*T} = \frac{A^* J_n^T}{J_n A^* J_n^T} = (\frac{2}{3}, \frac{1}{3}).$$

and

$$v_I = \frac{\det(A)}{J_n A^* J_n^T} = 0, \quad v_{II} = \frac{\det(B)}{J_n B^* J_n^T} = -\frac{1}{3}.$$

In this part of the problem A does not have an inverse and $\det(A) = 0$. There are also two pure Nash equilibria $X_1^* = (0,1), Y_1^* = (0,1)$ and $X_2^* = (1,0), Y_2^* = (1,0)$.

2.27 If $ab < 0$, there is exactly one strict Nash equilibrium: (i) $a > 0, b < 0 \implies$ the Nash point is $X^* = Y^* = (1,0)$; (ii) $a < 0, b > 0 \implies$ the Nash point is $X^* = Y^* = (0,1)$.

If $a > 0, b > 0$ there are three Nash equilibria $X_1 = Y_1 = (1,0), X_2 = Y_2 = (0,1)$, and the mixed Nash $X_3 = Y_3 = (\frac{b}{a+b}, \frac{a}{a+b})$. The mixed Nash is obtained from equality of payoffs:

$$xa = (1-x)b \implies x = \frac{b}{a+b}, \text{ and } ya = (1-y)b \implies y = \frac{b}{a+b}.$$

If $a < 0, b < 0$, there are three Nash equilibria $X_1 = (1,0), Y_1 = (0,1), X_2 = (0,1), Y_2 = (1,0)$, and the mixed Nash $X_3 = Y_3 = (\frac{b}{a+b}, \frac{a}{a+b})$.

2.29.a First we define $f(x, y_1, y_2) = XAY^T$ where $X = (x, 1-x), Y = (y_1, y_2, 1 - y_1 - y_2)$. We calculate

$$f(x, y_1, y_2) = 3 - 3y_1 + y_2 + x(1 + y_1 - 2y_2)$$

and

$$x = BR_1(y_1, y_2) = \begin{cases} 0 & \frac{1}{2} \leq y_2 \leq 1, \text{ and } 0 \leq y_1 < -1 + 2y_2, \\ [0,1] & \frac{1}{2} \leq y_2 \leq 1 \text{ and } y_1 = -1 + 2y_2, \\ 1 & 0 < 1 + y_1 - 2y_2. \end{cases}$$

Similarly,

$$g(x, y_1, y_2) = XBY^T = 1 + y_1 - y_2 + x(3y_2 - 2)$$

and the maximum of g is achieved at (y_1, y_2) where

$$y_1(x) = \begin{cases} 0 & \frac{2}{3} < x \leq 1 \\ [0,1] & x = \frac{2}{3} \\ 1 & 0 \leq x < \frac{2}{3} \end{cases} \text{ and } y_2(x) = \begin{cases} 0 & 0 \leq x < \frac{2}{3} \\ [0,1] & x = \frac{2}{3} \\ 1 & \frac{2}{3} < x \leq 1 \end{cases}$$

The best response set, $BR_2(x)$, for player II is the set of all pairs $(y_1(x), y_2(x))$.

This is obtained by hand or using the simple Mathematica commands:

$$\text{Maximize}\,[\{g[x,y_1,y_2],y_1 \geq 0,y_2 \geq 0,y_1+y_2 \leq 1\}, \{y_1,y_2\}].$$

2.29.b This is a little tricky because we don't know which columns are used by player II with positive probability. However, if we notice that column 3 is dominated by a convex combination of columns 1 and 2, we may drop column 3. Then the game is a 2×2 game and

$$2(1-x) = x \implies x = \frac{2}{3} \implies X = (\frac{2}{3}, \frac{1}{3})\ \text{and}\ v_{\text{II}} = \frac{2}{3}.$$

Then, for player II

$$2y + 3(1-y) = 4(1-y) \implies y = \frac{1}{3}\ \text{and}\ v_{\text{I}} = \frac{8}{3}.$$

The Nash equilibrium is $X^* = (\frac{2}{3}, \frac{1}{3}), Y^* = (\frac{1}{3}, \frac{2}{3}, 0)$. Then $y_1 = \frac{1}{3}, y_2 = \frac{2}{3}, x = \frac{2}{3}$, and by definition $\frac{2}{3} \in BR_1(\frac{1}{3}, \frac{2}{3})$. Also, $(\frac{1}{3}, \frac{2}{3}) \in BR_2(\frac{2}{3})$.

2.31 There are three Nash equilibria: $X_1 = (1,0), Y_1 = (0,1), X_2 = (0,1), Y_2 = (1,0), X_3 = (\frac{2}{3}, \frac{1}{3}) = Y_3$. The last Nash gives payoffs $(0,0)$. The best response functions are

$$x = BR_1(y) = \begin{cases} 0 & \frac{2}{3} < y \leq 1 \\ [0,1] & y = \frac{2}{3} \\ 1 & 0 \leq y < \frac{2}{3} \end{cases} \quad \text{and} \quad y = BR_2(x) = \begin{cases} 0 & \frac{2}{3} < x \leq 1 \\ [0,1] & x = \frac{2}{3} \\ 1 & 0 \leq x < \frac{2}{3} \end{cases}$$

The Mathematica commands to solve this and all the 2×2 games is the following

```
A = {{-1, 2}, {0, 0}}
f[x_, y_] = {x, 1 - x}.A.{y, 1 - y}
Maximize[{f[x, y], 0 <= x <= 1}, {x}]
B = {{-1, 0}, {2, 0}}
g[x_, y_] = {x, 1 - x}.B.{y, 1 - y}
Maximize[{g[x, y], 0 <= y <= 1}, {y}]
Show[{Graphics[Line[{{0, 1}, {0, 2/3}, {1, 2/3}, {1, 0}}]],
  Graphics[Line[{{0, 1}, {2/3, 1}, {2/3, 0}, {1, 0}}]]}, Axes -> True]
```

The last command produces the graph:

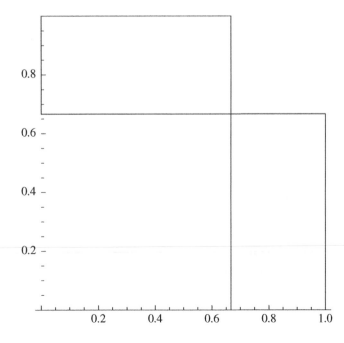

2.33.a First note that there is no pure Nash equilibrium. We could use Problem (2.25) to solve this problem or do it directly by equality of payoffs. The equations to solve are

$$50y_2 + 40y_3 = v$$
$$40y_1 + 50y_3 = v$$
$$50y_1 + 40y_2 = v$$
$$y_1 + y_2 + y_3 = 1$$

Adding the first three equations and using the fourth gives us $v = 30$. Then solving the first three equations gives $Y^* = (\frac{1}{3}, \frac{1}{3}, \frac{1}{3})$. Since the game is symmetric, $X^* = Y^*$. The expected payoff to each player is 30.

2.33.b If player II uses $Y = (y_1, y_2, y_3)$ then the expected payoff to player I is

$$E_1(X, Y) = XAY^T = x_1(50y_2 + 40y_3) + x_2(50y_3 + 40y_1) + x_3(50y_1 + 40y_2).$$

The best response for player I will be to choose $x_i = 1$ corresponding to the largest coefficient in E_1. The only way to get a mixed strategy as a best response for player I is if the coefficients are all equal, and then they must all be 30, as we have seen. Thus, at least one coefficient must be greater than 30. Assume $50y_2 + 40y_3 > 30$. Then $x_1 = 1, x_2 = x_3 = 0$ and player I's expected payoff will be $50y_2 + 40y_3 > 30$. Thus, no matter what, if player II does not use the mixed Nash, then player I's expected payoff using the best response will be more than 30. This also applies for player II. Thus, if any player deviates from the Nash equilibrium, the best response gives a larger expected payoff to the other player.

2.35 Since this is a 2×2 game, all we need to do is match up the coefficients in the constraints $AY^T \le pJ_2^T$ and $XB \le qJ_2$. $A = \begin{bmatrix} 50 & 80 \\ 90 & 20 \end{bmatrix}$, $B = \begin{bmatrix} 50 & 20 \\ 10 & 80 \end{bmatrix}$. Then use Mathematica to get $X^* = (0.7, 0.3), Y^* = (0.6, 0.4), p = 62, q = 38$.

2.37 Take $B = -A$. The algorithm is then

$$\max_{X, Y, p, q} \quad XAY^T - XAY^T - p - q = -p - q$$

subject to

$$AY^T \le pJ_n^T$$
$$-A^TX^T \le qJ_m^T \quad \text{(equivalently } XA \ge -qJ_m)$$
$$x_i \ge 0, y_j \ge 0, \quad XJ_n = 1 = YJ_m^T$$

where $J_k = (1\ 1\ 1\ \cdots 1)$ is the $1 \times k$ row vector consisting of all 1's. In addition, $p^* = E_1(X^*, Y^*) = X^*AY^{*T}$, and $q^* = E_{II}(X^*, Y^*) = -X^*AY^{*T} = -p^*$. You can see that this problem is the same as Method 2 of linear programming for solving a game. Thus Lemke-Howson reduces to Method 2.

The Nash equilibrium found using Lemke-Howson is $X^* = (\frac{5}{8}, \frac{3}{8}, 0), Y^* = (0, \frac{5}{8}, \frac{3}{8})$, and the value of the game is $v(A) = \frac{1}{8}$.

2.39 The Nash equilibria are

X^*	Y^*
$(0, 0, 0.333333, 0.666667)$	$(0.25, 0, 0.75, 0)$
$(0, 0.66667, 0.111108, 0.222221)$	$(0.341463, 0.146341, 0.512195, 0)$
$(0.020202, 0.777778, 0, 0.20202)$	$(0, 1, 0, 0)$
$(0.0896435, 0, 0.00464034, 0.905716)$	$(0.25, 0, 0.75, 0)$

The corresponding respective payoffs are

E_I	E_{II}
−0.25	1
−0.560975	0.3333
2	0.525253
−0.25	2.34465

2.41 Suppose first that (i^*, j^*) is a pure Nash equilibrium. Then $P = (p_{ij})$ defined by $p_{i^*j^*} = 1$ and $p_{ij} = 0$ if $i \neq i^*, j \neq j^*$ is a correlated equilibrium. Indeed, as a Nash equilibrium we know

$$a_{i^*j^*} \geq a_{ij^*} \quad \text{and} \quad b_{i^*j^*} \geq b_{i^*j}, \quad \forall \, i = 1, 2, \ldots, n, j = 1, 2, \ldots, m.$$

The condition for P to be a correlated equilibrium is

$$\sum_{j=1}^{m} a_{ij} p_{ij} = a_{i^*j^*} \geq \sum_{j=1}^{m} a_{qj} p_{ij} = a_{qj^*}$$

for all rows $i = 1, 2, \ldots, n, q = 1, 2, \ldots, n$, and

$$\sum_{i=1}^{n} b_{ij} p_{ij} = b_{i^*j^*} \geq \sum_{i=1}^{n} b_{ir} p_{ij} = b_{i^*r}$$

for all columns $j = 1, 2, \ldots, m, r = 1, 2, \ldots, m$. But those conditions hold if i^*, j^* is a Nash equilibrium. Thus P is a correlated equilibrium.

Now suppose $X^* = (x_1, \ldots, x_n), Y^* = (y_1, \ldots, y_m)$ is a mixed Nash equilibrium with component $x_k > 0$. We show that $P = (p_{ij})$, with $p_{ij} = x_i y_j$ is a correlated equilibrium. Since (X^*, Y^*) is a Nash equilibrium we know that

$$v_I = E_I(X^*, Y^*) = E_I(k, Y^*)$$

for every k for which $x_k > 0$. Thus,

$$E_I(k, Y^*) \geq E_I(q, Y^*), \text{ for all } q = 1, 2, \ldots, n,$$

which says

$$\sum_{j=1}^{m} a_{kj} y_j \geq \sum_{j=1}^{m} a_{qj} y_j, \quad q = 1, 2, \ldots, n.$$

Multiplying both sides by $x_k > 0$ gives

$$\sum_{j=1}^{m} a_{kj} x_k y_j \geq \sum_{j=1}^{m} a_{qj} x_k y_j, \quad q = 1, 2, \ldots, n.$$

This is true for any component of X^* which is positive, but it is also true for any component which is zero (because then both sides are zero). Similarly,

$$\sum_{i=1}^{n} b_{ij} x_i y_j \geq \sum_{i=1}^{n} b_{ir} x_i y_j, \quad r = 1, 2, \ldots, m.$$

This says that $P = (x_i, y_j)$ is indeed a correlated equilibrium.

2.43.a The Nash equilibria are $(X_1 = (0, 1), Y_1 = (1, 0))$, $(X_2 = (1, 0), Y_2 = (0, 1))$, and $X_3 = (\frac{1-\varepsilon}{2-\varepsilon}, \frac{1}{2-\varepsilon}) = Y_3$. The corresponding correlated equilibria are

$$P_1 = \begin{bmatrix} 0 & 0 \\ 1 & 0 \end{bmatrix}, \quad P_2 = \begin{bmatrix} 0 & 1 \\ 0 & 0 \end{bmatrix}, \quad P_3 = \frac{1}{(2-\varepsilon)^2} \begin{bmatrix} (1-\varepsilon)^2 & 1-\varepsilon \\ 1-\varepsilon & 1 \end{bmatrix}$$

2.43.b The linear program we have to solve is

Maximize $2p_{11} + (p_{12} + p_{21})(3 - \varepsilon)$
Subject to
$p_{12}(1 - \varepsilon) \geq p_{11}$
$p_{21} \geq (1 - \varepsilon)p_{22}$
$p_{21}(1 - \varepsilon) \geq p_{11}$
$p_{12} \geq (1 - \varepsilon)p_{22}$
$p_{11} + p_{12} + p_{21} + p_{22} = 1$
$p_{ij} \geq 0.$

Considering the objective function, since $2 \leq 3 - \varepsilon$, the maximum is achieved at $p_{11} = 0, p_{12} + p_{21} = 1$, if the rest of the constraints are satisfied.

If $p_{11} = 0, p_{12} + p_{21} = 1$, then $p_{22} = 0$. The constraints simply reduce to $p_{ij} \geq 0$. Thus, there is a collection of correlated equilibria

$$P = \begin{bmatrix} 0 & p_{12} \\ p_{21} & 0 \end{bmatrix}, \quad p_{12} + p_{21} = 1,$$

all giving the maximum social welfare $3 - \varepsilon$. In particular, P_1 and P_2 are both correlated equilibria maximizing the social welfare. One red light should do the trick.

Solutions for Chapter 3

3.1 Player I has two pure strategies: 1=take action 1 at node 1:1, and 2=take action 2 at node 1:1. Player II has three pure strategies: 1=take action 1 at information set 2:1, 2=take action 2 at 2:1, and 3=take action 3 at 2:1.

The game matrix is

I/II	1	2	3
1	1, 1	0, 3	−1, 0
2	0, 0	−3, −2	4, 1

There is a pure Nash at $(4, 1)$.

3.3.a The extensive form along with the solution is shown in the figure.

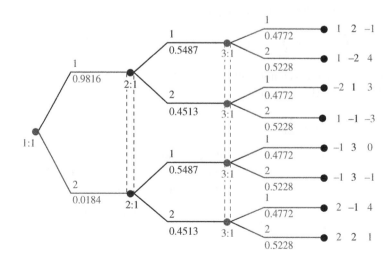

The probability a player takes a branch is indicated below the branch.

3.5.a The game tree follows the description of the problem in a straightforward way. It is considerably simplified if you use the fact that if a player fires the pie and misses, all the opponent has to do is wait until zero paces and winning is certain. Here is the tree.

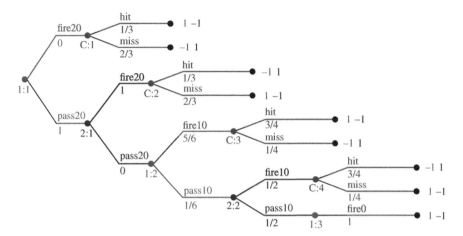

Aggie has 3 pure strategies:

(a1) Fire at 20 paces;

(a2) Pass at 20 paces; If Baggie passes at 20, Fire at 10 paces;

(a3) Pass at 20 paces; If Baggie passes at 20, Pass at 10 paces; If Baggie passes at 10, Fire at 0 paces.

Baggie has 3 pure strategies:

(b1) If Aggie passes, Fire at 20 paces;

(b2) If Aggie passes, Pass at 20 paces; If Aggie passes at 10, Fire at 10 paces;

(b3) If Aggie passes, Pass at 20 paces; If Aggie passes at 10, Pass at 10 paces.

The game matrix is then

Aggie/Baggie	(b1)	(b2)	(b3)
(a1)	$-\frac{1}{3}$	$-\frac{1}{3}$	$-\frac{1}{3}$
(a2)	$\frac{1}{3}$	$\frac{1}{2}$	$\frac{1}{2}$
(a3)	$\frac{1}{3}$	$-\frac{1}{2}$	1

To see where the entries come from, consider (a1) vs (b1). Aggie is going to throw her pie and if Baggie doesn't get hit, then Baggie will throw her pie. Aggie's expected payoff will be

$$(+1)\frac{1}{3} + (-1)\frac{2}{3} = -\frac{1}{3}$$

because Aggie hits Baggie with probability $\frac{1}{3}$ and gets $+1$, but if she misses, then she is going to get a pie in the face.

If Aggie plays (a2) and Baggie plays (b3), then Aggie will pass at 20 paces; Baggie will also pass at 20; then Aggie will fire at 10 paces. Aggies' expected payoff is then

$$(+1)\frac{3}{4} + (-1)\frac{1}{4} = \frac{1}{2}.$$

3.5.b The solution of the game can be obtained from the matrix using dominance. The saddle point is for Aggie to play (a2) and Baggie plays (b1). In words, Aggie will pass at 20 paces, Baggie will fire at 20 paces, with value $v = \frac{1}{3}$ to Aggie. Since Baggie moves second, her best strategy is to take her best shot as soon as she gets it.

Here is a slightly modified version of this game in which both players can miss their shot with a resulting score of 0. The value of the game is still $\frac{1}{3}$ to Aggie. It is still optimal for Aggie to pass at 20, and for Baggie to take her shot at 20. If Baggie misses, she is certain to get a pie in the face.

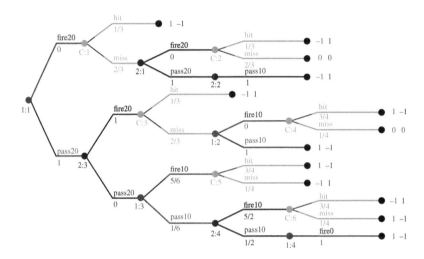

3.7.a

	aa	ab	ba	bb
$AA\_$	-3	-3	2	2
$AB\_$	4	4	-2	-2
B_A	5	0	5	0
B_B	4	0	4	0
B_C	-2	0	-2	0

3.9

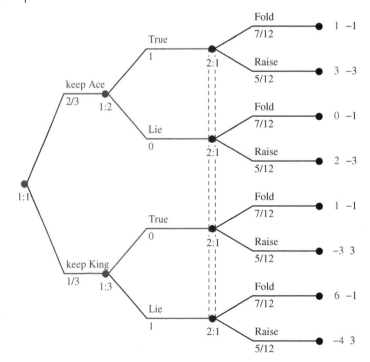

PI should announce I kept an Ace with probability $\frac{2}{3}$ and say it truthfully. With probability $\frac{1}{3}$ PI should announce she kept the King and always lie in the announcement. PII should fold with probability $\frac{7}{12}$ and raise with probability $\frac{5}{12}$. The payoffs are $\frac{11}{6}$ to PI and -1 to PII.

3.11 The backward Nash is : player 1 plays branch 1; player 2 (from node 2:1) plays branch 2. The payoff is $(4, 5)$.

3.17 If we use Gambit to solve this game we get the game matrix

BAT/PM	1	2
111	$-3, -2$	$1, -4$
112	$-3, -2$	$2, 3$
113	$-3, -2$	$-2, 5$
121	$-4, 2$	$1, -4$
122	$-4, 2$	$2, 3$
123	$-4, 2$	$-2, 5$
131	$-2, 4$	$1, -4$
132	$-2, 4$	$2, 3$
133	$-2, 4$	$-2, 5$
2__	$0, 6$	$0, 6$

By dominance, the matrix reduces to the simple game

BAT/PM	1	2
132	$-2, 4$	$2, 3$
2__	$0, 6$	$0, 6$

This game has Nash equilibria at (1) $(0, 6)$ and (2) $(2, 4)$.

The subgame perfect equilibrium is found by backward induction. At node 1:2 BAT will play Out with payoff $(-2, 4)$ for each player. At 1:3, BAT plays Passive with payoffs $(2, 3)$. At 2:1, PM has the choice of either payoff 4 or payoff 3 and hence plays Tough at 2:1. Now player BAT at 1:1 compares $(0, 6)$ with $(-2, 4)$ and chooses Out. The subgame perfect equilibrium is thus BAT plays Out, and PM has no choices to make. The Nash equilibrium at $(0, 6)$ is subgame perfect.

3.19 The game tree is given in the figure.

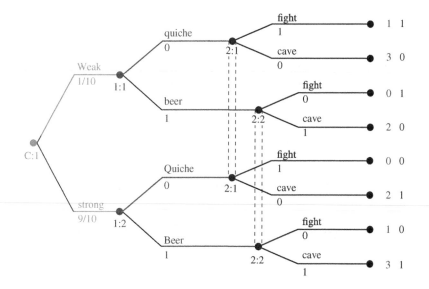

Let's first consider the equivalent strategic form of this game. The game matrix is

Curly/Moe	11	12	21	22
11	$0.1, 0.1$	$0.1, 0.1$	$2.1, 0.9$	$2.1, 0.9$
12	$1, 0.1$	$2.8, 1$	$1.2, 0$	$3, 0.9$
21	$0, 0.1$	$0.2, 0$	$1.8, 1$	$2, 0.9$
22	$0.9, 0.1$	$2.9, 0.9$	$0.9, 0.1$	$2.9, 0.9$

First notice that the first column for player II (Moe) is dominated strictly and hence can be dropped. But that's it for domination.

There are four Nash equilibria for this game. In the first Nash, Curly always chooses to order quiche, while Moe chooses to cave. This is the dull game giving an expected payoff of 2.1 to Curly and 0.9 to Moe. An alternative Nash is Curly always chooses to order a beer and Moe chooses to cave, giving a payoff of 2.9 to Curly and 0.9 to Moe. It seems that Moe is a wimp.

3.21.a The game tree is in the figure.

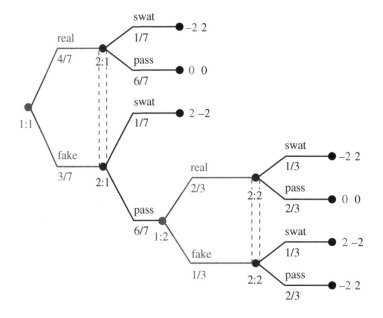

There are 3 pure strategies for Jack: (1) Real, (2) Fake, then Real, and (3) Fake, then Fake. Jill also has 3 pure strategies: (1) Swat, (2) Pass, then Swat, and (3) Pass, then Pass.

Jack/Jill	(1)	(2)	(3)
(1)	-2	0	0
(2)	2	-2	0
(3)	2	2	-2

We can use Gambit to solve the problem, but we will use the invertible matrix theorem instead. We get

$$v = \frac{1}{J_3 A^{-1} J_3^T} = -\frac{2}{7}, X^* = v\, J_3 A^{-1} = \left(\frac{4}{7}, \frac{2}{7}, \frac{1}{7}\right), Y^* = v \cdot A^{-1} J_3^T = \left(\frac{1}{7}, \frac{2}{7}, \frac{4}{7}\right).$$

3.21.b Jack and Jill have the same pure strategies but the solution of the game is quite different. The matrix is

Jack/Jill	(1)	(2)	(3)
(1)	−1	0	0
(2)	2	−1	0
(3)	2	2	−2

For example, (1) vs (1) means Jack will put down the real fly and Jill will try to swat the fly. Jack's expected payoff is

$$(-2)\frac{3}{4} + (2)\frac{1}{4} = -1.$$

The game tree from the first part is easily modified noting that the probability of hitting or missing the fly only applies to the real fly. Here is the figure with Gambit's solution below the branches.

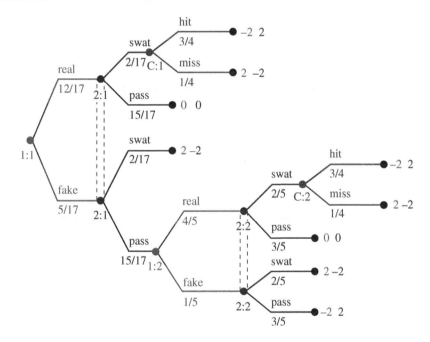

The solution of the game is

$$v = \frac{1}{J_3 A^{-1} J_3^T} = -\frac{2}{17}, X^* = v\, J_3 A^{-1} = \left(\frac{12}{17}, \frac{4}{17}, \frac{1}{17}\right), Y^* = v \cdot A^{-1} J_3^T = \left(\frac{2}{17}, \frac{6}{17}, \frac{9}{17}\right).$$

3.23 The result is $p_{1,1} = 1$ and $p_{i,j} = 0$ otherwise. The correlated equilibrium also gives the action that each player Stops.

3.25 This problem can be tricky because it quickly becomes unmanageable if you try to solve it by considering all possibilities. The key is to recognize that after the first round, if all three stooges survive it is as though the game starts over. If only two stooges survive, it is like a duel between the survivors with only one shot for each player and using the order of play Larry, then Moe, then Curly (depending on who is alive).

According to the instructions in the problem we take the payoffs for each player to be 2, if they are sole survivor, $\frac{1}{2}$ if there are two survivors, 0 if all three stooges survive, and -1 for the particular stooge who is killed. The payoffs at the end of the first round are -1 to any killed stooge in round 1, and the expected payoffs if there are two survivors, determined by analyzing the two person duels first. This is very similar to using backward induction and subgame perfect equilibria.

Begin by analyzing the two person duels Larry vs. Moe, Moe vs. Curly, and Curly vs. Larry. We assume that round one is over and only the two participants in the duel survive. We need that information in order to determine the payoffs in the two person duels. Thus, if Larry kills Curly, Larry is sole survivor and he gets 2, while Curly gets -1. If Larry misses, he's killed by Curly and Curly is sole survivor.

1. Curly versus Larry:

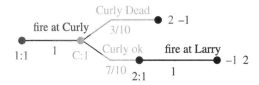

In this simple game we easily calculate the expected payoff to Larry is $-\frac{1}{10}$ and to Curly is $\frac{11}{10}$.

2. Moe versus Curly:

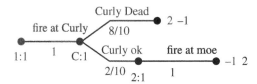

The expected payoff to Moe is $\frac{7}{5}$, and to Curly is $-\frac{2}{5}$.

3. Larry versus Moe:

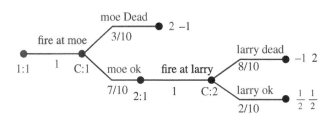

In this case, if Larry shoots Moe, Larry gets 2 and Moe gets -1, but if Larry misses Moe, then Moe gets his second shot. If Moe misses Larry, then there are two survivors of the true, and each gets $\frac{1}{2}$. The expected payoff to Moe is $\frac{89}{100}$, and to Curly is $-\frac{11}{100}$.

Next consider the tree for the first round incorporating the expected payoffs for the second round.

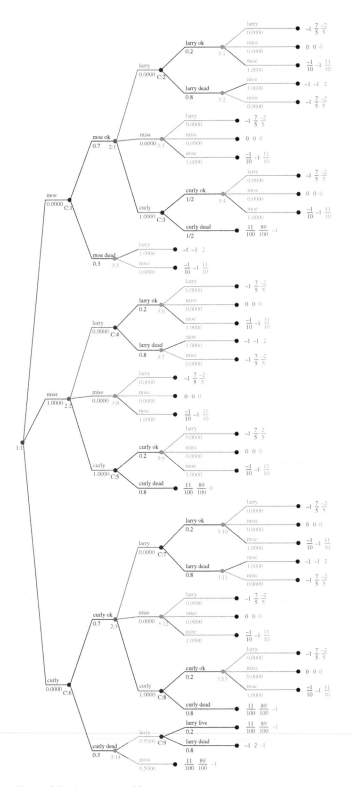

Figure 1.1 Larry versus Moe

The expected payoff to Larry is 0.68, Moe is 0.512, and Curly is 0.22. The Nash equilibrium says that Larry should deliberately miss with his first shot. Then Moe should fire at Curly; if Curly lives, then Curly should fire at Moe–and Moe dies. In the second round, Larry has no choice but to fire at Curly, and if Curly lives, then Curly will kill Larry. Does it make sense that Larry will deliberately miss in the first round? Think about it. If Larry manages to kill Moe in the first round, then surely Curly kills Larry when he gets his turn. If Larry manages to kill Curly, then with 80% probability, Moe kills Larry. His best chance of survival is to give Moe and Curly a chance to shoot at each other. Since Larry should deliberately miss, when Moe gets his shot he should try to kill Curly because when it's Curly's turn, Moe is a dead man. To find the probability of survival by each player assuming the Nash equilibrium is being played we calculate:

1. For Larry:

$Prob$(Larry Survives) = $Prob$(Moe Kills Curly in round 1 ∩ Larry Kills Moe in round 2)

$\quad = Prob$(Moe Kills Curly in round 1)

$\quad \times Prob$(Larry Kills Moe in round 2)

$\quad = 0.8 \times 0.3 = 0.24.$

2. For Moe:

$Prob$(Moe Survives) = $Prob$(Moe Kills Curly in round 1 ∩ Moe Kills Larry in round 2)

$\quad = Prob$(Moe Kills Curly in round 1)

$\quad \times Prob$(Moe Kills Larry in round 2)

$\quad = 0.8 \times 0.8 = 0.64.$

3. For Curly:

$Prob$(Curly Survives) = $Prob$(Moe misses Curly in round 1 ∩ Curly Kills Larry in round 2)

$\quad = Prob$(Moe misses Curly in round 1)

$\quad \times Prob$(Curly Kills Larry in round 2)

$\quad = 0.7 \times 1 = 0.7.$

3.27 Here's the figure.

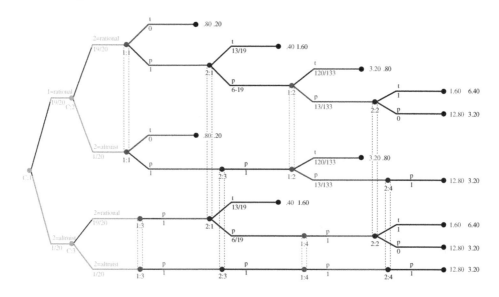

The expected payoffs to each player are $\frac{69}{50}$ for I and $\frac{276}{175}$ for II. The Nash equilibrium is the following:

If both players are rational, I should Pass, then II should Pass with probability $\frac{6}{19}$, then I should Pass with probability $\frac{13}{133}$, then player II should Pass with probability 0.

If I is rational and II is an altruist, then I should Pass with probability 1, then II Passes, then I Passes with probability $\frac{13}{133}$, then II Passes.

If I is an altruist and II is rational, then I Passes, II Passes with probability $\frac{6}{19}$, I Passes, and II Passes with probability 0.

If they are both altruists, they each Pass at each step and go directly to the end payoff of 12.80 for I and 3.20 for II.

3.31 The value of this zero sum game is $v = \frac{4}{9}$. The optimal strategies are player I always chooses wheel 1, and player II always chooses wheel 2. The game matrix is

I/II	111	112	121	122	211	212	221	222
1	$\frac{4}{9}$	$\frac{4}{9}$	$\frac{4}{9}$	$\frac{4}{9}$	$\frac{5}{9}$	$\frac{5}{9}$	$\frac{5}{9}$	$\frac{5}{9}$
2	$-\frac{4}{9}$	$-\frac{4}{9}$	$-\frac{14}{9}$	$-\frac{14}{9}$	$-\frac{4}{9}$	$-\frac{4}{9}$	$-\frac{14}{9}$	$-\frac{14}{9}$
3	$-\frac{5}{9}$	$\frac{14}{9}$	$-\frac{5}{9}$	$\frac{14}{9}$	$-\frac{5}{9}$	$\frac{14}{9}$	$-\frac{5}{9}$	$\frac{14}{9}$

It is easy to check that the numbers in the matrix are the actual expected payoffs. For instance, if player I uses wheel 1 and player II chooses wheel 2, the expected payoff is $(-3 - 5 - 7 + 4 - 5 - 7 + 9 + 9 + 9)/9 = \frac{4}{9}$.

3.35 The extensive form and matrix are

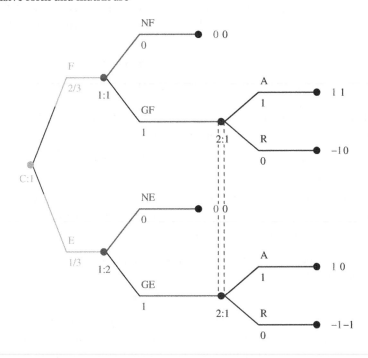

The strategic form of this game is

	A	R
NF NE	$(0,0)$	$\boxed{(0,0)}$
NF GE	$\left(\frac{1}{3},0\right)$	$\left(-\frac{1}{3},-\frac{1}{3}\right)$
GF NE	$\left(\frac{2}{3},\frac{2}{3}\right)$	$\left(-\frac{2}{3},0\right)$
GF GE	$\boxed{\left(1,\frac{2}{3}\right)}$	$\left(-1,-\frac{1}{3}\right)$

Let p =probability I plays NF if friend, and q =probability I plays NE if I is an enemy. Let r =probability II accepts. Then

$$\mu_{2:1} = \frac{2/3(1-p)}{2/3(1-p)+1/3(1-q)} = \frac{2(1-p)}{2(1-p)+1-q}$$

$$E_{II}(r,2:1) = \frac{2(1-p)}{2(1-p)+1-q}r + \frac{1-q}{2(1-p)+1-q}(r-1) = r - \frac{1-q}{2(1-p)+1-q}$$

which is maximized when $r = 1$, i.e., always Accept a gift. Next

$$E_I(p,1:1) = 1-p \implies p = 0, \text{ and } E_I(q,1:2) = (1-q)(r(1)+(1-r)(-1)) = 1-q$$

which is maximized when $q = 0$. Player I should always give a gift if I is a friend and player II should always accept the gift. We have that $p = 0, q = 0, r = 1$ is a PBE, i.e., a sequential equilibrium consistent with beliefs. The NEs $((NF,NE),R)$ and $((GF,GE),A)$ are indeed Nash equilibria but only the latter is a PBE.

3.37 Begin by calculating the expected payoff to player II at info sets 2:1 and 2:2. At info set 2:1 let p =probability of playing fight. We have the beliefs for 2:1 as $\mu_1 = (1/2)(2/3)/((1/2)(2/3) + 0(1/3)) = 1$ and $\mu_2 = 0$.

$$E_{II}(p,2:1) = \mu_1(p(1)+(1-p)0) + \mu_2(p(0)+(1-p)(1)) = p$$

which is maximized when $p = 1$ meaning that player II should always fight if player I is spotted with a Quiche.

For 2:2 the beliefs are $\mu_1 = (2/3)(1/2)/((2/3)(1/2) + (1/3)1) = 1/2$ and $\mu_2 = 1/2$. If q is the probability player II plays Fight we have

$$E_{II}(q,2:2) = \tfrac{1}{2}(q(1)+(1-q)0) + \tfrac{1}{2}(q(0)+(1-q)1) = \tfrac{1}{2},$$

which is independent of q. This means that taking $q = \frac{1}{2}$ is consistent and so playing Fight and Cave with equal probability if player I has a Beer is part of a sequential equilibrium. Moving next to 1:1, let a =the probability player I plays Quiche. We have

$$E_I(a,1:1) = a(1) + (1-a)(1) = 1,$$

which is independent of a and so $a = \frac{1}{2}$ is a legitimate equilibrium. Finally, if b = probability player I plays Quiche at 1:2, we have

$$E_I(b,1;2) = b(0) + (1-b)(\tfrac{1}{2} + \tfrac{3}{2}) = (1-b)2$$

which is maximized when $b = 0$. Consequently player I will never play Quiche if Strong. Concluding, we have shown that the NE is indeed a sequential equilibrium

3.39 If we model this as an extensive game in which Nature is a player which chooses which game is played, with the information given we have

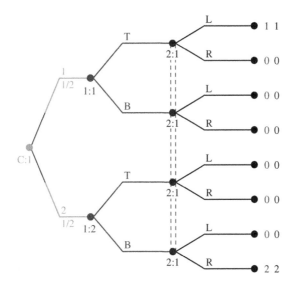

The strategic form of this game is

	L	R
TT	$\left(\frac{1}{2},\frac{1}{2}\right)$	$(0,0)$
TB	$\left(\frac{1}{2},\frac{1}{2}\right)$	$(1,1)$
BT	$(0,0)$	$(0,0)$
BB	$(0,0)$	$(1,1)$

Let p =probability PI plays T in game 1 and q =probability PI plays T in game 2. Also, let r =probability PII plays L. Since PII does not know which game is played r is the only unknown for PII. The beliefs at 2:1 for PII are

$$\mu_1 = \frac{p}{2}, \mu_2 = \frac{1-p}{2}, \mu_3 = \frac{q}{2}, \mu_4 = \frac{1-q}{2}.$$

Then

$$E_{II}(r, 2:1) = r(\frac{p}{2} + q - 1) + 1 - q \implies r^* = \begin{cases} 1, & \frac{p}{2} + q > 1 \\ 0, & \frac{p}{2} + q < 1, \\ \text{arb.,} & \frac{p}{2} + q = 1 \end{cases}$$

$$E_I(p, 1:1) = rp \implies p^* = \begin{cases} 1, & r > 0 \\ \text{arb.,} & r = 0 \end{cases}$$

$$E_I(q, 1:1) = 2(1-q)(1-r) \implies q^* = \begin{cases} 0, & r < 1 \\ \text{arb.,} & r = 1 \end{cases}$$

We consider cases.

1. If $0 < r^* < 1 \implies q^* = 0$. If also $p^* = 1$ then $\frac{p^*}{2} + q^* = \frac{1}{2} \neq 1$ so this case cannot give a PBE.

2. If $r^* = 1 \implies q^* = $ arb., $p^* = 1 \implies \frac{p^*}{2} + q^* = \frac{1}{2} + q^* > 1 \implies q^* < \frac{1}{2}$. In addition, if $q^* = \frac{1}{2}$, then $\frac{p^*}{2} + q^* = 1$ and this is consistent with $r^* = 1$. Therefore, we conclude that $r^* = 1, p^* = 1, q^* \geq \frac{1}{2}$ all give PBEs. In particular, PI plays T always in game 1 and T in game 2 with any probability $q^* \geq \frac{1}{2}$, and PII plays L no matter what.

3. If $r^* = 0$, then $q^* = 0$ and $\frac{p^*}{2} < 1$ always. Therefore, $r=0, q^* = 0, 0 \le p^* \le 1$, also results in a PBE.

3.41 The game tree with the optimal strategies is

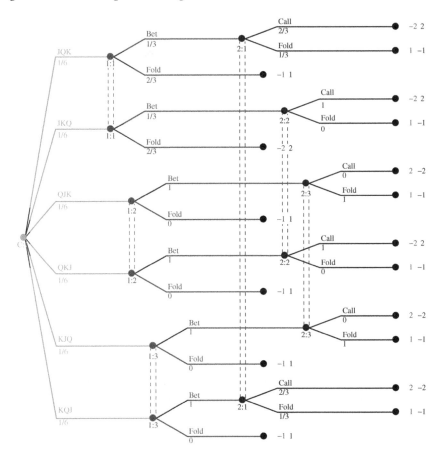

At each info set for PII the belief at the top node is $\frac{1}{4}$. At each info set for PI the belief is $\frac{1}{2}$ at the top node.

3.43.a We will use backward induction. Hitler in the final moves decides on Peace or War.

If Hitler tells the Truth, his best response to Chamberlain Caving is Peace. The best response to Chamberlain standing Firm is War.

If Hitler is Lying, his best response to Chamberlain Caving is War. His best response to Chamberlain standing Firm is also War.

Chamberlain is unaware of Hitler's type so we need to calculate his expected utility for selecting Caving and standing Firm.

If Chamberlain Caves, his expected payoff is $.6 \times 3 + .4 \times 1 = 2.2$. If Chamberlain stands Firm, his expected payoff is $.6 \times 2 + .4 \times 2 = 2$. This means Chamberlain's best strategy is to Cave.

Hitler's best strategy depends on if Nature chooses for him to Lie or tell the Truth.

If he tells the Truth, he should play Peace if Chamberlain Caves and War if Chamberlain Stands Firm.

If he Lies, then he should play War after Chamberlain Caves, and War if Chamberlain stands Firm.

What Actually Happened? The meeting took place in Munich on September 30, 1938. Chamberlain Caved and Hitler played War (when he invaded Poland September 1, 1939). It turns out that Hitler was lying throughout the Munich meeting.

3.43.b The extensive for of this game is now

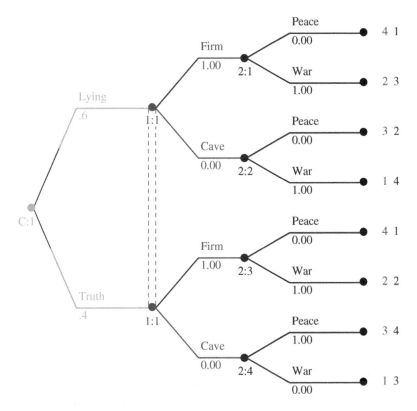

The Bayesian Equilibrium is for Chamberlain to stand Firm always and for Hitler (who is aware of his type, i.e., whether he is lying or not), to play War if he is Lying or if Chamberlain plays Firm. Chamberlain's payoff is 2 and Hitler's payoff is 2.6.

3.45 We have the connection $\mu = \frac{b}{b+c}$.

The expected payoff to player II if player II is then

$$E_{II}(p, 2:1) = \mu(p \cdot 1 + (1-p)2) + (1-\mu)(p \cdot 0 + (1-p)(-1)) = p(1-2\mu) + 3\mu - 1.$$

This is maximized when $p = p^* = \begin{cases} 1, & \text{if } \mu < \frac{1}{2}, \\ 0, & \text{if } \mu > \frac{1}{2}, \\ \text{arb.}, & \text{if } \mu = \frac{1}{2}. \end{cases}$ Equivalently, since $\mu = \frac{b}{b+c}$, we have

$$p^* = \begin{cases} 1, & \text{if } b < c, \\ 0, & \text{if } b > c, \\ \text{arb.}, & \text{if } b = c. \end{cases}$$

The expected payoff to player I is then

$$E_I(a, b, c) = 0 \cdot a + b((-1)p^* + 1(1-p^*)) + c(2p^* + (-2)(1-p^*)) = b(1-2p^*) + c(4p^* - 2).$$

This is maximized when

$$b = \begin{cases} 1, & p^* < \frac{1}{2}, \\ 0, & p^* > \frac{1}{2} \\ \text{arb.}, & p^* = \frac{1}{2}. \end{cases}$$

Since $a + b + c = 1$ this forces $b = 1 \implies a = c = 0$, and $b = 0 \implies c = 1, a = 0$ and $p^* = \frac{1}{2} \implies b = c$. Also with $p^* = \frac{1}{2}, E_I(a, b, b) = 0$.

Combining these results we have

$$b = 1 \implies p^* < \frac{1}{2} \Leftrightarrow b > c \qquad\qquad \implies b = 1, c = 0, a = 0, p^* = 0$$

$$b = 0 \implies p^* > \frac{1}{2} \Leftrightarrow b < c \qquad\qquad \implies b = 0, c = 1, a = 0, p^* = 1$$

$$0 < b < 1 \implies p^* = \frac{1}{2} \Leftrightarrow b = c.$$

3.47 The NEs are (i)(A, r), (ii)(B, l), (iii)(C, R, a). The PBE is (C, R, a) with payoff 8 to player I and $7 = \frac{1}{4}4 + \frac{3}{4}8$ to player II.

3.49 There are three BEs.

1. If nature chooses y, player 2 plays F with probability 1 and player 1 chooses F with probability 1. If nature chooses n, player 2 plays B with probability 1 and player 1 chooses F with probability 1. The beliefs at each node of 1:1 are $\frac{1}{2}, 0, 0, \frac{1}{2}$. The payoff to 1 is 1 and to player 2 is $\frac{3}{2}$.

2. If nature chooses y, player 2 plays F with probability $\frac{2}{3}$ and player 1 chooses F with probability $\frac{2}{3}$. If nature chooses n, player 2 plays B with probability 1 and player 1 chooses F with probability $\frac{2}{3}$. The beliefs at each node of 1:1 are $\frac{1}{3}, \frac{1}{6}, 0, \frac{1}{2}$. The payoff to 1 is $\frac{2}{3}$ and to player 2 is 1.

3. If nature chooses y, player 2 plays B with probability 1 and player 1 chooses F with probability $\frac{1}{3}$. If nature chooses n, player 2 plays F with probability $\frac{2}{3}$ and player 1 chooses F with probability $\frac{1}{3}$. The beliefs at each node of 1:1 are $0, \frac{1}{2}, \frac{1}{3}, \frac{1}{6}$. The payoff to 1 is $\frac{2}{3}$ and to player 2 is 1.

For instance, to check the beliefs of BE(3), if $p_i = $ probability of being at node $i = 1, \dots, 4$ in info set 1:1, the belief at node i consistent with the equilibrium is $\mu_i = \frac{p_i}{p_1 + p_2 + p_3 + p_4}$ so

$$\mu_1 = \frac{0}{0 + 1/2 + 1/3 + 1/6} = 0, \mu_2 = \frac{1/2}{1} = \frac{1}{2}, \text{ and } \mu_3 = \frac{1}{3}, \mu_4 = \frac{1}{6}.$$

3.51.b

$$\mu_{2:1} = \frac{\frac{1}{4}p}{\frac{1}{4}p + \frac{1}{4}1} = \frac{p}{p+1}.$$

Also, $\mu_{2:2} = \frac{p}{p+1}$. Similarly, for player 1, $\mu_{1:1} = \frac{1/4}{(1/4+1/4)} = \frac{1}{2} = \mu_{1:2}$.

3.51.c

$$E_{II}(q, 2 : 1) = \mu_{2:1}(-4(1 - q)) + (1 - \mu_{2:1})(-4(1 - q) - 10q) = q(10\mu_{2:1} - 6) - 4,$$

which is maximized by $q = \begin{cases} 1, & \mu_{2:1} > \frac{3}{5} \\ 0, & \mu_{2:1} < \frac{3}{5} \\ \text{arb.}, & \mu_{2:1} = \frac{3}{5}. \end{cases}$ But $\mu_{2:1} = \frac{p}{p+1} > \frac{3}{5}$ if and only if $p > \frac{3}{2}$ which

is impossible. Therefore the maximizing q is $q = 0$ which means player 2 will fold at 2:1. Similarly,

$$E_{II}(q, 2 : 2) = \mu_{2:2}(10) + (1 - \mu_{2:2})(0) = 10\mu_{2:2},$$

which is independent of q so player 2 always folds unless she holds an Ace and she must bet. For player 1, we have

$$E_I(p, 1 : 1) = \tfrac{1}{2}((1 - p)(-4) + 4p) + \tfrac{1}{2}(-4(1 - p) - 10p) = p - 4$$

which is maximized for $p = 1$ which means player 1 should always bet even with a King.

3.51.d The PBE is player I should always bet and player II should fold with a King and bet with an Ace. The expected payoffs to each player are $-\tfrac{1}{2}$ to player I and $\tfrac{1}{2}$ to player II.

3.51.e (a) If PI gets an Ace she should always Bet and PII should Call with an Ace, Call with probability $\tfrac{5}{7}$ with a King, and Fold with a Queen.

If PI gets a King she should always Bet and PII should Call with an Ace, Call with probability $\tfrac{5}{7}$ with a King, and Fold with a Queen.

If PI gets a Queen she should Bet with probability $\tfrac{1}{7}$ and PII should Call with an Ace, Call with probability $\tfrac{5}{7}$ with a King, and Fold with a Queen.

The payoff to PI is $-\tfrac{32}{63}$.

3.51.f Player I should always Bet with a King. Player II will always Fold with a King. The payoff to PI is $-\tfrac{2}{3}$.

3.53 For the 2 players in a dispute recall that in the case $p = \tfrac{1}{3}$ the BE was (F, FY). Now we take a general p. We know that player II knows her own type (strong or weak) but player I does not know II's type. Let

- β = the probability player I will fight.
- α = the probability player II will fight if he is strong
- γ = the probability player II will fight if he is weak.

		fight α	yield $1 - \alpha$
II is strong prob p	fight β	$(-1, 1)$	$(1, 0)$
	yield $1 - \beta$	$(0, 1)$	$(0, 0)$

		fight γ	yield $1 - \gamma$
II is weak prob $1 - p$	fight β	$(1, -1)$	$(1, 0)$
	yield $1 - \beta$	$(0, 1)$	$(0, 0)$

The beliefs of PI at 1:1 are given by

$$\mu_1 = p\,\alpha, \mu_2 = p(1 - \alpha), \mu_3 = (1 - p)\gamma, \mu_4 = (1 - p)(1 - \gamma)$$

Then,

$$E_{II}(\alpha, 2 : 1) = \alpha(\beta + 1(1 - \beta)) = \alpha \implies \alpha = 1.$$

$$E_{II}(\gamma, 2 : 2) = \gamma(-\beta + 1 - \beta) = \gamma(1 - 2\beta) \implies \gamma = \begin{cases} 1, & \beta < \tfrac{1}{2} \\ 0, & \beta > \tfrac{1}{2}, \\ \text{arb.}, & \beta = \tfrac{1}{2} \end{cases}$$

and

$$E_I(\beta, 1 : 1) = \mu_1(-\beta) + \mu_2\beta + \mu_3\beta + \mu_4\beta = \beta(1 - 2\mu_1) = \beta(1 - 2p).$$

The fact that $\alpha = 1$ means II will always fight if strong. Now, if $1 - 2p > 0$, then $\beta = 1$ and so PI will always fight if the chances PII is strong is less than 50%. If $p > \tfrac{1}{2}$, then $\beta = 0$ and PI will always yield if the chance PII is strong is more than 50%.

If $p < \frac{1}{2}$ then $\beta = 1$ and so $\gamma = 0$. If the chances PI is strong is less than 50% then PII will always fight if strong but yield if weak and PI will always fight. If $p > \frac{1}{2}$, then $\beta = 0$ and $\gamma = 1$ so PII will fight if strong or if weak, PI will yield. If $p = \frac{1}{2}$ the game matrix is

	FF	FY	YF	YY
F	$(0,0)$	$(0,\frac{1}{2})$	$(1,-\frac{1}{2})$	$(1,0)$
Y	$(0,1)$	$(0,\frac{1}{2})$	$(0,\frac{1}{2})$	$(0,0)$

The strategy is for PII to always fight and PI to fight or yield with equal probability.

Solutions for Chapter 4

4.1.a Since (q_1^*, q_2^*) maximizes both u_1 and u_2 we have

$$u_1(q_1^*, q_2^*) = \max_{(q_1, q_2)} u_1(q_1, q_2), \quad \text{and} \quad u_2(q_1^*, q_2^*) = \max_{(q_1, q_2)} u_2(q_1, q_2).$$

Thus,

$$u_1(q_1^*, q_2^*) \geq u_1(q_1, q_2^*) \quad \text{and} \quad u_2(q_1^*, q_2^*) \geq u_2(q_1^*, q_2),$$

for every $q_1 \neq q_1^*, q_2 \neq q_2^*$. Thus (q_1^*, q_2^*) automatically satisfies the definition of a Nash equilibrium. A maximum of both payoffs is a much stronger requirement than a Nash point.

4.1.b Consider $u_1(q_1, q_2) = q_2^2 - q_1^2$ and $u_2(q_1, q_2) = q_1^2 - q - 2^2)$ and $-1 \leq q_1 \leq 1, -1 \leq q_2 \leq 1$.

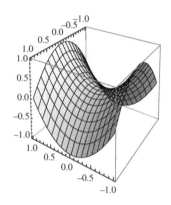

There is a unique Nash equilibrium at $(q_1^*, q_2^*) = (0, 0)$ since

$$\frac{\partial u_1(q_1, q_2)}{\partial q_1} = -2q_1 = 0, \quad \text{and} \quad \frac{\partial u_2(q_1, q_2)}{\partial q_2} = -2q_2 = 0$$

gives $q_1^* = q_2^* = 0$ and these points provide a maximum of each function with the other variable fixed because the second partial derivatives are $-2 < 0$. On the other hand,

$$u_1(q_1^*, q_2^*) = 0, \quad \max_{(q_1, q_2)} u_1(q_1, q_2) = 1, \quad \text{and} \quad u_2(q_1^*, q_2^*) = 0, \quad \max_{(q_1, q_2)} u_2(q_1, q_2) = 1.$$

Thus $(q_1^*, q_2^*) = (0, 0)$ maximizes **neither** of the payoff functions. Furthermore, there does not exist a point which maximizes **both** u_1, u_2 at the same point (q_1, q_2).

4.3.a By taking derivatives we get the best response functions

$$BR_c(y) = \begin{cases} 1, & \text{if } 1 \leq y \leq \sqrt{2}; \\ \frac{y^2}{2}, & \text{if } \sqrt{2} \leq y \leq \sqrt{6}; \\ 3, & \text{if } \sqrt{6} \leq y \leq 3. \end{cases},$$

$$BR_s(x) = \begin{cases} 3, & \text{if } 1 \leq x \leq \frac{4}{3}; \\ \frac{4}{x}, & \text{if } \frac{4}{3} \leq x \leq 3; \end{cases}$$

The graphs of the rational reaction sets are

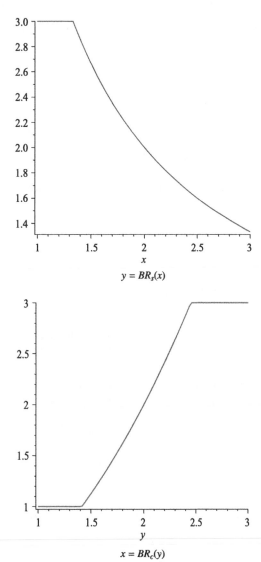

$$y = BR_s(x)$$

$$x = BR_c(y)$$

4.3.b The rational reaction sets intersect at the unique point $(x^*, y^*) = (2, 2)$, which is the unique pure Nash equilibrium.

4.5.a The payoff function for each player $i = 1, 2, \ldots, N$, is

$$u_i(r_1, \ldots, r_N) = h(r_i) - (f(r_i) + g(R - r_i)) = \sqrt{r_i} - 2r_i^2 - (R - r_i)^2,$$

where $R = r_1 + r_2 + \cdots + r_N$.

4.5.b Taking a partial derivative of u_i with respect to r_i gives

$$\frac{\partial u_i}{\partial r_i} = \frac{1}{2}\frac{1}{\sqrt{r_i}} - 4r_i = 0$$

which implies $r_i = \frac{1}{4}$. Since $\frac{\partial^2 u_i}{\partial r_i^2} < 0$ we conclude that $(r_1, \ldots, r_N) = (\frac{1}{4}, \ldots, \frac{1}{4})$ is the Nash equilibrium. The total amount of resources used by all the players is then $R = \frac{N}{4}$.

When $N = 12$, the total resources used will be $R = 3$. The payoff to each player when $N = 12$ is

$$u_i(\frac{1}{4}, \ldots, \frac{1}{4}) = \frac{1}{2} - \frac{1}{8} - \left(\frac{11}{4}\right)^2 = -7.396.$$

4.5.c Set

$$F(R) = N(h(\frac{R}{N}) - f(\frac{R}{N}) - g(R - \frac{R}{N})) = \sum_{i=1}^{N} u_i\left(\frac{R}{N}, \ldots, \frac{R}{N}\right)$$

We have to find the maximum of F. Take a derivative with respect to R and set to zero:

$$F'(R) = \left(4 - \frac{6}{N} - 2N\right)R + \frac{1}{2\sqrt{\frac{R}{N}}} = 0$$

After some algebra we get

$$R^s = \frac{N}{\left(2(4 + 2(N-1))^2\right)^{2/3}}.$$

Since $F''(R) = 4 - \frac{6}{N} - 2N - \frac{\sqrt{N}}{4R^{3/2}} < 0$ we know that R^s provides a maximum.

When $N = 12$ we get $R^s = 0.192547$ and the value of the maximum social welfare is $F(R^s) = 1.14004$.

4.7 According to the solution of the median voter problem in Example 4.1 we need to find the median of X. To do that we solve for γ in the equation

$$\int_0^\gamma f(x) \, dx = \int_0^\gamma -1.23x^2 + 2x + 0.41 \, dx = \frac{1}{2}.$$

This equation becomes $-0.41\gamma^3 + \gamma^2 + 0.41\gamma - 0.5 = 0$ which implies $\gamma = 0.585$. The Nash equilibrium position for each candidate is $(\gamma^*, \gamma^*) = (0.585, 0.585)$. With that position, each candidate will get 50% of the vote.

4.9 The best responses are easily found by solving $\frac{\partial u_1}{\partial x} = 0$ and $\frac{\partial u_2}{\partial y} = 0$. The result is

$$x(y) = \frac{4+y}{2}, \quad y(x) = \frac{4+x}{2}$$

and solving for x and y results in $x = 4, y = 4$ as the Nash equilibrium. The payoffs are $u_1(4, 4) = u_2(4, 4) = 16$.

4.11 Calculate the derivatives and set to zero: $\frac{\partial u_1}{\partial q_1} = w_1 - 2q_1 - q_2 = 0 \implies q_1(q_2) = \frac{w_1 - q_2}{2}$ is the best response. Similarly, $q_2(q_1) = \frac{w_2 - q_1}{2}$. The NE is then $q_1^* = \frac{2w_1 - w_2}{3}, q_2^* = \frac{2w_2 - w_1}{3}$, assuming that $2w_1 > w_2, 2w_2 > w_1$. If $w_1 = w_2 = w$, then $q_1 = q_2 = \frac{w}{3}$ and each citizen should donate one-third of their wealth.

4.13

$$\int_0^5 \left(500 - \frac{t^3}{3}\right) p(t)dt - (800 + \frac{5^3}{3}) \int_5^\infty p(t)dt.$$

4.13.a The payoff to C1 is

$$\int_0^5 \left(500 - \frac{t^3}{3}\right) 2e^{-2t}dt - (800 + \frac{5^3}{3}) \int_5^\infty 2e^{-2t}dt = 499.692.$$

4.13.b If C2 quits at time $s > 0$,

$$\int_0^s \left(500 - \frac{t^3}{3}\right) p(t)dt - (800 + \frac{s^3}{3}) \int_s^\infty p(t)dt = \text{Constant}$$

4.13.c The first differentiation gives after simplifying

$$1300p(s) - s^2 \int_s^\infty p(t)\,dt = 0.$$

The second differentiation gives,

$$-2s \int_s^\infty p(t)\,dt + s^2p(s) + 1300p'(s) = -\frac{(t^3 - 2600)}{t}p(t) + 1300p'(t) = 0.$$

4.13.d The equation is $p'(t) + \frac{t^3 - 2600}{1300t}p(t) = 0$ which is a linear first order equation which may be solved using integrating factors. The solution is $p(t) = Ct^2 e^{-t^3/3900}$. Using the condition $\int_0^\infty p(t)\,dt = 1$ we get $C = 1/1300$ and so $p(t) = \frac{1}{1300}t^2 e^{-t^3/3900}$.

4.15 The payoffs are

$$u_1(t_1, t_2) = \begin{cases} v_1 - c_1 t_2 & \text{if } t_1 > t_2; \\ -c_1 t_1 & \text{if } t_1 < t_2; \\ \frac{v_1}{2} - c_1 t_1 & \text{if } t_1 = t_2. \end{cases} \qquad u_2(t_1, t_2) = \begin{cases} v_2 - c_2 t_1 & \text{if } t_2 > t_1; \\ -c_2 t_2 & \text{if } t_2 < t_1; \\ \frac{v_2}{2} - c_2 t_2 & \text{if } t_1 = t_2. \end{cases}$$

First, calculate $\max_{t_1 \geq 0} u_1(t_1, t_2)$, and $\max_{t_2 \geq 0} u_2(t_1, t_2)$. For example

$$\max_{t_1 \geq 0} u_1(t_1, t_2) = \begin{cases} v_1 - c_1 t_2 & \text{if } t_2 < \frac{v_1}{c_1}; \\ 0 & \text{if } t_2 > \frac{v_1}{c_1}; \\ 0 & \text{if } t_2 = \frac{v_1}{c_1}. \end{cases}$$

and the maximum is achieved at the set valued function

$$t_1^*(t_2) = \begin{cases} (t_2, \infty) & \text{if } t_2 < \frac{v_1}{c_1}; \\ 0 & \text{if } t_2 > \frac{v_1}{c_1}; \\ 0 \cup (q_2, \infty), & \text{if } t_2 = \frac{v_1}{c_1}. \end{cases}$$

This is the best response of country 1 to t_2. Next calculate the best response to t_1 for country 2, $t_2^*(t_1)$:

$$t_2^*(t_1) = \begin{cases} (t_1, \infty) & \text{if } t_1 < \frac{v_2}{c_2}; \\ 0 & \text{if } t_1 > \frac{v_2}{c_2}; \\ 0 \cup (t_1, \infty) & \text{if } t_1 = \frac{v_2}{c_2}. \end{cases}$$

If you now graph these sets on the same set of axes, the Nash equilibrium are points of intersection of the sets. The result is

$$(t_1^*, t_2^*) = \begin{cases} \text{either } t_1^* = 0 \text{ and } t_2^* \geq \frac{v_1}{c_1} \\ \text{or } t_2^* = 0 \text{ and } t_1^* \geq \frac{v_2}{c_2} \end{cases}$$

For example, this says that either (1) country 1 should concede immediately and country 2 should wait until first time $\frac{v_1}{c_1}$, or (2) country 2 should concede immediately and country 1 should wait until first time $\frac{v_2}{c_2}$.

4.17.a If three players take $A \to B \to C$, their travel time is 51. No player has an incentive to switch. If 4 players take this path, their travel time is 57 while the two players on $A \to D \to C$ travel only 45, clearly giving an incentive for a player on $A \to B \to C$ to switch.

4.17.b If 3 cars take the path $A \to B \to D \to C$ their travel time is 32 while if the other 3 cars stick with $A \to D \to C$ or $A \to B \to C$, the travel time for them is still 51, so they have an incentive to switch. If they all switch to $A \to B \to D \to C$, their travel time is 62. If 5 cars take $A \to B \to D \to C$ their travel time is 52 while the one car that takes $A \to D \to C$ or $A \to B \to C$ has travel time $33 + 31 = 64$. Notice that no player would want to take $A \to D \to B \to C$ because the travel time for even one car is 66. Continuing this way we see that the only path that gives no player an incentive to switch is $A \to B \to D \to C$. That is the Nash equilibrium and it results in a travel time which is greater than before the new road was added.

4.19 We have the derivatives

$$\frac{\partial u_1}{\partial x} = -C_1 - V\frac{x}{(x+y)^2} + V\frac{1}{x+y}, \quad \frac{\partial u_2}{\partial y} = -C_2 - V\frac{y}{(x+y)^2} + V\frac{1}{x+y},$$

and

$$\frac{\partial^2 u_1}{\partial x^2} = -2V\frac{y}{(x+y)^3} < 0, \quad \frac{\partial^2 u_2}{\partial y^2} = -2V\frac{x}{(x+y)^3} < 0.$$

and hence concavity in the appropriate variable for each payoff function. Solving the first derivatives set to zero gives

$$x^* = \frac{C_2 V}{(C_1 + C_2)^2}, \quad y^* = \frac{C_1 V}{(C_1 + C_2)^2}.$$

Then

$$u_1(x^*, y^*) = \frac{C_2^2 V}{(C_1 + C_2)^2} \quad \text{and} \quad u_2(x^*, y^*) = \frac{C_1^2 V}{(C_1 + C_2)^2}.$$

4.21 The payoff function can be written as

$$u_i(x_1, \ldots, x_{100}) = \begin{cases} 1, & \text{if } x_i \text{ closest to } \frac{2}{3}\bar{x}; \\ \frac{1}{2}, & \text{if } x_i = x_j, i \neq j, \ x_i \text{ closest to } \frac{2}{3}\bar{x}; \\ \frac{1}{3}, & \text{if } x_i = x_j = x_k, i \neq j \neq k, \ x_i \text{ closest to } \frac{2}{3}\bar{x}; \\ \vdots, & \\ \frac{1}{100}, & \text{if } x_1 = x_2 = \cdots = x_{100}. \end{cases}$$

We need to show that

$$u_i(1, 1, 1, \cdots, 1) \geq u_i(x_i, 1_{-i}), \quad x_i = 2, \ldots, 100.$$

Recall that 1_{-i} denotes that all the players except player i are playing 1. Note that since $\bar{x} = 1, u_i(1, 1, \ldots, 1) = \frac{1}{100}$ and $|1 - \frac{2}{3}\bar{x}| = \frac{1}{3}$. What is a better x_i for player i to switch to? In order to be better it must satisfy

$$|x_i - \frac{2}{3}\bar{x}| = |x_i - \frac{2}{3}\frac{99 + x_i}{100}| < \frac{1}{3}.$$

Note that $x_i - \frac{2}{3}\frac{1}{100}x_i - \frac{2}{3}\frac{99}{100} = x_i\frac{298}{300} - \frac{198}{300} > 0$. Hence we must have

$$0 < x_i\frac{298}{300} - \frac{198}{300} < \frac{1}{3} \implies x_i\frac{298}{300} < \frac{298}{300} \implies x_i < 1.$$

That's impossible. Hence $(1, 1, \cdots, 1)$ is a Nash equilibrium.

4.23 The price subsidy kicks in if $p = 15 - \frac{q_1+q_2+q_3}{150000} \leq 2$, which implies that the total quantity shipped, $Q = q_1 + q_2 + q_3 \leq 1,950,000$. The price per bushel is given by

$$p(q_1, q_2, q_3) = \begin{cases} 15 - \frac{q_1+q_2+q_3}{150000}, & \text{if } Q < 1950000; \\ 2, & \text{if } Q \geq 1950000. \end{cases}$$

Each farmer has the payoff function

$$u_i(q_1, q_2, q_3) = p(q_1, q_2, q_3)q_i, \quad i = 1, 2, 3.$$

Assume that $Q < 1950000$. Take a partial derivative and set to zero to get $q_i = 562500$ bushels each. Then $Q = 1687500 < 1950000$. So an interior pure Nash equilibrium consists of each farmer sending 562500 bushels each to market and using 437500 bushels for feed. The price per bushel will be $p = 15 - \frac{3 \times 562500}{150000} = 3.75$, which is greater than the government-guaranteed price.

If each producer ships 562500 bushels, the profit for each producer is

$$u_i(562500, 562500, 562500) = 3.75 \times 562500 = 2,109,375.$$

Now suppose producer 1 ships his entire crop of $q_1 = 10^6$ bushels. Assuming the other producers are not aware that producer 1 is shipping 10^6 bushels, if the other producers still ship 562500 bushels the total shipped will be $2,125,000 > 1,950,000$ and the price per bushel would be subsidized at 2. The revenue to each producer is then

$$u_1(10^6, 562500, 562500) = 2000000,$$
$$u_2(10^6, 562500, 562500) = 1125000,$$
$$u_3(10^6, 562500, 562500) = 1125000.$$

This means all the producers make less money but 2 and 3 make significantly less.

It's very unlikely that 2 and 3 would be happy with this solution. We need to find 2 and 3's best responses to 1's shipment of 10^6. Assume first that $10^6 + q_2 + q_3 \leq 1950000$. We calculate

$$\max_{q_2} u_2(10^6, q_2, q_3) = \max_{q_2} q_2(15 - \frac{1}{150000}(10^6 + q_2 + q_3))$$

The problem for q_3 is similar. Taking a derivative, setting to zero, and solving for q_2, q_3 results in $q_2 = q_3 = (1.25 \times 10^6)/3 = 416,667$. That is the best response of producers 2 and 3 to producer 1 shipping his entire crop assuming $Q \leq 1950000$. The price per bushel at total output $1,833,334$ bushels will be \$2.78 and the profit for each producer will be $u_1 = 2,780,000, u_2 = u_3 = 1,158,334$.

Now assume that $Q \geq 1950000$. In this case $p = 2$ and the profit for each producer is $u_1 = 2000000, u_2 = 2 \times q_2, u_3 = 2 \times q_3$. The maximum for producers 2 and 3 is achieved with $q_2 = q_3 = 10^6$ and all 3 producers ship their entire crop to market. This is clearly the best response of producers 2 and 3 to $q_1 = 10^6$.

4.25.a If we take the first derivatives we get

$$\frac{\partial u_C(x,y)}{\partial x} = -\frac{1}{I-x} + \frac{\delta(1+r)}{y+(1+r)x}, \quad \frac{\partial u_C(x,y)}{\partial y} = -\frac{1}{T-y} + \frac{\alpha\delta}{y+(1+r)x}$$

The second partials give us

$$\frac{\partial^2 u_C(x,y)}{\partial x^2} = -\frac{1}{(I-x)^2} - \frac{\delta(1+r)^2}{(y+(1+r)x)^2}, \quad \frac{\partial^2 u_C(x,y)}{\partial y^2} = -\frac{1}{(T-y)^2} - \frac{\alpha\delta}{(y+(1+r)x)^2}$$

both of which are < 0. That means that we can find the Nash equilibrium where the first partials are zero. Solving, we get the general solution

$$x^* = \frac{(1+\alpha\delta)I(1+r) - \alpha T}{(1+\alpha+\alpha\delta)(1+r)}$$

$$y^* = \frac{-I(1+r) + \alpha(1+\delta)T}{1+\alpha+\alpha\delta},$$

and that is the Nash equilibrium as long as $0 \le x^* < I, 0 \le y^* < T$.

4.25.b For the data given in the problem $x^* = 338.66, y^* = 357.85$.

4.27.a

$$u_2(x,y) = \begin{cases} -y, & \text{if } x > y; \\ \frac{1}{2} - y, & \text{if } x = y; \\ 1 - y, & \text{if } x < y. \end{cases}$$

4.27.b If $y < 1$, then investor I would like to choose an $x \in (y, 1)$ (in particular we want to choose $x = y$ but $y \notin (y, 1)$) and therefore no best response exists. If $x < 1$, then investor II would like to choose a $y \in (x, 1)$ and II has no best response either. The last case is $x = y = 1$. In this case both investors have best responses $x = 0$ or $y = 0$.

4.27.c First observe that $u_1(x,y) = u_2(y,x)$ which means that if X is a NE for player I, then it is also a NE for player II. From the definition of v_I, $v_I = \int_0^x (1-x)g(y)\,dy + \int_x^1 (-x)g(y)\,dy$ so that taking derivatives,

$$0 = -\int_0^x g(y)\,dy + (1-x)g(x) - \int_x^1 g(y)\,dy + xg(x) = g(x) - \int_0^1 g(y)\,dy.$$

Now since $\int_0^1 g(y)\,dy = 1$, we conclude that $g(x) = 1, 0 \le x \le 1$. This is the density for a uniformly distributed random variable on $[0, 1]$. Thus the Nash equilibrium is that each investor chooses an investment level at random.

Finally, we will see that $v_I = v_{II} = 0$. But that is immediate from

$$v_I = \int_0^x (1-x)\,dy + \int_x^1 (-x)\,dy = 0.$$

4.31.a If they have the same unit costs, the two firms together should act as a monopolist. This means the optimal production quantities for each firm should be

$$q_1^* = q_2^* = \frac{q_m^*}{2} = \frac{1}{4}(\Gamma - c).$$

The payoffs to each firm is then $u_i(q_1^*, q_2^*) = \frac{1}{8}(\Gamma - c)^2$. which is greater than the profit under competition. Furthermore, the cartel price will be the same as the monopoly price since the total quantity produced is the same.

4.31.b The best response of firm 1 to the quantity q_2^* is given by

$$q_1(q_2^*) = \frac{1}{2}(\Gamma - c - q_2^*) = \frac{3}{8}(\Gamma - c) > \frac{1}{4}(\Gamma - c) = q_2^*$$

Thus firm 1 should produce *more* than firm 2 as a best response. Similarly, firm 2 has an incentive to produce more than firm 1. The cartel collapses.

4.33 The profit function for firm i is $u_i(q_1, \ldots, q_i, \ldots, q_N) = q_i((\Gamma - \sum_{j=1}^{N} q_j)^+ - c_i)$. If we take derivatives and set to zero we get

$$\frac{\partial u_i}{\partial q_i} = \Gamma - \sum_{j=1}^{N} q_j - c_i - q_i = 0, \ i = 1, 2, \ldots, N.$$

Since $\frac{\partial^2 u_i}{\partial q_i^2} = -2 < 0$, any critical point will provide a maximum. Set $\alpha_i = \Gamma - c_i$. We have to solve the system of equations $q_i + \sum_{j=1}^{N} q_j = \alpha_i$. In matrix form we get the system

$$A\vec{q} = \vec{\alpha}, \quad A = \begin{bmatrix} 2 & 1 & 1 & \cdots & 1 \\ 1 & 2 & 1 & \cdots & 1 \\ \vdots & \vdots & \vdots & \vdots & \vdots \\ 1 & 1 & 1 & \cdots & 2 \end{bmatrix}, \quad \vec{q} = \begin{bmatrix} q_1 \\ q_2 \\ \vdots \\ q_N \end{bmatrix}, \quad \vec{\alpha} = \begin{bmatrix} \alpha_1 \\ \alpha_2 \\ \vdots \\ \alpha_N \end{bmatrix},$$

It is easy to check that

$$A^{-1} = \frac{1}{N+1} \begin{bmatrix} N & -1 & -1 & \cdots & -1 \\ -1 & N & -1 & \cdots & -1 \\ \vdots & \vdots & \vdots & \vdots & \vdots \\ -1 & -1 & -1 & \cdots & N \end{bmatrix}$$

and then

$$\vec{q} = A^{-1}\vec{\alpha}$$

After a little algebra we see that the optimal quantities that each firm should produce is

$$q_i = \frac{1}{N+1} \left(\Gamma - Nc_i + \sum_{j=1, j \neq i}^{N} c_j \right).$$

If $c_i = c, \ i = 1, 2, \ldots, N$, then $q_i = \frac{\Gamma - c}{N+1} \to 0, N \to \infty$. The profits then also approach zero.

4.35.a No firm would produce more than 10 because the cost of producing and selling one gadget would essentially be the same as the return on that gadget,

4.35.b The profit function for the monopolist is

$$u(q) = q P(q) - q = \frac{1}{4}q^3 - 5q^2 + 25q.$$

Taking a derivative and setting to zero gives

$$u'(q) = \frac{3}{4}q^2 - 10q + 25 = 0 \implies q = 10 \text{ or } q = \frac{10}{3}$$

and it is easy to check that $q = \frac{10}{3}$ provides the maximum. Also $u(\frac{10}{3}) = \frac{1000}{27}$

4.35.c Now we have

$$u_i(q_1, q_2) = q_i P(q_1 + q_2) - q_i, \ i = 1, 2.$$

We can take advantage of the fact that there is symmetry between the two firms since they have the same price function and costs. That means, for firm 1,

$$u_1(q_1, q_2) = u_1(q_1) = q_1 P(2q_1) - q_1$$

The maximum is achieved at $q_1 = \frac{5}{3}$. Thus the Nash equilibrium is $q_1 = \frac{5}{3}, q_2 = \frac{5}{3}$ and $u_1 = u_2 = \frac{500}{27}$.

4.37.a We have to solve the system

$$q_1 = \frac{1}{2}[(\Gamma - q_2^1 - c_1)p_1 + (\Gamma - q_2^2 - c_1)p_2 + (\Gamma - q_2^3 - c_1)p_3],$$

$$q_2^i = \frac{1}{2}[\Gamma - q_1 - c^i], i = 1, 2, 3,$$

Observe that q_1 is the expected value of the best response production quantities when the costs for firm 2 are $c^i, i = 1, 2, 3$ with probability $p_i, i = 1, 2, 3$. The system of best response equations has solution

$$q_1 = \frac{1}{3}\left[\Gamma - 2c_1 + p_1(c^1 - c^3) + p_2(c^2 - c^3)\right]$$

$$q_2^1 = \frac{1}{3}\left[\Gamma + c_1 - \frac{c^3}{2} + \frac{1}{2}p_1(c^3 - c^1) + \frac{p_2}{2}(c^3 - c^2)\right] - \frac{c^1}{2}$$

$$q_2^2 = \frac{1}{3}\left[\Gamma + c_1 - \frac{c^3}{2} + \frac{1}{2}p_1(c^3 - c^1) + \frac{p_2}{2}(c^3 - c^2)\right] - \frac{c^2}{2}$$

$$q_2^3 = \frac{1}{3}\left[\Gamma + c_1 + \frac{1}{2}p_1(c^3 - c^1) + \frac{p_2}{2}(c^3 - c^2)\right] - \frac{2c^3}{3}$$

4.37.b The optimal production quantities with the information given are $q_1 = \frac{749}{24}$ for firm 1, and $q_2^1 = \frac{529}{16}, q_2^2 = \frac{521}{16}$, and $q_2^3 = \frac{497}{16}$.

4.39 First we have to find the best response functions for firm 2 and firm 3. We have from the profit functions and setting derivatives to zero,

$$q_2(q_1) = \frac{\Gamma - q_1 - q_3 - c_2}{2}, \quad \text{and} \quad q_3(q_1) = \frac{\Gamma - q_1 - q_2 - c_3}{2}.$$

Solving these two equation results in

$$q_2(q_1) = \frac{\Gamma - q_1 - 2c_2 + c_3}{3}, \quad q_3(q_1) = \frac{\Gamma - q_1 - 2c_3 + c_2}{3}.$$

Next firm 1 will choose q_1 to maximize

$$u_1(q_1, q_2(q_1), q_3(q_1)) = q_1(\Gamma - q_1 - q_2(q_1) - q_3(q_1) - c_1).$$

Taking a derivative and setting to zero we get $q_1 = \frac{\Gamma + c_2 + c_3 - 3c_1}{2}$. Finally, plugging this q_1 into the best response functions q_2, q_3, we get the optimal Stackelberg production quantities

$$q_1 = \frac{\Gamma + c_2 + c_3 - 3c_1}{2}$$

$$q_2 = \frac{\Gamma - 5c_2 + c_3}{6} + \frac{c_1}{2}$$

$$q_3 = \frac{\Gamma + c_2 - 5c_3}{6} + \frac{c_1}{2}.$$

The profits for each firm are

$$u_1(q_1, q_2, q_3) = \frac{(\Gamma + c_2 + c_3 - 3c_1)^2}{12}$$

$$u_2(q_1, q_2, q_3) = \frac{(\Gamma - 5c_2 + c_3 + 3c_1)^2}{36}$$

$$u_2(q_1, q_2, q_3) = \frac{(\Gamma + c_2 - 5c_3 + 3c_1)^2}{36}$$

4.41 If $c_1 = c_2 = c$ we have the profit function for firm 2

$$u_2(p_1, p_2) = \begin{cases} p_2(\Gamma - p_2) - c(\Gamma - p_2) & \text{if } p_2 < p_1; \\ \dfrac{(p - c)(\Gamma - p)}{2} & \text{if } p_1 = p_2 = p \geq c; \\ 0 & \text{if } p_2 > p_1. \end{cases}$$

We will show that $u_2(c, c) \geq u_2(c, p_2)$, for any $p_2 \neq c$. Now, $u_2(c, c) = 0$ and

$$u_2(c, p_2) = \begin{cases} (p_2 - c)(\Gamma - p_2) < 0 & \text{if } p_2 < c; \\ \dfrac{(p - c)(\Gamma - p)}{2} = 0 & \text{if } p_1 = p_2 = p \geq c; \\ 0 & \text{if } p_2 > c. \end{cases}$$

In every case $u_2(c, c) = 0 \geq u_2(c, p_2)$. Similarly $u_1(c, c) \geq u_1(p_1, c)$ for every $p_1 \neq c$. This says (c, c) is a Nash equilibrium and both firms make zero profit.

4.43 The profit functions for each firm are

$$u_1(p_1, p_2) = \begin{cases} (p_1 - c_1) \min\{(\Gamma - p_1), K\} & \text{if } p_1 < p_2; \\ \dfrac{((p - c_1)(\Gamma - p))}{2} & \text{if } p_1 = p_2 = p \geq c_1; \\ 0 & \text{if } p_1 > p_2, p_2 \geq \Gamma - K. \\ (p_1 - c_1)(\Gamma - p_1 - K) & \text{if } p_1 > p_2, p_2 < \Gamma - K. \end{cases}$$

and

$$u_2(p_1, p_2) = \begin{cases} (p_2 - c_2) \min\{(\Gamma - p_2), K\} & \text{if } p_2 < p_1; \\ \dfrac{((p - c_2)(\Gamma - p))}{2} & \text{if } p_1 = p_2 = p \geq c_2; \\ 0 & \text{if } p_2 > p_1, p_1 \geq \Gamma - K. \\ (p_2 - c_2)(\Gamma - p_2 - K) & \text{if } p_2 > p_1, p_2 < \Gamma - K. \end{cases}$$

To explain the profit function for firm 1, if $p_1 < p_2$, firm 1 will be able to sell the quantity of gadgets $q = \min\{\Gamma - p_1, K\}$, the smaller of the demand quantity at price p_1 and the capacity of production for firm 1.

If $p_1 = p_2$, so that both firms are charging the same price, they split the market and since $K \geq \frac{\Gamma}{2}$ there is enough capacity to fill the demand.

If $p_1 > p_2$ in the standard Bertrand model, firm 1 loses all the business since they charge a higher price for gadgets. In the limited capacity model, firm 1 loses all the business only if, in addition, $p_2 \geq \Gamma - K$; that is, if $K \geq \Gamma - p_2$, which is the amount that firm 2 will be able to sell at price p_2, and this quantity is less than the production capacity.

Finally, if $p_1 > p_2$, and $K < \Gamma - p_2$, firm 1 will be able to sell the amount of gadgets that exceed the capacity of firm 2. That is, if $K < \Gamma - p_2$, then the quantity demanded from firm 2 at price p_2 is greater than the production capacity of firm 2 so the residual amount of gadgets can be sold to consumers by firm 1 at price p_1. But notice that in this case, the number of gadgets that are demanded at price p_1 is $\Gamma - p_1$, so firm 1 can sell at most $\Gamma - p_1 - K < \Gamma - p_2 - K$.

What about Nash equilibria? Even in the case $c_1 = c_2 = 0$ there is no pure Nash equilibrium even at $p_1^* = p_2^* = 0$. This is known as the Edgeworth paradox.

4.45 Start with profit functions

$$u_1(p_1, p_2) = (\Gamma - p_1 + bp_2)(p_1 - c_1), \quad \text{and} \quad u_2(p_1, p_2) = (\Gamma - p_2 + bp_1)(p_2 - c_2).$$

Solve $\frac{\partial u_2}{\partial p_2} = 0$ to get the best response function

$$p_2(p_1) = \frac{1}{2}(c_2 + \Gamma + bp_1).$$

Next solve $\frac{\partial u_1(p_1, p_2(p_1))}{\partial p_1} = 0$, to get

$$p_1^* = \frac{2c_1 + 2\Gamma - c_1 b^2 + bc_2 + b\Gamma}{4 - 2b^2}$$

and then the optimal price for firm 2 is

$$p_2^* = p_2(p_1^*) = \frac{-4c - 2 + c_2 b^2 - 4\Gamma + \Gamma b^2 - 2bc_1 + c_1 b^3 - 2b\Gamma}{4b^2 - 8}.$$

The profit functions become

$$u_1(p_1^*, p_2^*) = \frac{1}{8} \frac{(-2c_1 + b^2 c_1 + 2\Gamma + bc_2 + b\Gamma)^2}{2 - b^2}$$

and

$$u_2(p_1^*, p_2^*) = \frac{1}{16} \frac{\left(-4\Gamma + \Gamma b^2 + 4c_2 - 3c_2 b^2 - 2bc_1 + c_1 b^3 - 2b\Gamma\right)^2}{\left(-2 + b^2\right)^2}$$

Notice that with the Stackelberg formulation and the formulation in the preceding problem the Nash equilibrium has positive prices and profits.

For the data of the problem we have

$$p_1^* = 74.07, \quad p_2^* = 69.02, \quad u_1 = 4174.50, \quad u_2 = 4626.43$$

4.47.a Profit is $P = (80 - N)N - 10N - 25 = 70N - N^2 - 25$. To optimize, set the derivative equal to zero, which gives $70 - 2N = 0$, so $N = 35$.

The resulting profit is $45 \times 35 - 350 - 25 = 1200$ thousands of dollars, so 1.2 million.

4.47.b A second firm entering the market has a profit function: $P_2 = (80 - N_1 - N_2)N_2 - 10N_2 - 25$. To optimize, set the derivative equal to zero, which gives $70 - N_1 - 2N_2 = 0$, so $N_2 = 35 - \frac{N_1}{2}$.

Thus, firm 1 knows that an entering firm's potential profit is

$$(70 - N_1 - 35 + \frac{N_1}{2})(35 - \frac{N_1}{2}) - 25 = (35 - \frac{N_1}{2})^2 - 25.$$

Firm one sets this to zero, meaning firm 2 can't make a profit. So, they solve:

$$(35 - \frac{N_1}{2})^2 - 25 =$$

giving $N_1 = 60$.

The profit for this value (assuming firm 2 decides not to enter the market) is

$$20 \times 60 - 10 \times 60 - 25 = 575.$$

So, quite a bit less than the monopoly value.

4.49.a $P_1(x, y) = \begin{cases} x - y - b, & x > y \\ 0, & x = y \\ y - x + b, & y > x. \end{cases}$ $P_2(x, y) = -P_1(x, y).$

4.49.b

$$E[P_1(X, y)] = \int_0^1 P_1(x, y) X(x)\, dx = \int_0^y (y - x - b) X(x)\, dx + \int_y^1 (y - x + b) X(x)\, dx$$

Simplifying, this becomes $f(y) = E[P_1(X, y)] = y - \int_0^1 x X(x)\, dx - b \int_0^y X(x)\, dx + b \int_y^1 X(x)\, dx$.

4.49.c

$$\frac{\partial f(y)}{\partial y} = 1 - bX(y) - bX(y) = 1 - 2bX(y) = 0 \implies X(y) = \frac{1}{2b}.$$

This is a constant which says that X should be a uniform density on the interval $0 \leq y \leq 2b \leq 1$. This says, player 1 should choose a random amount of the resource in $[0, 2b]$.

4.49.d Since the game as set up is symmetric with respect to the payoffs, the optimal strategy for player 2 is also uniform on $[0, 2b]$

4.49.e The expected payoff to each player in a symmetric zero-sum game is zero. To verify by calculation,

$$f(y) = y - \int_0^1 x X(x)\, dx - b \int_0^y X(x)\, dx + b \int_y^1 X(x)\, dx$$

$$E[f(Y)] = b - b - b\frac{E(Y)}{2b} + b\frac{2b - EY}{2b}$$

$$= -\frac{1}{2}b + \frac{1}{2}(2b - b) = 0.$$

4.55 The payoff functions are

$$u_i(b_1, \ldots, b_N) = \begin{cases} 0 & \text{if } b_i < M, \text{ she is not a high bidder;} \\ v_i - b_i & \text{if } b_i = M, \text{ she is the sole high bidder;} \\ \dfrac{v_i - b_i}{k} & \text{if } i \in \{k\}, \text{ she is one of } k \text{ high bidders.} \end{cases}$$

and recall that $v_1 \geq v_2 \cdots \geq v_N$. We have to show that $u_i(v_1, \ldots, v_N)$ gives a larger payoff to player i if player i makes any other bid $b_i \neq v_i$. We assume that $v_1 = v_2$ so the two highest valuations are the same. Now for any player i if she bids less than $v_1 = v_2$ she does not win the object and her payoff is zero. If she bids $b_i > v_1$ she wins the object with payoff $v_i - b_i < v_1 - b_i < 0$. If she bids $b_i = v_1$, her payoff is $\frac{v_i - b_i}{3} = \frac{v_i - v_1}{3} < 0$. In all cases she is worse off if she deviates from the bid $b_i = v_i$ as long as the other players stick with their valuation bids.

4.57 In the first case it will not matter if she uses a first or second price auction. Either way she will sell it for $100,000. In the second case the winning bid for player 1 is between $95,000 < b_1 < 100,000$ whether it is a first or second price auction. However, in the second price auction the house will sell for $95,000. Thus, a first price auction is better for the seller.

4.59 The expected payoff of a bidder with valuation v who makes a bid of b is given by

$$u(b) = v\,Prob(b \text{ is high bid}) - b = vF(\beta^{-1}(b))^{N-1} - b = v\beta^{-1}(b)^{N-1} - b.$$

Differentiate, set to zero, and solve to get $\beta(v) = \frac{(N-1)}{N})v^N$.

Since all bidders will actually pay their own bids and each bid is $\beta(v) = (\frac{N-1}{N})v^N$, the expected payment from each bidder is

$$E[\beta(V)] = \frac{N-1}{N} \int_0^1 v^N\, dv = \frac{N-1}{N(N+1)}.$$

Since there are N bidders, the total expected payment to the seller will be $\frac{N-1}{N+1}$.

4.61.a The stable matching from the Gale-Shapley algorithm is $(m1, w1), (m2, w2), (m3, w3)$. All the men are matched with their first choice and all the women are matched with their worst choice.

4.61.b The stable matching from the Gale-Shapley algorithm is $(w1, m3), (w2, m1), (w3, m2)$. All the women get matched with their first choice and all the men are matched with their second choice.

4.61.c Consider the pair $(m3, w1)$. Man m3 ranks w1 higher than w2, and woman w1 ranks m3 higher than m1. But, in the matching, m3 is paired with w2 and w1 is paired with m1. This means m3 and w1 could run off with each other.

4.63.a A: 1 5 7 6 2

B: 2 3 4 6 1 5

C: 3 7 2 5

5 fails to match.

4.63.b 1: B

2: A

3: A C

4: B C

5:

6: A C

7: B C

Solutions for Chapter 5

5.1 The grim-trigger used by both players:

- Play C to start the game and play C as long as both players are doing so.
- If any player ever chooses to play D, then switch to D for every game thereafter.

If both players use the Grim-Trigger strategy the payoff to each player is 2 and they play (C, C) at every stage. We have to show that if a player deviates from C to D at stage k that player gets less.

$$(1 - \gamma) \sum_{i=0}^{\infty} \gamma^i 2 = 2 \geq (1 - \gamma) \left(\sum_{i=0}^{k-1} \gamma^i 2 + \gamma^k 3 + \sum_{i=k+1}^{\infty} \gamma^i (-2) \right) \Leftrightarrow$$

$$2 \geq (1 - \gamma) \left(2 \frac{1 - \gamma^k}{1 - \gamma} + 3\gamma^k - 2 \frac{\gamma^{k+1}}{1 - \gamma} \right)$$

$$= 2 - 2\gamma^k + 3\gamma^k - 3\gamma^{k+1} - 2\gamma^{k+1} \Leftrightarrow$$

$$2 \geq 2 + \gamma^k - 5\gamma^{k+1} \Leftrightarrow$$

$$\gamma \geq \frac{1}{5}.$$

This shows that if the probability another game will be played after the current game is more than 20% a player would prefer to stick with C at every stage.

5.3.a $\gamma^* = \frac{3}{8}$.

5.3.b The payoff to player I is $(1 - \gamma)(-3 - 3\gamma^2 - 3\gamma^4 \cdots + 8\gamma + 8\gamma^3 + 8\gamma^5 \cdots) = \frac{8\gamma - 3}{1 + \gamma}$. If player I deviates in some round and plays D then they both play D forever. Observe that using the strategies of the problem, player I's payoff in the first round is -3 so that is the best round for player I to deviate and play D. In this case the payoff to player I is 0 and remains 0. Therefore, the pair of strategies will be a NE if $\frac{8\gamma-3}{1+\gamma} \geq 0$. This results in $\gamma^* = \frac{3}{8}$.

5.5.a $v_1 = 2, v_2 = 5$.

5.5.b The largest payoff to player I by switching to U, assuming player II is playing the given strategy is $(1 - \gamma)(4 + 2\gamma$, since II switches to Y forever. Since then I's response is X, the resulting payoff is 2 on each round. This largest payoff is ≤ 3 if and only if $\gamma \geq \frac{1}{2}$. In a similar way, the largest payoff to II from deviating to R, assuming I is still playing the giving strategy, is $8(1 - \gamma) + 5\gamma$. This is ≤ 6 if and only if $\gamma \geq \frac{2}{3}$. The conclusion is that the strategies is a subgame perfect NE if $\gamma \geq \frac{2}{3}$.

5.7 If a firm deviates in round k then the profit to the firm is

$$(1 - \gamma)\left(\sum_{j=1}^{k-1} \gamma^{k-1}\frac{(\Gamma - c)^2}{8} + \gamma^{k-1}\frac{(\Gamma - c)^2}{4} + 0 \right).$$

In order to be subgame perfect NE we need to make sure there is no gain from deviating, which means

$$(1 - \gamma)\sum_{j=1}^{\infty} \gamma^{j-1}\frac{(G - c)^2}{8} \geq (1 - \gamma)\left(\sum_{j=1}^{k-1} \gamma^{k-1}\frac{(\Gamma - c)^2}{8} + \gamma^{k-1}\frac{(\Gamma - c)^2}{4} + 0 \right)$$

$$\Leftrightarrow \gamma^{k-1}\frac{(\Gamma - c)^2}{8} \geq (1 - \gamma)\gamma^{k-1}\frac{(\Gamma - c)^2}{4}$$

$$\Leftrightarrow \gamma \geq \frac{1}{2}.$$

5.9 Suppose that player 2 sticks with tit-for-tat.

Suppose the players choose (C, C) in the any stage. If player 1 sticks with tit-for-tat the play is (C, C) in every period, so that her discounted average payoff in the subgame is x. If player 1 chooses D in the first period of the subgame, then sticks to tit-for-tat, the plays alternates between (D, C) and (C, D), and her discounted average payoff is $(1 - \gamma)\sum_{k=1}^{\infty} (\gamma^2)^{k-1}y = y/(1 + \gamma)$. Thus we need $x \geq y/(1 + \gamma)$, or $\gamma \geq (y - x)/x$, for a one-period deviation from tit-for-tat not to be profitable for player 1.

Suppose the players choose (C, D). If player 1 sticks to tit-for-tat the plays alternate between (D, C) and (C, D), so that her discounted average payoff is $y/(1 + \gamma)$. If she deviates to C in the first period of the subgame, then sticks to tit-for-tat, the play is (C, C) in every period, and her discounted average payoff is x. Thus we need $y/(1 + \gamma) \geq x$, or $\gamma \leq (y - x)/x$, for a one-period deviation from tit-for-tat not to be profitable for player 1.

Suppose the players choose (D, C). If player 1 sticks to tit-for-tat the plays alternate between (C, D) and (D, C), so that her discounted average payoff is $\gamma y/(1 + \gamma)$. If she deviates to D in the first period of the subgame, then adheres to tit-for-tat, the outcome is (D, D) in every period, and her discounted average payoff is 1. Thus we need $\gamma y/(1 + \gamma) \geq 1$, or $\gamma \geq 1/(y - 1)$, for a one-period deviation from tit-for-tat not to be profitable for player 1.

Suppose the players choose (D, D). If player 1 sticks to tit-for-tat the play is (D, D) in every period, so that her discounted average payoff is 1. If she deviates to C in the first period of the subgame, then adheres to tit-for-tat, the plays alternate between (C, D) and (D, C), and her discounted average payoff is $\gamma y/(1 + \gamma)$. Thus we need $1 \geq \gamma y/(1 + \gamma)$, or $\gamma \leq 1/(y - 1)$, for a one-period deviation from tit-for-tat not to be profitable for player 1.

The same arguments apply to deviations by player 2, so we conclude that (tit-for-tat, tit-for-tat) is a subgame perfect equilibrium if and only if $\gamma = (y - x)/x$ and $\gamma = 1/(y - 1)$, or $y - x = 1$ and $\gamma = 1/x$.

5.11.a $(P - c)q_i$

5.11.b $Profit = (P - c)q_i = (300 - 5(\sum_{j \neq i} q_j + q_i) - c)q_i$. Set $q_{-i} = \sum_{j \neq i} q_j$. The best response for country i is $q_i = (300 - c)/10 - q_{-i}/2$. To solve this we assume all q's are equal. We get $q_i = (300 - c)/25 = 11.2$ and then $P = 300 - 5Q = 76$. The profit to each country is then $(76 - 20)11.2 = 627.2$.

5.13 When cooperating, each country earns 800 with a payoff $800 + 800\gamma/(1 - \gamma)$. When deviating (optimally), a country earns 845 for one day, and 627.2 /day ever after with a payoff $845 + 667\gamma/(1 - \gamma)$. In order to sustain $q = 10$, it must be that $800 + 800\gamma/(1 - \gamma) \geq 845 + 667\gamma/(1 - \gamma)$, which holds for $\gamma \approx 1$. Hence the last statement is wrong.

5.15 The minimax values of both players is 0 by charging 4% as can easily be checked, $(m_1, m_2) = (0, 0)$. The cooperative values they want are $(v_1, v_2) = (230, 230)$ so that each player charges 8%. The maximum payoff to each player is $d_1 = d_2 = 320$. The grim-trigger strategy is for each player to charge 8% to begin with and as long as they both go with the program. If any player deviates in some round it will be in order to reap the largest possible payoff of $d_1 = d_2 = 320$ by charging 6% while the opponent sticks with 8%. But, if that happens they will both switch to 4% giving payoffs of $(P_1, P_2) = (100, 100)$ which is the NE strategy. Therefore, this grim-trigger strategy for both players will be a subgame perfect NE as long as the discount factor satisfies

$$\gamma > \frac{d_i - v_i}{d_i - P_i} = \frac{320 - 230}{320 - 100}(i = 1, 2) = \frac{90}{220} = 0.4091 = \gamma^*$$

On the other hand, if the grim-trigger with punishment strategy is used, the lower bound on γ becomes

$$\gamma > \frac{d_i - v_i}{d_i - m_i} = \frac{320 - 230}{320 - 0}(i = 1, 2) = \frac{90}{320} = 0.2813 = \gamma^*.$$

One way to think of this is that if the chances of a game being played again is at least 29% cooperation will be played forever and they will both charge 8%.

Now suppose that $\gamma < \frac{9}{32}$ so that the discount rate does not support cooperation at the 8% commission rate. Is it possible that collusion will work if they both charge 6%? Suppose they start out cooperating at 6% but then someone deviates so they switch to 4%. Notice that we may drop the fourth row and column of the game matrix because 8% is eliminated.

How large does the discount factor have to be to support this strategy? Since $d_i = 220$, $v_i = 180$ and $P_i = 100$, the analysis of grim-trigger gives

$$\gamma > \frac{220 - 180}{220 - 100} = \frac{1}{3} = \gamma^*.$$

If we use grim-trigger with punishment we get the better bound

$$\gamma > \frac{220 - 180}{220 - 0} = \frac{4}{22} = 0.1818 = \gamma^*.$$

If we use the regular grim-trigger to NE strategy, we may conclude that if $\frac{1}{3} \leq \gamma < \frac{9}{22}$ then cooperation will work to charge 6% instead of 4% but not 8%. With this grim-trigger strategy if the probability that there will be another item up for auction pitting these two houses against each other for the business is at least 33% but less than 41%, then they should agree to charge 6% to maximize their payoffs. If we use grim-trigger with punishment, then the range of γ we need is $\frac{4}{22} \leq \gamma < \frac{9}{32}$, or equivalently, from 19% to 28%.

Solutions for Chapter 6

6.1.a The normalized function is $v'(S) = \frac{v(S) - \sum_{i \in S} v(i)}{v(N) - \sum_{i \in N} v(i)} = 0, S \subsetneq N$, and $v'(N) = 1$.

6.1.b The core using v' is

$$C(0) = \{\vec{x}' \mid x_i' \geq 0, \sum_{i=1}^{n} x_i' = 1\}$$

The unnormalized allocations satisfy

$$x_i = x_i' \left(v(N) - \sum_{i=1}^{n} v(i) \right) + v(i) = x_i' \left(1 - \sum_{i=1}^{n} \alpha \right) + \alpha = x_i'(1 - n\alpha) + \alpha.$$

Since $1 - n\alpha > 0, x_i - \alpha \geq 0$ and $\sum_{i=1}^{n} x_i = (1 - n\alpha) + n\alpha = 1$. In terms of the unnormalized allocations, $C(0) = \{\vec{x} \mid x_i \geq \alpha, \sum_{i=1}^{n} x_i = 1\}$.

6.3 Let $\vec{x} = (-b, \ldots, -b)$. Calculate for $S \subsetneq N$,

$$e(S, \vec{x}) = v(S) - \vec{x}(S) = \max\{|S|(-nc), |S|(-(n - |S|)c) - |S|b\} + b|S|$$
$$= \max\{|S|(b - nc), |S|(-(n - |S|)c)\}.$$

Since $b < nc$ and $n > |S|$ both terms in the max are negative so $e(S, \vec{x}) \leq 0, S \subsetneq N$. If $S = N$,

$$e(N, \vec{x}) = \max\{-n^2 c, -nb\} + nb = \max\{n(b - nc), 0\} = 0$$

since $b < nc$. By definition we conclude $\vec{x} \in C(0)$.

6.5 The characteristic function is

$$v(1) = 1, v(2) = 2, v(3) = 3$$
$$v(12) = 4, v(13) = 5, v(23) = 6, v(123) = 8$$

The normalized characteristic function is

$$v'(1) = 0, v'(2) = 0, v'(3) = 0$$
$$v'(12) = \frac{1}{2}, v'(13) = \frac{1}{2}, v'(23) = \frac{1}{2}, v'(123) = 1$$

The normalized least core is $C(-\frac{1}{6}) = \{(\frac{1}{3}, \frac{1}{3}, \frac{1}{3})\}$. The unnormalized least core is $C(-\frac{1}{3}) = \{(\frac{5}{3}, \frac{8}{3}, \frac{11}{3})\}$.

The normalized least core comes from the inequalities

$$\frac{1}{2} - \varepsilon \leq x_1 + x_2, \ x_1 + x_2 \leq 1 + 2\varepsilon \implies -\frac{1}{2} \leq 3\varepsilon \implies -\frac{1}{6} \leq \varepsilon,$$

which decides $\varepsilon = -\frac{1}{6}$ is the first ε. Then $x_1 + x_2 = \frac{2}{3} \implies x_1 = \frac{1}{3}$. Then $x_1 \leq \frac{1}{2} - \frac{1}{6} = \frac{1}{3}, x_2 \leq \frac{1}{2} - \frac{1}{6}$ implies $0 = x_1 - \frac{1}{3} + x_2 - \frac{1}{3}$ and hence $x_1 = x_2 = \frac{1}{3}$.

6.7 In normalized form since $v(i) = 0$, simply divide each allocation value by $v(1234) = 13 : \vec{x}$ unnormalized=$\left\{ \left(\frac{13}{4}, \frac{33}{8}, \frac{33}{8}, \frac{3}{2} \right) \right\}$.

6.9 $v(\emptyset) = 0, v(1) = \frac{3}{5}, v(2) = 2, v(3) = 1, v(12) = 5, v(13) = 4, v(23) = 3, v(123) = 16$.

6.11 Writing out the inequalities for $C(\varepsilon)$ we have the inequalities

$$e(23, \vec{x}) - x_2 - x_3 \leq \varepsilon, \ e(12, \vec{x}) - x_1 - x_2 \leq \varepsilon$$

and $x_1 + x_2 + x_3 = 4$. These imply that $3 - \varepsilon \leq x_1 + x_2 \leq 5 + \varepsilon$ and then $-2 \leq 2\varepsilon$. Consequently, we derive that $\varepsilon^1 = -1$ and the least core is $C(-1) = \{\vec{x} = (2, 2, 0)\}$.

6.13 Suppose $\vec{x} \in C(0)$ so that $e(S, \vec{x}) \leq 0, \forall S \subsetneq N$. Take the single player coalition $S = \{i\}$ so $v(i) + v(N - i) = v(N)$. Since the game is essential, $v(N) > \sum_{i=1}^{n} v(i)$. Since $\vec{x}$ is in the core, we have

$$v(N) > \sum_{i=1}^{n} v(i) = \sum_{i=1}^{n} v(N) - v(N - i) = nv(N) - \sum_{i=1}^{n} v(N - i),$$

$$\Longrightarrow$$

$$v(N)(n - 1) < \sum_{i=1}^{n} v(N - i) \leq \sum_{i=1}^{n} \sum_{j \neq i} x_j = \sum_{i=1}^{n} v(N) - x_i$$

$$= nv(N) - \sum_{i=1}^{n} x_i = (n - 1)v(N).$$

That's a contradiction and hence $C(0) = \emptyset$.

6.15 Suppose $i = 1$. Then

$$x_1 + \sum_{j \neq 1} x_j = v(N) = v(N - 1) \leq \sum_{j \neq 1} x_j,$$

and so $x_1 \leq 0$. But since $-x_1 = v(1) - x_1 \leq 0$, we have $x_1 = 0$.

6.17 Let $\vec{x} \in C(0)$. Since $v(N - 1) \leq x_2 + \cdots + x_n = v(N) - x_1$, we have $x_1 \leq v(N) - v(N - 1)$. In general, $x_i \leq v(N) - v(N - i), 1 \leq i \leq n$. Now add these up to get $v(N) = \sum_i x_i \leq \sum_i \delta_i < v(N)$, which says $C(0) = \emptyset$.

6.19 If we remove, say player 1 from N, then $N - 1$ has 2^{n-1} coalitions (including the empty coalition), none of which contain player 1. But N has 2^n coalitions and so the number of coalitions which do not contain player 1 is $2^n - 2^{n-1} = 2^{n-1}$ and consequently there are 2^{n-1} coalitions which do contain player 1.

6.21.a The characteristic function for this game is

$$v(1) = 15, v(2) = 30, v(3) = 45, v(12) = 75, v(13) = 90, v(23) = 105, v(123) = 150.$$

For example, for $S = 1$, players 2 and 3 get their capacity first which is 135; the 15 left over go to player 1.

6.21.b If we find the least core and show that $\varepsilon^1 < 0$ then it must be true that $C(\varepsilon^1) \subset C(0) \neq \emptyset$. The least core inequalities we must solve are

$$15 - x_1 \leq \varepsilon, 30 - x_2 \leq \varepsilon, 45 - x_3 \leq \varepsilon$$

$$75 - \varepsilon \leq x_1 + x_2, 90 - \varepsilon \leq x_1 + x_3 = 150 - x_2, 105 - \varepsilon \leq x_2 + x_3 = 150 - x_1$$

$$x_1 + x_2 + x_3 = 150$$

Hence $x_2 \leq 60 + \varepsilon, x_1 \leq 45 + \varepsilon \implies 75 - \varepsilon \leq x_1 + x_2 \leq 105 + 2\varepsilon$ which gives $\varepsilon^1 = -10$. Then $x_1 + x_2 = 85, x_3 = 65$, and since $x_1 \leq 35, x_2 \leq 50$, it must be true that $x_1 = 35, x_2 = 50$. The least core is $C(-10) = \{(35, 50, 65)\}$ the exact same allocation as before but with a nonempty core. Every player is ok with this allocation.

6.23.a The payoff for player i is the number of bags of garbage dumped in his yard times -1. The payoff for a coalition is the number of bags of garbage dumped on all the yards of members of S times -1. If there are n players, there are $n - |S|$ bags of garbage that will be dumped on the yards of the coalition S.

6.23.b Suppose $n > 2$ and $C(0) \neq \emptyset$. Since $v(S) = |S| - n \leq \sum_{j \in S} x_j$ we have for any $k = 1, 2, \ldots,$ $n, -1 = v(N - k) \leq \sum_{j \in N - k} x_j$. This says

$$x_2 + x_3 + \cdots + x_n \geq -1$$
$$x_1 + x_3 + \cdots + x_n \geq -1$$
$$x_1 + x_2 + \cdots + x_n \geq -1$$
$$\vdots$$
$$x_1 + x_2 + \cdots + x_{n-1} \geq -1$$

Add these up and use the fact that $\sum_{i=1}^{n} x_i = v(N) = -n$ to get

$$(n-1)(x_1 + x_2 + \cdots + x_n) = -n(n-1) \geq -n \implies n \leq 2.$$

With this contradiction we see that $n > 2 \implies C(0) = \emptyset$.

Another way to see this is to use a result from an earlier Problem (6.17) which gives us a criterion to use to determine when the core is empty. The criterion says

$$\sum_{i=1}^{n} \delta_i = \sum_{i=1}^{n} (v(N) - v(N-i)) < v(N) \implies C(0) = \emptyset.$$

In the garbage game we have

$$\sum_{i=1}^{n} [-n - (-(n - |N - i|))] = \sum_{i=1}^{n} [-n + n - (n-1)]$$
$$= -n(n-1) < v(N) = -n$$

implies that $n - 1 > 1$ or $n > 2$. We conclude that when $n > 2$, from Problem (6.17), the core of the garbage game is empty.

6.23.c A coalition that works is $S = \{12\}$.

6.25.a $v(\emptyset) = 0, v(1) = -100, v(2) = -150, v(3) = -400, v(12) = -150, v(13) = -400, v(23) = -400, v(123) = -400.$

6.25.b The least ε−core is given by

$$C(\varepsilon) = \{x_1 \geq -100 - \varepsilon, x_2 \geq -150 - \varepsilon, x_3 \geq -400 - \varepsilon$$
$$= x_1 + x_2 \geq -150 - \varepsilon, x_1 + x_3 \geq -400 - \varepsilon,$$
$$x_2 + x_3 \geq -400 - \varepsilon, x_1 + x_2 + x_3 = -400\}.$$

Next,

$$-400 - \varepsilon \leq x_1 + x_3 = -400 - x_2 \implies x_2 \leq \varepsilon$$
$$-400 - \varepsilon \leq x_2 + x_3 = -400 - x_1 \implies x_1 \leq \varepsilon$$
$$-150 - \varepsilon \leq x_1 + x_2 \leq 2\varepsilon \implies \varepsilon \geq -50$$

and that is the first $\varepsilon^1 = -50$. This implies

$$x_1 \leq -50, x_2 \leq -50, x_1 + x_2 = -100 \implies x_1 = -50, x_2 = -50, x_3 = -300.$$

6.27 We have

$$C(\varepsilon) = \{(x_1, x_2, x_3) \mid x_i \geq -\varepsilon, i = 1, 2, 3, x_1 + x_2 \geq \frac{4}{5} - \varepsilon,$$
$$x_1 + x_3 \geq \frac{2}{5} - \varepsilon, x_2 + x_3 \geq \frac{1}{5} - \varepsilon\}$$

Checking all possible combinations of inequalities results in the first ε from the inequalities

$$\frac{4}{5} - \varepsilon \leq x_1 + x_2 = 1 - x_3 \implies -\varepsilon \leq x_3 \leq \frac{1}{5} + \varepsilon \implies \varepsilon \geq -\frac{1}{10}$$

and $\varepsilon^1 = -\frac{1}{10}$. Then, the first least core is

$$X^1 = C(-\frac{1}{10}) = \{x_3 = \frac{1}{10}, x_1 + x_2 = \frac{9}{10}, \frac{4}{10} \leq x_1, \frac{2}{10} \leq x_2\}.$$

Next we calculate the excesses for $\vec{x} \in X^1$.

$$e(1, \vec{x}) = v(1) - x_1 = -x_1 \qquad\qquad e(12, \vec{x}) = \frac{4}{5} - x_1 - x_2 = -\frac{1}{10}$$

$$e(2, \vec{x}) = v(2) - x_2 = -x_2 \qquad\qquad e(13, \vec{x}) = \frac{2}{5} - x_1 - x_3 = \frac{3}{10} - x_1$$

$$e(3, \vec{x}) = v(3) - x_3 = -x_3 = -\frac{1}{10} \qquad e(23, \vec{x}) = \frac{1}{5} - x_2 - x_3 = \frac{1}{10} - x_2$$

Thus we may eliminate coalitions 3 and 12 since their excesses cannot be reduced further. The next least core is

$$X^2 = C(\varepsilon) = \{(x_1, x_2, \frac{1}{10}) \mid x_i \geq -\varepsilon, i = 1, 2, e(13, \vec{x}) \leq \varepsilon, e(23, \vec{x}) \leq \varepsilon, x_1 + x_2 = \frac{9}{10}\}$$

$$= \{(x_1, x_2, \frac{1}{10}) \mid x_i \geq -\varepsilon, i = 1, 2, \frac{3}{10} - x_1 \leq \varepsilon, \frac{1}{10} - x_2 \leq \varepsilon, x_1 + x_2 = \frac{9}{10}\}$$

We get

$$\frac{3}{10} - \varepsilon \leq x_1, \frac{1}{10} - \varepsilon \leq x_2 \implies \frac{4}{10} - 2\varepsilon \leq x_1 + x_2 = \frac{9}{10} \implies \varepsilon \geq -\frac{1}{4}.$$

The first ε making $C(\varepsilon) \neq \emptyset$ is $\varepsilon^2 = -\frac{1}{4}$. Then we calculate the next least core is $X^2 = C(-\frac{1}{4}) = \{(\frac{11}{20}, \frac{7}{20}, \frac{2}{20})\}$. This contains only one point and is the nucleolus.

6.29 The characteristic function can be taken to be the interest earned on the investment:

$$v(1) = 90,000, \ v(2) = 36,000, \ v(3) = 16,000$$

$$v(12) = 135,000, \ v(13) = 110,000 \ v(23) = 65,000$$

$$v(123) = 170,500$$

To get the least core we have

$$C(\varepsilon) = \{90 - \varepsilon \leq x_1, 36 - \varepsilon \leq x_2, 16 - \varepsilon \leq x_3,$$

$$135 - \varepsilon \leq x_1 + x_2, 110 - \varepsilon \leq x_1 + x_3, 65 - \varepsilon \leq x_2 + x_3$$

$$x_1 + x_2 + x_3 = 170.5\}$$

Using the inequalities $90 - \varepsilon \leq x_1 \leq 105.5 + \varepsilon$ we get the first $\varepsilon^1 = -\frac{15.5}{2}$. Then $x_1 = 97.75$ and $x_2 + x_3 = 72.75$.

We need the next least core. We calculate the excesses for $\vec{x} \in X^1 = C(\varepsilon^1)$ to get

$$e(1, \vec{x}) = 90 - 97.75 = -7.75, \quad e(12, \vec{x}) = 135 - 97.75 - x_2$$
$$e(2, \vec{x}) = 36 - x_2, \qquad\qquad\qquad e(13, \vec{x}) = 110 - 97.75 - x_3$$
$$e(3, \vec{x}) = 16 - x_3, \qquad\qquad\qquad e(23, \vec{x}) = 65 - 72.75 = -7.75$$

The next least core will work with coalitions $\{2, 3, 12, 13\}$ since only those coalitions have excesses which may be lowered. We have

$$X^2 = \{36 - x_2 \leq \varepsilon, 16 - x_3 \leq \varepsilon, 37.25 - x_2 \leq \varepsilon, 12.25 - x_3 \leq \varepsilon, x_2 + x_3 = 72.75\}$$

The two inequalities $16 - x_3 \leq \varepsilon, 37.25 - x_2 \leq \varepsilon$ lead to $53.25 - 2\varepsilon \leq x_2 + x_3 = 72.75$ and $\varepsilon^2 = -9.75$ as the first ε making $X^2 \neq \emptyset$. Then $37.25 + 9.75 = 47 \leq x_2$, and $16 + 9.75 = 25.75 \leq x_3$ imply that $x_2 = 47, x_3 = 25.75$.

The nucleolus is $x_1 = \frac{391}{4} = 97750, x_2 = 47000, x_3 = \frac{103}{4} = 25750$.

It is interesting to compare this with the common assumption that the fair allocation should be that each player will get the amount of 170500 proportional to the amount they invest. That would lead to the allocation $y_1 = \frac{90}{142}170500 = 108063.38, y_2 = \frac{36}{142}170500 = 43225.35$ and $y_3 = \frac{16}{142}170500 = 19211.27$. But, this does not take into account that player 3 is a very important investor. It is her money that pushes the grand coalition into the 5.5% rate of return. Without player 3 the most they could get is 5%. Consequently player 3 has to be compensated for this power. The proportional allocation doesn't do that.

6.31.a Let's draw an appointment schedule:

9	10	11	12	1	2	3
			Shemp	Shemp	Shemp	Shemp
Curly	Curly	Curly	Curly	Curly	Curly	Curly
	Larry	Larry	Larry	Larry	Larry	Larry
					Moe	Moe

The characteristic function is the number of hours saved by a coalition. Single coalitions save nothing $v(i) = 0, i = 1, 2, 3, 4$, and

$$v(12) = 4, v(13) = 4, v(14) = 3, v(23) = 6, v(24) = 2, v(34) = 2,$$

$$v(123) = 10, v(124) = 7, v(134) = 7, v(234) = 8, v(1234) = 13.$$

For instance $v(124) = 7$ since Shemp and Curly overlap 4 hours and Shemp and Moe overlap 3 hours.

6.31.b We will use Mathematica to solve this problem. We show that the Nucleolus $=\{(\frac{13}{4}, \frac{33}{8}, \frac{33}{8}, \frac{3}{2})\}$ with units in hours.

The first least core is determined from $\varepsilon^1 = -\frac{3}{2}$,

$$X^1 = C\left(-\frac{3}{2}\right) = \left\{ x_1 + x_2 + x_3 = \frac{23}{2}, x_4 = \frac{3}{2}, \right.$$

$$x_1 + x_2 + x_3 + x_4 = 13, x_1 + x_2 + x_4 \geq \frac{17}{2},$$

$$x_2 + x_3 + x_4 \geq \frac{19}{2}, x_1 \geq \frac{3}{2}, x_2 \geq \frac{3}{2},$$

$$x_1 + x_2 \geq \frac{11}{2}, x_3 \geq \frac{3}{2}, x_1 + x_3 \geq \frac{11}{2},$$

$$x_2 + x_3 \geq \frac{15}{2}, x_1 + x_4 \geq \frac{9}{2}, x_2 + x_4 \geq \frac{7}{2},$$

$$\left. x_3 + x_4 \geq \frac{7}{2}, x_1 + x_3 + x_4 \geq \frac{17}{2} \right\}$$

A decided allocation is $x_4 = \frac{3}{2}$. Next, to get X^2 first calculate the excesses $e(S, \vec{x})$ for $\vec{x} \in X^1$ and we see that we may eliminate coalitions $S = 4, S = 123$.

We then obtain $\varepsilon^2 = -\frac{7}{4}$, and $x_1 = \frac{13}{4}$. Again we calculate the excesses $e(S, \vec{x}), \vec{x} \in X^2 = C(-\frac{7}{4})$, and we determine we may eliminate coalitions $S = 1, S = 14, S = 1234$.

Finally we calculate $\varepsilon^3 = -\frac{15}{8}, X^3 = C(\varepsilon^3)$ using only the coalitions $S = 2, 3, 124, 134$ and we get $X^3 = \{(\frac{13}{4}, \frac{33}{8}, \frac{33}{8}, \frac{3}{2})\}$. That is the nucleolus.

6.31.c The schedule is set up as follows:

1. Since Curly(2) starts at 9am, and works 7 hours, he saves 4.125 hours and works 2.875 hours. That means he works from 9 to 11:52.5 am.
2. Larry(3) comes in after Curly. He saves 4.125 hours from the 6 he worked before cooperation and so he works 1.875 hours. So he starts at 11:52.5 am and leaves at 1:45pm.
3. Shemp(1) comes in after Larry. He saves 3.25 hours from the 5 he worked before. He must work 1.75 hours, which means he starts at 1:45pm and leaves at 3:30pm.
4. Moe(4) is the last to arrive. He saves 1.5 hours from the 3 hours he worked. He arrives at 3:30pm and leaves at 5pm, closing the office and turning out the lights.

6.33 The characteristic function for the **savings game** is $v(\emptyset) = 0, v(i) = 0, v(1234) = 22 - 8.5$, and

$$v(12) = 13 - 7.5, \quad v(13) = 11 - 7, \quad v(14) = 12 - 7.5,$$
$$v(23) = 10 - 6.5, \quad v(24) = 11 - 6.5, \quad v(34) = 9 - 5.5,$$
$$v(123) = 17 - 7.5, \quad v(124) = 18 - 8, \quad v(134) = 16 - 7.5,$$
$$v(234) = 15 - 7.$$

The least core is

$$X^1 = C(-1.125) = \{x_1 + x_2 + x_3 + x_4 = 13.5, x_1 + x_2 + x_3 \geq 10.625,$$
$$x_1 + x_2 + x_4 \geq 11.125, x_1 + x_3 + x_4 \geq 9.625,$$
$$x_2 + x_3 + x_4 \geq 9.125, x_1 \geq 1.125, x_2 \geq 1.125, x_1 + x_2 \geq 6.625,$$
$$x_3 \geq 1.125, x_1 + x_3 \geq 5.125, x_2 + x_3 \geq 4.625,$$
$$x_4 \geq 1.125, x_1 + x_4 \geq 5.625, x_2 + x_4 \geq 5.625, x_3 + x_4 \geq 4.625\}$$

This reduces to $X^1 = \{(\frac{35}{8}, \frac{31}{8}, \frac{19}{8}, \frac{23}{8})\} = \{(4.375, 3.875, 2.375, 2.875)\}$, so this is the nucleolus.

In the original terms of costs, we have

$$y_1 = c(1) - \frac{35}{8} = 2.625$$
$$y_2 = c(2) - \frac{31}{8} = 2.125$$
$$y_3 = c(3) - \frac{19}{8} = 1.625$$
$$y_4 = c(4) - \frac{23}{8} = 2.125.$$

6.37 We have $T \in 2^N \setminus \Pi^i$ if and only if $T \cup i \in \Pi^i$. Also, $|T \cup i| = |T| + 1$. Hence

$$x_i = \sum_{S \in \Pi^i} [v(S) - v(S - i)] \frac{(|S| - 1)!(n - |S|)!}{n!}$$
$$= \sum_{T \in 2^N \cdot \Pi^i} [v(T \cup i) - v(T)] \frac{(|T| + 1 - 1)!(n - |T| - 1)!}{n!}$$
$$= \sum_{T \in 2^N \cdot \Pi^i} [v(T \cup i) - v(T)] \frac{|T|!(n - |T| - 1)!}{n!}.$$

6.39.a A reasonable characteristic function is

$$v(M) = 155 - 80 = 75, \ v(L) = 160 - 75 = 85, \ v(C) = 140 - 78 = 62,$$
$$v(ML) = 240 - 40 = 200, \ v(MC) = 217 - 40 = 177, \ v(LC) = 227 - 40 = 187,$$
$$v(MLC) = 115 + 125 + 102 - 40 = 302.$$

6.39.b The Mathematica statements to find the least core are very simple:

$$c = \{75 - x_M \le \varepsilon, 85 - x_L \le \varepsilon, 62 - x_C \le \varepsilon,$$
$$200 - x_M - x_L \le \varepsilon, 177 - x_M - x_C \le \varepsilon, 187 - x_L - x_C \le \varepsilon,$$
$$x_M + x_L + x_C = 302\}$$

Then

$$\text{Minimize}\,[\{\varepsilon, c\}, \{x_M, x_L, x_C\}]$$

Of course this can also be done tediously by hand. The result is

$$\varepsilon^1 = -\frac{40}{3}, C(\varepsilon^1) = \left\{ \left(\frac{305}{3}, \frac{335}{3}, \frac{266}{3} \right) \right\}$$

and that's one fair way to divide the total of 302.

6.39.c We calculate the Shapley value from the table:

Order of Arrival	Moe	Larry	Curly
MLC	75	125	102
MCL	75	125	102
LMC	115	85	102
LCM	115	85	102
CML	115	125	62
CLM	115	125	62
Shapley	101.66	111.66	88.66

The Shapley allocation is the same as the nucleolus. Looks like waxing is lucrative.

6.41 Take the characteristic function representing the hours saved by a coalition,

$$v(M) = v(C) = v(L) = 0, v(MC) = 0, v(ML) = 2, v(CL) = 2, v(MCL) = 4.$$

Then the nucleolus is $C(-1) = \{(1, 1, 2)\}$. Thus Moe and Curly both save 1 hour while Larry saves 2. Since Curly arrives first, he can now work 9-12. Then Larry works 12-3, and Moe works 3-5.

The Shapley allocation is same as least core. For example,

$$x_M = (v(M) - v(\emptyset))\frac{(3-1)!}{(1-1)!}3!$$
$$+ (v(MC) - v(C) + v(ML) - v(L))\frac{(3-2)!}{(2-1)!}3!$$
$$+ (v(MCL) - v(CL))\frac{(3-3)!}{(3-1)!}3!$$
$$= 0 + 2\frac{1}{6} + 2\frac{2}{6} = 1$$

Here is the table giving the Shapley allocation:

Order of Arrival	Moe	Larry	Curly
MLC	0	2	2
MCL	0	4	0
LMC	2	0	2
LCM	2	0	2
CLM	2	2	0
CML	0	4	0
Shapley	1	2	1

6.43.a The characteristic function is $v(i) = 0, v(12) = 100, v(13) = 130, v(23) = 0, v(123) = 130$. We calculate the core,

$$C(0) = \{x_i \geq 0, i = 1, 2, 3, 100 \leq x_1 + x_2, 130 \leq x_1 + x_3, 0 \leq x_2 + x_3,$$

$$x_1 + x_2 + x_3 = 130\}$$

$$= \{x_2 = 0, x_1 + x_3 = 130, x_1 \geq 0, x_2 \geq 0\}.$$

The Shapley value is $\vec{x} = \{(\frac{245}{3}, \frac{50}{3}, \frac{95}{3})\}$, and this point is not in $C(0)$ since $x_2 = \frac{50}{3} > 0$ and

$$x_1 + x_3 = \frac{340}{3} \neq 130.$$

The nucleolus of this game is $\{(115, 0, 15)\}$ which is quite different from the Shapley value. Under the nucleolus, of the 130 available if the players cooperate, the seller gets 115 for her object, and player 3 gets to keep 15 of her 130. She pays the seller 115. Under Shapley, the seller gets 81.67, player 3 pays a total of 130-31.67=98.33 of which 81.67 goes to the seller to buy the object and 16.67 goes to player 2 to go away. Do you think this is what would actually happen?

6.43.b To be individually rational we need $v(i) \leq x_i, i = 1, 2, 3$ which is clearly satisfied for the Shapley value. The group rational condition requires $\sum_{i=1}^{3} x_i = v(N)$. In our case,

$$x_1 + x_2 + x_3 = \frac{245 + 50 + 95}{3} = 130 = v(N).$$

6.45 Shapley value=$(\frac{4}{3}, \frac{16}{3}, \frac{16}{3})$.

6.47 The Shapley value for the game is $(-1, -1, -1, -1)$. To see why, since the game is symmetric clearly the formula for the Shapley value will give $x_1 = x_2 = x_3 = x_4$. Since $x_1 + x_2 + x_3 + x_4 = v(N) = -4$, we must have $x_1 = x_2 = x_3 = x_4 = -1$.

6.49 Shapley value=$\{(3.91667, 3.66667, 2.75, 3.16667)\}$.

6.51.a The characteristic function is

$$v(i) = 0, i = 1, 2, 3, 4, v(12) = v(13) = v(14) = 1, v(23) = v(24) = v(34) = 0,$$

$$v(123) = v(124) = v(134) = 2, v(234) = 0, v(1234) = 2.$$

6.51.b The core is $C(0) = \{(2, 0, 0, 0)\}$ To see this,

$$C(0) = \{x_i \geq 0, i = 1, 2, 3, 1 \leq x_1 + x_2, 1 \leq x_1 + x_3, 1 \leq x_1 + x_4$$

$$2 \leq x_1 + x_2 + x_3, 2 \leq x_1 + x_2 + x_4, 2 \leq x_1 + x_3 + x_4,$$

$$x_1 + x_2 + x_3 + x_4 = 2\}$$

Adding the 3 player coalitions we see that

$$3x_1 + 2x_2 + 2x_3 + 2x_4 \geq 6 \implies x_1 \geq 2.$$

Then $x_1 + x_2 + x_3 + x_4 = 2 \implies x_1 = 2, x_2 = x_3 = x_4 = 0$. Thus $C(0) = \{(2,0,0,0)\}$. Since the core contains only one point, it is the nucleolus. Player 1 is allocated everything while players 2, 3 and 4 get nothing. Doesn't seem fair.

6.51.c Players 2, 3 and 4 are symmetric so the Shapley allocation is of the form (x, y, y, y), where $x + 3y = 2$. By the formula for the Shapley value,

$$x = \sum_{S \in \Pi^1} [v(S) - v(S-1)] \frac{(|S|-1)!(4-|S|)!}{4!}$$

$$= \frac{(2-1)!(4-2)!}{4!} \times 3 + \frac{(3-1)!(4-3)!}{4!} \times 6 + \frac{(|4|-1)!(4-|4|)!}{4!} \times 2$$

$$= \frac{1}{4} + \frac{1}{2} + \frac{1}{4} = \frac{5}{4}.$$

Then $y = \frac{2-x}{3} = \frac{1}{4}$. The Shapley value is then $\vec{x} = (\frac{5}{4}, \frac{1}{4}, \frac{1}{4}, \frac{1}{4})$.

6.51.d The characteristic function is $v(1) = v(2) = 10, v(12) = 30$. The core is

$$C(0) = \{x_1 \geq 10, x_2 \geq 10, x_1 + x_2 = 30\}$$

which has more than one point. That means we have to calculate the least core.

$$C(\varepsilon) = \{x_1 \geq 10 - \varepsilon, x_2 \geq 10 - \varepsilon, x_1 + x_2 = 30\}$$

$$= \{30 = x_1 + x_2 \geq 20 - 2\varepsilon, x_1 \geq 10 - \varepsilon, x_2 \geq 10 - \varepsilon\}.$$

Thus $\varepsilon \geq -5 \implies \varepsilon^1 = -5 \implies x_1 = x_2 = 15$ and $C(-5) = \{(15, 15)\}$, an even split. The Shapley allocation is

$$x_i = \frac{1}{2}[10 - 0] + \frac{1}{2}[30 - 10] = 15, \quad i = 1, 2$$

which is the same as the nucleolus.

6.53 If $n = 5, b = 3, c = 2$ we have

$$v(S) = \begin{cases} \max\{|S|(-10), |S|(-(5-|S|)2) - |S|3\}, & \text{if } S \subset N; \\ -15, & \text{if } S = N. \end{cases}$$

A direct calculation shows that

$$v(S) = \begin{cases} -10, & \text{if } |S| = 1; \\ -18, & \text{if } |S| = 2; \\ -21, & \text{if } |S| = 3; \\ -20, & \text{if } |S| = 4; \\ -15, & \text{if } S = N, |N| = 5. \end{cases}$$

Plugging into Shapley's formula gives $\vec{x} = (-3, -3, -3, -3, -3)$.

6.55.a Since

$$c(1) = 16, c(2) = 15, c(3) = 15$$

$$c(12) = 18, c(13) = 16, c(23) = 18, c(123) = 19$$

we compute

$$v(i) = 0, v(12) = 13, v(13) = 15, v(23) = 12, v(123) = 27.$$

This characteristic function is superadditive and they all have an incentive to join the carpool.

6.55.b The inequalities we need to solve are

$$C(\varepsilon) = \{0 \leq x_i \leq -\varepsilon, 13 - \varepsilon \leq x_1 + x_2, 15 - \varepsilon \leq x_1 + x_3, 12 - \varepsilon \leq x_2 + x_3$$

$$x_1 + x_2 + x_3 = 27\}$$

Eliminating x_3 we get $x_2 \leq 12 + \varepsilon, x_1 \leq 15 + \varepsilon \implies 13 - \varepsilon \leq x_1 + x_2 \leq 27 + 2\varepsilon$. This leads to $\varepsilon^1 = -\frac{14}{3}$ and then $x_1 + x_2 = \frac{53}{3}, x_3 = \frac{28}{3}$.

Then $x_2 \leq 12 + \varepsilon^1 = \frac{22}{3}, x_1 \leq 15 + \varepsilon^1 = \frac{31}{3}$. Since $31 + 22 = 53$, we conclude $x_2 = \frac{22}{3}, x_1 = \frac{31}{3}$. Thus

$$X^1 = C(-\frac{14}{3}) = \left\{\left(\frac{31}{3}, \frac{22}{3}, \frac{28}{3}\right)\right\}.$$

6.55.c It seems reasonable that Larry and Curly should pay Moe the difference in the amount it costs them to drive on their own and the amount of the savings attributable to that player. We get Larry should pay Moe $c(2) - \frac{22}{3} = 15 - \frac{22}{3} = 7.66$ and Curly should pay Moe $c(3) - \frac{28}{3} = 5.66$.

6.55.d The table for calculation of the Shapley value is given below.

Arrival/Player	1	2	3
123	0	13	14
132	0	12	15
213	13	0	14
231	15	0	12
312	15	12	0
321	15	12	0
Total	58	49	55

The Shapley allocation is therefore $x_1 = \frac{58}{6}, x_2 = \frac{49}{6}, x_3 = \frac{55}{6}$. We get Larry should pay Moe $c(2) - \frac{49}{6} = 6.83$ and Curly should pay Moe $c(3) - \frac{55}{6} = 5.83$.

6.55.e Let Shemp be player 4. The costs are

$$c(1) = 16, c(2) = 15, c(3) = 15, c(4) = 15$$

$$c(12) = 18, c(13) = 16, c(14) = 16, c(23) = 18, c(24) = 17, c(34) = 16,$$

$$c(123) = 19, c(124) = 18, c(134) = 17, c(234) = 18,$$

$$c(1234) = 19.$$

The corresponding cost savings are

$$v(i) = 0, i = 1, 2, 3, 4$$

$$v(12) = 13, v(13) = 15, v(14) = 15, v(23) = 12, v(24) = 13, v(34) = 14,$$

$$v(123) = 27, v(124) = 28, v(134) = 29, v(234) = 27,$$

$$v(1234) = 42.$$

By going through similar calculations, we get

$$\text{Nucleolus} = C\left(-\frac{15}{4}\right) = \left\{\left(x_1 = \frac{45}{4}, x_2 = \frac{37}{4}, x_3 = \frac{41}{4}, x_4 = \frac{45}{4}\right)\right\}$$

The Shapley value is

$$x_1 = \frac{133}{12}, x_2 = \frac{115}{12}, x_3 = \frac{125}{12}, x_4 = \frac{131}{12}.$$

Larry, Curly and Shemp, should pay Moe the difference in the amount it costs them to drive on their own and the amount of the savings attributable to that player.

1. Larry should pay Moe $c(2) - \frac{115}{12} = 5.42$ under Shapley and $c(2) - \frac{37}{4} = 5.75$ under nucleolus.

2. Curly should pay Moe $c(3) - \frac{125}{12} = 4.58$ under Shapley and $c(3) - \frac{41}{4} = 4.75$ under nucleolus.

3. Shemp should pay Moe $c(4) - \frac{131}{12} = 4.08$ under Shapley and $c(4) - \frac{45}{4} = 3.75$ under nucleolus.

6.57 One simple way to calculate the Banzhaf-Coleman index is to list for each coalition, the players who are critical for that coalition (critical means win with the player, lose without her). The table contains the results

Coalition	Votes	Critical Players
12	6	1, 2
13	5	1, 3
14	5	1, 4
123	7	1
124	7	1
134	6	1
1234	8	1

Next, count up the number of times each player is critical for a coalition. For player 1 it is 7 times; player 2, 3, and 4, exactly 1 time each. The total number of winning coalitions for each critical player is 10. Hence,

$$b_1 = \frac{7}{10}, \ b_2 = b_3 = b_4 = \frac{1}{10}.$$

Player 1's power index is 7 times that of the other players even though she has only twice as many votes (as player 2).

6.59.a

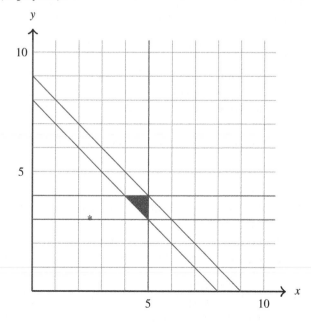

6.59.b The least core is at $(4\frac{2}{3}, 3\frac{2}{3}, 2\frac{2}{3})$

6.59.c

	1	2	3
132	0	4	7
213	5	3	3
231	5	3	3
312	5	4	2
321	5	4	2
	20	26	20

payoff: $(3\frac{1}{3}, 4\frac{1}{3}, 3\frac{1}{3})$

This is interesting because it is not stable! Also you can note Shapley is much worse for PI.

Solutions for Chapter 7

7.1.a The matrices are $A = \begin{bmatrix} \frac{1}{2} & 0 \\ 1 & 0 \end{bmatrix}, B^T = A$. Calculate easily that $value(A) = value(B^T) = 0$ and hence the security point is $(0, 0)$. The Nash bargaining problem is then

$$\text{Maximize } uv, \quad \text{subject to } u + v \leq 1, 0 \leq u \leq 1, 0 \leq v \leq 1.$$

By calculus, the solution is $\bar{u} = \frac{1}{2}, \bar{v} = \frac{1}{2}$. The bargained solution is that each contestant should Split. That seems fair and natural.

7.1.b First we calculate the threat security point. Since the Pareto optimal boundary is $u + v = 1$ we have $v = -u + 1$ and so $m_p = -1, b = 1$. Then we calculate $-m_p A - B = A - B = \begin{bmatrix} 0 & -1 \\ 1 & 0 \end{bmatrix}$, and $value(A - B) = 0$, with a saddle point $X_t = (0, 1), Y_t = (0, 1)$. The threat security point is therefore $u^t = X_t A Y_t^T = 0$ and $v^t = X_t B Y_t^T = 0$, exactly the same as the security point we found before. Immediately we see that the threat bargaining solution is

$$\bar{u} = \frac{1}{-2m_p}(b + value(-m_p A - B)) = \frac{1}{2}, \bar{v} = \frac{1}{2}(b - value(-m_p A - B)) = \frac{1}{2},$$

the same solution as before. This means that in order to achieve the threat bargained solution, they should agree to always play row 1, column 1, or if they play many times, half the time they should play row 2, column 1, and half the time row 1, column 2. The expected payoffs are the same.

7.1.c If player 1 announces he will Claim the prize and split it after the show is over, player 2 has two choices.

1. Player 2 can believe that player 1 will actually split the prize. In this case player 2 and player 1 both receive $\frac{1}{2}$.

2. Player 2 does not believe player 1 will split the prize claimed. In that case, player 2 should threaten to claim the prize as well if player 1 does not agree to Split. If player 1 does not agree to Split, player 2 should carry out the threat. Either way, since player 2 doesn't believe player 1, player 2 ends up with either 0 or $\frac{1}{2}$, but if 0, then player 1 also gets 0, not 1.

7.3 The matrices are $A = \begin{bmatrix} 4 & 2 \\ -1 & 2 \end{bmatrix}$ and $B = \begin{bmatrix} 2 & -1 \\ 2 & 4 \end{bmatrix}$. A computation shows $value(A) = 2, vAlue(B^T) = 2$ and the security point is $(2, 2)$. With $(2, 2)$ security point, the bargaining solution is (3, 3) since that is the solution of the problem

$$\text{Maximize } (u - 2)(v - 2), \text{ subject to } u + v \le 6, 2 \le u \le 4, 2 \le v \le 4.$$

In fact take a derivative of $(u - 2)(6 - u - 2)$ and set to zero to see that $u = 3$.

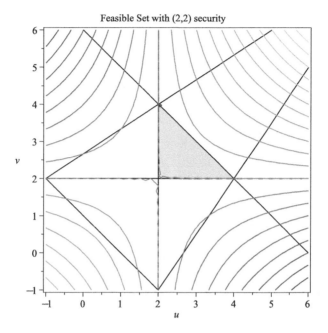

Feasible Set with (2,2) security

Next, since the Pareto optimal boundary is $u + v = 6$, we have $m_p = -1, b = 6$. We calculate $value(-m_p A - B)$ and the saddle strategies for that game to get $value(A - B) = 2, X_t = (1, 0), Y_t = (1, 0)$.

The threat security point is therefore $(u^t, v^t) = (4, 2)$ with threat strategies $X_t = (1, 0) = Y_t$.
Next we use the formulas

$$\bar{u} = \frac{(m_p u^t + v^t - b)}{2m_p} \quad \text{and } \bar{v} = \frac{(m_p u^t + v^t + b)}{2}$$

with $b = 6$. The threat solution is $(\bar{u}, \bar{v}) = (4, 2)$ with both players threatening to play row 1, column 1.

The KS line for $(u^*, v^*) = (2, 2)$ is $v - 2 = k(u - 2), k = \frac{\max v - 2}{\max u - 2} = 1$, or simplified to $v = u$.
This intersects $u + v = 6$, the Pareto-optimal boundary, at $(3, 3)$. Therefore $(3, 3)$ is the KS solution.

There is no KS line for $(u^*, v^*) = (4, 2)$ because it is on the edge of the Pareto line.
If we consider the cooperative game, the characteristic function is $v(1) = v(2) = 2, v(12) = 6$, which gives nucleolus and Shapley value $(3, 3)$ and matches with the solution for the $(2, 2)$ security point.

7.5 The matrices are

$$A = \begin{bmatrix} -3 & 0 & 1 \\ 2 & \frac{5}{2} & -1 \end{bmatrix} \quad B = \begin{bmatrix} -1 & 5 & \frac{19}{4} \\ \frac{7}{2} & \frac{3}{2} & -3 \end{bmatrix}.$$

We have $value(A) = -\frac{1}{7}, value(B^T) = \frac{19}{8}$. That is our safety point. The Pareto optimal boundary has three line segments:

$$\begin{cases} \frac{1}{4}u + v = 5, & \text{if } 0 \le u \le 1; \\ \frac{5}{4}u + v = 6, & \text{if } 1 \le u \le 2; \\ 2u + \frac{1}{2}v = \frac{23}{4}, & \text{if } 2 \le u \le \frac{5}{2}. \end{cases}$$

The Nash problem is

$$\text{Maximize } (u + \frac{1}{7})(v - \frac{19}{8})$$

subject to $(u, v) \in S$. The part of the Pareto optimal boundary for this problem is the line segment $\frac{5}{4}u + v = 6, 1 \le u \le 2$. Using calculus we find

$$\bar{u} = \frac{193}{140}, \quad \bar{v} = \frac{479}{112}.$$

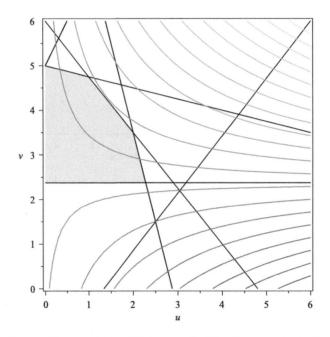

Next we consider the threat solution. We have to find the threat strategies for all three line segments.

1. $v = -\frac{1}{4}u + 5, 0 \le u \le 1$. Then $m_p = -\frac{1}{4}, b = 5$, and

$$value(\frac{1}{4}A - B) = -\frac{487}{236}, \ X_t = (\frac{17}{59}, \frac{42}{59}), Y_t = (\frac{33}{59}, \frac{26}{59}, 0),$$

and

$$u^t = X_t A Y_t^T = 1.097, \ v^t = X_t B Y_t^T = 2.34, \implies \bar{u} = 5.87, \bar{v} = 3.53.$$

Since $5.87 \notin [0, 1]$, this is not the threat solution.

2. $v = -\frac{5}{4}u + 6, 1 \le u \le 2$. Then $m_p = -\frac{5}{4}, b = 6$, and

$$value(\frac{5}{4}A - B) = -1, \ X_t = (0, 1), Y_t = (1, 0, 0).$$

Then

$$u^t = X_t A Y_t^T = 2, \quad v^t = X_t B Y_t^T = \frac{7}{2} \implies \bar{u} = 2, \ \bar{v} = \frac{7}{2}.$$

This is in the range. Let's check the final segment.

3. $v = -4u + \frac{23}{2}, 2 \le u \le \frac{5}{2}$. In this case $m_p = -4, b = \frac{23}{2}$, and

$$value(4A - B) = -\frac{115}{126}, \ X_t = (\frac{22}{63}, \frac{41}{63}), Y_t = (\frac{1}{63}, 0, \frac{62}{63}).$$

The safety point is then

$$u^t = X_t A Y_t^T = -0.294, \quad v^t = -0.258 \implies \bar{u} = 1.32, \ \bar{v} = 6.206.$$

Since $1.32 \notin [2, \frac{5}{2}]$, this too is not the threat solution.

We conclude that the threat solution is $\bar{u} = 2, \ \bar{v} = \frac{7}{2}$ and player 1 threatens to always play the second row; player 2 threatens to use the third column unless they both come to their senses.

Finally we determine the KS solution for the safety point $u^* = -\frac{1}{7}, v^* = \frac{19}{8}$.
We have $a = \max_{(u,v)\in S} u = \frac{73}{32}, b = \max_{(u,v)\in S} v = 5$. Then

$$v = v^* + k(u - u^*) = \frac{19}{8} + \frac{196}{181}(u + \frac{1}{7})$$

is the KS line, and this intersects the Pareto optimal boundary through the segment $v = -\frac{5}{4}u + 6$ at the point $\bar{u} = 1.487, \bar{v} = 4.140$, and that is the KS solution.

7.7 The spread is $350 - 305 = 45K$. If the process ends with at most 3 stages, player 1 should offer $305 + .99(45) = 349.55$. If the process could go on indefinitely, player 1 should offer $305 + .5025(45) = 327.625$, because $x_1 = 1 - 0.99 + 0.99^2 \frac{1}{1+0.99}$.

7.9.a The Nash problem is

$$\text{Maximize } (u - u^*)(v - v^*) = (f(w) - pw)(pw + (W - w)p_0 - Wp_0)$$
$$= (f(w) - pw)(p - p_0)w$$

with $(p, w) \in S$, where

$$S = \{(u, v) \mid u \ge u^*, v \ge v^*\}$$
$$= \{(u, v) \mid u \ge 0, v \ge Wp_0\}$$
$$= \{(p, w) \mid f(w) - pw \ge 0, p \ge p_0, 0 \le w \le W\}$$

7.9.b Set $h(p, w) = (f(w) - pw)(p - p_0)w$. We have

$$\frac{\partial h}{\partial p} = w((p_0 - 2p)w + f(w)) = 0$$
$$\frac{\partial h}{\partial w} = (p - p_0)(f(w) + w(f'(w) - 2p)) = 0.$$

Solving the first equation for p gives

$$p = \frac{wp_0 + f(w)}{2w}.$$

Substitute this p into the equation $h_w = 0$ to get

$$f(w) + w(f'(w) - \frac{wp_0 + f(w)}{w}) = f(w) + wf'(w) - p_0 w - f(w)$$
$$= w(f'(w) - p_0) = 0,$$

which implies $f'(w) = p_0$.

7.9.c Using the formulas from the previous part, $p_0 = f'(w^*) = \frac{1}{w^*+a}$ which implies $w^* = \frac{1}{p_0} - a > 0$ and

$$p^* = \frac{wp_0 + f(w)}{2w} = \frac{p_0(ap_0 - \ln(\frac{1}{p_0} - a) - b - 1)}{2ap_0 - 2}.$$

7.11 1. $(-4, 1)$: $(0, 3)$
    2. $(-2, 2)$: $(0, 3)$
    3. $(-3, -1)$: $(2, 2)$
    4. $(-2, -2)$: $(2, 2)$
    5. $(2, -3)$: $(\frac{17}{6}, -\frac{1}{2})$
    6. $(4, -4)$: $(4, -4)$

Solutions for Chapter 8

8.1 Let's look at the equilibrium $X_1 = (1, 0)$. We need to show that for $x \neq 1$, $u(1, px + (1 - p)) > u(x, px + (1 - p))$ for some p_x, and for all $0 < p < p_x$. Now $u(1, px + (1 - p)) = 1 - p + px$, and $u(x, px + (1 - p)) = p + x - 3px + 2px^2$. In order for X_1 to be an ESS, we need $1 > 2p(x - 1)^2$, which implies $0 < p < 1/(2(x - 1)^2)$. So, for $0 \leq x < 1$, we can take $p_x = 1/(2(x - 1)^2)$ and the ESS requirement will be satisfied. Similarly, the equilibrium $X_2 = (0, 1)$, can be shown to be an ESS. For $X_3 = (\frac{1}{2}, \frac{1}{2})$, we have

$$u(\frac{1}{2}, px + (1 - p)/2) = \frac{1}{2} \text{ and } u(x, px + (1 - p)/2) = \frac{1}{2} + \frac{p}{2} - 2px + 2px^2.$$

In order for X_3 to be an ESS, we need

$$\frac{1}{2} > \frac{1}{2} + \frac{p}{2} - 2px + 2px^2,$$

which becomes $0 > 2p(x - \frac{1}{2})^2$, for $0 < p < p_x$. This is clearly impossible, so X_3 is not an ESS.

8.3 The pure Nash equilibria $(X_2, Y_2), (X_3, Y_3)$ are not symmetric and hence cannot be ESS's. We determine whether or not X_1 is an ESS. We calculate for $x \neq \frac{1}{7}$,

$$u(\frac{1}{7}, px + (1 - p)\frac{1}{7}) = \frac{8}{7} + \frac{p}{28}(5 - 35x)$$

and

$$u(x, px + (1 - p)\frac{1}{7}) = \frac{8}{7} + \frac{p}{28}(3 - 7x - 98x^2).$$

The question is whether (for small enough $0 < p < 1$) we have

$$\frac{8}{7} + \frac{p}{28}(5 - 35x) > \frac{8}{7} + \frac{p}{28}(3 - 7x - 98x^2) \Leftrightarrow 98x^2 - 28x + 2 > 0.$$

Since $98x^2 - 28x + 2 = 2(7x - 1)^2 > 0$ for any $x \neq \frac{1}{7}$, we conclude that $X_1 = (\frac{1}{7}, \frac{6}{7})$ is indeed an ESS. Consequently, in a series of naval battles between the British and French, strong navies will always attack, while weak navies will attack with probability $\frac{1}{7}$.

8.5 The Nash equilibria and their payoffs are shown in the following table; they are all symmetric.

X^*	$u(X^*, X^*)$
$(1, 0, 0)$	2
$(0, 1, 0)$	2
$(0, 0, 1)$	2
$(\frac{3}{4}, \frac{1}{4}, 0)$	$\frac{5}{4}$
$(\frac{1}{4}, 0, \frac{3}{4})$	$\frac{5}{4}$
$(0, \frac{3}{4}, \frac{1}{4})$	$\frac{5}{4}$
$(\frac{1}{3}, \frac{1}{3}, \frac{1}{3})$	$\frac{2}{3}$

For $X^* = (1, 0, 0)$ you can see this is an ESS because it is strict. Consider next $X^* = (\frac{3}{4}, \frac{1}{4}, 0)$. Since $u(Y, X^*) = \frac{5}{4}(y_1 + y_2) - y_3/2$, the set of best response strategies is $Y = (y, 1 - y, 0)$. Then $u(Y, Y) = 4y^2 - 4y + 2$, and $u(X^*, Y) = -\frac{1}{4} + 2y$. Since it is **not** true that $u(Y, Y) < u(X^*, Y)$, for all best responses $Y \neq X^*$, X^* is not an ESS.

8.7.a There is a unique Nash, strict and symmetric ESS $X^* = (0, 1)$ if $a < 0, b > 0$, and $X^* = (1, 0)$ if $b < 0, a > 0$.

8.7.b There are three Nash equilibria, all symmetric, Nash equilibria $= (1, 0), (0, 1), X$, where $X = \left(\frac{b}{(a+b)}, \frac{a}{(a+b)} \right)$. Both $(1, 0), (0, 1)$ are strict, $(1, 0), (0, 1)$ are both evolutionary stable. The mixed X is not an ESS since

$$E(1, 1) = a > \frac{ab}{(a + b)} = E(X, 1).$$

8.7.c There are two strict asymmetric Nash Equilibria, and one symmetric Nash Equilibrium given by $X = (\frac{b}{(a+b)}, \frac{a}{(a+b)})$. However now X is an ESS since

$$E(X, Y) = cay_1 + (1 - c)by_2 = \frac{ab}{(a + b)}, \quad \text{where } c = \frac{b}{(a + b)},$$

and for every strategy

$$Y \neq X, \quad E(Y, Y) = ay_1^2 + by_2^2 < \frac{ab}{(a + b)} = E(X, Y),$$

so X is an ESS.

8.9.a The replicator equation in simplified form is

$$\frac{dp}{dt} = p(1 - p)(1 - 2p).$$

8.9.b The three Nash equilibria are $X_1 = (\frac{1}{2}, \frac{1}{2}) = Y_1$, and the two nonsymmetric Nash points $((0, 1), (1, 0))$ and $((1, 0), (0, 1))$. So only X_1 is a possible ESS. It is not hard to show directly that X_1 is an ESS but we can also use the fact from (8.1.9) that if $a_{11} \neq a_{21}$ and $a_{22} \neq a_{12}$, then there must be an ESS. Since it can't be the nonsymmetric Nash's, X_1 must be an ESS.

8.9.c From the following figure you can see that $(p_1(t), p_2(t)) \to (\frac{1}{2}, \frac{1}{2})$ as $t \to \infty$ and conclude that (X_1, X_1) is an ESS. Verify directly using the stability theorem that it is asymptotically stable.

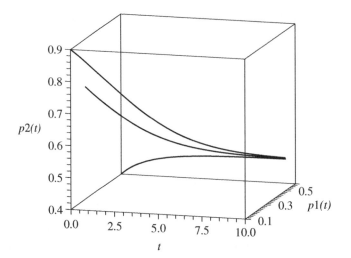

t

The figure shows trajectories starting from three different initial points. In the three-dimensional figure you can see that the trajectories remain in the plane $p_1 + p_2 = 1$. For the problem, Determinant $(a1) = 5 > 0$, and $b1 = -6 > 0$, so the stability Theorem 8.2.3 allows us to conclude that $(\frac{1}{2}, \frac{1}{2})$ is asymptotically stable.

8.11.a The unique Nash equilibrium is $X = (\frac{15}{44}, \frac{20}{44}, \frac{9}{44}) = Y$, and it is symmetric. We obtain this from the equality of payoffs theorem or Problem (2.25). We calculate using

$$Y^{*T} = \frac{A^{-1} J_3^T}{J_3 A^{-1} J_3^T}$$

and get $Y^* = \left(\frac{15}{44}, \frac{20}{44}, \frac{9}{44}\right)$. Observe that if we want to use the formula for X^* in Problem (2.25), we must use $B = A^T$.

8.11.b We check that with $u(X, Y) = X A Y^T$

$$u(X^*, X^*) = \frac{95}{44} = u(X, X^*)$$

for any $X \in S_3$. Now we check if $u(X^*, X) > u(X, X)$ for all $X \in S_3$. We have

$$u(X^*, X) - u(X, X) = \frac{75}{44} - \frac{50}{11}x_1 + 4x_1^2 - \frac{45}{11}x_2 + 4x_1 x_2 + 3x_2^2$$

The easiest way to check if this is > 0 is to graph the function, or to minimize the function. We'll do both. First, we have

$$f(x_1, x_2) = \frac{75}{44} - \frac{50}{11}x_1 + 4x_1^2 - \frac{45}{11}x_2 + 4x_1 x_2 + 3x_2^2$$

and the problem becomes

$$\text{Minimize} f(x_1, x_2), 0 \le x_1 \le 1, 0 \le x_2 \le 1, x_1 + x_2 \le 1$$

We can do this by calculus or with Mathematica. The minimum is 0 attained at $x_1 = \frac{15}{44}, x_2 = \frac{20}{44}$. Thus $f(x_1, x_2) > 0$ at any point except at the Nash equilibrium X^*. This tells us that X^* is indeed an ESS. Here is the graph of f showing it is positive except for the minimum point.

8.11.c The replicator equations are

$$\frac{dp_i}{dt} = p_i(E(i, \pi) - E(\pi, \pi)), \quad i = 1, 2, 3.$$

Using $p_3 = 1 - p_1 - p_2$, and a lot of algebra we get the replicator equations for p_1, p_2 are

$$\frac{dp_1(t)}{dt} = p_1(t)(-9p_1(t) - 8p_2(t) + 5 + 4p_1(t)^2 + 4p_1(t)p_2(t) + 3p_2(t))^2$$

$$\frac{dp_2(t)}{dt} = p_2(t)(-p_1(t) - 3p_2(t) + 4p_1(t)^2 + 4p_1(t)p_2(t) + 3p_2(t))^2$$

Then $p_1 = \frac{15}{44}, p_2 = \frac{20}{44}$ is a stationary solution. To check stability we use Theorem 8.2.3. These calculations may be done by hand but mathematica is way easier. Here are the commands to do this.

$$f[p_1, p_2] = p_1(-9p_1 - 8p_2 + 5 + 4p_1^2 + 4p_1p_2 + 3p_2)^2$$

$$g[p_1, p_2] = p_2(-p_1 - 3p_2 + 4p_1^2 + 4p_1p_2 + 3p_2)^2$$

$$s[p_1, p_2] = D[f[p_1, p_2], p_1] + D[g[p_1, p_2], p_2]$$

$$s[15/44, 20/44] = -0.230517 < 0$$

$$t[p_1, p_2] = D[\{f[p_1, p_2], g[p_1, p_2]\}, \{\{p_1, p_2\}\}]$$

$$Det[t[15/44, 20/44]] = 1.90447 > 0$$

The third line calculates $f_{p_1} + g_{p_2}$, and the fifth line calculates the determinant of the Jacobian matrix given in the fourth line. By the Theorem 8.2.3 we conclude that $p_1 = \frac{15}{44}, p_2 = \frac{20}{44}$ is asymptotically stable.

The remaining stationary solutions are $(p_1 = 0, p_2 = 1)$, $(p_1 = 0, p_2 = 0)$, $(p_1 = 1, p_2 = 0)$. The convergence to $(\frac{15}{44}, \frac{20}{44}, \frac{9}{44})$ is shown in the figure.

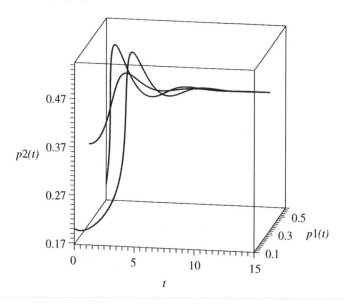

8.13.a $X^* = (\frac{1}{3}, \frac{1}{3}, \frac{1}{3})$ is the unique symmetric Nash equilibrium. But since $u(Y, Y) = 0$, $u(Y, X^*) = 0$, and $u(X^*, Y) = 0$, it will **not** be true that $u(Y, Y) < u(X^*, Y)$ for all Y that is a best response to X^*.

8.13.b It still is true that $X^* = (\frac{1}{3}, \frac{1}{3}, \frac{1}{3})$ is the unique symmetric Nash equilibrium. Now we calculate for any $Y = (y_1, y_2, 1 - y_1 - y_2)$

$$u(Y, Y) = YAY^T = a(1 + 2y_1^2 + (2y_1 + 2y_2)(-1 + y_2))$$

$$u(Y, X^*) = \frac{a}{3}$$

$$u(X^*, Y) = \frac{a}{3}$$

$$u(X^*, X^*) = \frac{a}{3}$$

Hence, we are in the condition $u(X^*, X^*) = u(Y, X^*)$ and we have to check if $u(X^*, Y) > u(Y, Y)$ for every $Y \neq X^*$. However,

$$\text{Minimum}\,[u(X^*, Y) - u(Y, Y)]$$

is $-\frac{8a}{3} < 0$. Thus, it is not true that X^* is an ESS.

8.13.c The reduced replicator equations are

$$\frac{dp_1}{dt} = f(p_1, p_2) = -p_1(-1 + p_1 + 2p_2 + a(1 + 2p_1^2 + 2p_2(-1 + p_2) + p_1(-3 + 2p_2)))$$

$$\frac{dp_2}{dt} = g(p_1, p_2) = -p_2(1 - 2p_1 - p_2 + a(1 - 3p_2 + 2(p_1^2 + p_1(-1 + p_2) + p_2^2)))$$

Then

$$(f_{p_1} + g_{p_2})(\frac{1}{3}, \frac{1}{3}) = \frac{2a}{3} > 0$$

and $\det(J(f, g))(\frac{1}{3}, \frac{1}{3}) = \frac{1}{3} + \frac{a^2}{9} > 0$. According to Theorem 8.2.3, X^* is unstable. This is illustrated in the following figure in the case when $a = 0$.

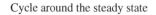

Cycle around the steady state

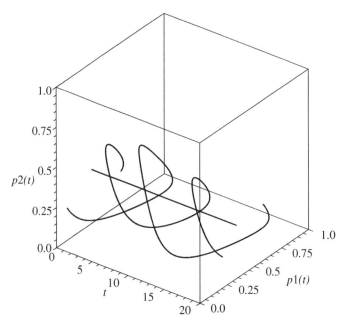

The case when $a = 2$ is shown in the following figure.

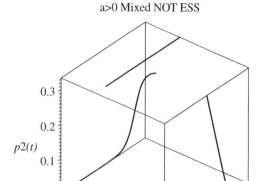

a>0 Mixed NOT ESS

You can see that unless you start exactly at X^*, the trajectories converge to one of the pure Nash equilibria.

8.15 The difference between this problem and the matrix in Problem (8.13.b) is that we are taking the negative of the matrix. We consider

$$A = \begin{bmatrix} -a & 1 & -1 \\ -1 & -a & 1 \\ 1 & -1 & -a \end{bmatrix} = - \begin{bmatrix} a & -1 & 1 \\ 1 & a & -1 \\ -1 & 1 & a \end{bmatrix}.$$

It still is true that $X^* = (\frac{1}{3}, \frac{1}{3}, \frac{1}{3})$ is the unique symmetric Nash equilibrium. We see this by calculating

$$X^* = \frac{J_3 A^{-1}}{J_3 A^{-1} J_3^T} = (\frac{1}{3}, \frac{1}{3}, \frac{1}{3}).$$

Now we calculate for any $Y = (y_1, y_2, 1 - y_1 - y_2)$

$$u(Y, Y) = Y A Y^T = -a(1 + 2y_1^2 + (2y_1 + 2y_2)(-1 + y_2))$$
$$u(Y, X^*) = -\frac{a}{3}$$
$$u(X^*, Y) = -\frac{a}{3}$$
$$u(X^*, X^*) = -\frac{a}{3}$$

Again the condition $u(X^*, X^*) = u(Y, X^*)$ holds and we have to check if $u(X^*, Y) > u(Y, Y)$ for every $Y \neq X^*$. We consider,

$$\text{Minimum } [u(X^*, Y) - u(Y, Y)]$$

over all strategies Y and calculate that this minimum is zero, attained only when $Y = X^*$. Thus, in this case X^* is an ESS. Here is the figure for the replicator dynamics.

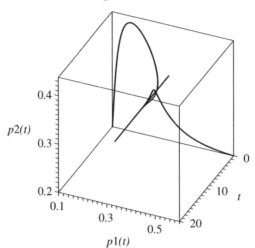

Diagonal<0 Mixed IS ESS

8.19.a At $(\frac{1}{3}, \frac{1}{12})$, $(0, \frac{3}{4})$ and $(1, 0)$. These are all fixed points of the dynamical system.

8.19.b At $(0, \frac{3}{4}, \frac{1}{4})$ and $(1, 0, 0)$. These are all fixed points of the dynamical system.

8.19.c 1. Will tend toward $(1, 0, 0)$.

2. Will tend toward $(0, \frac{3}{4}, \frac{1}{4})$.

3. Initially the system will head toward $(\frac{1}{3}, \frac{1}{12}, \frac{7}{12})$. But this is an unstable fixed point, so randomness will knock it off, and it will head toward either $(0, \frac{3}{4}, \frac{1}{4})$ or $(1, 0, 0)$. It is impossible to know which.

References

1 C. Aliprantis and S. Chakrabarti, *Games and Decision Making*, Oxford University Press, New York, 2000.

2 D. Austen-Smith and J. Banks, *Information aggregation, rationality, and the Condorcet jury theorem*, Am. Political Sci. Rev. 90 (1996), no. 1, 34–45.

3 K. Binmore, *Playing for Real: A Text on Game Theory*, Oxford University Press, 2007.

4 A. Devinatz, *Advanced Calculus*, Holt, Rinehart & Winston, New York, 1968.

5 M. Dresher, *The Mathematics of Games of Strategy*, Dover, New York, 1981.

6 T. Ferguson, *Game Theory*, Notes for a Course in Game Theory, available at www.gametheory .net.

7 D. Fudenberg and J. Tirole, *Game Theory*, MIT Press, 1991.

8 D. Fudenberg and J. Tirole, *Perfect Bayesian equilibrium and sequential equilibrium*, J. Econ. Theory 53 (1991), no. 2, 236–260.

9 H. Gintis, *Game Theory Evolving*, Princeton University Press, Princeton, NJ, 2000.

10 J. Goeree and C. Holt, *Stochastic game theory: For playing games, not just doing theory*, Proc. Natl. Acad. Sci. U.S.A. 96 (1999), 10564–10567.

11 J. Gonzalez-Diaz, I. Garcia-Jurado, and M. G. Fiestrar-Janeiro, *An Introductory Course on Mathematical Game Theory*, Graduate Studies in Mathematics, American Mathematical Society, 2010.

12 J. Harrington, *Games, Strategies, and Decision Making*, 2nd ed., Worth Publishers, 2014.

13 J. E. Harrington Jr., *The Theory of Collusion and Competition Policy*, MIT Press, Cambridge, MA, 2017.

14 J. Hofbauer, *Minimax via replicator dynamics*, Dyn. Games Appl. 8 (2018), 637–640.

15 J. Hofbauer and K. Sigmund, *Evolutionary game dynamics*, Bull. Am. Math. Soc. 40 (2003), no. 4, 479–519.

16 A. J. Jones, *Game Theory: Mathematical Models of Conflict*, Horwood Publishing, West Sussex, UK, 2000.

17 L. Jonker and P. D. Taylor, *Evolutionarily stable strategies and game dynamics*, Math. Biosci. 40 (1978), 145–156.

18 E. Kalai and M. Smorodinsky, *Other solutions to the Nash's bargaining problem*, Econometrica 43 (1975), 513–518.

19 E. L. Kaplan, *Mathematical Programming and Games*, Wiley, New York, 1982.

20 S. Karlin, *Mathematical Methods and Theory in Games, Programming, and Economics*, Vols. I, II, Dover, New York, 1992.

21 C. Lemke and J. Howson, *Equilibrium points of bimatrix games*, Soc. Ind. Appl. Math. 12 (1964), 413–423.

Game Theory: An Introduction, Third Edition. E. N. Barron.
© 2024 John Wiley & Sons, Inc. Published 2024 by John Wiley & Sons, Inc.

22 M. Leng and M. Parlar, *Analytic solution for the nucleolus of a three-player cooperative game*, *Nav. Res. Logist.* 57 (2010), 667–672.

23 W. Lucas (ed.), *Game Theory and its Applications*, Proceedings of Symposia in Applied Mathematics, vol. 24, American Mathematical Society, Providence, RI, 1981.

24 N. MacRae, *John Von Neumann: The Scientific Genius who Pioneered the Modern Computer, Game Theory, Nuclear Deterrence, and Much More*, American Mathematical Society, Providence, RI, 1999.

25 V. Mazalov, *Mathematical Game Theory and Applications*, Wiley, New York, 2014.

26 N. McCarty and A. Meirowitz, *Political Game Theory An Introduction*, Cambridge University Press, New York, 2007.

27 M. Mesterton-Gibbons, *Introduction to Game Theoretic Modelling*, 2nd ed., American Mathematical Society, 2000.

28 P. Milgrom, *Auctions and bidding: A primer*, *J. Econ. Perspect.* 3 (1989), no. 3, 3–22.

29 R. Myerson, *Game Theory, Analysis of Conflict*, Harvard University Press, Cambridge, MA, 1991.

30 S. Nasar, *A Beautiful Mind: The Life of Mathematical Genius and Nobel Laureate John Nash*, Touchstone, New York, 1998.

31 J. von Neumann and O. Morgenstern, *Theory of Games and Economic Behavior*, 3rd ed., Princeton University Press, Princeton, NJ, 1953 (1st ed., 1944).

32 M. Osborne, *An Introduction to Game Theory*, Oxford University Press, New York, 2004.

33 G. Owen, *Game Theory*, 3rd ed., Academic Press, San Diego, CA, 2001.

34 T. Parthasarathy and T. E. S. Raghavan, *Some Topics in Two-Person Games*, vol. 22, Elsevier, New York, 1971.

35 E. Rasmussen, *Games and Information*, Basil Blackwell, Cambridge, UK, 1989.

36 J. Roemer, *Kantian equilibrium*, *Scand. J. Econ.* 112 (2010), 1–24.

37 S. Ross, *A First Course in Probability*, Prentice-Hall, Englewood Cliffs, NJ, 2002.

38 E. Scheinerman, *Introduction to Dynamical Systems*, Prentice-Hall, Englewood Cliffs, NJ, 1996.

39 T. Vincent and J. Brown, *Evolutionary Game Theory, Natural Selection, and Darwinian Dynamics*, Cambridge University Press, Cambridge, UK, 2005.

40 N. Vorobev, *Game Theory, Lectures for Economists and Systems Scientists*, Springer-Verlag, New York, 1977.

41 J. Wang, *The Theory of Games*, Oxford University Press, New York, 1988.

42 J. Webb, *Game Theory: Decisions, Interaction and Evolution*, Springer Undergraduate Mathematics Series, Springer, 2006.

43 J. Weibull, *Evolutionary Game Theory*, MIT University Press, Cambridge, MA, 1995.

44 W. Winston, *Operations Research, Applications and Algorithms*, 2nd ed., PWS-Kent, Boston, MA, 1991.

45 E. Wolfstetter, *Auctions: An introduction*, *J. Econ. Surv.* 10 (1996), 367–420.

Index

Game Theory: An Introduction, Third Edition. E. N. Barron.
© 2024 John Wiley & Sons, Inc. Published 2024 by John Wiley & Sons, Inc.

Printed and bound by CPI Group (UK) Ltd, Croydon, CR0 4YY

16/04/2025

14658372-0004